OPTOELECTRONIC DEVICES AND PROPERTIES

Edited by **Oleg Sergiyenko**

Optoelectronic Devices and Properties
Edited by Oleg Sergiyenko

Published by InTech
Janeza Trdine 9, 51000 Rijeka, Croatia

Publishing Process Manager Ivana Lorkovic

Technical Editor Teodora Smiljanic

Cover Designer InTech Design Team

Additional hard copies can be obtained from orders@intechopen.com

Optoelectronic Devices and Properties, Edited by Oleg Sergiyenko
p. cm.
ISBN 978-953-307-204-3

For EU product safety concerns:
IN TECH d.o.o., Prolaz Marije Krucifikse Kozulić 3, 51000 Rijeka, Croatia,
info@intechopen.com or visit our website at intechopen.com.

Meet the editor

Oleg Yu. Sergiyenko received the B.S., and M.S., degrees in Kharkiv National University of Automobiles and Highways, Kharkiv, Ukraine, in 1991, 1993, respectively. He received the Ph.D. degree in Kharkiv National Polytechnic University on specialty "Tools and methods of non-destructive control" in 1997.

In March 1997, he joined the Kharkiv National University of Automobiles and Highways. He has written 67 papers and holds 1 patent of Ukraine; currently he is reviewer for IEEE Transaction on Industrial Electronics, IEEE Transaction on Mechatronics, etc.; he participates as reviewer and section chair in several IEEE conferences in Japan, USA, UK, Italy, etc.

Contents

Preface

Optoelectronic devices impact many areas of society, from simple household appliances and multimedia systems to communications, computing, spatial scanning, optical monitoring, 3D measurements and medical instruments. This is the most complete book about optoelectromechanic systems and semiconductor optoelectronic devices; it provides an accessible, well-organized overview of optoelectronic devices and properties that emphasizes basic principles. Coverage combines an optional review from key concepts such as properties of compound semiconductors, semiconductor statistics, carrier transport properties, optical processes, etc., up to gradual progress through more advanced topics. This book includes the recent developments in the field, emphasizes fundamental concepts and analytical techniques, rather than a comprehensive coverage of different devices, so readers can apply them to all current, and even future, devices.

In this book are introduced novel materials and physico-chemical phenomena useful for new tasks solution. It discusses important properties for different types of application, such as analog or digital links, the formation and analysis of optical waveguides; channel waveguide components; guided wave interactions; electrooptical effects; time dependence, bandwidth and electrical circuits.

Given the demand for ever more compact and powerful systems, there is growing interest in the development of nanoscale devices that could enable new functions and greatly enhanced performance. Semiconductor nanowires are emerging as a powerful class of materials that, through controlled growth and organization, are opening up substantial opportunities for novel photonic and electronic nanodevices.

Also progress in the area of nanowires growth is reviewed, as well as the fundamental electronic and optoelectronic properties of semiconductor nanowires and nanowire heterostructures, as well as strategies for and emerging results demonstrating their promise for nanoscale device arrays. Nanowires made could be ideal building blocks for making nano-optoelectronic devices; the nanowires sometimes show periodic defect structures along their lengths, which may be crucial for determining the optical properties of the material, so nanostructures may lead to further novel properties and promising applications such as point defects and stacking faults.

A significant part of optoelectronic methods are contributed in various geometric measurements like rangefinders, various 2D and 3D vision systems, with several applications in robot navigation, structural health monitoring, medical and body scanners.

Optoelectronic measurements are still among of the most attractive tools in a both spatial and frequency domains.

Independently a review of a wide range of optical fiber communication and optoelectronic systems is presented. In such networks, the electrical and the optical characteristics of guided-wave devices have a profound effect on the system design and overall performance. This book generally combines both the optical and electrical behavior of optoelectronic devices so that the interwoven properties, including interconnections to external components. It also shows the impact of material properties on various optoelectronic devices, and emphasizes the importance of time-dependent interactions between electrical and optical signals. It provides the key concepts and analytical techniques that readers can apply to current and future devices.

This is an ideal reference for graduate students and researchers in electrical engineering and applied physics departments, as well as practitioners in the optoelectronics industry.

Oleg Sergiyenko
The Engineering Institute,
Autonomous University of Baja California,
Mexicali, Mexico

Part 1

New Materials in Optoelectronics

1

Organic-Organic Semiconductor Interfaces for Molecular Electronic Devices

Ji-Seon Kim[1] and Craig Murphy[2]
[1]*Department of Physics & Centre for Plastic Electronics, Imperial College London,*
[2]*National Physical Laboratory (NPL)*
United Kingdom

1. Introduction

Molecular (Plastic) electronics encompasses the materials science, chemistry and physics of molecular electronic materials and the application of such materials to displays, lighting, flexible thin film electronics, solar energy conversion and sensors. The field is a growth area, nationally and globally, evidenced by the rapidly expanding organic display and printed electronics industries. Organic semiconductors combine the semiconductor properties traditionally associated with inorganic materials with the more desirable properties of plastics. Moreover, the organic syntheses of these materials allow for great flexibility in the tuning of their electronic and optical properties. By combining these properties, organic semiconductors such as conjugated polymers have been demonstrated as the active material in light-emitting diodes (LEDs), transistors, and photovoltaic (PV) cells. Furthermore, these conjugated polymers provide a new way of looking at many of the broad fundamental scientific issues related to using molecules for electronics. A great deal of the physics which governs the behaviour of molecules for electronics occurs at the organic-organic interfaces (heterojunctions). For example, the nature of organic interfaces determines the fate of excitons to be either stabilised (for efficient LEDs) or destabilised (for efficient PV cells) at the interfaces. Therefore, by selecting semiconductors with proper band-edge offsets between their conduction and valence bands, different device characteristics can be readily achieved. While significant progress has been made in developing the materials and high performance organic devices, many fundamental aspects of organic-organic semiconductor interfaces remain to be understood. In particular, fundamental understanding of the correlation between nanostructures and interfaces of organic semiconductors in thin films and multilayers and associated device performance still remain to be fully explored. In this Chapter, we will introduce how to control and characterise various length-scale organic-organic interfaces facilitating the rational design of materials, device architectures and fabrication methods via increased understanding of fundamental properties of organic-organic interfaces and their modification due to processing. In particular, we will address the distinctive optoelectronic and charge transport properties which have been observed across different organic-organic interfaces depending on their length-scale (micron-scale in the blends down to molecular-scale in the copolymers) and nature (interchain vs intrachain), providing the deeper understanding of organic interfaces and their vital roles in various optoelectronic devices. The key advances in organic semiconductor interfaces achieved so far will provide important insight into a design rule of organic semiconductors which is essential for future development in molecular electronic devices.

2. The main aim and contents of this chapter

This chapter aims to review fundamental scientific issues related to using molecules for electronics down to the single-molecule scale by studying a range of well-controlled organic-organic interfaces, with a particular focus on their role on electronic structures and electronic processes of organic semiconductors and their devices. Specific topics were:
1. Control and characterisation of various length-scale organic interfaces (Section 3)
2. Photophysical dynamics of electronic species at the organic interfaces (Section 4)
3. Electronic processes of charge carriers across the organic interfaces (Section 5)
4. Charge-carrier operational dynamics across the organic interfaces (Section 6)

3. Control and characterisation of various length-scale organic interfaces

3.1 Interfaces controlled by polymer molecular weight variation

Polymer molecular weight (MW) (i.e. chain length) variation was used as a tool to control the phase separation laterally and/or vertically in blend films (Yim et al., 2010). The conjugated polymers studied are poly(9,9-di-*n*-octylfluorene-*alt*-benzothiadiazole) F8BT (M_n = 9 - 255 kg/mol) and poly(9,9-di-*n*-octylfluorene-*alt*-(1,4-phenylene-((4-*sec*-butylphenyl) imino)-1,4-phenylene) TFB (M_n = 3 - 102 kg/mol) (Chemical structures in Table 1). Micron-scale lateral phase separation is observed in blend films that consist of high MW of both F8BT and TFB (M_n > 60 kg/mol), in which domain sizes increase with MW of each homopolymer. For these blend films, the micro-Raman spectroscopy study indicates that the higher-lying domains are F8BT-rich and the lower-lying domains are TFB-rich. In contrast, the blend films that consist of at least one low MW homopolymer (M_n < 10 kg/mol) show relatively smooth surface with sub-micron or no measurable lateral phase separation.

Using the surface-sensitive X-ray photoelectron spectroscopy (XPS) technique, it is observed first that, for blend films that consist of at least one low MW polymer (M_n < 10 kg/mol), there is a significant enrichment of the short polymer chains at the film-air interface. This reveals that the vertical phase segregation at the film-air interface is driven by the contrast of MW between the two homopolymers. On the other hand, for blend films that show micron-scale lateral phase separation, the film-air interface is always enriched with TFB, suggesting the presence of TFB capping layer apart from the exposed TFB-rich domains. Second, for all the blend films at the film-substrate interface, there is an enrichment of the lower surface energy material (TFB). The extent of TFB enrichment is found to increase with the MW of both polymers, possibly due to increased thickness or purity of the TFB wetting layer in these blend films. These observations suggest that surface energy contrast (as oppose to MW contrast) might be the dominant driving force in determining the vertical phase segregation at the film-substrate interface. Based on the morphological and compositional analyses of these blend films, we proposed two different models of the final phase separated structures (Fig 1a and 1b) for blend films without and with micron-scale lateral phase separation, respectively.

For the blend films with no visible lateral phase separation (in which a large MW contrast exists between the two homopolymers), the film-air interface is enriched with the low MW polymer, while the film-substrate interface is always enriched with the lower surface energy polymer TFB. For the blend films with obvious micron-scale lateral phase separated structures, additional nanoscale vertical phase segregation also occurs resulting in a continuous TFB wetting layer at the film-substrate interface and a discontinuous TFB capping layer at the film-air interface (aKim et al., 2004). These remarkably different lateral and vertical phase separation observed in the F8BT:TFB blend films has important implications on LED performance.

Material	Chemical Structure	HOMO [eV]	PL efficiency
P3HT	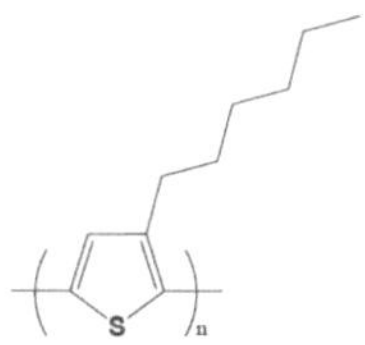	-4.8	0.1 0[a]
PFB	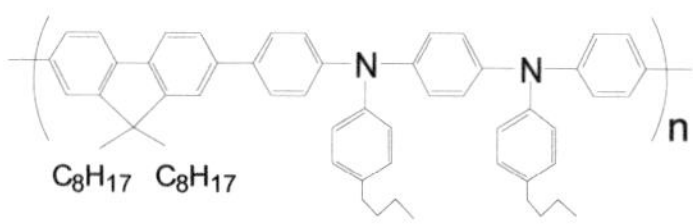	-5.1	0.65 0[a] 0.35[b]
TFB	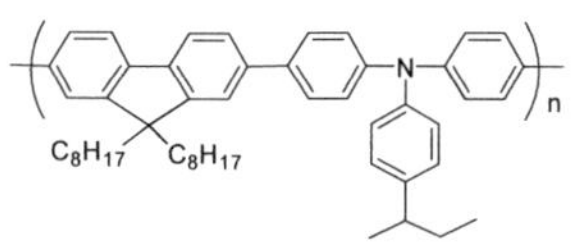	-5.3	0.4 0.05[a] 0.1[b]
F8BT	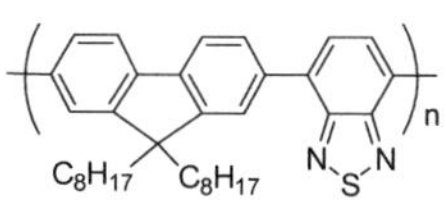	-5.9	0.6 0.05[a]
F_4-TCNQ	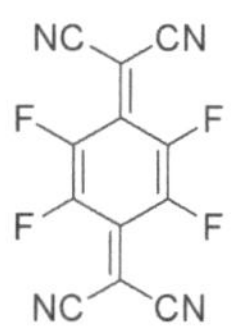	-5.2[c]	-

Table 1. Chemical structures and optoelectronic properties of conjugated polymers and F_4TCNQ. [a] PL efficiency of 5 % F4TCNQ-doped polymer, [b] PL efficiency of 5 % F4TCNQ-doped polymer after annealing (N_2 environment, 200 °C, 1 hr), [c] LUMO level of F_4TCNQ

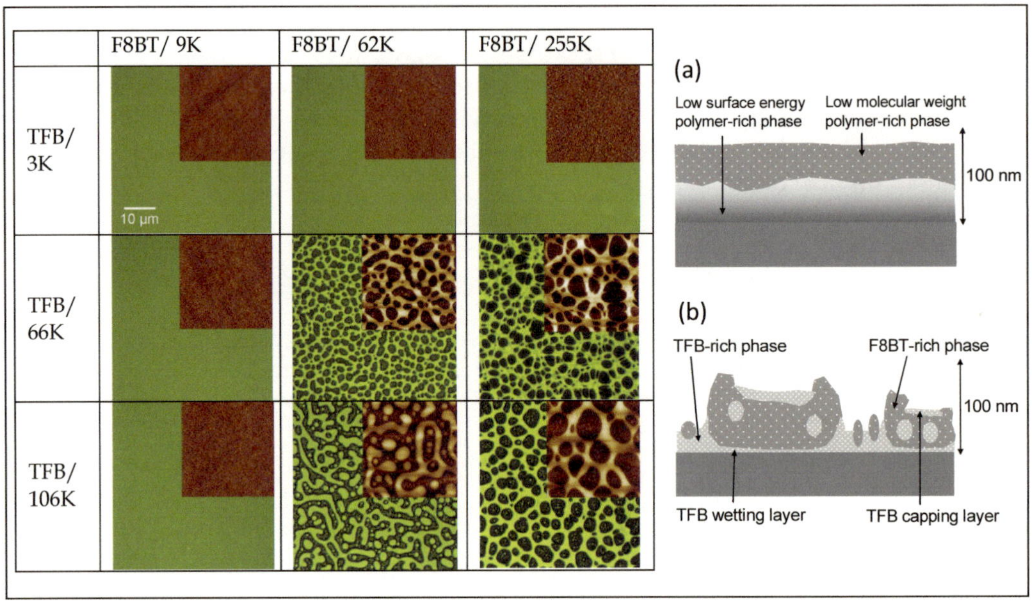

Fig. 1. Left: PL images of F8BT:TFB blend films (100nm, 1:1 by weight) with different MWs under blue excitation. The bright regions correspond to F8BT-rich phases while the dark regions TFB-rich phases. Inset: AFM images (20μmX20μm). Right: Proposed cross sections (a) at least one low MW homopolymers and (b) high MW of both homopolymers

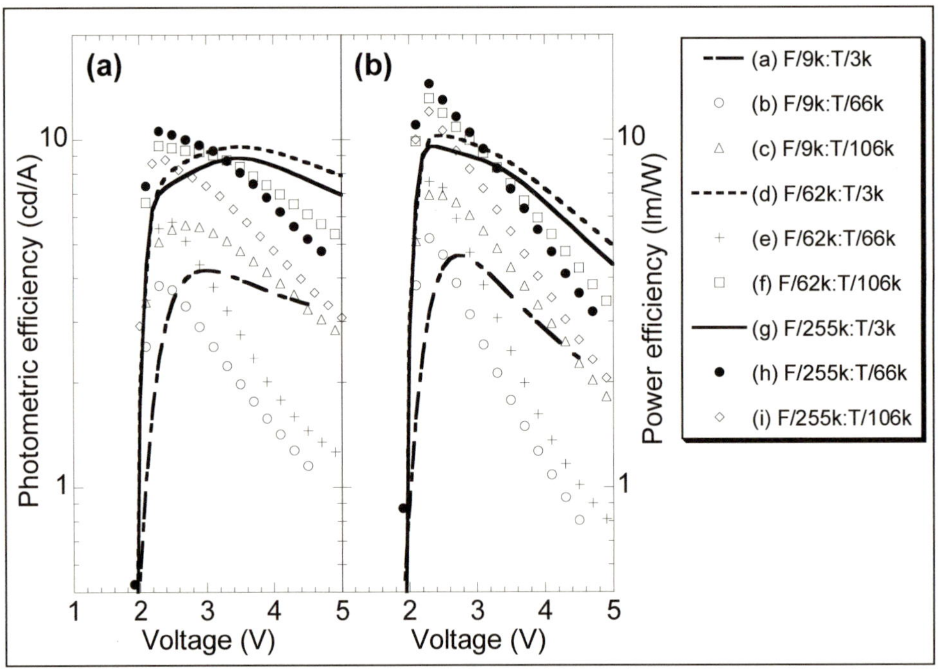

Fig. 2. EL efficiency-voltage characteristics of LEDs fabricated with F8BT:TFB blend films with different molecular weights of each copolymer, in (a) cd/A and (b) lm/W

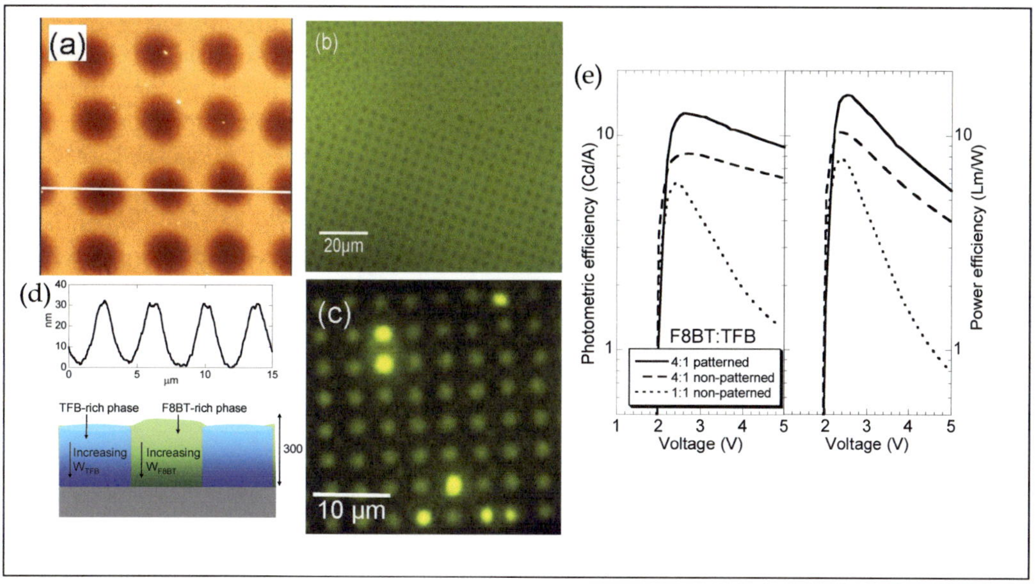

Fig. 3. (a) AFM image (15 μm X 15 μm, on 70 nm height scale) of the patterned F8BT:TFB blend film. Surface line scan (below) indicates that the height difference between the higher-lying F8BT-rich and lower-lying TFB-rich phases is ~30 nm. (b) PL image of ~300 nm-thick patterned F8BT:TFB blend film under blue excitation (2.85 eV). The bright and dark regions in the PL image correspond to F8BT-rich and TFB-rich phases, respectively. Note the contrast between the areas with and without the 2D pattern. (c) EL image of the patterned LED at 7V showing EL from enclosed TFB-rich domains. Differences in brightness between the TFB-rich domains might be due to thickness variation across the blend film. (d) Schematic drawing summarizes the proposed cross section of the patterned F8BT:TFB blend film based on micro-Raman compositional analysis. Both domains show increased purity of the corresponding polymer nearer to the patterned substrate. (e) EL efficiency-voltage characteristics of LED fabricated with the patterned blend film expressed in Cd/A and Lm/W. F8BT:TFB blend devices (4:1 and 1:1 by weight) prepared by spin-coating are included for comparison

The performance of LEDs fabricated with these blend films is found to be closely related to the blend thin film morphology, which varies remarkably with the molecular weight of both polymers (Figure 2). All the devices fabricated with the blend films exhibit sharp turn-on in both current and luminance at ~2 V. Two distinctive efficiency-voltage characteristics are observed in these blend devices. First, blend films that exhibit micron-scale lateral phase separation show high initial efficiencies just after turn-on, but decreases rapidly at high voltages. Such device characteristics are closely related to the blend film morphology. While the continuous TFB wetting layer might assist hole injection/transport and act as electron blocking layer at the anode interface, the discontinuous TFB capping layer might localise electron injection from the cathode, resulting in a high degree of spatial confinement of charge carriers. This then leads to high electron-hole recombination efficiency at organic-organic interfaces (Morteani et al., 2003), which may explain the observed high initial EL efficiencies in these blend films. However, at high voltages, the presence of very thin lower-lying TFB-rich domains provide a pathway for holes to punch through the blend film without undergoing radiative recombination with electrons, causing imbalance of charge carriers and hence an increase in leakage current and the rapid decay in EL efficiencies. Second, among the blend films with no observable micron-scale lateral phase separation, those that consist of TFB/3K show relatively lower peak initial efficiency, comparing to those with micron-scale lateral phase separation. This is attributed to the lack of phase separated features that can assist spatial confinement of charge carriers, as discussed earlier. However, improved film thickness uniformity and balance of charge carriers do contribute to reduce leakage current at high voltages, explaining the observed slower decay in EL efficiencies. Furthermore, the amount of surface out-coupling of light in the forward direction observed in blend devices is found to be positively correlated to the distribution of periodicity of the phase-separated structures in the active layer.

3.2 Interfaces controlled by patterned substrate

The phase separation in organic blend thin films can be controlled via chemical modification of the substrate with a periodic contrast of the substrate surface energy by microcontact printing ([a]Yim et al., 2008). With appropriate choice of polymer molecular weight and blend ratio, the phase-separated structures in the blend film closely replicate the underlying 2D pattern since the low surface energy component TFB preferentially migrate away from regions of higher surface energy (Figure 3). Micro-Raman analysis revealed nanometer-scale vertical segregation of the polymers within both lateral domains, with regions closer to the substrate being substantially pure with each of the two polymers. This indicates the absence of a continuous TFB wetting layer typically formed in blend films spin-coated on non-patterned surfaces, and has important implications on device performance. It also implies the formation of periodic TFB/F8BT (and reversed) heterojunctions structures which favour (and suppress) charge carrier injection from both electrodes in the TFB-rich (F8BT-rich) domains. As a result, charge carrier injection is confined in the well-defined enclosed TFB-rich domains, leading to high EL efficiency. The overall reduction in the patterned blend film roughness as compared to reference spin-coated blend (1:1 by weight) leads to slower decay in EL efficiency at high voltages. The amount of surface out-coupling of light in the forward direction observed in blend devices is also found to be positively correlated to the distribution of periodicity of the phase-separated structures in the active layer.

3.3 Interfaces controlled by thin film transfer printing technique

The fabrication of functional multilayered conjugated polymer structures with well-defined organic-organic interfaces for optoelectronic device applications is constrained by the common solubility of many polymers in most organic solvents ([b]Yim et al., 2008). A simple, low-cost, large-area transfer printing technique for the deposition and patterning of conjugated polymer thin films has been demonstrated. This method utilises a planar poly(dimethylsiloxane) (PDMS) stamp, along with a water-soluble sacrificial layer, to pick up an organic thin film (~20nm-1μm) from a substrate and subsequently deliver this film to a target substrate. The versatility of this transfer printing technique and its applicability to optoelectronic devices have been shown by fabricating bilayer structures of TFB/F8BT and poly(3-hexylthiophene) (P3HT)/ methanofullerene([6,6]-phenyl C_{61} butyric acid methyl ester) (PCBM), and incorporating them into light-emitting diodes and photovoltaic cells, respectively (Figure 4). For both types of devices, bilayer devices fabricated with this transfer printing technique showed equal, if not superior performance to either blend devices or bilayer devices fabricated by other techniques. This indicates well-controlled organic-organic interfaces achieved by the transfer printing technique.

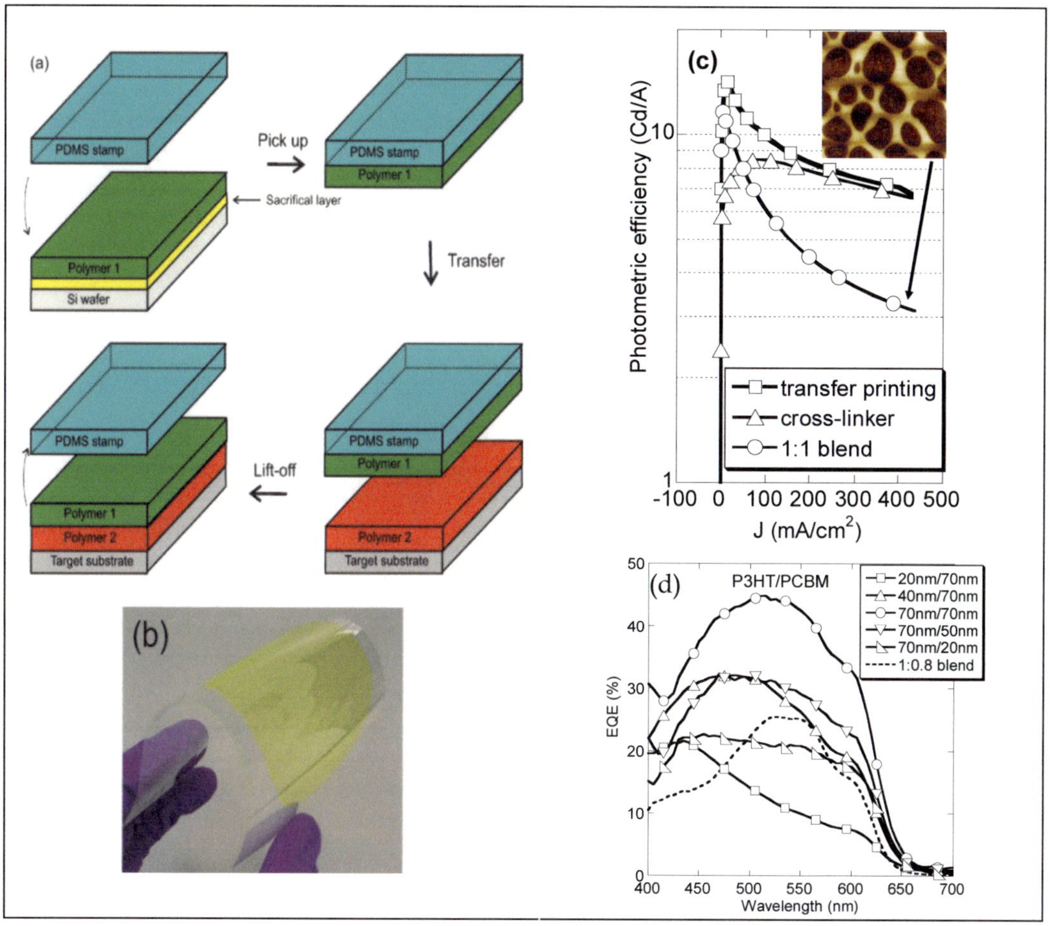

Fig. 4. (a) Schematic depiction of the thin film transfer printing process to form a bilayered conjugated polymer structure. (b) A photograph of a 4-inch transfer printed F8BT thin film (~ 70 nm) on a flexible PET sheet. (c) EL efficiency (Cd/A) as a function of diode current (J) of LEDs with TFB/F8BT bilayered films (20nm/80nm) fabricated by transfer printing technique (square) and cross-linker (triangle), and TFB:F8BT blend film (1:1 by weight, circle). (d) External quantum efficiency (EQE) of P3HT/PCBM bilayered solar cells fabricated by transfer printing technique, with different thicknesses of either P3HT or PCBM (20-70 nm), while keeping the thickness of the other material (70 nm) and other processing conditions constant. An EQE spectrum of a P3HT:PCBM blend device (1:0.8 by weight) is included for comparison

3.4 Interfaces controlled by copolymerization of electron donor and acceptor units

The molecular-scale intrachain interfaces (heterojunctions) can be created by copolymerization of TFB (electron donor) and F8BT (electron acceptor) polymers by covalently attaching them to a main conjugated polymer backbone (Kim et al., 2008) and also by adding different amounts of the strongly electron accepting BT unit into the F8 polymer backbone, $F8_{1-x}BT_x$ copolymers ($0 \leq x \leq 0.5$) (Winfield et al., 2009). We have observed that not only are excitons in $F8_{1-x}BT_x$ copolymers ($0 \leq x \leq 0.5$) strongly charge transfer (CT) in character but that the strength of this CT character increases when the proportion of BT units in the polymer chains is small. Incorporation of the strongly electron-accepting BT unit, even in small proportions, into the F8 polymer chain results in localization of the lowest unoccupied molecular orbital (LUMO) on the BT units. There are no significant changes in the highest occupied molecular orbital (HOMO) and LUMO energies with BT content leading to the same shape of the F8BT emission in the PL spectrum for all $F8_{1-x}BT_x$ copolymers. Increased coupling of the excited state to the ground state at higher BT content results in shorter exciton lifetime and higher PL efficiency. The increased CT character of the excitons in lower-BT content copolymers is also seen in the stronger solvent dependence of the emission spectra and excited state lifetimes of these copolymers.

4. Photophysical dynamics of electronic species at the organic interfaces

4.1 Photoinduced intrachain charge transfer state in copolymer

The optoelectronic properties at the organic-organic semiconductor interfaces formed between polymer chains (interchain) and within a polymer chain (intrachain) are studied (Kim et al, 2008). These interfaces are fabricated using (TFB [F8-tfb]) (electron-donor) and (F8BT [F8-BT]) (electron-acceptor) conjugated polymers, by blending them together or by covalently attaching them via a main polymer backbone (copolymer) . When a bulky and twisted tfb molecule is incorporated into a rigid F8BT conjugated backbone, it disturbs the conjugation of F8BT polymer, leading to a blue-shift in the lowest absorption transition. However, by acting as an effective electron donor, it assists the formation of an intrachain singlet exciton that has a strong CT character, leading to a red-shifted and longer-lived emission than that of F8BT. An extremely efficient and fast energy transfer from tfb donor to BT acceptor is observed in the copolymer (<1 ps) compared to transfer from TFB to F8BT in the blend (tens of ps). This efficient energy transfer in the copolymer is found to be associated with its low fluorescence efficiency (40-45% vs 60-65% for blend) because of the migration of radiative singlet excitons to low-energy states such as triplet and exciplex states that are nonemissive or weakly emissive.

4.2 Dielectric switching of the nature of excited state in copolymer

In a conjugated random copolymer (RC) composed of electron donor (TFB) and electron acceptor (F8BT) units, the spectral evolution of an intrachain neutral singlet exciton toward a CT state in solvents of increasing polarity has been monitored by time-resolved photoluminescence and ultrafast transient absorption spectroscopy (Petrozza et al., 2010). The PL spectra of the RC solutions in different polarity of solvents are shown and compared to the F8BT spectra (Figure 5). Very diluted solutions in o-xylene, the RC emission shows ~16 nm red-shift with respect to that of F8BT, with its lifetime slightly longer than that of F8BT (2.6 ns for RC vs. 2 ns for F8BT). This suggests that as the TFB unit is covalently linked to the F8BT backbone, the emission in RC occurs from an electronic state of different nature compared to F8BT. The F8BT itself shows a red shift of ~30 nm in PL spectra as the polarity of the medium is increased from the o-xylene (dielectric constant, ε = 2.57) to the o-dichlorobenzene (ε = 4.81) solvents underlying the polar nature of the relaxed emissive state resulted from the localisation of the electron wavefunction on the BT unit. In the RC, this solvatochromic effect is strongly amplified. There is a significant increase in the red-shift, ~114 nm shift going from the cyclohexane (ε = 2.02) to the *o*-dichlorobenzene (ε = 9.93) solution, and spectral changes in emission spectra. In addition, these spectral changes are accompanied by a concomitant enhancement of the luminescence lifetime (from 2.6 ns to 5.7 ns) and reduction of the PL quantum yield when the solvent polarity increases. The sensitivity of the emissive state to the solvent dielectric constant supports its strong CT character in the relaxed configuration.

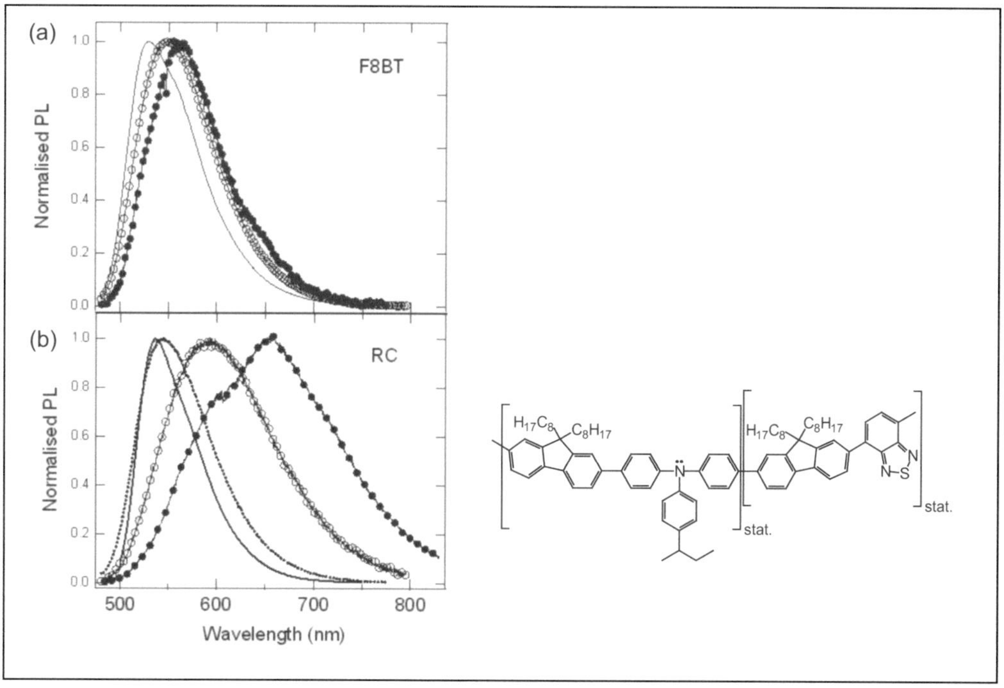

Fig. 5. PL spectra of diluted solutions of (a) F8BT and (b) RC in various solvents with different polarity, cyclohexane (solid line), *o*-xylene (dot line), chloroform (empty circles), *o*-DCB (filled circles). Excitation at 470 nm

In order to obtain further insight into the photophysical dynamics of the relaxation pathways, ultrafast transient absorption (TA) experiments have been performed. The TA spectra of F8BT in diluted *o*-xylene solution (Figure 6a) present two main features. For λ < 600 nm, a positive differential transmission signal ($\Delta T/T$) was observed. It is assigned to stimulated emission (SE) originating from the first excited singlet state, since its spectral position matches the fluorescence of the polymer. The SE could be observed over the entire detection time (about 2 ns). At λ > 600 nm, a photoinduced absorption (PA) band occurred which extended into the near infrared. This arises from the $S_1 \rightarrow S_n$ excited state absorption, since the SE and PA bands exhibited the same lifetimes and decay dynamics. A spectral broadening of the SE band was observed for *o*-xylene solution, whereas in the polar solvents (chloroform and *o*-DCB) a distinct dynamic red-shift of the band was observed (not shown). This is in agreement with the solvatochromic effect observed in the steady-state PL spectra and confirms the presence of a relaxed emissive state, which has a small polar character and is intrinsically sensitive to the environment.

The TA spectra of RC in *o*-xylene solution (Figure 6b) were found to be similar to the spectra of F8BT exhibiting a SE band below 620 nm and a PA band at longer wavelength. However, compared to the TA spectra of F8BT the SE band showed a dynamic red-shift and spectral broadening in the first 22 ps before decaying, indicating a stronger nuclear relaxation after photoexcitation. Fig. 6c shows the TA spectra of RC in chloroform solution. The SE and PA bands were observed at very early times (hundreds of fs). However, no spectral diffusion could be traced and within 3 ps the SE band evolved into a broad absorption band, which decayed slowly over the timescale of the experiment. The broad band shows a reduced PA response at 590 nm, which corresponds to the spectral region where the RC PL spectrum in chloroform solution peaks. This is ascribed to the competition between the SE and PA signal of the CT state. Similar spectral features have been observed in the *o*-DCB solution (Figure 6d), although they evolve considerably more slowly. Here, a complete relaxation of the SE band and a red-shift to about 590 nm in the first 25 ps have been observed followed by quenching of the SE band and the emergence of a broad absorption band to the infrared. It indicates that in polar solvents a charge-like absorption superimposes the region of stimulated emission and leads to a dramatic reduction in gain implying that CT states in the RC can be detrimental for light amplification and lasing.

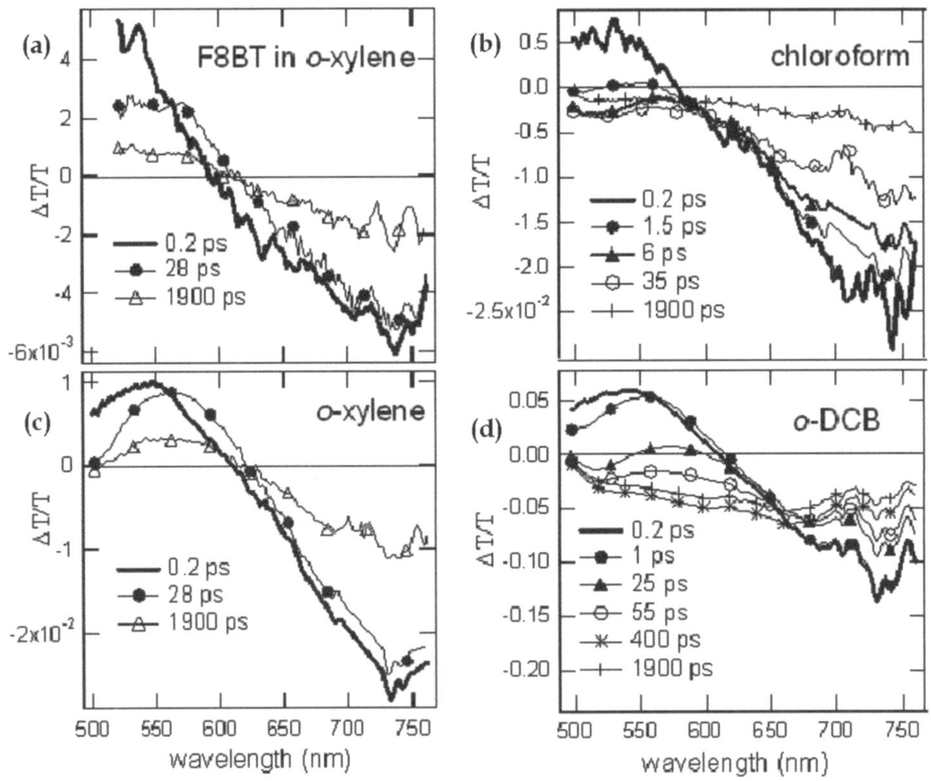

Fig. 6. Femto- to picosecond TA spectra of (a) F8BT in *o*-xylene solution and RC in (b) *o*-xylene, (c) chloroform, and (d) *o*-DCB solutions. TA spectroscopy was used to study the evolution and dynamics of the excited states on a timescale of hundreds of femtoseconds to milliseconds by probing the relative change in transmission (ΔT/T) after photoexcitation at 490 nm

4.3 Intrachain versus intermolecular interactions at high pressure

The optical spectroscopy studies of F8BT polymers at high pressure have been performaed in order to disentangle the intramolecular and intermolecular effects of hydrostatic pressure (Schmidtke et al., 2007). The PL spectrum of F8BT in a dilute solid state solution in polystyrene redshifts by ~320 meV over 7.4 GPa, while that of a F8BT thin film redshifts ~460 meV over a comparable pressure range. We attributed the redshift in solution to intrachain pressure effects, principally conformational planarization (i.e. a reduction in the torsional angle between the F8 and BT subunits of F8BT at high pressure). The additional contribution from interchain π-electron interactions accounts for the larger redshift of films.

4.4 Dynamic emission polarization anisotropy for aligned polymer films

Time-integrated and femtosecond time-resolved PL spectroscopy has been used to study the dynamic emission polarization anisotropy for aligned F8BT thin films (Schmid et al., 2008). The results indicate a high degree of chain alignment, with the presence of a small fraction of unaligned chain domains in film regions far from the imprinted surface. The time-averaged emission from aligned domains is found to be slightly shifted to higher photon energies compared to that from more disordered film regions. This effect is attributed to a subtly different chain packing geometry in the more aligned regions of the film, which leads to a reduced exciton diffusivity and inhibits energetic relaxation of the exciton in the inhomogeneously broadened density of states. While for an unaligned reference film, exciton migration results in a nearly complete depolarization of the emission over the first 300 ps. For the aligned films, interchain exciton hopping from unaligned to aligned domains is found to increase the anisotropy over the same time scale. In addition, excitons generated in aligned film domains were found to be slightly more susceptible to nonradiative quenching effects than those in disordered regions deeper inside the film, suggesting a marginally higher defect density near the nanoimprinted surface of the aligned film.

5. Electronic processes of charge carriers across the organic interfaces

5.1 Effects of polymer packing structure on optoelectronic properties

Spin-coated F8BT films of different molecular weights (M_n = 9 - 255 kg/mol), both in the pristine and annealed state, were studied in an effort to elucidate changes in the polymer packing structure and the effects this structure has on the optoelectronic and charge transport properties of these films (Donley et al., 2005; Zaumseil et al., 2006). A model based on quantum chemical calculations, wide-angle X-ray scattering, atomic force microscopy, Raman spectroscopy, photoluminescence, and electron mobility measurements was developed to describe the restructuring of the polymer film as a function of polymer chain length and annealing. In pristine high molecular weight films, the polymer chains exhibit a significant torsion angle between the F8 and BT units, and the BT units in neighboring chains are close to one another. Annealing films to sufficiently high transition temperatures allows the polymers to adopt a lower energy configuration in which the BT units in one polymer chain are adjacent to F8 units in a neighboring chain ("alternating structure"), and the torsion angle between F8 and BT units is reduced. This restructuring, dictated by the strong dipole on the BT unit, subsequently affects the efficiencies of interchain electron transfer and exciton migration. Films exhibiting the alternating structure show significantly lower electron mobilities than those of the pristine high molecular weight films, due to a decrease in the efficiency of interchain electron transport in this structure (Figure 7a). In addition, interchain exciton migration to low energy weakly emissive states is also reduced for these alternating structure films, as observed in their photoluminescence spectra and efficiencies (Figure 7b).

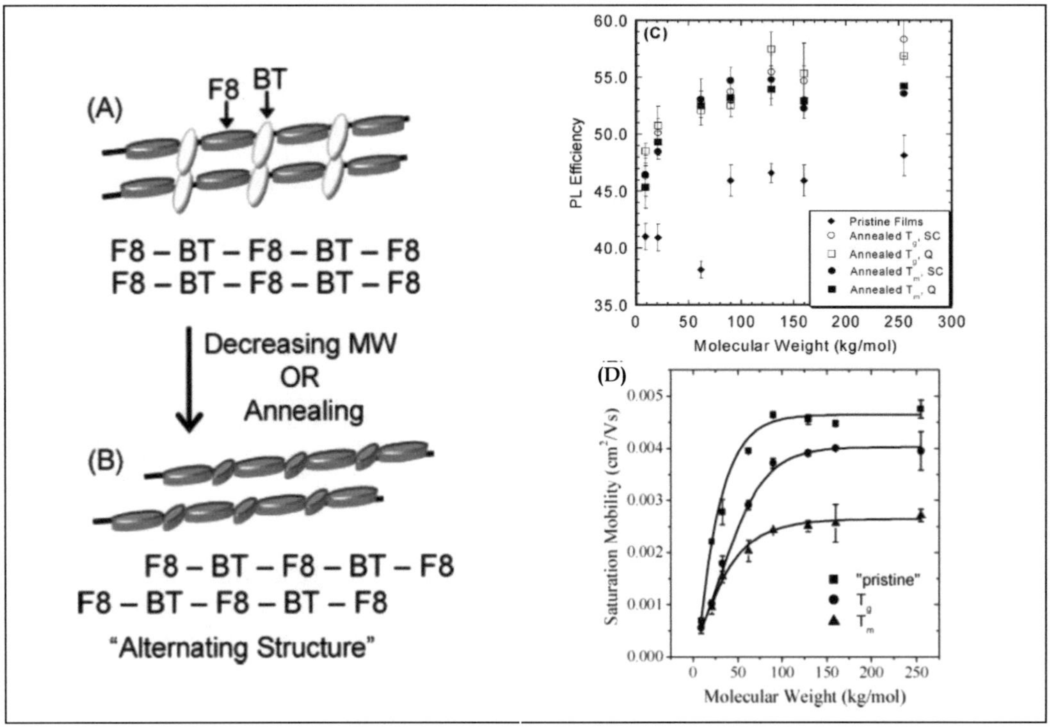

Fig. 7. Schematic illustrations showing (A) the initial packing structure of the high molecular weight pristine films. The BT units exhibit a relatively high torsion angle with respect to the F8 units, and in neighboring polymer chains, the BT units are adjacent to each other. (B) The packing structure for the low molecular weight pristine films or annealed films. Adjacent polymer chains have been translated with respect to one another, so that the BT units in one chain are adjacent to the F8 units in the neighboring chain (termed "alternating structure"). This structure forces the BT units into a geometry that is more planar with the F8 units. (C) PL efficiencies for pristine and annealed films. The sample with a molecular weight of 62 kg/mol is known to have some inorganic impurities and shows a low PL efficiency. (D) Electron mobilities calculated in the saturation region for F8BT films annealed to different temperatures. Decreases in mobility were observed upon annealing and for the shorter molecular weights. "Pristine" films are those that were heated only to 100 °C to remove residual solvent trapped in the films before further processing steps

5.2 Enhanced charge transport properties of aligned F8BT films by nanoconfinement

The uniaxial alignment of a liquid-crystalline conjugated polymer F8BT, by means of nanoconfinement during nanoimprinting has been demonstrated (Zheng et al., 2007). The orientation of the conjugated backbones was parallel to the nanolines imprinted into the polymer film. Polarized UV-vis absorption and photoluminescence spectra were measured to quantify the degree of alignment, showing that the polarization ratio and uniaxial molecular order parameter were as high as 66 and 0.97, respectively. The aligned F8BT film was used as the active layer in a PLED, which resulted in polarized electroluminescence with a polarization ratio of 11. Ambipolar polymer FET in a top-gate configuration with aligned F8BT as the active semiconducting layer showed mobility enhancement when the chains were aligned parallel to the transport direction. Mobility anisotropies for hole and electron transports were 10-15 and 5-7 respectively, for current flow parallel and perpendicular to the alignment direction.

5.3 Intrachain versus interchain electron transport

F8BT displays very different charge-transport properties for holes versus electrons when comparing annealed and pristine thin films and transport parallel (intrachain) and perpendicular (interchain) to the polymer axes. We have performed a quantum-chemical calculation focusing on the electron-transport properties of F8BT chains and compared the efficiency of intrachain versus interchain transport in the hopping regime (Van Vooren et al., 2008). The theoretical results rationalize significantly lowered electron mobility in annealed F8BT thin films and the smaller mobility anisotropy ($\mu_{\parallel}/\mu_{\perp}$) measured for electrons in aligned films (i.e. 5–7 compared to 10–15 for holes).

5.4 Controlled electrical properties via a solution-based *p*-type doping

We have controlled *p*-doping of P3HT, PFB, TFB and F8BT conjugated polymers by co-blending with F_4TCNQ in a common organic solvent ([c]Yim et al., 2008). Doping leads to significant increase in the bulk conductivity and hole current of the polymers with gradual disappearance of turn-on voltage. The effectiveness of doping increases as the HOMO level of the polymers becomes smaller, from F8BT (5.9eV) to P3HT (4.8eV), indicating that *p*-doping occurs via electron transfer from the HOMO level of the polymers to the LUMO level of F_4TCNQ (Figure 8). F_4TCNQ appears as a promising candidate to *p*-dope a wide range of conjugated polymers. This solution-based doping process will be one of the most effective and desirable ways to control the electrical properties of organic materials, in particular for solution processable organic semiconductors and their associated devices. In particular, a single polymer material can be used as both semiconductor and conductor in a single device.

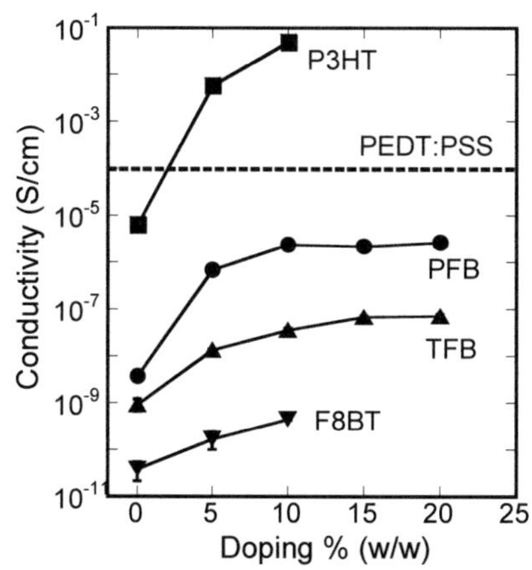

Fig. 8. Measured conductivity of the conjugated polymer films doped with different concentrations of F_4TCNQ. Polymer films were deposited on substrates with inter-digitated ITO structures. The applied electric field was $\leq 4 \times 10^{-4}$ V/nm. The conductivity of PEDT:PSS film typically used in organic devices is included for comparison

5.5 Improved PLED efficiency by inserting a thin polymer interlayer

It is demonstrated that adding a thin (10 nm) conjugated polymer interlayer between PEDT:PSS and an emissive semiconductor prevents the exciton quenching at the PEDT:PSS interface, resulting in a significant improvement in the device efficiency of polymer LEDs (Kim et al., 2005). For PLEDs with the TFB interlayer, the external quantum efficiency (EQE) increases from 0.7% (0.4 CdA at 3.7 V) to 1.9% (1.0 CdA at 3.3 V) at 100 Cd/m^2 for red LEDs and from 1.9% (6.2 Cd/A at 3.4 V) to 3.0% (10.1 Cd/A at 3.0 V) at 1000 Cd/m^2 for green LEDs. An EQE of 4.0% is also observed in blue LEDs (35% increase). The increase in the efficiency is accompanied by a large increase in the device lifetime (up to five times for red LEDs and four times for green LEDs). This thin-conjugated polymer interlayer is spin-coated from TFB solution directly on top of the PEDT:PSS layer. TFB is a triarylamine-based large-band-gap semiconductor (3.0eV) often used as a hole transporter due to its low ionization potential (5.33 eV) and high hole mobility. One of the main roles of this TFB interlayer is considered to be a blocking layer that prevents the radiative excitons from direct quenching by PEDT:PSS and thus to remove a nonradiative decay channel introduced by PEDT:PSS. This exciton blocking property of the TFB interlayer contributes to improvement of the device performance. We demonstrate this by directly measuring the exciton lifetime, the time taken for an excited state to decay radiatively, of F8BT emissive semiconductor in direct contact with PEDT:PSS and TFB interlayer, using time-correlated single-photon counting technique (Figure 9).

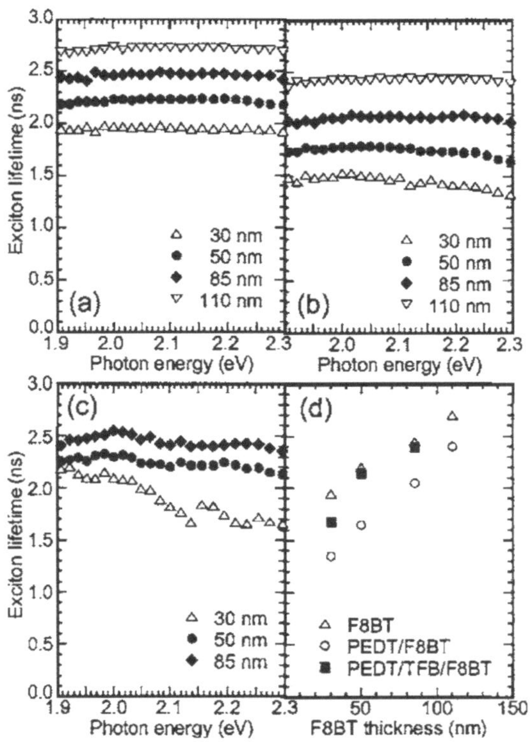

Fig. 9. Exciton lifetime of (a) F8BT, (b) PEDT:PSS/F8BT, (c) PEDT:PSS/TFB/F8BT films as a function of emitted photon energy, and (d) as a function of F8BT film thickness at 2.27 eV. PEDT:PSS is a poly(3,4-ethylenedioxythiophene) doped with poly(styrenesulphonate)

6. Charge-carrier operational dynamics across the organic interfaces

6.1 Effects of charge balance on transient electroluminescence

Charge balance inside the active layer has been studied by applying a pulse-mode electrical excitation to the device. The transient response of the devices under pulsed excitation yields important information related to charge injection and transport pathways (Seeley et al., 2004), in addition to its obvious use in high-brightness pulsed devices. In this study, a constant voltage pulse (4 μs pulses with a repeat rate of 30 Hz) was sent to the devices and their transient EL characteristics were monitored (Kim et al., 2008). The device fabricated with a TFB (50nm)/ F8BT (50nm) bilayer was added for the pulse measurements in order to address the important role of different length-scale polymer-polymer interfaces in the charge-carrier transport and recombination processes. The pulsed measurements revealed a remarkably clear trend through the appearance of a "turn-on" spike in the EL output of the devices (Figure 10a). This "turn-on" spike was strongest in the bilayer device and gradually decreases as the length-scale of the organic interfaces decreases, starting with micron-scale in the blend and ending with molecular-scale in the RC. Therefore, no turn-on spike was observed in the RC. The turn-on spikes were not observed either in the neat F8BT or neat TFB devices.

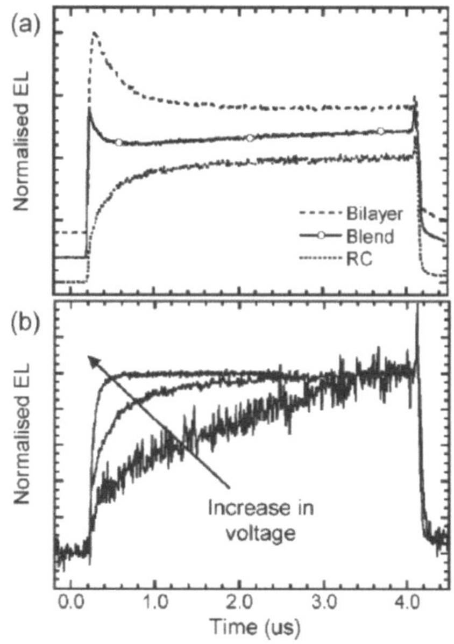

Fig. 10. (a) Pulsed measurements of TFB:F8BT LEDs fabricated with bilayer (50 nm TFB/50 nm F8BT), polymer blend, and RC. A constant voltage pulse (5.5 V for the blend and 11-12 V for others) with the duration of 4 μs and 30 Hz repeat rate was used. (b) Pulsed measurements of the RC LED with an increase in the voltage from 6.5 V to 14.5 V

This turn-on spike can be understood by associating it with charge balance inside the active layer after an initial electrical excitation, although its exact origin has not yet been clearly established. An ohmic contact is formed at the cathode interface between F8BT and Ca electrode since the F8BT energy level for electron injection (i.e. LUMO level, ~2.95 eV) lies very closely to the work function of Ca electrode (~2.9 eV). Hence once a contact is made, there is no injection barrier for electrons and this provides a barrier-free injection of electrons into the F8BT layer even before the driving voltage is applied. Inside the bilayer device, this would lead to an accumulation of electrons at the interface between TFB and F8BT layers due to the energy barrier for electrons induced by the relatively low LUMO level of TFB (~2.25 eV). When a voltage pulse is applied, holes are injected into the TFB layer and meet the high density of electrons already accumulated at the TFB/F8BT interface to recombine and give rise to light emission, thus a sudden spike can appear in the EL. As time passes, the accumulated electrons quickly run out and the flow of opposite charges becomes more balanced and thus produces more constant EL.

In the blend device, μm length-scale TFB-rich phases are dispersed in a F8BT-rich matrix. This F8BT-rich matrix may provide a reasonable pathway for the electron injection and transport, leading to accumulation of barrier-free injected electrons upon contact with Ca cathode. Once the voltage pulse is applied, the flow of holes into the active layer would cause a similar but smaller turn-on spike as that seen in the bilayer. The smaller turn-on spike in the blend device can be understood since freely injected electrons are distributed more evenly throughout the whole active layer, differently from the bilayer device in which high density accumulation of electrons occurs at the abrupt TFB/F8BT interface. As the length scale of the TFB/F8BT interfaces decreases in the copolymer, no accumulation of electrons is expected. In the RC device, where no continuous pathway for the electron transport is present due to randomly distributed TFB and F8BT interfaces at a molecular-scale, no turn-on spike is observed. Note that the absence of the turn-on spike in the RC does not depend on the voltage applied (Figure 10b). This observation agrees well with the single-carrier device data, confirming a better charge balance in the RC which produces more stable characteristics of LEDs in terms of device efficiencies.

6.2 Intrinsic and extrinsic degradation mechanisms

Understanding the diversity of steady-state operational mechanisms focusing on the electrochemical reaction pathways of injected charges across the organic semiconductor interfaces is very important to improve device efficiency and stability. However, there has been little work to address intrinsic and extrinsic mechanisms governing the electrical stability of these devices primarily due to the challenge of tracking minute chemical reactions *in-situ* in the 100-nm-thick buried active layers of polymer LEDs during operation. Work with micro-Raman spectroscopy has given very encouraging results in this field (Winfield et al., 2010; Ballantyne et al., 2010), in particular for degradation of the active layer of PLED devices (Kim et al., 2002; [b]Kim et al., 2004).

We have also monitored in-situ changes to the chemistry of the polymers and their interfaces in the devices using non-destructive Raman spectroscopy. Fig. 11(a) shows an optical micrograph of an encapsulated polymer LED. It shows a 2-μm-wide pinhole defect surrounded by a diffusion disk on the ITO area. The pinhole and associated disk (which appears black under EL and therefore termed "black spot") are marked by an optically distinct boundary. Fig. 11(b) shows the Raman spectra taken across this black spot. Our results confirm that the black spots are associated with cathode pinhole defects and caused by electrochemical activity between the cathode and hole-transport conducting polymer PEDT:PSS layer. We have also performed in-situ Raman measurements in hole-only poly(9,9-dioctylfluorene-co-bis-N,N-(4-butylphenyl)-bis-N,N-phenyl-1,4-phenylenediamine) (PFB) diodes to study a hole mediated interfacial reaction pathway during electrical operation. The results suggest that a primary reaction pathway of the electrical properties of these diodes is the slow quasi-irreversible interfacial electrochemical oxidation (doping) of PFB adjacent to the PEDT:PSS layer (Figure 11c). Understanding such failure mechanisms under device operational conditions leads to the development of materials and devices that are intrinsically more resistant to degradation.

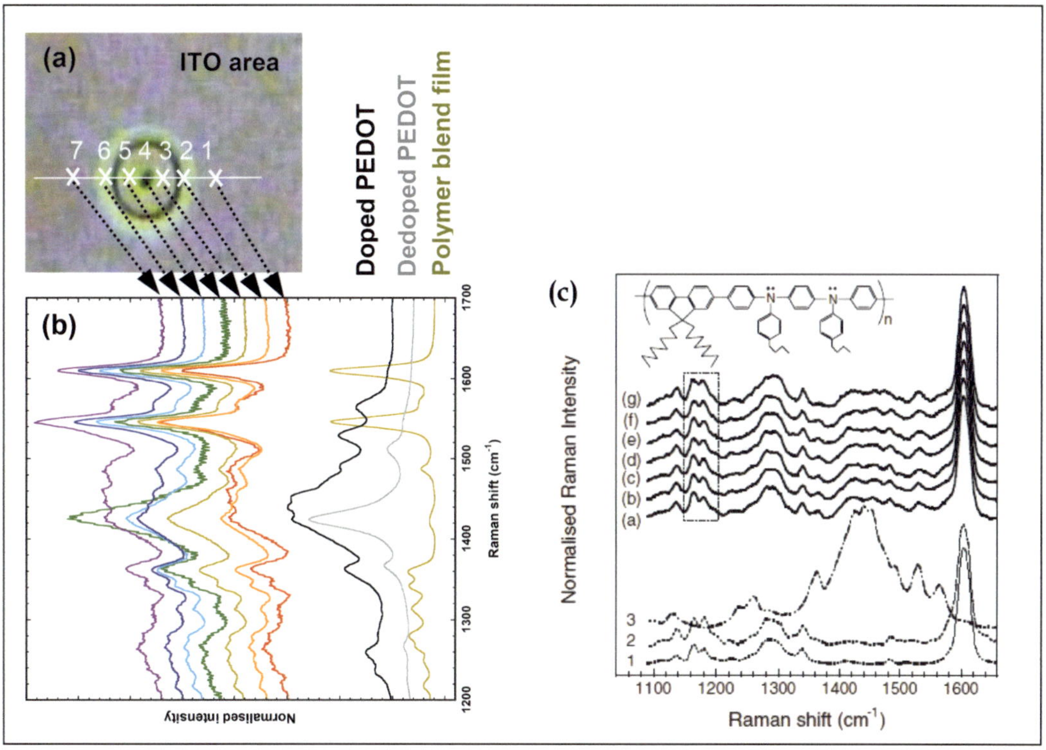

Fig. 11. (a) Enlarged optical micrograph of the black spot (ITO area), (b) Raman spectra taken across this black spot along the dotted line. All the spectra were normalised to the 1609 cm^{-1} peak of the EL polymer blend. Reference spectra for the EL polymer blend, doped-PEDOT:PSS and dedoped-PEDT:PSS, and (c) Raman spectra of the ITO/PEDT:PSS (60 nm)/PFB (100 nm)/Al (400 nm) diode after the J–V test at each point in Fig. 1A (excitation, 633 nm; objective NA, 0.7; spectra accumulation, 10_20 s). Reference spectra of (1) PFB, (2) electrochemically oxidised PFB, and (3) doped PEDT:PSS thin films are also shown. All the spectra except PEDT:PSS are normalised to the PFB Raman intensity at 1602 cm_1. The chemical structure of PFB is also shown

6.3 In-situ identification of a luminescence quencher in OLEDs

In-situ Raman spectroscopy was also used to identify a luminescence quencher formed during OLED operation. Raman spectroscopy revealed that oxo-bridged dimerization occurs during the operation of $[Ru(bpy)_3]^{2+}(PF_6^-)_2$ devices, where bpy is 2,2′-bipyridine. Photoluminescence spectroscopy showed that oxo-bridged dimers such as $[Ru(bpy)_2(H_2O)]_2O^{4+}(PF_6^-)_4$ effectively quench photoluminescence. Comparison of the Raman spectra from devices with the spectra from prepared blended films of $[Ru(bpy)_3]^{2+}$ $(PF_6^-)_2$ and

$[Ru(bpy)_2(H_2O)]_2O^{4+}(PF_6^-)_4$ demonstrated that sufficient dimerization occurs in the device to account for the luminescence quenching observed upon device driving. Dimerization occurred particularly where oxygen and moisture could penetrate the organic film. Dimerization could be a general failure mode of organic electroluminescent devices that incorporate metal complexes. Understanding failure under device-relevant conditions can lead to the development of materials and devices that are intrinsically more resistant to degradation (Slinker et al., 2007; Soltzberg et al., 2006).

7. Conclusion

In this chapter, we have reviewed our recent work carried out to understand the role of different length-scale organic-organic interfaces with an important focus on molecular-scale electronic structures and electronic processes across these interfaces. We have fabricated various length-scale interchain and intrachain organic interfaces using polymer blends with different molecular weight homopolymers and copolymers. We have also used surface patterning and transfer printing techniques to control length-scale of the organic interfaces. At these various interfaces, we have studied in depth the photophysical processes and dynamics of electronic species including intrachain charge transfer states and the transport and recombination of charge carriers. The distinctive optoelectronic and charge transport properties have been observed across different organic-organic interfaces depending on their length-scale (micron-scale in the blends down to molecular-scale in the copolymers) and nature (interchain vs intrachain), providing the fundamental understanding of these interfaces and their vital roles in various optoelectronic devices. Furthermore, based on the studies of charge carrier operational dynamics at these organic interfaces, we have been able to identify failure mechanisms of organic devices including intrinsic and extrinsic degradation mechanisms. The key advances in organic-organic semiconductor interfaces achieved so far will provide important insight into a design rule of organic semiconductors which is essential for future development in molecular electronic devices.

8. Acknowledgements

We thank Dr K. H. Yim, Dr A. Petrozza, Dr C. Donley, Dr. J. Winfield, Prof. H. Sirringhaus , Prof. Sir R. H. Friend, Dr. Z. Zheng, Prof. W. T. S. Huck, Prof. W. R. Salaneck, Dr. J. Cornil, Dr. D. Beljonne, Prof. G. G. Malliaras, Dr J. H. Burroughes for collaborations, and Cambridge Display Technology (CDT) Ltd. for materials supply. We thank the EPSRC-NPL Post-Doctoral Research Partnership Grant (EP/G062056/1) and the World Class University (WCU) Program through the National Research Foundation of Korea funded by the Ministry of Education, Science and Technology (Grant No. R32-10051).

9. References

Ballantyne, A.; Ferenczi, T.; Campoy-Quiles, M.; Clarke, T.; Maurano, A.; Wong, K.; Stingelin, N.; Kim, J. S.; Bradley, D.; Durrant, J.; McCulloch, I.; Zhang, W.; Heeney, M.; Nelson, J.; Mueller, C.; Smith, P.; Tierney, S.; Duffy, W. (2010). Towards Understanding The Influence of Morphology on Poly(3-hexylselenothiophene): PCBM Solar Cells. *Macromolecules*, 43, 1169-1174

Donley, C. L., Zaumseil, J., Andreasen, J. W., Nielsen, M. M., Sirringhaus, H., Friend, R. H. and Kim, J. S. (2005). Effects of Packing Structure on the Optoelectronic and Charge Transport Properties in Poly (9,9-Dioctylfluorene-co-Benzothiadiazole). *J. Am. Chem. Soc.*, 127, 12890-12899

[a]Kim, J. S., Ho, P. K. H., Murphy, C. E. and Friend, R. H. (2004). Phase separation in polyfluorene-based conjugated polymer blends: Lateral and vertical compositional analysis of the blend thin films. *Macromolecules*, 37, 2861-2871

[b]Kim, J. S., Ho, P. K. H., Murphy, C. E., Seeley, A. J. A. B., Grizzi, I., Burroughes, J. H. and Friend, R. H. (2004). Electrical degradation of triarylamine-based LEDs monitored by micro-Raman spectroscopy. *Chem. Phys. Lett.*, 386, 2-7

Kim, J. S., Grizzi, I., Burroughes, J. H., and Friend, R. H. (2005). Spin-cast thin semiconducting polymer interlayer for improving device efficiency of polymer light-emitting diodes. *Appl. Phys. Lett.*, 87, 023506

Kim, J. S., Lu, L., Sreearunothai, P., Seeley, A., Yim, K. H., Petrozza, A., Murphy, C. E., Beljonne, D., Cornil, J., Friend, R. H. (2008). Optoelectronic and Charge Transport Properties at Organic/Organic Semiconductor Interfaces: Comparison Between Polyfluorene-Based Polymer Blend and Copolymer. *J. Am. Chem. Soc.*, 130, 13120-13131

Morteani, A. C., Dhoot, A. S., Kim, J. S., Silva, C., Greenham, N. C., Friend, R. H., Murphy, C., Moons, E., Cina, S., Burroughes, J. (2003). Barrier-free electron-hole capture in polymer blend heterojunction light-emitting diodes. *Adv. Mater.* 15, 1708-1712

Petrozza, A., Laquai, F., Kim, J. S. and Friend, R. H. (2010). Dielectric Switching of the Nature of Excited Singlet State in a Donor-Acceptor-type Polyfluorene Copolymer. *Phys. Rev. B*, 81, 205421

Schmid, S. A., Yim, K.H., Chang, M. H., Kim, J. S., Friend, R. H. and Herz, L. M. (2008). Polarization anisotropy dynamics for thin films of a conjugated polymer aligned by nanoimprinting. *Phys. Rev. B*, 77, 115338

Schmidtke, J. P., Kim, J. S., Gierschner, J., Silva, C. and Friend, R. H. (2007). Optical spectroscopy of a polyfluorene copolymer at high pressure: intra- and inter-molecular interactions. *Phys. Rev. Lett.*, 99, 167401

Seeley, A. J. A. B., Friend, R. H., Burroughes, J. H. and Kim, J. S. (2004). Trap-assisted hole injection and quantum efficiency enhancement in poly(9,9′ dioctylfluorene-co-benzothiadia zole) polymer light-emitting diodes. *J. Appl. Phys.*, 96, 7643-7649

Slinker, J. D., Kim, J. S., Flores-Torres, S., Delcamp, J. H., Abruna, H. D., Friend, R. H. and Malliaras, G. G. (2007). In situ identification of a luminescence quencher in an organic light-emitting device. *J. Mater. Chem.*, 17, 76-81

Soltzberg, L.J., Slinker, J.D., Flores-Torres, S., Bernards, D.A., Malliaras, G.G., Abruna, H.D., Kim, J.S., Friend, R.H., Kaplan, M.D., Goldberg, V. (2006). Identification of a Quenching Species in Ruthenium Tris-Bipyridine Electroluminescent Devices. *J. Am. Chem. Soc.* 128, 7761-7764

Van Vooren, A., Kim, J. S., Cornil, J. (2008). Intrachain versus Interchain Electron Transport in Poly(fluorine-*alt*-benzothiadiazole): Quantum-Chemical Insight. *ChemPhysChem*, 9, 908-993

Winfield, J. M., Van Vooren, A., Park, M. J., Hwang, D. H., Cornil, J., Kim, J. S., and Friend, R. H. (2009). Charge-transfer character of excitons in poly[2,7-(9,9-di-n-octylfluorene)$_{(1-x)}$-co-4,7-(2,1,3- benzothiadiazole)$_{(x)}$]. *J Chem. Phys.*, 131, 035104

Winfield, J. M., Donley, C. L., Friend, R. H. and Kim, J. S. (2010). Probing thin-film morphology of conjugated polymers by Raman spectroscopy *J. Appl. Phys.*, 107, ISSN: 0021-8979

[a]Yim, K. H., Zheng, Z., Friend, R. H., Huck, W. T. S., and Kim, J. S. (2008). Surface-directed phase separation of conjugated polymer blends for efficient light-emitting diodes. *Adv. Funct. Mater.* 18, 2897

[b]Yim, K. H., Zheng, Z., Liang, Z., Friend, R. H., Huck, W. T. S. and Kim, J. S. (2008). Efficient Conjugated Polymer Optoelectronic Devices Fabricated by Thin Film Transfer Printing. *Adv. Funct. Mater.*, 18, 1012-1019

[c]Yim, K. H., Whiting, G., Murphy, C.E., Halls, J. J.M., Burroughes, J. H., Friend, R. H., and Kim, J. S. (2008). Controlling electrical properties of conjugated polymers via a solution-based p-type doping. *Adv. Mater.*, 20, 3319-3324

Yim, K. H., Doherty, W. J., Salaneck, W. R., Murphy, C. E., Friend, R. H. and Kim, J. S. (2010) Controlling Phase Separation for Efficient Polymer Blend Light-Emitting Diodes, *Nano Lett.*, 10, 385

Zaumseil, J., Donley, C. L., Kim, J. S., Friend R. H. and Sirringhaus, H. (2006). Efficient Top-Gate, Ambipolar, Light-Emitting Field-Effect Transistors Based on a Green-Light-Emitting Polyfluorene. *Adv. Mater.*, 18, (2006), 2708 – 2712

Zheng, Z., Yim, K. H., Saifullah, M. S. M., Welland, M. E., Friend, R. H., Kim, J. S. and Huck, W. T. S. (2007). Uniaxial Alignment of Liquid-Crystalline Conjugated Polymers by Nanoconfinement. *Nano Lett.*, 7, 987-992

2

A Study of Adhesion of Silicon Dioxide on Polymeric Substrates for Optoelectronic Applications

E. Amendola[1,2], A. Cammarano[2] and D. Acierno[3]
[1]*Institute of Composite and Biomedical Materials, National Research Council, Piazzale E. Fermi 1, 80055 Portici (NA),*
[2]*Technological District of Polymer and Composite Materials Engineering and Structures, IMAST S.c.a.r.l, Piazzale E.Fermi 1, 80055 Portici (NA),*
[3]*Department of Materials and Production Engineering, University of Naples "Federico II", Piazzale Tecchio 80, 80125 Naples*
[1,2,3]*Italy*

1. Introduction

The use of plastic film substrates for organic electronic devices promises to enable new applications.
Plastic substrates have several advantages, such as ruggedness, robustness, ultra lightness, conformability and impact resistance over glass substrates, which are primarily used in flat panel displays (FPDs) today (Imparato et al., 2005). However, high transparency, proper surface roughness, low gas permeability and highly transparent electrode conductivity of the plastic substrate are required for commercial applications (Choi et al., 2008) (Mannifacier et al., 1979) (Adhikari & Majumdar, 2004).
Polyesters, both amorphous and semicrystalline, are a promising class of commercial polymers for optoelectronic applications.
Despite the best premises, the adoption of polymers for electronic applications has been slowed by their limited compatibility with semiconductor fabrication processes, at least during the first stage of the transition towards all-polymeric functional devices. In particular, the relatively high linear expansion coefficient, α, and low glass transition temperature, Tg, of most polymers limit their use to temperatures above 250°C. Therefore, the high-temperature process leads to considerable mechanical stress and difficulties in maintaining accurate alignment of features on the plastic substrate.
The availability of suitable polymeric functional materials, with reliable and durable performances, will eventually results in development of fully polymeric devices, with milder processing requirements in term of high temperature exposure.
At the present stage, inorganic materials are used as buffer, conductive and protective layers for functional organics and high performance polymer substrates.
Several high-Tg polymers (Tg >220°C) with optical transparency, good chemical resistance and barrier properties have recently been developed for applications in organic display technology, and these latest developments have motivated the present research.

Ferrania Imaging Technologies, has developed amorphous polyester material, AryLite™, with high glass transition temperature (Tg ≈ 320°C) and good optical transparency (Angiolini & Avidano, 2001).
Substrates for flexible organic electronic devices are multilayer composite structures comprising a polymer-based substrate on which are deposited a number of functional coatings, with specific roles:

- chemical protection from the hostile environment during processing;
- mechanical protection, such as improvement of the scratch resistance;
- a diffusion (or permeation) barrier. A polymer based permeation barrier may be sufficient for protection during, for instance, processing during display manufacturing;
- electrical connections.

Taking into account that for a number of these functions transparent coatings are required, silicon dioxide (SiO_2) layer has been deposited on AryLite™ substrate at temperatures below 50°C in an Electron Cyclotron Resonance (ECR) plasma reactor from H_2, SiH_4, and N_2O gas mixture. Silicon dioxide possesses excellent physical and chemical properties, such as transparency from ultraviolet to infrared, good thermal stability, chemical inertness, wear and corrosion resistance and low gas permeation.
In a multilayer structure, the adhesion between organic/inorganic layer plays an important role in determining the reliability of the optoelectronic devices.
As a matter of fact, the effort is focused on the improvement of adhesion between organic-inorganic materials, and the use of nanocomposite (hybrid) substrates (Amendola et al., 2009).
Adhesion properties can be varied by modifying the surface, by means of several chemical and/or physical processes (Goddard & Hotchkiss, 2007).
The most common techniques include plasma-ion beam treatment, electric discharge, surface grafting, chemical reaction, metal vapour deposition, flame treatment, and chemical oxidation. In this way it's possible to change hydrophobic polymer surface into a hydrophilic one without affecting the bulk properties.
Adhesion can be improved also by using an adhesion promoter such as a silane on the polymer surface. In this work the surface of polyester films was modified via chemical solution. Afterward, samples have been treated with (3-Aminopropyl)triethoxysilane (APTEOS) that function as an adhesion promoter between organic substrate and SiO_2 layer. In particular, SiOH silane functional groups are suitable for coupling with SiO_2 layer.
Contact angle and roughness measurements, surface free energy calculation and attenuated total reflectance Fourier transform infrared spectroscopy (ATR-FTIR) were used to monitor the effects of silane treatments on the physical and chemical characteristics of pristine and modified polyester surfaces. Infrared spectroscopic analysis has been performed in order to study the reaction between amino group present on the organosilane backbone and carbonilic group of polyester substrate.
Conventional characterization techniques are not appropriate for the measurement of mechanical and adhesion properties of thin functional layers on substrate. Nano-indentation and nano-scratch testing are alternative approaching methods. Both techniques have become important tools for probing the mechanical properties of small volumes of material at the nano-scale.
Indentation measurements has been used to evaluate the hardness and Young's modulus of films. The film adhesion was determined by the nano-scratch test.

2. Materials

AryLite™ (supplied by Ferrania Imaging Technologies S.p.A.) characterised by very high glass transition temperature, has been selected due to its outstanding thermo-mechanical and optical properties. Polymer films of 10 cm x 10 cm and of 100 μm in thickness have been used.

Silicon dioxide (SiO_2) layers were deposited at temperatures below 50 °C in an electron cyclotron resonance (ECR) plasma reactor from N_2O, SiH_4, and H_2 gas mixture.

Coupling agent with amino functional group (3-Aminopropyl)triethoxysilane (APTEOS) has been supplied by Aldrich and used without further purification.

3. Method

3.1 Thermo-Mechanical properties of substrates

Thermal properties of substrates under investigation have been evaluated in order to determine glass transition temperature Tg and degradation temperature by differential scanning calorimetry (DSC) and thermogravimetric analysis (TGA) respectively.

The glass transition (Tg) was investigated by DSC-Q1000 (TA Instruments). The DSC thermal analysis technique measures heat flows and phase changes on a sample under thermal cycles. Since the Tg of AryLite™ is overlaid by an enthalpic relaxation phenomenon, deeper investigations were performed with Modulated DSC (MDSC).

Enthalpic relaxation is an endothermic process that can vary in magnitude depending on the thermal history of the material. Traditional DSC measures the sum of all thermal events in the sample. When multiple transitions occur in the same temperature range, results are often confusing and misinterpreted. MDSC eliminates this problem by separating the total heat flow signal into two separated contribution, namely "Reversing" and "Non Reversing". The reversing signal provides information on heat capacity and melting, while the non reversing signal shows the kinetic process of enthalpic recovery and cold crystallization.

In MDSC analysis, the samples were heated from 150 °C to 400 °C, at heating rate of 2.5 °C/min, with a modulated temperature amplitude of 0.5 °C and a period of 60 sec under a nitrogen flow.

The degradation temperature and thermal stability were investigated by thermogravimetric analysis TGA-Q5000 (TA Instruments). The weight loss due to the formation of volatile products caused by the degradation at high temperature was monitored as a function of temperature. The heating occurred both under a nitrogen and oxygen flow, from room temperature up to 900°C with a heating rate of 10 °C/min.

Elastic modulus and ultimate properties were investigated according to UNI EN ISO 527-3 on rectangular specimens with 150 mm length, 25 mm width and 0.1 mm thick using a mechanical dynamometer SANS 4023 with a 30 kN loading cell and a traverse speed of 20mm/min.

3.2 Surface treatments

3.2.1 Surface modification by coupling reactions

Polymer films were preliminary immersed in an alcohol/water (1/1, v/v) solution for 2 h in order to clean the surface and then rinsed with a large amount of distilled water. They were dried under reduced pressure for 12 h at 25 °C.

AryLite™ samples have been functionalized with (3-Aminopropyl)triethoxysilane. Untreated AryLite™ samples have been used as substrate for the sake of comparison.

Prior to AryLite™ surface treatments, the SiOR groups of the silane were transformed to active SiOH groups for the subsequent condensation reactions. The transformation is realized by hydrolyzing the silane in a aqueous solution. 7.5 wt % silane solution were prepared by adding the silane to a mixture of 70:30 ethanol and distilled water. The pH of the solution was adjusted to 5.5 by inclusion of a few droplet of acetic acid. The solution was stirred for 10 minutes and the system was kept 1 h at room temperature for hydrolysis reaction and silanol formation. Subsequently the films were dipped into the solution for 30 minutes at room temperature.
These silane-treated specimens were rinsed with distilled water to eliminate the unreacted silane and dried under reduced pressure at 25°C overnight.
Reaction path is reported in figure 1.
The reaction proceeds through a nucleophilic attack of NH_2 nitrogen atom to the carbon atom of carbonilic group generating an amide group.

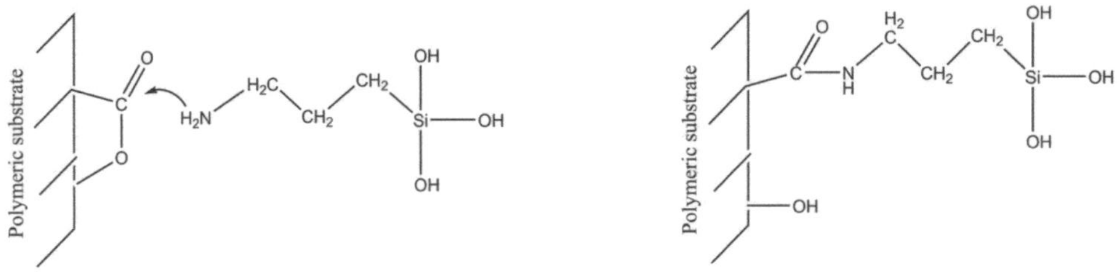

Fig. 1. Scheme of nucleophilic addition of NH_2 group to the polyester carbonilic group

3.2.2 Electron Cyclotron Resonance (ECR) deposition

The deposition process was performed by ENEA Portici research centre (Naples) using Multichamber System MC5000, a Ultra High Vacuum Multichamber for Plasma Enhanced Chemical Vapour Deposition.
Thousand nm thick SiO_2 layer was deposited by Electron Cyclotron Resonance (ECR) on a single face of AryLite™ substrate. During deposition process gas flows are kept constant at 2 sccm (standard cubic centimeters per minute) for SiH_4, 70 sccm for H_2 and 40 sccm for N_2O.
Deposition was performed for 13 minutes setting magnetron power to 400 W. Samples were heated at 50°C under hydrogen flow for 5 minutes before SiO_2 deposition. Films were purged under nitrogen flow for 5 minutes at the end of the treatment.

3.3 Spectroscopic analysis FTIR-ATR

Infrared spectroscopic analysis has been performed by Nicolet Nexus 670 FTIR equipped with attenuated total reflection (ATR) smart ARK HATR accessory.
In ATR, the sample is placed in optical contact on a zinc selenide (ZnSe) crystal. The IR beam penetrate a short distance into the sample. This penetration is termed the evanescent wave. The sample interacts with the evanescent wave, resulting in the absorption of radiation by the sample, which closely resembles the transmission spectrum for the same sample. However, the ATR spectrum will depend upon several parameters, including the angle of incidence (θ) for the incoming radiation, the wavelength of the radiation (λ), and the refractive indices of the sample (n_2) and the ATR crystal (n_1). The penetration depth (d_p) of the evanescent wave, is defined by equation 1.

$$d_p = \frac{\lambda}{2\pi(n_1^2 sen^2\theta - n_2^2)^{1/2}} \tag{1}$$

In the 400 - 4800cm^{-1} wavenumber investigated range, d_p varies from 5 µm to 15 µm for measured substrates (Zuwei et al., 2007).

A spectroscopic investigation has been performed, also, by using a transmitted infrared analysis to verify the kind of chemical reaction that occurs between polymer substrate and organosilane.

The sample for FTIR analysis has been prepared by adding the amminosilane in a polyarilate solution in dichloromethane solvent. In this way After treatment, films have been obtained by solvent casting technique. They have been heated at 100 °C for 1 h in order to remove the whole solvent. Treated films finely divided, were ground and dispersed in a matrix of KBr (300 mg), followed by compression at 700 MPa to consolidate the formation of the pellet for FTIR measurements.

All spectra were recorded in the range of 4000–800 cm^{-1}.

3.4 Contact angle measurements

Contact angle measurements have been used to verify chemical surface modification. The surface wettability was evaluated by contact angle measurements using the sessile drop method (Mack, 1936) considering the shape of the small liquid drop to be a truncated sphere. Prior to contact angle measurement, samples were washed in ethanol and deionised water.

Contact angles were obtained using a Dataphysics OCA-20 contact angle analyzer with 1 µL of liquid. A digital drop image has been processed by an image analysis system, which calculated both the left and right contact angles from the shape of the drop with an accuracy of ±0.1°. Drop contact angle were used to assess efficiency of surface modification suffered by the polymer films. Each solid sample was measured ten times with liquid at room temperature. The contact angle data were obtained at room temperature using two different liquids: water and ethylene glycol (Ozcan & Hasirci, 2008). The contact angle is a method for evaluation of the solid surface free energy (SFE) (van Oss et al., 1988).

3.5 Evaluation of surface free energy (SFE)

The values of SFE were obtained at room temperature using two liquids (Cantin et al., 2006) with known surface tension (table 1). Two liquids with different polarity (P) have been selected: water and ethylene glycol. The liquids, supplied by Aldrich, were used without further purification.

Surface free energy of the polymer substrates was calculated using the methods proposed by Owens and Wendt (1969) which divide the total surface free energy (γ) in 2 parts: dispersive (γ_s^d) and polar (γ_s^p). The dispersive component accounts for all the London forces such as dispersion (London-van der Waals), orientation (Keesom-van der Waals), induction (Debye-van der Waals) and Lifshitz-van der Waals (LW) forces. The polar component is affected by hydrogen bonding components.

The theory of contact angle of pure liquids on a solid was developed nearly 200 years ago in terms of the Young equation (1805):

$$\gamma_L \cos\theta = \gamma_S + \gamma_{SL} \tag{2}$$

	Surface tension [mJ/m²] and Polarity			
	γ	$\gamma_L{}^d$	$\gamma_L{}^p$	P
Distilled Water	72.8	21.8	51	0.70
Ethylene Glycol	48	29	19	0.40

Table 1. Surface tension data of test liquids (Ozcan & Hasirci, 2008)

where γ_L is the experimentally determined surface tension of the liquid, θ is the contact angle, γ_S is the surface free energy of the solid and γ_{SL} is the solid-liquid interfacial energy. In order to obtain the solid surface free energy γ_S an estimate of γ_{SL} has to be done. Fowkes (1962) pioneered a surface free energy component approach. He divided the total surface free energy in 2 parts: dispersive part and non-dispersive (or polar) part. Owens and Wendt (1969) extended the Fowkes equation and included the hydrogen bonding term. They used geometric mean to combine the dispersion force and hydrogen bonding components:

$$\gamma_{SL} = \gamma_S + \gamma_L - 2\sqrt{\gamma_S^d \gamma_L^d} - 2\sqrt{\gamma_S^p \gamma_L^p} \tag{3}$$

Dispersion force and polar components are indicated respectively by superscript *d* and *p*. From the Young equation it follows that:

$$\gamma_L(1+\cos\theta) = 2\sqrt{\gamma_S^d \gamma_L^d} + 2\sqrt{\gamma_S^p \gamma_L^p} \tag{4}$$

In order to obtain $\gamma_S{}^d$ and $\gamma_S{}^p$ of a solid, contact angle data for a minimum of two known liquids are required. If two liquids are used, then, one must be polar and the other is non-polar.

3.6 Topography measurements

Topography measurements were performed in this studies in order to investigate the roughness of polyester films before and after surface treatment. The apparatus used in this work was NanoTest™ Platform by Micro Materials Ltd. This instrument monitors and records the load and displacement of a diamond three-sided pyramidal indenter tip with a radius of curvature of about 100 nm. A constant load of 10 μN has been applied. Scans (200 μm) were collected with the tip in close proximity to the surface, but not in contact. Nanoindenter is able to achieve sub-nanometric depth resolution in the horizontal plane. This resolution allows for the detection of changes to topography and providing valuable information on contribution of surface roughness to adhesion strength.

Two specimens of each film were randomly selected for recordings 20 measurements per sample. Average surface root means squared roughness (R_{RMS}) was calculated from equation 5 (Faibish et al., 2002).

$$R_{RMS} = \sqrt{\frac{\int_0^L (x_n - \bar{x})^2 \, dy}{L}} \tag{5}$$

Where x_n is the height of a random location on the scanned profile, x is the mean height of all measured heights and L is the sampling scan length.

3.7 Nano-indentation

Nano-indentation and nano-scratch techniques have been used in order to investigate the adhesion between organic substrate and inorganic layer.

Hardness and elastic modulus are calculated from the load vs. displacement data obtained by nano-indentation on coating at twenty different indentation depths ranging from 20m to 300 nm. NanoTest™ Platform, already described in section 3.6, monitors and records the dynamic load of a three-sided pyramidal diamond indenter. Berkovich tip with a radius of about 100 nm has been used.

In a indentation test, a Berkovich diamond tip is driven to indent the surface of the coating from the SiO_2 side. Tests were carried out in depth controlled mode, selecting a minimum and a maximum depth of 20 nm and 300 nm respectively. The experiments were performed with an initial load of 10 µN at loading and unloading rate of 10 µN/sec. The hold time of 30 sec at peak load was kept constant. Additional hold at 90% unload in all tests was set for thermal drift correction.

The indenter needs to be held at the constant indentation load for a certain amount of time in order to eliminate the dynamic effect and reach the quasi-steady flow state.

Each successive indent was displaced by 50 µm in order to avoid overlapping of plastic deformation zone onto neighbouring indents.

All data were corrected for thermal drift and instrument compliance and subsequently analysed with the Oliver and Pharr method (1992).

During the course of the instrumented indentation process, a record of the depth of penetration is made, and then the area of the indent is determined using the known geometry of the indentation tip. Indenting parameters, such as load and depth of penetration, can be measured. A record of these values can be plotted on a graph to create a load-displacement curve.

Figure 2 shows a typical loading and unloading displacement curves during a nanoindentation cycle on AryLite™. A power law curve was fit to the unload data points.

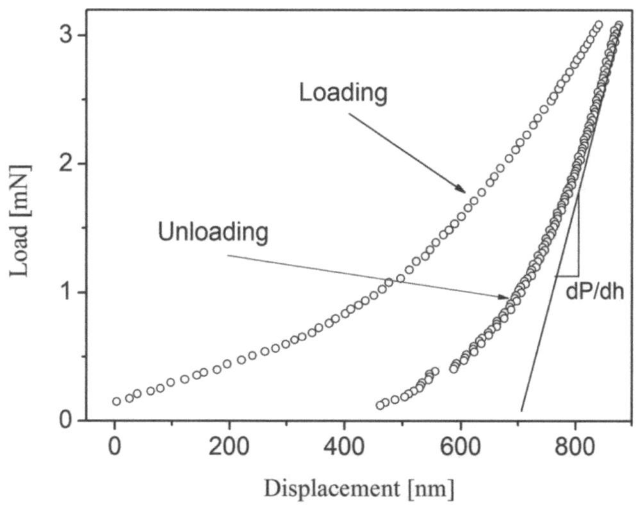

Fig. 2. Typical loading and unloading displacement curves during a nanoindentation cycle on AryLite™ substrate

The slope dP/dh at the maximum load data point is used to calculate the elastic modulus (Er). Hardness is calculated by dividing the loading force by the projected residual area of the indentation. Hardness and the Young's modulus of elasticity can be obtained from the slope of the unloading curve. The hysteresis indicates that the deformation is not fully elastic and partially inelastic.

3.8 Nano-Scratch test

In nano-scratch studies a conical indenter is drawn over the sample surface with ramping up of the load until damage occurs. The load corresponding to this damage provides a measure of scratch resistance or adhesive strength of a coating and is called the "critical load" (Park & Kwon, 1997). The definition of damage can be the onset of cracking around the scratching tip, spalling of the coating, or the formation of a channel in which all of the coating has been removed from the substrate. The critical loads are indicators of the scratch resistance of these samples.

Scratches have been made by translating the sample while ramping the loads on the conical diamond tip (1μm tip radius) over different load ranges from 0 mN to 10 mN. A sudden increase in the scratch load was related to coating damage.

Multipass test experiments have been performed. They consist of three sequential scans over the same 250 μm track, all at 2 μm/sec scan speed. In the first topography scan the applied load was constant at 100 μN. Surface roughness was measured from this scan. In the second scratch scan, the load applied after 50 μm was ramped at a constant rate of 0.1 mN/sec to a maximum load reached of 10 mN. In the final scan the resultant topography was observed by using a low applied load of 100 μN. Five repeat tests were performed on each sample.

4. Results and discussion

4.1 Thermo-Mechanical properties of AryLite™

AryLite™ exhibits excellent Tg (324°C) (figure 3) and good optical transparency. The lack of crystalline phase is a consequence of the aromatic and rigid nature of the polymer backbone which hinders conformational rearrangements into a regular crystalline structure. On the other hand the rigid backbone is responsible of reduced elongation at break and the lack of crystalline reinforcement results in poor elastic and ultimate properties (table 2).

The initial degradation temperatures (T_{id}) have been measured by using thermogravimetric analyser at a scanning rate of 10°C/min in N_2 purging flow. T_{id} is associated to 3 % weight loss.

Thermal and mechanical behavior of polyester indicate that AryLite™ is a good candidate for optoelectronic application.

Thermal and mechanical properties of AryLite™ are shown in table 2.

	AryLite™
Glass transition temperature T_g [°C]	324
Initial degradation temperature [°C]	488
Young's Modulus E_s [GPa]	2.82±0.24
Elongation at break ε_r [%]	10.24±2.56

Table 2. Thermo-mechanical properties of Arylite™

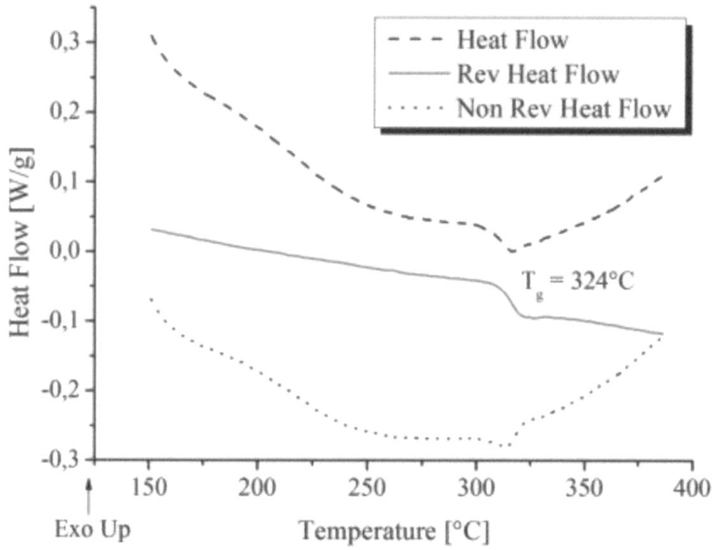

Fig. 3. Modulated Differential Scanning Calorimetry of AryLite™.

4.2 Spectroscopic analysis FTIR-ATR

Infrared spectroscopic analysis has been performed to verify the reaction between amino group located on the organo-silane and carboxylic group of polymeric substrate.

In figure 4 normalized ATR spectra of untreated AryLite™ and AryLite™ treated with APTEOS solution have been plotted.

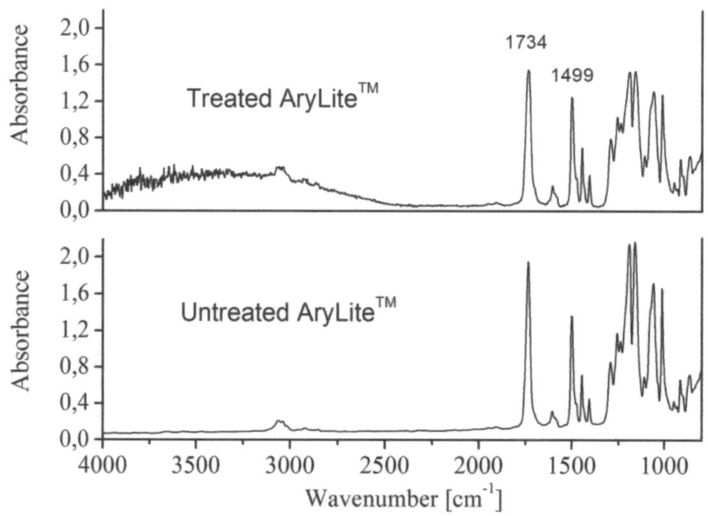

Fig. 4. ATR spectra of AryLite™ before and after surface silane treatment in the range 4000-700 cm^{-1}

In order to make a comparison between the treated and untreated samples the intensity of each peak was rationed (Gu et al., 2001) against the absorbance of the invariant band at 1499 cm^{-1}. These peaks are due to C-H in-plane bending of the benzene ring.

In the FTIR-ATR spectra, silane-treated specimens show a wide peak between about 2500 - 4000 cm^{-1}. This is attributable to the presence of SiOH groups (Anderson & Smith, 1974).

Treated sample spectra show a decrease of C=O ester linkage at 1734 cm^{-1} and a decrease of C-O ester stretching vibrations in the region 1300 - 1000 cm^{-1} after silane treatment (Bellamy, 1975) (Colthup et al., 1990).

This is attributable to nucleophilic addition of the amine group (NH_2) to the carbonilic group. Following this reaction path, the organosilane APTEOS has been grafted on the AryLite™ substrate.

In fact, a chemical reaction could happen between amine group (NH_2) of silane and carbonilic group of polymer substrate. The reaction produce an amide group (O=C-NH). This hypothesis has been supported by a spectroscopic study. Figure 5 shows double peak of amide group located at about 1640-1540 cm^{-1}.

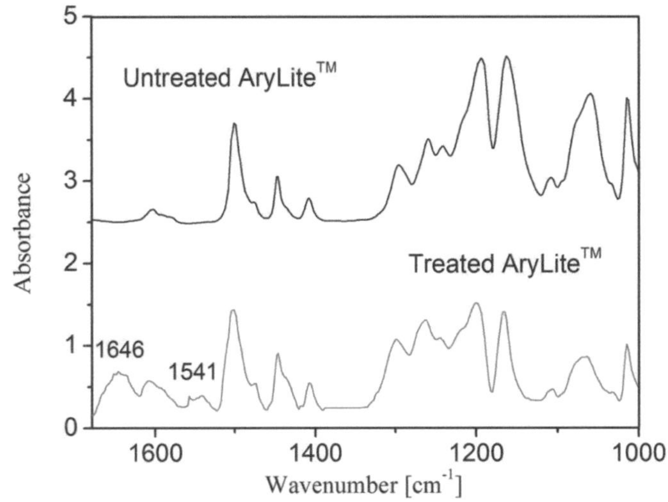

Fig. 5. FTIR spectra of polyarilate before and after silane treatment in a dichloromethane solution

4.3 Contact angle measurements

AryLite™ is an hydrophobic film with a water contact angle of 92 degrees. Samples treated with silane show a decrease of water contact angle of about 20 degrees. It has been demonstrated that silane treatment is effective in increasing the hydrophilicity of samples.

4.4 Evaluation of surface free energy (SFE)

In order to obtain γ_S^d and γ_S^p of a solid, contact angle data for a minimum of two known liquids are required. If two liquids are used, then, one must be polar and other is non-polar.

In this study two test liquids were used as a probe for surface free energy calculations: distilled water and ethylene glycol. The data for surface tension components of the test liquids at 20 °C are given in table 1.

Silane treatment binds OH groups on the polymeric surface. This phenomenon increase polyester surface polarity and surface free energy (Clint, 2001) (table 3). The increase in surface polarity causes an increase in molecular forces between substrates and hence an increase in adhesion strength (Burnett et al., 2007) (Comyn, 1992) (Lee & Wool, 2002).
In table 3 are reported SFE of untreated AryLite™ and treated AryLite™ calculated by using geometric mean method (Spelt et al., 1996).

	Surface Free Energy [mJ/m²]			
	γ	γ_S^d	γ_S^p	P
Untreated AryLite™	29.97	27.80	2.17	0.07
Treated AryLite™	30.19	17.10	13.09	0.43

Table 3. Surface free energy and polarity of AryLite™ films before and after surface treatment

4.5 Topography measurements

The topography scan technique was used to measure the roughness of samples before and after the silane treatments.
The changes of surface roughness produce a surface topography variation and provide a contribution to substrates adhesion strength.
Untreated AryLite™ films are smooth and they possess the lower roughness value of about 2.64±0.57 nm. Treated and untreated samples exhibit a root means squared roughness (R_{RMS}) lower than 20 nm. All materials can be considered as totally flat from a topographic point of view (Ponsonnet et al., 2003).
Clearly, surface roughness of the treated polymers was higher than the untreated polyester. After the chemical treatment AryLite™, exhibit a R_{RMS} value 2.7 times higher than unmodified material.
As expected, the silane treatment increase the surface roughness of polyester film.

4.6 Nano-indentation test

Nano-mechanical tests on SiO_2 layers deposited on AryLite™ have been carried out to further investigate the interface between organic substrate and inorganic layers.
Since coating of SiO_2 is less than 1 μm thick, small volume testing methods such as depth-sensing nano-indentation and nano-scratch were used as a indicator of film adhesion strength. Nano-indentation tests has been conducted with Berkovich tip, on 100 nm thick SiO_2 side deposited on AryLite™ substrates. Results are shown in figure 6.
The presence of rigid hard coating on soft substrate results in an depth sensitive properties. The variation of young's modulus (E) and hardness (H) with depth were plotted in figure 6(a) and 6(b).
As reported in figure 6(a) Young's modulus aren't greatly affected by compatibilizing treatment and the elastic modulus decrease with increase depth. Both polymer substrates exhibit the same elastic properties approximately 3.5 GPa for depth higher than 300 nm. Beyond this limit the elastic modulus is not influenced from the presence of the SiO_2 coating. This value is slightly higher than 2.9 GPa calculated by tensile test and reported in literature (Abdallah et al., 2008). According to Zheng et al. (2005) elastic modulus measured by depth sensing indentation are invariantly higher than tensile test by a value of 5-20%. The

difference between uniaxial and indentation results is probably attributable to the data reduction procedure used in analyzing the indentation data. If the polymer was creeping while unloading, this would tend to increase the slope of the unloading curve and hence the calculated elastic modulus.

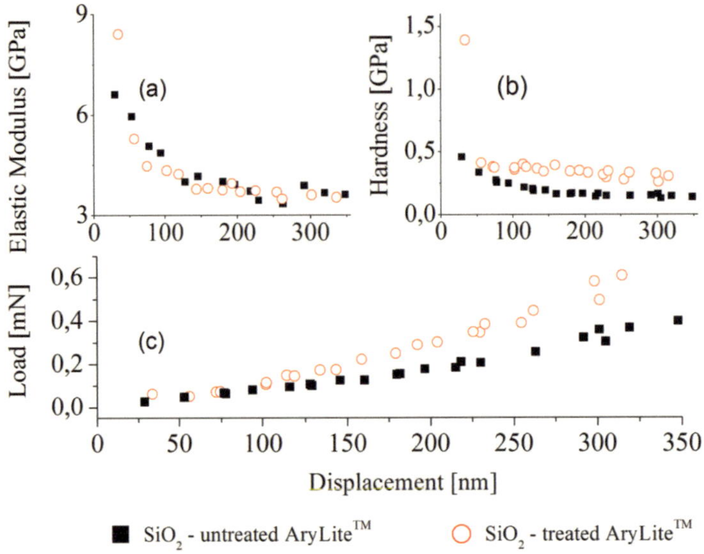

Fig. 6. Young's modulus (a), hardness (b) and load vs. depth (c) of untreated and treated AryLite™ coated with SiO_2 layer

The load applied to obtain depth higher than 100 nm was much greater for treated samples than for untreated substrate (Malzbender et al., 2002) (figure 6(c).

4.7 Nano-scratch test and Work of Adhesion

The polymer surface of AryLite™ has been modified to improve the adhesion between organic-inorganic layer and to reduce the inorganic layer cracking surface (Park & Jin, 2001) (Lian et al., 1995).

Depth of scratches with increasing normal load were measured in situ by topography scan of the film before and after the scratch event. Length of the test has been 250 µm. The load for initial and post scratch scan was 100 µN. In order to make effective use of the displacement data for production of a profile, it was assumed that region associated with the pre-scratch scan was unaffected by the deformation. Data from these regions were used to account for both the slope and curvature of the sample surface so that the entire scratch could be viewed with the surface of the sample as the baseline for deformation.

Five scratches were made at each load at different areas of specimen.

The load corresponding to the damage provides a measure of scratch resistance or adhesive strength of a coating and is called "critical load" (Lc) (Beake & Lau, 2005) (Beake et al., 2006) (Rats et al., 1999) (Zheng & Ashcroft, 2005) (Charitidis et al., 2000). Untreated AryLite™ exhibit a Lc value of 3.4±0.2 mN (figure 7); a 1 µm conical indenter has been used resulting in evident and reproducible data. The failure begins abruptly by brittle fragmentation and spallation in the coating. Spalling coating failure modes (Burnett & Rickerby, 1987) occurs as a result of the compressive stress field preceding the moving tip. Spallation is the result of total delamination and adhesive failure.

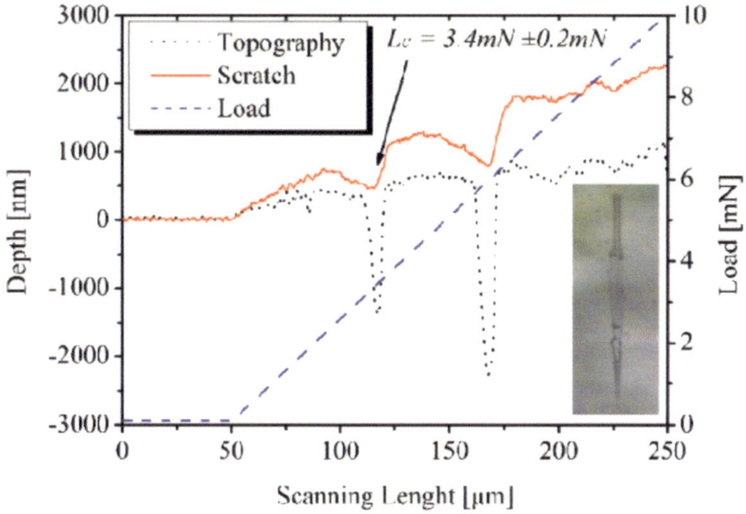

Fig. 7. Scratch test curve of SiO_2 layer of untreated AryLite™ with optical image in the plot inset (magnification 20x)

The scratch curve test of SiO_2 coating reveal that no inflexion can be found in depth curves of silane-treated samples which means that the films cannot be delaminated in the scratch process. Indeed, critical load disappeared after surface modification treatment, there is no abrupt change in the displacement curves, implying that the coatings did not peel off during the scratch ramping load (figure 8).

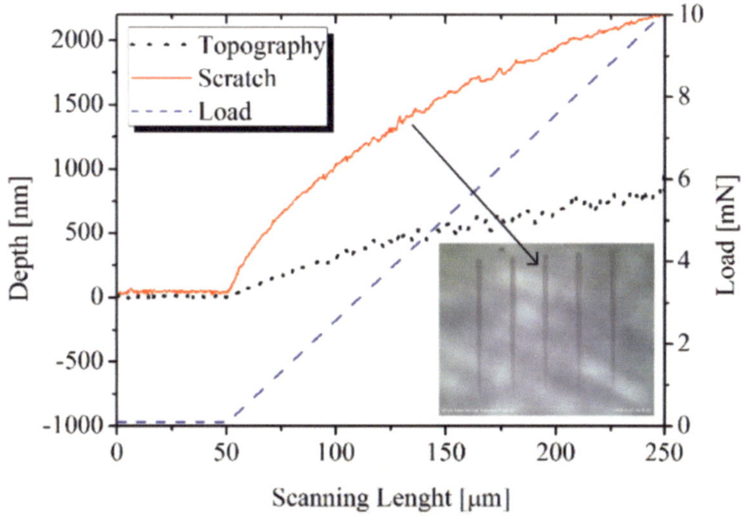

Fig. 8. Scratch test curve of SiO_2 layer of treated AryLite™ with optical image in the plot inset (magnification 20x). Topography and scratch scan refers to the third wear

A careful observation by optical microscopy has been made by comparing the topography pre-scan and post-scan to the scratch-scan for silane-treated and coated sample. The optical images reveal that the scratch is extremely smooth and shallow.

Similar observations performed at higher load confirm the same failure mechanisms. The scratch surface remains smooth until the wear exposes the polymer substrate. This is a conformal crack coating failure mode (Burnett & Rickerby, 1987)that consist of cracking within the scratch only and it occur when the coating remains fully adherent.

Evaluation of the adhesion strength between the coating and the substrate is complicated since it depends on a combination of many factors. In the first approximation, the adhesion can be modeled in terms of the strain energy released during fracture of the coating.

Griffith (1920) cast the problem of fracture in terms of energy balance. Griffith's study was based on the idea that all materials contain imperfections on a very small scale. Griffith's idea was to model a static crack as a reversible thermo-dynamical system. Equilibrium (i.e. no crack extension or contraction) is attained when over an infinitesimally small increase in crack length, dC, there is no overall change in energy of the system (U).

This can be expressed by Griffith energy balance concept:

$$\frac{dU}{dC} = 0 \tag{6}$$

Using a Griffith energy balance approach, the strain energy released provides the surface energy for a crack to form at the coating-substrate interface. The stress responsible for interface failure is related to the work of adhesion, W. Burnett and Rickerby (1987) have identified three contributions to the stress responsible for coating detachment: elastic-plastic indentation stress, internal stress and tangential frictional stress. In this model, the elastic-plastic indentation stress is considered to be dominant; the shear (frictional) stress is small compared with the ploughing (indentation) stress, they established the following relation:

$$W = \frac{32}{\pi^2} \frac{t \cdot L_c^2}{E_c \cdot w_c^4} \tag{7}$$

where w_c represents the width of the scratch track at the critical load, t and E_c is respectively the thickness and elastic modulus of inorganic coating.

In the case of a coating on polymer substrate, W increases with the film thickness. For unmodified and coated sample it has been found that W is about 0.18 J/m^2.

The proposed approach is not useful for the evaluation of W in the case of very strong interface.

The absence of critical failure for modified and coated sample has not allowed the experimental evaluation of Lc value, thus preventing the use of equation 7.

This is clearly evidenced by figure 7, where a scratch depth of 2000 nm have been reached without spallation.

5. Conclusion

In this paper polymer surface has been modified by chemical treatment in order to improve the adhesion properties. Coupling agent with amino functional group (3-Aminopropyl)triethoxysilane (APTEOS) has been grafted on polyester surface. A significant decrease of water contact angle have been measured for treated sample resulting in increased wettability and surface free energy of polymeric substrate. The increase of surface polarity enhance the adhesion of silicon dioxide subsequently deposited with Electron Cyclotron Resonance (ECR).

The improvement of adhesion is associated with the presence of SiO_x grafted on the surface. Small volume testing method, such as nano-indenter and nano-scratch, have been used to characterize interfaces of multilayer composite.

Nano-scratch test of SiO_2 layer of untreated AryLite™ exhibit a critical load (Lc) coupled with surface fractures, delamination and blistering. This is a clear evidence of a poor adhesion at the interface between substrate and inorganic films.

Critical load for SiO_2 detachment from treated sample was not observed, as a consequence of a different failure mechanism, due to surface modification. In fact, there is no abrupt change in the displacement curves, implying that the coatings did not peel off during the scratch ramping load scratch. The optical images reveal that the scratch is extremely smooth and shallow, implying that the coatings strongly adheres on substrate. The results showed that adhesion of SiO_2 on AryLite™ has been improved by substrate silane surface treatment.

6. Acknowledgements

The activities was performed in the framework of the project FIRB "Poliflex" (RBIP06SH3W) granted to IMAST S.c.a.r.l. The authors gratefully acknowledge Ferrania Imaging Technologies S.p.A. for providing AryLite™ substrates. The authors also wish to thank Mrs Marcedula M. and Mr De Angioletti M. for experimental tests.

7. References

Abdallah A.A., Bouten P.C.P., den Toonder J.M.J., de With G. (2008). The effect of moisture on buckle delamination of thin inorganic layers on a polymer substrate. *Thin Solid Films*, 516, 1063–1073.

Adhikari B., Majumdar S. (2004). Polymers in sensor applications. *Prog. Polym. Sci*, 29, 699–766.

Amendola E., Cammarano A., Pezzuto M., Acierno D. (2009). Adhesion of functional layer on polymeric substrates for optoelectronic applications. *Journal of the European Optical Society - Rapid publications*, 09027, 4, 1-6.

Anderson D.R., Lee Smith A. (1974). *Analysis of Silicones*. Wiley-Interscience editor, Chapter 10, New York.

Angiolini S., Avidano M. (2001). P-27: Polyarylite Films for Optical Applications with Improved UV-Resistance. *Organic Synthesis Lab. Polymers, Ferrania Imaging Technologies, Technical Information.*

Beake B.D., Lau S.P. (2005). Nanotribological and nanomechanical properties of 5-80 nm tetrahedral amorphous carbon films on silicon. *Diamond & Related Materials*, 14, 1535-1542.

Beake B.D., Ogwu A.A., Wagner T. (2006). Influence of experimental factors and film thickness on the measured critical load in the nanoscratch test. *Materials Science and Engineering A* 42, 70-73.

Bellamy L.J. (1975). The infrared spectra of complex molecules. John Wiley & Sons, 3rd ed., p.49, New York.

Burnett P.J. and Rickerby D.S. (1987). The relationship between hardness and scratch adhesion. *Thin Solid Films*, 154, 403-416.

Burnett D., Thielmann F., Ryntz R. (2007). Correlating thermodynamic and mechanical adhesion phenomena for thermoplastic polyolefins. *Journal of Coatings Technology and Research*, 4, 2, 211-215l.

Cantin S., Bouteau M., Benhabib F. and Perrot F. (2006). Surface free energy evaluation of well-ordered Langmuir–Blodgett surfaces Comparison of different approaches. *Colloids and Surfaces A: Physicochem Eng. Aspects*, 276, 107-112.

Charitidis C., Logothetidis S., Gioti M. (2000). A comparative study of the nanoscratching behavior of amorphous carbon films grown under various deposition conditions. *Surface and Coatings Technology*, 125, 201–206.

Choi M.C., Kim Y., Ha C. S. (2008). Polymers for flexible displays: From material selection to device applications. *Prog. Polym. Sci*, 33, 581–630.

Clint J. H. (2001). Adhesion and components of solid surface energies. *Current Opinion in Colloid & Interface Science*. 6, 28-33.

Colthup N.B., Daly L.H., Wiberley S.E., Introp P. (1990). Introduction to infrared and Raman spectroscopy. Academic, 3rd ed., p 94.

Comyn J. (1992). Contact Angles and Adhesive Bonding. *International Journal of Adhesion and Adhesives*, 12, 3, 145–9.

Faibish R.S., Yoshida W., Cohen Y. (2002). Contact Angle Study on Polymer-Grafted Silicon Wafers. *Journal of Colloid and Interface Science*, 256, 341 - 350.

Fowkes F.M. (1962). Determination of interfacial tensions, contact angles, and dispersion forces in surfaces by assuming additivity of intermolecular interactions in surfaces. *Journal of Physical Chemistry*. 66, 2, 382.

Goddard J.M., Hotchkiss J.H. (2007). Polymer surface modification for the attachment of bioactive compounds. *Progress in Polymer Science*, 32, 698-725.

Griffith (1920). The phenomenon of rupture and flow in solids. *Philosophical Transactions of the Royal Society of London*, 221, 163.

Gu X., Raghavan D., Nguyen T., VanLandingham M.R., Yebassa D. (2001). Characterization of polyester degradation using tapping mode atomic force microscopy: exposure to alkaline solution at room temperature. *Polymer Degradation and Stability*, 74, 139–149.

Imparato A., Minarini C., Rubino A., Tassini P., Villani F., Guerra A., Amendola E., Della Sala D. (2005). Thin silicon films on polymeric substrates. *Macromol. Symp*, 228, 167-176.

Lee I., Wool R.P. (2002). Thermodynamic Analysis of Polymer-Solid Adhesion: Sticker and Receptor Group Effects. *Journal of Polymer Science Part B: Polymer Physics*, 40, 20, 2343–53.

Lian Y.M., Leu K.W., Liao S.L., Tsai W.H. (1995). Effects of surface treatments and deposition conditions on the adhesion of silicon dioxide thin film on polymethylmethacrylate. *Surface and Coatings Technology*, 71, 142-150.

Ma Z., Mao Z., Gao C. (2007). Surface modification and property analysis of biomedical polymers used for tissue engineering. *Colloids and Surfaces B: Biointerfaces*, 60, 137–157.

Mack G.L. (1936). Determination of contact angles from measurements of the dimension of small bubbles and drops. I. The spheroidal segment methods for acute angles. *J Phys Chem*, 40, 159-167.

Malzbender J., den Toonder J.M.J., Balkenende A.R., de With G. (2002). Measuring mechanical properties of coatings: a methodology applied to nano-particle-filled sol-gel coatings on glass. *Materials Science and Engineering*, 36, 47-103.

Mannifacier J.C., Szepessy L., Bresse J.F., Perotin M., Stuck R. (1979). In_20_3:(Sn) and SnO_2:(F) films–Application to solar energy conversion; part 1 - preparation and characterization. *Mater. Res. Bull*, 14, 163.

Oliver W.C., Pharr G.M. (1992). An improved technique for determining hardness and elastic-modulus using load and displacement sensing indentation experiments. *Journal of Materials Research*. 7, 1564-83.

Owens D.K., Wendt R.C. (1969). Estimation of the surface free energy of polymers. *J. Appl. Polym. Sci.*, 13, 1741–1747.

Ozcan C., Hasirci N. (2008). Evaluation of Surface Free Energy for PMMA Films. *Journal of Applied Polymer Science*, 108, 438–446.

Park H.S., Kwon D. (1997). An energy approach to quantification of adhesion strength from critical loads in scratch tests. *Thin Solid Films*. 307, 156-162.

Park S.J. and Jin J.S. (2001). Effect of Silane Coupling Agent on Interphase and Performance of Glass Fibers/Unsaturated Polyester Composites. *Journal of Colloid and Interface Science*, 242, 174–179.

Ponsonnet L., Reybier K., Jaffrezic N., Comte V., Lagneau C., Lissac M., Martelet C., (2003). Relationship between surface properties (roughness, wettability) of titanium and titanium alloys and cell behaviour*Materials Science and Engineering C*, 23, 551.

Rats D., Hajek V., Martinu L. (1999). Micro-scratch analysis and mechanical properties of plasma-deposited silicon-based coatings on polymer substrates. *Thin Solid Films*, 340, 33-39.

van Oss C.J., Good R.J., Chaudhurys M.K. (1988). Additive and Nonadditive Surface Tension Components and the Interpretation of Contact Angles. *Langmuir*, 4, 884-891.

Young T. (1805). An Essay on the Cohesion of Fluids. *Philosophical transactions Royal Society London*. 95, 65-87.

Zhao Q., Liu Y., Abel E.W. (2004). Effect of temperature on the surface free energy of amorphous carbon films. *Journal of Colloid and Interface Science*, 280, 174–183.

Zheng S., Ashcroft I.A. (2005). A depth sensing indentation study of the hardness and modulus of adhesives. *International Journal of Adhesion & Adhesives*, 25, 67–76.

3

Organic Semiconductor Based Heterostructures for Optoelectronic Devices

Anca Stanculescu[1] and Florin Stanculescu[2]
[1]*National Institute of Materials Physics, Bucharest-Magurele,*
[2]*University of Bucharest, Bucharest-Magurele*
Romania

1. Introduction

In the last decades organic materials are considered a very important alternative to the inorganic semiconductors in the manufacturing of a large variety of devices because of the great diversity of the organic compounds, their remarkable properties, low production cost, compatibility and complementarity with the very well known silicon technology. Lately the organic semiconductors started to represent a new class of materials very attractive for opto-electronic applications (Huang, 2002; Tang, 1988; Forrest, 2000; Kalinowski, 2003) and electronic applications (Dodabalapur, 1995; Karl, 2000; Katz, 2000 a; Horowitz, 1999; Torsi, 2000; Schön, 2000 a; Inoue, 2005) in particular light emitting devices and lighting sources (Yersin, 2007; Van Slyke, 1996; Wang, 2005; Dini, 2005), solar cells (Lane, 2005; Ameri, 2009; Troshin, 2008; Duggal, 2005), photodetectors (Hofmann, 2005; Troshin, 2008), field effect transistors (Molinari, 2007; Mas-Torrent, 2008) and lasers (Duarte, 2009; Liu, 2009) being considered candidates to replace the inorganic semiconductors.
The electroluminescence (EL) of the organic molecules is a well-known phenomenon for more than 50 years, but only in the late '80th has become utile for practical applications. The successful application of the organic materials luminescence in Light Emitting Devices (OLEDs) requires adequate device structures to overcome the problems associated with the high resistivity of the organic materials and with the difficulty to obtain a good charge injection from the electrodes in the organic layers. These devices have a thin film architecture, which includes both organic/organic and organic/inorganic interfaces.
Starting with the first prototype of OLED (Vincett, 1982), lately has been proposed the first efficient OLED based on heterostructures containing thin films from small molecule organic compounds (Tang, 1988) that have shown an attractive efficiency (1% external quantum efficiency, 1.5 lm/W luminous efficiency, higher than 1000 cd/m^2 brightness) and driving voltage below 10 V.
From this moment, different types of OLEDs have been realised, based on both fluorescence and phosphorescence, that have started to be used in mobile, small dimension displays. Probably, the most spectacular application of the Organic Light Emitting Devices is in the flat panel displays for TV and monitor technology because they are characterized by high brightness and wide viewing angle, which are the most important advantages over the liquid crystals. Significant effort is devoted now to increase the quantum efficiency, lifetime and thermal stability of these devices. Devices with high efficiencies, low working voltages

and long lifetime involve low electrical resistivity and good chemical stability of the contact organic semiconductor/metal to assure a good charge carriers injection and transport. On the other hand the particularities of the charge injection and transport across the inorganic/organic and organic/organic heterojunctions are determinant for the properties of the devices. The improvement of the organic electroluminescent diode performances involves the separate optimisation of the two processes implied in the radiative emission phenomena: 1. the injection and transportation of the charge carriers and 2. their radiative recombination in different layers. The injection of the charge carriers must be not only efficient but also stable under operation conditions.

In this chapter we summarize some results of our work in the field of applied research concerning the charge injection and transport through inorganic/single and multilayer organic heterostructures.

2. OLED structure. General considerations

The basic p-n structure is formed from two organic thin layers, one a holes transporting layer (HTL) and the other an electrons transporting layer (ETL), which can be also the emitting layer where the recombination takes place. The organic layers are disposed between an anode and a cathode and are offering an adequate medium for the charge transfer through the interface situated between the two organic layers. In the OLED structures the relevant mechanism of electroluminescence involves several steps: injection, transport, capture and radiative recombination of the positive and negative charge carriers inside an organic semiconductor characterized by an energy band gap suitable for yielding visible light.

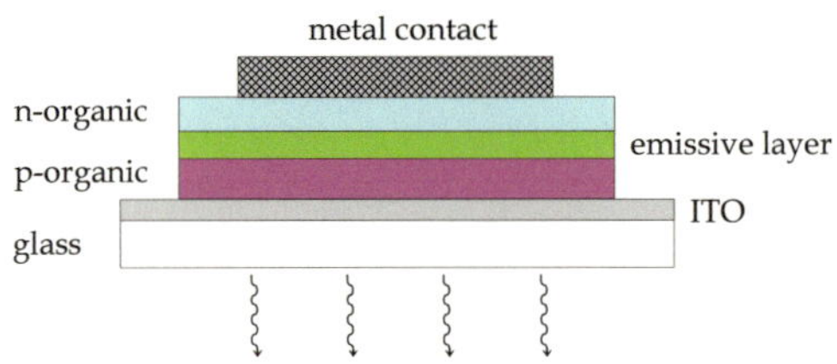

Fig. 2.1. Basic p-n junction for OLED application

The OLED structure is delimited by the metallic layer and the transparent conductor such as indium tin oxide (ITO)/glass interfaces. While the electrical properties are controlled by the mobility of the charge carriers and the energy levels offset between the layers that compose the heterostructure, the optical properties are controlled by the refractive index mismatches at the interfaces glass/air and organic layer/transparent conductor. These mismatches generate the trapping of a large fraction of the incident light by the total internal reflection into glass and transparent conductor (ITO) (Lu, 2002). An improved electroluminescence can be obtained by the individual optimization of the steps mentioned above (injection and recombination), in the configuration of organic multilayer light emitting devices based on hetero-interfaces between different organic materials.

Multiple layers can be used to increase the quantum efficiency or avoid the degradation of the devices. This way can be assured a balanced injection of the charge carriers. Since recombination of the electrons and holes within an organic layer gives rise to luminescence, it is necessary that both charge carriers are simultaneously injected in equal numbers from both electrodes, in the organic layer. This can be achieved by the optimization of both electrons and holes transporting layers or by the use of intermediate organic layers acting as blocking layers that assure a good separation of the charge carriers.

The first critical factor in determining the OLED efficiency is the charge injection being correlated with the lifetime of the device. The electrodes are used to inject charge into or to extract charge from the organic semiconductor layers, more precisely to inject electrons and holes into opposite sides of the emissive organic layers: (1) the electrons are injected from the Fermi level of the cathode into the lowest unoccupied molecular orbital (LUMO) of the organic, overcoming the contact barrier; (2) the hole are injected from the Fermi level of the anode into the highest occupied molecular orbitals (HOMO) of the organic. To improve the carrier injection, the energy level for the charge carriers in organic should match the electrode work function. On the other hand the HTL and the ETL should be characterized by high mobility carriers to assure the transport of the injected carriers to the emission zone.

The second critical factor is the radiative recombination of the charge carriers in the bulk of the organic layer. The electrons and holes accumulated at the HTL and ETL interface are creating pairs, known as excitons, that release the energy as light. Beside the improvement in the fluorescence yield of the emitting material, which can be obtained by doping, a well-balanced injection of positive and negative charge carriers is necessary.

The inorganic semiconductor/organic and/or metal/organic junction, as the key element(s) of any organic device, can offer the possibility to overcome the disadvantages of the conventional inorganic heterostructures, but the process of injection of the charge carriers across the interfaces of the structure and the transport of the hole and electrons inside the structure influence the performances of the organic devices.

Other important processes dominating the electrical properties of the interfaces and affecting the performance and lifetime of the devices, are the diffusion and chemical reaction of the metal deeper into the organic layer at the metal/organic interface and the molecular interdiffusion at the organic/organic interface.

In some molecular compounds, the intermolecular coupling is supposed to be stronger because of the significant overlapping between the π electrons clouds supporting the charge carriers generation and transport, leading to a significant delocalization of electrons and, as a consequence, a relatively significant intrinsic conductivity.

The limitation in performance of the device realized with organic semiconductor characterized by low charge mobility (μ<<1 cm^2/Vs) is determined by the current limitation due to the space charge. This process is favoured if one contact is able to inject more carriers than would be present in semiconductor at thermal equilibrium. The electrical transport properties are determined by the low mobility of the charge carriers in organic solids (2 or 3 orders of magnitude lower than in silicon) that induces a limitation associated with the space charge and by the grain boundaries, defects and imperfections in the organic films acting as trap centres that induce a limitation associated with the trap charge.

Another source for the generation of interfacial space charge in hetero or multilayer devices can be the energy level offset at the organic/organic interfaces. Even the energy barriers are not present, the differences in the charge carriers mobility between the layers can generate a mobility barrier at hetero-interfaces, which can also be a source for the generation of interfacial space charge. Trapped and interfacial charges have an important effect on the performances of the OLEDs (Riess, 2001).

A stronger emission can be obtained introducing a supplementary luminescent layer between the holes transporting layer and the electrons transporting layer, as it is presented in Figure 2.1, but this way the manufacturing of the OLED more is complicated.
If we use doping to improve the fluorescence, the electrical properties of the resulted "host-guest" material can be controlled through the selection of the dopant (Makinen, 2002). The electronic transfer direction and the doping polarity are strongly depending of the relative position between HOMO and LUMO for dopant (guest) and matrix (host) molecules. So, a good concordance between the ionisation energy (IE) corresponding to HOMO for the host matrix and the electron affinity (EA) corresponding to LUMO of the dopant molecule is necessary to produce a „p" type doping while the alignment between the LUMO level of the host matrix and the HOMO level of the dopant molecule leads to a „n" type doping. Fermi level will freely move inside the band gap HOMO-LUMO of the host matrix and dopant, supposing that this movement is not limited by the band edges (no pinning phenomenon).
The role of the dopants is very complex. On one hand the doping of the "p" and "n" transport layers has two important transport effects: 1. generates narrow space charge regions at the contacts and an improved charge carriers' injection through an efficient tunnelling; 2. assures a low voltages across the transport layers because of the high conductivity. On the other hand, the "p" doping of the "n" conduction layer has a stabilization effect by trapping the positive charge carriers partially responsible for the degradation of the device (Zhi-Lin, 1998; Cheng, G, 2006).

3. Theoretical approach

Many theories have been proposed in the last years to describe charge injection into materials characterised by a hopping transport, such as organic compounds. The electrical transport theories in organics try to describe the charge injection process as thermally-assisted tunnelling from the metal to localized states (Abkowitz, 1995), tunnelling into polaron levels in polymers (Conwell, 1997), thermally-assisted injection into an energetically disordered dielectric (Arkhipov, 1998), or as diffusion-limited thermionic emission (Emtage, 1966; Scott, 1999). The most important factors playing a role in the injection of the charge carriers from metal to organic are the charge mobility in the organic layer (Emtage, 1966; Scott, 1999), the dependence of the mobility of the electric field intensity (Borsenberger, 1998) and of charge density (Roichman, 2002), the trapping of injected charge at the interface due to the image potential (Gartstein, 1996), the interface dipoles arising from the charge transfer (Crispin, 2002) or the interfacial chemistry (Abkowitz, 1998) and disorder in these interface dipoles (Baldo, 2001).
The simplest case to analyse can be a metal electrode that injects electrons in a trap free organic semiconductor. At lower voltages and neglecting the diffusion, the current is determined by the motion of the free electrons that are present in the semiconductor and the current density is given by the Ohm's law (Lambert, 1970):

$$J_{\Omega} = q\mu_0 N_0 \frac{V}{d} \tag{3.1}$$

where q is charge of the electron, N_0 is the number of free electrons per unit volume, μ is the mobility of electron, V is the applied voltage and d is the length of the sample (film thickness). In the case of purely injection-limited currents, neglecting the specific limiting mechanisms, the current at constant intensity field, E, has no explicit thickness dependence and is given by:

$$J=J(E) \tag{3.2}$$

The transport over an interfacial barrier at a metal/semiconductor interface can be analysed with one of the following theories:

1. For the process thermally activated at room temperature, is used the thermionic, Richardson-Schottky (RS) emission (Sze, 1981; Ashcroft, 1976) over the barriers to describe the transport characteristics.
2. For tight barriers at low temperatures, electron tunnelling or Fowler-Nordheim (FN) tunnelling (Sze, 1981) dominates the transport process.

The (RS) model is based on the lowering of the energetic barrier height by the image charge potential under the effect of an external electric field of intensity $E=V/d$. In this model, the current density J_{RS}, as a function of field intensity, is given by the equation:

$$J_{RS} = A^{*}T^{2}\exp\left(-\frac{\phi_B - \beta_{RS}\sqrt{E}}{k_B T}\right) \tag{3.3}$$

where: $A^{*} = 4\pi q m^{*} k_B^2 / h^3$ [$A^*=120$ A/cm^{2}K^2 for $m^*=m_0$ (electron mass)] is the Richardson constant (q=electron charge, k_B=Boltzmann constant; h=Planck constant, T= Thermodynamic temperature); $\beta_{RS} = \sqrt{q^3/4\pi\varepsilon\varepsilon_0}$ (ε is the relative permittivity and ε_0 is the vacuum permittivity) and Φ_B is the zero-field injection barrier.

In the case of FN tunnelling model, the Columbic effects are neglected. It is considered the tunnelling through a triangular barrier into a continuum of states. In this model the current density is given by the following equation (Gao, 2002):

$$J_{FN} = \frac{A^{*}q^{2}E^{2}}{\Phi_B \alpha^2 k_B^2}\exp\left(-\frac{2\alpha\Phi_B^{3/2}}{3qE}\right) \tag{3.4}$$

Under certain conditions, the (RS) and (FN) models can be applied to explain the conduction also in inorganic semiconductors characterized by extended band states and large mean free path.

At low voltages, and neglecting the diffusion, the current is determined by the motion of the free electrons and the current density is given by the Ohm's law. As the voltage increases, electrons injected from the contact begin to exceed the electrons that are initially present inside the semiconductor. This happens when the charge corresponding to the number of electrons becomes approximately equal to $C \cdot V$, where C is the capacitance of the sample and V is the applied voltage. This is the space charge limited (SCL) regime and the Mott-Gurney (Mott, 1940; Shen, 2004) equation can be applied for SCLC for field-independent mobility and the current density expression becomes:

$$J_{SCLC} = \frac{9}{8}\varepsilon\varepsilon_0 \mu \frac{V^2}{d^3} \tag{3.6}$$

where $\varepsilon\varepsilon_0$ is the permittivity of the organic. The threshold voltage, V_0, at which the current turns from ohmic to SCL can be found by the relation:

$$V_0 = \frac{8}{9} e N_0 \frac{d^2}{\varepsilon\varepsilon_0} \tag{3.7}$$

The SCL conduction relies on a phenomenological theory developed in the 1950s and subsequently improved with more specific details (Lampert, 1970; Kao, 2004).
Considering a trap free semiconductor with a limitation of the current imposed only by the space charge (SCLC), with the mobility depending or being independent of the field intensity, the current at constant intensity field becomes

$$J = J^{*}(E)/d \tag{3.8}$$

where d is the thickness of the layer as mentioned above.
Space charge limited transport appears in undoped, wide-gap semiconductors in which the density of charge carriers at equilibrium is very small (de Boer, 2005). In these materials the current is generated by the charge injected from the contacts. The density of the charge carriers is determined (limited) by the electrostatic phenomena.
Especially in amorphous molecular materials, molecularly doped polymers and most of the conjugated polymers, the mobility is characterized by a field-dependence and an approximate analytical expression has been deduced (Murgatroyd, 1970). This is the Poole-Frenkel (PF) field dependence of the mobility:

$$\mu(E) = \mu_0 \exp\left(\beta\sqrt{E}\right) \tag{3.9}$$

where μ_0 is the zero-field mobility, E is the electric field strength and β is associated with the Poole-Frenkel-like field dependence of the mobility in the organic (Silveira, 2004). For field dependent mobility the SCL current density becomes:

$$J_{SCLC} = \frac{9}{8}\varepsilon\varepsilon_0\mu_0\frac{V^2}{d^3}\exp\left(0.89\beta\sqrt{V/d}\right) \tag{3.10}$$

Measuring the current as a function of voltage it was remarked a transition from „ohmic" conduction J_Ω at low voltage to a SCL conduction at high voltages, with the corresponding, low voltage, current densities J_Ω given by:

$$J_\Omega = \mu_0 N_0 \frac{V}{d}\exp(\beta\sqrt{V/d}) \tag{3.11}$$

where N_0 is the density of free carriers and J_{SCL} is given by Eq. (3.10).
The conduction in organic semiconductors can be injection-limited or transport-limited (Boer, 2005). For injection-limited conduction the models for bulk-limited transport, such as trap charge limited transport (Burrows, 1996; Lampert, 1970; Campbell, 1998; Campbell, 1997; Hung, 1997; Parker, 1994; Sze, 1981) and space charge limited transport with a field and temperature dependence of the mobility (Bloom, 1997) can not be applied.
For a bulk-limited transport such as the trap charge limited conduction (TCLC), with an exponential trap distribution and a mobility independent of field, the current at constant field varies with d^{-l} with l>1 (Lampert, 1970; Kao, 1981):

$$J = J^{*}(E)/d^{l} \tag{3.12}$$

The bulk-limited transport explains both the thickness and temperature dependences of current-voltage characteristics over many order of magnitude and the variations in injection

for different cathode materials. This model can not explain the dependence of current-voltage characteristics on the composition of the injection contact (Baldo, 2001).
The shape of the I-V curves can be the result of the conjugated effect of the surface and bulk traps and not of different physical mechanism (de Boer, 2005).
In the case of Poole-Frenkel field dependence of the mobility, SCLC can show exactly like TCLC, for the current versus voltage curve, a power law behaviour, but the power coefficient is higher than 2. The distinction between these two mechanisms can be done by the difference in the thickness dependence of these mechanisms given by Eq. (3.8) and Eq. (3.12) respectively.

4. Experimental

4.1 Types of substrates

To investigate the electrical conduction properties of the interface inorganic/organic and organic/organic have been prepared sandwich type heterostructures on different substrates: ITO with different sheet resistance and single crystal silicon with different types of conductibility, p and n, and different resistivity, and a given degree of surface processing [double side etched (E/E)]. Single silicon wafers (Si) could be used as an ohmic charge carrier injection contact for organic compounds of high ionization energy (high value of HOMO) because of the promising properties such as the high mobility of the charge carriers, the possibility to be p and n doped and because of reduced surface reactivity and interfacial interactions obtained by chemical processing by etching.
Theoretically, the hole/electron injecting contacts must be metals with high/low work function to reduce the height of the injection barriers (Hill, 1998). An efficient injection of holes into a p-type conduction organic semiconductor is possible from an ITO semiconductor layer characterized by a work function situated between 4.6 eV and 5 eV. If the effective electrode work function is higher than the ionisation energy of the organic semiconductor, the contact has an ohmic behaviour and the charge carriers are injected from electrode to organic. The main difficulty is associated with the high value for the ionisation energy (HOMO), higher than 5.5 eV, in some p type conduction organic semiconductors.

4.2 Organic thin films preparation

Beside the wide band gap organic semiconductors [benzil, meta-dinitrobenzene (m-DNB)] studied for the first time from the point of view of the films' electrical properties, other organic semiconductors, which have attracted a great research interest, were:
perylene: IE=5.1 eV (Kang, 2005), Eg=2.5 eV (Kang, 2005); **3,4,9,10-perylenetetracarboxylic dianhydride (PTCDA):** IE=6.2 eV (Gao, 2003), IE=6.8 eV (Gao, 2001), Eg=2.2 eV (Rajagopal, 1998); **tris (8-hydroxyquinoline) aluminium (Alq3):** IE=5.7 eV (Hirose, 1996), IE=5.95 (Gao, 2001), EA=3.25 eV (Rajagopal, 1998); **zinc phthalocyanine (ZnPc):** IE=5.28 eV (Gao, 2002), Eg=1.94 eV (Gao, 2001; Gao, 2003); **N,N'-di-[(1-naphtalenyl)-N,N'-diaphenyl]-(1,1'-biphenyl)-4,4'diamine (α-NPD):** IE=5.52 eV (Gao, 2003), Eg=3.1 eV (Rajagopal, 1998); **5,10,15,20-tetra(4-pyridyl)-21H,23H-porphine synthetic (TPyP):** EA=4.1 eV (Antohe, 2008), Eg=2.7 eV, evaluated from the UV-VIS transmission spectra of the thin films (Rasoga, 2009).
Two different methods have been used to prepare organic molecular thin films: 1. the rapid directional solidification method and 2. the vacuum evaporation and deposition method.
It is adequate to grow films from organic compounds with a melting point $T_{melting}$<100°C between two silicon wafers using a temperature gradient for solidification ΔT>50°C to

counteract the supercooling that characterises the large molecule organic compounds (Stanculescu, 2004; Stanculescu, 2006b; Stanculescu, 2006c). This method has been also used for the preparation of doped organic semiconductor layer. As dopants have been used oxine (8-hydroxyquinoline) or resorcinol (1, 3 dihydroxybenzene) in m-DNB and Na or Ag or m-DNB in benzil (Stanculescu, 2005; Stanculescu, 2006 b).
Thin films of organic molecular compounds with $T_{melting}$>100 °C have been prepared by vacuum evaporation and deposition processes that allows the preparation of stable, homogeneous films with a specified geometry and a good adhesion to the substrate (Stanculescu, 2006 a; Stanculescu, 2008; Rasoga, 2009).

4.3 Organic multilayer heterostructures manufacturing

Organic multilayer heterostructures have been realised using two different procedures. The successive organic layers have been prepared by vacuum evaporation and deposition using aluminium foil masks. The metallic contact has been obtained by sputtering or vacuum evaporation of the metal.

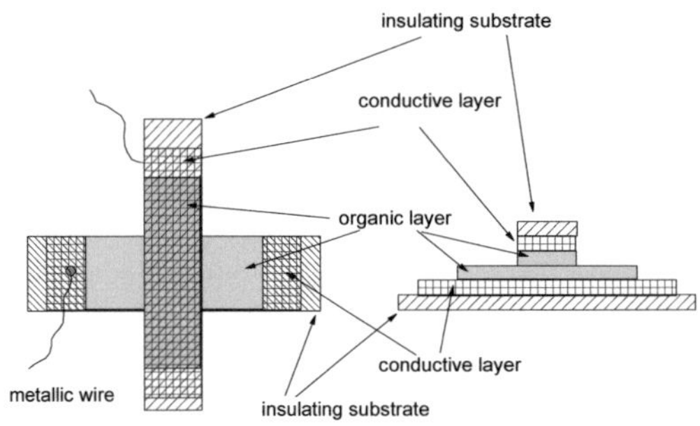

Fig. 4.1. Schematic cross section through the following heterostructures: insulating substrate (glass, organic insulator), conductive layer (Al, Cu, Si, ITO), organic single or double layers, insulating substrate (Stanculescu, 2007 b)

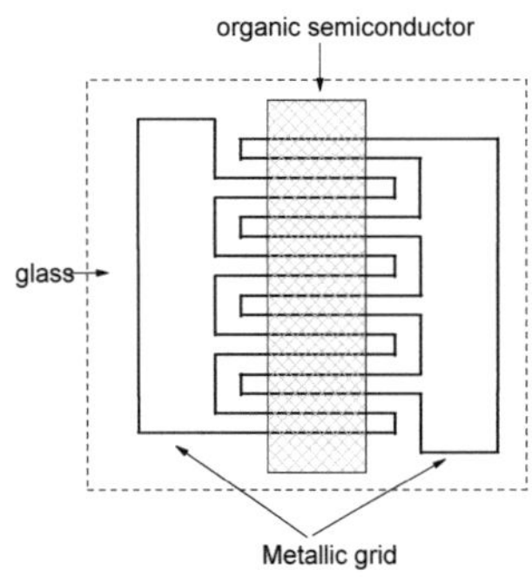

Fig. 4.2. Planar grid contact geometry (Stanculescu, 2007 b)

Semiconductor/Insulator/Semiconductor (SIS) sandwich type Si/single or double organic layer/Si structures and Metal/Insulator/Semiconductor (MIS) sandwich type (Cu or Al)/ single or double organic layer/Si or ITO structures have been prepared by a process in two steps.

Firstly have been manufactured the inorganic/organic (SI) junctions and metal/organic (MI) junctions and subsequently the SIS and MIS heterostructures putting into direct contact the organic film of two (SI) or, one (SI) and one (MI) individual junction as it is shown in Figure 4.1 (Stanculescu, 2007 b; Stanculescu, 2008).
Metal Insulator Metal (MIM) heterostructures have been prepared by the deposition of Al electrodes with a special grid configuration, presented in Figure 4.2, on glass substrate followed by the deposition by vacuum evaporation of the organic thin film.

5. Electrical conduction in organic molecular compounds heterostructures

5.1 SIS type structures: inorganic /organic /inorganic (Si/organic/Si)

When the order is induced by the substrate, the molecular planes of PTCDA have a preferred orientation and the (102) planes form stacks with small interplanar spacing, perpendicular to the substrate (Wu, 1997). This arrangement in the solid state generates a significant overlap of the π orbitals of the aromatic rings belonging to molecules situated in different planes. In Alq3 the conduction properties depend on the dominant type of stereoisomer, meridian or facial (Cölle, 2003), and are generated by the overlapping degree of the π orbitals of the hydroxyquinoline ligands belonging to the neighbouring molecules.

5.1.1 Si/wide band gap organic semiconductor/Si heterostructures

The wide band gap organic semiconductors used for the thin film deposition have different particularities of the molecular configuration: 1. one aromatic ring and two NO_2 substituent groups, meta-dinitrobenzene (m-DNB); 2. two aromatic rings coupled by two CO substituent groups, benzil (known also as diphenyl-α,β-diketone, diphenylethanedione, dibenzoyl). In Figure 5.1 and Figure 5.2 can be seen that the value of the current is comparable for the both types of structures prepared with non-purified compounds: $\sim 10^{-6}$ A for an applied voltage of 1 V.
For the structures based on organic layers prepared by rapid solidification in the experimental conditions mentioned above in paragraph 4, it has been remarked a non-ohmic behaviour at low voltages. Some differences have been evidenced in the shape of the I-V curves for the structures prepared with m-DNB and benzil and presented in Figure 5.1 and Figure 5.2. The origin of these differences could be in the different molecular structures of the two compounds, which are reflected in different crystalline quality of the layers.
Despite the fact that a linear behaviour has been expected at low voltages for an organic molecular film grown between two Si electrodes a deviation from the ohmic dependence has been obtained, which can be explained by two counteracting mechanisms. The high value of the ionization energy in some of the benzene derivatives [higher than 5.5 eV (Hersberg, 1959)] means deep HOMO levels. This position of the energetic level favours the accumulation of the charge carriers near the surface and as consequence an ohmic dependence. But this effect is dominated by the very low mobility of the charge carriers in these organic molecular compounds that favours the generation of the space charge and a limitation of the current by this mechanism, leading to non-linear I-V dependences.
In m-DNB and benzil the conduction is favoured by the strong interaction between the electrons of different substituent groups (NO_2 or CO). In the solid state these groups are close one to another and assure a higher mobility of the charge carriers and a weaker influence on the current of the space charge limitation.

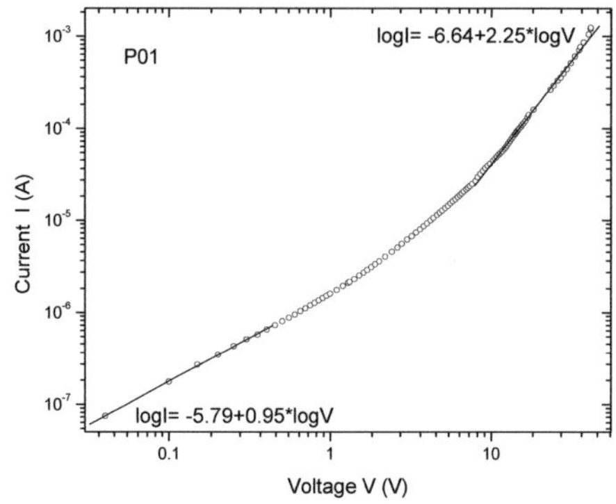

Fig. 5.1. I-V plot for as-synthesised m-DNB layer grown between two Si wafers, Si(p), E/E, ρ=12-15 Ωcm, (P01 in Table 1) (Stanculescu, 2006 b)

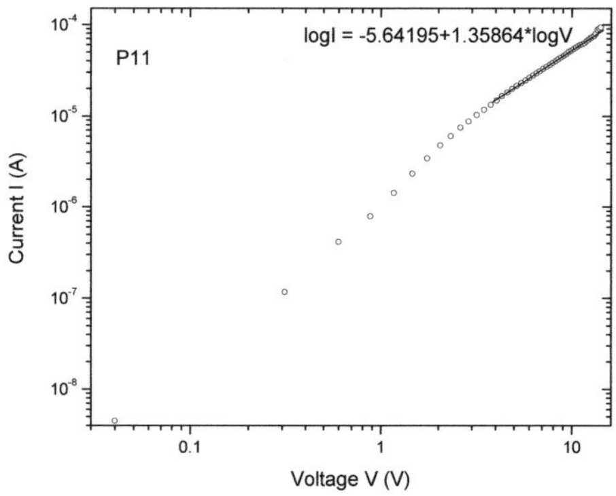

Fig. 5.2. I-V plot for as-synthesised benzil layer grown between two Si wafers, Si(p), E/E, ρ=5.5-27 Ωcm, (P11 in Table 1) (Stanculescu, 2006 b)

The interactions between the molecules are stronger in m-DNB than in benzil because the group NO_2 in m-DNB has a stronger polar character compared to the group CO in benzil and favours the overlapping between the π orbitals of the neighbour molecules. As a consequence, m-DNB is a compound characterized by a narrow electronic band structure with a high delocalization for electrons and holes that assures an increase in the electrical conduction (Stanculescu, 2006 b). The moderate polar character of the CO groups in benzil is correlated with a lower overlapping of the π orbitals and a higher degree of localization of the charge carriers. This generates a slightly wider electronic band structure of the compound (benzil) and a slightly lower conductivity.

5.1.2 Si/ molecular organic/Si

The type of molecules stacking in the organic solid state is very important for the electrical behaviour of the organic semiconductor because it could favour the significant overlapping between the π electrons clouds supporting the charge carriers generation and transport. This leads to a significant delocalisation of the charge carriers and, as consequence, to a relatively significant intrinsic conductivity.

Perylene and 3,4,9,10-perylenetetracarboxylic dianhydride (PTCDA) are characterised by a special stacked arrangement with distances between the parallel planes of molecules of 3.46 Å (Hirose, 1996) and 3.21 Å (Wu, 1997) respectively. This stacking assures significant overlapping between the π electrons clouds of the neighbouring molecules from successive planes and sustains a stronger intermolecular coupling in the direction perpendicularly to the substrate. This way are favoured highly anisotropic electrical transport properties. The charge carriers transport in the direction parallel to the molecular plane of perylene and PTCDA is limited by the distance between the adjacent molecules and by the polar binding of the hydrogen and oxygen atoms.

To investigate the injection properties of these organics at the contact with different types of single crystal Si electrodes, the position of the energy level in Si and in organic must be compared: the energy of the holes in the valence band of the p type silicon (E_{VB}~5.1 eV) must be compared with the energy of the HOMO level of the organic compound, and the energy of the electrons in the conduction band of the n type silicon (E_{CB}~4 eV) must be compared with the energy of the LUMO level of the organic compound.

It can be anticipated from theoretical considerations that a higher number of holes is injected from the p type silicon in Si/perylene/Si heterostructure because of the small difference between the energy of the valence band level in Si situated around 5 eV and the energy of the HOMO level in perylene, an organic compound characterized by an ionisation energy of IE=5.1 eV (Hirose, 1996).

Experimentally data obtained for low applied voltages (~ 0.1 V) presented in Figure 5.3 a and Figure 5.3 b show a slightly higher value of the current passing through the Si/PTCDA junction (10^{-8} A) compared to Si/perylene junction (4×10^{-9} A), when the both structures are prepared with the same type of Si electrode (p type conduction, resistivity of 1.3 Ωcm). This confirms that a lower number of charge carriers are injected in perylene because of the surface states. The specific interaction of the organic molecules with the Si surface, such as surface adsorption, is important in the generation of the injection barrier and can significantly change the electronic properties of the Si surface.

For the structure realised with n type Si electrodes we have anticipated an increase in the number of electrons injected in PTCDA determined by the lower position of the energetic level associated with the electron affinity in PTCDA, calculated using IE given in paragraph 4.2, EA=4 eV or EA=4.6 eV compared to perylene, EA=2.6 eV. The band gap energy was considered Eg=2.2 eV in PTCDA (Rajagopal, 1998) and Eg=2.5 eV in perylene (Kang, 2005).

For n type of Si electrodes and an applied voltage of 1 V, the current obtained experimentally in Si/perylene/Si structures, drawn in Figure 5.3 b, is one order of magnitude higher than the current obtained in Si/PTCDA/Si structures drawn in Figure 5.3 a.

A steeper increase in the current at voltages higher than 0.1 V, presented in Figure 5.3 b, has been obtained using as electrodes p type, low resistivity silicon wafers, ρ=0.015 Ωcm. This behaviour can be explained by the high density of defects, which characterizes the low resistivity (highly doped) silicon, that could be induced in the organic layer.

For the heterostructure with two n type Si electrodes, the difference between the current, at forward and reverse bias, at a given voltage, is very small and the contact Si/perylene is practically ohmic at voltages lower than 1 V, as shown in Figure 5.4. At low voltages (0.1-1 V), the slope of the logarithmic I-V characteristics is n=1, meaning that the behaviour of the above mentioned structures is ohmic and the current is not limited. In this case, the accumulation of the charge carriers is not so significant and the injection of the charge carriers is not reduced by the created electrostatic field.

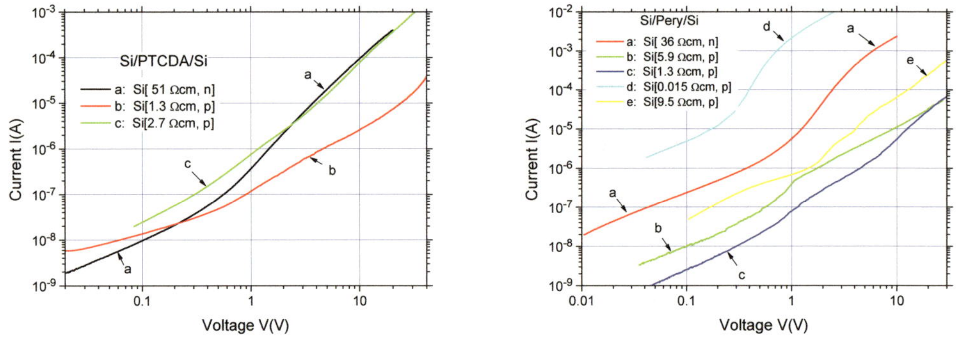

(a) organic-PTCDA; (b) organic-perylene

Fig. 5.3. I-V characteristics for Si/organic/Si heterostructures for different Si electrodes (Stanculescu, 2008)

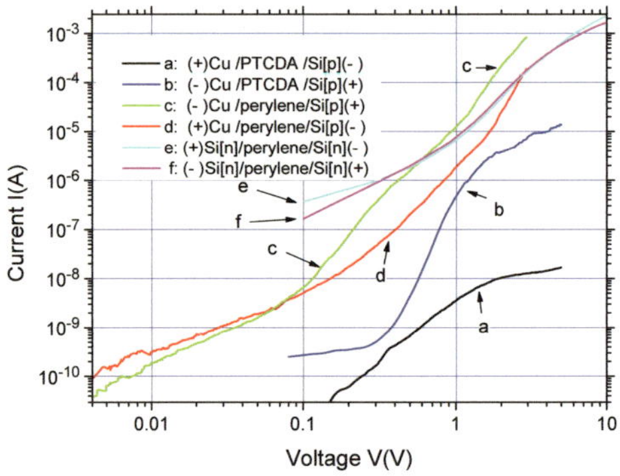

Fig. 5.4. I-V characteristics of Cu(Si)/perylene(PTCDA)/Si heterostructures at forward and reverse bias (Stanculescu, 2008).

Because a Poole-Frenkel (PF) mechanism characterised by n=1/2, which involves the field dependence of the mobility, has not been evidenced, the conduction could be space charge limited (without electric field dependence of the mobility) and trap charge limited (Gao, 2002). In the higher applied voltages region (> 1 V) the charge transport process appears to be space charge limited, SCLC (n=2) when the charge injected from the inorganic electrodes (Si) is larger than the charge existing in material in equilibrium, and trap charge limited, TCLC (n>2) when exist discrete trapping levels associated with defects in the organic layer. A transition region situated around 1V has been evidenced in the region where the conduction process is trap charge limited, being associated with a mechanism of trap filling.

The general shape of the I-V characteristics is not very different for structures realized with n type Si and n type conduction organic molecular compounds, n type Si/TPyP or n type Si/Alq3, as can be seen in Figure 5.5. The current is less than one order of magnitude higher in TPyP based heterostructure compared to Alq3 based heterostructure for an applied voltage lower than 1 V.

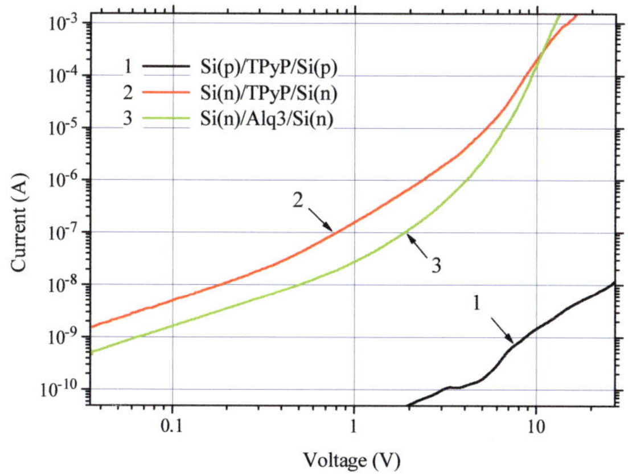

Fig. 5.5. I-V characteristics for SIS symmetrical heterostructures based on single n type organic thin film (Rasoga, 2009):
1. Si(Cz, E/E, p: 1.34 Ωcm)/TPyP/ Si(Cz, E/E, p: 1.34 Ωcm);
2. Si(Cz, E/E, n: 0.008 Ωcm)/TPyP/ Si(Cz, E/E, n: 0.008 Ωcm);
3. Si(Cz, E/E, n: 0.008 Ωcm)/Alq3/ Si(Cz, E/E, n: 0.008 Ωcm)

A higher number of injected electrons was obtained in TPyP with E_{LUMO}~4.1 eV (Antohe, 2008) compared to Alq3 with E_{LUMO}=3.25 eV (Rajagopal, 1998). This behaviour is sustained by the position of the electron affinity (LUMO) level in TPyP compared to the position of the conduction band level in silicon. The difference between the energy of the conduction band in n type Si and the energy of the electron affinity is higher in Alq3 (ΔE~0.75 eV) compared to TPyP (ΔE~0.10 eV). The height of the energetic barrier favours the injection of the electron at the contact n type Si/TPyP. As it is shown in Figure 5.5, for an applied voltage of 1 V, the interface between n type Si wafer and n type conduction organic semiconductor (TPyP, Alq3) favours the injection of a much higher number of charge carriers, $I_{TPyP}=1.5\times10^{-7}$ A and $I_{Alq3}=2.5\times10^{-8}$ A, than the interface between p type Si wafer and the same n type conduction organics (I_{TPyP} is lower than 10^{-10} A).
The lower injection at the contact n type Si/Alq3 compared to the contact n type Si/TPyP could be explained by the lower efficiency of the coupling between the n type Si substrate and the π electrons system of Alq3 compared to TpyP (Rasoga, 2009).
The number of charges injected at the contact p type Si and p conduction organic, Si/ZnPc, is higher (6×10^{-4} A at an applied voltage of 1 V) because ZnPc (as PTCDA as well) is a planar aromatic molecule characterised by a preferred orientation of the molecular planes and small distances between the intermolecular planes (Wu, 1997).

5.1.3 Si/organic multilayer/Si heterostructures
The order of preparation of organic/organic structure in multilayer heterostructures is not critical because the system is not submitted to high thermal variations during the vacuum evaporation process. The behaviour of the interface organic/organic has an important influence on the shape of the I-V characteristics by a charge accumulation process. Figure 5.6 shows the I-V characteristics for the structures with double organic layer, n type organic semiconductor (TPyP or Alq3) on p type organic semiconductor (PTCDA or perylene). The

highest value of the current, was obtained for an applied voltage of 1 V at reverse polarization (n type Si electrode positively biased and p type Si electrode negatively biased) in the heterostructures Si/TPyP/PTCDA/Si ($I=8\times10^{-6}$ A) and Si/Alq3/perylene/Si ($I=5\times10^{-6}$ A).
The energetic barriers for the charge carriers at the p-n junction results from the value of the ionisation energy and electron affinity for each organic compound:

$\Delta E_{TPyP/PTCDA}$~0 eV; $\Delta E_{Alq3/PTCDA}$=0.85 eV; $\Delta E_{TPyP/perylene}$=1.7 eV; $\Delta E_{Alq3/perylene}$=0.85 eV. (holes)

$\Delta E_{TPyP/PTCDA}$~0 eV; $\Delta E_{Alq3/PTCDA}$=1.35 eV; $\Delta E_{TPyP/perylene}$=1.5 eV; $\Delta E_{Alq3/perylene}$=0.65 eV (electrons)

The position of the HOMO (IE) level in TPyP can be estimated from the position of the LUMO level (EA=4.1 eV) and the optical band gap (Eg=2.7 eV). The optical band gap has been evaluated from the optical transmission spectra of TPyP thin films deposited on quartz substrate.

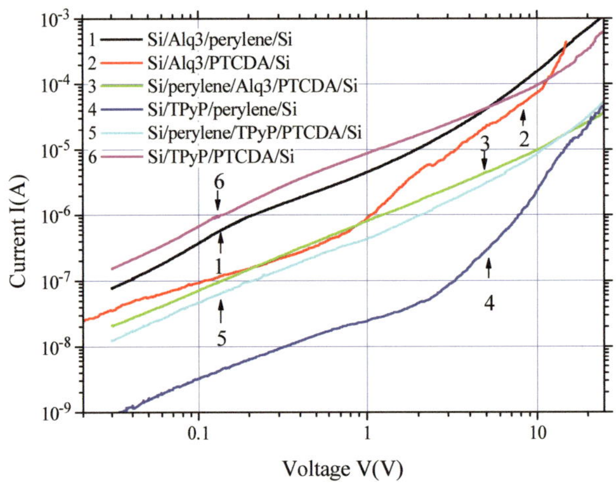

Fig. 5.6. I-V characteristics for SIS heterostructures based on double n-p organic thin films (Rasoga, 2009):
1. Si(Cz, E/E, n: 0.008 Ωcm)/Alq3/ perylene/Si(Cz, E/E, p: 6.78 Ωcm);
2. Si(Cz, E/E, n: 0.008 Ωcm)/Alq3/ PTCDA/Si(Cz, E/E, p: 1.34 Ωcm);
3. Si(Cz, E/E, p: 1.34 Ωcm)/perylene/ Alq3/PTCDA/Si(Cz, E/E, p: 6.78 Ωcm);
4. Si(Cz, E/E, n: 0.008 Ωcm)/TPyP/ perylene/Si(Cz, E/E, p: 1.34 Ωcm);
5. Si(Cz, E/E, p: 1.34 Ωcm)/perylene/ TPyP/PTCDA/Si(Cz, E/E, p: 1.34 Ωcm);
6. Si(Cz, E/E, n: 0.008 Ωcm)/TPyP/ PTCDA/Si(Cz, E/E, p: 1.34 Ωcm)

In the structures with three organic layer perylene/Alq3/PTCDA, presented in Figure 5.6, the currents are slightly different at the reverse bias (p type, ρ=1.34 Ωcm Si electrode positively biased and p type ρ=6.78 Ωcm Si electrode negatively biased) compared to the structure with only two organic layers, Alq3/PTCDA at reverse bias (p type, ρ=1.3 Ωcm Si electrode, negatively biased and n type, ρ=0.008 Ωcm, positively biased).
The supplementary perylene layer decreases with more than one order of magnitude the value of the current at an applied voltage of 1 V in the heterostructure Si/TPyP/PTCDA/Si.

This is a result of the high energetic barrier at the contact perylene/TPyP (ΔE=1.7 eV). In the structure Si/Alq3/PTCDA/Si, a supplementary perylene layer has no influence on the current for applied voltages lower than 1 V. The effect of the energetic barrier height at the contact perylene/Alq3 (ΔE=0.85 eV) becomes important in the heterostructure Si/perylene/Alq3/PTCDA/Si at applied voltages higher than 1 V (Rasoga, 2009).

5.2 Effect of the dopant

The doping process, which refers to the introduction of a strong electron donor or acceptor in the host organic matrix, could have a strong impact on the material properties, increasing the electrical conductivity of these host compounds and involving or not charge transfer between the organic host matrix and the guest dopant (organic or inorganic). For example, by the intercalation of benzil matrix with monovalent alkali (sodium) or non-alkali (silver) metallic atoms can be induced delocalised charge carriers, electrons and holes, respectively. Similarly results are expected for organic dopants containing substituent groups on the aromatic nucleus with strong electron donors (such as OH) or acceptors (such as NO_2 and CO), properties.

The organic molecular films show some particularities of the transport mechanism of the charge carriers, small amount of impurity or dopant strongly affecting or even masking or hiding the intrinsic properties by a trapping mechanism.

It has been evidenced that the relatively low intrinsic conductivity of optical wide-gap organic semiconductor, benzil [Eg=2.84 eV (Stanculescu, 2004; Stanculescu, 2006 a] and m-DNB [Eg=2.92 eV (Stanculescu, 2004; Stanculescu, 2006 a)] has been improved by doping: in benzil doped with m-DNB and, in m-DNB doped with 8-hydroxyquinoline (oxine) or 1,3 dihydroxybenzene (resorcinol). In these situations the resistivity of the organic layer decreases to 10^6-10^7 Ωcm. New states have been generated within the band gap by doping, which can be involved in the conduction process.

Benzil can be easily doped with electrons donors such as alkali metals (sodium) because it is characterized by a high electron affinity. At low concentration, c=1 wt %, it was evidenced an increased conductivity compared to the pure compound.

Another factor with an important effect on the electrical properties of the heterostructures is the crystalline quality of the intermediate organic layer. An increase in the resistivity is obtained for higher concentration of sodium, c=6 wt %, or by p type doping with silver. A high concentration of dopants can induce a higher concentration of defects. Microstructured patterns of the organic crystalline layers are created either by dopants or by the grain boundaries whose generation is controlled by the thermal regime of the solidification process. The effect of dopant on the resistivity of the organic layer is presented in Table 1.

The heterostructures prepared with doped intermediate benzil layer show a non-linear behaviour of the I-V characteristics, as illustrated in Figure 5.7.

We have emphasised a special behaviour for the organic layer doped with organic molecule. In benzil doped with m-DNB (3 wt %) we have obtained an important decrease in resistivity and a higher conductibility. m-DNB doped with resorcinol shows a different I-V characteristic. The exponential dependence can be correlated with a field intensity dependence of the mobility of the charge carriers.

An even stronger effect of the dopant over the resistivity of the organic layer is evidenced in m-DNB. The resistivity of the m-DNB layer can be reduced with three orders using, as dopants, oxine (c=1 wt %) or resorcinol (10 wt %) as mentioned in Table 1.

An exponential dependence between I and $\sqrt{V}$ has been evidenced in Figure 5.8 for m-DNB layer highly doped with resorcinol (10 %), in the low voltage range, around 1 V, suggesting a Poole-Frank conduction mechanism involving the field dependence of the charge carrier mobility (Silveira, 2005) :

$$I \sim V \cdot e^{C\sqrt{V}} \tag{5.1}$$

Sample	Organic material	Film thickness (μm)	Si conductivity type	Si surface processing	Resistivity of the organic film ρ (Ωcm)
P01	m-DNB as synthesized	21	p	E/E	2.1×10^{9}
P03	m-DNB purified by melting zone	16	p	E/E	6.6×10^{9}
P07	m-DNB+oxine	75	n	E/E	0.49×10^{6}
P08	m-DNB+resorcinol	20	n	E/E	5.3×10^{6}
P11	benzil as synthesized	52	p	E/E	0.76×10^{9}
P12	benzil purified by melting zone	17	p	E/E	5.0×10^{9}
P15	benzil+m-DNB	30	p	E/E	2.3×10^{7}
P17	benzil+1 wt % Na	44	p	E/E	$0.79\text{x}10^{9}$
P18	benzil +6 wt % Na	43	p	E/E	1.9×10^{10}
P19	benzil+2.4 wt % Ag	72	p	E/E	4.5×10^{10}

Table 1. The properties of prepared SIS heterostructures (Stanculescu, 2006 b)

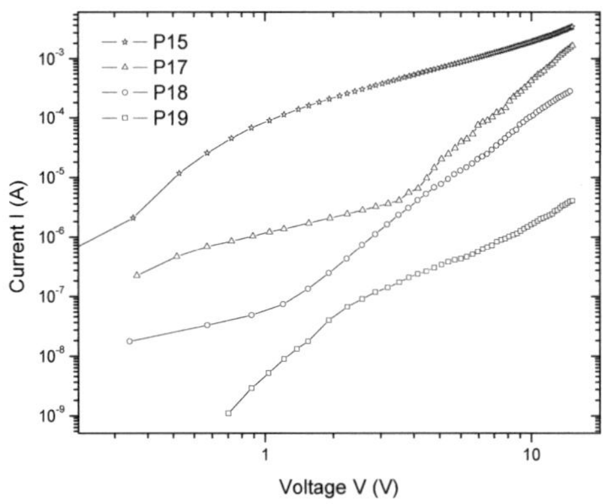

Fig. 5.7. I-V plots (Stanculescu, 2006 b) for:
a) benzil doped with 1 wt% Na, Si(p), (P17);
b) benzil doped with 6 wt% Na, Si(p), (P18);
c) benzil doped with 2.4 wt% Ag, Si(p), (P19);
d) benzil doped with 3 wt% m-DNB, Si(p), (P15)

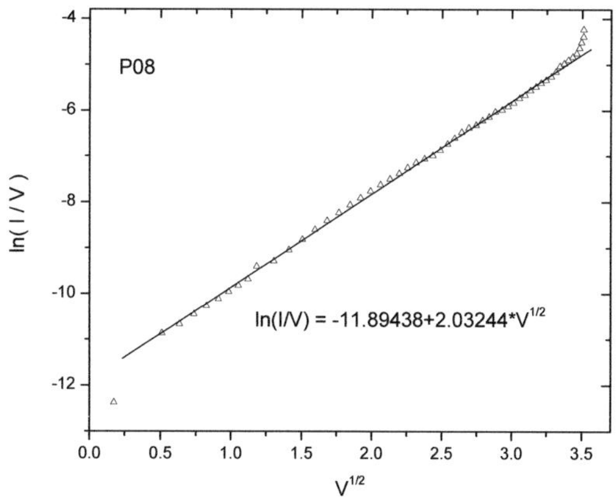

Fig. 5.8. The plot $\ln(I/V) - \sqrt{V}$ for a layer of m-DNB highly doped (10 wt%) with resorcinol grown between two Si wafers, Si (n), E/E, ρ=0.90-1.9 Ωcm, (P08 in Table 1) (Stanculescu, 2006 b)

6. Effect of metallic contact on the electrical properties of organic semiconductor thin films

The charge carrier injection at this type of interface is more complicated than at the interface metal/inorganic semiconductor because in this case, the contact properties can be influenced by the defects induced on the organic surface by the metallic contact and by the space charge effect that characterises the organic semiconductors, effect that can obscure or alter the real contact properties (Stanculescu, 2007 a).

The injection efficiency at the interface metal/organic semiconductor depends on many factors such as the energetic barriers, which must to be overcome by the charge carriers, the morphology and composition of the interface. Diffusion processes or reaction tacking place at the interface could also influence the behaviour of the metal/organic contact (Hirose, 1996). When the physical contact between metal and organic solid is realized, interface states are induced by metallization or interdiffusion (Hirose, 1996).

6.1 MIS type structures: metal/organic semiconductor/inorganic semiconductor

Different types of structures have been investigated from the point of view of the electrical conduction properties. In the structures based on perylene and PTCDA the organic layer has been obtained by vacuum deposition, while in the structure based on wide band gap organic semiconductor, with $T_{melting}$ < 100 °C, Cu/m-DNB/Si, by rapid directional solidification between two substrates one of Si and the other of an organic insulator substrate covered with Cu (Stanculescu, 2007 a).

The order of the steps for the deposition of the layers that compose the heterostructure (metal on organic or organic on metal) can be very important for the behaviour of the heterostructure (Hill, 1998). The chemisorption of the organic molecules in determined sites controlled by the extended π-electrons system irrespective of the functional groups appear when the organic layer is deposited on a metallic layer. When a metal is deposited on an organic layer an important modification of the interface may take place as a result of the

chemical reaction between the metal and the organic molecules and/or of diffusion of a large number of metallic atoms into the organic layer. These processes generate a high density of states in the organic semiconductor band gap (Hill, 1998).
The diffusion into the organic layer depends on the type of metal and plays an important role in the behaviour of the metal/organic contact. The main difference between different metals is generated by the degree of diffusion of the metal atoms in the organic film. The diffusion process depends on the reactivity of the metal atom with the organic compound. The reactivity of the organic molecule is strongly dependent on the molecular structure. The groups that substitute the hydrogen to the aromatic nucleus, and sometimes the ends groups like anhydride in the case of PTCDA, control the interaction of the metal with the organic molecule.
For perylene or perylene derivative (such as PTCDA) the transport of the charge carriers at the contact metal/organic is influenced by the properties of the single silicon electrode and by the polarization of the heterostructure. A low height injection barrier is sustained by a value of the metallic cathode work function (Φ) close to the value of the electron affinity of the organic. This assures a high level of electrons injection from metal to the organic layer.
Some studies have mentioned In with the work function, Φ_{In} =4.2 eV (Hirose, 1996) and Al with Φ_{Al}=4.25 eV (Hirose, 1996), as the most adequate metals for contacting PTCDA (EA=4 eV or 4.6 eV depending on the reference).
For example, the distance between the molecular planes parallel to the substrate in PTCDA films is large compared to the metal atomic and ionic radius and assures a natural path for the motion of the metallic atom/ion in the organic crystalline layer of PTCDA. The motion is also facilitated by the local deformation of the organic crystalline lattice characterised by weak van der Waals bonds (Hirose, 1996). The motion of the metal atoms can also be influenced by the presence of defects, such as grain boundaries, which have influence on the structural quality of the film.
The chemical reaction between In or Al on one hand and PTCDA on the other hand involves predominantly the anhydride group of the organic molecule generating the oxidation of the adatoms and the reduction of C atoms from the carbonyl group, as a consequence of the high affinity of these metals for oxygen. The heat of formation of the reactive metals oxides is with orders of magnitude higher (ΔH_{298}=-1676 KJ/mol for Al_2O_3 and ΔH_{298}=-926 KJ/mol for In_2O_3) than the heat necessary for Ag_2O formation (ΔH_{298}=-31 KJ/mol) and the heat necessary for the formation of ordinary metal oxides. This means that the generation of In, and Al oxides take place in the detriment of metal-metal bonding, because the more negative is the heat of formation the more stable is the compound. This is due to the fact that more energy is lost to the surroundings when the compound is formed and the compound has a lower energy being more stable (Stanculescu, 2007 b).
For PTCDA, the attachment of the electronegative anhydride group to the perylene core causes the increase of the ionisation energy of perylene, IE=5.1 eV (Hirose, 1996). PTCDA has higher ionisation energy, IE=6.2 eV (Hirose, 1996) or IE= 6.8 eV (Gao, 2001) and electron affinity EA=4 eV or EA=4.6 eV respectively and its contact with metals are characterised by the transfer of the negative charge carriers (electrons) from metal (In, Al, Ag) to organic molecule (Hill, 1998).
The charged metallic ions created at the interface are driven through the materials by the electric field and can react with the organic molecule, depending on its chemical structure and reactivity. Considering Cu and Al metallic electrodes, the low atomic radius of Cu (1.57

Å) and Al (1.82 Å) compared to the distance of separation between the molecular planes (for compounds characterised by a nearly planar molecular configuration such as PTCDA and ZnPc) assures a structural way for the motion of metal atoms (ions) in this type of compounds.

This high diffusivity of metal atoms/ions can be attributed to a relatively low value of the first ionisation energy that favours the transfer of an electron to the host organic matrix, this ionisation leading to a Coulomb type repulsion between positively charged In ions situated in interstitials sites (organic crystalline materials are characterised by large dimension of the interstitials zones), acting as driving forces for diffusion. The diffusion rate is also substantial for the other selected metals (Al) but slowly decreases in the succession In>Al because the first ionisation energy for these metals increases in the following succession: 5.786 eV (In) < 5.985 eV (Al) (Stanculescu, 2007 b).

The reactivity of Al with perylene is not significant (there are not substituent groups to the aromatic nucleus), Al atoms diffuses through the perylene film characterised by a separation of 3.46 Å between the parallel plans of molecules (Hirose, 1996). At an applied voltage of 0.1-0.2 V has been remarked a region of transition though a region characterised by a power dependence between I and V. This power dependence with an exponent n=2 characterises the presence of the space-charge limited currents (SCLC) and with n>2 characterises the presence of the trap-charge limited currents (TCLC).

As can be seen from Figure 6.1, the I-V characteristics of the Si/perylene/Al heterostructure are not linear at low applied voltages (< 1 V) , but tend to an ohmic behaviour through a weak superlinear dependence at forward (direct) polarization at very low voltages (<0.03 eV). At reverse polarization, the I-V dependence is weak sublinear at voltages <1 V (Stanculescu, 2007a). As can be seen in Figure 6.1 for direct bias there is a region between 0.1 and 1 V characterised by power dependence with an exponent n>2. The region with power dependence (n>2) between I and V is present also for reverse bias but it is shifted between 1 and 10 V.

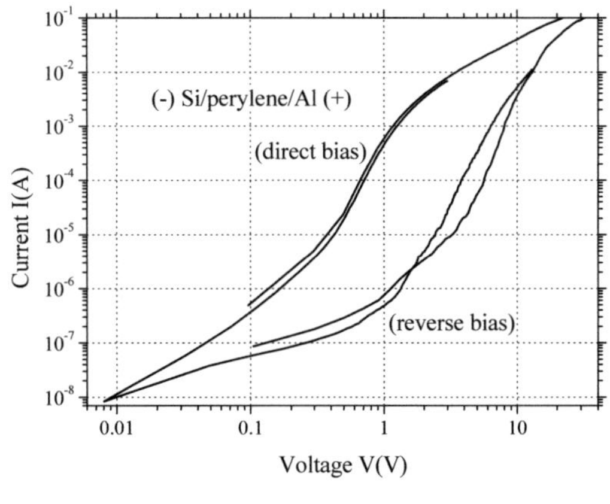

Fig. 6.1. I-V plot of Si/perylene/Al heterostructure (Stanculescu, 2007 a)

These experimental results confirm that the charge carriers transport by hopping and tunnelling mechanisms in the metal/organic semiconductor structure are favoured by the energetic states generated in the band gap by the diffusion of the metallic (Al; In) atoms and/or by the reaction taking place at the PTCDA (perylene)/Al(In) interface (Hirose, 1996).

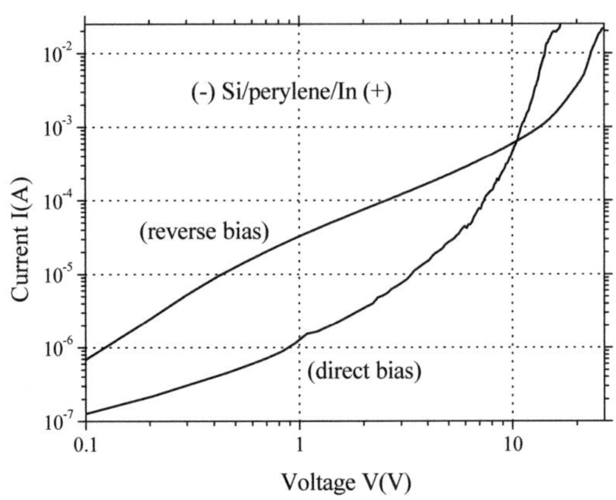

Fig. 6.2. I-V plot of Si/perylene/In heterostructure (Stanculescu, 2007 a)

The behaviour of the perylene/In contact is shown in Figure 6.2. In this case the reactivity of In is very weak, but the diffusion of In through the perylene layer is significant and induces states under the Fermi level. The ohmic character of this contact, associated with the charge carrier injection mechanism, evidenced at voltage lower than 2 V at direct bias and between 1 V and 10 V at reverse bias, can be explained by the conduction involving these states situated in the band gap, in the interface region, just below the Fermi level.

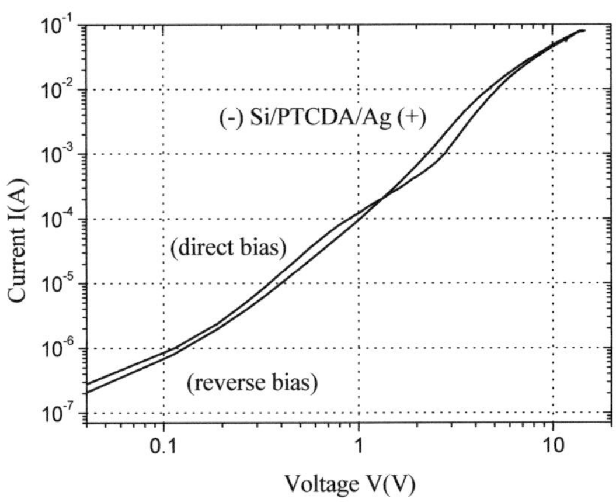

Fig. 6.3. I-V plot of Si/PTCDA/Ag heterostructure (Stanculescu, 2007 a)

The I-V characteristics of the PTCDA/Ag contact presented in Figure 6.3, are ohmic only at low voltages (<0.1 V), both at direct or reverse polarization. This can be explained by the reduced reactivity of silver with PTCDA molecule. The silver atoms are not inducing enough interface states in the band gap to be involved in the conduction process. The space charge limitation of the current (n=2) become important at slightly higher voltages (~ 0.1-1 V) both for forward and reverse polarization.

The quasi-linear I-V dependence of perylene/Ag contact shown in Figure 6.4, at applied voltages lower than 0.7 V, becomes a power dependence with n>2 suggesting the presence of defects (traps) in the organic semiconductor band gap and the appearance of the TCLC phenomenon at applied higher voltages. In the region of quasi linear dependence between I and V the current has high value, between 10^{-6} and 10^{-5} A, that favours, in time, the degradation of the organic layer of the structure.

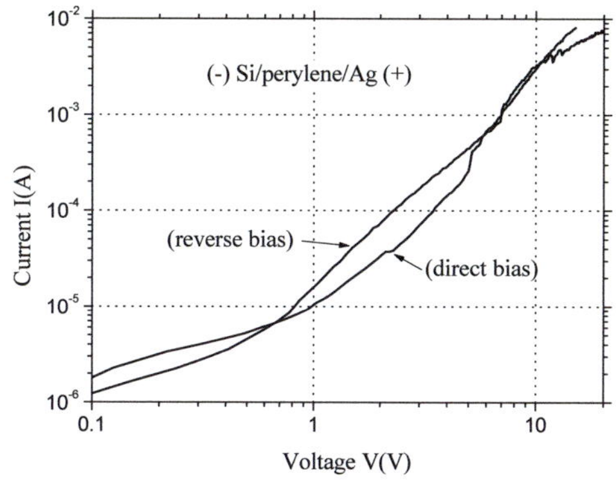

Fig. 6.4. I-V plot of Si/perylene/Ag heterostructure (Stanculescu, 2007 a)

The most important characteristic of In, Al and Ag diffusion process in organic layer is the high level of incorporation of isolated atoms and the absence of clustering process (Hirose, 1996).

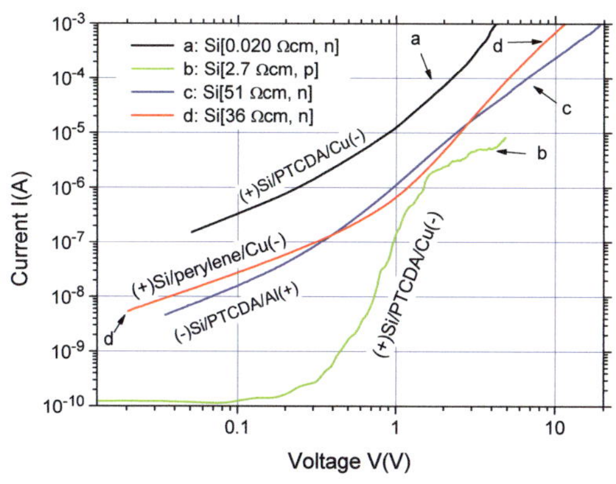

Fig. 6.5. I-V characteristics of metal/organic/Si heterostructures for different metallic electrodes and polarizations: metal-Cu, Al; organic-perylene, PTCDA (Stanculescu, 2008)

For silver and copper the situation is different because the relatively high first ionisation energy, 7.57 eV for Ag and 7.72 eV for Cu are not favourable to the diffusion in PTCDA as individual atoms, but are favourable to the clustering process. Another process that is favoured by these energetic considerations is the immobilization of the Cu atoms at the surface of the organic. The atoms of metals with high values of the first ionisation energy, such as Ag and Cu, remain neutral at the surface of the organic film and form cluster. The steeper increase in the current, that suggest the presence due to the effect due to the

interface beside the effect of the energetic barrier, has been evidenced in Figure 6.5 for the contact Cu/PTCDA [Φ_{Cu}=4.5 eV (Li, 2005)] at forward bias, (-) applied on Cu electrode and (+) on p type Si electrode.
For the heterostructure Cu/PTCDA/Si presented in Figure 6.5, with a semiconductor electrode of p type Si for an applied voltage of 0.4 eV, the current varied between 2×10^{-9} A at forward bias and 5×10^{-10} A at reverse bias, and for an applied voltage of 0.7 V, the current varied between 10^{-7} A at direct bias and 2×10^{-9} A at reverse bias showing a steep increase in current around 1 V.

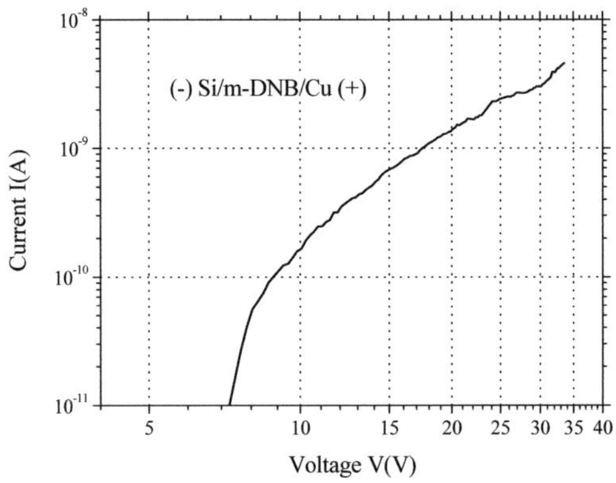

Fig. 6.6. I-V plot of Si/m-DNB/Cu heterostructure (Stanculescu, 2007 a)

For the heterostructure Cu/perylene/Si, for an applied voltage of 0.1 V no significant difference has been remarked between the current at direct and reverse bias (4-5×10^{-9} A). For an applied voltage of 0.4 V, the current varied between 2×10^{-6} A at forward bias and 2×10^{-7} A at reverse bias, suggesting a limiting behaviour of the metallic contact for applied voltages higher than 0.1 V. For n type Si electrode at voltages higher than 0.1 V, the current is higher, I=3×10^{-7} A in the heterostructure containing the contact Cu/PTCDA at forward bias, than in the heterostructure containing the contact Al/PTCDA [Φ_{Al}=4.28 eV (Eastment, 1973)], I=2×10^{-8} A, at reverse bias, as it is presented in Figure 6.5.
A distinct type of heterostructure Si/m-DNB/Cu presented in Figure 6.6 is characterised by a strong increase in the value of the current for voltages between 7 V and 10 V. For higher voltages situated between 10 and 20 V the behaviour of the I-V characteristic is close to linear.
Modifications in the I-V characteristics can be determined by modifications in the morphology of the film which are correlated with the preparation conditions. The preparation conditions are critical especially for the heterostructure based on rapid directional solidified films of m-DNB.

6.2 MIM type structures: metal/organic semiconductor/metal- (metal/organic/metal)

As it was already mentioned, there are two counteracting factors that affect the electrical behaviour of the organic/metal contact: the motion of the metallic atoms through the organic film determining the penetration depth correlated with the Coulomb repulsion and the chemical reaction with the substituent group that limits the motion of the atoms. The metal atoms diffusion stops when the saturation of the organic film takes place or when the front of molecules , which have not reacted, are too deep below the surface of the organic film.

As can be seen in Figure 6.7, the structure shows a characteristic close to ohmic for voltages between 1 V and 10 V. An exponential dependence between I and V , previously observed in m-DNB films doped with resorcinol, has been evidenced for PTCDA films at voltages < 1 V suggesting a Poole-Frenkel conduction mechanism (Stanculescu, 2006 b; Silveira, 2004) with a field dependent mobility of the charge carriers. At voltages > 10 V the effect of the space-charge limited current became significant.

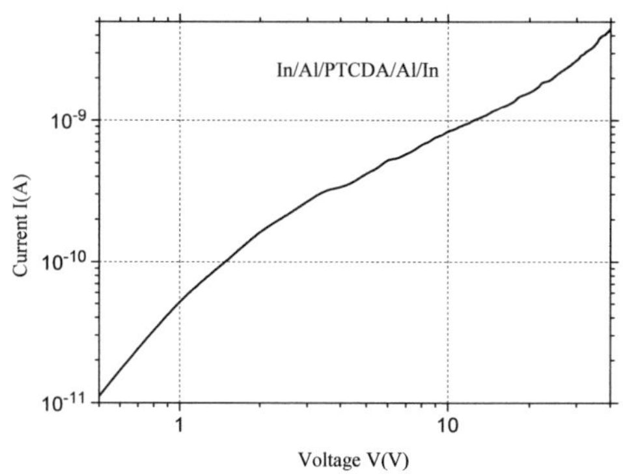

Fig. 6.7. I-V characteristic for Al/PTCDA/Al structure with planar grid electrodes (the contact with the electrodes was realised with In) (Stanculescu, 2007 a)

7. Organic heterostructures with buffer layer

A way to control the conduction properties through multilayer organic heterostructures is to introduce a buffer organic layer with special structural and energetical properties between the electrode and the first deposited organic layer. The quality of the initial organic layer deposited on a substrate can also be controlled by the morphology of this intermediate layer situated between the substrate and the organic layer being a way to minimize the influence of the deposition substrate.

A supplementary organic layer could facilitate the injection of the holes in the Si/organic/organic/Si structures, if the condition of a small difference between the ionisation energy of the supplementary organic layer and the valence band level in semiconductor (Si) is satisfied. The electrons injection could be facilitated in metal/organic/organic/Si structures for small differences between the work function of the metal and the electron affinity of the supplementary organic layer. The effect of the electrode can be eliminated using as contact electrodes the same type of Si wafers.

The injection of holes from the Si electrode to perylene or PTCDA could be improved with an intermediary organic layer having a higher level of the ionisation energy than perylene or PTCDA. An improvement in the injection of the charge carriers from Si to PTCDA has been obtained introducing an intermediate organic layer of ZnPc [IE=5.28 eV (Eastment, 1973); Eg=1.94 eV (Gao, 2003; Gao, 2001)] and generating a ZnPc/PTCDA, organic/organic heterojunction. The difference between the valence band energy in p type Si and the ionisation energy is smaller for ZnPc than for PTCDA and therefore the intercalation of the ZnPc layer leads to a significant increase (more than one order of magnitude) in the current that passes through the Si/PTCDA/Si structure, as can be seen in Figure 7.1. No significant change has been observed in Figure 7.1 for the structure Si/ZnPc/perylene/Si because ZnPc and perylene have close ionisation energy.

The injection properties in the structure Cu/organic/n type Si, for a given polarization can be influenced by the introduction of an intermediate layer of ZnPc. Because the electron affinity of ZnPc (EA=3.34 eV) is lower than that of PTCDA (EA=4 eV), an intermediate layer of ZnPc is not significantly improving the electron injection from Cu cathode to PTCDA. The current has increased from 10^{-5} A to 3×10^{-4} A, for an applied voltage of 1 V. An increase in injection could be obtained at forward bias with an intermediate organic layer with an electron affinity lower than PTCDA.

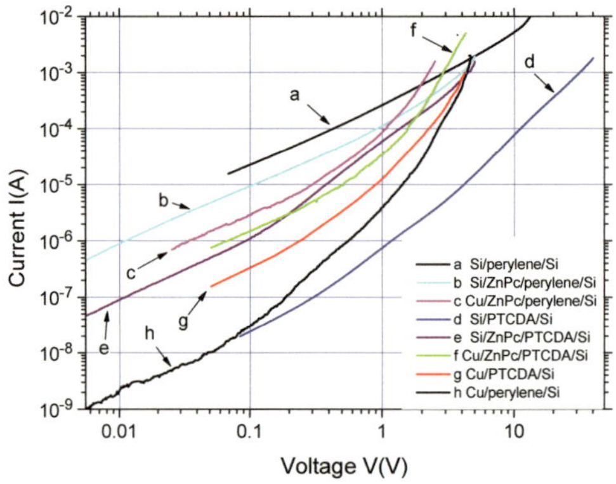

Fig. 7.1. The effect of the intermediate ZnPc layer on the I-V characteristics of the Si(Cu)/perylene (PTCDA)/Si heterostructures (Stanculescu, 2008)

Introducing a supplementary layer of ZnPc, for an applied voltage of 1 V, the current has increased from 8×10^{-7} A to 6×10^{-5} A in Si/PTCDA/Si heterostructure because the ionisation energy of ZnPc is situated between the hole energy in Si and the ionisation energy in PTCDA and favours the injection of the holes from p type Si to PTCDA.

A supplementary layer of ZnPc has not significantly influenced the conduction in the heterostructure Si/perylene/Si, because the ionisation energy of ZnPc and perylene have close value, 5.25 eV and 5.1 eV respectively. A significant increase in the electrons injection at forward bias in Cu/ZnPc/perylene/Si(p) structure is evidenced in Figure 7.1 and has been obtained because the difference between the work function of Cu and the electron affinity of the organic is smaller for ZnPc than for perylene. The current has increased from 3×10^{-6} A to 10^{-4} A, with ~ two orders of magnitude, for an applied voltage of 1 V.

8. SIS and MIS type structures on ITO substrate

The electrical conduction properties of different types of heterostructures realised on ITO/glass substrate, with single or multi organic layer and different types of metallic or p type Si contacts, at forward or reverse polarization, have been investigated. The electrode positively biased was name anode. High purity (5N) In was used to contact the electrodes.

The structures prepared with double organic layer, Alq3/PTCDA have shown similar behaviour revealed in Figure 8.1, for the same anode, Si(p)/In, and different cathodes (ITO/In; Al/In).

If the succession of the organic layer is changed from Alq3/PTCDA to PTCDA/Alq3 and if it is used the Al/In anode instead of Si(p)/In anode, no significant changes have appeared in the shape of the I-V characteristics presented in Figure 8.1, for voltages higher than 0.1 V. The only change is in the value of current. For an applied voltage of 1 V the highest current, $I=3x10^{-4}$ A ,was obtained in the structure consisting of a glass substrate, In contact, Al cathode, electron transporting layer Alq3, hole transporting layer PTCDA, Si(p) anode and In contact, In/Al/Alq3/PTCDA/Si(p)/In, presented in Figure 8.1.

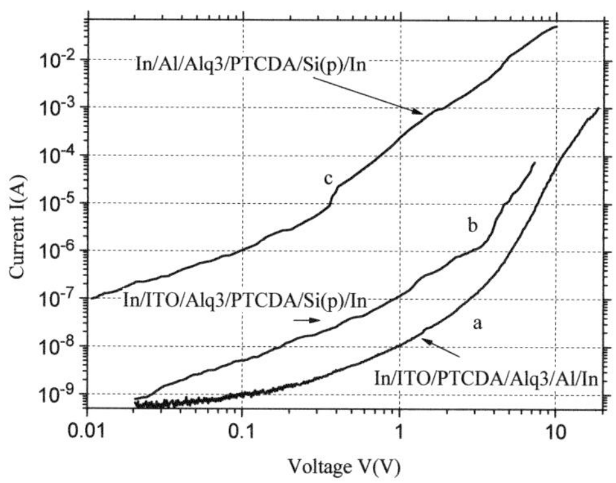

Fig. 8.1. I-V characteristics in double-organic layer structures: (Stanculescu, 2007 b)
(a) In/ITO/PTCDA/Alq3/Al/In;
(b) In/ITO/Alq3/PTCDA/Si(p)/In;
(c) In/Al/Alq3/PTCDA/Si(p)/In

In the structures ITO/organic/Si at reverse polarization (ITO is cathode), the barrier for holes is higher at the contact with α-NPD [IE=5.52 eV (Hirose, 1996), Eg=3.1 eV (Rajagopal, 1998)] compared to the contact with ZnPc [IE=5.28 eV (Gao, 2002), Eg=1.94 eV (Gao, 2001; Gao, 2003)]. In consequence, the injected current is two orders of magnitude lower in ITO/α-NPD/Si heterostructure than in ITO/ZnPc/Si heterostructure for voltages lower than 7 V, as it is presented in Figure 8.2.

This behaviour can be explained by the higher ionisation energy of α-NPD compared to ZnPc that determines better injection properties for the contact ITO/ZnPc compared to ITO/α-NPD.

Introducing a supplementary layer of perylene characterised by IE=5.1 eV (Kang, 2005) and Eg=2.5 eV (Kang, 2005) or PTCDA characterised by IE=6.8 eV (Rajagopal, 1998) and Eg=2.2 eV (Rajagopal, 1998), the current passing through the structure ITO/ZnPc/Si decreases from $I=2\times10^{-4}$ A to $I=3\times10^{-6}$ A in the structure with a supplementary layer of perylene, and from $I=2\times10^{-4}$ A to $I=6\times10^{-8}$ A in the structure with a supplementary layer of PTCDA, for an applied voltage of 1 V.

In Figure 8.2 can also be seen that, at voltages lower than 5 V the supplementary layer of α-NPD has as effect a decrease in the value of the current in the structure ITO/perylene/Si while the supplementary layer of ZnPc has as effect an increase in the value of the current.

This behaviour is determined by the height of the energetic barrier: $\Delta E_{\alpha\text{-NPD/perylene}}$=0.42 eV and $\Delta E_{ZnPc/perylene}$=0.18 eV. The current is predominantly a hole current resulting from the efficient hole injection from the low resistivity (6.7 Ωcm), p type Si anode.

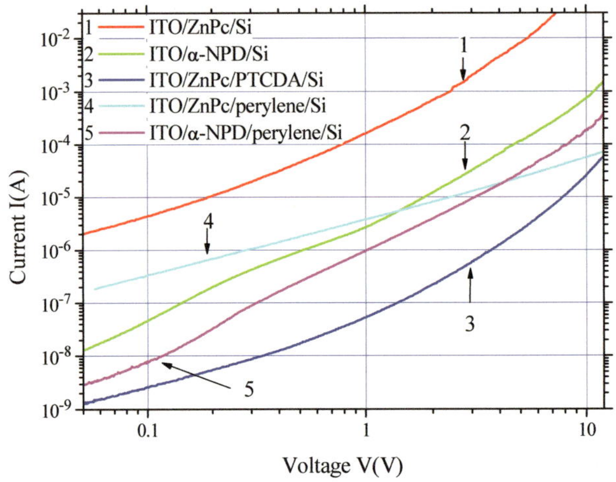

Fig. 8.2. I-V characteristics for ITO (Si)/single (double) p type organic thin film(s)/Si heterostructures (Rasoga, 2009):
1. ITO/ZnPc/Si(Cz, E/E, p: 0.018 Ωcm);
2. ITO/ α-NPD/Si(Cz, E/E, p: 0.64 Ωcm);
3. ITO/ZnPc/PTCDA/Si(Cz, E/E, p: 0.015 Ωcm);
4. ITO/ZnPc/perylene/Si(Cz, E/E, p: 0.014 Ωcm);
5. ITO/α-NPD/perylene/Si(Cz, E/E, p: 6.77 Ωcm)

9. Conclusion

The purpose of this chapter is to enlarge the knowledge about the electrical conduction processes in organic thin films and organic multilayer heterojunctions. The conduction properties of the following types of organic heterostructures have been analysed:

i. inorganic semiconductor [ITO, single silicon (Si)]/organic compound (Semiconductor/Insulator heterostructure=SI);
ii. organic compound/metal (Metal/Insulator heterostructure=MI);
iii. inorganic semiconductor (ITO, Si)/single (multi) organic layer(s)/metal, Si (Semiconductor/Insulator/Semiconductor heterostructure=SIS; Metal/Insulator/Semiconductor heterostructure=MIS);
iv. metal/organic layer/metal (Metal/Insulator/Metal structure=MIM).

Two basic processes are controlling the electrical transport reflected by the I-V characteristics in the inorganic semiconductor/organic/metal (semiconductor) heterostructures: (a) the injection of the charge carrier from electrodes into the semiconductor layer and (b) the transport of charge carriers in the bulk of the layer. The electrical transport is either injection limited by the energetic barrier at the interface or bulk transport limited by the low mobility of the charge carriers in organics (which usually is 2 or 3 orders of magnitude lower than in silicon) that induces a limitation associated with the space charge and space charge limited currents (SCLC). The electrical conduction properties can also be determined by the defects (grain boundaries and imperfections) generated during the deposition in the organic films and acting as trap centres that induce a limitation associated with the trap charge and trap charge limited current (TCLC).

Therefore, charge carrier injection at the interface between metal and organic is more complicated than the charge carrier injection at the contact metal/inorganic because of the defects than can be induced in the organic layer during the deposition of the metallic contact.

The efficient injection of the holes into a p type conduction organic semiconductor is possible from an ITO semiconductor layer with an effective electrode work function higher than the ionization potential of the organic that favours an ohmic contact and the injection of the charge carriers from electrode to organic.

In this chapter have been presented some comparative investigations on the electrical properties of the interface inorganic/organic and organic/organic in different single and multi organic layer based structures and the electrical conduction through different types of heterojunctions {examples: ITO, Si/wide-gap n type or/and p type organic semiconductors [5,10,15,20-tetra(4-pyridyl)-21H,23H-porphine synthetic (TPyP); tris(8-hydroxyquinoline) aluminium (Alq3) or/and zinc phthalocyanine (ZnPc); perylene; 3,4,9,10 perylenetetracarboxylic dianhydride (PTCDA), N,N'-di[(1-naphtalenyl)-N,N'-diaphenyl]-(1,1'-biphenyl)-4,4'-diamine (α-NPD); benzil, meta-dinitrobenzene (m-DNB)]/Si(Cu)}.

The effect of the types of electrodes, preparation conditions, succession order of the layers and molecular packing on electrical conduction in organic compound based heterostructures have been analysed.

10. References

Abkowitz, M. A. & Mizes, H. A. (1995). Emission limited injection by thermally assisted tunnelling into a trap free transport polymer. *Appl. Phys. Lett.*, Vol. 66, No. 10, (March 1995) 1288-1290, ISSN 0003-6951

Abkowitz, M.; Facci, J. S. & Rehm, J. (1998). Direct evaluation of contact injection efficiency into small molecule based transport layers: Influence of extrinsic factors. *J. Appl. Phys.*, Vol. 83, No. 5, (March 1998) 2670-2676, ISSN 0021-8979

Ameri, T.; G. Dennler, G.; Lungenschmied, C. & Brabec, C. J. (2009) Organic tandem solar cells: A review. *Energy Environ. Sci.*, Vol.2, No.4, 347-363, print: ISSN. 1754-5692, on line: ISSN 1754-5706

Antohe, S. (2008). Chapter 11 „Electronic and Optoelectronic Devices Based on Organic Thin Films" In *Handbook of ORGANIC ELECTRONICS and PHOTONICS (Electronic Materials and Devices)*, Hari Singh Nalwa, (Ed.), 435-446, American Scientific Publishers, ISBN: 1-58883-096-9, Los Angeles, California, USA

Arkhipov, V. I.; Emelianova, E. V.; Tak, Y. H. & H. Bässler, H. (1998). Charge injection into light-emitting diodes: theory and experiment. *J. Appl. Phys.*, Vol. 84, No. 2., (July 1998) 848-856, ISSN 0021-8979

Ashcroft, N. W. & Mermin, N. D. (1976). *Solid State Physics*, Thomson Learning Inc., ISBN-10: 0030839939; ISBN-13: 978-0030839931, London, UK

Baldo, M. A. & Forrest, S. R. (2001). Interface-limited injection in amorphous organic semiconductors. *Phys. Rev. B*, Vol. 64, No. 8, (August 2001) 085201/1-085201/17, ISSN 0163-1829

Borsenberger, P. M. & Weiss, D. S. (1998). Organic Photoreceptors for Xerography, In *Optical Engineering*, Borsenberger, P. M. (Ed.), Marcel Dekker Inc, ISBN-10: 0824701739. ISBN-13: 978-0824701734 , New York

Burrow, P. E.; Shen, Z.; Bulovic,V.; McCarty, D. M.; Forrest, S. R.; Cronin, J. A. & Thompson, M. E. (1996). Relationship between electroluminescence in organic heterojunction light-emitting devices. *J. Appl. Phys.*, Vol. 79, No. 10 (May 1996) 7991-8006, ISSN 0021-8979

Campbell, A. J.; Bradley, D. D. C. & Lidzey, D. G. (1997). Space-charge limited conduction with traps in poly(phenylene vinylene) light emitting dides. *J. Appl. Phys.*, Vol. 82, No. 12, (November 1997) 6326-6342, ISSN 0021-8979

Campbell, A. J.; Weaver, M. S.; Lidzey, D. G. & Bradley, D. D. C. (1998). Bulk limited conduction in electroluminescent polymer devices. *J. Appl. Phys.*, Vol. 84, No. 12, (December 1998) 6737-6746, ISSN 0021-8979

Cheng, G.; Zhang, Y.; Zhao, Y. & Liu, S. (2006). Improved efficiency for white organic light-emitting devices based on phosphor sensitized fluorescence. *Appl. Phys. Lett.*, Vol. 88, No.8, (February 2006) 083512/1-083512/3, ISSN 0003-6951

Cölle, M.; Gmeiner, J.; Milius, W.; Hillebrecht, H. & Brütting, W.(2003). Preparation and characterization of blue-luminescent tris(8-hydroxyquinoline)-aluminum (Alq3), *Adv. Funct. Mater.*, Vol. 13, No. 2, (February 2003) 108-112, print ISSN: 1616-301X; online ISSN: 1616-3028

Conwell, E. M. & Wu, M. W. (1997). Contact injection into polymer light-emitting diodes. *Appl. Phys. Lett.* Vol. 70, No.14, (April 1997) 1867-1869, ISSN 0003-6951.

Crispin, X.; Geskin, V.; Crispin, A.; Cornil,J.; Lazzaroni, R.; Salaneck, W. R. & Bredas, J. L. (2002). Characterization of the interface dipole at organic/ metal interfaces. *J. Am. Chem. Soc.*, Vol. 124, No. 27, (July 2002) 8131-8141, ISSN 0002-7863

De Boer, R. W. I. & Morpurgo, A. F. (2005). Influence of surface traps on space charge limited current. *Phys. Rev. B*, vol.72,No.7, 073207/1-073207/4, ISSN 0163-1829

Dini, D. (2005). Electrochemiluminescence from Organic Emitters. *Chem. Mater.*, vol.17, No.8, (April 2005)1933-1945, ISSN 0897-4756

Dodabalapur, A.; Torsi, L. & Katz, H. E. (1995). Organic transistors: two-dimensional transport and improved electrical characteristics. *Science*, Vol. 268, No. 5208, (April 1995) 270-271, ISSN 0036-8075

Duarte, F. J. & R. O. James (2009). Chapter 4 Tunable Lasers Based on Dye-Doped Polymer Gain Media Incorporating Homogeneous Distributions of Functional Nanoparticles, In *Tunable Laser Applications*, F. J. Duarte (Ed.), 121-142, 2nd Ed., CRC Press, print ISBN ISBN-10: 1420060090, ISBN-13: 978-1420060096, New York.

Duggal A. R. (2005). Chapter 10 Organic electroluminescent devices for solid state lightning, In *Organic Electroluminescence*, Z. Kafafi (Ed), 437-466, CRC Press, , print ISBN: 9780824759063, ISBN 10: 0824759060 , eBook ISBN: 978-1-4200-2820-1, New York

Eastment, R. M. & Mee, C. H. B. (1973). Work function measurements on (100), (110) and (111) surfaces of aluminium. *J. Phys. F: Metal Phys.*, Vol.3, No.9, (September 1973) 1738-1745, ISSN 0305-4608

Emtage, P. R. & O'Dwyer, J. J. (1966). Richardson-Schottky effect in insulators. *Phys. Rev. Lett.*, Vol.16, No.9, (February 1966) 356-358

Gao, W. & Kahn, A. (2001). Controlled p-type doping of an organic molecular semiconductor. *Appl. Phys. Lett.*, 79, 4040-4042, ISSN 0003-6951

Gao, W. & Kahn, A. (2002). "Electronic structure and current injection in zinc phthalocyanine doped with tetrafluorotetracyanoquinodimethane: Interface versus bulk effects", *Org. Electr.*, Vol.3, No.2, (June 2002) 53-63, ISSN 1566-1199

Gao, W. & Kahn, A. (2003). Controlled p doping of the hole-transport molecular material N,N-diphenyl-N,N-bis(1-naphthyl)-1,1-biphenyl-4,4-diamine with tetrafluorotetracyanoquinodimethane. *J. Appl. Phys.*, Vol.94, No.1, (July 2003) 359-366, ISSN 0021-8979

Gartstein, Y & Conwell, N. E. M. (1996). Field-dependent thermal injection into a disordered molecular insulator. *Chem. Phys. Lett.*, 255, 93-98, ISSN 0009-2614

Hill, I. G.; Rajagopal, A.; Kahn, A. & Hu, Y. (1998). Energy-level alignment at interfaces between metals and the organic semiconductor 4,4'-N,N'-dicarbazolyl-biphenyl. *J. Appl. Phys.*, Vol. 84, No. 6, (September 1998) 3236-3241, ISSN 0021-8979

Hirose, Y.; Kahn, A.; Aristov,V.; Soukiassian, P.; Bulovic, V. & Forrest, S. R. (1996). Chemistry and electronic properties of metal-organic semiconductor interfaces: Al, Ti, In, Sn, Ag, and Au on PTCDA. *Phys. Rev. B*, Vol.54, No.19, (November 1996) 13748-13758, ISSN 0163-1829

Hofmann, O.; Miller, P.; Sullivan, P.; Jones, T. S.; deMello, J. C.; Bradley, D. D. C. & deMello, A. J. (2005). Thin-film organic photodiodes as integrated detectors for microscale chemiluminescence arrays. *Sen. And Act. B*, Vol.106, No.2, (May 2005) 878-884, ISSN 0925-4005

Horowitz, G.; Hajlaoui, R.; Bourgouiga, R. & Hajlaoui, M. (1999). Theory of organic field-effect transistors. *Synth. Met.*, Vol.101, No.1-3, (May 1999) 401-404, ISSN 0379-6779

Huang, J.; Pfeiffer, M.; Werner, A.; Blockwitz, J.; Liu, S. & Leo, K. (2002). Low-voltage organic electroluminescent devices using pin structures. *Appl. Phys. Lett.*, Vol.80, No.1, (January 2002) 139-141, ISSN 0003-6951

Hung, L. S.; Tang, C. W. & Mason, M. G.(1997). Enhanced electron injection in organic electroluminescence devices using an Al/LiF electrode. *Appl. Phys. Lett.*, Vol.70, No.2, (January 1997) 152-154, ISSN 0003-6951

Inoue, Y.; Sakamoto, Y.; Suzuki, T.; Kobayashi, M.; Gao, Y. & Tokito, S. (2005). Organic thin-film transistors with high electron mobility based on perfluoropentacene. *Jap. J. Appl. Phys.*, 44, No. 6A, (June 2005) 3663-3668, print ISSN 0021-4922, online ISSN 1347-4065

Kalinowski, J.; Palilis, L. C.; Kim, W. H. & Kafafi, Z. H.(2003). Determination of the width of the carrier recombination zone in organic light-emitting diodes. *J. Appl. Phys.*, Vol.94, No.12, (December 2003) 7764-7767, ISSN 0021-8979

Kang, S. J.; Yi, Y.; Cho, K.; Jeong, K.; Yoo, K. -H. & Whang, C. N. (2005). Electronic structure of perylene on Au studied by ultraviolet photoelectron spectroscopy and density functional theory. *Synth. Metals*, Vol.151, No.2, (June 2005) 120-123, ISSN 0379-6779

Kao, K. C. & Hwang, W. (1981). *Electrical Transport in Solids*, Pergamon Press, ISBN-10: 0080239730; ISBN-13: 978-0080239736 , Oxford, UK

Kao, K. C. (2004). *Dielectric Phenomena in Solids*, Elsevier Academic Press, ISBN-13: 9780123965615, San Diego, California, U. S. A

Karl, N. (2000). Chapter 3, Charge carrier mobility in organic molecular crystals In *Organic Electronic Materials, Part II, Low Molecular Weight Organic Solids*, R. Farchioni, G. Groosso, (Ed.), 215-241, Springer-Verlag, ISSN 0933-033x; ISBN 3-540-66721-0, Berlin

Katz, H. E.; Lovinger, A. J.; Johnson, J.; Kloc, C.; Siergist, T.; Li, W.; Lin, Y.-Y.; & Dodabalapur, A. (2000 a). A soluble and air-stable organic semiconductor with high electron mobility. *Nature,* Vol.404, No. 6777, (March 2000) 478-481 ISSN 0028-0836

Kommandeur J. (1965) Conductivity, In: *Physics and Chemistryof the Organic Solid State,* D. Fox, M. M. Labes, A. Weissberger (Ed.), vol. 2, 1-66, Interscience, New York

Lambert, M. A. & Mark, P. (1970). *Current Injection in Solids,* Academic Press, ISBN-10: 0124353509; ISBN-13: 978-0124353503, New York

Lane, P. A. & Kafafi, Z. H. (2005). Solid-State Organic Photovoltaics: A Review of Molecular and Polymeric Devices, In: *Organic Photovoltaics: Mechanisms, Materials and Devices,* Sam-Shajing Sun, Niyazi Sedar Sariciftci (Ed.), CRC Press, 49-106ISBN 9780824759636; ISBN 10: 082475963X, Boca Raton, Florida, U.S.A.

Lee J.; Lee J.-Ik; Lee J.; Lee J. Y.; Kangc D. M.; Yuanc W.; Kwon S.-Ki & Chu H. Y. (2009). Efficient white phosphorescent organic light-emitting diodes for solid-state lighting applications using an exciton-confining emissive-layer structure", *Journal of Information Display,* 10, 92-, ISSN 1598-0316

Li, W. & Li, D. Y. (2005). On the correlation between surface roughness and work function in copper. *J. Chem. Phys.,* Vol.122, No.6, (February 2005) 064708-064714, ISSN 0022-3654

Liu X.; Li, H.; Song, C.; Liao, Y. & Tian, M. (2009). Microcavity organic laser device under electrical pumping. *Opt. Lett.,* Vol.34, No.4, (February 2009) 503-505, ISSN 0146-9592

Lu, M. –H. & Sturm, J. C. (2002). Optimization of external coupling and light emission in organic light-emitting devices: modelling and experiment. *J. Appl. Phys.,* Vol.91, No.2, (January 2002) 595-604, ISSN 0021-8979

Makinen, A. J.; Hill, I. G. & Kafafi, Z. H. (2002). Vacuum level alignment in organic guest-host systems. *J. Appl. Phys.,* Vol.92, No.3, (August 2002) 1598-1603, ISSN 0021-8979

Mas-Torrent, M. & Rovira, C. (2008). Novel small molecules for organic field-effect transistors: towards processability and high performance. *Chem. Soc. Rev.,* 37, (February 2008) 827-838, ISSN 0306-0012

Molinari, A.; Gutiérrez, I.; Hulea, I, N.; Russo. S. & Morpurgo, A. F. (2007). Bias-dependent contact resistance in rubrene single-crystal field effect transistors", *Appl. Phys. Lett.,* Vol.90, No. 21., (May 2007) 212103/1-212103/3, ISSN 0003-6951

Mott, N. F. & Gurney, R. W. (1940). *Electronic Processes in Ionic Crystals,* Oxford University Press, New York

Murgatroyd, P. N. (1970). Theory of space-charge-limited current enhanced by Frenkel effect. *J. Phys. D,* Vol.3, No.2, (February 1970) 151-156, ISSN 0022-3727

Parker, I. D (1994). Carrier tunneling and device characteristics in polymer light-emitting diodes. *J. Appl. Phys.,* Vol.75, No.3, (February 1994) 1656-1666, ISSN 0021-8979

Rajagopal, A.; Wu, C. I. & Kahn, A. (1998). Energy level offset at organic semiconductor heterojunctions. *J. Appl. Phys.,* Vol.83, No. 5, (March 1998) 2649-2655, ISSN 0021-8979

Rasoga, O.-L.; Socol, M. & Stanculescu, F. (2009). Charge carrier transport phenomena in some organic heterojunctions *J. Opt. Adv. Materials,* Vol.11, No.4, (April 2009) 509-514, print ISSN 1454-4164; on line ISSN 1841-7132

Riess, W.; H. Riel, H.; Beierlein, T.; Brütting, W.; Müller, P. & Seidler, P. F. (2001). Influence of trapped and interfacial charges in organic multilayer light-emitting devices. *IBM J. Res. & Dev.*, Vol.45, No.1 (January 2001) 77-88, ISSN 0018-8646.

Roichman, Y. & Tessler, N. (2002). Generalized Einstein relation for disordered semiconductors—implications for device performance *Appl. Phys. Lett.*, Vol.80, No.11, (March 2002) 1948-1950, ISSN 0003-6951

Schön, J. H.; S. Berg, S.; Kloc, Ch. & Batlogg, B. (2000 a). Ambipolar pentacene field-effect transistors and inverters. *Science*, Vol.287, No. 5455, (February 2000) 1022-1023, ISSN 0036-8075

Scott, J. C. & Malliaras, G. G. (1999). Charge injection and recombination at the metal-organic interface. *Chem. Phys. Lett.*, Vol.299, No.2, (January 1999) 115-119, ISSN 0009-2614

Silveira, W. R. & Marohn, J. A. (2004). Microscopic view of charge injection in an organic semiconductor. *Phys. Rev. Lett.*, Vol.93, No.11, (September 2004) 116104/1-116104/4, ISSN 0031-9007

Stanculescu, A.; Antohe, S.; Alexandru, H. V.; Tugulea, L.; Stanculescu, F. & Socol, M. (2004). Effect of dopant on the intrinsic properties of some multifunctional aromatic compounds films for target applications. *Synth. Met.*, Vol.147, No.1-3, (December 2004) 215-220, ISSN 0379-6779

Stanculescu, A.; Tugulea, L.; Alexandru, H. V.; Stanculescu, F. & Socol, M. (2005). Molecular organic crystalline matrix for hybrid organic-inorganic (nano) composite materials. *J. Cryst. Growth*, Vol.275, No.1-2, (February 2005) e1779-e1786, ISSN 0022-0248

Stanculescu, F.; Stanculescu, A. & Socol, M. (2006 a). Light absorption in meta-dinitrobenzene and benzil crystalline films. *J. Opt. Adv. Materials*, Vol8, No.3, (June 2006) 1053-1056, print ISSN 1454-4164; on line ISSN 1841-7132

Stanculescu, A.; Stanculescu, F.; Alexandru, H. & Socol, M. (2006 b). Doped aromatic derivatives wide-gap crystalline semiconductor structured layers for electronic applications. *Thin Solid Films*, Vol.495, No.1-2, (January 2006) 389-393, ISSN 0040-6090

Stanculescu A.; Stanculescu, F.; Tugulea, L. & Socol, M. (2006 c). Optical properties of 3,4,9,10-perylenetetracarboxylic dianhydride and 8-hydroxyquinoline aluminium salt films prepared by vacuum deposition. *Mat. Sci. Forum*, Vol.514-516, Part 2, 956-960, ISSN 0255-5476

Stanculescu, F.; Stanculescu, A. & Socol, M. (2007 a). Effect of metallic contact on the electrical properties of organic semiconductor film. *J. Opt. Adv. Materials*, Vol.9, No.5, (May 2007) 1352-1357, print ISSN 1454-4164, on line ISSN 1841-7132

Stanculescu, A. & Stanculescu, F. (2007 b). Investigation of the properties of indium tin oxide-organic contacts for optoelectronic applications. *Thin Solid Films*, Vol.515, No.24, (October 2007) 8733-8737, ISSN 0040-6090

Stanculescu, A.; Stanculescu, F.; Socol, M. & Grigorescu, O. (2008). Electrical transport in crystalline perylene derivatives films for electronic devices. *Solid State Science*, Vol.10, No.12, (December 2008) 1762-1767, ISSN 1293-2558

Sze, S. M. (1981). *Physics of Semiconductor Devices*, Wiley, ISBN-10: 0471056618; ISBN-13: 978-0471056614, New York

Tang, C. W. & Van Slyke, S. A. (1988). Organic electroluminescent diodes. *Appl. Phys. Lett.*, Vol.51, No.12, (September 1987) 913-915, ISSN 0003-6951

Torsi, L.; Dodabalapur, A.; Sabbatini, L.; Zambonoin, P. (2000). Multi-parameter gas sensors based on organic thin-film-transistors. *Sens. And Act. B*, Vol.67, No.3, (September 2000) 312-316, ISSN 0925-4005

Troshin, P. A; Lyubovskaya, R. N. & Razumo, V. F. (2008). Organic solar cells: structure, materials, critical characteristics, and outlook. *Nanotechnologies in Russia,* Vol.3, No.5-6, 242-271, print ISSN: 1995-0780, online ISSN: 1995-0799

Van Slyke, S. A.; Chen; C. H. & Tang, C. W. (1996). Organic electroluminescent devices with improved stability. *Appl. Phys. Lett.*, Vol.69, No.15, 2160-2162, ISSN 0003-6951

Wang, Y.; Gao, W.; Braun, S.; Salaneck, W. R.; Amy, F.; Chan, C. & Kahn, A. (2005). Enhancement of iridium-based organic light-emitting diodes by spatial doping of the hole transport layer. *Appl. Phys. Lett.*, Vol.87, No.19, (October 2005) 193501-193503, ISSN 0003-6951

Wu, C. I.; Hirose, Y.; Sirringhaus, H. & Kahn, A. (1997). Electron-hole interaction energy in the organic molecular semiconductor PTCDA. *Chem. Phys. Lett.*, Vol.272, No.1-2, (June 1997) 43-47, ISSN 0009-2614

Yersin H. (Ed.) (2007). *Highly Efficient OLEDs with Phosphorescent Materials,* Wiley-VCH Verlag GmbH & Co, KGaA, ISBN 978-3-527-40594-7, Weinheim

Zhi-lin, Z.; Xue-Jin, J.; Shao-Hong, X.; Nagatomo, T. & Omoto, O. (1998). The effect of rubrene as a dopant on the efficiency and stability of organic thin film electroluminescent devices. *J. Phys. D. Appl. Phys.*, Vol.31, No.1, (June 1998) 32-35, ISSN 0022-3727

4

Thin-Film Diamond Phototransistors

Linjun Wang, Jian Huang, Ke Tang, Jijun Zhang and Yiben Xia
School of Materials Science and Engineering, Shanghai University, Shanghai 200072, China

1. Introduction

In 21st century, with optoelectronic integration technology fast developing, the new technology has put forward higher demand for devices operating in high-power, high-frequency, and high-temperature environment. Under the circumstances, the devices based on common used semiconductor meterials (silicon, GaAs, etc.) more and more show the limitation of properties. The researches of new generation semiconductors suitable for application in severe environments (high-power, high-temperature, high radiation flux, etc.) focus on wide bandgap semiconductors such as SiC[1-2] 、GaN[3-4] and diamond .

The comparison of main properties between diamond, GaN, SiC and commonly used semiconductors are shown in table 1 [5-8], from which we can see that diamond is one the most promising candidate for new generation semiconductor material due to its unique properties, including wide bandgap, high carrier mobility, high hole-saturation velocity, highest thermal conductivity, high electric breakdown field, and chemical inertness, etc. In the past decades, the development of diamond-based devices are hampered by several problems, a big issue is the high price and rare resource of single diamond. However, in 1980s, the success of diamond film synthesis by chemical vapor deposition(CVD) have opened up the possibilities for a wide range of applications, such as high-temperature, high-power microelectronics device, and ultraviolet light emitting optoelectronics. Over the past few years a variety of state-of-the-art diamond film devices have been fabricated, analysed and simulated including field effect transistors (FET).

Two concepts have been developed concerning p-channel diamond film FETs: boron-doped surface channel FET and hydrogen-induced surface channel FET. However, boron is not an ideal dopant as at moderate concentration levels it displays a deep acceptor level of about 0.37eV resulting in low carrier densities at room temperature [9-10]. In 1989, Ravi and Landstrass[11] reported a substantial surface conductivity of hydrogenated diamond surfaces (p-type semiconducting layer), both of single crystals and of films prepared by chemical vapor deposition, respectively. This surface conductivity is unique among semiconductors and can be a promising candidate for the application in electronic devices due to the smaller acceptor activation energy less than 50meV [12-13].

While boron-doped channel FETs still encounter serious technological problems, hydrogen-terminated surface channels have been successfully used for the fabrication of Schottky diodes[14], metal semiconductor gate FETs (MESFETs) and metal–insulator gate FETs (MISFETs) [15]. To date, although some valuable information about this hydrogen-terminated surface conductivity of diamond has been obtained, it is still far from sufficient and more detailed research is indispensable.

Properties	diamond	β-SiC	GaAs	CdZnTe	GaN	Si
Atomic number / Z	6	14/6	31/33	48/30/52	31/7	14
Hardness/(GPa)	100	3.43	0.59			0.98
Bandgap/(ev)	5.5	3.0	1.43	1.5-2.2	3.45	1.12
Thermal expansion coefficient / (10-6/°C)	1.1	4.7	5.9		5.6	2.6
Dielectric constant	5.7	9.7	12.5	10.9	9	11.8
Resistivity/(Ω·cm)	>1013	150	108	1011	>1010	105
Electron mobility /(cm^2/V·s)	2200	400	8500	1350	1250	1500
Hole mobility /(cm^2/ V·s)	1800	50	400	120	850	600
Breakdown field/(104V/cm)	1000	400	40	0.15	>100	30
Thermal conductivity /(W·cm^{-1}·K^{-1})	20	5	0.46		1.3	1.5
Electron sacturated velocity/(107cm/s)	2.7	2.5	1		2.2	1
Working temperature /(°C)	<800		130	300	>300	77

Table 1. Comparison of properties between diamond, GaN, SiC and commonly used semiconductors

In this paper, high quality freestanding diamond (FSD) films and formation of H-terminated p-type channel on the diamond film surface were investigated and the origin of this H-terminated high-conductivity layer was discussed. The realization and properties of the optically activated MESFETs were also described here.

2. Preparation and characterization of FSD films

In this work, a microwave plasma chemical vapour deposition (MPCVD) technique at 2.45 GHz using a gaseous mixture of methane, hydrogen was applied to deposit FSD films on p-type low resistivity single crystalline silicon substrates. All three gases were metered into the chamber using mass flow controllers. The deposition parameters for FSD films were shown in table 2.

Flux of hydrogen (sccm)	Flux of methane (sccm)	Chamber pressure (KPa)	Substrate temperature (°C)
100	1	4.5	800

Table 2. Parameters for FSD deposition

After deposition, the silicon substrates were chemically etched to obtain FSD films with a smooth surface at the nucleation side. The FSD films were disposed in mixed solution of H_2O_2 and H_2SO_4 for 15 min to eliminate non-diamond surface layer of the diamond films.

These samples were then cleaned by ultrasonic vibration in deionized water. A thermal annealing treatment in nitrogen atmosphere at 650°C for an hour was performed to further improve the quality of the diamond films. The thickness of the film was about 110μm, as shown in Fig.1.

Fig. 1. Cross section image of the FSD film

Figure 2 and figure 3 showed SEM images of growth surface and nucleation surface of FSD films. The mean grain size of growth surface of the FSD film range from a few micrometers to tens of micrometers and the growth surface was very rough. Whereas, the nucleation surface was very smooth

Fig. 2. SEM image of growth surface of FSD

The typical AFM images of the nucleation side of FSD film are shown in Fig. 4, from which it could be seen that the nucleation side was very smooth with a mean surface roughness of about 10 nm in a scanning area of 1.5×1.5 μm^2. The result was consistent with that obtained from SEM images.

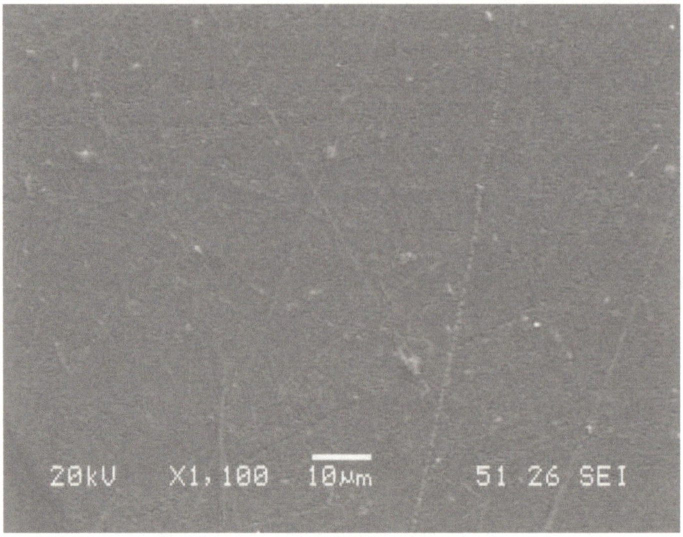

Fig. 3. SEM image of nucleation surface of FSD

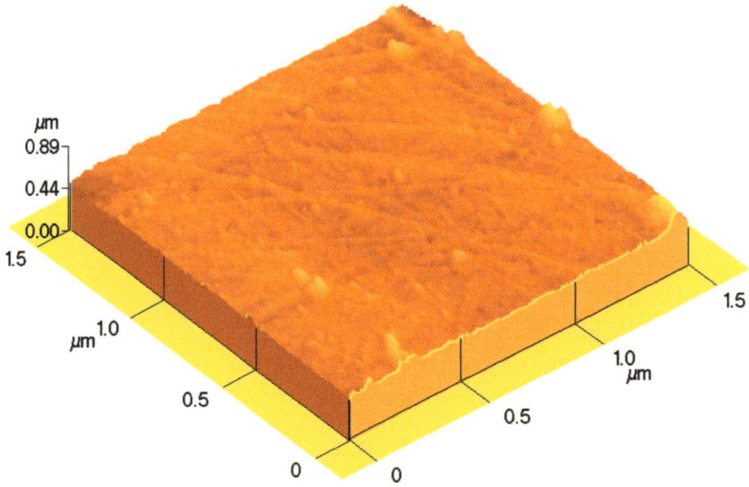

Fig. 4. AFM image of nucleation surface of FSD film

The Raman spectrum of the nucleation side of freestanding diamond film was shown in Fig.5. For both FSD films with and without post-treatment, a strong Raman scattering peak, located at about 1332cm^{-1}, is the characteristic of diamond and a weaker Raman scattering band, existed in range of 1400~1600cm^{-1}, is the characteristic of non-diamond carbon [16]. It is well known that the Raman signal for non-diamond carbon phase is about 75 times of that for diamond. So Raman spectroscopy is also used to estimate the non-diamond carbon content (C_{nd}) [17]: Cnd =1/ [1+75(I_{dia}/I_{nd})], where Idia is Raman peak intensity for diamond crystals and Ind is Raman peak intensity for non-diamond carbon phase. Therefore, the Raman results from Fig.5 indicated a high quality diamond of nucleation side of FSD diamond films with low content of non-diamond carbon. The Fig.5 also revealed that post-treatment (wet chemical etch, annealing process) was helpful to improve the quality of FSD films.

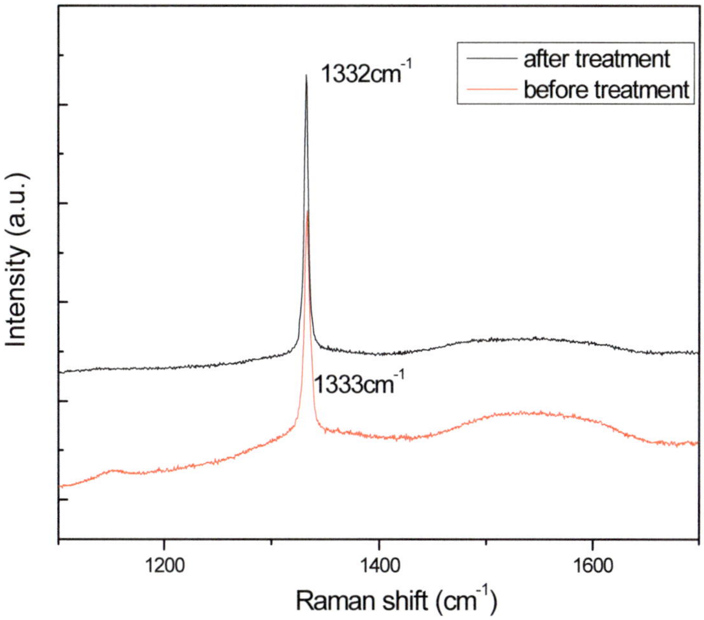

Fig. 5. Raman spectrum of nucleation side of freestanding diamond film

3. Preparation and characterization of H-terminated p-type channel on FSD films

The FSD films prepared above were exposed to hydrogen plasma at 750°C using a MPCVD apparatus. The time of hydrogen plasma treatment on the p-type behavior of undoped FSD nucleation surfaces were investigated by Hall Effect measurement. The electrical properties of the FSD nucleation surfaces following different annealing temperature were also measured.

Figure 6 shows the sheet carrier density and sheet resistivity of p-type FSD films as a function of time of hydrogen plasma treatment. The sheet carrier density rises with the time of hydrogen plasma treatment and a stable value is achieved after about 30 min, whereas the sheet resistivity reduces with time, with stable value being achieved after the same period of time.

The sheet carrier density and sheet resistivity of p-type FSD films as a function of annealing temperature in air and in vacuum were shown in Fig.7 and Fig.8, respectively. All the FSD films are exposed to hydrogen plasma treatment for 30 min at 750°C before a 180 min annealing process. The values of sheet carrier density and sheet resistivity remained in a relative stable range but change dramatically after annealing at temperature above 250 °C in air, whereas the sheet carrier density and sheet resistivity kept a stable value up to 600°C in vacuum.

Figure 9 and Fig.10 showed the values of sheet carrier density and sheet resistivity obtained as a function of time that a sample had been annealed at various temperatures in air. The data presented are typical of that obtained for many samples. Following annealing at 100 °C, little variation was apparent in measurements taken over a prolonged period. However, if the film was annealed at higher temperatures of 200 °C and 250°C the sheet carrier density was seen to decrease with time, although a stable value was reached after a given period, whereas the film resistivity showed a reversed trend, gradually increasing to a stable resistivity with the increase of annealing.

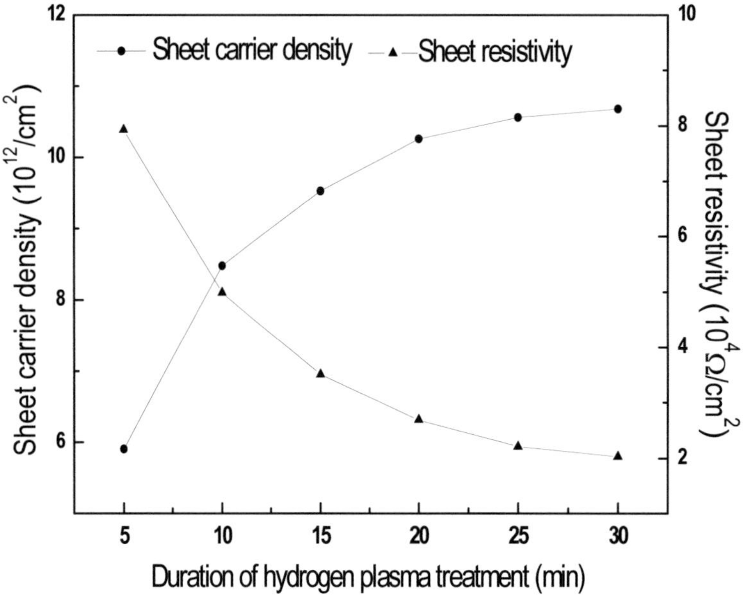

Fig. 6. Sheet carrier density and sheet resistivity of FSD against duration of hydrogen plasma treatment

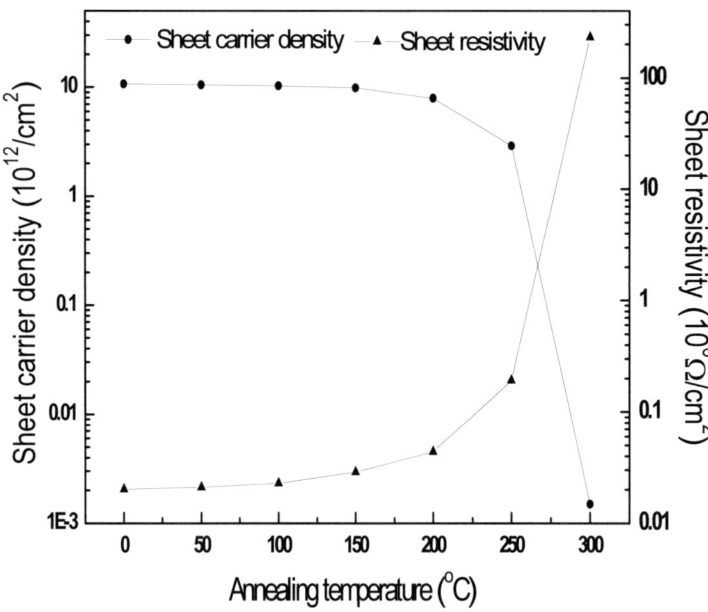

Fig. 7. Sheet carrier density and sheet resistivity of FSD against annealing temperature in air

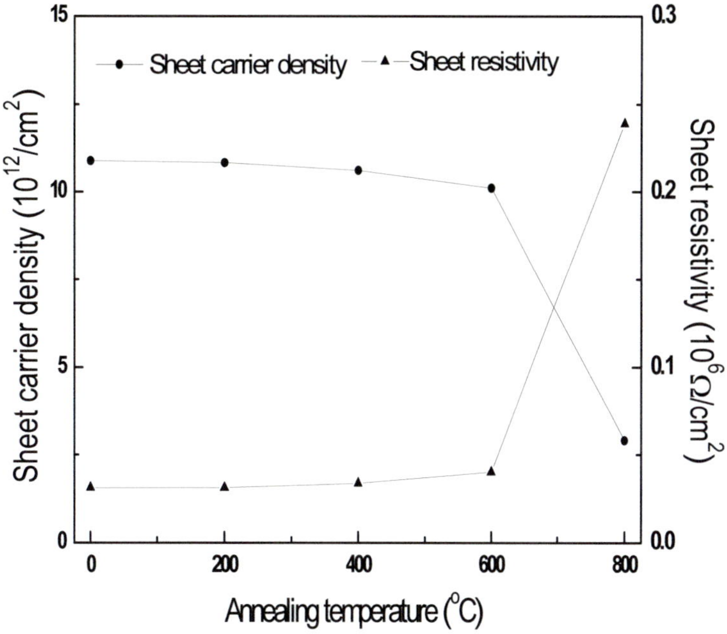

Fig. 8. Sheet carrier density and sheet resistivity of FSD against annealing temperature in vacuum

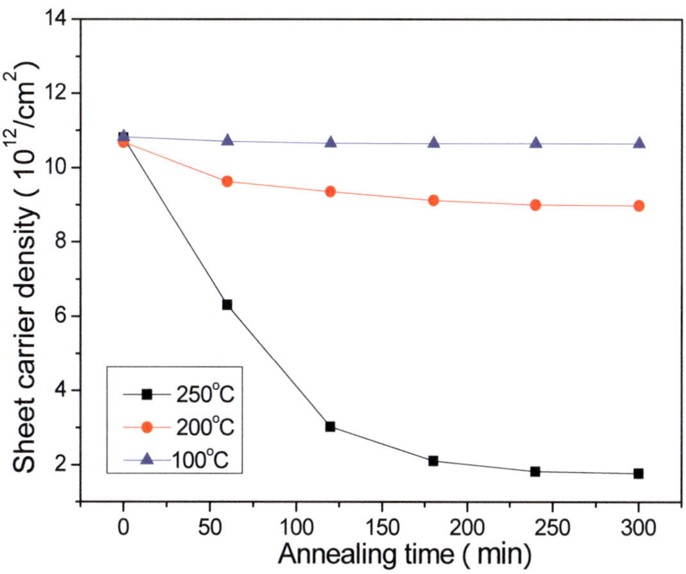

Fig. 9. Sheet carrier density against time of annealing

It's well known that the growth of diamond films at low pressure is a kinetic nonequilibrium process because of the thermodynamic instability of diamond under these conditions [18]. During such a nonequilibrium growth process, hydrogen plasma or atomic hydrogen radicals get rid of the graphitic phase, and sp3 species are rearranged to form crystalline diamond by gradually adjusting their positions and orientations near the growth face region [19]. The incompletely grown subsurface layer, where vacancies and dangling bonds were concentrated, was continuously transformed into the "perfect" bulk diamond [20]. Thus an imperfect thin layer will always exist on the diamond growth face. Hayashi et al. also reported that high density hydrogen was incorporated into the subsurface region rather than in bulk of as-grown diamond films by secondary ion mass spectroscopy (SIMS) [21]. Therefore, the diamond surface conductivity may be related to the complexes of absorbed hydrogen atoms with carbon dangling bonds. For example, if two adjacent carbon-dangling bonds share one hydrogen atom, an acceptor state should be generated in the band gap, since each hydrogen atom has only one electron.

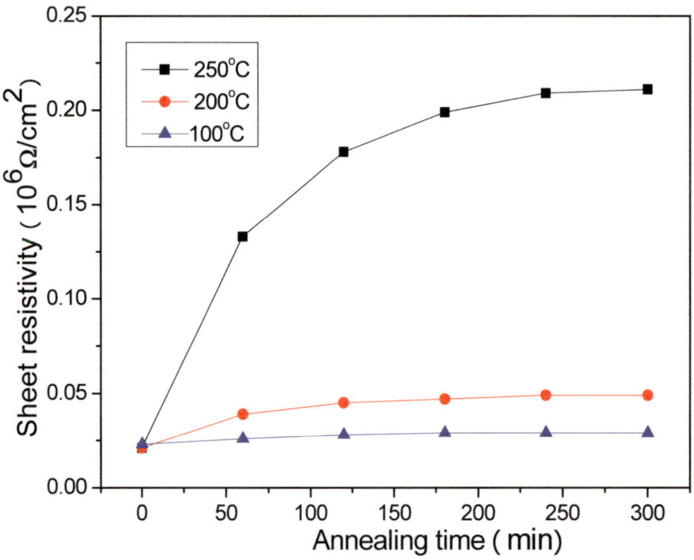

Fig. 10. Sheet resistivity against time of annealing

Starting from this point, the above experimental results can be well understood. The nucleation surface of the CVD diamond film is full of defects (e.g., vacancies, dangling bonds). The hydrogen plasma treatment may promote the complexes of hydrogen atoms with vacancies and dangling bonds. After a period of time, a stable value of the sheet carrier density is achieved when almost all the vacancies and dangling bonds are hydrogenated. The sheet carrier density reduces after annealing at a temperature high enough, which is due to desorption of hydrogen from the surface.

The loss of chemisorbed hydrogen from diamond surfaces requires temperatures of ~700 °C to occur with any significant rate[22], so the simple loss of surface hydrogen would not appear to account for the observations made here at temperature lower than 600°C in vacuum. However, the loss of hydrogen from diamond could occur at temperatures lower than 300 °C in air due to oxidation[23].

The fact that the sheet carrier density remains in the range of 10^{12} -10^{13} cm^2 following a annealing process below 250 °C in air and 600°C in vacuum suggests that the fabrication of devices which would operate up to this temperature using hydrogenation as a source of carriers appears viable.

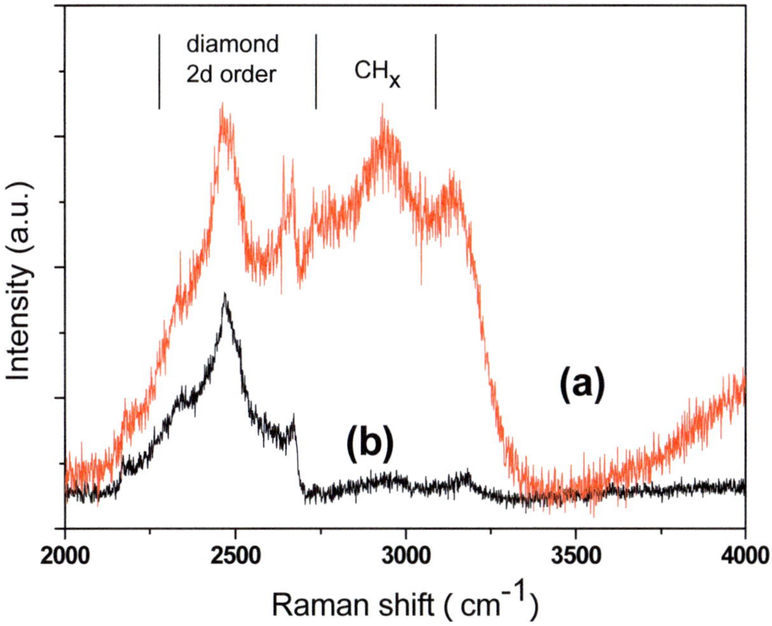

Fig. 11. Ultraviolet Raman scattering spectra for hydrogenated diamond nucleation surface sample (a) and annealed diamond surface sample (b)

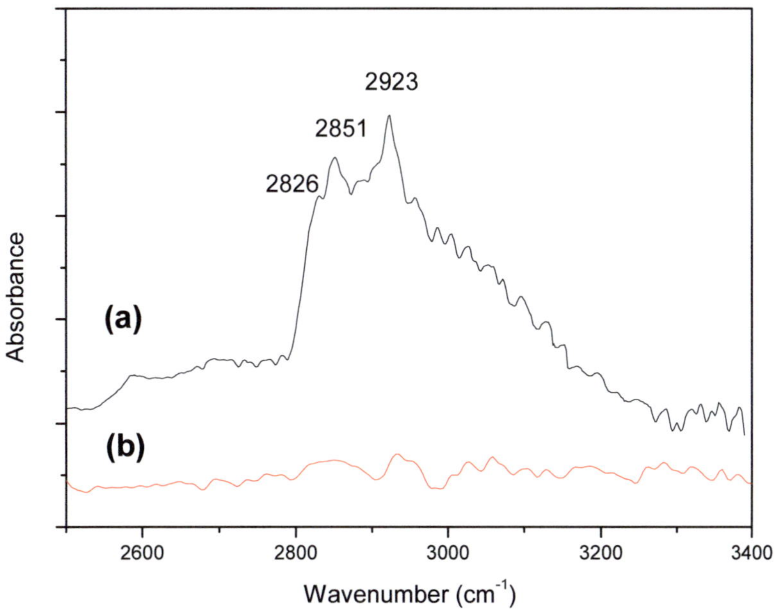

Fig. 12. Infrared spectra for hydrogenated diamond nucleation surface sample (a) and annealed diamond surface sample (b)

In order to take a further insight into surface conductivity of the nucleation surface of diamond films, the ultraviolet Raman scattering spectroscopy was used to characterize hydrogenated nucleation surface of diamond sample (a) and 500°C annealed (in air) diamond surface sample (b). The magnified profiles of ultraviolet Raman scattering spectra scanned in the 2000 -4000 cm^{-1} region were shown in Fig. 11. From the figure, sample (a) and sample (b) both had a strong peak at about 2468.49 cm^{-1}, representing the second order of the diamond peak, and a weaker peak at about 2669.52 cm^{-1} and at 3148.26 cm^{-1}, representing the second order of the D band and G band of graphine respectively. However, sample (a) had a stronger peak at 2930.93 cm^{-1} which indicates sp3 CH_x [24], in comparison with annealed nucleation surface of sample (b). It meant that, after annealed at a temperature of 500 °C in air, hydrogen desorbed from the nucleation surface of FSD films.
The internal reflection infrared spectrum obtained from sample (a) hydrogenated nucleation surface of diamond and sample (b) 500°C annealed (in air) diamond surface were shown in Fig. 6, from which the symmetric C-H stretching modes s at 2826 cm^{-1}, symmetric stretching mode of CH_2 at 2851 cm^{-1} and the antisymmetric stretching mode of CH_2 at 2923 cm^{-1} can be observed in films after hydrogen plasma treatment[23,25]. However, there was no obvious hydrocarbon adsorbates from the spectrum of Fig. 11(b) obtained after annealing at 500 °C in air, which indicated that the hydrogen desorbed from the surface of diamond film after annealing. All the above results confirmed that the diamond surface conductivity was related to the complexes of absorbed hydrogen atoms with carbon dangling bonds.

4. Fabrication and characterization of phototransistor based on diamond MESFETs

The FSD films were used to fabricate devices of MESFET. The fabrication of surface devices using the nucleation surface of the FSD films solved the problem of the surface roughness without the need of any kind of polishing. The smoothness of the nucleation surface allows a higher control of the electrodes. And the problem of the high resistivity can be easily overcome by a proper exposure of the surface to hydrogen plasma, as described above.
FSD films using for device fabrication were prepared as decribed in paragraph 2. Then, FSD films were exposed to hydrogen plasma at 750°C for 30min using a MPCVD apparatus.

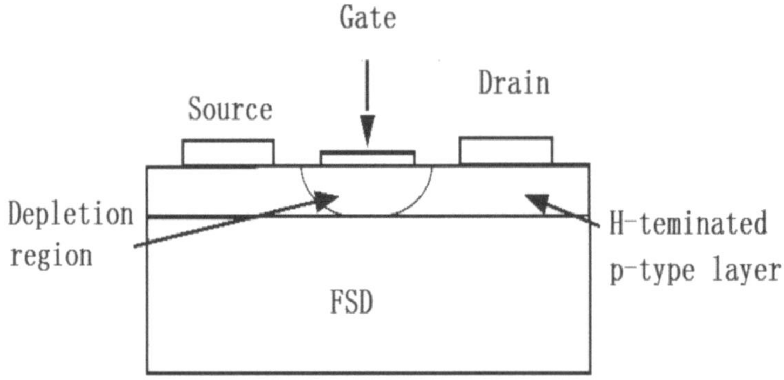

Fig. 13. A schematic the diamond MESFET device structure

Gold Ohmic contacts were evaporated as source (S), drain (D) and Aluminum contacts as gate (G) by standard lithographic procedures. The thickness of the drain/source and gate contacts was 300nm and 200nm, respectively. The channel length and width were 10 μm and 5 mm, respectively. The distance between Al electrode and Au electrode is about 10μm. A schematic picture of the device structure and optical micrograph of the device were shown in Fig. 12 and Fig.13, respectively. All devices were packaged and wire bonded prior to testing.

Fig. 14. Optical micrograph of the MESFET device

Current-Voltage (I-V)behaviour of the gold-gold electrodes and aluminum-gold electrodes on Hydrogen-teminated nucleation surface of FSD films were shown in Fig.14 (a) and Fig.14 (b). The I-V characteristics of adjacent Au contacts were near to linear, indicating Ohmic-like behavior, however Al-Au electrodes showed a strongly asymmetric I-V behavior, due to the presence of a Schottky barrier at the Al-diamond interface.

Figure 15 showed the Current-Voltage (I-V) behaviour of the source and drain electrodes without any applied gate voltage. The gold electrodes, evaporated directly on the Hydrogen-teminated nucleation surface of FSD films, behave as ohmic contacts. The small asymmetry in the characteristics is related probably with a slight heating of the electrode when the current starts to flow, increasing the resistivity of the material and producing an extra decrease of the current [26].

Hydrogen-teminated FSD film MESFET structures, with an Al gate and Au source and drain contacts, showed clear modulation of channel current as a gate bias was applied. The drain current as a function of drain-source voltage (V_{DS}) plotted for differing gate bias (V_{GS}) was shown in Fig.16. Field effect was seen for negative V_{GS}, revealing a p-channel. There was no current for V_{GS}= 0 V, and channel current considerably increased as V_{GS} was increased, indicating the device was an enhancement-mode MESFET. For all gate bias values, I_{DS} saturated for higher V_{DS}, indicating channel pinch-off.

Figure17 showed I_{DS} against V_{DS} with gate voltages of -0.1V which was illuminated with 200 nm light with varying intensity. The effect of the light is clearly to enhance the channel current level, with increasing optical powers giving higher saturated I_{DS} values. The results suggest that phototrasistors based on hydrogenated diamond MESFETs may be ideally suited for UV switching applications. The devices are not "visible blind" in the way that photoconductive structures can be, but they do offer the potential of high switching speed allied to high sensitivity.

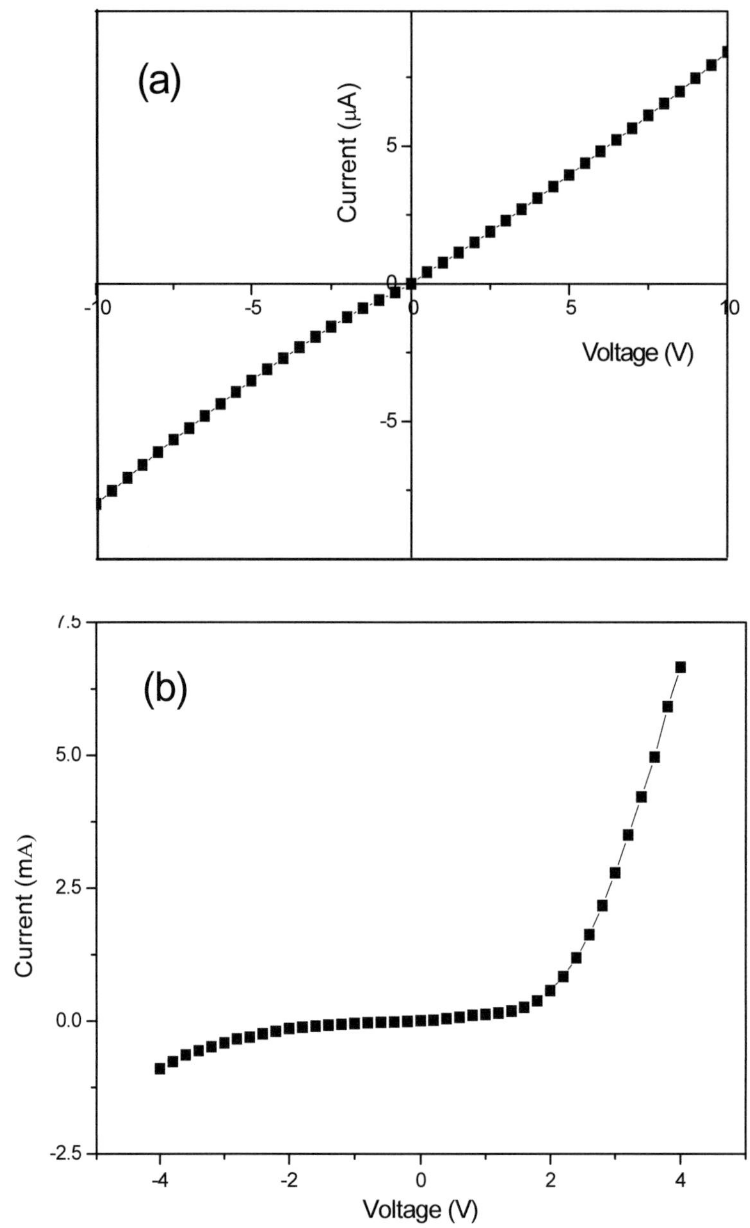

Fig. 15. Current-Voltage behaviour of the gold-gold electrodes (a) and aluminum-gold electrodes (b)

A PTI optical system and monochromator combination was used to investigate the response of the device across the spectral range 200–350 nm. Responsivity of diamond phototransistors as a function of illuminating wavelength with V_{DS} of -12V was shown in Fig.18. For the phototransistor, a pronounced increase in responsivity could be seen at around 230 nm, which corresponds to the band-gap energy of diamond. The response at the longer wavelengths (>300nm) was much smaller.

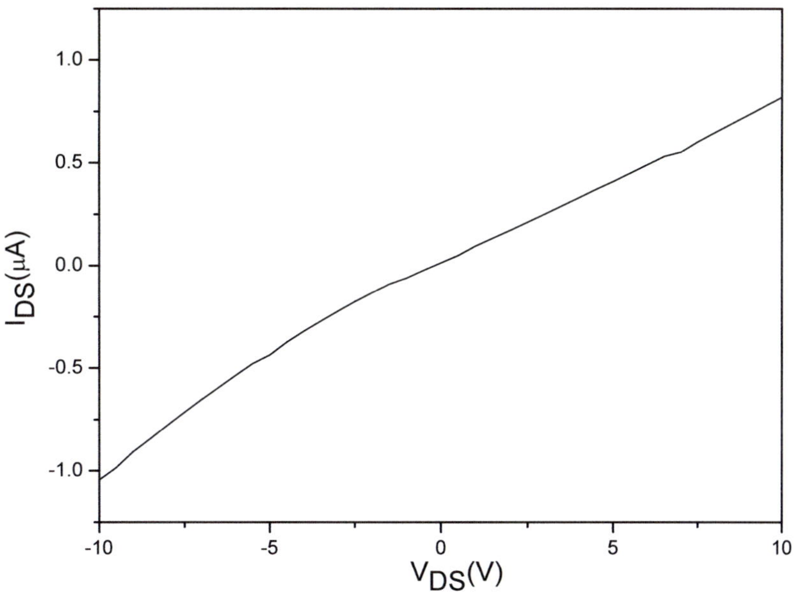

Fig. 16. Current-Voltage behaviour of the source and drain electrodes without any applied gate voltage

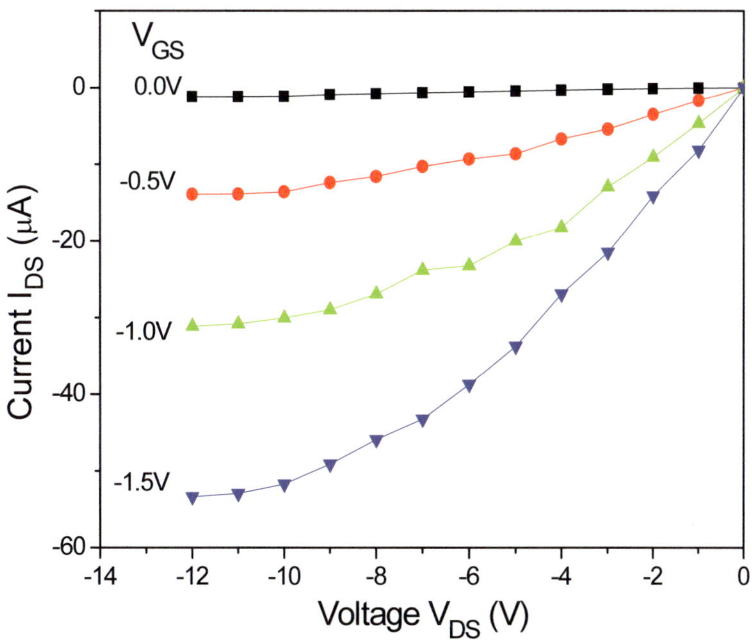

Fig. 17. Output characteristics of Hydrogen-teminated FSD film MESFET structures for negative bias. Drain-source voltage (V_{DS}) swept between 0 and -12 V; gate voltages (V_{GS}) swept between 0 and -1.5 V

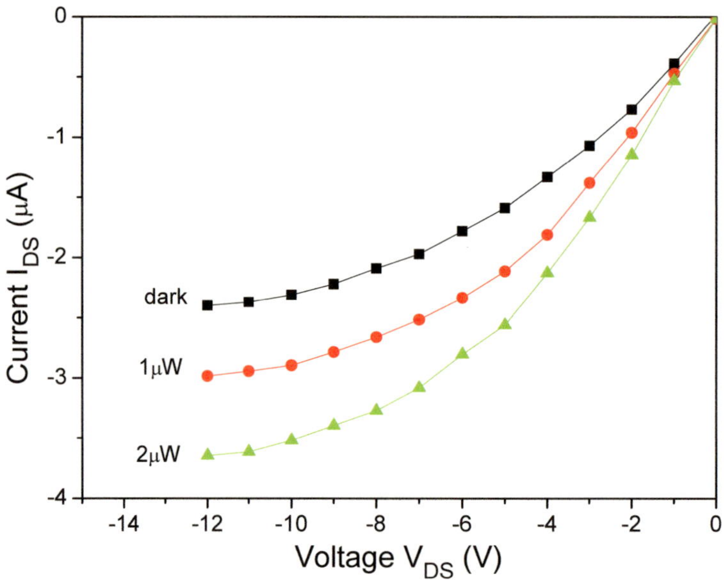

Fig. 18. Drain current against drain-source voltage for differing illumination power (λ= 200 nm) with gate voltages of -0.1V

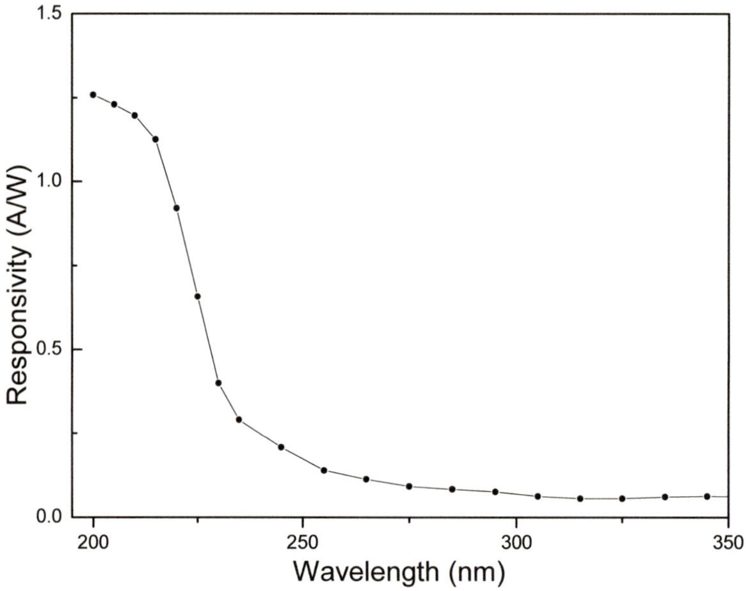

Fig. 19. Responsivity of diamond phototransistors as a function of illuminating wavelength with V_{DS} of -12V

5. Conclusions

In this work, high quality freestanding diamond (FSD) films were grown by microwave plasma chemical vapor deposition (MPCVD) method. The effects of hydrogen plasma treatment and annealing process on the p-type behavior of FSD films were investigated. The origin of this high-conductivity layer of FSD films was also discussed. The fabrication and properties of phototrasistors based on hydrogenated diamond MESFETs were studied. The main conclusions of the work were as below:

1. The thickness of FSD films, the structure, and morphology of the FSD nucleation surface were analyzed by scanning electron microscopy (SEM), atomic force microscopy (AFM), and Raman spectroscopy. The results indicated that the nucleation sides of FSD films prepared by both methods were very smooth with a mean surface roughness of about 10 nm in a scanning area of 1.5x1.5 μm^2. The thickness of the film was about 110μm. The Raman results showed a high quality diamond of nucleation side of FSD diamond films with low content of non-diamond carbon. The post-treatment (wet chemical etch, annealing process) was helpful to improve the quality of FSD films.
2. The nucleation sides of FSD films prepared by MPCVD method were exposed to hydrogen plasma treatment. The effects of hydrogen plasma treatment and annealing process on the p-type behavior of FSD films were investigated. The origin of this high-conductivity layer of FSD films was also discussed by using ultraviolet (UV) Raman spectroscopy, Fourier-transform infrared spectroscopy (FTIR) and secondary ion mass spectrometry (SIMS). The nucleation side of FSD films with a p-type conductivity layer (~50nm thick) could be obtained by hydrogen plasma treatment; The origin of this conductivity layer may be related to the complexes of absorbed hydrogen atoms with carbon dangling bonds; The sheet carrier concentration of the FSD film increased and sheet resistivity decreased with the time (5-30min) of plasma treatment; Surface conductivity of hydrogenated diamond surfaces disappeared gradually after annealing at temperature above 200 °C in the air, or above 600 °C in the vacuum.
3. The properties of metal contacts on hydrogenated p-type diamond surfaces were discussed. The results suggested that ohmic contacts could be realized between the p-type diamond and the Au electrodes. However, it's easy to form schottky contacts between Al electrodes and the p-type diamond. Preparation and properties of phototransistors based on hydrogen-terminated diamond film p-type channel metal–semiconductor field effect transistors (MESFETs) were reported. The results showed a typical characteristic of enhancement-mode MESFET and the phototrasistor may be ideally suited for UV switching applications. The phototransistor showed a pronounced increase in responsivity at around 230 nm, which corresponds to the band-gap energy of diamond.

6. Acknowledgments

This work was supported by National Natural Science Foundation of China (60877017), Program for Changjiang Scholars and Innovative Research Team in University (No:IRT0739), Shanghai Leading Academic Disciplines (S30107) and Innovation Program of Shanghai Municipal Education Commission (08YZ04).

7.References

H. L. Lu, Y. M. Zhang, Y. M. Zhang, et al. A comprehensive model of frequency dispersion in 4H-SiC MESFET [J]. *Solid-State Electronics*, 2009, 53: 285–291.

Yasuhiro Ueda, Yasuhiko Nomura, Seiji Akita, et al. Characteristics of 4H-SiC Pt-gate metal-semiconductor field-effect transistor for use at high temperatures [J]. *Thin Solid Films*, 2008, 517: 1468–1470.

Yuki Niiyama, Hiroshi Kambayashi, Shinya Ootomo, et al. 250 °C operation normally-off GaN MOSFETs [J]. *Solid-State Electronics*, 2007, 51:784–787.

E. Tschumak, M. P. F. deGodoy, D. J. As, et al. Insulating substrates for cubic GaN-based HFETs [J]. *Microelectronics Journal*, 2009, 40: 367–369.

M. Bavdaz, A. Peacock, A. Owens. Future space applications of compound semiconductor X-ray detectors [J]. *Nucl. Instr. and Meth.* A, 2001, 458: 123-131.

M. Schieber, H. Hermon, A. Zuck, et al. Theoretical and experimental sensitivity to X-rays of single and polycrystalline HgI2 compared with different single-crystal detectors [J]. *Nucl. Instr. and Meth.* A, 2001, 458: 41-46.

E. Rossa, C. Bovet, D. Meier, et al. CdTe photoconductors for LHC luminosity monitoring [J]. *Nucl. Instr. and Meth.* A, 2002, 480: 488-493.

I. M. Buckley-Golder, R. Bullough, M.R. Hayns, et al. Post-processing of diamond and diamond films: a review of some Harwell work [J]. *Diam. Relat. Mater.*, 1991, 1 (1) : 43-50.

Yi Zhang, Linjun Wang, Qi Xiao, et al. The research on the surface structure and conductivity of free-standing diamond films for photo-transistor applications. *Proc. of SPIE*, 2009, 7381: 73811M-1--73811M -4.

Qian Fang, Linjun Wang, Qingkai Zeng, et al. Preparation and photoelectric properties of a thin-film nanocrystalline diamond phototransistor, *Proc. of SPIE*, 2009, 7381: 73811N-1-73811N -4.

K. Hayashi, S. Yamanaka, H. Okushi and K. Kajimura. Appl. Phys. Lett. , 1996,68:376-379.

A. Aleksov, A. Denisenko, N. Kab, W. Ebert, E. Kohn, Diamond power FET concept. *Proceedings of the 2000 IEEE/ Cornell Conference on High Performance Devices*, Cornell, USA, 2000: 266.

P. Gluche, A. Aleksov, A. Vescan, W. Ebert, M. Birk, E. Kohn, First diamond power FET structure. *Proceedings of the Fifty- Fifth Device Research Conference, Book of Abstracts*, Fort Collins, CO, USA, 1997: 42.

M. Silveira, M. Becucci, E. Castellucci, et al. Non-diamond carbon phases in plasma-assisted deposition of crystalline diamond films: a Raman study [J]. *Diamond Relat Mater*, 1993, 2 (9): 1257-1262.

Y. Bar-Yam and T. D. Moustakas. Defect-induced stabilization of diamond films [J]. *Nature*, 1989, 342: 786 - 787.

J. Nan and I. Toshimichi. Electrical properties of surface conductive layers of homoepitaxial diamond films [J]. *JOURNAL OF APPLIED PHYSIC*, 1999, 85(12): 8267-8273.

J. Nan, H. Akimitsu and I. Toshimichi. Nitrogen Doping Effects on Electrical Properties of Diamond Films [J]. *Jpn. J. Appl. Phys.*, 1998, 37: L1175-L1177.

C. Su, J.-C. Lin. Thermal desorption of hydrogen from the diamond C(100) surface [J]. *Surface Science*, 1998, 406: 149–166.

F. Maier, M. Riedel, B. Mantel, et al. Origin of Surface Conductivity in Diamond [J]. *PHYSICAL REVIEW LETTERS*, 2000, 85 (16): 3472-3475.

D. Ballutaud, T. Kociniewski, J. Vigneron, et al. Hydrogen incorporation, bonding and stability in nanocrystalline diamond films [J]. *Diamond & Related Materials*, 2008, 17: 1127–1131.

C. L. Cheng, C F. Chen, W. C. Shaio, et al. The CH stretching features on diamonds of different origins [J]. *Diamond & Related Materials*, 2005, 14: 1455 - 1462.

J. C. Madaleno, L. Pereira, G. Lavareda, , et al. A MIS transistor using the nucleation surface of polycrystalline diamond [J]. *Diamond & Related Materials*, 2008,17 : 768 - 771.

5

Multilayer Approach in Light-Emitting Transistors

Gianluca Generali, Stefano Toffanin and Raffaella Capelli
National Research Council – Institute for the study of nanostructured materials (CNR-ISMN)
Italy

1. Introduction

1.1 OLED Vs. OLET

Plastic electronics, i.e. electronics based on organic polymer and molecular semiconductors, is a key low-cost technology for new vast application areas. The scope of applications in plastic electronics is virtually endless. It ranges from opto-electronic devices and flexible organic light-emitting diodes (OLEDs) displays, to ultra-cheap, radio-frequency identification tags that can replace the now ubiquitous bar-codes, and to wearable computing and electronic bio-interfaces.

The first demonstration of the OLED dates back to the 1960s when electrically driven light emission from non-crystalline organic materials was first observed. After that, several studies were carried out by academic groups and companies (Kodak, Pioneer, Motorola, NEC, etc…) both for fundamental physics comprehension and application purposes. Nowadays electronic products containing displays are becoming more and more portable. Therefore, they need some peculiarities like lightweightness, flexibility, brightness, etc… These, with many others, are the strong points of the OLEDs. In fact they are thinner, lighter and more flexible with respect to their inorganic counterpart. Moreover, OLEDs can be as bright as LEDs and they consume much less power. Due to the organic processability, they are easier to produce and can be made on larger area. Finally OLEDs have large fields of view, about 170 degrees, a significantly advantage over, for example, liquid crystal displays. Of course, these devices present also some disadvantages: they have typically shorter lifetime (in particular lifetime of the blue emitter is critical, about 1.000 hours), they are not very stable and can easily be contaminated by water or oxygen.

In general, with respect to OLEDs, organic light-emitting transistors (OLETs) present some fascinating characteristics which overcome many physical and technical drawbacks in the realization of nano-scale integrated electroluminescent devices.

The main difference between the vertical (OLED) and planar (OLET) device geometry is a direct consequence of the different device structures. In OLED, charge transport occurs perpendicular to the organic layers (bulk charge transport) while in OLET the transport occurs horizontally (field-effect charge transport).

Under the typical biasing conditions, the electron and hole mobility can be about four orders of magnitude higher in OLETs than in OLEDs, thus affecting directly the material lifetime and exciton emission. In a typical OLED structure, the minority carriers travel only few tens of nanometers to encounter the opposite sign carriers and to recombine radiatively. In OLETs, instead, both carriers must travel longer distances (typically tens of micrometers), which impose more requirements on the organic semiconductors charge transport properties.

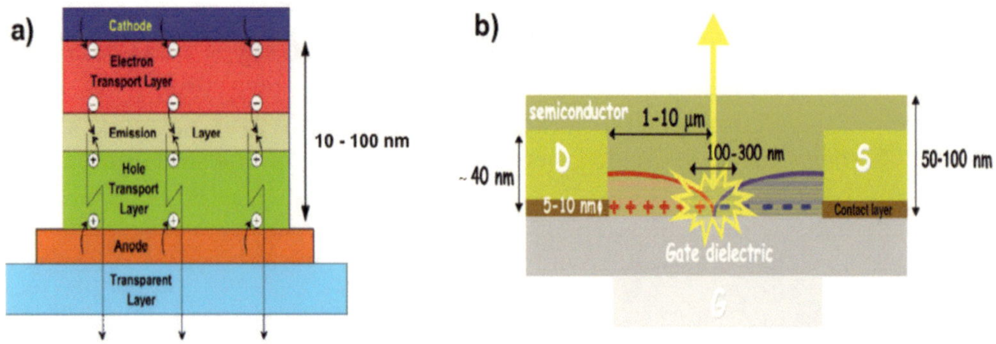

Fig. 1. Schemes of an OLED (a) and an OLET (b) showing for each the typical dimensions

A clear advantage of the OLETs is the virtually higher electroluminescence quantum efficiency inherent to the device structure. Indeed, in OLETs it is possible to drastically reduce the exciton quenching due to the interaction with charge carriers, with the externally applied electric field and with metal contacts.

For what concerns the exciton-charge interaction, even though the current density in an OLET is expected to be higher (1-10 A/cm^2, for a 1-nm-thick accumulation layer) compared to OLEDs (10^{-3}-10^{-2} A/cm^2), the spatial localization of exciton formation an OLET could favor an effective separation between the exciton population and the charge carriers thus avoiding any quenching for that kind of interaction. The presence of a third electrode helps to achieve a balanced charge carriers current, therefore further reducing exciton-charge quenching. In OLET, compared to OLED, all the operational requirements of the basic electronic and optoelectronic elements in active matrix displays, are satisfied in a single device structure. Indeed in conventional OLED electronics, such a high degree of integration cannot be achieved and, for each pixel, an electrical switch and a separate light source must be combined. Lastly, in OLETs is possible to control the position of the emission zone inside the channel length.

2. Approaches to OLETs

According to the previous section, it turns out how the OLET structure presents more intriguing potentiality compared to OLED. However, the actual degree of exploitation of this technology depends on the development of new organic materials combining multiple functionalities and high performances. Moreover, many structure-related issues are to be considered in order to get good electrical properties and intense light emission from the device.

Indeed, since FETs (and thus OLETs) are considered as truly interface devices, processes that take place in the device active region, like charge transport, energy/charge transfer, exciton formation, charge trapping, are strongly dependent on interface. In recent years it has been demonstrated that the chemical structure of the organic semiconductor is not the only factor that determines whether an organic FET exhibits predominantly p-channel or n-channel behaviour. Processing and characterization conditions, device architecture, and choice of electrodes are important as well.

To get an insight into the main phenomena that happen at the interfaces, many aspects must be taken into account: the chemical and electronic interaction taking place at the different interfaces (dielectric/organic, organic/organic and organic/metal), the modulation of the molecular electronic structure in the active material due to the proximity with other functional layers (polarization, relaxation), the bonding and ordering of molecules at the interface (which strongly determine the growth morphology). For example, crucial processes of charge accumulation and transport in OLETs take place at the interface between the gate dielectric and the semiconductor. Thus, the properties of this interface and the dielectric have a huge influence on device opto-electronic performances.

Device parameters such as mobility, threshold voltage, sub-threshold swing, etc. depend not only on the nature of the semiconductor but also on the chemical structure and dielectric properties of the insulator. The requirements for gate dielectrics in OFETs are rigorous. They should show high dielectric breakdown strength, contain only minimal concentrations of impurities, that could act as traps, and must be easily processable and be environmentally stable.

From the point of view of light emission, the requirements depend on the approach used for fabricating the OLET, but regardless the implemented architecture, it is absolutely important the use of high emission efficiency organic materials. In the next sections we will discuss some issues related to the emission in different OLET approaches, each with its points of strength and weaknesses.

2.1 Unipolar OLETs

Historically the first demonstrated OLET was achieved in unipolar charge carriers transistors made by tetracene by Hepp et al. in 2003 (Hepp et al., 2003). The device was fabricated on Si/SiO_2 substrates in bottom-gate bottom-contact configuration, with gold electrodes. The material was chosen because of its good charge carrier transport and EL properties. The electrical characteristic of the OLET were typically of a unipolar p-type transistor and can be seen in Fig. 2, along with a picture of the illuminated channel (located near the drain contact). Despite the electrical unipolar behaviour, the observation of light emission from tetracene proved an undoubted presence of opposite charges recombining inside the material. Given the inhomogeneous illuminated channel area, Hepp et al. introduced then an empirical model in which they supposed different injection mechanisms at the source or drain electrodes as a consequence of the thin-film physical imperfections. In particular they assumed that, due to under-etching problems of the electrodes that act as a shadow mask, a thinner tetracene layer is formed at the organic/electrode interface resulting in a poor electrical contact.

During device operation, the channel is filled with holes. However due to imperfections of the electrode, holes cannot directly reach the drain through the channel, but they must travel a certain distance in the tetracene bulk film. On the other side, since the gate field is screened by holes at the dielectric/organic interface, there is a high electric field at the drain electrode. This strong field, magnified by the local spikes due to the imperfections on the contact, could be intense enough to allow injection of electrons into the organic (Murata et al., 2001). For more details, see fig. 3.

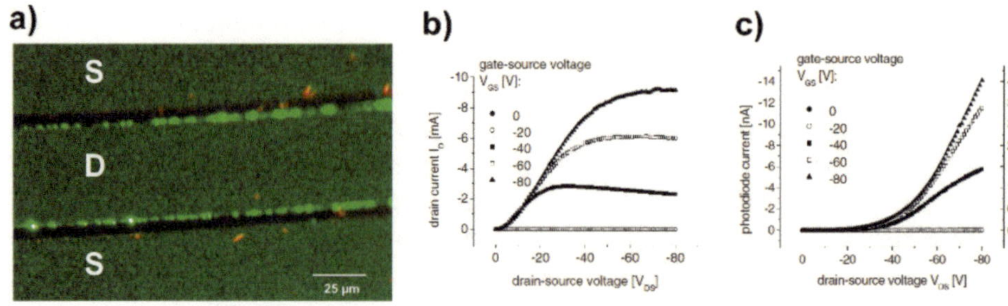

Fig. 2. Image of the illuminated channel with the typical green-light emission of tetracene (a), I-V characteristic (b) and electro-luminescence (c) measurement of the device, in the negative bias region

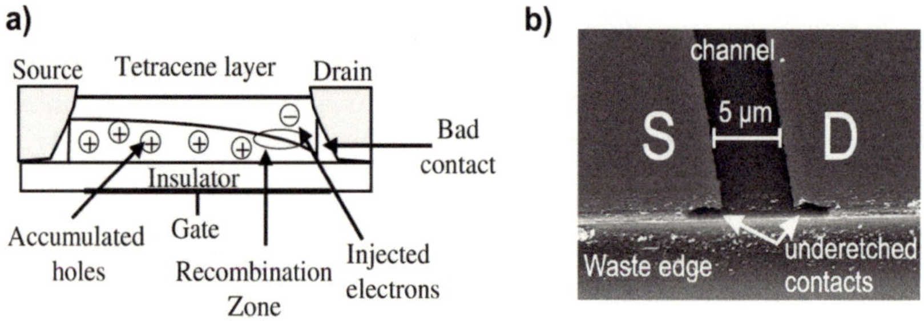

Fig. 3. Scheme of the device with the edge effect at the contacts proposed to demonstrate the electrons injection from the drain contact (a), example of a SEM image showing the underetching effect (b)

A successive study proposed a phenomenological model to explain the working mechanism of the OLET. The proposed model assumed that the voltage drop at drain electrode, caused by a contact barrier, induces a distortion of the Highest Occupied (HOMO) and Lowest Unoccupied (LUMO) Molecular Orbital levels of tetracene near the contact, thus determining the conditions for the tunnelling of electrons from the drain to the LUMO of the organic (Santato et al., 2004).

In the proposed model, the external quantum efficiency (EQE) is proportional to the tunneling probability and thus proportional to the drain-source voltage (V_{ds}), but it is independent from gate voltage (V_g). Indeed, increasing V_g leads to an overall increase of the electroluminescence (EL) but leaves unaffected the EQE.

Several other approaches have been performed on unipolar OLETs (Oyamada et al., 2005a). In particular, Oyamada et al. succeeded in demonstrating how the channel length could affect EQE of thiophene-derivative devices and they obtained, for a channel length of 0.8 µm, an EQE of 6.4×10^{-3}%. Although the noticeable EL improvement, it is worth of notice that in a configuration with a very small channel the contact resistance effects at the metal/organic interface are predominant with respect to the FET working mode. In the same year, Omayada et al. showed also a new single layer unipolar OLET with an EQE of almost 0.8%, made with a blend of 1wt%-rubrene doped tetraphenylpyrene (TPPy) (Omayada et al., 2005b).

Other studies have been done using spin-coated or drop-casted polymers like poly-phenylenevinylene, poly-fluorene or poly-arylenevinylene derivatives (Sakanoue et al.,

2004; Ahles et al., 2004; Swensen et al., 2005a) as active layer. These experiments, besides showing a possible extension of the concept of unipolar OLET, demonstrated the feasibility of wet-technique fabrication process that could open up many possibilities of large-area and low cost devices. Within the same studies, it was demonstrated also that the use of different metals for drain-source contacts could lead to a considerable increase of EL.

Generally speaking, despite the enhancement in EL in unipolar devices, the EQE achieved is still too low for any practical application. Indeed, some problems arisen from the type of architecture, limit drastically its potentiality. First of all the EL emission takes place in proximity to the metal electrode with a consequent quenching due to the interaction between exciton and metal surface plasmons. Moreover, the exciton formation and recombination occur in the same spatial region where charges flow, so leading to a significant exciton-charge quenching. In practice, unipolar OLETs suffer from the same negative effects of OLEDs.

2.2 Ambipolar OLETs

So far we showed a possible approach to fabricate OLETs, using a unipolar organic material (polymeric or small molecule) as the active single layer. In this case the emission is localized in a small region underneath the metal contacts.

Indeed, most of the scientifically and technologically remarkable properties that make light-emitting transistors desirable are, however, only present in ambipolar OLETs since they can provide an effective pn-junction within the device channel that allows exciton formation and radiative recombination.

The most simple structures are composed by a single organic material capable of transporting electrons as well as holes (single component approach) or by a combination of two unipolar transport materials (multi-component approach).

2.2.1 Ideal single layer ambipolar OLETs

In an ideal ambipolar transistor with just one polymeric or small-molecule semiconductor layer, the ambipolar regime is characterized by a hole and an electron accumulation layers, next to the respective electrode, that meet at some point within the transistor channel. There, oppositely charged carriers recombine. In electroluminescent materials, this leads to light generation within the channel. The length of each channel and thus position of the recombination zone depends on the applied gate and source-drain voltage and mobility ratio. The behaviour of an ambipolar field-effect transistor in the ambipolar regime can roughly be imagined as that of a saturated hole and electron channels in series within the field-effect transistors.

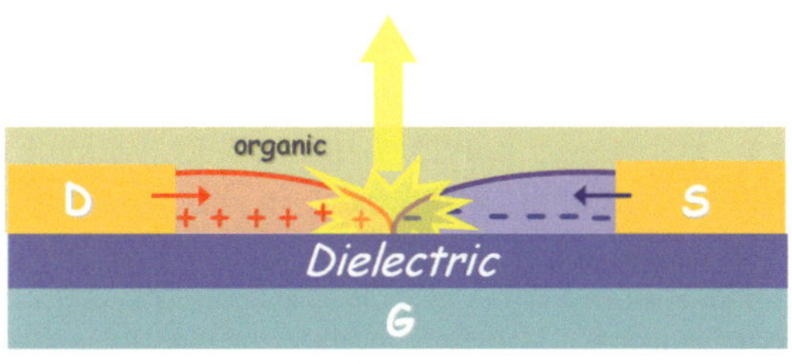

Fig. 4. Schematic of an ideal ambipolar OLET, when the two opposite charge carriers meet, there is exciton formation and consequently light emission

In general, ambipolarity in FETs is an attractive characteristic since it enables the fabrication of complementary logic circuits like CMOS transistors and permits a higher light emission efficiency due to the maximization of the exciton recombination through a better electron-hole charge balance. In addition, with an opportune tuning of the gate voltage, it is possible to move the emission through the channel length. On the other hand, one of the limiting negative aspect of this class of devices remain the poor mobility of charge carriers, usually of the order of 10^{-4} cm^2/Vs.

2.2.1.1 Ambipolar polymer OLET

The first research showing the feasibility of an ambipolar single layer OLET, was done by Swensen et al. (Swensen et al., 2005b). The material used was a poly-phenylenevinylene (PPV) derivative polymer, called SuperYellow (SY) and the device was done on a Si/SiO_2 substrate.

Previous studies on the same class of material/substrate FETs showed only p-type mobility. It was just after the discovery of the hydroxyl groups trapping effect for electrons and the consequent SiO_2 surface passivation process implementation (Chua et al., 2005), that it was possible the realization of electroluminescent ambipolar polymer transistors. In particular, the presence of a so called "two-color" geometry of the device was the key feature for enabling ambipolar transport from SY. In this reported geometry, they implemented the use of a low work-function metal (Ca) for the n-type injection electrode and a high work-function metal (Ag) for the p-type injection electrode. A spatially resolved recombination zone was then observed under ambipolar conditions controlled by the gate bias. The emission zone moved across the channel as the gate bias swept (see. Fig.5) and in case of balanced electron and hole currents the recombination was located at the channel centre.

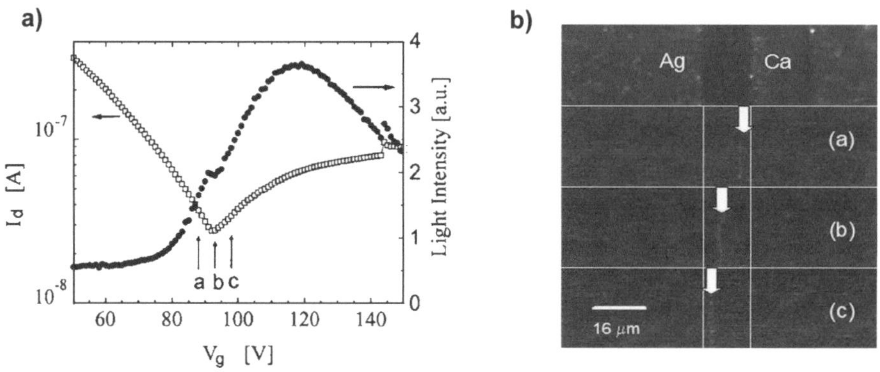

Fig. 5. Transfer characteristic of the SY OLET along with EL intensity (a), image of EL inside the channel at different spatial positions depending on gate bias (b)

Almost contemporary, another group achieved the same results using a spin-coated layer of a different conjugated polymer, poly(2-methoxy-5-(3,7-dimethyloctoxy)-p-phenylene-vinylene) (OC_1C_{10}-PPV)(Zaumseil et al., 2006). Also in this case, to obtain a good ambipolar behaviour, with balanced electron-hole charge densities, two different work function metals were used. For this material, the reported EQE was 0,35%, thus similar to the EQE of a bulk LED based on the same polymer.

Recently in 2008, again Zaumseil et al. fabricated ambipolar OLETs with poly(9,9-di-*n*-octylfluorene-*alt*-benzothiadiazole) (F8BT) and poly((9,9-dioctylfluorene)-2,7-diyl-*alt*-[4,7-

bis(3-hexylthien-5-yl)-2,1,3-benzothiadiazole]-2′,2″-diyl) (F8TBT) (see Fig.6) (Zaumseil et al., 2008).

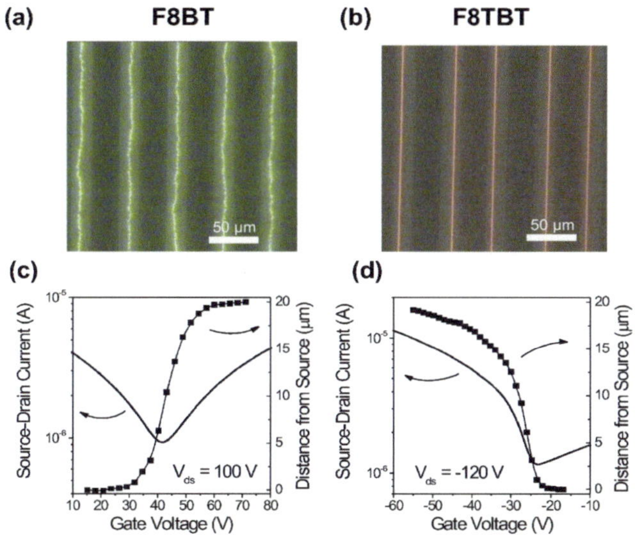

Fig. 6. Image of light emission from an interdigitated source-drain OLET made with F8BT (a) and F8TBT (b). Transfer curves and position of the light with respect to source contact for F8BT (c) and F8TBT (d)

In their study by using theoretical models they showed that the ambipolar regime can be thought of as a saturated electron channel and a saturated hole channel in series within the OLET channel. Within the model, EQE was expected to be constant, depending only on singlet-triplet ratio and not from, for example, voltage conditions. The increase of EQE with current density up to saturation, measured during the experiments, was considered as a consequence of trap-assisted non-radiative decay mechanisms at the semiconductor-dielectric interface. Furthermore, they demonstrated that, when complete recombination of all charges happened the maximum saturated EQE of F8BT, in top gate OLET configuration, is 0,8%. This conclusion signed a very important step in OLET fabrication showing clearly, for single layer devices, that an EQE of 0,8% is the highest limit.

2.2.1.2 Ambipolar small molecule OLET

Another approach in the realization of ambipolar single layer OLETs, using an intrinsic ambipolar light-emitting small molecule, was explored by Capelli et al., describing the realization of α,ω-dihexyl-carbonyl-quaterthiophene (DHCO4T) based device. The advantage in using a physical vapour deposited (PVD) small molecule instead of a polymer deposited by solution-processes is that, in the first case, the resulting film presents a more ordered and crystalline structure and thus, in principle, higher hole-electron mobilities due to larger charges delocalization.
Unfortunately electro-luminescence was present mainly in unipolar region (see Fig. 7). In order to have a better understanding of the phenomena, they tested the material in different conditions, changing dielectrics and metal contacts. Through this study, they showed the strong dependence of DHCO4T-based OLETs performances from the dielectric-organic interface. This aspect profoundly affects the electrical properties of both charge carriers in

terms of mobility and threshold voltage and thus electro-luminescence. They identified this main issue as the first cause of limited optoelectronic performances of their device (Capelli et al., 2008).

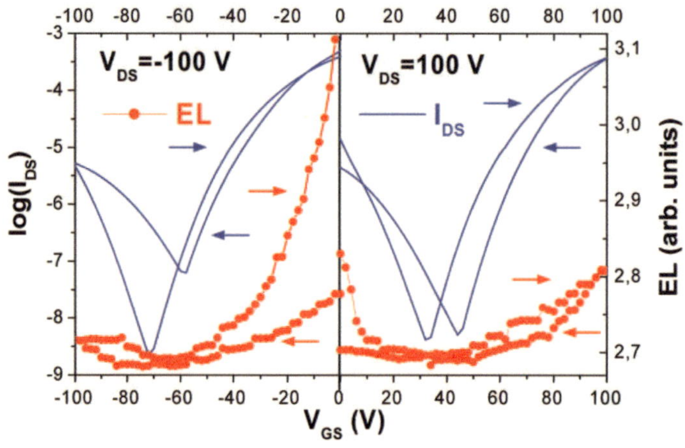

Fig. 7. Example of transfer characteristics of a DHCO4T OLET on SiO_2/poly-methyl-metacrylate (PMMA) substrate. The left plot is in p-type bias mode, while the right one is in n-type bias mode

2.2.2 Multi-layer ambipolar OLETs

In multi-component approach, OLETs can be obtained either by simultaneous co-evaporation of two unipolar materials realizing a bulk hetero-junction or by superposing two layers of unipolar materials in a bi-layer structure.

2.2.2.1 Bulk hetero-junction OLETs

In bulk organic hetero-junction approach, exciton formation and charge transport are competitive processes due to the dispersed interface between the p-type and n-type transport materials. Clearly, the wider the interface surface is, the higher the probability that electrons and holes recombine forming excitons. Nevertheless, connected percolative paths are needed for the charges to migrate by hopping so that interface can represent a physical obstacle for efficient charge transport. Furthermore, even if the interface morphology is precisely controlled during vacuum sublimation, well-balanced ambipolar behaviour has not yet been achieved.

The first documented fabrication of an ambipolar device belonged to this class of OLETs. In 2004, for the first time, Rost et al. proposed a new structure made by a co-evaporation with 1:1 ratio of *N,N'-ditridecylperylene-3,4,9,10-tetracarboxylic diimide* (P13) for n-type transport and α-quinquethiophene (T5) for p-type transport (see fig. 8) (Rost et al., 2004).These two materials are known for their good hole-transporting (T5, with mobility of 10^{-2} cm^2/Vs) and electron-transporting (P13, with mobility of 10^{-1} cm^2/Vs) properties. Of course, one of the most important prerequisite for having exciton formation and thus light formation, is the relative positions of the materials energy levels of highest occupied (HOMO) and lowest unoccupied molecular orbital (LUMO). Indeed, there must be the conditions to allow recombination in the material with the smaller energy gap.

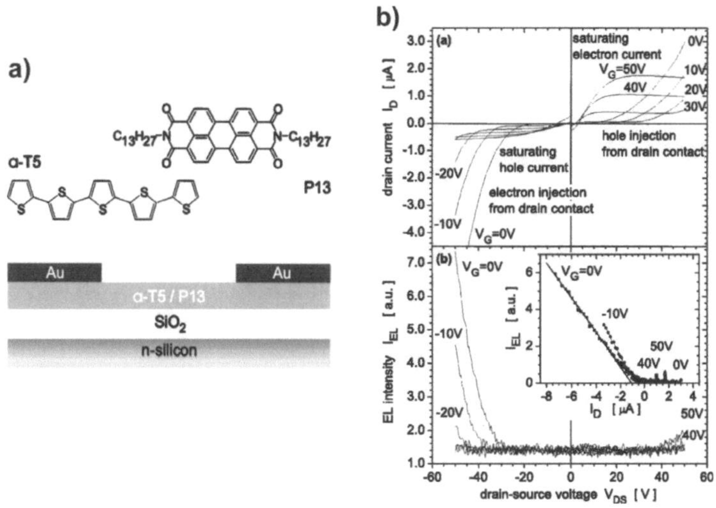

Fig. 8. Device and molecules chemical structures (a), I-V characteristic with EL emission at different gate biases

Through a fine control of the two materials co-evaporation, it is possible to obtain a good tuning of both charge carriers mobilities and quite good EL. However, in general, in this kind of device structure, the absolute mobility values are definitely lower, if compared to other fabrication strategies. In particular, Rost et al., after a deep investigation, found the best deposition parameter tuning in order to obtain very high mobility in both materials, though with low EL emission. Mobility values achieved were, respectively, 10^{-4} cm^2/Vs for hole transport and 10^{-3} cm^2/Vs for electron transport.

2.2.2.2 Bi-layer vertical hetero-junction OLETs

In this structure the organic layers are deposited in a vertical stack. Each layer is devoted to a single functionality and can be optimized by controlling the growth conditions of the different organic/organic, organic/metal and organic/dielectric interfaces. This approach, compared to others, presents the advantage of enhanced charge transport and mobility values. It is known that in OFETs the charge transport is confined to the first few layers next to the dielectric. Thus, electron and hole paths are confined at the interface between the first layer and the substrate and at the interface between the two organic films. If the two films are continuous, the charge transport should be uniform in both films over all the device channel area and therefore good transport properties are expected.

Dinelli et al. reported on a bi-layer of α,ω-dihexyl-quaterthiophene (DH4T) and P13 OLET that showed good ambipolar behaviour and light emission (Dinelli et al., 2006). They demonstrated, studying two different possible organic configurations (DH4T-P13 or P13-DH4T), that the device, in which DH4T was evaporated directly on the dielectric surface, had the best balanced mobility in ambipolar region (10^{-2} cm^2/Vs for both charge carriers). Through analysis of the interfaces in the two cases, they underlined the importance of having the best growth compatibility between the hole and electron transport materials in order to form continuous films and thus enhancing the optoelectronic performance. They observed also that EL occurs only when the device is biased with FET transport in the bottom layer and that the light emission originates, from P13, as expected from energetic considerations.

In the bi-layer approach, the spatial separation between charge carriers mainly prevents excitons formation inside the device channel, and so, EL is present only in unipolar bias region. This means that the pn-junction forms only underneath the electrodes as in the case of single layer unipolar OLETs. However, this structure has, up to date, the highest balanced ambipolar mobility ever obtained in OLETs.
Instead of implementing n-transport and p-transport materials in the bi-layer device realization, Heeger et al. utilized a structure comprising a hole transporting polymer, poly(2,5-bis(3-tetradecylthiophen-2-yl)thieno[3,2-b]thiophene) (PBTTT-C14) and a light emitting polymer, Super Yellow (SY), a polyphenylenevinylene derivative showing good output characteristics and brightness. Both materials were deposited by wet techniques (see. Fig.9) (Namdas et al., 2008).

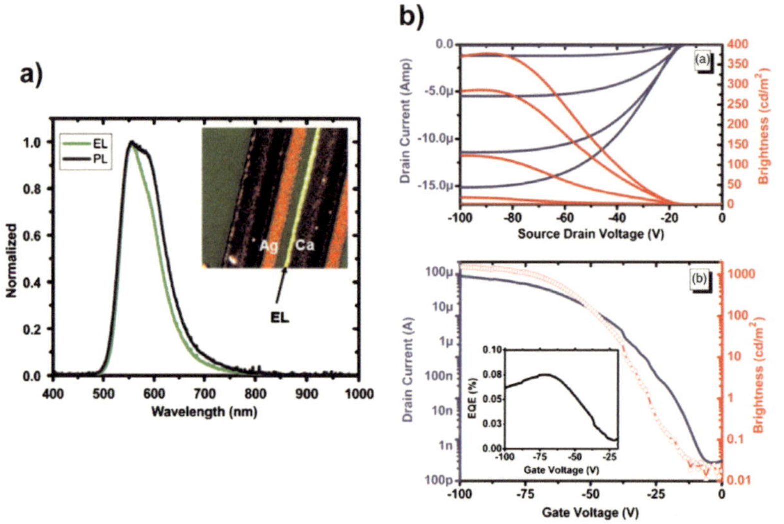

Fig. 9. Image of the EL emission spectrum of the device compared to PL emission of SY, along with a picture of the luminescent channel (a), I-V characteristic, transfer characteristic and EL curve of the bi-layer device(b)

It is well known that, in OLET devices, light emission is quite low due to the fact that the organic materials present either low carrier mobility with high photoluminescence (PL), i.e. amorphous materials or high mobility with weak PL, i.e. crystalline materials. In order to obtain good performances, materials should be capable of good ambipolar behaviour and have an high PL efficiency in thin film. In this case, although they did not achieve good ambipolar behaviour since electron transport was significantly lower than hole one, using two different metals as drain-source electrodes (Ag for hole injection and Ca for electron injection), they obtained a device showing intense EL, independent from the gate bias, with an efficiency of 0,35% and located under the electrode.

2.2.3 The tri-layer OLET approach

Lastly, a novel strategy in OLET realization, the tri-layer vertical hetero-junction, is presented. So far, we have seen OLETs based on unipolar single layer which reached high brightness but EQE as low as 0,2% due to exciton-charge and exciton-metal quenching effects. Then we reported on ambipolar single layer OLETs that enable, under proper bias conditions, the spatial localization of the EL far from the electrodes, but since charge carrier and exciton coexists in the same region, large exciton-charge quenching happens. Finally we reported ambipolar bi-layer approaches for OLETs, but in each of the two cases proposed (superposition of either two unipolar semiconductors of different charge transport or a highly efficient luminescent layer over a unipolar conductive layer) the device architecture does not offer any control on the exciton quenching due to charge and metal electrodes interactions.

In this new tri-layer hetero-structure approach, proposed by Capelli et al. in 2010, it is reported an OLET enabling simultaneous control of the electrode induced photon losses, the exciton-metal and the exciton-charge interactions. In this condition they showed that devices could reach the considerable EQE value of 5%, out-performing the OLED state of the art based on the same emitting layer with an EQE of 2.2% (Capelli et al., 2010).

In the trilayer configuration the first organic thin-film in contact with the device dielectric layer is devoted to the unipolar field-effect transport. The second layer deposited onto is the recombination layer which presents high emission quantum efficiency and OLED-like vertical bulk mobility value. In the specific case a host-guest system with a dye doped matrix was used. The third layer is devoted to unipolar charge transport complementary to that of the first layer (see Fig. 10a).

The key idea of the vertical tri-layer heterojunction approach in realizing OLET is that each layer has to be optimized according to its specific function (charge transport, energy transfer, radiative exciton recombination). Clearly, matching the overall device characteristics with the functional properties of the single materials composing the active region of the OFET, is a great challenge that requires a deep investigation of the morphological, optical and electrical features of the system.

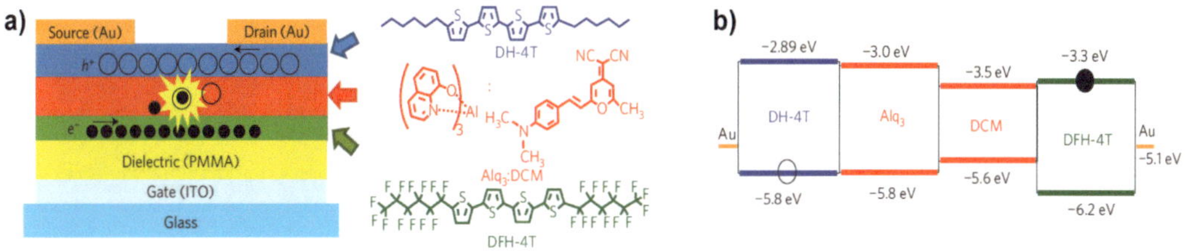

Fig. 10. Tri-layer device structure along with chemical structures of the organic materials used for the device fabrication (a), HOMO and LUMO energetic levels of the materials (b)

In this structure, the main aim is to enable charges to percolate into the middle emitting layer. In order to do so, first of all, material energetic levels must be considered. The LUMO level of the n-type transport layer should be equal or higher than the LUMO level of the guest matrix of the host-guest system consisting the middle layer, while the HOMO level of the p-type transport layer should be less or equal than the guest matrix HOMO level (see fig. 10b). In addition to these considerations, attention must be paid to the control over the interfaces morphologies, in order to allow the formation of a continuous stack.

Indeed, in this approach more than in others, functional interfaces play the predominant role in determining the performance of vertical tri-layer heterojunction. As in the case of the bilayer-based OLETs, it is clear that the interfaces between the dielectric and the bottom transport layer and between the recombination and the top transport layers are crucial for guaranteeing good ambipolar field-effect electrical characteristics. Moreover interfaces between the bottom transport and the recombination layer and between the recombination and the top transport layer should provide the favourable conditions for the charge percolation to happen in the recombination layer and to form excitons.

Through the research, Capelli et al. found that the materials that best fitted these conditions were α,ω-dihexyl-quaterthiophene (DH4T) for p-type transport, α,ω-diperfluorohexyl-quaterthiophene (DFH4T) for n-type transport and tris-(8-hydroxyquinoline)aluminum (Alq3):DCM doped host-guest system.

For what concerns light emission, in ambipolar region EL is located inside the device channel far from electrodes, thus preventing photon losses due to exciton-metal quenching. Moreover, since the emission layer is separated from charge flows, the exciton-charge quenching is also prevented. The light generation process is based on the percolation of the opposite charges from the transport layers into the recombination layer. This percolation is made easier due to the transverse electric field generated by electrons and holes in the respective transport layers. In the tri-layer structure charges recombine in the middle layer because they cannot travel through several microns of opposite charge accumulation layer without recombining, similar to what happens in ambipolar single-layer OLETs in which charges recombine in the middle of the channel. Indeed, a self-regulated equilibrium exists between the amount of charges located in the transport layers and those entering the recombination one that prevent any possible exciton-charge interaction.

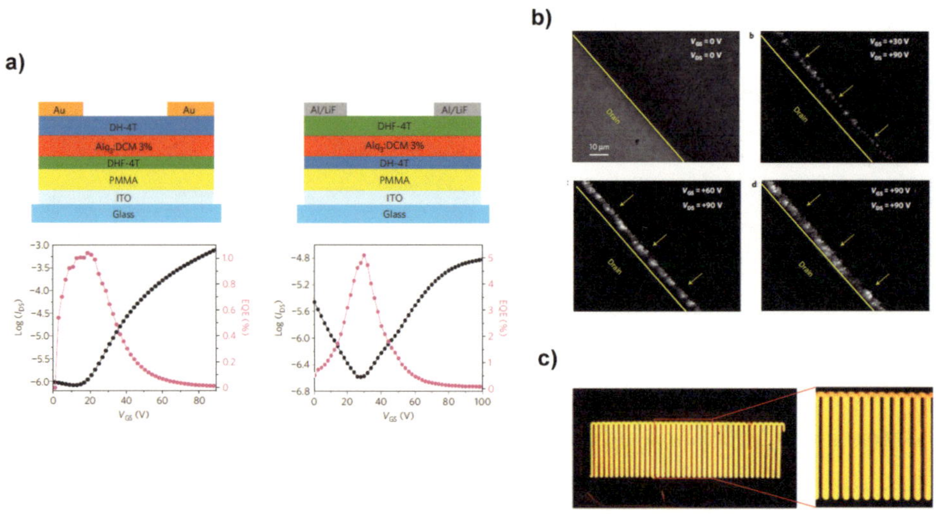

Fig. 11. Schemes of two tri-layer structure configurations and their EQE as a function of the applied voltage in a transfer characteristic (a), images of the EL channel moving toward the drain contact with increasing gate voltage (b), image of a tri-layer fabricated with an interdigitated structure (c)

This new OLET structure showed clearly the full potential of field-effect devices compared to OLED devices. The actual limits of this structure mainly depend on the materials performances. Future investigations on this structure will lead, surely, to the study of the implementation of a triplet emitter for the recombination layer, thus, potentially improving further the emission brightness.

An issue that actually must be addressed is the device operating voltage. Usually in OLED devices one of the most significant parameter is considered the power efficiency, that represents the power emitted at a certain voltage. Unfortunately in the tri-layer OLET device, the maximum light power emission is found at a gate voltage of 70 V. This issue could be overcome by developing an high capacitance gate insulator.

3. Future developments

After the demonstration of the first OLET in 2004 made by Rost et al., many great advancements have been achieved. These advancements opened up the possibility of many future developments of OLETs, tightly connected to their potential market application, especially for solid-state lighting.

Indeed, many interests are focused around OLETs, since their higher brightness and light emission efficiency with respect to OLEDs. The small spatial localization of the illuminated channel, prerogative of OLET devices (hardly achievable in OLEDs), could lead to the development of new high definition displays. The planar technology used for OLETs, will be surely exploited in future, to develop a new generation of devices, in which organic photonic field-effect transistors are used for both light generation and detection. One of the possible applications of this integration between light sensing and light generation effects in one device could be used for the fabrication of the so called "lab-on-a-chip", a miniaturized cheap and disposable device for bio-sensing that permits quantitative diagnostic tests, that up to date, are limited to laboratory or hospital facilities.

Moreover OLETs planar architecture is considered the ideal platform for the realization of resonant micro-cavity where the active region is separated by few microns from the injecting metal electrodes. This device configuration which prevents exciton quenching, will make OLETs suitable for the development of electrically-pumped organic lasers.

In addition to these promising technological scenarios of OLETs developments, the latest results, together with the continuous development of understanding of the chemical and physical properties of the device interfaces and the synthesis of new molecules by chemical tailoring, will surely open, new perspectives for the full exploitation of OLETs potentialities.

4. References

Ahles, M. et al. (2004). Light emission from a polymer transistor. *Appl. Phys. Lett.*, 84, 3, 2004, 428-430.

Capelli, R. et al. (2008). Investigation of the Optoelectronic Properties of Organic Light-Emitting Transistors Based on an Intrinsically Ambipolar Material. *J. Phys. Chem. C*, 112, 33, 12993-12999.

Capelli, R. et al. (2010). Organic light-emitting transistors with an efficiency that outperforms the equivalent light-emitting diodes. *Nat. Mater.*, 9, 6, 2010, 496-503.

Chua, L.-L. et al. (2005). General observation of n-type field-effect behavior in organic semiconductors. *Nature*, 434, 2005, 194-199.

Dinelli, F. et al. (2006). High-Mobility Ambipolar Transport in Organic Light-Emitting Transistors. *Adv. Mater.*, 18, 11, 2006, 1416-1420.

Hepp, A. (2003). Light-Emitting Field-Effect Transistor Based on a Tetracene Thin Film. *Phys. Rev. Lett.*, 91, 15, 2003, 7406-1 7406-4.

Murata, K. et al. (2001). Barriers to electron extraction in polymer light-emitting diodes. *Appl. Phys. Lett.*, 79, 8, August 2001, 1193-1195.

Namdas, E. B. et al. (2008). High performance light emitting transistors. *Appl. Phys. Lett.*, 92, 18, 2008, 3304-1 3304-3.

Oyamada, T. et al. (2005a). Electroluminescence of 2,4-bis(4-(2′-thiophene-yl)phenyl) thiophenes in organic light-emitting field-effect transistors. *Appl. Phys. Lett.*, 86, 09, 2005, 3505-1 3505-3.

Oyamada, T. et al. (2005b). Lateral organic light-emitting diode with field-effect transistor characteristics. *J. Appl. Phys.*, 98, 07, 2005, 4506-1 4506-7.

Rost, C. et al. (2004). Ambipolar light-emitting organic field-effect transistor. *Appl. Phys. Lett.*, 85, 9, August 2004, 1613-1615.

Sakanoue, T. et al. (2004). Visible light emission from polymer-based field-effect transistors. *Appl. Phys. Lett.*, 84, 16, 2004, 3037-3039.

Santato, C. et al. (2004). Tetracene-based organic light-emitting transistors: optoelectronic properties and electron injection mechanism. *Synth. Met.*, 146, 3, 2004, 329-334.

Swensen, J. ; Moses, D. & Heeger, A. J. (2005a). Light emission in the channel region of a polymer thin-film transistor fabricated with gold and aluminium for the source and drain electrodes. *Synth. Met.*, 153, 1-3, 2005, 53-56.

Swensen, J.; Soci, C. & Heeger, A. J. (2005b). Light emission from an ambipolar semiconducting polymer field-effect transistor. *Appl. Phys. Lett.*, 87, 25, 2005, 3511-1 3511-3.

Zaumseil, J.; Friend, R. H. & Sirringhaus, H. (2006). Spatial control of the recombination zone in an ambipolar light-emitting organic transistor. *Nat. Mater.*, 5, 1, 2006, 69-74.

Zaumseil, J. et al. (2008). Quantum efficiency of ambipolar light-emitting polymer field-effect transistors. *J. Appl. Phys.*, 103, 6, 2008, 4517-1 4517-10.

6

Effects of Ionizing Radiation on Optoelectronic Devices

V. Th. Tsakiri[1], A. P. Skountzos[1], P. H. Yannakopoulos[1] and E. Verrelli[2]
[1]TEI Piraeus – P.Ralli and Thivon 250, 12244 Aigaleo, Attica,
[2]National Technical University of Athens, Dept. of Applied Physics,
Zografou Campus, 15780
Greece

1. Introduction

The devices, whose function is based on the interaction of electronic processes with either light or optical processes, are called optoelectronic devices. This interaction is usually accompanied by an energy conversion process (such as electrical to optical and vice versa). Optoelectronic devices are made from semiconducting materials. Research and development of optoelectronic devices and optoelectronic integrated circuits have received a tremendous boost with the development of low-loss optical fibers for long distance communication and the development of new immune to radiation optoelectronic devices used in the aerospace industry. In the following paragraphs we describe the semiconductor materials used in optoelectronic applications in combination with the electromagnetic theory. We focused on the OLEDs and IRLEDs and we present some of the SRIM –TRIM and CASINO simulation results.

2. Optoelectronic devices

2.1 Semiconductor materials

The elemental semiconductors, especially Si, which is proved to be very useful for microelectronics, have some important drawbacks. Their band gap is indirect, resulting in poor light emission and low absorption coefficients. On the other hand compound semiconductor materials offered many of the desired properties and could be synthesized relatively easy. These materials are made from elements of different columns of the periodic table resulting in III-V, II-VI or IV-IV compounds. III-V compounds have been the first and most widely used semiconducting materials

Before referring to the advantages of some characteristic compounds, we should point out the structural difference between direct and indirect semiconductors. The top of the valence band of most semiconductors occurs at a value of effective momentum (k), equal to zero. Semiconductors in which the bottom of the conduction band is also at $k = 0$ are direct band gap materials. Semiconductors in which the bottom of the conduction band occurs at other points in momentum space are indirect band gap materials. According to the atomic structure of the semiconductors, their outermost valence electrons are in s- or p-type orbitals. Although true only for elements in their atomic form, the s- or p-like character is also retained in the crystalline semiconductors. The bands of a semiconductor are a result of the crystal potential that originates from the equilibrium arrangement of atoms in the lattice. If the edge of the conduction band is made up of s-type states, the semiconductor is direct band gap. If, on the other hand, the lowest conduction band edge is made up of p-type states, then the semiconductor is indirect band gap.

One characteristic example of semiconductor compound is indium antimonite (InSb), the first to be discovered in 1950, presenting low bandgap, E_g = 0.17 eV, and consequently applied in far infrared detector technology. Moreover, the invention of the semiconductor laser and the discovery of the Gunn Effect, turned the interest to other III-V compounds such as GaAs (E_g = 1.43 eV) and InP (E_g = 1.35 eV). GaP, which has its band gap (E_g = 2.1 eV) in the visible part of the spectrum is important for the development of the light-emitting diode (LED). GaP band gap is indirect, but by certain doping techniques, the radiative efficiency can be improved.

One main advantage of the Binary compounds is that they can be combined or "alloyed" to form Ternary or Quaternary compounds. These compounds are made up of three or four group III and group V atoms and, by choosing different elements it is possible to create materials of different band gaps, with various emission energies for light sources. However, by alloying, it is possible to vary the band gap continuously and monotonically and together with it, the band structure and the electronic and optical properties. The varying band gap also allows the building of heterojunctions, which are important for high-performance electronic and optoelectronic devices. The quaternary compounds of $In_xGa_yAl_{1-x-y}As$, and $In_{1-x}Ga_xAs_yP_{1-y}$ present band gaps which correspond to the spectral window in which silica fibers have their lowest loss and dispersion, making their research and development very important for optical communications. (Bhattacharya, 1997)

2.2 Electron – hole formation and recombination

The creation or annihilation of electron-hole pairs rules the operation of almost all optoelectronic devices. Pair formation involves raising an electron in energy from the valence band to the conduction band, resulting with a hole behind in the valence band. In principle, any energetic particle incident on a semiconductor, which can impart energy at least equal to the bandgap energy to a valence band electron, will create pairs. The simplest way to create electron-hole pairs is to irradiate the semiconductor. Photons with sufficient energy are absorbed, and these impart their energy to the valence band electrons raising them to the conduction band. This process is, therefore, also called absorption. The reverse process, that of electron and hole recombination, is associated with the pair giving up its excess energy after recombination. This process may be radiative or non-radiative. In a non radiative transition, the excess energy due to recombination is usually imparted to phonons and dissipated as heat in the material. In a radiative transition, the excess energy is dissipated as photons, usually having energy equal to the bandgap. This is the luminescent process, which is classified in three cases according to the method by which the electron-hole pairs are created: a) Photoluminescence, involves the radiative recombination of electron-hole pairs created by injection of photons, b) cathodoluminescence, involves radiative recombination of electron-hole pairs created by electron bombardment and c) electroluminescence, involves radiative recombination following carriers' injection in a p-n junction.

In the case of a semiconductor in equilibrium (i.e., without any incident photons or injection of electrons), the carrier densities can be calculated from an equilibrium Fermi level using Fermi-Dirac or Boltzmann statistics. When excess carriers are created by one of the techniques described above, non-equilibrium conditions are generated and the concept of a Fermi level is no longer valid. One can, however, define non-equilibrium distribution functions for electrons and holes as:

$$f_n(E) = \frac{1}{1+\exp\left(\frac{E-E_{fn}}{k_\beta T}\right)} \quad (1)$$

$$1 - f_p(E) = \frac{1}{1+\exp\left(\frac{E-E_{fp}}{k_\beta T}\right)} \quad (2)$$

where E_{fn} and E_{fp} are the quasi-Fermi levels for electrons and holes respectively. For the non-degenerate case (the Fermi level is several kT below E_c), the distribution functions of equations 1, 2 can be written as:

$$f_n(E) \cong \exp\left(\frac{E_{fn}-E}{k_\beta T}\right) \quad (3)$$

$$f_p(E) \cong \exp\left(\frac{E-E_{fp}}{k_\beta T}\right) \quad (4)$$

and the non-equilibrium carrier concentrations are given by:

$$n = N_c \exp\left(\frac{E_{fn}-E_c}{k_\beta T}\right) \quad (5)$$

$$p = N_v \exp\left(\frac{E_v-E_{fp}}{k_\beta T}\right) \quad (6)$$

The quasi-Fermi levels, provide the proper tool for calculating the changes of carrier concentration as a function of position in a semiconductor.

Generation and recombination processes involve transition of electrons across the energy bandgap and differ for direct and indirect bandgap semiconductors. In a direct bandgap semiconductor, the valence band maximum and the conduction band minimum occur at the zone center ($k = 0$) so that by an upward or downward transition of electrons the momentum is conserved. Therefore, in direct bandgap semiconductors such as GaAs, an electron raised to the conduction band, (i.e. by photon absorption) will dwell there for a very short time and recombine again with a valence band hole to emit light of energy equal to the bandgap. In the case of an indirect bandgap semiconductor, where the conduction band minima are not at $k = 0$, upward or downward transition of carriers results in a momentum change so that the involvement of a phonon is needed for the conservation of momentum. Thus, an electron dwelling in the conduction band minimum, cannot recombine with a hole at $k = 0$ until a phonon with the right energy and momentum is available. Both phonon emission and absorption processes can assist the downward transition. In order for the right phonon collision to occur, the dwell time of the electron in the conduction band increases. Consequently the probability of radiative recombination is much higher in direct than indirect bandgap semiconductors justifying the use of them as light sources such as light-emitting diodes and lasers.

Moreover impurities and defects in the crystal lattice also serve as traps and recombination centers. It is most likely that an electron and a hole recombine non-radiatively through such a defect center and the excess energy is dissipated into the lattice as heat. As advanced epitaxial techniques are being developed, the purity of the crystals continues to improve. The simulation that is presented aims in the investigation of the defects generated under ionized radiation. .(Bhattacharya, 2003)

2.3 Radiative and non-radiative recombination

For continuous carrier generation by optical excitation or injection, a quasi-equilibrium or steady state is produced. Electrons and holes are created and annihilated in pairs and, depending on the injection level, a steady-state excess density is established in the crystal, also necessary for the overall charge neutrality. When the excitation source is removed, the density of excess carriers returns to the equilibrium values, n_0 and p_0. The excess carriers usually decay exponentially with respect to time following the -exp (-t/r), where r is defined as the lifetime of excess carriers. The lifetime is determined by a combination of intrinsic and extrinsic parameters, and affects the performance characteristics of most optoelectronic devices. It should be noted that, depending on the semiconductor sample and its surface, there can be a very strong surface recombination component which depends on the density of surface states.

Generally, the excess carriers decay by radiative and/or non-radiative recombination, in which the excess energy is dissipated by photons and phonons. The former is of importance for the operation of luminescent devices. Non-radiative recombination usually takes place via surface or bulk defects and traps and reduces the radiative efficiency of the material. Consequently the research in materials aims in reducing these non-radiative centers by optimizing the fabrication method or by applying passivation techniques on these centers.

2.4 Electromagnetic description of reflection and refraction

The function and the properties of the materials used in optoelectronic devices are strongly determined by the propagation of plane, single-frequency, electromagnetic waves. By the Maxwell's equations, useful expressions can be obtained for the dissipation, storage and transport of energy, resulting from the propagation of waves in material media. Moreover analyzing the incidence, reflection and transition of an electromagnetic wave on the border surface between two dielectric materials, can result in a better comprehension of optoelectronic devices.

2.4.1 Maxwell's equations

The following four laws constitute the basis for the electromagnetic theory:

Gauss's law for electric fields:
$$\vec{\nabla} \circ \vec{E} = {}^{\rho}\!/_{\varepsilon_0} \tag{7}$$

Gauss's law for magnetic fields:
$$\vec{\nabla} \circ \vec{B} = 0 \tag{8}$$

Faraday's law:
$$\vec{\nabla} \times \vec{E} = -\frac{\partial \vec{B}}{\partial t} \tag{9}$$

Ampere – Maxwell law:
$$\vec{\nabla} \times \vec{B} = \mu_0 \left(\vec{J} + \varepsilon_0 \frac{\partial \vec{E}}{\partial t} \right) \tag{10}$$

The laws above can be applied in combination with the constitutive equations:

$$\vec{D} = \varepsilon_0 \vec{E} + \vec{P} \text{ and } \vec{B} = \mu_0 \left(\vec{H} + \vec{M} \right) \tag{11}$$

where J, is the current density, E and H are the electric and magnetic fields respectively, D and B, are the electric and magnetic displacements and P and M, are the electric and magnetic polarizations respectively.

Considering that in the dielectric materials there are no free charge carriers and currents, the application of Maxwell's equations, in combination with the constitutive equations result in the boundary conditions standing at the boundary surface between two dielectric materials. These conditions denote that during the transition from the dielectric 1 to the dielectric 2, the tangential (at the boundary surface) components of E and H as well as the normal (at the boundary surface) components of D and B, are continuous. The mathematical expression can be written as follows:

$$E_{2t} = E_{1t}, \quad H_{2t} = H_{1t}, \quad D_{2n} = D_{1n} \quad \text{and} \quad B_{2n} = B_{1n}$$

where the indexes 1 and 2 denote the limiting values, in the case of approaching the boundary surface from the dielectric 1 or 2.

These boundary conditions will serve for the calculation of important coefficients in the case of reflection and transmission of light. (Yariv,1976)

2.4.2 Reflection and refraction

2.4.2.1 Angle of incidence = 0°

In figure 1, the vectors of the electric (**E**) and magnetic (**H**) fields are subscripted as "i" for incident wave, "r" for reflected wave and "t" for transmitted wave. In this case the respective equations in complex form, for the vectors of the fields E(electric) and H(magnetic) will be:

$$\begin{aligned}
E_{iy} &= E_{i0} e^{i(\omega t - k_1 x)} \quad H_{iz} = H_{i0} e^{i(\omega t - k_1 x)} \quad (x \prec 0) \\
E_{ry} &= E_{r0} e^{i(\omega t - k_1 x)} \quad H_{rz} = H_{r0} e^{i(\omega t - k_1 x)} \quad (x \prec 0) \\
E_{ty} &= E_{t0} e^{i(\omega t - k_2 x)} \quad H_{tz} = H_{t0} e^{i(\omega t - k_2 x)} \quad (x \succ 0)
\end{aligned} \tag{12}$$

In the above equations, the unknown quantities are the amplitudes: E_{r0} and E_{t0}. By these, are defined the coefficient of reflection $r=E_{r0}/E_{i0}$ and the coefficient of transmission $t=E_{t0}/E_{i0}$. These coefficients can be calculated by applying boundary conditions for the fields E and H, and finally the quantities of E_{r0} and E_{t0} can be calculated. Finally:

$$r = \frac{n_1\mu_2 - n_2\mu_2}{n_1\mu_2 + n_2\mu_1} \tag{13}$$

and

$$t = \frac{2n_1\mu_2}{n_1\mu_2 + n_2\mu_1} \tag{14}$$

where $n_{1,2}$ and $\mu_{1,2}$ are the refraction coefficient and magnetic permeability respectively, of the two dielectric materials. (Papakitsos,2002)

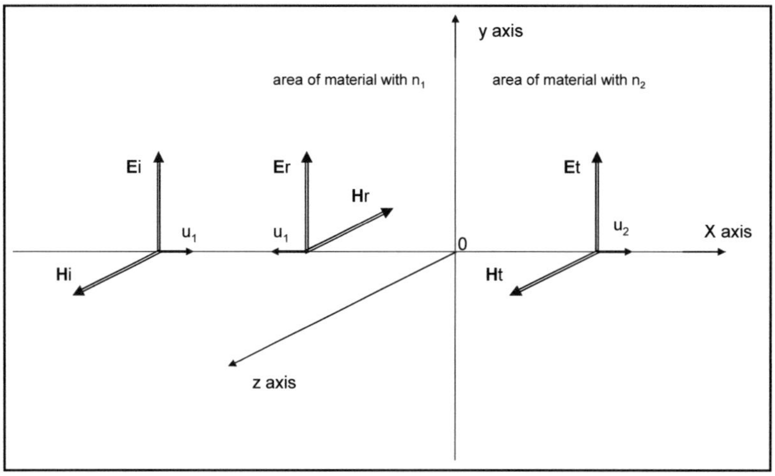

Fig. 1. EMF vectors presenting the transmission and reflection of an EM wave incident at a boundary surface between to dielectric materials

By the coefficients above, also the quantities of reflectance, R and transmittance, T can be calculated. These quantities are defined as: $R = \frac{I_r}{I_i}$ and $T = \frac{I_t}{I_i}$ where I_r is the intensity of the reflected wave, I_i is the intensity of the incident wave and I_t is the intensity of the transmitted wave. In the special case of angle of incidence = 0°,

$$R = \frac{n_1|E_{r0}|^2/2c\mu_1}{n_1|E_{i0}|^2/2c\mu_1} = \left|\frac{E_{r0}}{E_{i0}}\right|^2 = |r|^2 \tag{15}$$

and

$$T = \frac{n_2|E_{t0}|^2/2c\mu_2}{n_1|E_{r0}|^2/2c\mu_1} = \frac{n_2\mu_1}{n_1\mu_2}\left|\frac{E_{t0}}{E_{r0}}\right|^2 = \frac{n_2\mu_1}{n_1\mu_2}|t|^2 \tag{16}$$

From the above relations, the energy conservation is proved, since R+T=1!

In the case of transparent dielectrics, where the magnetic permeability is semi equal to the magnetic permeability of vacuum, the above relations are far simplified as follows:

$$r = \frac{n_1 - n_2}{n_1 + n_2},\ t = \frac{2n_1}{n_1 + n_2},\ R = \left(\frac{n_1 - n_2}{n_1 + n_2}\right)^2 \text{ and } T = \frac{4n_1 n_2}{(n_1 + n_2)^2}$$

It is important to note that $r = \begin{cases} \succ 0 & for \quad n_1 \succ n_2 \\ \prec 0 & for \quad n_1 \prec n_2 \end{cases}$ but $t \succ 0$ for every n_1, n_2.

The physical meaning hiding here is that when a wave is heading from a material with n_1 to a material with n_2 with $n_1 \succ n_2$ the phase of the reflected wave is unchanged in accordance to the phase of the incident wave. On the opposite case ($n_1 \prec n_2$) the phase of the reflected wave has a difference of π in accordance to the phase of the incident wave.

2.4.2.2 Angle of incidence = θ°

In the case of a random angle of incidence, we consider the transverse wave of the electric field (TE) and the transverse wave of the magnetic field (TM). Consequently, following the method of figure 1, the coefficients r and t are now written as:

For the transverse wave of the electric field $r_E = \frac{E_{r0}^E}{E_{i0}^E}$, $t_E = \frac{E_{t0}^E}{E_{i0}^E}$ and

for the transverse wave of the magnetic field $r_M = \frac{E_{r0}^M}{E_{i0}^M}$, $t_M = \frac{E_{t0}^M}{E_{i0}^M}$.

By applying again the boundary conditions for the fields, the coefficients r and t can be calculated:

$$r_E = \frac{\cos\theta - \sqrt{n^2 - \sin^2\theta}}{\cos\theta + \sqrt{n^2 - \sin^2\theta}},\ t_E = \frac{2\cos\theta}{\cos\theta + \sqrt{n^2 - \sin^2\theta}} \tag{17}$$

$$r_M = \frac{\sqrt{n^2 - \sin^2\theta} - n^2\cos\theta}{\sqrt{n^2 - \sin^2\theta} + n^2\cos\theta},\ t_M = \frac{2n\ \cos\theta}{\sqrt{n^2 - \sin^2\theta} + n^2\cos\theta} \tag{18}$$

where $n = {}^{n_2}\!/_{n_1}$ the relative refraction index between material 2 and material 1. The equations above, known as the "Fresnel equations", apply for transparent materials.
As described before two cases should be clarified again depending on the relation between n_1 and n_2.

a. $\boldsymbol{n_1 < n_2}$

In this case, both $r_{E,M}$ and $t_{E,M}$ have real values for all incident angles $\theta < \pi/2$. Also t_E and t_M have positive values which means that the refracting transverse waves of electric and magnetic field have the same phase with the respective incident ones. On the other hand the value of r_E is negative for every angle $\theta < \pi/2$, and the value of r_M is: negative for $\theta < \theta_B$, zero for $\theta = \theta_B$, and positive for $\theta_B < \theta < \pi/2$, where θ_B is the Brewster angle ($\tan\theta_B = n = n_2/n_1$). The negative values mean that the respective reflecting waves present a phase difference of π in accordance to the phase of the incident wave. In the special case of $\theta = \theta_B$, there is no reflectance for the transverse magnetic field (!) and when $\theta_B < \theta < \pi/2$ there is no phase difference for it.

b. $n_1 > n_2$

Both $r_{E,M}$ and $t_{E,M}$ have real values but only for incident angles $\theta < \theta_c$, where θ_c is the critical angle for total reflection ($\sin\theta_c = n = n_2/n_1$). For those values of θ, both t_E and t_M as well as r_E are positive indicating that these waves exhibit no phase difference with the respective incident ones. As before the value of r_M depends on the Brewster angle but in a different way, r_M is: positive for $\theta < \theta_B$, zero for $\theta = \theta_B$, and negative for $\theta_B < \theta < \theta_c$. So for the last case a phase difference of π is observed. Finally in the case of $\theta_c < \theta < \pi/2$, for all coefficients the values are complex numbers and the phenomenon of total reflection is observed.

2.4.3 Optical tunneling

The optical tunnelling is an interesting optical phenomenon which can be explained only by considering the light as an electromagnetic wave. This phenomenon arises in the case of total reflection where:

$$\sin\theta \succ \sin\theta_c = n \Rightarrow n^2 - \sin^2\theta \prec 0 \Rightarrow \sqrt{n^2 - \sin^2\theta} = \pm i\sqrt{\sin^2\theta - n^2}$$

By substitution of the negative values in the r_E and r_M relations:

$$r_E = \frac{\cos\theta + i\sqrt{\sin^2\theta - n^2}}{\cos\theta - i\sqrt{\sin^2\theta - n^2}}, \quad r_M = \frac{n^2\cos\theta + i\sqrt{\sin^2\theta - n^2}}{-n^2\cos\theta + i\sqrt{\sin^2\theta - n^2}} \tag{19}$$

The magnitude of the complex coefficients above equals zero, so they can also be written as: $r_E = e^{i\phi_E}$, $r_M = e^{i\phi_M}$. The arguments φ_E and φ_M can be calculated by the trigonometrical equations:

$$\tan\left(\frac{\varphi_E}{2}\right) = \frac{1}{\cos\theta}\sqrt{\sin^2\theta - n^2} \equiv \tan\beta \ , \ \tan\left(\frac{\varphi_M}{2} + \frac{\pi}{2}\right) = \frac{1}{n^2}\tan\beta \tag{20}$$

where $0 \prec \beta \prec \pi/2$.

The physical meaning of the relations above is that the amplitudes of the reflected waves of the transverse electric and magnetic fields remain the same. But the reflected wave of the electric field is leading the respective incident wave by a phase of φ_E and the reflecting wave of the magnetic field is following the respective incident wave by a phase of φ_M. Moreover since $|r_E| = |r_M| = 1$, also the $R_E = R_M = 1$ (Eq. 15). This means that during the internal total reflection, the total energy of the incident waves is reflected.

According to the geometric optic, one would expect the reflected wave to arise from the point of the incidence. However in total reflection phenomenon, this is not always the case! The reflected wave appears shifted by a distance z, relatively to the incidence point, known as "Goose-Haenchen shift" as shown in Fig. 2b.(reported in Isaac Newton's works, who could not explain it without Maxwell's equations).

This shift means that part of the incident wave penetrates the material 2 (with $n_2 < n_1$) and moves parallel (z axis) to the boarder surface with an amplitude that is diminishing exponentially as a function of the depth (y axis) in the material 2. This wave is called "evanescent" (Fig. 2a) and is described by the following equations:

$$E_e^E = E_{e0}^E(y)\cos\left(\omega t - k_e z + \varphi_E/2\right) \tag{21}$$

where: $E_{e0}^E(y) = A_e^E e^{-ay}$ and $A_e^E = 2A_i^E \cos\frac{\varphi_E}{2}$ (A_i^E is the amplitude of the incident transverse electric file wave)

$$a = \sqrt{k_1^2 \sin^2\theta - k_2^2} = \frac{2\pi n_2}{\lambda}\sqrt{\left(\frac{n_1}{n_2}\right)^2 \sin^2\theta - 1} \tag{22}$$

$k_e = k_1 \sin\theta$ ($k_1 = n_1 k$, $k_2 = n_2 k$, $k = \frac{2\pi}{\lambda}$, where λ is the wavelength in vacuum)

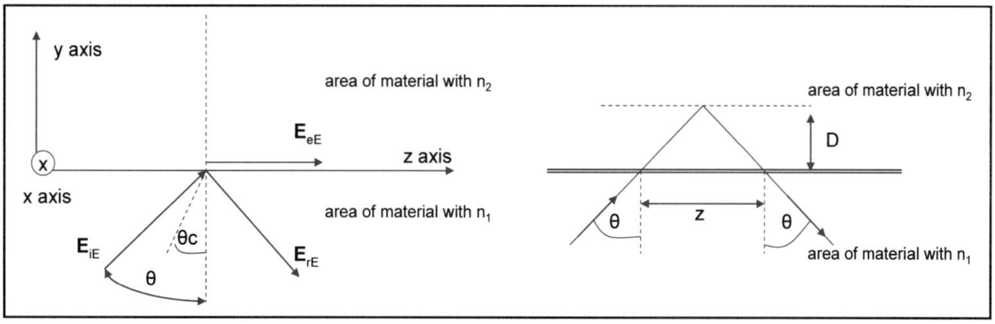

Fig. 2. a) Evanescent wave (Left) and b) Goose-Haenchen shift (right)

The "penetration depth" D, in material 2 is defined as the depth where the amplitude $E_{e0}^E(D) = A_e^E e^{-aD}$ decreases by the factor of e $E_e^E(D) = \frac{A_e^E}{e}$. D is given by:

$$D = \frac{1}{a} = \frac{\lambda}{2\pi n_2} \Bigg/ \sqrt{\left(\frac{n_1}{n_2}\right)^2 \sin^2\theta} \tag{23}$$

By geometry, it is found that the shift of the reflected wave equals: $z = 2D\tan\theta$. In the case of a thin enough layer of material 2 (thickness < D), not all of the evanescent wave rises as a reflected wave. Part of it, propagates through the material 2. In that case the energy of the reflected wave is smaller than the energy of the incident wave and depends on the thickness of material 2. This case, known as optical tunnelling, is one of the effects of the wave nature of light.

2.5 Light Emitting Diodes (LEDs)

Since 1904, these devices have found wide applications in the form of indicator lamps, display elements and sensors. In a junction LED, photons of near-band-gap energy are generated by the process of injection (luminescence or electroluminescence, as described above), in which a large population of electrons, injected into a normally empty conduction

band by forward bias, recombine with holes in the valence band. The device emits light by spontaneous emission. Some of the obvious advantages of the LED as a light source, are simpler fabrication procedures, lower cost, and simpler drive circuitry.
In the case of electroluminescence, the LED converts the input electrical energy into output optical radiation in the visible or infrared portion of the spectrum, depending on the semiconductor material. This conversion takes place in two consecutive events. Initially the energy of carriers in the semiconductor is raised above their equilibrium value by the electrical input energy. Finally most of these carriers, after having lived a mean lifetime in the higher energy state, they give up their energy as spontaneous emission of photons with energy nearly equal to the band gap of the semiconductor. In the case of a doped semiconductor (p- or n- type), the energy of the current carriers can be increased by applying a forward bias to a p-n junction. (Bhattacharya, 1997)
The internal quantum efficiency of the device is defined as the rate of the emission of photons divided by the rate of the supplied electrons. As known, all the injected electrons do not recombine radiatively with holes, leading to an efficiency always less than 100%.
The choice of the proper semiconductor material, depends on the wavelength of the light emission required for a specific application. As expected, lower band gap materials are required for infrared and far-infrared applications while higher band gap materials are required for applications in the visible part of the spectrum. It is important to note that higher band gap materials exhibit higher melting point and lower structural stability. Another important disadvantage of the higher band gap materials is their higher resistivity and the difficulty of being doped. The III-V compounds and especially the binary compounds of GaAs and GaP and the ternary $GaAs_{1-x}P_x$ are considered to be the most important materials for LED's fabrication due to their properties described before.

2.5.1 IR LEDs

For longer wavelengths in the near-infrared region of the spectrum, InP- based compounds are important (i.e. $In_{0.53}Ga_{0.47}As$, $In_xGa_{1-x}P_yAs_{1-y}$). Devices made of these materials operate in the 1.1-1.6 μm range, which overlaps with the spectral region of low loss and minimum dispersion in optical fibers, making them important for applications in optical fiber communication. In general, the use of IR leds offer great advantages in circuit design due to the:

a. low power requirements, making them ideal for personal digital equipment like laptops and telephones
b. low cost and simplicity of circuitry
c. lack of interference with other devices' signals

The advantages they exhibit make them appropriate for applications involving multi-touch screens, based on FTIR technolology. As an example, the designing of single display cockpits (with IR LEDs) is reported by Xalas et al.(2009).
On the other hand there are some design restrictions since signals are wirelessly transmitted. All transmitters and receivers should be almost directly aligned and any blocking material should be avoided. Moreover the transmission length is limited since the performance of such devices drops off by distance. Research is heading for the development of higher power IR LEDs in order to eliminate this problem and as referred, there are already IR LEDs at 940nm with an output power 2.5 times higher than that of a conventional 940 nm LED. (Kitabayashi et al., 2010)

2.5.2 OLEDs

The special case of OLEDs involves the use of organic materials, for example a conducting polymer layer (polyaniline PANI:PSS, Polyethylenedioxythiophene PDOT:PSS) and an emissive polymer layer (Polyohenylenevinylene R-PPV, Polyfluorene PF), between the anode and the cathode of the diode. These devices offer mainly the advantage of different emission colors by synthesizing a variety of chemical structures. Moreover the fabrication cost is less since the vacuum deposition steps required are less and these materials are compatible with printing techniques resulting in low cost for full color applications. The advantages mentioned, led the research to the semiconducting organic materials in order to fabricate electronic devices that are thin, flexible, low cost, and disposable.

The efficiency of an OLED device is defined as the ratio between the electrical power supplied to the device and the optical power that comes out through the glass substrate. It can be divided into the internal efficiency, that is the number of generated photons per injected electron, and the optical outcoupling efficiency which is the percentage of the generated light that is able to escape from the device through the glass substrate. Three factors affect mainly the outcoupling efficiency: a) the refractive index of the emitting material (n = 1.7), b) the refractive index of the substrate (n = 1.5) and c) the reflectivity of the cathode. In a thick device (relative to the wavelength of light), and in absence of scattering effects, the outcoupling efficiency (η) is given by the following analytical formula following from Snell's law:

$$n = \frac{1+R}{2}\left(1-\sqrt{1-\left(\frac{1}{n_e}\right)^2}\right) \tag{24}$$

But in real cases, the thickness of every layer is in the order of the wavelength of the generated light, yielding an angular distribution very different from the isotropic. Therefore, the thickness of every layer and the location of the emitter play also an important role on the angular distribution of the generated photons.

Optimization of OLEDs can be done by enhancing the amount of light that is emitted into air. This can be achieved by optimizing the thickness of the different layers and by inhibiting lateral propagation of the light inside the OLED. The presence of scattering centers in the substrate or especially in the active medium, where most of the light is trapped, could result in an increment of the outcoupling efficiency. Photonic crystals or nanostructured surfaces can also be applied to avoid lateral propagation of the light inside the OLED, bringing more light into the escape cone.

Research about improving the life time and the efficiency seems to be worth it due to the great advantages of OLEDs:

- High efficiency and large area sources.
- High brightness and wide viewing angle.
- Thin, flat and lightweight.
- Low voltage and fast switching technology.
- Form freedom and tunable emission.
- Flexible displays.
- Low cost production.

The great number of OLED applications led to the fabrication of different types according to special needs. Two different classifications can be made according to the matrix type and according to the emission of the OLED. (Andoniadis, 2003)

Matrix type:

a. **Passive matrix OLED (PMOLED)**
PMOLEDs have strips of cathode, organic layers and strips of anode. The anode strips are arranged perpendicular to the cathode strips. The pixels, emitting light, are formed by the intersections of the cathode and anode. Current is applied by external circuitry to specific strips of anode and cathode, depending on which pixels are chosen to emit light. PMOLEDs are easy to make, but they consume more power than other types of OLED, mainly due to the power needed for the external circuitry. PMOLEDs are most efficient for text and icons and are best suited for small screens such as cell phones, PDAs and, MP3 players. It should be noted that even with the external circuitry, passive-matrix OLEDs consume less battery power than the LCDs that currently power these devices.

b. **Active matrix OLED(AMOLED)**
AMOLEDs have full layers, instead of strips as PMOLEDs, of cathode, organic molecules and anode. In this case the pixels' matrix is formed by a thin film transistor (TFT) array overlaying the anode. The TFT array itself is the circuitry that determines which pixels get turned on to form an image. Considering that AMOLEDs consume less power than PMOLEDs, since the TFT array requires less power than external circuitry, and that they exhibit faster refresh rates, they are the appropriate choice for large displays such as computer monitors, large-screen TVs and electronic signs or billboards.

OLED emission:

a. **Transparent OLED**
All the components of the OLED, in this case are transparent (substrate, cathode and anode) and, when turned off, are up to 85 percent as transparent as their substrate. When a transparent OLED display is turned on, it allows light to pass in both directions. A transparent OLED display can be either active- or passive- matrix and can be used for heads-up displays.

b. **Top-emitting OLED**
In top-emitting OLEDs the substrate is either opaque or reflective and they are more appropriate for active-matrix design. They are usually applied in smart cards.

c. **Foldable OLED**
In this case the substrates are made of very flexible metallic foils or plastics. Foldable OLEDs are very lightweight and durable. Their use in devices such as cell phones and PDAs can reduce breakage, a major cause for return or repair. Potentially, foldable OLED displays can be attached to fabrics to create "smart" Clothing, with an integrated computer chip, cell phone, GPS receiver and OLED display sewn into it.

d. **White OLED**
White OLEDs emit white light that is brighter, more uniform and more energy efficient than that emitted by fluorescent lights. Since OLEDs can be made in large sheets, white OLEDs can replace fluorescent lights currently used in homes and buildings resulting in energy saving.

3. Experiment

Simulation is important in proton (or electron) testing because real testing is expensive and it is complicated by the strong dependence of proton damage on energy. The effects of shielding must also be taken into account when this calculation is done, because most of the low energy protons will be removed by even thin amounts of shielding. One important practical difficulty for proton testing is activation of circuits and test boards during irradiation.

We have simulated the damage induced by protons and beta particles of various energies using SRIM-TRIM and CASINO packages respectively. We have modeled an OLED using its cross section (Youtian Tao et al., 2010) which is shown in Table 1.

Layer Name	*Width (Å)*	*Density (g/cm^3)*
Al (aluminum)	1000	2.70
Lithium fluoride	10	2.64
Alq_3 tris(8-hydroxyquinoline) aluminum	300	1.39
BCP(2,9-dimethyl-4,7-diphenyl-1,10-phenanthroline)	100	1.29
NPB (emissive layer) ((1,4-bis(1-naphthylphenylamino)biphenyl)	800	1.326
MoO_3	100	4.69
ITO (indium tin oxide)	1500	7.1

Table 1. Cross Section of an OLED

Before starting the irradiation of the structure above, we performed simulations for protons of various energies on a single layer of aluminum of 1 cm thickness, considering it as a characteristic example of a craft's shield. It was calculated that only protons of energies grater than ~45 MeV could cross over the aluminum layer. Further more it was observed that protons with slightly higher energies than the above mentioned threshold, traverse the shield with residual energy in the MeV range. For example for 50 MeV protons the transmitted ions that emerge from the shield have a mean energy of 11MeV.

In this work protons were simulated with SRIM for the energies of 10keV, 30keV, 100keV, 300keV, 1MeV, 3MeV, 10MeV and 20MeV and around 3million ions considering these values as the protons' energies emerging from a possible shield.

SRIM-TRIM 2010.01 (Stopping and Range of Ions in Matter) can handle ions and neutrons. SRIM 2010.00 calculates the range of ions in the matter (10eV-2GeV/amu) using collisions of ions - atoms. TRIM (Transport of Ions in Matter) accepts up to 8 complex targets from various materials. It calculates the 3D spread of ions as well as the all kinetic phenomena that are related with the loss of their energy: damage of the target, sputtering, ionization and phonons production. With SRIM we obtained the plots of depth versus y-axis, ionization, phonons, collision events, atom distributions and energy to recoils. (Fig. 3, 4).

30keV

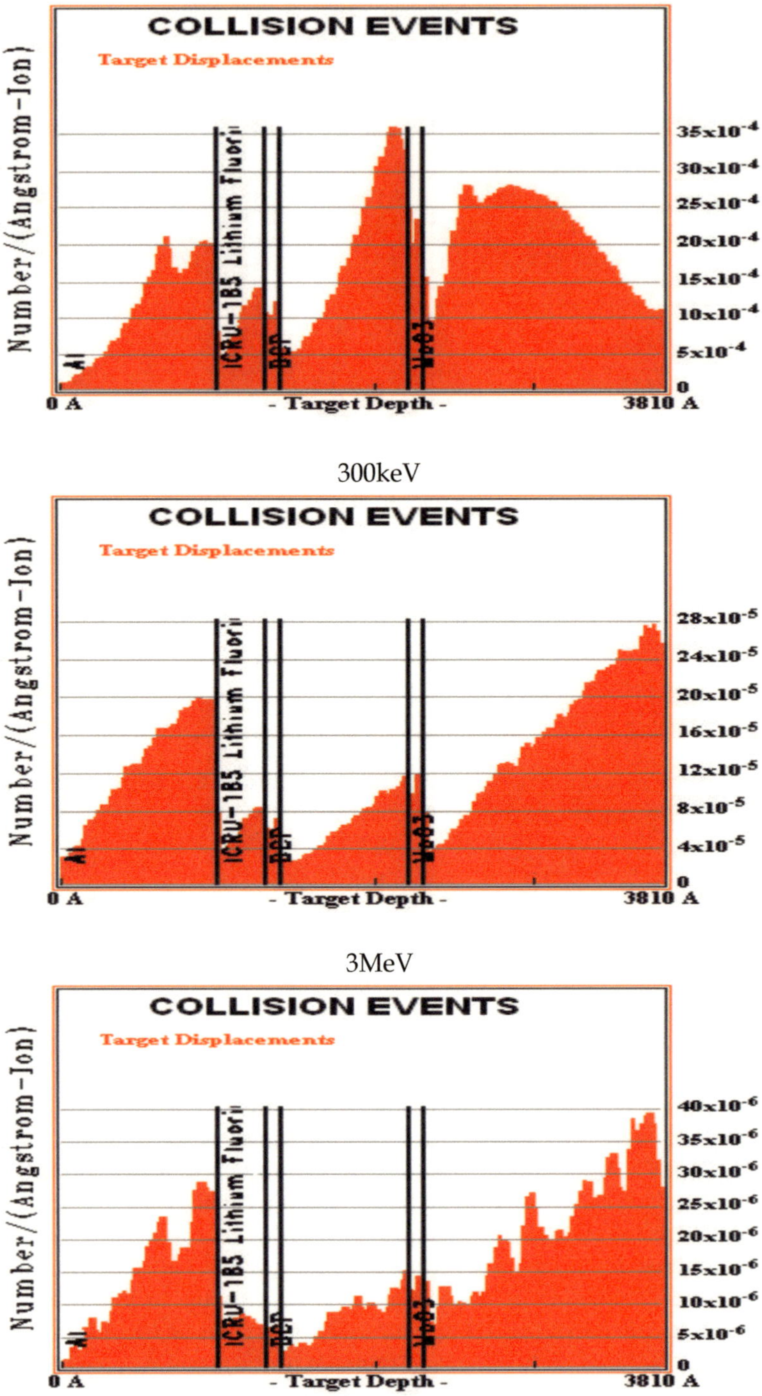

Fig. 3. Displacements distribution across the OLED

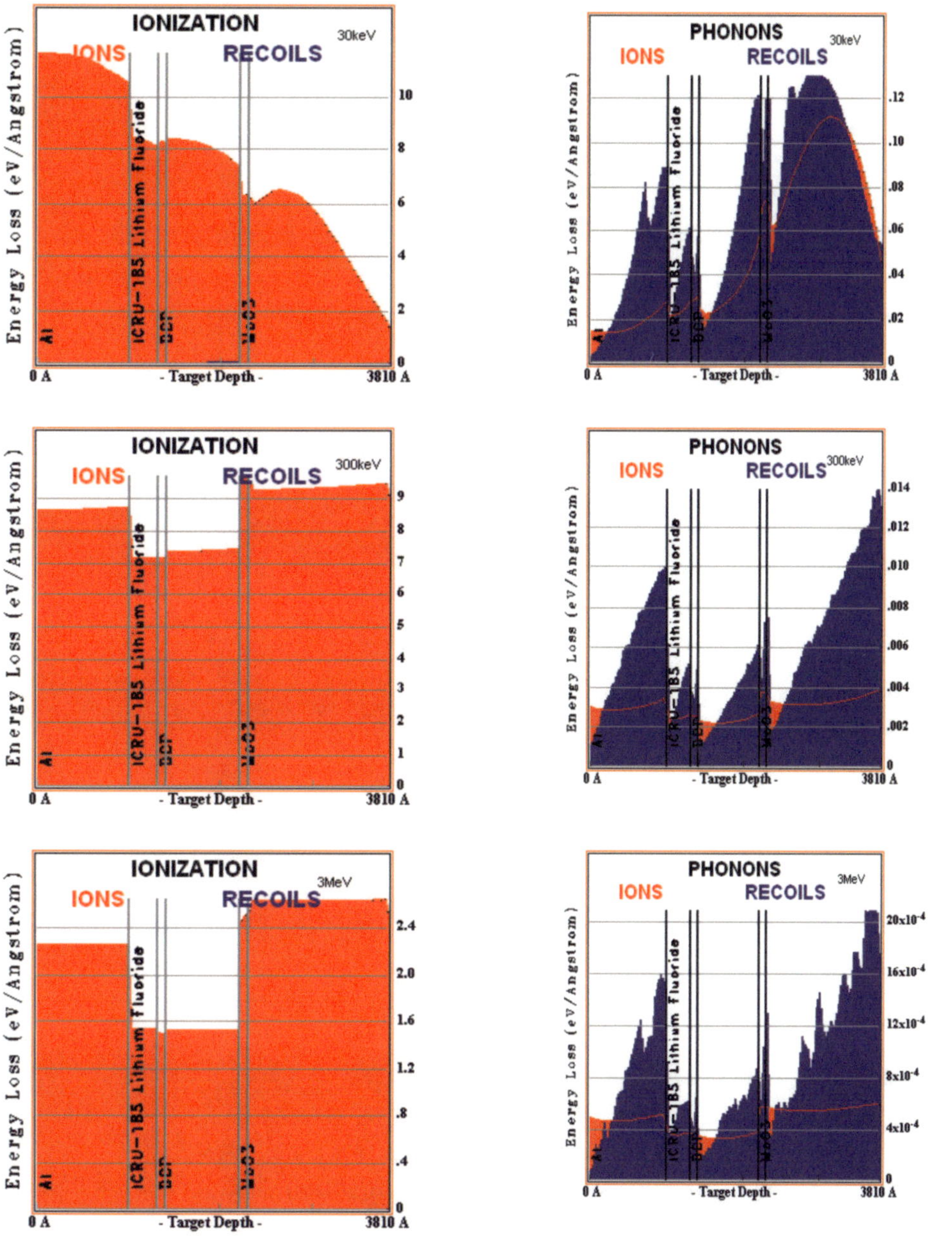

Fig. 4. Ionization (left column) and phonons (right column) distribution across the OLED

Beta particles were simulated with CASINO (Fig.5) with energies ranging from 3 to 30 keV. CASINO v.2.42 (Monte Carlo Simulation of electron trajectories in Solids) is a program with which we can see the trajectories of electrons in the matter as well as their results. When electrons pass through a material, four possible processes may occur: Ionization, Bremsstrahlung, Elastic scattering and nuclear excitation which is usually negligible.

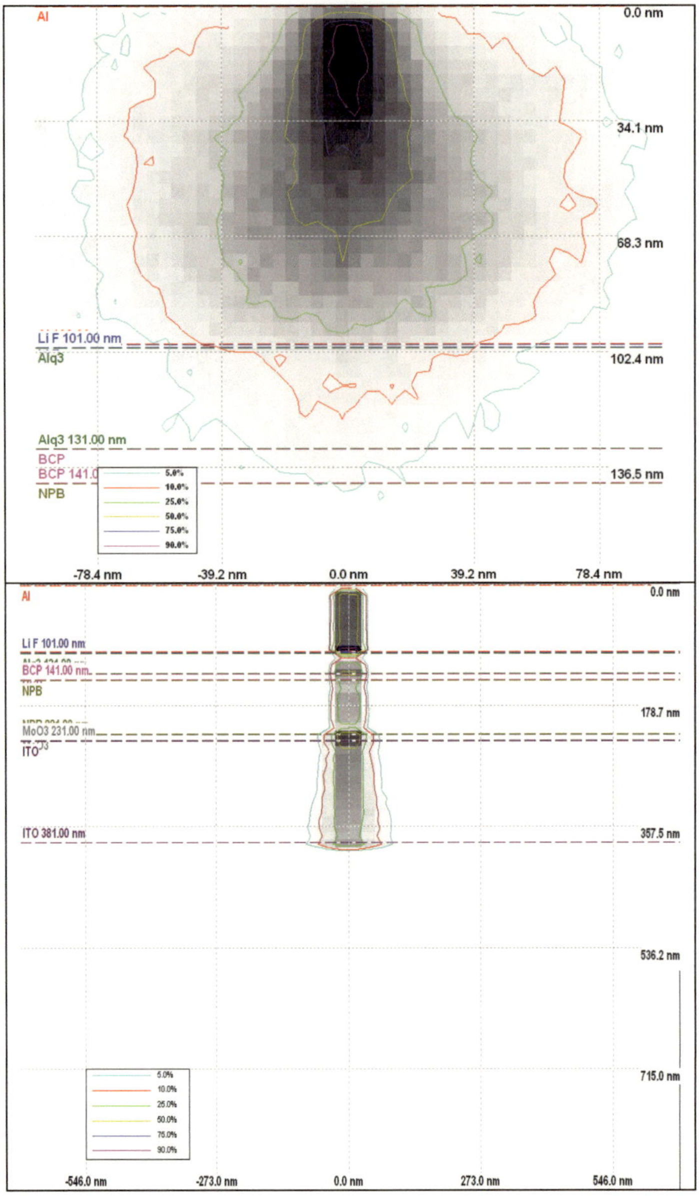

Fig. 5. Typical CASINO output at 3keV(top) and 30keV(bottom)

4. Results and discussion

The main types of radiation effects on materials can be grouped into the following categories:

a. *Impurity Production*: transmutation of nuclei into other nuclei which themselves may be radioactive; this mechanism is from neutrons only through fission and activation. Impurities can also be deposited from the creation of hydrogen or helium when a proton or an alpha particle becomes neutralised in the material of passage.

b. *Atom Displacement* from their normal position in the structure of the material; displacement atoms may leave lattice vacancies and lodge in interstitial locations or cause interchange of dissimilar atoms in the lattice structure.
c. *Ionization*: removal of electrons from atoms in the material and formation of ion pairs in the path of the charged particles.
d. Large *Energy Release* in a small volume, which can result in thermal heating of the material.

By the CASINO energy dispersion graphs it is shown that for energies less than 3keV, the emissive layers of the optoelectronic device remain practically unaffected. Most of the radiation is stopped in the substrate layers affecting their electronic properties. (Fig. 5)

Concerning the TRIM simulations, two major limitations that should be considered are: (a) There is no build-up of ions or damage in the target. Every ion is calculated with the assumption of *zero dose*, i.e. the target is perfect and previous ions have no effect on subsequent ions. (b) The target temperature is 0^oK, and there are no thermal effects changing the distribution of ions (thermal diffusion) or affecting the target damage (thermal annealing). Several experiments have been reported at very low temperatures ($15\text{-}40^oK$) which validate the TRIM results, but how that thermal effects can be quite substantial. For example, for a low dose of boron ions into crystalline silicon, more than 99% of the target damage anneals at room temperature.

A proton loses energy in small steps through interactions with the electrons in the material. It slows down almost entirely due to Coulomb interactions with the atomic electrons of the target material. Because of the large number of these interactions, the slowing down procedure is nearly continuous and along a straight-line path. As the particle slows down, it captures electron(s) to form a neutral atom and thus has an increased probability to have nuclear collisions that may induce displacements and vacancies in the target material lattice.

The overall behaviour of the OLED studied by SRIM, showed the reduction of ionization, phonons and displacements at higher proton energies justifying the fact that protons of higher energies can interact less with the structure. (Fig. 3, 4)

Moreover phonons produced by ions, exhibit an even distribution across the OLED's layers above 100keV showing no spatial dependence. On the other hand phonons produced by recoils exhibit a distribution similar to the distribution of displacements showing that the formation of displacements is related to the recoil phonons.

All the graphs show less irradiation affects in the emissive layers, which are of main concern in the OLED performance. For that reason we examined in more detail the behaviour of the NPB layer under the protons irradiation. Initially, the phonon dependence on ion energy was examined throughout the OLED after summing the total phonons produced due to ion and due to recoils throughout the device. They were found to follow an exponential decay $phonon = 10^{0.582}(ion\ energy)^{-0.900}$. (Fig. 6)

A similar decay was observed in the NPB emissive layer but with some interesting variations of the exponents. It was calculated that in this layer ion phonons decrease as the recoil phonons do, as the ion energy increases although the amount of recoil phonons dominates at each energy. More specifically $ion\ phonons = 10^{-0.354}(ion\ energy)^{-0.887}$ and $recoil\ phonons = 10^{0.0800}(ion\ energy)^{-0.910}$. (Fig. 7) Similar decay behaviour holds for the ionization distribution where, $ionization\ due\ to\ ions = 10^{2.53}(ion\ energy)^{-0.686}$ and $ionization\ due\ to\ recoils = 10^{-0.164}(ion\ energy)^{-0.935}$. (Fig. 8) These last relations demonstrate that in the emissive layer, ionization from ions dominates the ionization from recoils (~ 3 orders of magnitude) and also that the ionization due to recoils falls off much faster than the one from ions as the energy of the incident protons increases.

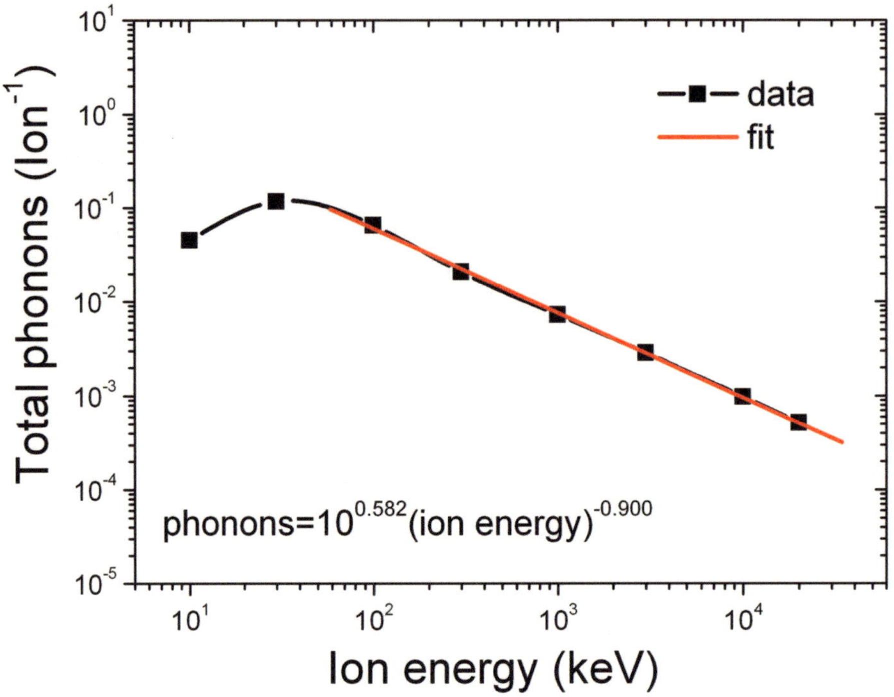

Fig. 6. Total phonons as a function of proton energy

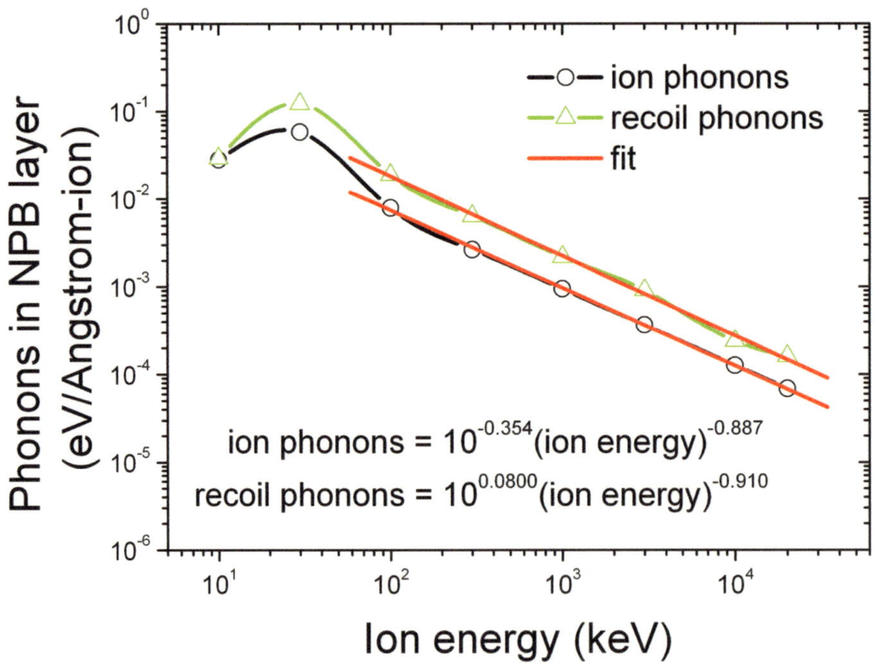

Fig. 7. Phonons generated by protons and recoils in NPB layer

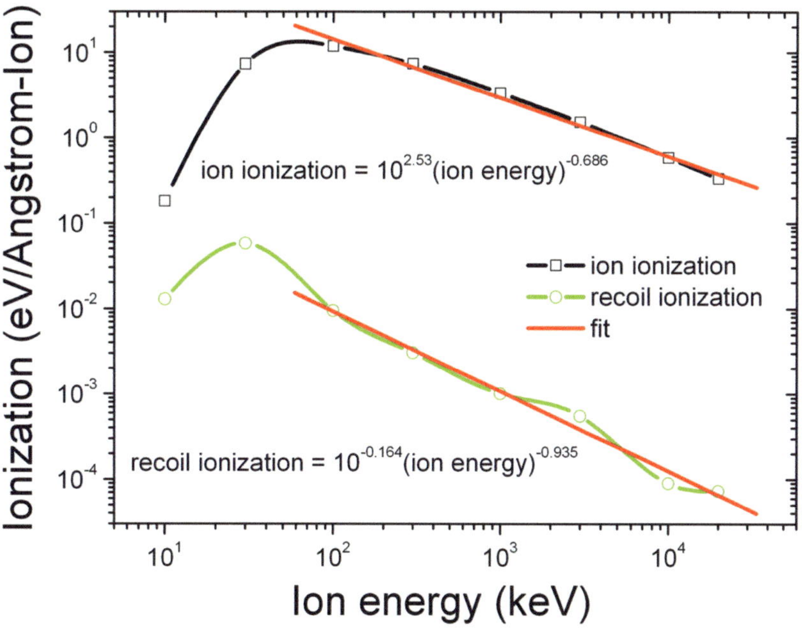

Fig. 8. Ionization generated by protons and recoils in NPB layer

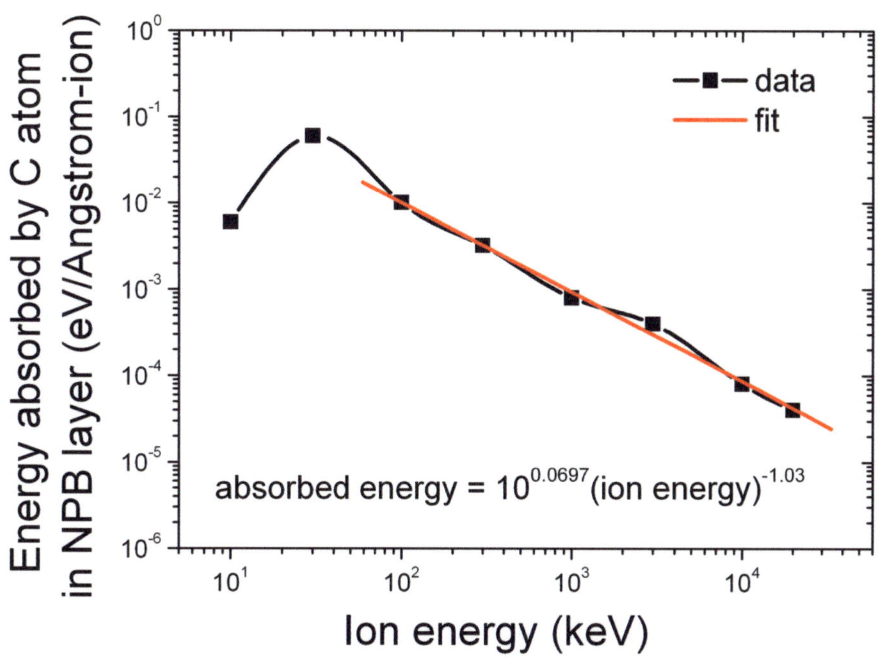

Fig. 9. Energy absorbed by C in NPB layer

By comparing the results for phonons and ionization, one can see that phonons due to ions decrease in a faster manner than the ionization due to ions as the ion energy increases. Considering that phonons are strictly connected to displacements and consequently to the formation of non radiative centres, the above statement indicates the possibility that in the MeV range the electrical behaviour is more affected due to excess of carriers, while the effects seem to be less for the optical behaviour which should be highly affected in the keV range.

Eventually, we studied the behaviour of carbon in the NPB layer since this element exhibited the higher energy absorption in the specific layer. We found a decay behaviour similar to those discussed above for the energy absorption (Fig. 9) and the vacancy formation (Fig. 10) in carbon that is: *energy absorbed*$=10^{0.0697}$*(ion energy)*$^{-1.03}$ and *vacancies*$=10^{-2.02}$*(ion energy)*$^{-0.985}$. Both of them present a dependence onto the ion energy very similar to that of the previous quantities with exception of the ionization due to ions.

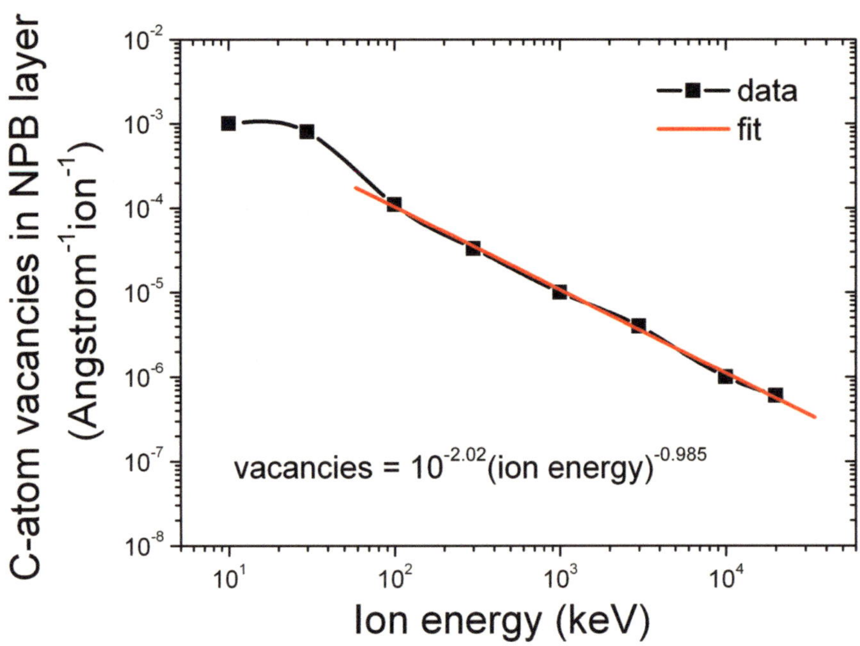

Fig. 10. Vacancies induced in C in NPB layer

5. Conclusions

Simulation of damages induced in optoelectronic devices carried out using SRIM and CASINO packages. The energies used for protons particles were from 10keV to 20MeV. The energies used at CASINO, for b- particles, were from 3keV to 30keV. We obtained the plots of depth versus y-axis, ionization, phonons, collision events, atom distributions and energy to recoils. Ionizing radiation is proved to affect the layers of the OLED structure simulated, as described above after the analysis of ionization, displacement and phonons induced by proton irradiation. As described before, in the theoretical part, changes in the material structure, strongly affect its optoelectronic properties and thus the device operation resulting in the whole circuit's malfunction.

We aim to investigate the appropriate elements that can be used in the substrate and in the emissive layers by a specific analysis layer by layer. We also aim to study the neutrons and alpha particles radiation effect.

There are additional permanent damage effects caused by heavy ions. These include microdose damage, and the problem of gate rupture in thin oxides which is less well understood. Recent work (Johnston A.H., 2000) on catastrophic damage in linear circuits shows that this remains a significant problem in space and more work needs to be done.

The optimization of the simulation results together with the modelling of novel devices which will be produced for us in Crete (FORTH) will assist the design of the prototype of a multi-touch screen based on FTIR technology for single display cockpits (IR LEDS) immune to any radiation effects.

It is critical that soft errors (or single event upset (SEU)) (neutrons density increases with height and neutrons with lower energy introduce SEUs) (Psarakis, 2008) which are induced by ionizing radiation must be eliminated from the optoelectronic devices as they are used in the aerospace industry (circuits with high fidelity requirements).

By comparing the effect of ionizing radiation on different material layers and thicknesses, we aim to minimize the operational faults due to structural changes. It should be noted that current research on packaging and device scaling has also improved the performance of ICs.

6. References

Andoniadis H. ,(2003) *Overview of OLED display technology*, OSRAM

Bhattacharya Pallab (1997) Semiconductor Optoelectronic Devices, Prentice Hall, Upper Saddle River, ISBN:0-13-495656-7, New Jersey

Drouin D. , Scanning, 29, Issue 3 , May 2007, 92 - 101

Freudenrich Craig (2004), *How OLEDs Work,*

Johnston A.H., (2000) *4th International Workshop on Radiation Effects on Semiconductor Devices for Space Application*, Tsukuba, Japan, (2000) Radiation Damage of Electronic and Optoelectronic Devices in Space.

Kitabayashi Hiroyuki , Kawabata Yoshisumi , Matsubara Hideki, Miyahara Kenichi and So Tanaka (2010). Development of High Power Infrared LED, *Sei Technical Review*, 70, (April, 2010), 71-74

Papakitsos A., (2002) *Lectures Notes on Applied Optoelectronics*, TEI of Pireaus, Greece.

Psarakis M., (2008) *Lecture Notes on Reliable embedded systems*, University of Piraeus, Greece.

Lin Shih-Yen, Tseng Chi-Che, Lin Wei-Hsun, Mai Shu-Cheng, Wu Shung-Yi, Chen Shu-Han, and Jen-Inn Chyi. (2010). Room-temperature operation type-II GaSb/GaAs quantum-dot infrared light-emitting diode. *Applied Physics Letters*, 96, (March, 2010), ISSN: 0-553-37783-3

Xalas A., Sgouros N., Kouros P., Ellinas J., .(2009). One Display for a Cockpit Interactive Solution, *era-4 Conference Proceedings*, ISSN:1791-1133, Spetses, September 2009, TEI Piraeus, Aigaleo

Yariv Amnon (1976) Introduction to Optical Electronics Holt, Rinehart and Winston, ISBN: 9780030898921, New York

Youtian Tao, Qiang Wang, Liang Ao, Cheng Zhong, Chuluo Yang, Jingui Qin, and Dongge Ma. Highly Efficient Phosphorescent Organic Light-Emitting Diodes Hosted by 1,2,4-Triazole-Cored Triphenylamine Derivatives: Relationship between Structure and Optoelectronic Properties, *J. Phys. Chem.* C, 114, 601–609, (2010)

Ziegler J., Biersack Jochen P., Ziegler Matthias D. (2010). SRIM - The Stopping and Range of Ions in Matter, Lulu Press Co., Morrisville, NC, 27560 USA

7

Identification of Emergent Research Issues: the Case of Optoelectronic Devices

Ivana Roche[1], Nathalie Vedovotto[1], Dominique Besagni[1], Claire François[1], Roger Mounet[1], Edgar Schiebel[2] and Marianne Hörlesberger[2]

[1]*Institut de l'Information Scientifique et Technique (INIST-CNRS) Allée du Parc de Brabois, CS 10310, 54519 Vandœuvre-lès-Nancy*

[2]*Austrian Institute of Technology (AIT) Tech Gate Vienna, Donau-City-Straße 1, 1220 Wien*

[1]*France*

[2]*Austria*

1. Introduction

The optoelectronic devices field is one of the last decade's most promising technological fields. Light emitting diodes gain more applications in cars and housing lighting, OLED displays are introduced in electronic devices and consumer electronics. Optimization of lighting power and tuning of light spectrum are well known important research topics.

Early recognition of new and alternative products and production processes is a strategic necessity contributing to appropriate assessment and decision-making. Emerging technologies are essential to advances in research, industry and society. But the detection of emerging technologies remains unsolved.

An interesting introduction to the use of bibliometrics for the identification of upcoming issues from online databases was published by Lancaster & Lee (1985). These authors analyzed the spread of an issue from pure sciences literature to popular press, through applied sciences literature. They measured the occurrence of a keyword over time in different online databases covering scientific literature. Additionally they suggested a procedure to identify growing issues in terms of time gradients of the number of published articles where single keywords occur.

The selection tree introduced by Armstrong & Green (2007) gives an additional picture of the landscape of the forecasting methods that could be employed to detect these emerging technologies. This tree illustrates the dichotomy between judgmental and quantitative forecasting methods and shows the great diversity of existing approaches like the Delphi or the Nominal Group Technique ones. Forecasts are obtained in a structured way from two or more experts. Other methods combine expert domain knowledge with statistical techniques and allow the identification of causal forces acting on trends.

Even though the number of available data is large enough to apply quantitative methods, the important question of the data type used for forecasting remains. The two data types mostly used to detect new topics with bibliometric analysis are provided by bibliographic databases covering scientific literature and patent databases. These methods consist of simple statistical techniques, such as growth curve analysis, or more sophisticated ones, such as clustering or network analysis. A third data type, related to state-funded research grants, is very interesting. The analysis of these three sources of information as a whole, and not as separate entities, allows gaining an understanding of the triple-helix interfaces between university, industry and government.

The framework for this study was the PROMTECH project (PROMTECH (2007)), financed by the European Commission. The main goal of the project was to elaborate a methodology enabling the identification of promising emerging technologies.

As the optoelectronic devices field turns out to be among the most promising ones, we focus on it by identifying its emerging research topics. We then study their evolution by considering data sets of bibliographic records related to two successive time periods and represented by their associated keywords. We apply this diachronic approach to the following analytical methodologies:

- a "Diffusion Model" using a bibliometric filter that distributes keywords in different diffusion stages in order to model the field terminology evolution (Schiebel & Hörlesberger 2007). The visualization of the results was operated by the software BibTechMon™, a bibliometric monitoring tool (Kopcsa & Schiebel 1998);
- a cluster analysis enabling the identification of emerging research topics by comparing the clusters related to each period (Roche et al. 2008). The clustering was operated by the software-tool Stanalyst (Polanco et al. 2001) that produces a graphical representation of the clustering results under the form of a network of clusters.

Firstly we refer to the data acquisition. Secondly the applied methodologies are illustrated. Thirdly the results of the applied methodologies to the field "optoelectronic devices" are presented. Finally we discuss the two approaches, summarize and conclude our discussion.

2. Data

The data framework is a sample recorded from PASCAL, a multidisciplinary bibliographic database produced by the French institute for the scientific and technological information (INIST-CNRS), that was specifically adapted to the purpose of our approach. PASCAL provides broad multidisciplinary coverage of scientific publications and contains, nowadays, about 20 million bibliographic records that are derived from the analysis of the scientific and technical international literature published predominantly in journals and conference proceedings. PASCAL is appropriate for the intended analysis, as each analyzed document is registered by a very fine classification, much more sophisticated than the category codes often employed in other sources of bibliographic data. The PASCAL classification categories, named also classification codes, belong to a structured classification scheme hugely detailed that is a taxonomy of every field and subfield of all the scientific disciplines covered in the database.

In addition, PASCAL covers international publications in physics, chemistry, engineering sciences, life sciences, and medicine and produces an indexing, on the one hand, by keywords and, on the other hand, by classification categories, both given in the database and assigned to the individual publications, either manually by scientific experts or automatically based on a content analysis. The query operated in this work was based on the information conveyed by these indexations. By extracting from the obtained corpus of bibliographic records the keyword indexing we produced a terminology related to the studied domain. After a verification step, done by a scientific expert, that terminology can be employed in our analysis.

The search focused on publications related, on the one hand, to Optoelectronics by means of classification codes or keyword indexing and, on the other hand, to peripheral scientific fields whose research issues, even if apparently far from the concerns of the researchers working on Optoelectronics, could have an impact on this domain. The main idea is to produce a finite space of exploration constituted by the information contained in the set of records answering the search strategy. The choice of these peripheral scientific fields is delicate. Indeed, if it is too large, the exploratory space could present a high level of noise with a great number of records that will be definitively uninteresting and that we would need to eliminate before analysis. But conversely, if the operated choice is too sharp, the exploratory space could not reflect in a suitable way the scientific environment richness of the studied domain and this silence would be very harmful for its analysis.

After some iterations, the list of peripheral fields converged towards: "Physics", "Chemical Industry", "Physico-chemistry of polymers" and "Polymer industry". The choice of these fields was motivated by their interest towards the synthesis and characterization of materials involved in optoelectronic devices. A search strategy has been detailed:

- firstly, we looked for the PASCAL classified simultaneously in the field "Optoelectronics" and in a field that is either "Physics" or "Chemical Industry" or "Physico-chemistry of polymers" or "Polymer industry":
- then we added the records indexed by the keywords *"LED"* or *"Electroluminescent device"* or *"Optoelectronic device"* and classified either in "Physics" or "Chemical Industry" or "Physico-chemistry of polymers" or "Polymer industry".

The result is a corpus of 8,169 bibliographic records all in all, divided into the periods from 2000 to 2004 and 2005 to 2010 delivers:

- 2,590 records for the first period (2000-2004)
- 5,219 records for the second period (2005-2009/2010[1])

The 8,169 bibliographic records (all years, 2000-2010) are assigned to more than 160 different PASCAL codes, whereas a reference can be assigned to more than one code.

The first 20 fields with the biggest number of records in the two considered periods (2000-2004 and 2005-2010) are showed, respectively, in Tables 2 and 3. Among the 16 fields matching in both periods, 12 confirm their position or rise in the second period ranking, like e.g. "Infrared, submillimeter wave, microwave and radiowave instruments, equipment and techniques", "Sensors (chemical, optical, electrical, movement, gas, etc)", "Optical elements, devices and systems", "Material sciences: Methods of deposition of films and coatings ; film growth and epitaxy" (see in table 2, the ranks 1 to 4, 6 to 9, and 12 to 15), and 4 fields lose places in the ranking of the second period (see in table 3 the fields written in italic): "Physico-chemistry of polymers: Organic polymers", "Polymer industry: Technology of polymers", "Electronics: Materials", "Electric, optical and optoelectronic circuits".

Four fields appear in the top 20 field list in the second period: "Nanoscale materials and structures: Fabrication and characterization", "Structure and nonelectronic properties of surfaces, interfaces and thin films", "Natural energy", "Materials science" and, conversely, four fields disappear from the top 20 field list: "Industrial chemicals", "Electrical engineering. Electrical power engineering", "General and physical chemistry", "Computer science: Software".

---

[1] 2010 can not be complete: the data, recorded on June 2010, contains around a half of the expected production of one entire year

ACOUSTICS.
ANALYTICAL CHEMISTRY.
ATOMIC AND MOLECULAR PHYSICS.
BUILDINGS. PUBLIC WORKS.
CHEMICAL ENGINEERING.
CHEMICAL INDUSTRY AND CHEMICALS.
COMPUTER SCIENCE. CONTROL THEORY. SYSTEMS.
CONDENSED MATTER: ELECTRONIC STRUCTURE, ELECTRICAL,
CONDENSED MATTER: STRUCTURE, MECHANICAL AND THERMAL
CROSS-DISCIPLINARY PHYSICS: MATERIALS SCIENCE, RHEOLOGY.
EARTH, OCEAN, SPACE.
ELECTRICAL ENGINEERING. ELECTRICAL POWER ENGINEERING.
ELECTROMAGNETISM, ELECTRON, AND ION OPTICS.
ELECTRONICS.
ENERGY.
FLUID DYNAMICS.
FUNDAMENTAL AND APPLIED BIOLOGICAL SCIENCES.
GENERAL AND PHYSICAL CHEMISTRY.
GROUND, AIR AND SEA TRANSPORTATION, MARINE
HEAT TRANSFER.
INFORMATION SCIENCE. DOCUMENTATION.
INFORMATION, SIGNAL AND COMMUNICATION THEORY.
INORGANIC CHEMISTRY AND ORIGINS OF LIFE.
INSTRUMENTS, APPARATUS, COMPONENTS AND TECHNIQUES
MATHEMATICS.
MECHANICAL ENGINEERING. MACHINE DESIGN.
MEDICAL SCIENCES.
METALS. METALLURGY.
NUCLEAR PHYSICS.
OPERATIONAL RESEARCH. MANAGEMENT SCIENCE.
OPTICS.
ORGANIC CHEMISTRY.
PHYSICOCHEMISTRY OF POLYMERS.
PHYSICS OF ELEMENTARY PARTICLES AND FIELDS.
PHYSICS OF GASES, PLASMAS AND ELECTRIC DISCHARGES.
POLLUTION.
POLYMER INDUSTRY, PAINTS, WOOD.
SOLID MECHANICS.
TELECOMMUNICATIONS.

Table 1. General scientific fields present in the corpus "Optoelectronic devices"

Rank	PASCAL classification scheme headings
1	Semiconductor electronics. Microelectronics. Optoelectronics. Solid state devices
2	Physico-chemistry of organic polymers
3	Polymer industry: Technology of polymers
4	Infrared, submillimeter wave, microwave and radiowave instruments, equipment and techniques
5	Electronics: Materials
6	Condensed matter: Optical properties and spectroscopy
7	Industrial chemicals
8	Electric, optical and optoelectronic circuits
9	Lasers
10	Electrical engineering. Electrical power engineering
11	Optical elements, devices and systems
12	Sensors (chemical, optical, electrical, movement, gas, etc)
13	Structure of solids and liquids; crystallography
14	Material sciences: Methods of deposition of films and coatings; film growth and epitaxy
15	General and physical chemistry
16	Condensed matter: Electronic structure and electrical properties of surfaces, interfaces and thin films
17	Computer science: Software
18	Optical instruments, equipments and techniques
19	Optical materials
20	Optical sources and standards

Table 2. The most important fields in the first period (2000-2004) expressed by means of the PASCAL classification scheme headings

At first sight, these evolutions suggest a decline in P2 of the number of records dealing with the characterization of materials, for the benefit of documents dedicated to the study of devices, as well as the emergence of new forms of materials as thin films and nanomaterials. More than 200 journals contributed to form each of the two studied corpus. In the first period, 9,666 authors working in institutions located in 69 different countries produce the 2,950 obtained records. In the second period, the number of authors is almost multiplied by 2 and the number of affiliation countries also rise to 81. In table 4, we show the top 10 affiliation countries for both periods. If the countries present in both lists are the same, their rankings are quite different. The USA remains in the first place with an almost unchanged rate, but we can see the remarkable ascent of China that doubles its affiliation rate coming from the sixth to the second place. With the exception of Canada, all other countries see their affiliation rate decreasing.

Rank	PASCAL classification scheme headings
1	Semiconductor electronics. Microelectronics. Optoelectronics. Solid state devices
2	Condensed matter: Optical properties and spectroscopy
3	Infrared, submillimeter wave, microwave and radiowave instruments, equipment and techniques
4	Sensors (chemical, optical, electrical, movement, gas, etc)
5	*Physico-chemistry of organic polymers*
6	Optical elements, devices and systems
7	Material sciences: Methods of deposition of films and coatings; film growth and epitaxy
8	Optical materials
9	Lasers
10	Nanoscale materials and structures: Fabrication and characterization
11	*Electric, optical and optoelectronic circuits*
12	Condensed matter: Electronic structure and electrical properties of surfaces, interfaces and thin films
13	Optical sources and standards
14	Structure and nonelectronic properties of surfaces, interfaces and thin films
15	Optical instruments, equipments and techniques
16	Structure of solids and liquids; crystallography
17	*Polymer industry: Technology of polymers*
18	Natural energy
19	Materials science
20	*Electronics: Materials*

Table 3. The most important fields in the second period (2005-2010) expressed by means of the PASCAL classification scheme headings

First period			Second period		
Rank	**Country**	**% of affiliations**	**Rank**	**Country**	**% of affiliations**
1	USA	28	1	USA	29
2	Japan	13	2	China	18
3	UK	10	3	Japan	10
4	Germany	10	4	Taïwan	9
5	South Korea	10	5	South Korea	8
6	China	9	6	Germany	7
7	France	7	7	UK	6
8	Italy	6	8	France	6
9	Taïwan	6	9	Canada	4
10	Canada	3	10	Italy	3
… 69			… 81		

Table 4. The top 10 countries in terms of affiliation rates in the first and second periods

This literature is of English expression in more than 99 % of documents and was published in 14 and 11 countries, respectively, in the first and second period. In table 5, we give the top 5 publishing countries for both periods. The USA strengthens its first place, the Netherlands rate grows up and the rates of Switzerland, United Kingdom and Germany go down.

First period			Second period		
Rank	**Country**	**% of publications**	**Rank**	**Country**	**% of publications**
1	USA	41	1	USA	58
2	Switzerland	16	2	Netherlands	15
3	UK	14	3	UK	12
4	Germany	12	4	Switzerland	9
5	Netherlands	11	5	Germany	4
... 14			... 11		

Table 5. The top 5 countries in terms of publishing rates in the first and second periods

3. Methodology

3.1 Diffusion model

New scientific and technological developments are presented and documented in scientific publications. These new ideas are described using specific words and expressions and they sometimes lead to the creation by scientists of new terminological representations. In a bibliographic record, the quintessence of a publication is represented by means of keywords, the investigation of which is meaningful for tracking the developments of new technologies or new ideas in science, especially if these keywords are given by the authors.

Our model is based on the consideration that technologies find their roots in an invention which was created by one person or a small group of scientists who use specific terms for describing it. After a specific invention is published, the respective knowledge is available to other scientists. If scientists are inspired by the new ideas, they make use of the new theories and technologies, publish them and add their own ideas. At that stage, the new terminology is used in few publications and appears therefore in few bibliographic records. New terms are created or applied coming from other disciplines with sometimes a new meaning. If the findings are of interest for other research groups, they pick them up and the number of publications, and of records, including these terms grows. In this second stage the findings are primarily interesting for researchers in the same discipline and the growth is essentially restricted to this area. If the findings are also considered as fundamental for the use in another discipline, the related terms diffuse over the scientific community and they occur in publications of many disciplines. By investigating the terminology of a suitable corpus of bibliographic records at a certain time span, we can divide the terminology into three stages, which expresses an invention or new findings, growth in a discipline and diffusion in the scientific community:

- Stage 1: New terms created or adapted from other fields in connection with a specific considered technology (or discipline, etc.) can be found in a few publications. These terms occur with a low frequency in this considered field where they form an exotic set of unusual terms.
- Stage 2: Already established specific terms in the considered technology (or discipline, etc.) occur in this stage 2. We refer to these terms as established terms.

- Stage 3: This is the stage where we consider the so called cross-section terms. Such terms are well established and applied in different technology fields. They are well understood not only in the considered field. They show a broad diffusion to other research or technology fields. One can detect a high number of articles in different technology fields using these terms.

We call this three-stage model the "Diffusion Model". Therefore, the diffusion model identifies, with the help of simple criteria, three categories of terms: unusual terms in a considered field, established terms in a considered field, and cross-section terms regarding to a considered field. The categorisation of a term is based on the following indicators and their related criteria:

	STAGE 1	STAGE 2	STAGE 3
Consider a specific term for the following indicators	**Unusual terms in a considered field**	**Established terms in a considered field**	**Cross-section terms regarding a considered field**
Total frequency	low	high	high
Number of fields where the term occurs (diffusion of a term)		in less than 1/3 of all fields	in more than 1/3 of all fields
Ratio of the frequency in the considered field to total frequency (which means that a term focuses in a specific technology – Field Focus)	more than 2/3	more than 2/3	less than 0.9
Age of the term	young	old	old

Table 6. Criteria to dispatch the terms of a considered field into the three stages

These criteria are linked to the corpus characteristics and, in the framework of a diachronic analysis as ours, their numerical value can differ within each period (see table 7 and 9).

The "total frequency" of a term represents the number of different bibliographic records in which it occurs at least once. Each bibliographic record is associated to at least one scientific field. The more specific a publication is, the fewer the scientific fields associated with its bibliographic record. The fewer scientific fields a term is associated with, the more the term is specific.

The "field focus" of a term is an indicator of its specificity and it lies between 0 and 1. The higher the field focus of a term, the more the term seems to focus on a specific field or the less it is spread to other fields.

The "age of the term" is defined in relation with the year when it occurs in a bibliographic record for the first time in the considered field for a given period. It is considered old if it appears during the first years of the period and young otherwise.

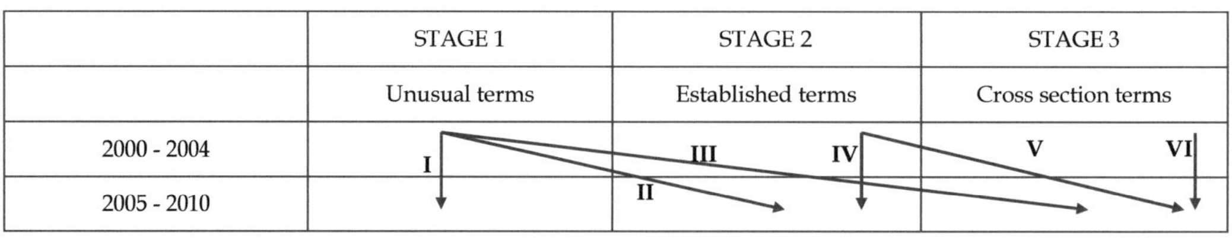

Fig. 1. Taxonomy of the migration pathways of terms through different stages of the diffusion model in two time periods

In order to dispatch the terminology between the three stages, each four indicators given in table 6 has to be determined for each term. The dynamics of emerging technologies can be described by pathways of terms through diffusion stages from one period to the following one. Figure 1 illustrates the pathways. The dynamics of emerging technologies can be revealed by identifying the terms that follow the mentioned pathways, see Besagni et al. (2009).

Following the diffusion approach it can be assumed that terms migrate from earlier stages in the first period to later stages in the second period like the following:

- terms from stage 1 in the first period stay in this stage or migrate in the second period to stage 2 or stage 3
- terms from stage 2 in the first period stay in stage 2 or migrate to stage 3
- terms from stage 3 stay in stage 3

We define pathways as follows:

- Pathway I: the same terms remain in stage one in both periods
- Pathway II: terms move from stage 1 in first period to stage 2 in second period
- Pathway III: terms move from stage 1 in first period to stage 3 in second period
- Pathway IV: the same terms remain in stage two in both periods
- Pathway V: terms move from stage 2 in first period to stage 3 in second period
- Pathway VI: the same terms remain in stage three in both periods

For the visualization and analysis we used the software BibTechMon™, a bibliometric monitoring tool (Kopcsa & Schiebel 1998). The core of the software is an algorithm to position related objects (in our case, terms) in a two dimensional space. The relation between terms is defined by their pairwise co-occurrence in bibliographic records. These co-occurrences are normalized by the often used Jaccard index.

The terms are arranged in a network structure by a spring model coming from mass point mechanics. Thus, a term is positioned with respect to all other ones. If a term in the centre of the network or of a subnetwork has a rather strong relationship to many other nodes in the network graph, that is called a high centrality.

In our case, established terms or cross-section terms are centred. New terms and unusual terms are positioned in the periphery. A bunch of terms that are cited often together crossing over many publications forms up a subnetwork.

3.2 Cluster analysis

The diachronic cluster analysis is based on the study of the evolution of cluster maps and contents obtained by a clustering method and considering data sets related to two or more successive time periods. In this work, we deal with two time periods and we apply a clustering algorithm on the corpus of each period, in which documents are represented by the keywords existing in the bibliographic records. The clustering tool we use (Lelu & François 1992; Lelu 1993) applies a non-hierarchical, non-supervised clustering algorithm, the axial K-means method that allows producing clusters presenting the following particular characteristics:

- they can overlap because the clustering method allows a document or a keyword to belong to more than one cluster,
- documents and keywords as the constituting elements of a cluster, are ranked by decreasing similarity with the cluster ideal type.

The clustering is operated by the software-tool Stanalyst (Polanco et al. 2001) that produces a graphical representation of the clustering results under the form of a network of clusters.

Each cluster is a homogeneous subset of the information contained in the analyzed corpus, corresponding to a topic. After the clustering, we represent graphically its results employing two methods: a principal component analysis and a connected component analysis. The first one produces a global map of the topics. The topics that are very far apart represent clusters very different with respect to the keywords defining them.

The connected component analysis is then employed to show the relationships that exist between the clusters. It is a method based on the graph theory and defining the connected components representing the relative proximity between clusters. The distance between two clusters is defined as the cosine of the angle between the two axes representing the clusters. A lower and an upper threshold are defined as respectively the minimum and the maximum value of the cosine between the clusters. Nine intermediate levels are defined in that interval. We consider that, at a given level, two clusters are connected if the value of the cosine associated to them is within the range of values of this level.

Finally, in order to analyse the evolution of the cluster vocabulary between the two considered periods, we build a comparison matrix (Roche et al. 2008) pointing out the percent of keywords belonging to the second-period clusters and already existing in the first-period clusters. The cumulated percentage is also calculated for each second-period cluster. Using this matrix, we can identify different cluster behaviours: stability, fusion or splitting. Using the cluster maps, we can also detect status change of the clusters in the global network. These detected phenomena need to be validated by experts from the technological domain.

The cluster behaviours are calculated in the following way. Let M be the comparison matrix:

$M = (m_{i,j})$ with:

$i = 1,\ldots,n_{P2}$ where n_{P2} = number of clusters obtained in the second period

$j = 1,\ldots,n_{P1}$ where n_{P1} = number of clusters obtained in the first period

m_{ij} = percent of keywords belonging to cluster i and existing in cluster j

and, for each line i, the marginal value is:

$$. \; m_{i.} = \sum_{j=1}^{j=n_{P1}} m_{ij} \; .$$

The analysis of the matrix M allows us to build the following hypotheses:

- Let (i,k) be a pair of homonymous clusters. The cluster i is supposed to be stable if:

$$m_{ij} < \frac{m_{ik}}{2}; \forall j \neq k \text{ and } j = 1, n_{P1} \tag{1}$$

- Let (i,k) be a pair of homonymous clusters. The cluster i is supposed to be unstable if there exists at least one j respecting the condition:

$$m_{ij} > \frac{m_{ik}}{2}; \forall j \neq k \text{ and } j = 1, n_{P1} \tag{2}$$

- Clusters having new titles (titles not existing in the first-period cluster list) and obeying the condition (1) are supposed to be stable;

- Clusters with the lowest marginal values have less inheritance from first-period clusters and we suppose they constitute new terminological notions;
- Clusters with the highest marginal values are supposed to aggregate a great number of notions already existing in the first-period clusters.

At this point, scientific expertise is needed. We ask scientific experts to validate our hypotheses regarding the analysis of the two cluster maps, taking into account the cluster contents.

4. Results

4.1 Diffusion model

The diffusion model has been applied to the 8,169 bibliographic records delivered by the PASCAL database query, split into two corpora according to their publication year: 2,950 references for P1 (2000 to 2004) and 5,219 references for P2 (2005 to 2010). In these bibliographic records:

- the technological fields are the PASCAL classification codes, reduced to the classification scheme headings,
- the terms are the PASCAL keywords,
- a term is considered old if it appears during the first 3 years of a period.

For P1, the terms have been dispatched into the three stages according to the following criteria:

	STAGE 1	STAGE 2	STAGE 3
Consider a specific term for the following indicators	**Unusual terms in a considered field**	**Established terms in a considered field**	**Cross-section terms regarding a considered field**
Total frequency	between 8 and 20	> 40	> 30
Number of fields where the term occurs (diffusion of a term)		< 17	≥ 17
Ratio of the frequency in the considered field to total frequency - Field Focus	> 0.7	> 0.7	< 0.9
Age of the term	young	old	old
Year of first occurrence	2003-2004	2000-2002	2000-2002

Table 7. Filter for the corpus of the first period (2000 – 2004) for 2,950 records

This filter does not allow for an overlap of common terms. Each term is assigned to only one of these three stages.

The criteria are determined as follows:

- Stage 1: The total frequency is defined higher than 7 in order to reject badly spelt terms. To exclude established terms, the total frequency upper bound is fixed to 20. As we obtain 2,590 records for the first period, a frequency of 20 means that such a term occurs in less than 0.8% of the documents of P1. Furthermore a term must be relevant and not used by chance. The value of field focus should then be high, at least 0.7, so that a focus on a field is ensured.
- Stage 2: Established terms must have a higher total frequency than in stage 1. Therefore we fixed the lower bound to 40 in P1 (which means a term frequency of 1.5%). An

established term should focus on a specific field. Therefore it has to occur in less than 17 fields, which corresponds to a third of the total number of fields of P1. The spread to other fields (the field focus) should be high, here 0.7.
- Stage 3: A cross-section term has a high total frequency : it is used in more than 40 records in P1. Furthermore, it must be used in a high number of fields, in P1 more than 17. The field focus should ensure a distribution to many fields and must not exceed 0.9.

This filtering results in the following table, presenting the top 10 terms of each stage:

Unusual terms	Established terms	Cross-section terms
CMOS image sensors	Binary compound	Photoluminescence
Thermal imaging	Semiconductor materials	Ternary compound
Sensor materials	Quantum well	Thin film
Imager	Gallium nitride	Conjugated copolymer
Chip	III-V semiconductors	Quantum yield
Test	Optical absorption	Phenylenevinylene derivative polymer
Spatial resolution	Molecular beam epitaxy	Measurement sensor
Digital image	Gallium arsenides	Phenylene derivative copolymer
Organometallic compound	Gallium Nitrides	Crystal growth from vapors
Diode array	Injection laser	Silicon

Table 8. Top 10 terms for the first period (2000 - 2004)

The P2 terms have been dispatched into three stages according to the following criteria:

	STAGE 1	STAGE 2	STAGE 3
Consider a specific term for the following indicators	**Unusual terms in a considered field**	**Established terms in a considered field**	**Cross-section terms regarding a considered field**
Total frequency	between 8 and 40	> 80	> 50
Number of fields where the term occurs (diffusion of a term)		< 16	≥ 16
Ratio of the frequency in the considered field to total frequency - Field Focus	> 0.7	> 0.7	< 0.9
Age of the term	young	old	old
Year of first occurrence	2008- 2010	2005-2007	2005-2007

Table 9. Filter for the corpus of the second period (2005 - 2010) for 5,219 records

The criteria are similar to the first period (see below Table 7) but adapted to the number of records of 5,219 records of this second period. Therefore the numbers of the total frequency differs from them in the first period.

This filtering results in table 10, presenting the top 10 terms of each stage.

Unusual terms	Established terms	Cross-section terms
Aluminium Nitrides	Voltage current curve	Electroluminescence
Electron transport	Binary compound	Photoluminescence
Oxine	Electrochemical properties	Thin film
Amplifier	Optical material	Quantum yield
Biomedical imaging	Brightness	III-V semiconductors
Computer software	Focal plane arrays	Nanostructure
Electronic conductivity	Luminescent material	Measurement sensor
Gallium Phosphides	Quantum well	Optical properties
Daylight	Conjugated polymer	Silicon
Holes	Thermal stability	Conjugated copolymer

Table 10. Top 10 terms for the second period (2005 - 2010)

4.1.1 Pathways

In that study dealing with the identification of emergent research issues, it appears that the Pathways II, III and V carry the bulk of the relevant information highlighting the potential scientific evolutions, or even breakthroughs. That is why we focus on these three pathways.

4.1.1.1 Pathway II

Terms which follow the Pathway II progress from "unusual term" in the first period to "established term" in the second period.

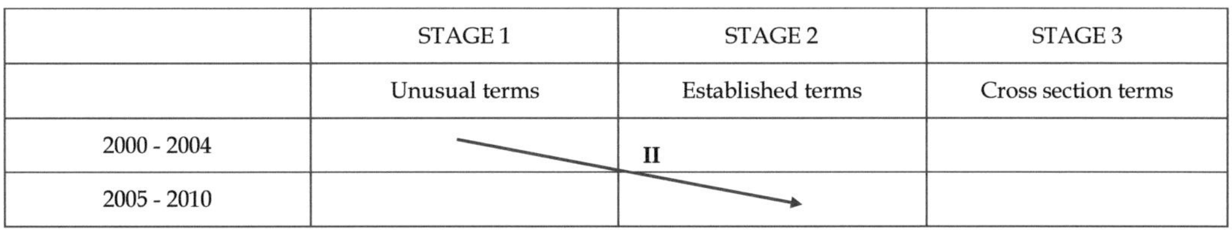

Fig. 2. Migration Pathway II

The terms which follow Pathway II are:

- *Sensor materials*
- *Thermal imaging*

These two keywords play a role in several articles as:

- Advanced pixel design for infrared 3D LADAR imaging
- Advanced infrared detectors for multimode active and passive imaging applications
- Dual-Band MW/LW IRFPAs made from HgCdTe grown by MOVPE
- Comparison of QWIP to HgCdTe detectors for gas imaging

4.1.1.2 Pathway V

Terms which follow the Pathway V progress from the "established stage" in P1 to the "cross-section stage" in P2.

These terms are:

- *Doping*
- *Gallium arsenides*
- *III-V semiconductors*
- *Injection laser*
- *Luminescence*
- *Quantum dot*

For example, the keywords "*Injection laser*", "*Luminescence*", or" *Quantum dot*" are used to index publications like:

- Carrier dynamics in active-region materials for diode laser applications.

The keywords "*Gallium arsenides*", "*Injection laser*" or "*Quantum dot*" are used in:

- Superluminescent Diodes versus Broadband Laser Diodes.

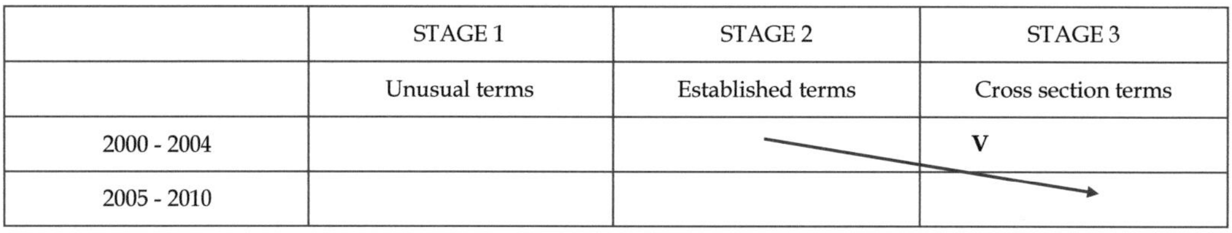

Fig. 3. Migration Pathway V

4.1.1.3 Pathway III

Terms which follow the Pathway III progress from the "unusual stage" in P1 to the "cross-section stage" in P2.

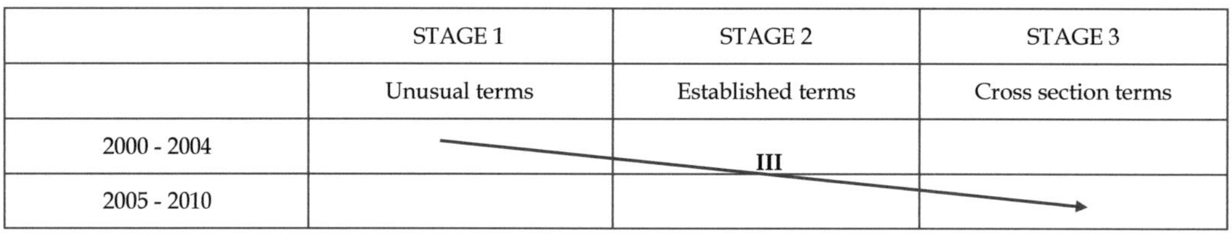

Fig. 4. Migration Pathway III

Pathway III results in the following terms:

- *Imager*
- *Nanoparticle*

The keyword "*Imager*" is used in publications as:

- Gated Diode Design to Mitigate Radiation Damage in X-Ray Imagers
- A Day and Night Vision MOS Imager With Robust Photonic-Crystal-Based RGB-and-IR.

The keyword "*Nanoparticle*" indexes publications as:

- Identification of unamplified genomic DNA sequences using gold nanoparticle probes and a novel thin film photodetector
- Direct gravure printing of indium tin oxide nanoparticle patterns on polymer foils
- Photoluminescence from ZnO Nanoparticles Embedded in an Amorphous Matrix
- Use of surface plasmon-coupled emission for enhancing light transmission through Top-Emitting Organic Light Emitting Diodes

4.1.2 BibTechMon™ visualization

For the calculation and the visualization of the term networks we used the software BibTechMon™. The algorithm in BibTechMon™ is based on a co-object analysis, which can be briefly described as follows. An object (in our case, a term) occurs with a certain frequency, defined by the number of bibliographic records in which the considered term occurs at least once. It is counted once per record. When a term occurs with another term in the same bibliographic record, we call that a co-occurrence. The co-occurrence strength is given by the number of co-occurrences pairwise. These co-occurrences are normalized via the so called Jaccard index. After that, the terms find their position inside the network by the application of a spring model, where the Jaccard indices are the measures for the interactions. Thus, a term is positioned with respect to all other terms and to their Jaccard indices.

The network consists of points which represent the terms and lines which show the intensity of the individual relation between two terms. The size of the points represents the total frequency of the term occurring in the corpus. The relative position of the objects compared to each other is important: objects which are positioned close to each other form subtopics and clusters because they occur in similar contexts.

Figure 5 shows the location of the terms of Pathway V in the network of the second period. In this figure, cross-section terms can be found in the subnetwork at the centre. They have connections to many other terms.

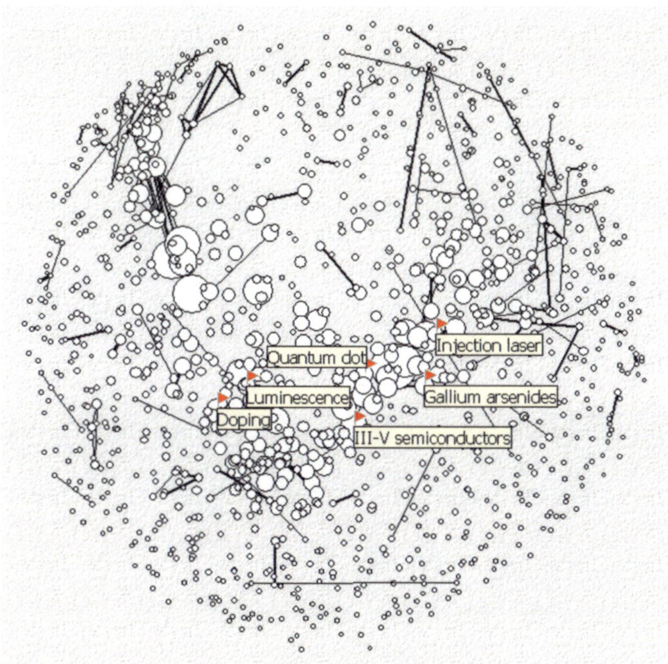

Fig. 5. Pathway V terms in the network of the second period

The study of Pathway III is even more interesting. Figures 6 and 7 show respectively the terms of Pathway III (*"Nanoparticle"*, *"Imager"*) in the network of the first period and the second period.

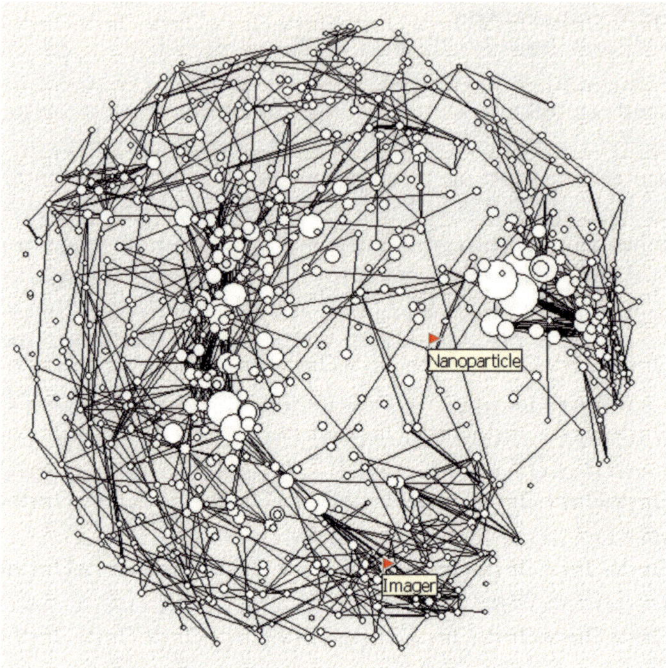

Fig. 6. Terms of Pathway III in the first period

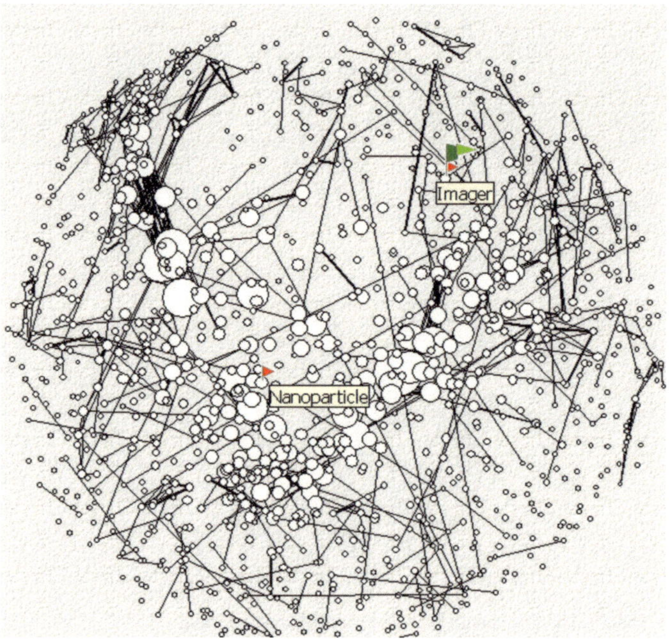

Fig. 7. Terms of Pathway III in the second period

In the first period, the terms "*Nanoparticle*" and "*Imager*" are unusual. Indeed, their position is marginal with respect to the strong subnetworks of the P1 graph (figure 6). Conversely, these terms play a more important role in the second period: they are located in the core of important and dense networks of the P2 graph (figure 7). Furthermore, as figure 8 shows, these two terms are connected to many other terms in the network of the second period.

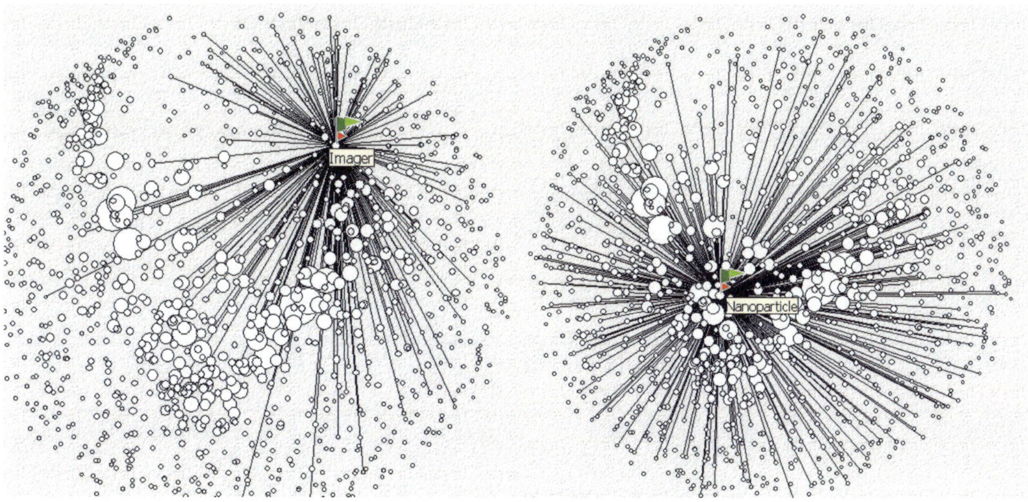

Fig. 8. Networks of "*Imager*" and "*Nanoparticle*" in the second period

The visualization obtained with BibTechMon™ thus supports the results of the diffusion model according to Pathway III and Pathway V that is the strong networks built by imaging techniques and nanotechnology in the optoelectronic technological field for the second period.

4.2 Cluster analysis

The 8,169 records delivered by the PASCAL database query have been split into two corpora according to their publication year: 2000 to 2004 (P1) and 2005 to 2010 (P2). In the first period, we obtained 2,950 documents and 6,735 different keywords among which 5,257 (78%) appear less than 4 times. The second period had 5,219 articles and 6,378 keywords among which 3,976 (62%) occur less than 4 times.

The clustering step was preceded by an expert revision of the keyword index of each period corpus in order to detect the presence of indexing artefacts, to homogenize the variational forms of common concepts and to eliminate keywords that are not discriminating in the context of our analysis. For instance, all the keywords employed in the query have been discarded. On the other hand, the expert decided to take into account only the keywords occurring at least 4 times in the two corpora. After this terminology validation step, we obtained for P1 and P2 two indexes formed, respectively, by 1,472 and 2,391 keywords that were the input data of the clustering step.

After that, we study the evolution of the clusters between P1 and P2 with the help of a comparison matrix before its validation by the experts.

4.2.1 Clustering step

We applied a clustering algorithm on each one of our two corpora in which the documents are represented by the keywords existing in the bibliographic records. In the table 11, we present, for the two periods, the name of the obtained clusters and the number of theirs keywords and documents. The homonymous clusters are highlighted.

	Nb keywords	Nb records	Second period	Nb keywords	Nb records	First period
Cluster 1	27	363	Organic light-emitting devices	90	124	Optical telecommunication
Cluster 2	21	725	Infrared detector	22	408	Infrared detector
Cluster 3	41	881	Electroluminescence	64	45	Image processing
Cluster 4	46	192	Single photon	142	162	Chemical synthesis
Cluster 5	80	125	Optical method	61	150	Semiconductor quantum wells
Cluster 6	67	136	Optical sensor	34	338	Semiconductor materials
Cluster 7	57	264	Semiconductor laser	14	454	Conjugated polymer
Cluster 8	73	483	Thin film	91	59	Printed circuit board
Cluster 9	144	134	Nonimaging optics	25	242	III-V semiconductors
Cluster 10	70	209	Lighting fitting	21	637	Electroluminescence
Cluster 11	42	448	III-V semiconductors	57	201	Thin film
Cluster 12	38	542	Conjugated copolymer	47	68	CCD image sensors
Cluster 13	38	379	Measurement sensor	31	458	Conjugated copolymer
Cluster 14	81	174	Integrated optics	96	66	Computer simulation
Cluster 15	94	172	Photonic crystal	94	89	Photons
Cluster 16	70	138	Radiation sources	20	583	Photoluminescence
Cluster 17	53	406	Nanostructure	45	156	Semiconductor laser
Cluster 18	55	249	Europium addition	58	148	Measurement sensor
Cluster 19	97	206	Photodiode	60	70	LED lamps
Cluster 20	22	1047	Photodetector	69	110	Optical sensor

Table 11. Clusters of the two periods: P1 (2000-2004) and P2 (2005-2010)

The clustering algorithm produced 20 clusters involving 2,938 documents and 823 keywords in P1, and 20 clusters relating 5,214 documents and 937 keywords in P2. Let us notice that the important reduction, in both periods, of the number of keywords associated to the clusters, around 85% with respect to the initial number of terms obtained directly from the query results, does not entail a significant loss of information at the level of the documents. Indeed, despite these huge reductions of terminology, we always had almost 100% of the initial information corpora in the clusters. The figures 9 and 10 show the cluster maps obtained for, respectively, P1 and P2.

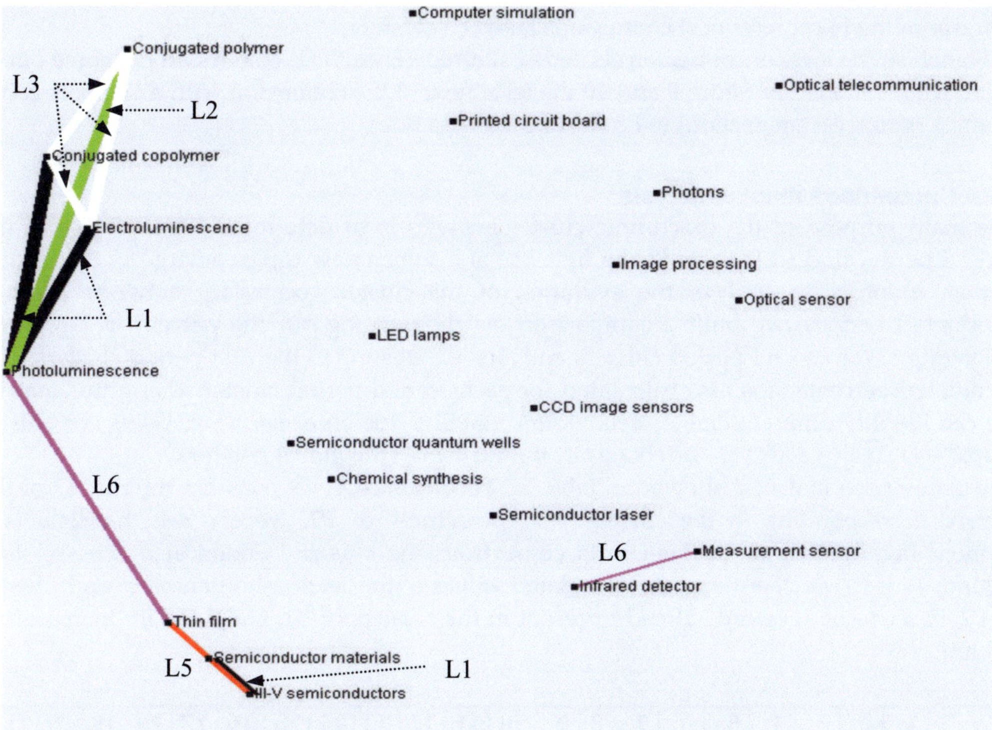

Fig. 9. Cluster map of P1 (2000-2004), with indication of the strength of the connections

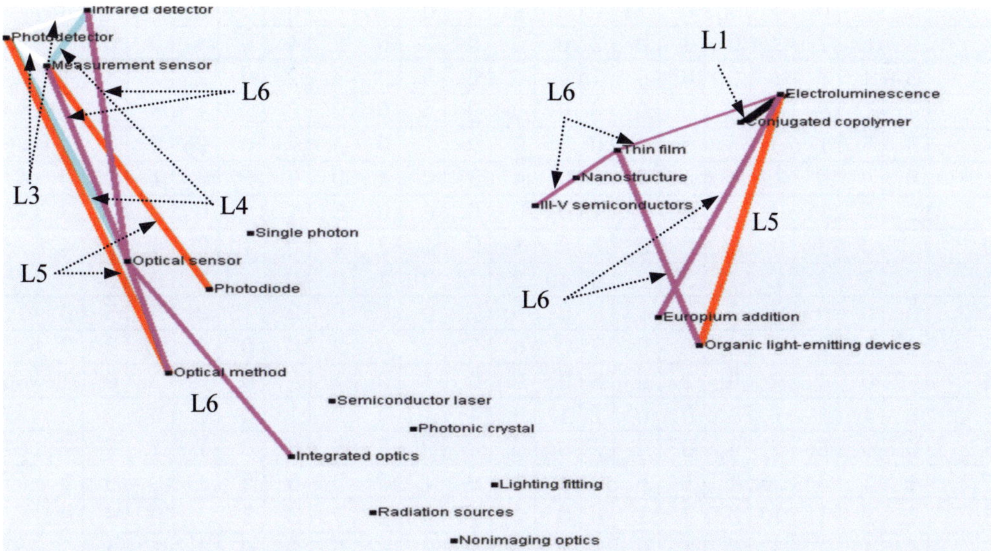

Fig. 10. Cluster map of P2 (2005-2010), with indication of the strength of the connections

In both maps produced by Stanalyst, each dot corresponds to a cluster and each line gives the connexion level between pairs of clusters. Note that we considered here only the 6

highest levels (out of the 10 defined by the connected component analysis step) corresponding to the strongest connexions between clusters.
In Stanalyst, the level of connexions is code-coloured. Since these colours do not come out in print, we mentioned in figure 9 and 10 the level next to the connexion with a symbols going from L1 (strongest connection) to L8 (weakest connection).

4.2.2 Comparison matrix analysis

The main purpose of the diachronic cluster analysis is to determine which topics of the second period find their roots in the first one and which new topics emerge in the second period. In order to analyze the evolution of the cluster vocabulary between the two considered periods, we built a comparison matrix pointing out the percent of keywords belonging to the second-period clusters and already existing in the first-period clusters. The cumulated percentage is also calculated for each second-period cluster. Using this matrix, we can identify different cluster behaviours: stability, fusion or splitting. Using the cluster maps, we can also detect status change of the clusters in the global network.
The comparison matrix is showed in table 12. For instance, if we consider the row 17 of the matrix, corresponding to the cluster "Nanostructure" of P2, we can see that 26% (see element highlighted) of its keywords come from the cluster "Chemical synthesis" (see column 4) of P1. Furthermore, the cumulated values of the set of inheritances of each cluster of P2, in terms of keywords already present in the clusters of P1, are given in the marginal column.

P2/P1	**1**	**2**	**3**	**4**	**5**	**6**	**7**	**8**	**9**	**10**	**11**	**12**	**13**	**14**	**15**	**16**	**17**	**18**	**19**	**20**	□
1	3	0	3	7	0	0	0	0	0	7	22	0	0	0	3	7	7	0	7	0	*66*
2	0	42	4	0	0	4	0	0	0	0	0	4	0	0	0	0	0	9	0	0	*63*
3	0	0	0	4	2	0	24	0	0	21	4	0	26	0	0	34	0	0	7	0	*122*
4	2	6	2	2	0	4	0	2	6	0	0	2	0	2	4	0	4	0	0	0	*36*
5	3	1	8	2	0	0	0	1	0	0	0	3	0	3	2	0	2	7	5	12	*49*
6	4	1	8	5	0	0	0	0	0	0	0	5	0	2	2	0	1	8	1	11	*48*
7	8	5	1	0	3	5	0	0	3	0	0	1	0	3	7	0	26	0	1	3	*66*
8	0	0	0	12	4	5	4	0	5	4	31	0	4	0	5	5	0	1	1	0	*81*
9	2	0	11	0	0	0	0	3	0	0	0	6	0	5	1	0	0	4	7	3	*42*
10	1	0	5	0	0	0	0	2	0	1	0	1	2	4	0	1	2	4	51	4	*78*
11	0	7	2	7	11	33	4	0	40	2	14	0	4	0	0	4	7	0	0	0	*135*
12	0	0	0	0	2	0	13	0	0	5	2	0	73	0	0	15	0	0	5	0	*115*
13	2	10	2	0	0	0	0	2	0	0	5	5	0	2	0	0	0	18	0	7	*53*
14	12	0	4	0	0	1	0	6	0	0	0	1	0	7	9	0	11	3	0	12	*66*
15	8	1	4	1	1	0	0	2	1	0	4	1	0	7	15	0	7	0	2	3	*57*
16	5	0	8	5	4	0	0	7	0	0	0	0	0	5	8	0	7	1	7	7	*64*
17	1	1	3	26	1	1	5	0	1	1	9	1	5	0	5	5	1	0	0	1	*67*
18	1	0	1	10	0	7	1	0	0	1	12	3	1	1	3	3	1	0	9	0	*54*
19	4	1	4	3	0	1	0	3	1	0	1	9	0	5	3	0	1	12	4	5	*57*
20	4	13	0	0	4	9	0	0	4	0	4	13	0	9	13	0	9	13	0	4	*99*

Table 12. Comparison matrix of the keywords of the two periods: P1 (2000-2004) and P2 (2005-2010)

The analysis of the comparison matrix allows us to singularize in the second period (see table 13):

- eight homonymous clusters, but only four of them seem to be stable
- three clusters with new titles but presenting characteristics of stability
- four clusters with a low (less than 50%) inheritance global rate from the clusters of P1 - *Are they new clusters?*
- three clusters with a strong (more than 100%) inheritance global rate from different clusters of P1 - *Do they convey "old", already known information?*

Type of singularization	Cluster name
Homonymy with P1, apparent stability (larger keywords inheritance coming from the homonymous cluster of P1)	Infrared detector, Semiconductor laser, Thin film, Conjugated copolymer
Homonymy with P1, less apparent stability (larger keywords inheritance coming from other clusters of P1 than the homonymous one)	Electroluminescence, Optical sensor, III-V semiconductors, Measurement sensor
New title with stability characteristics	Organic light-emitting devices, Lighting fitting, Nanostructure
Low inheritance global rate	Single photon, Nonimaging optics, Optical sensor, Optical method
High inheritance global rate	III-V semiconductors, Conjugated copolymer, Electroluminescence

Table 13. Clusters singularized in second period

4.2.3 Experts validation

Starting from these quantitative data the expert worked on the validation of our hypotheses by analysing the contents of the clusters and their relative position in the cluster network drew in the map of each period. Special attention has been paid on the detection of brutal changes of cluster relative position.

The expert's analysis confirms the stability of two homonymous clusters between P1 and P2: "Infrared detector" and "Semiconductor laser". The homonymous clusters "Thin film" and "Conjugated copolymer" are rather stable between P1 and P2, but the former shows a higher relative weight of organic towards inorganic materials in P2, and the latter a higher relative weight of devices towards characterization of the materials.

Despite their less apparent stability, due to a lower inheritance from P1 to P2, the homonymous clusters "Measurement sensor" and "III-V semiconductors" are stable between P1 and P2. The other clusters of P1 from which they inherit deal with similar topics.

The P2 cluster "Optical sensor" does not inherit predominantly from its P1 homonymous. Meanwhile, its stronger keywords inheritance comes from topically close clusters as "Image processing" and "Measurement sensors". This cluster is rather stable between P1 and P2, but it highlights the emergence in P2 of some applications related to biomedical imaging and security. The take off of these new applications of sensors can explain the relatively low inheritance level of the P2 clusters "Optical sensor" and "Optical method".

As the former one, the P2 cluster "Electroluminescence" does not inherit predominantly from its P1 homonymous. Indeed, three P1 clusters hand more keywords down to it than its homonymous: "Photoluminescence", "Conjugated polymers" and "Conjugated copolymers". This fact shows the strong inheritance relation between the studies of P1 related to synthesis of conjugated (co)polymers and characterization of their luminescent properties, and the studies of P2 related to the potential applications of these materials for the design of electroluminescent devices.

The strong inheritance of the P2 cluster "Lighting fitting" from the P1 cluster "LED lamps" shows the importance of the topic dealing with the use of electroluminescent devices as light sources through the two periods of time. In the second period, the weight of this topic is reinforced by the presence of the cluster "Nonimaging optics" and by a growing number of bibliographic records.

The P2 cluster "Organic light-emitting devices" inherits most of its keywords from the P1 cluster "Thin films". This relation shows the temporal evolution of the field, from the study of the potential use of thin films and heterostructures in optoelectronic devices in P1, to the fabrication and characterization of these devices in P2. The stronger implication of organic materials in P2, highlighted by the name of the P2 cluster, confirms the tendency formerly observed when analyzing the stability of the homonymous class "Thin films" between P1 and P2.

The P2 cluster "Nanostructure" inherits most of its keywords from the P1 cluster "Chemical synthesis", which deals essentially with the synthesis of nanomaterials (nanowires, nanocomposites, nanoboxes, nanospheres…). This inheritance, together with the strong relative growth of the number of documents between the two clusters, shows the growing importance of nanomaterials in the field of optoelectronic devices in P2.

Amongst the clusters of P2 showing a low marginal value, "Single photon" gathers bibliographic records dealings with the application of avalanche photodiodes to fields like quantum information, quantum communication or quantum cryptography. This low marginal value is thus closely related to the emergence of a new scientific field.

The analysis of the cluster maps obtained for the two periods completes and confirms these observations.

In the cluster map of the first period (see figure 11), the four cluster "Conjugated polymers", "Conjugated copolymers", "Electroluminescence" and "Photoluminescence" form a very strong network (N1), This result highlights the importance of organic semiconductors in the field of optoelectronic devices and confirms the conclusion of a former study, in which we analyzed the evolution of the same scientific field over an older time period (Schiebel et al. 2009). This network is still present in the second period, but it has included the cluster "Organic light-emitting device", illustrating the transition from materials to devices. Furthermore, the proximity of the cluster "Nanostructure" shows the growing implication of such materials in optoelectronic devices in P2.

The topic related to imaging devices is already present in the first period (see figure 11), but the corresponding clusters ("Measurement sensor", "Infrared detector", "Optical sensor", "Image processing") do not show strong relations. On the contrary, they form a much strong network (N2) in the second period (see figure 12). Though this topic is not new, the comparison of the two cluster maps give evidence of its growing importance in the second period. Close to this network, the appearance in the second period of the cluster "Single photon" illustrates the emergence of quantum information processing in the field.

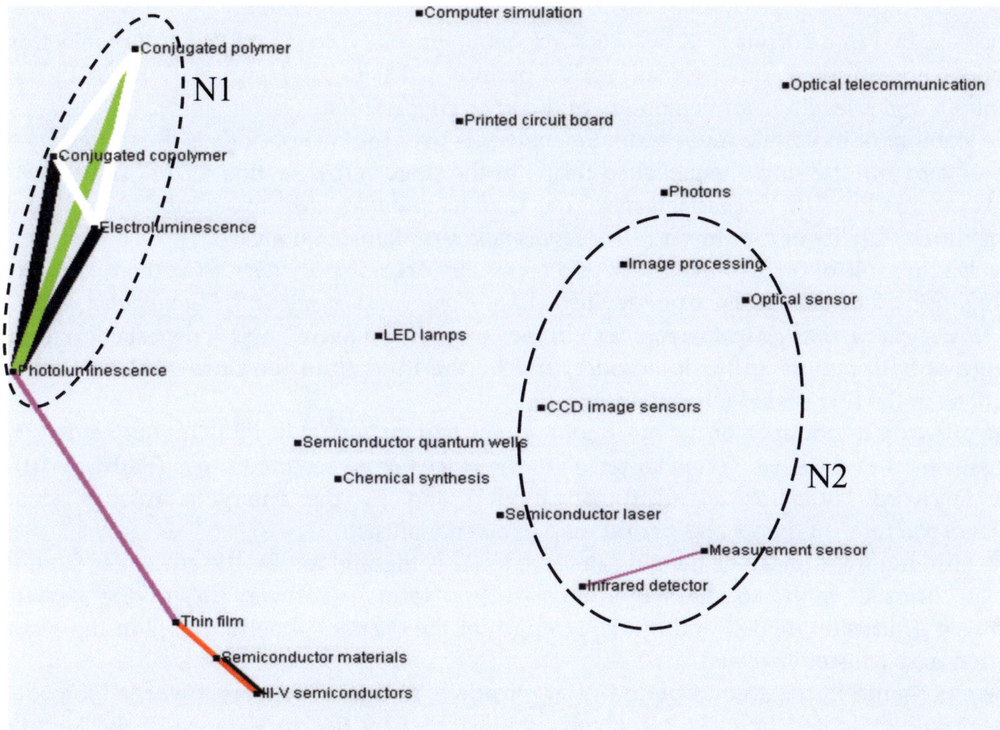

Fig. 11. Networks of clusters for P1

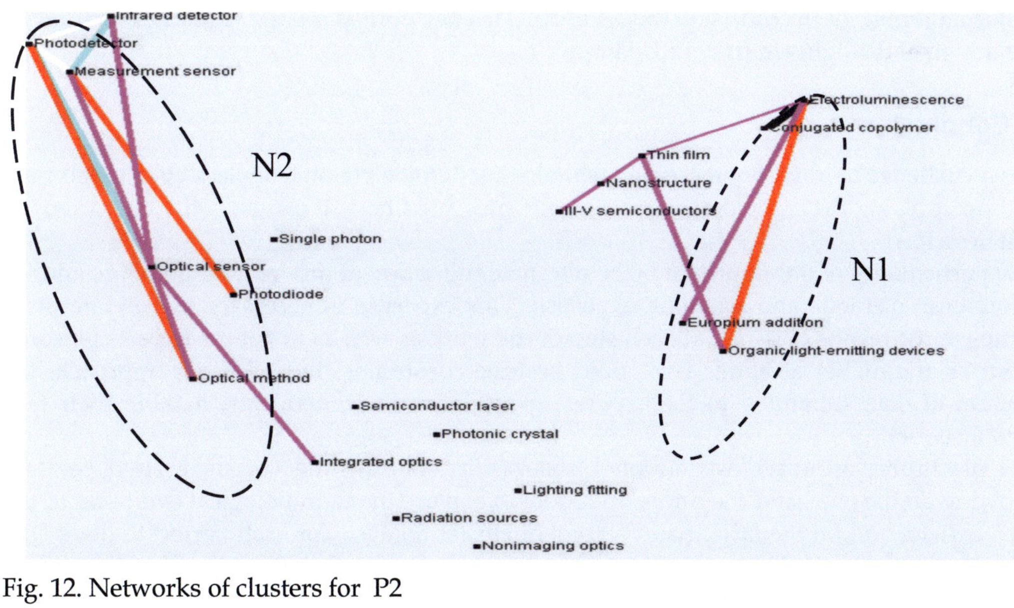

Fig. 12. Networks of clusters for P2

Now we study the complementarity of both methodologies by comparing their results.

5. Convergence between the two methodologies

Once applied to a corpus of 8,169 PASCAL bibliographic records related to optoelectronic devices and divided into two successive periods (2000 to 2004 and 2005 to 2010), both methodologies lead to complementary and converging results.
The stability of inorganic semiconductor materials over the two periods is illustrated by the crossover from the stage "established term" to the stage "cross-section terms" (Pathway V) of keywords like "*III-V semiconductors*" or "*Gallium arsenides*" (diffusion model) and by the diachronic stability of the cluster "*III-V semiconductors*" (cluster analysis).
The leading role of organic materials, and especially organic polymers, over the two periods is highlighted by the presence of keywords like "*Conjugated copolymer*", "*Conjugated polymer*" or "*Phenylenevinylene derivative polymer*" in the "established terms" and "cross-section terms" stages of both periods (diffusion model) and by the strength of the cluster subnetwork N1 right from the first period (cluster analysis).
The growing importance of nanotechnology and nanomaterials in P2 is highlighted by the crossover from the stage "unusual term" to the stage "cross-section terms" (Pathway III) of the keyword "*Nanoparticle*" (diffusion model) and by the emergent role of cluster "Nanostructure" in the second period map (cluster analysis).
The growing importance of imaging devices in P2 is highlighted by the crossover from the stage "unusual term" to the stage "cross-section terms" (Pathway III) of the keyword "*Imager*" (diffusion model) and by the strength of the cluster subnetwork N2 in the second period map (cluster analysis).
The emergence in the second period of applications of optical sensors towards biomedical imaging, highlighted by cluster analysis, is illustrated by the appearance of the keyword "*Biomedical imaging*" amongst the top 10 "unusual terms" of the diffusion model in P2.
The reinforcement of the applications of LED as light sources, highlighted by cluster analysis, is illustrated by the appearance of the keyword "*Daylight*" amongst the top 10 "unusual terms" of the diffusion model in P2. This keyword is related to studies focused on the design of daylight emitting LED lamps.

6. Conclusion

It is a challenge to identify emerging technologies. "Optoelectronic devices" is a broad field and by applying the complementary methods presented here, it can be better characterised and described.
One particularity of our approach is the alternate utilisation of different bibliometric and/or informetric methods and scientific expertise. This expertise is necessary to validate or to complete the results obtained at each step of the work as well as to get the experts personal input on the matter at hand. This could be time consuming, but with our approach, the amount of data submitted to the experts' appreciation is limited, thus making their task notably easier.
The diachronic approach we adopted consists in splitting the corpus in two periods, applying on the one hand the diffusion model to explore the terminological evolution of the subject field, and on the other hand a content cluster analysis for each period to detect the evolution of the topics by examination of the vocabulary related to the respective clusters.
The application of the diffusion model is a novel bibliometric approach giving a more in-depth view of the considered field. Associating terms with technologies allows the

development of a new interesting analysis methodology based on the notion of terminology diffusion. The indicators we used such as term frequency, relative term frequency and age of terms helped work out the different features of a field. Exploiting these results, we were able to understand the field specifications. The bibliometric filter assigning keywords to three stages gives an insight on the emergence of research topics in a technology. At a first glance it is a formal taxonomy used to disassemble what is called a technology in its "atoms" of research and traces the breakthroughs if they happen. Terms in the unusual phase show potentials of future developments in a technology although it is neither a prognosis, nor a prediction. To discuss the merit of emergences and research potentialities it is important to obtain a validation of the results from experts in the concerned field.
Our cluster analysis approach allows to have a global view of the field landscape at two successive time periods. The analysis of the cluster contents and their relative position on the cluster maps supplies indications about their similarity with respect to their respective associated keywords. The observation of cluster maps allows to detect exceptional topics and interesting topic sub-sets. Here also, the experts' input remains at each step absolutely necessary to validate and position the analysis results in the field context by giving them a scientific foundation.
Applying at the same time cluster analysis and diffusion model allows to confirm the results detected by each method and also to lead to a deeper understanding and characterization of the technological field. Furthermore, the diffusion model approach allows new interpretation of clustering results introducing external term categorizations.

7. Acknowledgement

This work was carried out thanks to a European Union funding: Project number 15615 NEST Programme launched within the framework of the 6th Research and Development Framework Plan. The project acronym was PROMTECH and the project full title was "Identification and Assessment of Promising and Emerging Technological Fields in Europe". The Consortium was composed by the Austrian Institute of Technology GmbH (AIT, Vienna, Austria), the Fraunhofer Institut für Systemtechnik und Innovationsforschung (ISI, Karlsruhe, Germany) and the Institut de l'Information Scientifique et Technique (INIST-CNRS, Vandœuvre-lès-Nancy, France).

8. References

Armstrong, J. S., Green, K. C. (2007): http://www.forecastingprinciples.com/selection_tree.html

Besagni D., François C., Hörlesberger M., Roche I. & Schiebel E. (2009): Les émergences technologiques dans le domaine des dispositifs optoélectroniques : identification et caractérisation, *paper presented at the wokshop VSST* (Veille Stratégique Scientifique et Économique), 30-31 mars 2009, Nancy.

Kopcsa A. & Schiebel E. (1998): Science and technology mapping. A new iteration model for representing multidimensional relationships. *Journal of the American Society for Information Science*, 49 (1), 7-17.

Lancaster F. W. & Lee J. L. (1985): Bibliometric techniques applied to issues management: A case study. *Journal of the American Society for Information Science*, 36(8), 389-397.

Lelu A. (1993): Modèles neuronaux pour l'analyse de données documentaires et textuelles. PhD Dissertation, Université de Paris 6.

Lelu A. & François C. (1992): Hypertext paradigm in the field of information retrieval: A neural approach. *4th ACM conference on hypertext*, Milano, November 30th–December 4th.

PROMTECH (2007): Identification and Assessment of Promising Emerging Technological Fields in Europe. Final report, July 2007, AIT, Vienna.

Roche I., Besagni D., François C., Hörlesberger M. & Schiebel E. (2010): Identification and characterisation of technological topics in the field of Molecular Biology. *Scientometrics*, 82(3), 663-676.

Schiebel E., Hörlesberger M., Roche I., François C. & Besagni D. (2009): An advanced diffusion model to identify emergent research issues: the case of optoelectronic devices. *12th International Conference on Scientometrics and Infometrics*, ISSI 2009, July 14th-17th, 2009, Rio de Janeiro, Brazil.

Schiebel E., Hörlesberger M., Holste, D. Roche I., François C. & Besagni D. (2010): Identification of emerging technologies through pathway-III analysis, *Proceedings of the 19th International Conference on Management of Technology*. Cairo, Egypt, March 8th-12th, 2010

Schiebel E., Hörlesberger M., Roche I., François C. & Besagni D. (2010): An advanced diffusion model to identify emergent research issues: the case of optoelectronic devices. *Scientometrics*, 83(3), 765-781.

8

Synchronous Vapor-Phase Coating of Conducting Polymers for Flexible Optoelectronic Applications

Keon-Soo Jang and Jae-Do Nam
Sungkyunkwan University
Republic of Korea

1. Introduction

Since conducting polymers (CP) were first reported, poly(3,4-ethylenedioxythiophene) (PEDOT) is arguably one of the most commercially useful and most studied CPs in the last 20 years (Shirakawa et all., 1977; Chiang et al., 1977; Winther-Jensen et al., 2007; Truong et al., 2007) . PEDOT has been studied extensively on account of its many advantageous properties, such as high electrical conductivity, good transmittance and thermal stability with a low optical bandgap and thermal stability (Winther-Jensen & West, 2004; Jonas et al., 1991). These properties make PEDOT very attractive for applications, such as electrochromic windows (Welsh et al., 1999), organic electrodes for organic photovoltaic cells (OPVs) (Admassie et al., 2006; Gadisa et al., 2006) and hole injection layers (HIL) in organic light emitting devices (OLEDs) (Wakizaka et al., 2004; Hatton et al., 2009) and dye-sensitized solar cells (Saito et al., 2005). In particular, PEDOT is commonly used as a hole extraction layer in OPVs (Colladet et al., 2007; Kim et al., 2005). In most optoelectronic applications as a buffer or electrode layer, the bandgap of the layer plays an important role in determining the operating characteristics, quantum efficiency and electron/hole transport. Therefore, the main issues for electronic device applications include both the electrical conductivity and bandgap.

Oxidized PEDOT can be produced in a variety of forms using different polymerization techniques. Solution processing is used most commonly in synthesizing PEDOT in the form of spin-coating, solvent-casting or ink-jet printing. However, these PEDOT systems are relatively insoluble in most solvents, making it necessary to attach soluble functional groups to the polymer or dope it with stabilizing polyelectrolytes (Terje & Skotheim, 1998). An aqueous dispersion of poly(3,4-ethylenedioxythiophene)-poly(styrenesulfonate) (PEDOT-PSS), commercially available as Baytron P, is a stable polymer system with a high transparency up to 80% (Groenendaal et al., 2000). However, the PEDOT-PSS film exhibits relatively low electrical conductivity, 10-500 S/cm (Groenendaal et al., 2000), which does not often meet the high conductivity required for most applications. In addition, scanning-tunneling microscopy, neutron reflectivity measurements, and x-ray photoelectron spectroscopy have revealed a PSS rich layer on the top of the spin-coated PEDOT-PSS films due to the phase separation (Lee & Chung, 2008; Kemerink et al., 2004; Higgins et al., 2003). Since PSS is an electrical insulator, excessive PSS can limit the film conductivity (Kemerink et al., 2004), and an acidic PEDOT-PSS dispersion can etch indium tin oxide (ITO) during the polymer spin-coating process. Moreover, the hydrolysis of the deposited PEDOT-PSS by moisture absorption can also etch ITO to cause indium and tin incorporation into the polymer (Lee et al., 2007).

For the PEDOT systems without using polyelectrolytes, PEDOT can be deposited directly on the substrate surface by several in-situ polymerization techniques. One of the options is electrochemical polymerization, which has been reported to have higher electrical conductivity (Groenendaal et al., 2001). However, electrochemical polymerization results in a poor transparency and requires conducting substrates, which limits its practical applications. As an alternative, oxidative chemical polymerization either in the liquid or vapor phase is more versatile because it is not restricted by the substrates. In particular, one way of achieving a clear thin film with a smooth surface is to apply the oxidant using solvent coating processes and expose the coated surface to a reactive monomer vapor. This process is often referred to as vapor phase polymerization (referred to herein as VPP) (Kim et al., 2003; Fabretto et al., 2008; Cho et al., 2008; Ha et al., 2004; Winther-Jensen et al., 2004). The PEDOT films by VPP have been reported to have conductivities of approximately 15 S/cm at a thickness of 300 nm without any additives (Kim et al., 2003). Recently, a PEDOT film with high conductivity, exceeding 1000 S/cm, was reported using a base-inhibited VPP (Winther-Jensen et al., 2007). However, it should be noted that VPP PEDOT has a high bandgap and relatively low transmittance (Fabretto et al., 2008; Cho et al., 2008).

As another thiophene-based conducting polymer, poly(3-hexylthiophene) (P3HT) is also one of the most indispensable materials in OLEDs, OPVs, field effect transistors (FETs) and thin film transistors (TFTs) (Hatton et al., 2009; Kim et al., 2005; Grecu et al., 2006; Bartic et al., 2002). Over the last decade, blending or copolymerization techniques of conducting polymers have been investigated not only to overcome the drawbacks of a pristine conductive polymer, such as inadequate bandgap, rough surface, low conductivity and poor transmittance, but also to tailor the properties for various applications (Kim et al., 2009; Xu et al., 2006; Sarac et al., 2003). This study investigated P3HT thin films deposited using a vapor-phase polymerization (VPP) technique (Ha et al., 2004; Winther-Jensen et al., 2004), which desirably ensures thin film formation in various substrate materials without the additional processes to liquefy polymers (Jang et al., 2009). It is believed that the VPP technique for P3HT will allow the fabrication of thin coatings over a large surface area of various substrate materials. We used a similar route to VPP of PEDOT choosing appropriate catalyst and solvent systems. Using the VPP technique, therefore, PEDOT and P3HT may be copolymerized in the state of vaporized monomers and subsequently polymerized to form, most probably, a PEDOT/P3HT copolymer structure. However, it should be mentioned that the traditional VPP route, where the monomer is maintained in the state of thermodynamic equilibrium, may not simply be applied because the vapor pressure and polymerization rate of the EDOT and 3HT are different and, thus, the relative composition of the VPP copolymer is not controllable (Jang et al., 2010).

Therefore, the PEDOT to P3HT ratio was kinetically controlled in this study by adjusting the relative feed amount of the evaporating monomers to the reaction chamber to fabricate the PEDOT/P3HT films containing different ratios of PEDOT to P3HT. The developed synchronous VPP technique successfully provided PEDOT/P3HT copolymer thin coatings with tunable bandgap and optoelectronic properties.

2. Conducting polymers

2.1 Summary of conducting polymers

Conducting polymers–plastics that conduct electricity–continue to find market niches in consumer electronics and antistatic textiles, some of which have military applications. Among the most exciting applications is the use of conducting polymers in light-emitting devices (LEDs), replacing silicon as the traditional substrate material for clock radios, audio equipment, televisions, cellular telephones, automotive dashboard displays, and aircraft cockpit displays. Conducting polymers provide benefits to industries such as electronics by shielding against electromagnetic interference (EMI). Conductive polymers are also already used in devices that detect environmentally hazardous chemicals, factory emissions, and flavors or aromas in food products. Currently, their conductivity is being explored in electrostatic materials, conducting adhesives, electromagnetic shielding, artificial nerves, aircraft structures, diodes, and transistors.

2.2 Correlation of chemical structure and electrical conductivity

In traditional polymers such as polyethylenes, the valence electrons are bound in sp^3 hybridized covalent bonds. Such "sigma-bonding electrons" have low mobility and do not contribute to the electrical conductivity of the material. The situation is completely different in conjugated materials. Conducting polymers have backbones of contiguous sp^2 hybridized carbon centers. One valence electron on each center resides in a p_z orbital, which is orthogonal to the other three sigma-bonds. The electrons in these delocalized orbitals have high mobility when the material is "doped" by oxidation, which removes some of these delocalized electrons. Thus the p-orbitals form a band, and the electrons within this band become mobile when it is partially emptied. In principle, these same materials can be doped by reduction, which adds electrons to an otherwise unfilled band. In practice, most organic conductors are doped oxidatively to give p-type materials. The redox doping of organic conductors is analogous to the doping of silicon semiconductors, whereby a small fraction silicon atoms are replaced by electron-rich (e.g., phosphorus) or electron-poor (e.g. boron) atoms to create n-type and p-type semiconductors, respectively.

Although typically "doping" conductive polymers involves oxidizing or reducing the material, conductive organic polymers associated with a protic solvent may also be "self-doped."

The most notable difference between conductive polymers and inorganic semiconductors is the mobility, which until very recently was dramatically lower in conductive polymers than their inorganic counterparts. This difference is diminishing with the invention of new polymers and the development of new processing techniques. Low charge carrier mobility is related to structural disorder. In fact, as with inorganic amorphous semiconductors, conduction in such relatively disordered materials is mostly a function of "mobility gaps" with phonon-assisted hopping, polaron-assisted tunneling, etc., between localized states.

The conjugated polymers in their undoped, pristine state are semiconductors or insulators. As such, the energy gap can be > 2 eV, which is too great for thermally activated conduction. Therefore, undoped conjugated polymers, such as polythiophenes, polyacetylenes only have a low electrical conductivity of around 10^{-10} to 10^{-8} S/cm. Even at a very low level of doping (< 1 %), electrical conductivity of increases several orders of magnitude up to values of around 0.1 S/cm. Subsequent doping of the conducting polymers will result in a saturation of the conductivity at values around 0.1–10 kS/cm for different polymers. Highest values reported up to now are for the conductivity of stretch oriented polyacetylene with confirmed values of about 80 kS/cm. Although the pi-electrons in polyactetylene are delocalized along the chain, pristine polyacetylene is not a metal. Polyacetylene has alternating single and double bonds which have lengths of 1.44 and 1.36 Å, respectively. Upon doping, the bond alteration is diminished in conductivity increases. Non-doping increases in conductivity can also be accomplished in a field effect transistor (organic FET or OFET) and by irradiation. Some materials also exhibit negative differential resistance and voltage-controlled "switching" analogous to that seen in inorganic amorphous semiconductors.

2.3 Synthesis of conducting polymer

Chemical, oxidative coupling, electrochemical, chemical oxidative, Grignard poly-condensation, and quasi-living polymerization are commonly used. In chemical oxidative polymerization, a monomer, oxidizing agent, and a dopant are reacted on the dielectric surfaces to form the conductive polymer. In electochemical polymerization, conducting polymers such as Polypyrrole and Polyaniline films of uniform morphology and good electrical conductivity are formed on Indium Tin Oxide glass electrodes in supercritical CO_2 where the only added ingredient is electrolyte dopant. Furthermore, conducting polymers are chemically polymerized in an aqueous medium to which an oxidant is added. Sometimes, in addition to the ingredients, a protonic acid is added to the aqueous polymerization mixture that renders the final polymers conductive. The oxidative polymerization of monomers to obtain conductive polymer is affected by the oxidation potential level of monomer as well as the oxidizing capability of oxidizing agent. In the example of Fe (III) oxidizing agent, the Fe (III) ion would form a complex with a specific compound having lone pair electrons, which alters the oxidizing strength of Fe (III) ion. The stronger the complex bonding is, the lower the oxidizing capability of the oxidizing agent is.

3. Fabrication of the VPP conducting polymers

3.1 VPP-P3HT

The substrates were washed and rinsed with DI-water and acetone while being sonicated for 10 minutes to remove any organic contaminants. The glass substrates plasma treated (KSC Korea switching, Korea) for 10 minutes (10 kHz & 10 V & 7 A at a speed of 50 cc Helium/min), and the ITO glass substrates were UV-treated for 20 min. The catalyst was a mixture of MeOH and EtOH at 1:1 ratio with $FeCl_3 \bullet 6H_2O$. After sonicating the catalyst solution for 2 minutes at 40°C, it was spin-coated onto the substrates at a speed of 500 rpm for 5 seconds and then at 1400 rpm for 5 seconds. Subsequently, the catalyzed substrates were placed in the vapor-phase-polymerization (VPP) chamber containing 3-hexylthiophene (3HT) monomer to evaporate and fill therein under a nitrogen purge, which is similar to that reported elsewhere (Truong et al., 2008). The 3HT monomers in the VPP chamber were polymerized for one hour at 50°C. The sample was soaked and washed sequentially with MeOH to eliminate the monomers remaining on the substrate. The washed P3HT film was further dried to remove the residual solvents for 10 minutes in ambient atmosphere.

3.2 VPP-P3HT/PEDOT copolymers

The substrates were washed and rinsed with DI-water and acetone and sonicated for 10 minutes to remove any organic contaminants. The glass substrates were plasma-treated

(PLAMAX, SPS Co., Ltd., Korea) for 30 seconds, and the ITO glass substrates were UV-treated for 20 minutes. The oxidant was a mixture of MeOH and EtOH at a 1:1 ratio with iron (III) chloride hexahydrate. The wt% ratio between the alcohol mixture and $FeCl_3{\cdot}6H_2O$ was 10:1, 5:1, and 3:1. A ratio of 10:1 was chosen to fabricate most thin films in this study. After sonicating the oxidant solution for 2 minutes at 40°C, it was spin-coated onto the substrates at a speed of 500 rpm for 5 seconds and then at 1400 rpm for 5 seconds. The oxidized substrates were placed in the vapor-phase-polymerization (VPP) chamber, where the containers of 3HT and EDOT were also placed. The vaporized monomers were supplied into the chamber through the feed inlets, which were made at the center of the container ceiling with different inlet diameters of D_{3HT} and D_{EDOT} (Scheme 1). The flow rate of flowing nitrogen was kept constant during polymerization, and the feed ratio of 3HT and EDOT monomer vapors was adjusted by the inlet diameters of D_{3HT} (0, 2, 4, 6, and 8 mm) and D_{EDOT} (fixed at 20 mm). The EDOT and 3HT monomers in the VPP chamber were polymerized for 20 min, 30 min, or 1 hour at 60°C. After polymerization, the sample was soaked and washed sequentially with MeOH to eliminate the monomers and Fe(III) solution remaining on the substrate. The washed PEDOT/P3HT copolymer film was dried further using a hot-air gun for 1 minute in an ambient atmosphere to remove the residual MeOH .

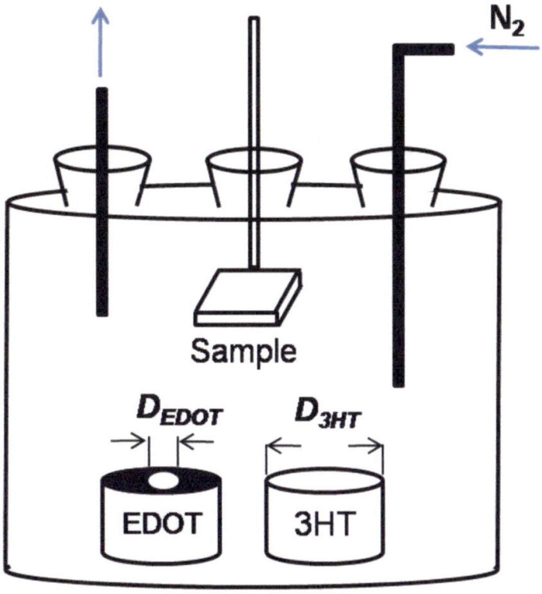

Fig. 1. Schematic of experimental setup for synchronous polymerization of EDOT and 3HT monomers, where the monomer concentrations are controlled by the inlet sizes of monomers (D_{EDOT} and D_{3HT}) to the reaction chamber under the flowing inert gas

4. Measurement

The thickness of the VPP PEDOT/P3HT was measured using an alpha step IQ (KLA Tencor corporate, the Yield Management company, San Jose CA, U.S.A.) and FE-SEM (Field emission scanning electron microscope, 1.0nm guaranteed at 15kV, JSM6700F, JEOL, Japan). The bandgap was determined by UV-vis-spectrophotometry (UV-3600, SHIMADZU,

Japan). AFM (SPA-300HV, SII Nano Technology Inc. Tokyo Japan) was used to examine the surface morphology of the VPP-PEDOT/P3HT coating, where a 5 μm x 5 μm area was analyzed to determine the surface roughness. The AFM 3D images revealed the surface roughness of the copolymer films with a z-axis of 100 nm. X-ray photoelectron spectroscopy (XPS, ESCA2000, VG MICROTECH) equipped with an Al Kα radiation source (hv = 1486.6 eV) was used to analyze the components. Attenuated total reflection-Fourier transform infrared (ATR-FTIR) spectroscopy (Bruker IFS-66/S, Bruker, Massachusetts U.S.A.) was also used for component analysis of the VPP-PEDOT/P3HT copolymers. Contact angle measurements were performed using a Digi-drop contact angle goniometer (JinTech, Korea). A water droplet was dropped onto the surface of a small sample. The droplet shape was recorded by a camera and analyzed to determine the contact angle. The electrical conductivity of the synthesized films was measured using a four-point probe method (Keithley 236 current source and Keithley 617 electrometer). The film thickness for the measurement of electrical conductivities ranged 100 – 500 nm coated on a glass substrate in the area of 25 x 25 mm^2. The type of the four-probe point was a linear type with 1 mm tip spacing. The HOMO (highest occupied molecular orbital) was measured using an ultraviolet photoelectron spectrometer (UPS, Surface Analyzer, Model AC-2, Riken-Keiki Co., Japan). The molecular weight of VPP-P3HT was measured by the matrix-assisted laser desorption ionization mass spectrometer (MALDI-MS, Voyager-DE STR biospectrometry Workstation, Applied Biosystems Inc.) equipped with N_2 UV-laser radiating at 337 nm wavelength using THF as a solvent and dithranol as a matrix.

5. Optical properties - bandgap

5.1 VPP-P3HT

Fig. 2 shows the UV-vis transmittance spectra of a VPP-P3HT film formed on a glass substrate after being polymerized at 50 °C for 1 hour. Taking the isobestic point at 650 nm as the energy gap, the band gap was estimated to be 1.91 eV, which is similar to that of conventional polythiophenes corresponding to the 70% regioregular head-to-tail (HT) P3HT

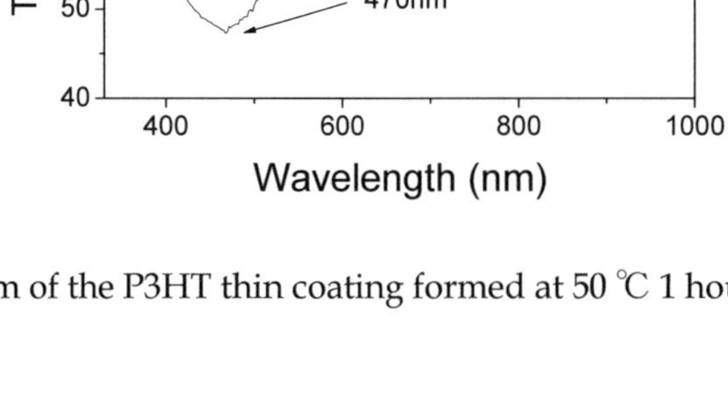

Fig. 2. UV-vis spectrum of the P3HT thin coating formed at 50 °C 1 hour

case (Chen et al., 1995; Chan et al., 1998). The measured value in this study is believed to be a proper band gap in OLED and photovoltaic cells (Li et al., 2004; Du Pasquier et al., 2005). The regioregularity of P3AT is also considered to be extremely important for the manifold properties, such as conductivity and the extent of crystal growth (McCullough, 1998; Skotheim & Reynolds, 2007; Chan et al., 1998). The regioregular head-to-tail P3ATs form well-defined and well-organized 3D structures in the form of π-stacks, which usually lead to both better material characteristics and enhanced device performance in almost all categories ranging from electrical conductivity to stability (Li et al., 2004; Du Pasquier et al., 2005). The minimum peak (λmin) at 470 nm in Fig. 2 indicates the regioregularity of P3HT corresponding to a 70% rr-P3HT of the π-π* transition (Chen et al., 1995; Chan et al., 1998).

5.2 VPP-P3HT/PEDOT copolymers

Fig. 3 presents the UV-vis absorbance spectra of VPP-PEDOT/P3HT films on glass substrates leading to changes of the bandgap of the VPP-PEDOT/P3HT copolymer at different feed ratio of **r** values. The bandgap was estimated to be 1.91, 1.98, 2.02, 2.07, and 2.16 eV for the specimens prepared with 0, 2, 4, 6 mm of feed diameters and the pristine VPP-PEDOT film, respectively, by taking the isobestic points at 650, 625, 615, 600, and 575 nm, respectively. The bandgap decreases gradually with decreasing EDOT monomer supply. Accordingly, the bandgap of the VPP-PEDOT/P3HT copolymer film can be desirably adjusted by changing the relative composition of PEDOT and P3HT, especially when a low bandgap is required in such applications as OLEDs and OPVs (Colladet et al., 2007). The inset in Fig. 3 shows the pristine PEDOT, P3HT and PEDOT/P3HT copolymers at different compositions coated on glass substrates. The color of the specimens clearly indicates that the compositions of PEDOT and P3HT vary with the fabrication conditions to give tunable bandgap.

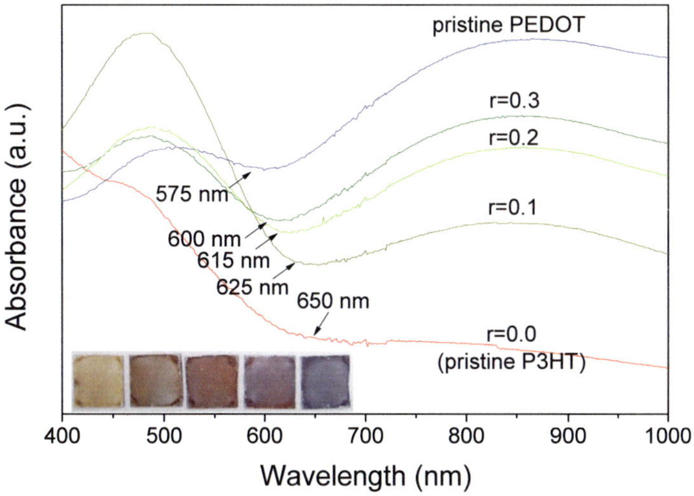

Fig. 3. UV-vis. absorbance spectra of the various PEDOT/P3HT copolymer films. The inset shows pictures of the PEDOT/P3HT copolymer films coated on ITO-coated glass using an oxidant at a 10:1 ratio of mixture to ferric chloride hexahydrate, representing r = 0, 0.1, 0.2, 0.3, 0.4, and the pristine PEDOT film from left to right, respectively, with a thickness of 300 nm

6. HOMO of VPP-P3HT

The highest occupied molecular orbital (HOMO) of the VPP-P3HT coating was approximately 5.07 eV (Fig. 4), which is similar to previous studies on organic light-emitting diodes and solar cells. Since the HOMO of P3HT can be adjusted using techniques, such as thermal annealing, the HOMO of the VPP-P3HT in this study can be adjusted for the applications in a variety of optoelectronic devices to meet different electron donor materials.

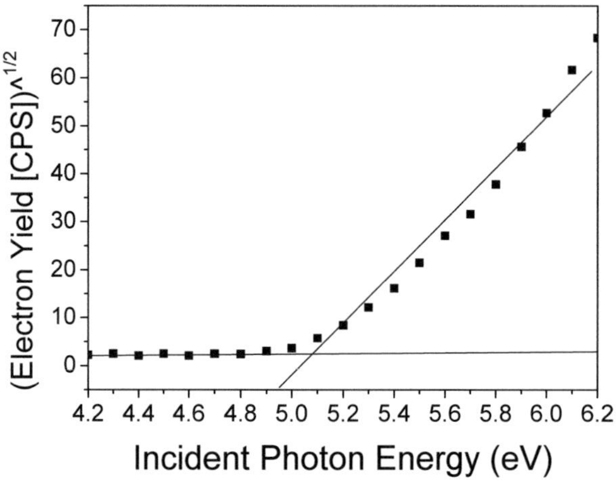

Fig. 4. HOMO (highest occupied molecular orbital) of P3HT thin coating on a ITO-glass substrate

7. Molecular weight and solubility test of VPP-P3HT

7.1 Molecular weight of VPP-P3HT

Determination of the molecular weights of polymers is almost invariably found by gel permeation chromatography (GPC) using polystyrene standards, which indicates relative molecular weights. Conjugation systems such as polythiophenes, however, are known to have a more rod-like conformation in solution, which tends to overestimate the molecular weights of such systems (Holdcroft, 1991). More specifically, in P3HT systems, the molecular weight by GPC is a factor of 1.2 - 2.3 times higher than that by a matrix-assisted laser desorption ionization mass spectrometer (MALDI-MS) (Liu, 1999). Consequently, MALDI-MS was used in this study in order to determine the molecular weight of VPP P3HT. Fig. 5 shows the molecular weight of VPP P3HT measured by MALDI-MS providing peaks at around 994, 1166, and 1333 corresponding to 6 through 8 repeat units of 3-HT (0.74 nm) (Goodman et al., 2009) giving a chain length of 4.44 - 5.92 nm. It is within the effective conjugation length of P3HT (Xu et al., 2007) reported differently in different applications as 9 - 10 thiophene units in polymer light emitting diodes (PLEDs) (Perepichka et al., 2005) and 5 – 7 thiophene units in field effect transistors (FETs) (Johansson & Larsson, 2004). To apply the thin VPP P3HT coatings to many optoelectronic devices, further work will be needed on controlling the molecular weights of VPP P3HT.

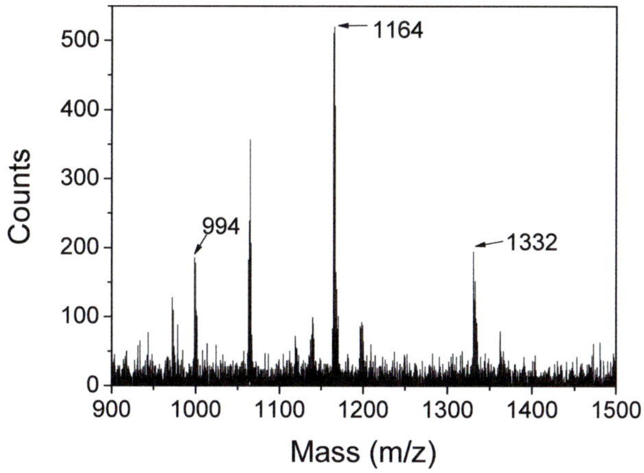

Fig. 5. MALDI-MS spectrum of P3HT vapor-phase polymerized at 55 ℃ for 1 hour

7.2 Solubility test of VPP-P3HT

Fig. 6 shows the solubility tests of P3HT, monomer (3-HT), and Fe(III) solution demonstrating that the vapor-phase polymerization was successfully carried out in this study. Fig. 6(a) through (c) shows that the VPP P3HT coating (200 nm of thickness in blue color) still remains on the glass as a red-brown layer after being washed with methanol for 3 hours, 1 hour, and 5 min, respectively. As THF is known to be a good solvent of P3HT, Fig. 6(d) shows that the VPP P3HT coating is completely dissolved in THF after 1 min of washing leaving a yellow-colored solution. Although the P3HT is not dissolved in methanol, the Fe(III) or 3-HT monomer is entirely dissolved in methanol. In order to confirm the solubility of the Fe(III) and 3-HT in methanol, the Fe(III) and monomer coatings are shown in Fig. 6(e) and (f) after being washing in methanol for 1 min. As can be seen, these coatings are immediately dissolved in methanol. Consequently, the solubility test proves that P3HT was successfully synthesized by the vapor-phase polymerization in this study because the synthesized P3HT coating remains intact on the glass substrate after thoroughly washed with methanol, which may very well remove unreacted monomers and the Fe(III) from the coating layer.

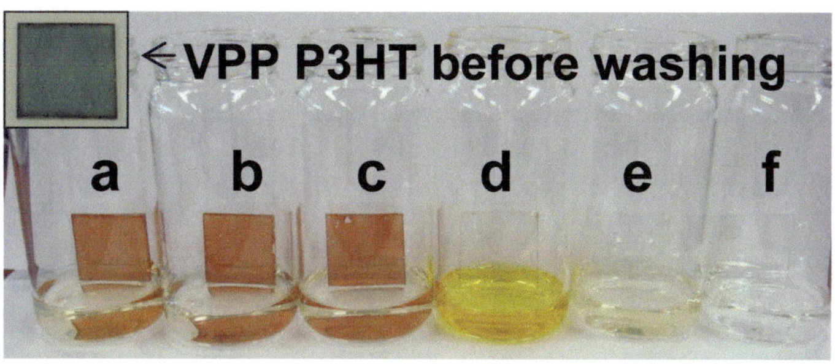

Fig. 6. Solubility test of VPP P3HT coating on glass after being washed with MeOH for 3 hours (a), 1 hour (b), and 5 min (c), and washed with THF for 1 min (d), also exhibiting Fe(III)Cl solution coating (e) and monomers coating (f) on glass after being washed with methanol for 1 min

8. Solvent selection as Fe(III)Cl•$6H_2O$ solution for fabrication of VPP-P3HT

The surface morphology is important when P3HT is used for thin coating applications because it influences a large number of properties, such as double layer capacitance and adhesion (Truong et al., 2008). It was reported that the polymerization rate of conjugated polymers, surface treatment and Fe(III) solution removal conditions substantially affect the surface morphology of thin film coatings (Truong et al., 2008). For example, B. Winther-Jensen (Winther-Jensen et al., 2004) made a variety of attempts to remove the residual Fe(III)

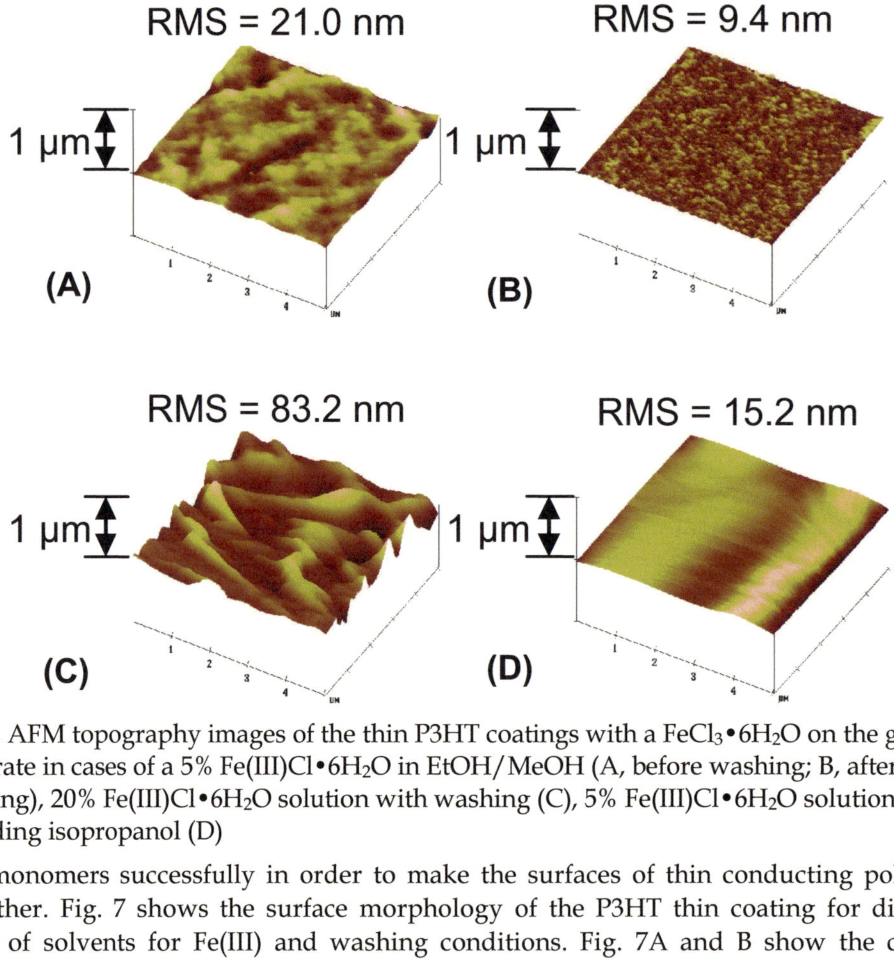

Fig. 7. AFM topography images of the thin P3HT coatings with a $FeCl_3$•$6H_2O$ on the glass substrate in cases of a 5% Fe(III)Cl•$6H_2O$ in EtOH/MeOH (A, before washing; B, after washing), 20% Fe(III)Cl•$6H_2O$ solution with washing (C), 5% Fe(III)Cl•$6H_2O$ solution including isopropanol (D)

and monomers successfully in order to make the surfaces of thin conducting polymers smoother. Fig. 7 shows the surface morphology of the P3HT thin coating for different types of solvents for Fe(III) and washing conditions. Fig. 7A and B show the coating surfaces before (A) and after (B) the washing steps for the 5% $FeCl_3$•$6H_2O$ in EtOH/MeOH, respectively. The washing process improved the RMS roughness from 21.0 nm to 9.4 nm. In general, the surface roughness of P3HT thin film coatings should not exceed ca. 10 nm for most optoelectronic applications (Truong et al., 2008), which was satisfied in this study. Fig. 7C shows the coating surface after removing the Fe(III) with EtOH/MeOH. The surface roughness of the coating increased with increasing Fe(III) concentration, as shown in Fig. 7B and C, giving a RMS roughness as 9.47 nm and 83.2 nm for the 5% and the 20% Fe(III)Cl•$6H_2O$ respectively. Although the result is not included here, the RMS roughness is increased with increasing polymerization time or temperature, mainly due to the increased polymerization rate. It should be noted that other Fe(III)Cl solutions incorporating isopropanol, propanol, or butanol provided produced rougher surfaces than the methanol/ethanol mixture, one of which is shown in Fig. 7D for the isopropanol case. On the other hand, thicker films tend to have higher electric conductivity, which is strongly dependant on the surface roughness. When the thickness of P3HT was more than 300 nm, the surface appeared rougher making it quite difficult to remove the remaining Fe(III) from the coating formed at high concentrations. Moreover, a long polymerization time at low concentration produced a better surface morphology than a short polymerization time at high concentrations, even though the film thicknesses are the same.

9. Experimental demonstration of copolymerization of P3HT and PEDOT

9.1 ATR-FTIR spectra of the PEDOT/P3HT copolymers

The thin copolymers were fabricated by adding a small amount of EDOT monomer in the VPP chamber with excessive 3HT monomers because the polymerization rate of the EDOT monomers was faster than that of the 3HT monomers. More specifically, the composition was controlled by adjusting the feed amount of EDOT to the catalyst-coated layer placed in the reaction chamber, which complies with the typical steady-state plug-flow reactor system (Levenspiel, 1962). In this case, the feed ratio of EDOT to 3HT may be controlled by the inlet size of the vaporized monomers in the stream of purging nitrogen. Simply quantifying the feed composition, the feed ratio (r) may be defined by the feed inlet sizes of EDOT and 3HT, viz: $r = D_{EDOT}/D_{3HT}$, where D_{EDOT} and D_{3HT} are the cross-sectional diameters of the feeding hole on the ceiling of the EDOT and 3HT containers, respectively.

The ATR-FTIR spectroscopy results of the pristine PEDOT, pristine P3HT and PEDOT/P3HT copolymer films are compared in Fig. 8 at different feed ratios. As the characteristic peaks of each pristine polymer, the peak intensity of the aliphatic C-H bond stretching vibration corresponds to the 3-hexyl group in P3HT, and the C-O-C bond stretching vibration corresponds to the ethylenedioxy group in PEDOT. As seen in Fig. 8B, the aliphatic C-H bond stretching in the 3-hexyl group in P3HT appears at 2854, 2924, and 2954 cm^{-1}, and its peak intensity decreases with the feed ratio of r (Chen et al., 1995; Singh et al., 2007). Assigning the bands at approximately 1058 and 1141 cm^{-1} to the stretching modes of the ethylenedioxy group (Zhan et al., 2008; Li et al., 2009), the intensity of these peaks increases with the increasing supply of the EDOT monomer. Consequently, the vapor phase polymerization ratio of PEDOT appears to reflect the controlled supply of the EDOT monomer via the feed ratio of r. Fig. 8C compares the bands of the pristine PEDOT (doped with $FeCl_3{\cdot}6H_2O$) with the copolymers at different feed ratios of r, seemingly exhibiting a redshift due to copolymerization of 3HT and EDOT. The inset in Fig. 8A shows the camera images of pristine PEDOT, P3HT, and PEDOT/P3HT at different compositions coated on the ITO substrates. Since the colors of the pristine PEDOT and P3HT are blue and orange, respectively, the color of the PEDOT/P3HT copolymers changes gradually from orange to dark blue with the increasing EDOT feed ratio of r demonstrating that the ratio was successfully controlled by the feed ratio in our experiments.

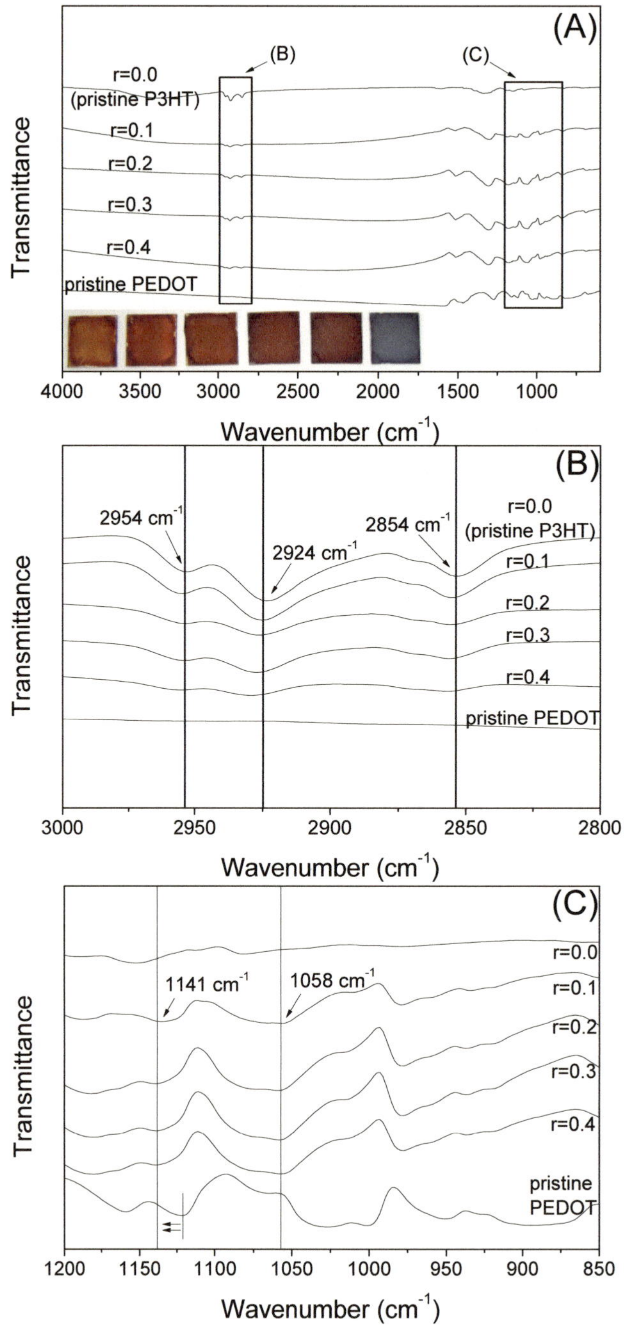

Fig. 8. Comparison of ATR-FTIR spectra of the PEDOT/P3HT copolymers at different feed ratio of ***r***. (A) full spectral region: 4000-600 cm^{-1}, (B) enlargement of the region of 3000-2800 cm^{-1}, (C) enlargement of the region of 1200-850 cm^{-1}. The inset in (A) shows pictures of the PEDOT/P3HT copolymer films coated on ITO-coated glass using an oxidant at a 5:1 ratio of the alcohol mixture to ferric chloride hexahydrate, representing **r** = 0, 0.1, 0.2, 0.3, 0.4, and the pristine PEDOT film from left to right, respectively, with a thickness of 500 nm

9.2 XPS spectra of the PEDOT/P3HT copolymers

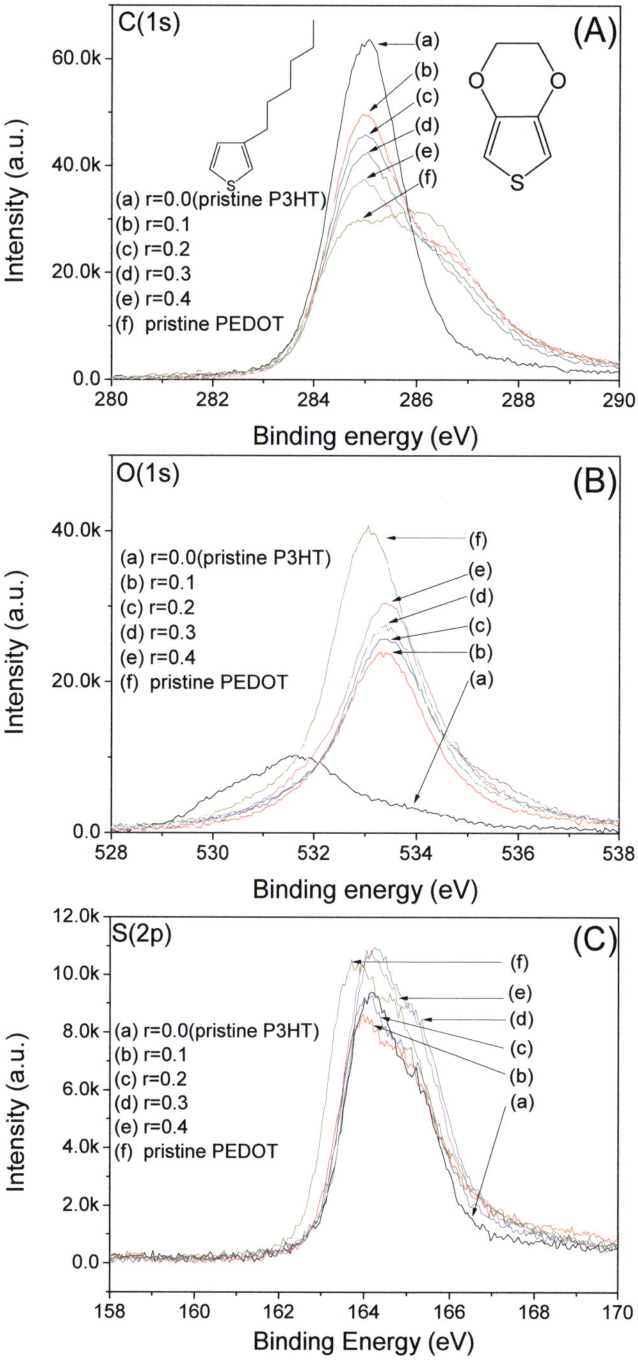

Fig. 9. Deconvolution of XPS spectra of (A) $C_{(1s)}$, (B) $O_{(1s)}$ core-level, and (C) $S_{(2p)}$ for PEDOT/P3HT copolymers at various feed ratio of **r**. The inset in (A) is chemical structures of PEDOT and P3HT, respectively

For copolymers, the surface composition may be quantified using several surface-sensitive techniques, such as XPS and time-of-flight secondary ion mass spectrometry (ToF-SIMS) (Vanden Eynde & Bertrand, 1999), XPS exhibiting somewhat better agreement with the theoretical compositions than ToF-SIMS (Cossement et al., 2006). Therefore, in this study, XPS was used to quantify the surface concentrations. The formation of PEDOT/P3HT copolymer films can be demonstrated by deconvoluting the XPS peaks at 285.0, 286.5, and 533.0 eV, as shown in Fig. 9. The XPS peaks at 285.0 and 286.5 eV in Fig. 9A can be assigned to C-C/C=C and C-O-C, respectively, corresponding to the spectra of the C(1s) core level (Marciniak et al., 2004; Hsiao et al., 2008). Fig. 9A and Table 1 indicate that the intensity of the peak at 285.0 eV decreases with increasing r, whereas that at 286.5eV increases, which agrees well with the chemical structures of PEDOT and P3HT, as shown in the inset in Fig. 9A. The O photoelectron spectrum in Fig. 9B and Table 1 shows that the O concentration increases with increasing r, as represented by the intensities of the XPS peak at 533 eV, which can be assigned to the oxygen-ether group (C-O-C) (Marciniak et al., 2004; Jonsson et al., 2003). The S photoelectron spectrum is shown in Fig. 9C and analyzed in Table 2.

Samples	C(1s)		
	C atom% at		h/g
	(g) 285.0 eV (C-C/C=C)	(h) 286.5 eV (C-O-C)	
a	78.4	0	0
b	50.0	24.6	0.49
c	48.1	25.9	0.54
d	42.4	28.4	0.67
e	35.1	28.9	0.82
f	29.2	32.1	1.10

Table 1. XPS atomic concentrations (%) of C and O for PEDOT/P3HT copolymers with (a) **r** = 0.0, (b) 0.1, (c) 0.2, (d) 0.3, (e) 0.4, and (f) pristine PEDOT

Samples	C(1s) atom%	S(2p) atom%	C/S theoretical	C/S experimental	PEDOT:P3HT experimental ratio(mol%)
a	84.1	8.2	10.0	10.3	0:100
b	80.4	8.7	-	9.2	31:69
c	78.4	8.8	-	8.9	39:61
d	75.4	9.7	-	7.8	69:31
e	74.8	9.9	-	7.5	78:22
f	69.5	10.4	6.0	6.7	100:0

Table 2. XPS atomic concentrations (%) of C and S comparing C/S ratios between theoretical and experimental results for (a) **r** = 0.0, (b) 0.1, (c) 0.2, (d) 0.3, (e) 0.4, and (f) pristine PEDOT

The S(2p) peak positions of pristine P3HT and PEDOT appear at 164.1 and 163.8 eV, respectively, and the peaks of their blend systems at different r values seem to appear between the two pristine polymers. Table 2 shows the elemental composition of the fabricated thin films demonstrating that the PEDOT to P3HT ratio increases with the EDOT feed ratio of r. As seen in Table 2, the concentration of S and the PEDOT to P3HT elemental ratio apparently increase with increasing r. Comparing the experimental and theoretical ratios of C/S, which may represent the ratios of P3HT to PEDOT, there are good agreements between the theoretical and experimental values. The slight discrepancy might be due to adventitious hydrocarbon contamination coming from the contact of the samples with the ambient atmosphere (Cossement et al., 2006).

10. Electrical and surface properties of the pristine PEDOT, P3HT, and PEDOT/P3HT copolymer films

10.1 Electrical conductivities

Fig. 10 shows the electrical conductivity of the copolymers plotted as a function of **r**. The electrical conductivity increases from 3.0 x 10^{-2} S/cm for the pristine P3HT up to 6.28 S/cm for the copolymer with **r** at 0.4. The conductivity of PEDOT/P3HT copolymer can be increased up to 200 times higher than that of the pristine P3HT, which can be achieved by adjusting the EDOT feed ratio of **r**. In this study, the conductivity of the pristine VPP-PEDOT was similar to that reported elsewhere (Kim et al., 2003), where a $FeCl_3$-based oxidant solution was used without any additives. However, the addition of basic inhibitors, such as pyridine (Winther-Jensen et al., 2004) and imidazole (Ha et al., 2004), has been reported to reduce the polymerization reaction kinetics and enhance the conductivity and transparency of the PEDOT film. Therefore, a weak base can be used to the PEDOT/P3HT copolymer systems to achieve high conductivities, which is desirable for electrode systems in various electronic devices.

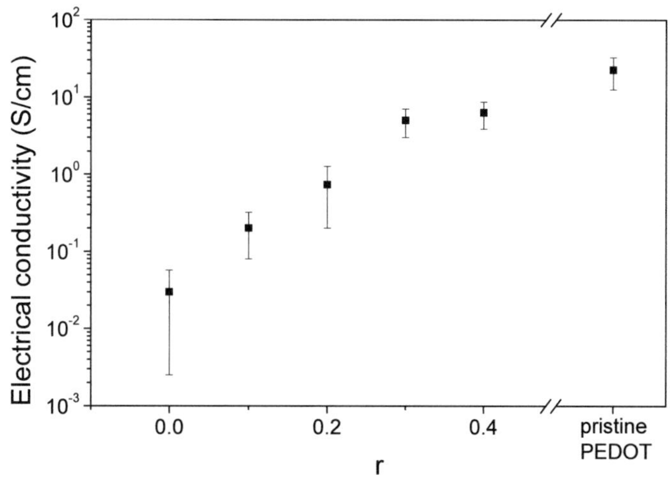

Fig. 10. Electrical conductivities of the pristine PEDOT, P3HT, and PEDOT/P3HT copolymer films

10.2 Water contact angles and RMS surface roughness

Fig. 11 shows the water contact angle and RMS surface roughness of the copolymerized films coated on the glass substrates. Since a liquid makes contact with the outermost molecular layer of a surface, the contact angles represent the chemical and structural differences at the coating surface. Furthermore, the wettability of the thin coating is important in the fabrication of many optoelectronic devices, where multilayers should be fabricated in contact with heterogeneous layers. Since the contact angle of the pristine

PEDOT at 36° is lower than that of the pristine P3HT at 54°, the PEDOT/P3HT copolymer films show a decreasing feature of the contact angle with a higher **r**, which is likely affected by the roughness of the coating since the water contact angle usually decreases with increasing RMS roughness. At a feed ratio of **r** = 0.4, however, the contact angle is slightly lower than the pristine PEDOT. Exhibited by the camera images of coating in the insets of Fig. 11, the color changes gradually from yellow to blue with the increasing EDOT feed ratio of **r**.

Overall, the developed VPP copolymerizing technique clearly demonstrates that the physicochemical properties of PEDOT/P3HT copolymers were successfully adjusted. Since the vapor pressure and polymerization rates of EDOT and 3HT monomers are different, the variation of the copolymer composition was ensured by adjusting the feed ratio of the monomers to the reaction chamber, which resulted in different reactant concentrations of EDOT and 3HT monomers. Different concentrations of reactants may well give different kinetic polymerization rates of EDOT and 3HT to provide a controllable synthesis route of PEDOT/P3HT copolymer coatings. Further study should be performed to identify the detailed copolymerization reaction and chemical structures of the VPP PEDOT/P3HT copolymer system.

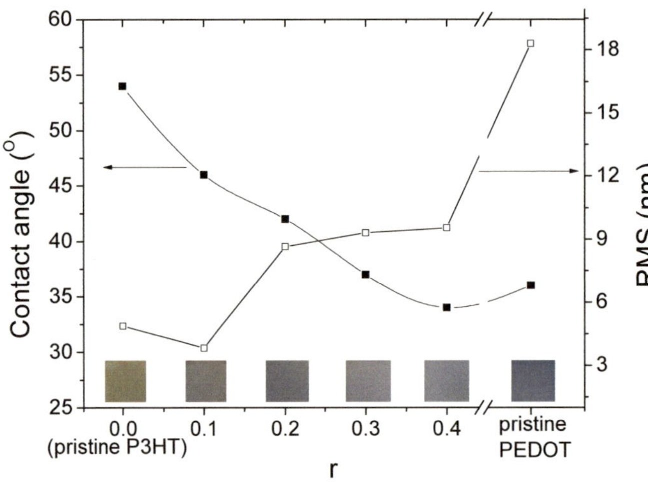

Fig. 11. Water contact angles and RMS surface roughness for differentPEDOT/P3HT films. The inset shows the camera images of PEDOT/P3HT copolymer films with thickness of ca. 100 nm

11. Conclusion

A conducting thin P3HT coating was successfully fabricated on various substrates with a thickness ranging from 50 nm to 1 μm by vapor phase polymerization of a 3HT monomer. The most appropriate Fe(III) solution was found to be 5 % $FeCl_3 \cdot 6H_2O$ in MeOH/EtOH (1:1) providing good conductivity (10^{-2}~10^{-4} S/cm) and transmittance (< 91 %). The RMS roughness of VPP-P3HT was 9.4 nm. The band gap and HOMO of VPP-P3HT were found to be 1.91 eV (650 nm) and 5.07 eV, respectively, which may be suitable for applications in optoelectronic devices. For further study, PEDOT/P3HT copolymer films were also successfully fabricated as thin films on the glass, Si wafer, ITO-covered glass and PET film substrates using a vapor phase polymerization technique. The ratios were kinetically controlled to ensure the tunable properties of bandgap, electrical conductivity, surface morphology and water contact angle of the thin copolymer films. The P3HT thin film incorporated with PEDOT may improve the mobility in OTFTs and charge extraction efficiency in OPVs. In addition to being used as a semiconductor with higher mobility, adding dopants to increase the electrical conductivity, P3HT/PEDOT copolymer films can also be used as a HIL in OLEDs with modified energy levels and properties.

12. References

Admassie, S.; Zhang, F.; Manoj, A.; Svensson, M.; Andersson, M.; Inganas, O. (2006). A polymer photodiode using vapour-phase polymerized PEDOT as an anode *Sol. Energ. Mat. Sol. C.* 90, 133-141.

Bartic, C.; Jansen, H.; Campitelli, A.; Borghs, S. (2002). Ta2O5 as gate dielectric material for low-voltage organic thin-film transistors *Org. Electron.* 3, 65-72.

Chan, H. S. O.; Ng, S. C. (1998). Synthesis, characterization and applications of thiophene-based functional polymers *Prog. Polym. Sci.* 23, 1167.

Chen, T. A.; Wu, X.; Rieke, R. D. (1995). Regiocontrolled Synthesis of Poly(3-alkylthiophenes) Mediated by Rieke Zinc: Their Characterization and Solid-State Properties *J. Am. Chem. Soc.* 117, 233.

Chen, T.; Wu, X.; Rieke, R. (1995). Regiocontrolled Synthesis of Poly(3-alkylthiophenes) Mediated by Rieke Zinc: Their Characterization and Solid-State Properties *J. Am. Chem. Soc.* 117, 233-244.

Chiang, C.K.; Fincher, C.R.; Park, Y.W.; Heeger, A.; Shirakawa, H.; Louis, E.J.; Gau, S.C.; MacDiarmid A.G. (1977). Electrical Conductivity in Doped Polyacetylene *Phys. Rev. Lett.* 39, 1098-1101

Cho, M.; Kim, S.; Nam, J.; Lee, Y. (2008). Preparation of PEDOT/Cu composite film by in situ redox reaction between EDOT and copper(II) chloride *Synthetic Met.* 158, 865-869.

Colladet, K.; Fourier, S.; Cleij, T.; Lutsen, L.; Gelan, J.; Vanderzande, D.; Neugebauer, H.; Sariciftci, S.; Aguirre, A.; Janssen, G. (2007). Low Band Gap Donor−Acceptor Conjugated Polymers toward Organic Solar Cells Applications *Macromolecules* 40, 65-72.

Du Pasquier, A.; Unalan, H. E.; Kanwal, A.; Miller, S.; Chhowalla, M. (2005). Conducting and transparent single-wall carbon nanotube electrodes for polymer-fullerene solar cells *Appl. Phys. Lett.* 87, 203511.

Fabretto, M.; Zuber, K.; Hall, C.; Murphy, P. (2008). High Conductivity PEDOT Using Humidity Facilitated Vacuum Vapour Phase Polymerisation *Macromol. Rapid Comm.* 29, 1403-1409.

Gadisa, A.; Tvingstedt, K.; Admassie, S.; Lindell, L.; Crispin, X.; Andersson, M.; Salaneck, W.; Inganas, O. (2006). Transparent polymer cathode for organic photovoltaic devices *Synthetic Met.* 156, 1102-1107.

Goodman, M. D.; Xu, J.; Wang, J.; Lin, Z. (2009). Semiconductor Conjugated Polymer−Quantum Dot Nanocomposites at the Air/Water Interface and Their Photovoltaic Performance *Chem. Mater.* 21, 934

Grecu, S.; Roggenbuck, M.; Opitz, A.; Brutting, W. (2006). Differences of interface and bulk transport properties in polymer field-effect devices *Org. Electron.* 7, 276-286.

Groenendaal, L.; Jonas, F.; Freitag, D.; Pielartzik, H.; Reynolds, J. (2000). Poly(3,4-ethylenedioxythiophene) and Its Derivatives: Past, Present, and Future *Adv. Mater.* 12, 481-494.

Groenendaal, L.; Zotti, G.; Jonas, G. (2001). Optical, conductive and magnetic properties of electrochemically prepared alkylated poly(3,4-ethylenedioxythiophene)s *Synthetic Met.* 118, 105-109.

Ha, Y.; Nikolov, N.; Pollack, S.; Mastrangelo, J.; Martin, B.; Shashidhar, R. (2004). Towards a Transparent, Highly Conductive Poly(3,4-ethylenedioxythiophene) *Adv. Funct. Mater.* 14, 615-622.

Hatton, R.; Blanchard, N.; Tan, L.; Latini, G.; Cacialli, F.; Silva, S. (2009). Oxidised carbon nanotubes as solution processable, high work function hole-extraction layers for organic solar cells *Org. Electron.* 10, 388-395.

Higgins, A.; Martin, S.; Jukes, P.; Geoghegan, M.; Jones, R.; Langridge, S.; Cubitt, R.; Kirchmeyer, S.; Wehrum, A.; Grizzi, I. (2003). Interfacial structure in semiconducting polymer devices *J. Mater. Chem.* 13, 2814-2818.

Holdcroft, S. J. (1991). Determination of molecular weights and Mark–Houwink constants for soluble electronically conducting polymers *Polym. Sci., Part B.* 29, 1585.

Jang, K.; Eom, Y.; Lee, T.; Kim, D.; Oh, Y.; Jung, H.; Nam, J. (2009). Fabrication of Poly(3-hexylthiophene) Thin Films by Vapor-Phase Polymerization for Optoelectronic Device Applications *ACS Appl. Mater. Interfaces* 1, 1567-1571.

Jang, K.; Kim, D.; Lee, J.; Hong, S.; Lee, T.; Lee, Y.; Nam, J. (2010). Synchronous vapor-phase polymerization of poly(3,4-ethylenedioxythiophene) and poly(3-hexylthiophene) copolymer systems for tunable optoelectronic properties *Org. Electron.* 11, 1668-1675.

Johansson, E. & Larsson, S. (2004). Electronic structure and mechanism for conductivity in thiophene oligomers and regioregular polymer *Synthetic Met.* 144, 183

Jonas, F.; Schrader, L. (1991). Conductive modifications of polymers with polypyrroles and polythiophenes *Synthetic Met.* 41, 831-836.

Kemerink, M.; Timpanaro, S.; De Kok, M.; Meulenkamp, E.; Touwslager, F. (2004). Three-Dimensional Inhomogeneities in PEDOT:PSS Films *J. Phys. Chem. B* 108, 18820-18825.

Kim, D.; Lee, L.; Kang, S.; Jang, K.; Lee, J.; Cho, M.; Nam, J. (2009). In-situ blends of polypyrrole/poly(3,4-ethylenedioxythiopene) using vapor phase polymerization technique *Thin Solid Films* 517, 4156-4160.

Kim, J.; Kim, E.; Won, Y.; Lee, H.; Suh, K. (2003). The preparation and characteristics of conductive poly(3,4-ethylenedioxythiophene) thin film by vapor-phase polymerization *Synthetic Met.* 139, 485-489.

Kim, Y.; Choulis, S.; Nelson, J.; Bradley, D.; Cook, S.; Durrant, J. (2005). Device annealing effect in organic solar cells with blends of regioregular poly(3-hexylthiophene) and soluble fullerene *Appl. Phys. Lett.* 86, 063502.

Lee, T. & Chung, Y. (2008). Control of the Surface Composition of a Conducting-Polymer Complex Film to Tune the Work Function *Adv. Funct. Mater.* 18, 2246-2252.

Lee, T.; Chung, Y.; Kwon, O.; Park, J. (2007). Self-Organized Gradient Hole Injection to Improve the Performance of Polymer Electroluminescent Devices *Adv. Funct. Mater.* 17, 390-396.

Levenspiel, O. (1962). Chemical reaction engineering, John Wiley and Sons, New York, 1962

Li, L.; Huang, Y.; Yan, G.; Liu, F.; Huang, Z.; Ma, Z. (2009). Poly(3,4-ethylenedioxythiophene) nanospheres synthesized in magnetic ionic liquid *Mater. Lett.* 63, 8-10.

Li, Z. L.; Yang, S. C.; Meng, H. F.; Chen, Y. S.; Yang, Y. Z.; Liu, C. H.; Horng, S. F.; Hsu, C. S.; Chen, L. C.; Hu, J. P. (2004). Patterning-free integration of polymer light-emitting diode and polymer transistor *Appl. Phys. Lett.* 84, 3558

Liu, J.; Loewe, R. S.; McCullough, R. D. (1999). Employing MALDI-MS on Poly(alkylthiophenes): Analysis of Molecular Weights, Molecular Weight Distributions, End-Group Structures, and End-Group Modifications *Macromolecules* 32, 5777

McCullough, R. D. (1998). The Chemistry of Conducting Polythiophenes *Adv. Mat.* 10, 93.

Perepichka, I. F.; Perepichka, D. F.; Meng, H.; Wudl, F. (2005). Light-Emitting Polythiophenes *Adv. Mat.* 17, 2281.

Saito, Y.; Kitamura, T.; Wada, Y.; Yanagida, S. (2002). Poly(3,4-ethylenedioxythiophene) as a hole conductor in solid state dye sensitized solar cells *Synthetic Met.* 131, 185-187.

Sarac, A.; Sonmez, G.; Cebeci, F. (2003). Electrochemical synthesis and structural studies of polypyrroles, poly(3,4-ethylene-dioxythiophene)s and copolymers of pyrrole and 3,4-ethylenedioxythiophene on carbon fibre microelectrodes *J. Appl. Electrochem.* 33, 295-301.

Shirakawa, H.; Louis E.; MacDiarmid, A.; Chiang, C.; Heeger, A. (1977). Synthesis of electrically conducting organic polymers: halogen derivatives of polyacetylene, (CH)x *J. S. C., Chem. Commun.* 1977, 578-580

Singh, R.; Kumar, J.; Singh, R.; Kant, R.; Chand, S.; Kumar, V. (2007). Micromorphology, photophysical and electrical properties of pristine and ferric chloride doped poly(3-hexylthiophene) films *Mater. Chem. Phys.* 104, 390-396.

Skotheim, T. A.; Reynolds, J. R. Eds.; Handbook of conducting polymers, CRS press 2007.

Terje A. Skotheim, Handbook of Conducting Polymer Marcel Dekker, New york, 1998.

Truong, T.; Kim, D.; Lee, Y.; Lee, T.; Park, J.; Pu, L.; Nam, J. (2008). Surface smoothness and conductivity control of vapor-phase polymerized poly(3,4-ethylenedioxythiophene) thin coating for flexible optoelectronic applications *Thin Solid Films* 516, 6020-6027.

Wakizaka, D.; Fushimi, T.; Ohkita, H.; Ito, S. (2004). Hole transport in conducting ultrathin films of PEDOT/PSS prepared by layer-by-layer deposition technique *Polymer* 45, 8561-8565.

Welsh, D.; Kumar, A.; Meijer, E.; Reynolds, J. (1999). Enhanced Contrast Ratios and Rapid Switching in Electrochromics Based on Poly(3,4-propylenedioxythiophene) Derivatives *Adv. Mater.* 11. 1379-1382.

Winther-Jensen, B.; Chen, J.; West, K.; Wallace, G. (2004). Vapor Phase Polymerization of Pyrrole and Thiophene Using Iron(III) Sulfonates as Oxidizing Agents *Macromolecules* 37, 5930.

Winther-Jensen, B.; Chen, J.; West, K.; Wallaces, G. (2004). Vapor Phase Polymerization of Pyrrole and Thiophene Using Iron(III) Sulfonates as Oxidizing Agents *Macromolecules* 37, 5930-5935.

Winther-Jensen, B.; Forsyth, M.; West, K.; Andreasen, J.; Wallace, G.; MacFarlane, D. (2007). High current density and drift velocity in templated conducting polymers *Org. Electron.* 8, 796-800.

Winther-Jensen, B.; West, K. (2004). Vapor-Phase Polymerization of 3,4-Ethylenedioxythiophene: A Route to Highly Conducting Polymer Surface Layers *Macromolecules* 37, 4538-4543.

Xu, J.; Wang, J.; Mitchell, M.; Mukherjee, P.; Jeffries-EL, M.; Petrich, J. W.; Lin, Z. (2007). Organic−Inorganic Nanocomposites via Directly Grafting Conjugated Polymers onto Quantum Dots *J. Am. Chem. Soc.* 129, 12828

Xu, Y.; Wang, J.; Sun, W.; Wang, S. (2006). Capacitance properties of poly(3,4-ethylenedioxythiophene)/polypyrrole composites *J. Power Sources* 159, 370-373.

Zhan, L.; Song, Z.; Zhang, J.; Tang, J.; Zhan, H.; Zhou, Y.; Zhan, C. (2008). PEDOT: Cathode active material with high specific capacity in novel electrolyte system *Electrochim. Acta* 53, 8319-8323.

Part 2

Nanostructures: Properties and Applications

9

ZnO Nanostructures for Optoelectronic Applications

Ashok K. Sood[1], Zhong Lin Wang[2], Dennis L. Polla[3], Nibir K. Dhar[3], Tariq Manzur[4] and A.F.M. Anwar[5]

[1]*Magnolia Optical Technologies Inc, 52-B Cummings Park, Suite 314, Woburn, MA 01801*

[2]*School of Materials Science and Engineering, Georgia Institute of Technology, 771 Ferst Drive, Atlanta, GA 30332*

[3]*DARPA/MTO, 3701 North Fairfax Drive, Arlington, VA 22203*

[4]*Naval Underwater Warfare Center, 1176 Howell Street, Newport, RI 02841*

[5]*Department of Electrical Engineering, University of Connecticut, Storrs, CT 06269 U.S.A.*

1. Introduction

ZnO is a unique material that exhibits both semiconducting and piezoelectric properties. ZnO is a unique material that exhibits both Semi conducting and piezoelectric properties. ZnO devices have been demonstrated for applications in piezoelectric pressure sensors and Pyroelectric infrared detectors [1] and Spintronic devices [2]. More recently, there has been significant effort underway for design and development of ZnO nanostructures such as ZnO nanowires for a variety of applications [3-7].

The ZnO nanostructures can be implemented in Optoelectronic, Sensors, Transducers and Biomedical applications [1, 2, 3, 4]. Use of these nanostructures, will allow building of Nanoscale nanosensors, nanocantilevers, field-effect transistors and nanoresonators for a variety of Military, Homeland Security and Commercial Applications. Due to the advancement of materials technology over the past decade, wide-band gap semiconductors such as SiC, GaN and ZnO have emerged as UV sensitive materials that have applications for UV lasers, UV Photodetector, switches, Bio-Sensors and solar cells.

ZnO wide-band gap semiconductor is promising for sensor applications in the UV range. The band-gap is 3.2 eV for ZnO. Therefore, GaN and ZnO, as well as SiC are potentially good materials to cover the UV spectral band (240-280 nm), when solar radiation is completely absorbed by the ozone layer of the earth atmosphere, so the background of solar radiation at the earth surface is essentially zero.

ZnO is transparent to visible light and can be made highly conductive by doping. ZnO is a versatile functional material that has a diverse group of growth morphologies. These growth morphologies have been demonstrated for nanowires (1), nanobelts (2), nanocages (3), nanocombs (4), nanosprings (5), nanorings (6), nanohelixes (7). The objective of this chapter is to review the unique ZnO nanostructure devices and characterized for optoelectronic applications.

2. Crystal structure of ZnO and its polar surfaces

Table 1 lists the basic physical properties of bulk ZnO. It is worth noting that as the dimension of the semiconductor materials continuously shrinks down to nanometer or even smaller scale, some of their physical properties undergo changes known as the "quantum size effects" [8]. For example, quantum confinement increases the band gap energy of quasi-one-dimensional (Q1D) ZnO, which has been confirmed by photoluminescence [8].
Bandgap of ZnO nanoparticles also demonstrates such size dependence. Understanding the fundamental physical properties is crucial to the rational design of functional devices. Investigation of the properties of individual ZnO nanostructures is essential for developing their potential as the building blocks for future nanoscale devices.

Properties	Value
Lattice constants (T = 300 K)	
a_0	0.32469 nm
c_0	0.52069 nm
Density	5.606 g/cm^3
Melting point	2248 K
Relative dielectric constant	8.66
Gap Energy	3.4 eV, direct
Intrinsic carrier concentration	$< 10^6$ cm^{-3}
Exciton binding Energy	60 meV
Electron effective mass	0.24
Electron mobility (T = 300 K)	200 cm^2/V s
Hole effective mass	0.59
Hole mobility (T = 300 K)	5-50 cm^2/V s

Table 1. Physical properties of wurtzite ZnO [Fan et al; Ref 8]

Zinc oxide has a hexagonal structure (space group $P6_3mc$) with lattice parameters a = 0.3296, and c = 0.52065 nm. The structure of ZnO can be simply described as a number of alternating planes composed of tetrahedral coordinated O^{2-} and Zn^{2+} ions, stacked alternatively along the c-axis (Figure 1a). The tetrahedral coordination in ZnO results in non-central symmetric structure and piezoelectricity. Another important characteristic of ZnO is the polar surfaces. The most common polar surface is the basal plane. The oppositely charged ions produce positively charged Zn-(0001) and negatively charged O-($000\bar{1}$) polar surfaces (Figure 1b), resulting in a normal dipole moment and spontaneous polarization along the c-axis.
Another polar surface is the {$01\bar{1}1$}. By projecting the structure along [$1\bar{2}10$], as shown in Figure 1b, beside the most typical ±(0001) polar surfaces that are terminated with Zn and oxygen, respectively, ±($10\bar{1}1$) and ±($10\bar{1}\bar{1}$) are also polar surfaces. The {$10\bar{1}1$} type surfaces are not common for ZnO, but they have been observed in a nanohelical structure found recently [9]. The charges on the polar surfaces are ionic charges, which are non-transferable and non-flowable. Because the interaction energy among the charges depends

on the distribution of the charges, the structure is arranged in such a configuration to minimize the electrostatic energy. This is the main driving force for growing the polar surface dominated nanostructures.

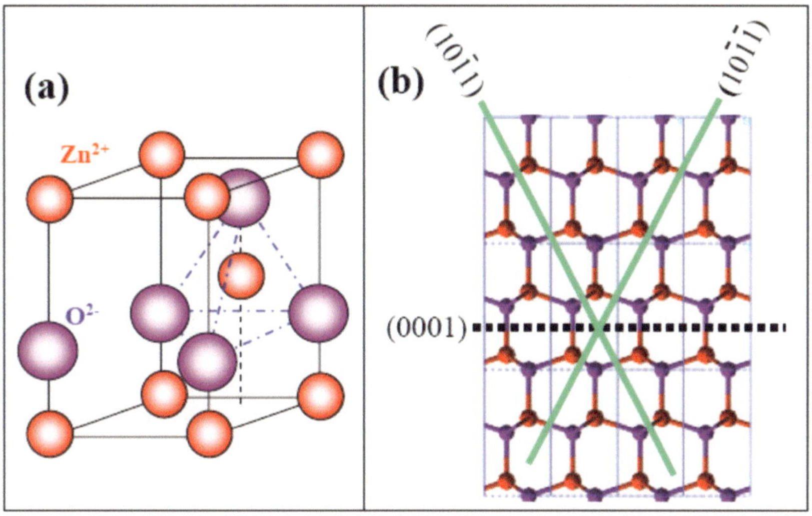

Fig. 1. (a) Wurtzite structure model of ZnO. The tetrahedral coordination of Zn-O is shown. (b) The structure model of ZnO projected along $[2\bar{1}\bar{1}0]$, displaying the ±(0001), $\pm(01\bar{1}\bar{1})$ and $\pm(01\bar{1}1)$ polar surfaces

3. Growth of ZnO nanostructures

Growth of ZnO nanostructures has been carried out by several groups using the techniques that are still being developed for their reliability and throughput. These approaches are listed in three broad categories. These include:

1. Vapor Phase Deposition Technique (Solid-Vapor Process) [10]
2. Chemical Synthesis Method [10]
3. Induction Heating Method [11]
4. Metal-Organic Chemical Vapor Deposition (MOCVD) [12,13]

There are several processing parameters such as temperature, pressure, carrier gas (including gas species and its flow rate), substrate and evaporation period, which can be controlled and need to be selected properly before and/or during the thermal vaporization [10]. The source temperature selection mainly depends on the volatility of the source material(s). Usually, it is slightly lower than the melting point of the source material. The pressure is determined according to the evaporation rate or vapor pressure of the source material(s). The schematic representation of the Vapor Phase Deposition technique is shown in figure 2.

The substrate temperature usually drops with the distance of its location from the position of the source material(s). The local temperature determines the type of product that will be obtained. It is also noted that the thermal evaporation process is very sensitive to the concentration of oxygen in the growth system. Oxygen influences not only the volatility of the source material(s) and the stoichiometry of the vapor phase, but also the formation of the product(s).

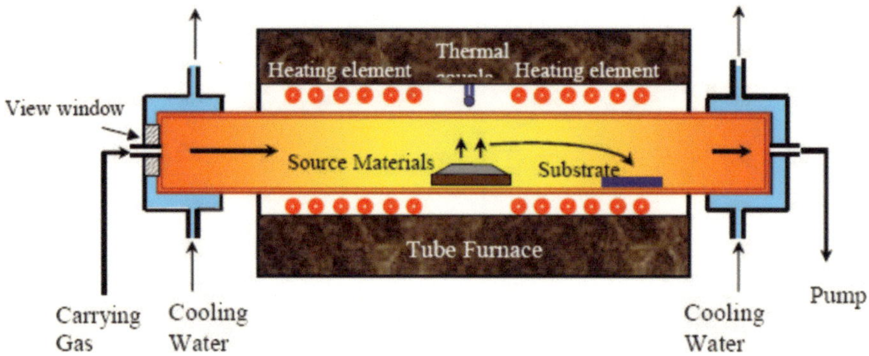

Fig. 2. A schematic diagram of the experimental apparatus for growth of oxides nanostructures by the solid-vapor phase process [Wang et al; Ref 10]

The Second Growth approach for ZnO nanowires uses the chemical synthesis method [10]. This is a hydrothermal method, where the growth procedure for growth of ZnO nanowires is as follows:1). Suspend the surface modified substrate in a Pyrex glass bottle filled with an equal molar aqueous solution of zinc nitrate hydrate (Zn(NO3)2.6H2O) and hexamethylenetetramine (C6H12N4) at temperatures between 60-90 C. 2). The temperature, PH, solution concentration, reaction time (a range of 1-72 hours) and substrate surface status were optimized for growing Nanowire arrays with controlled dimensions and orientation. 3). after completion of the reaction, the substrates were removed from the solution, and rinsed by de-ionized water, and dried in air at 65 C.

The third approach uses the inductive heating assisted fast synthesis of ZnO nanowires using ZnO/graphite solid source powder in a room temperature environment [13]. The internal heat generation induced by the alternating magnetic field at the synthesis specimen enables the fast temperature transition for ZnO nanowire growth, with a synthesis time less than 5 min compared to conventional methods. Furthermore, this demonstration illustrates the feasibility of a simple and fast nanoscale synthesis using inductive heating for nanomaterials synthesis [13].

Figure 3a shows the schematic illustration for the ZnO nanowire synthesis setup. The synthesis specimen is placed inside a quartz tube underneath the center of an eight-turn inductive coil with a pitch of 3.25 mm and an inner/outer diameter of 12.7/ 19.2 mm. The distance between the coil base and the synthesis specimen is - 6 mm.

The cross sectional view of the synthesis specimen, including a nickel coated heating chip, source powder, and growth chip inside a ceramic boat, is shown in Figure 3 b. When an alternating current is applied in the coil, an alternating magnetic field is generated, which induces eddy currents in the nickel layer and provides rapid Joule heating for nanowire synthesis. Analytically, one can derive each coil's magnetic field intensity generated at any point with a scalar distance of d with respect to the coil loop center from Biot-Savart's law [Lin et al 2007; Ref 11].

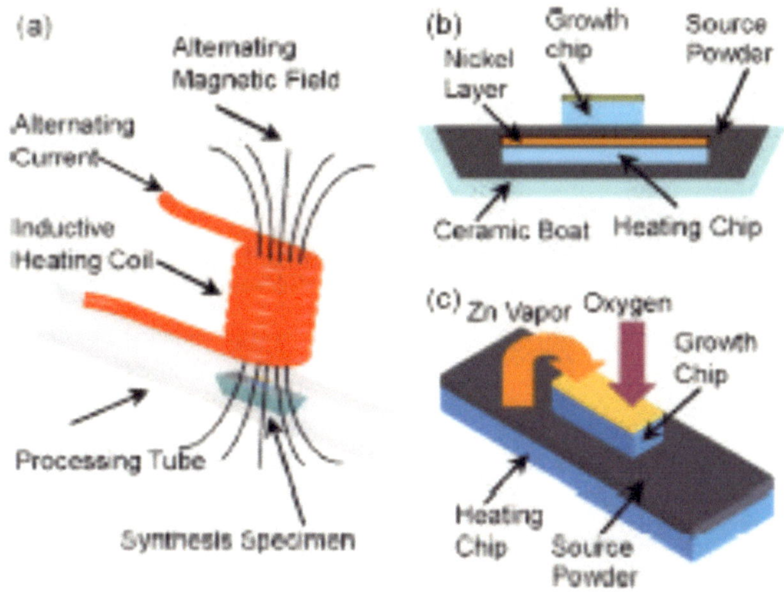

Fig. 3. Schematic illustration for ZnO nanowire synthesis setup using induction heating method (a). Cross sectional view of the synthesis specimen consisting of a heating chip, source powder, and growth chips inside a ceramic boat (b). Three-dimensional sketch of the heating chip/source powder/growth (c) [Lin et.al; Ref 11]

Several groups are have used Metal-Organic Chemical Vapor Deposition (MOCVD) successfully to grow ZnO nanostructures and have shown good results [12, 13]. Further work is underway to develop this technology for oriented ZnO nanowires for optoelectronic applications.

4. Oriented nanostructures growth

For the first two growth approaches, selectively patterned substrates have been used to control the density and selectivity of the aligned growth. The patterned materials can be catalysts (Au, Ni, etc.). The patternings methods are Electron-beam Lithography and Focused ion beam Microscopy.

Figure 4 presents a standard process for E-beam lithography. First step is to spin-coat E-beam resist PMMA of desired thickness layer on the substrate, and then using electron beam to selectively expose the patterned area of PMMA coating. Next step is to develop the exposed pattern using MIBK solution. After this step, the PMMA coating on the Si substrate is left with a pattern, which is subsequently filled by metal deposition, and the remaining PMMA can be lifted off in acetone.

This patterned substrate can be used for growth of ZnO nanowires. They are currently developing the lift-off process to fabricate the substrates. Once the process is refined on Silicon substrates, it can be further implemented on other substrates [10].

Figure 5 presents Au-pattern on Si substrate. We have recently achieved the Au-pattern using E-beam lithography. The patterned gold dot array is 2 um in diameter, 2 um in spacing, and 200 nm in height.

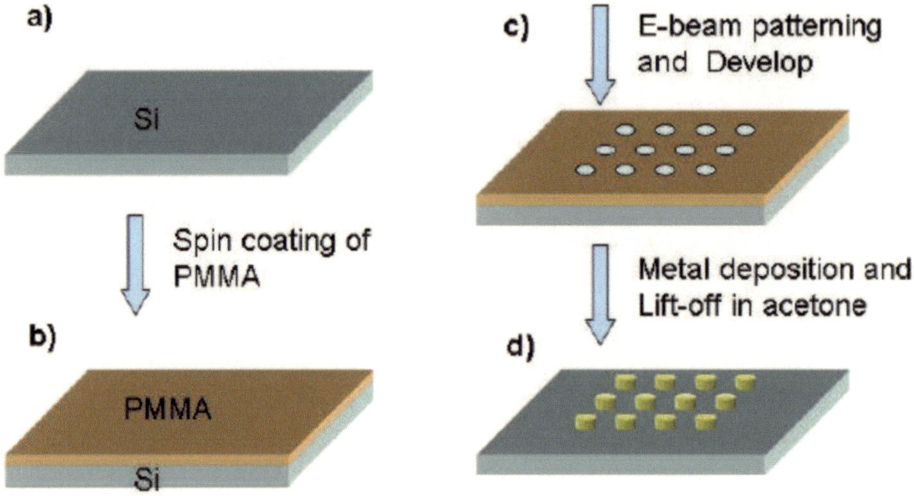

Fig. 4. E-beam Lithography Process Flow for growth of ZnO Nanowires [Wang et al; Ref 10]

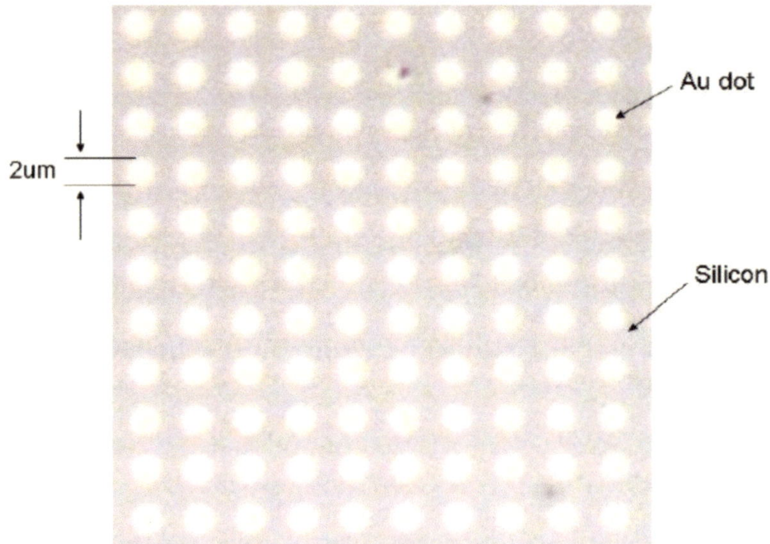

Fig. 5. Au Pattern on Silicon substrate with 20 micron size [Wang et al; Ref 10]

The hydrothermal synthesis method has been very successful in terms of synthesizing large scale nanowires arrays on Silicon substrates. As shown in figure 6a is a side view SEM image of the as grown ZnO nanowires arrays on Au-coated Si substrate, the alignment is very good. The zoom-in image in the inset displayed the dimensions of the grown nanowire, ~200 nm in diameter, ~ 5 um in lengths.

Figure 6b shows the top view of aligned nanowire arrays; the top-right inset enlarged picture clearly illustrated the hexagonal cross-section of the nanowires. The bottom-right CCD image is showing a 2-inch diameter sized area of nanowires arrays grown on an Au-coated substrate.

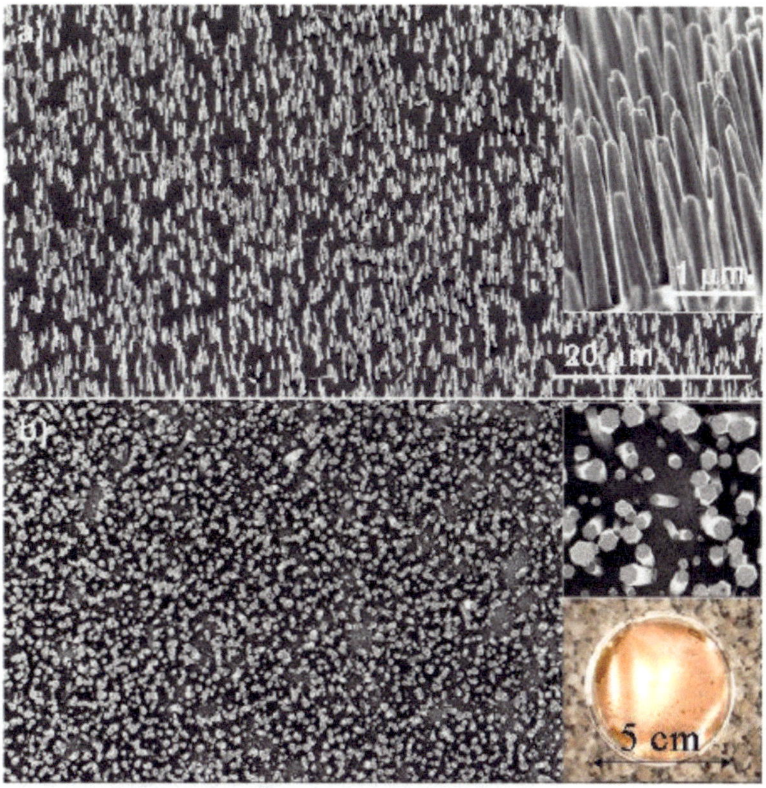

Fig. 6. ZnO Nanowires grown on Silicon Substrate [Wang et al; Ref 10]

Besides Au, they have also evaluated the ZnO seeds for promoting growth of ZnO nanowires using hydrothermal synthesis. As shown in figure 6a, a low-magnification SEM image showing the growth of ZnO nanowires on ZnO seeds coated (30 nm in thickness) PMMA film. The wavy feature is due to the non-uniform PMMA coating surface. The enlarged top view was revealed in figure 6b, witnessing the randomly grown nanowires with sharp tips ~20 nm in diameter. The EDS analysis in figures 6c-d revealed the main composition of ZnO with some impurities. They have identified the impurity to be Chromium (Figure 6d).

Density controls of nanowires growth are also being studied as an important parameter for the UV detector and Sensor application. As shown in figure 7, a typical dispersive nanowires array on Au-coated Si substrate. The wire diameter is about 1-2 um (figure 7 b), the length can be as long as 30 um, and the spacing in between nanowires are ~5-15 um. Figure 7c revealed the additional growth of nanoscale ZnO dots on the Si substrate corresponding to the densely distributed nanodots in figure 7b, which might be the seeds (nuclei) of ZnO nanowires growth.

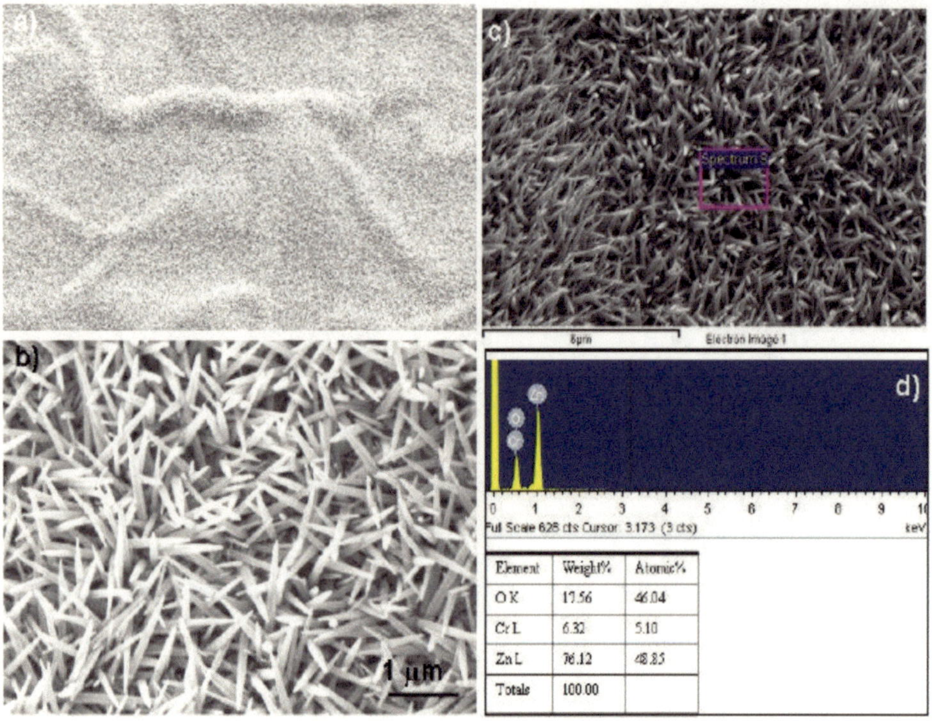

Fig. 7. ZnO nanowire growth on ZnO coated substrates [Wang et al; Ref 10]

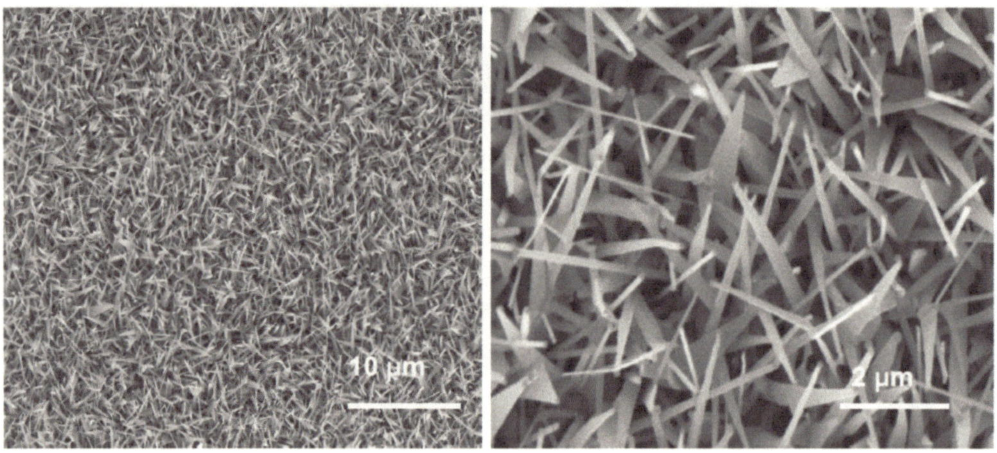

Fig. 8. ZnO nanowires grown on SiC (0001) substrates with 2 nm Au at 960 C [Wang et al; Ref 10]

They have also evaluated growth of ZnO nanowires on SiC substrates. SiC being a wide bandgap semiconductor can be used for active device fabrication. For this effort, we have used SiC (0001) substrates. We have used thermal evaporation method using ZnO and C as source material. Au, 2-4 nm thick was used as a catalyst for the growth. We also used Argon and Oxygen as the carrier gases. Figure 8 presents the ZnO nanowires and nanosheets on SiC (0001) coated with 2 nm gold. The growth was carried out at 960 C, using Argon and Oxygen. The growth was carried out for 30 minutes.

The results in figures 8 show their attempts to grow the ZnO nanowires. The process is still being optimized for oriented growth with different growth temperatures and gas flow along with various catalysts. These nanowires show that high density growth of ZnO nanowires is quite feasible and need further growth optimization for oriented growth. Figure 9 presents the TEM characterization of ZnO nanowires grown on SiC (0001) substrate with Au catalyst. This can open a variety of device configurations for optical and other applications using ZnO nanowires.

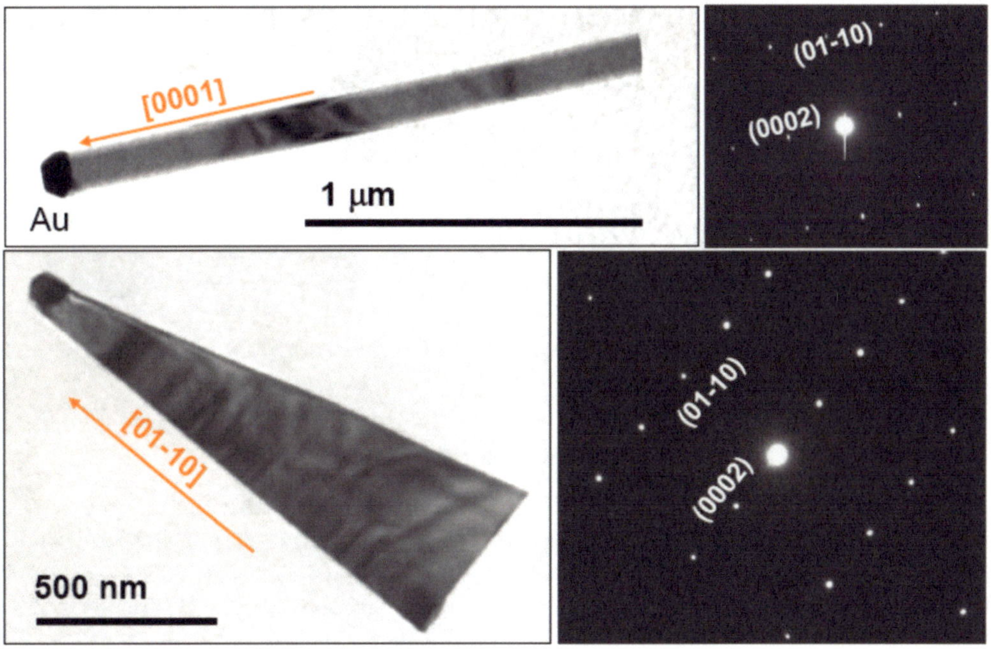

Fig. 9. TEM characterization of ZnO nanowires grown on SiC (0001) substrates shown in figure 8.The orientation of ZnO nanowires with two orientations are illustrated [Wang et al; Ref 10]

Their work has shown that Sn can also be an effective catalyst for growing ZnO nanorods [14]. Using the epitaxial growth of ZnO nanorods on ZnO crystal, aligned nanorods have been grown as shown in figure 10. In this case, the growth direction of the nanorods is led by the Sn catalyst and the epitaxial orientation is defined by the substrate that determines the aligned growth. The nanorods are also aligned in orientation because they tend to take the least mismatch orientation on the substrate, to reduce the interface mismatch energy.

The choice of substrate is important for the epitaxial growth. One must consider the crystallographic structure as well as the surfaces to be used for the growth. We have found recently that the atomic termination and charge status of the substrate can strongly affect the morphology of the grown nanorods [14]. The nanorods growing out of Zn-terminated (0001) ZnO substrates are very different from those grown out of the oxygen-terminated ($000\bar{1}$) ZnO substrates although the same type of Sn catalyst was used.

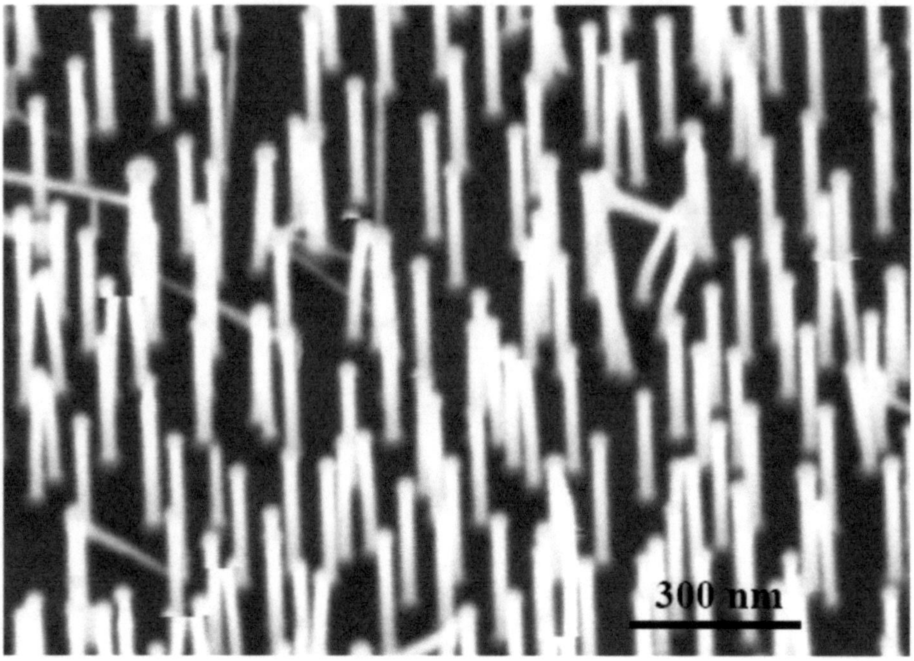

Fig. 10. Aligned ZnO nanorods/nanowires epitaxially grown on ZnO substrate using Sn as a catalyst [Gao et. al; Ref 15]

5. ZnO nanowires properties and optical device applications

ZnO exhibits a direct band-gap of 3.37 eV at room temperature with large exciton energy of 60 meV. The strong exciton binding energy, which is much larger than that of GaN (25 meV), and the thermal energy at room temperature (26 meV) can ensure an efficient exciton emission at room temperature under low excitation energy. As a consequence, ZnO is recognized as a promising photonic material in the blue–UV region.

Single-crystalline ZnO nanowires have been synthesized using high temperature VLS growth methods discussed above [10, 16]. Room temperature UV lasing in ZnO nanowires has been demonstrated [17]. PL spectra of the ZnO nanorods were measured with a fluorescence spectrophotometer using a Xe lamp with an excitation wavelength of 325 nm at room temperature. Figure 11 shows the PL spectrum of the ZnO nanorods with a diameter in the range 60 ± 80 nm. Three emitting bands, including a strong ultraviolet emission at around 386 nm and a very weak blue band (440 ± 480 nm) as well as an almost negligible green band (510 ± 580 nm), were observed.

The UV emission must be contributing to the near band edge emission of the wide band-gap ZnO. It has been suggested that the green band emission corresponds to the singly ionized oxygen vacancy in ZnO and results from the recombination of a photo-generated hole with the singly ionized charge state of this defect. The stronger the intensity of the green luminescence, the more singly ionized oxygen vacancies there are. Thus the almost negligible green band in figure 11 shows that there is a very low concentration of oxygen vacancies in the well-aligned ZnO nanorods. The observation of blue band emission (440 ± 480 nm) of ZnO film has also been reported, using cathodoluminescence.

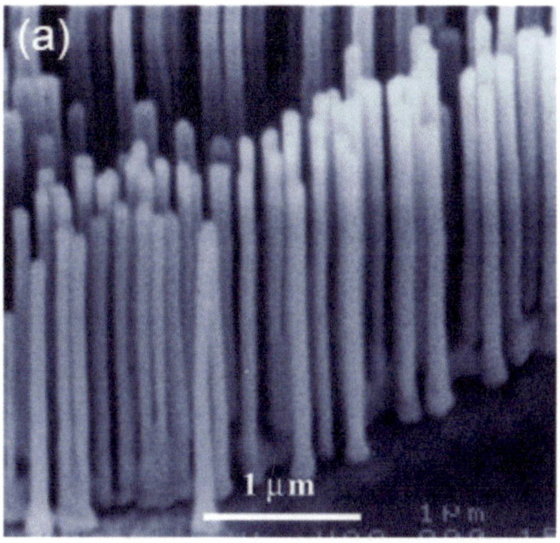

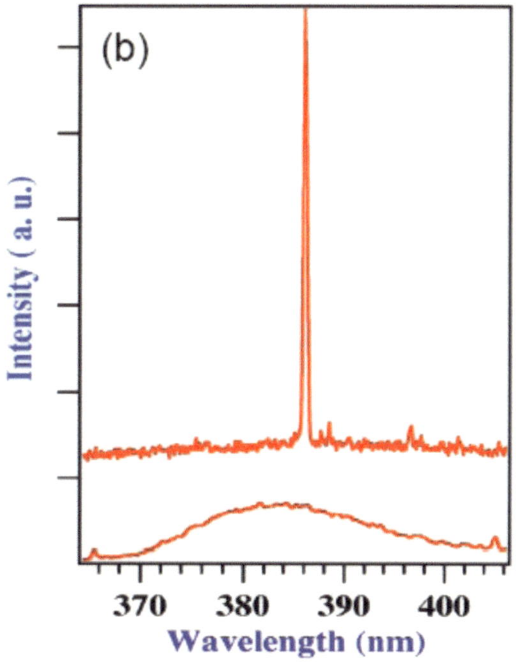

Fig. 11. (a) An SEM image of aligned ZnO nanowire arrays grown on sapphire substrate coated with a 3 nm thick layer of Au. (b) Emission spectra from nanowire arrays at pump powers of 20 and 100 kW cm–2. The spectra are offset for easy comparison. Stimulated emission from the nanowires was collected in the direction along the nanowire end-plane normal (the symmetric axis) with a monochromator. All experiments were carried out at room temperature [P.D, Yang et al.; Ref 17]

6. ZnO nanowire field effect transistor

Field effect transistors have been fabricated using individual nanobelts [18]. Large bundles of ZnO nanobelts were dispersed in ethanol by ultrasonication until mostly individual nanobelts were isolated. These ethanol dispersions were dried onto a SiO_2/Si substrate for imaging by non-contact mode atomic force microscopy (AFM). Field effect transistors were fabricated by depositing dispersed ZnO nanobelts on predefined gold electrode arrays. The SiO_2 gate dielectric thickness was 120 nm and the back gate electrode was fabricated by evaporation of gold on the Si (p+) side of the substrate.

In addition, in both fabrication schemes, the electrode arrays were variably spaced. They included electrode gaps as small as 100 nm and as large as 6 μm. An AFM image of the field effect transistor (FET) and a schematic diagram are shown in figure 12. The principle of this device is that controlling the gate voltage would control the current flowing from the source to the drain.

By forming metal electrode/nanostructure electrical contacts and capacitively coupling the nanostructure to a nearby gate electrode, an FET is produced using a nanobelt that allows the exploration of new aspects of the physical and chemical properties of the nanostructures [19, 20. 21].

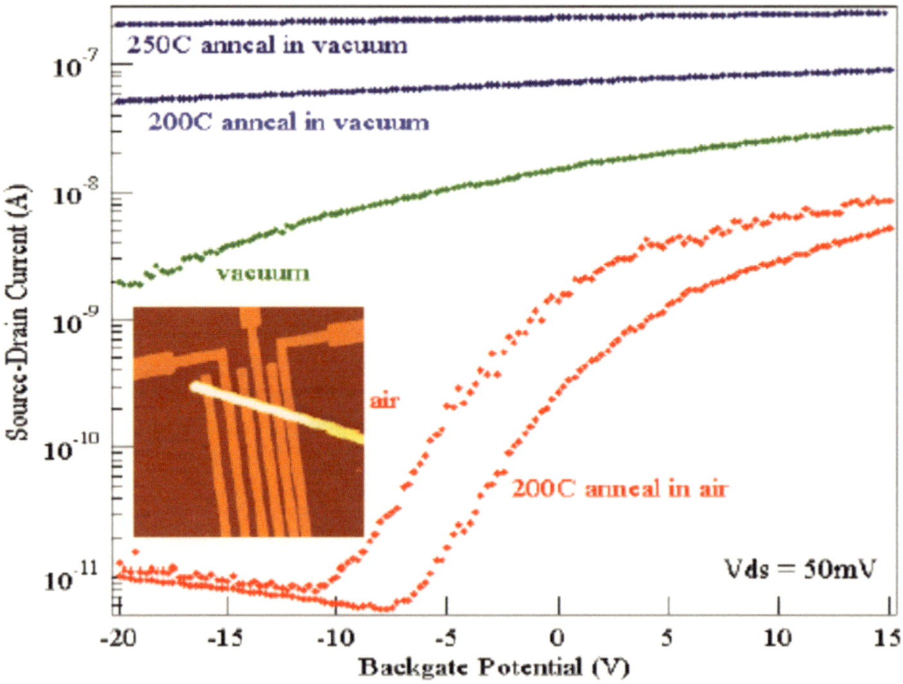

Fig. 12. ZnO nanobelt FET. Source–drain current versus gate bias for a SnO_2 FET after various treatments measured in this order: air, vacuum, 200 C vacuum anneal, 250 C vacuum anneal, 200 C air anneal. The inset shows a field effect transistor (FET) device made using a single nanobelt. [Wang et al; Ref 22]

The alternative way of contacting the nanobelts by simply depositing them on top of pre-fabricated gold electrodes led to very resistive contacts. A typical ZnO field effect transistor showed a gate threshold voltage of −15 V, a switching ratio of nearly 100 and a peak conductivity of $1.25 \times 10^{-3}\ \Omega\ cm^{-1}$. A completely analogous behavior has been observed in the case of carbon nanotubes deposited on top of Au electrodes or covered by Ti electrodes [22].

7. UV photoconductivity measurements on ZnO nanostructures

Ultraviolet light irradiation of the nanobelt diode of ZnO in air is observed to result in a significant increase of the conductivity [Figure. 13]. Light with a wavelength of 350 nm (E_λ = 3.54eV) was used, exceeding the direct band-gap of ZnO. The increase in the conductivity results from both photo- generation of electron-hole pairs as well as doping by UV light-induced surface desorption. These processes could be observed by introducing a shutter between the light source and the ZnO nanobelt so that the flux of UV photons could be turned 'ON' and 'OFF'.

The unique geometrical shape of nanobelts is ideal for field emission. MoO_3 nanobelts have been shown to exhibit superior performance. The work function at the tips of individual ZnO nanobelts has been measured by a novel technique [23].

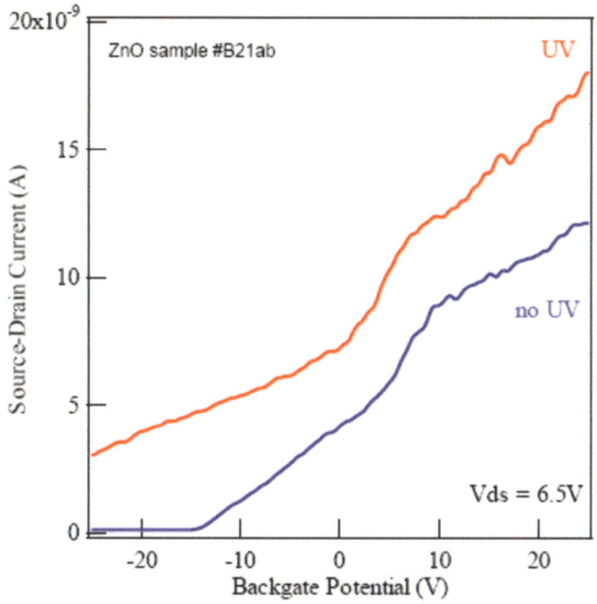

Fig. 13. Source-drain current measured from a ZnO nanobelt FET, showing a strong dependence on UV illumination [Wang et al; Ref 22]

8. ZnO nanowire piezoelectric nanogenerator

Wireless devices may allow in-situ, real-time biomedical monitoring and detection, but such devices still requires a power source. Ideally, such devices should be self-powered rather use a battery. The body provides numerous potential power sources - mechanical energy, vibrational energy, chemical energy (glucose), and hydraulic energy, but the challenge is their efficient conversion into electric energy. If accomplished on the nanoscale such power sources could have greatly reduced the size of the integrated nanosystems for optoelectronics [24], biosensors [25], resonators [26] and more. They have demonstrated an approach of converting mechanical energy into electric power using aligned ZnO NWs [27]. Their study is based on aligned ZnO NWs grown on c-plane oriented -Al_2O_3 substrate covered by a layer of ZnO film (Figure 14 a). The measurements were performed by AFM using a Si tip coated with Pt film. In the AFM contact mode, a constant normal force of 5 nN was maintained between the tip and sample surface. The tip scanned over the top of the ZnO NW, and the tip's height was adjusted according to the surface morphology and local contacting force. The output voltage across an outside load of resistance R_L = 500 MΩ was continuously monitored as the tip scans over the NWs (note the defined polarity of the voltage signal) (Figure 14b). No external voltage was applied in any stage of the experiment. Experimentally, both the topography (feedback signal from the scanner) and the corresponding output voltage (V_L) images across the load were recorded simultaneously when the AFM tip was scanned over the aligned NW arrays. In the voltage output image, many sharp output peaks (like discharge peaks) were observed, which are typically about 4 to 50 times higher than the noise level (Figure 14c). The location of the voltage peak is directly registered at the site of the NW.

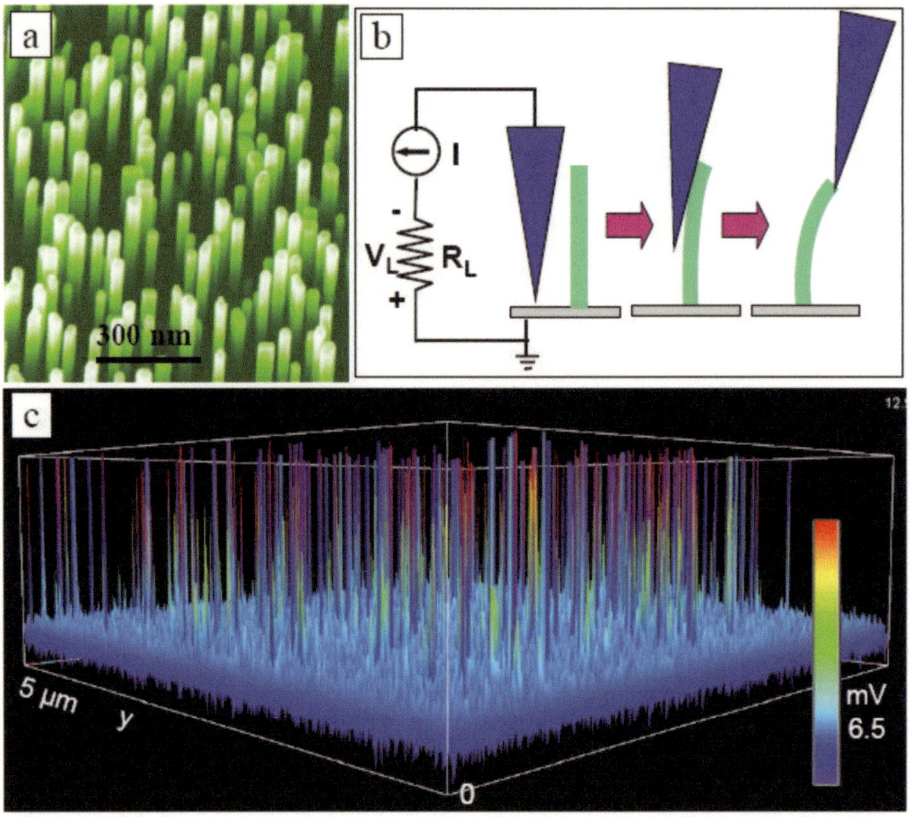

Fig. 14. (a) SEM images of aligned ZnO NWs grown on α-Al_2O_3 substrate. (b) Experimental set up and procedures for generating electricity by deforming a piezoelectric NW using a conductive AFM tip. The root of the NW is grounded, and an external load of R_L = 500 MΩ is applied, which is much larger than the inner resistance R_I of the NW. The AFM scans across the NW arrays in contact mode. (c) Output voltage image of the NW arrays when the AFM tip scans across the NW arrays. The discharging process is so quick that each discharge event is characterized by only a couple of data points. This gives a difficulty of displaying the data by rainbow color. [Wang et al; Ref 27]

The physical principle for creating, separating and preserving the PZ charges in the NW is a coupling of piezoelectric (PZ) and semiconducting properties [27, 28]. For a vertical, straight ZnO NW (Figures 15a), the deflection of the NW by AFM tip creates a strain field, with the outer surface being stretched (positive strain) and inner surface compressed (negative strain) (Figures 15b).

An electric field E_z along the NW (z direction) is then created inside the NW volume through the PZ effect, $E_z = \square_z / d$, where d is the PZ coefficient[29] along the NW direction that is normally the positive c-axis of ZnO, with the Zn atomic layer being the front terminating layer [30]. The PZ field direction is closely parallel to the z-axis (NW direction) at the outer surface and anti-parallel to the z-axis at the inner surface (Figures 15c). Under the first-order approximation, across the width of the NW at the top end, the electric potential distribution from the compressed to the stretched side surface is approximately between V_s^- (negative) to V_s^+ (positive).

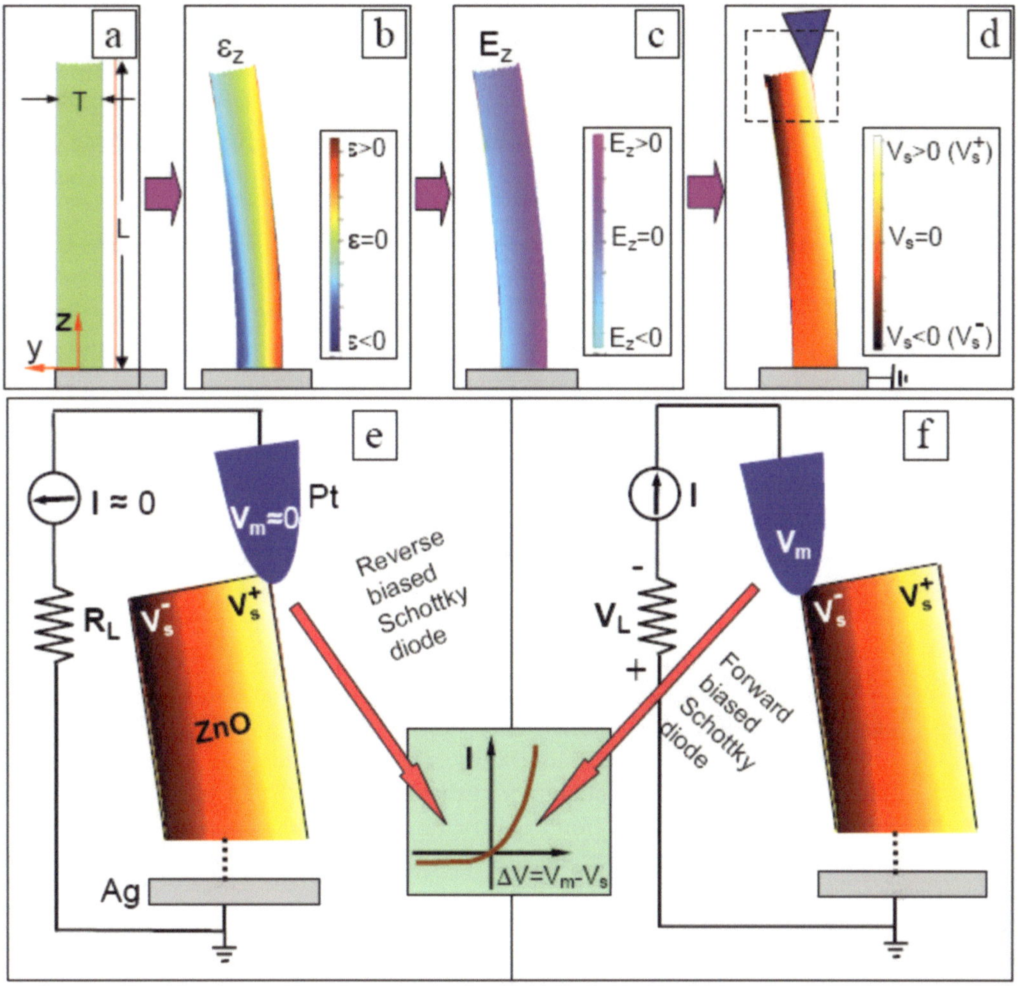

Fig. 15. Physical principle of the observed power generation process of a piezoelectric ZnO NW, showing a unique coupling of piezoelectric and semiconducting properties in this metal-semiconductor Schottky barrier governed transport process. (a) Schematic definition of a NW and the coordination system. (b) Longitudinal strain ε_z distribution in the NW after being deflected by an AFM tip from the side. The data was simulated by FEMLAB for a ZnO NW of length 1 μm, aspect ratio of 10. (c) The corresponding longitudinal piezoelectric induced electric field E_z distribution in the NW. (d) Potential distribution in the NW as a result of piezoelectric effect. The dashed box indicates the area to be examined in details in (e, f). (e, f) Metal and semiconductor contacts between the AFM tip and the semiconductor ZnO NW at two reversed local contact potentials (positive and negative), showing reverse and forward biased Schottky rectifying behavior, respectively (see text). It is this oppositely biased Schottky barrier across the NW makes it possible to preserve the piezoelectric charges and later produce the discharge output. Inset is a typical I-V characteristic of a metal-semiconductor (n-type) Schottky barrier. The process in (e) is built up the potential; the process in (f) is to discharge the potential. [Wang et al; Ref 27, 28]

The electrode at the base of the NW is grounded. Note the V_s^+ and V_s^- are the voltages produced by PZ effect, respectively. The potential is created by the relative displacement of the Zn^{2+} cations with respect to the O^{2-} anions due to PZ effect in the wurtzite crystal structure; thus, these ionic charges cannot freely move and cannot recombine without releasing the strain (Figures 15d). The potential difference is maintained as long as the deformation is in place and no foreign free charges (such as from the metal contacts) are injected.

We now consider the discharge process. In the first step, the AFM conductive tip that induces the deformation is in contact with the stretched surface of positive potential V_s^+ (Figure 15d and e). The Pt metal tip has a potential of nearly zero, $V_m = 0$, so the metal tip - ZnO interface is negatively biased for $\square V = V_m - V_s^+ < 0$. With consideration the n-type semiconductor characteristic of the as-synthesized ZnO NWs, the Pt metal- ZnO semiconductor (M-S) interface in this case is a reverse-biased Schottky diode (Fig. 40e), and little current flows across the interface. In the second step, when the AFM tip is in contact with the compressed side of the NW (Figure 415f), the metal tip - ZnO interface is positively biased for $\square V = V_L = V_m - V_s^- > 0$.

The M-S interface in this case is a positively biased Schottky diode, and it produces a sudden increase in the output electric current. The current is the result of $\square V$ driven flow of electrons from the semiconductor ZnO NW to the metal tip. The flow of the free electrons from the loop through the NW to the tip will neutralize the ionic charges distributed in the volume of the NW and thus reduce the magnitudes of the potential V_s^- and V_s^+.

The principle and technology demonstrated here have the potential of converting mechanical movement energy (such as body movement, muscle stretching, blood pressure), vibration energy (such as acoustic/ultrasonic wave) [74], and hydraulic energy (such as flow of body fluid, blood flow, contraction of blood vessel, dynamic fluid in nature) into electric energy that may be sufficient for self-powering nanodevices and nanosystems.

The technology could have important applications in wireless self-powered nanodevices by harvesting energy from the environment. It also provides a method for indirectly charging of a battery. It is also possible to fabricate large-power output electric generator by using arrays of ZnO wires/belts, which can be grown on substrates such as metal foils, flexible organic plastic substrates [31], ceramic substrates (such as alumina) and compound semiconductors (such as GaN and AlN). The nano-generator could be the basis for exploring new self-powering technology for in-situ, real-time and implantable Biosensing, biomedical monitoring and biodetection. It could have the potential of solving key energy requirement for remote sensing and actuating.

9. Ultrasonic wave driven nanogenerators using ZnO nanowires

For practical applications, we must make an innovative design to drastically improve the performance of the Nanogenerators in following aspects. First, we must eliminate the use of AFM for making the mechanical deformation of the NWs so that the power generation can be achieved by an adaptable, mobile and cost-effective approach over a larger scale. Secondly, all of the NWs are required to generate electricity simultaneously and continuously, and all the electricity can be effectively collected and output.

Finally, the energy to be converted into electricity has to be provided in a form of wave/vibration from the environment, so the nanogenerator can operate "independently" and wirelessly. An innovative approach has been developed by Wang's group to use ultrasonic waves to drive a nanogenerator built from an array of vertically aligned ZnO NWs [32, 33]. The design is novel and cost-effective and has the potential to meet the requirements outlined above. The approach is the basic platform for optimizing and improving the performance of the nanogenerator for its applications in nanotechnology.

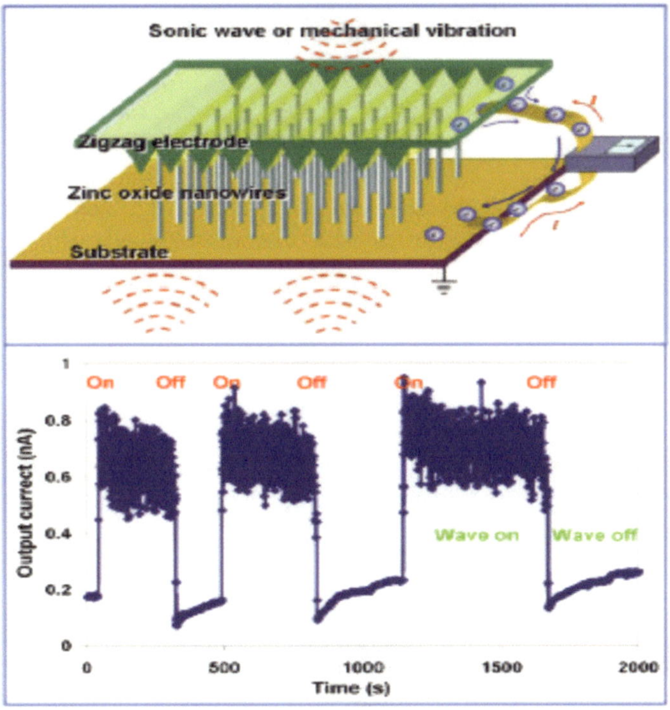

Fig. 16. Schematic diagram showing the direct current nanogenerator built using aligned ZnO nanowire arrays with a Zigzag top electrode. The nanogenerator is driven by an external ultrasonic wave or mechanical vibration and the output current is continuous. The lower plot is the output from a nanogenerator when the ultrasonic wave was on and off. The nanowire that simultaneously contributed to the DC current was estimated about 500

The experimental set up is schematically shown in Figure 16, in which an array of aligned ZnO NWs is covered by a zigzag Si electrode coated with Pt. The Pt coating is not only for enhancing the conductivity of the electrode but also for creating a Schottky contact at the interface with ZnO. The NWs were grown on either GaN substrates, which served as a common electrode for directly connecting with an external circuit. The asymmetric piezoelectric-potential across the width of a ZnO NW and the Schottky contact between the metal electrode and the NW are the two key processes for creating, separating, preserving/accumulating, and outputting the charges.

A top electrode is designed to achieve the coupling process and to replace the role played by the AFM tip, and its zigzag trenches are to act as an array of aligned AFM tips. The DC nanogenerator is driven by ultrasonic wave. Once the wave is on, about 500 nanowires within an area of 2 mm^2 gives 1 nA DC current. We propose to fabricate a nanogenerator that is expected to output the electricity generated by all of the nanowires continuously.

10. Thin film and nanostructure solar cells

ZnO thin films and nanostructures are being developed for design and development of high efficiency solar cells [34, 35, 36]. The use of ZnO thin films and nanostructures in solar cells will provide an inexpensive option for large area solar cells. The Zinc oxide nanorods are attractive for photo-voltaic applications and this claim is based on their properties. Typical ZnO nanorods have a wurtzite structure with lattice spacing a = 0.32469 nm and c = 0.52069 nm [37].

Significant progress has been made in thin film solar cells that have demonstrated efficiency in excess of 30 percent for GaAs based multijunction solar cells, which are being used for space mission [39]. There are several different configurations that would allow high efficiency solar cells. More recently, there is interest in using GaN / InGaN and ZnO / MgZnO for development of high efficiency multijunction solar cells [40].

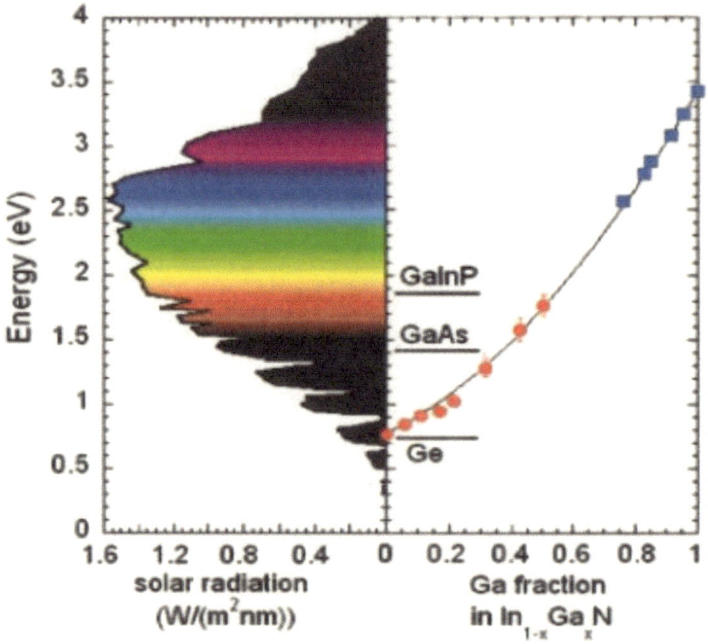

Fig. 17. Band gap energy for solar spectrum. Several material systems are being developed to provide high efficiency multijunction solar cells. These include material components Ge, GaAs and GaInP, InGaN and ZnO. [Wu et al; Reference 40]

Figure 17 presents complete solar spectrum for AM-1.5 solar cells. Different material systems can be utilized to accomplish high efficiency multijunction solar cells. Most of the multijunction solar cells have been developed for space missions where availability of high efficiency is of paramount importance and issue of radiation hardness is critical success for the space missions. There has been significant effort for development of low cost technologies that will allow high efficiency thin film solar cells for terrestrial applications.

ZnO is being explored as one of the key components for a variety of low cost thin film and nanostructure based solar cells. Recent work reported by Ganguly and his coworkers [41] has explored the use of amorphous silicon solar cells with ZnO with insertion of Germanium layers. They have shown that with insertion of a thin amorphous germanium layer at the ZnO-p-layer interface improves the cell performance and diode ideality by suppression of Oxygen and Zinc incorporation in the silicon layers. One of the advantages of ZnO is is its resistance to hydrogen plasma induced darkening and higher transmission thereby improving the solar cell efficiency.

Recent work by Peiro and his coworkers [36] at Imperial college and Manchester in UK have shown that solar cells fabricated from composites of conjugated polymers with nanostructure ZnO are of great interest for their stability, low cost and electron transport properties. ZnO polymer solar cells are promising for low temperature synthesis.

Zinc oxide (ZnO)/polymer solar cells are promising compared to other metal oxide/polymer combinations, on account of the possibility of low temperature synthesis, as well as the potential for controlling interface morphology through simple processing from solution. They have focused on the effect of surface morphology of ZnO films on photovoltaic device performance. They have successfully grown ZnO nanorods standing almost perpendicular to the electrodes on a flat, dense ZnO "backing" layer.

They have studied [36, 37] structures consisting of a conjugated polymer in contact with three different types of ZnO layer: a flat ZnO backing layer alone; ZnO nanorods on a ZnO backing layer; and ZnO nanoparticles on a ZnO backing layer. They also used scanning electron microscopy, steady state and transient absorption spectroscopy and photovoltaic device measurements to study the morphology, charge separation and recombination behavior and device performance of the three types of structures.

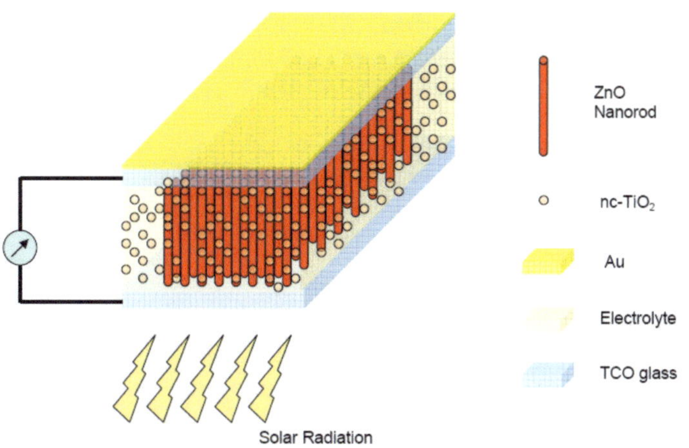

Fig. 18. Schematic diagram of ZnO based nanostructure solar cell [Sridhar et al; Ref 34]

They found that charge recombination in the structures containing vertically aligned ZnO nanorods is remarkably slow, with a half life of over 1 ms, over two orders of magnitude slower than for randomly oriented ZnO nanoparticles. A photovoltaic device based on the nanorods structure which has been treated with an ambiphilic dye before deposition of poly(3-hexyl thiophene) (P3HT) polymer shows a power conversion efficiency over four times greater than for a similar device based on the nanoparticles structure.

The best ZnO nanorods: P3HT device yields a short circuit current density of 2 mAcm-2 under AM1.5 illumination (100mWcm-2) and peak external quantum efficiency over 14%, resulting in a power conversion efficiency of 0.20%. Their effort demonstrates the potential of ZnO nanostructure based solar cells have the potential of low cost and further improvements can yield higher efficiency solar cells.

Several other groups in US, China and England [34, 36, 41, 42] are working on ZnO nanostructure solar cells. They are all trying to develop the necessary understanding for growth and characterization of ZnO nanostructures on variety of substrates.

Investigation of ZnO nanorods based solar cells is being conducted towards developing alternative, lightweight, flexible devices for commercial applications. A schematic view is shown in figure 18 .A lot of research has been done in the area of dye sensitized solar cells [37, 42] in particular, which is currently the most stable and efficient excitonic solar cell. Aligned ZnO nanorods, with their high carrier mobilities serve as the conduction pathways for the excitons. These approaches provide a glimpse of what is being done on using ZnO nanostructures fro solar cell applications. Further effort is needed in developing ZnO based solar cells that can be implemented in terrestrial applications.

11. UV atmospheric transmission model

There has been recent effort to study the atmospheric transmission models in UV, Visible and Infrared (IR) bands [43]. Modeling of atmospheric effects on transmission at UV wavelengths (250-350 nm range) was performed using MODTRAN. MODTRAN (MODerate resolution atmospheric TRANsmission) is an atmospheric radiative transfer model that may be used to determine the effects of various atmospheric layers and weather conditions on wavelength dependent free-space transmission. Six different atmospheric models were simulated: clear maritime (23 km visibility), desert extinction, drizzle (2 mm/hr), light rain (5 mm/hr), moderate rain (12.5 mm/hr), and heavy rain (25 mm/hr).

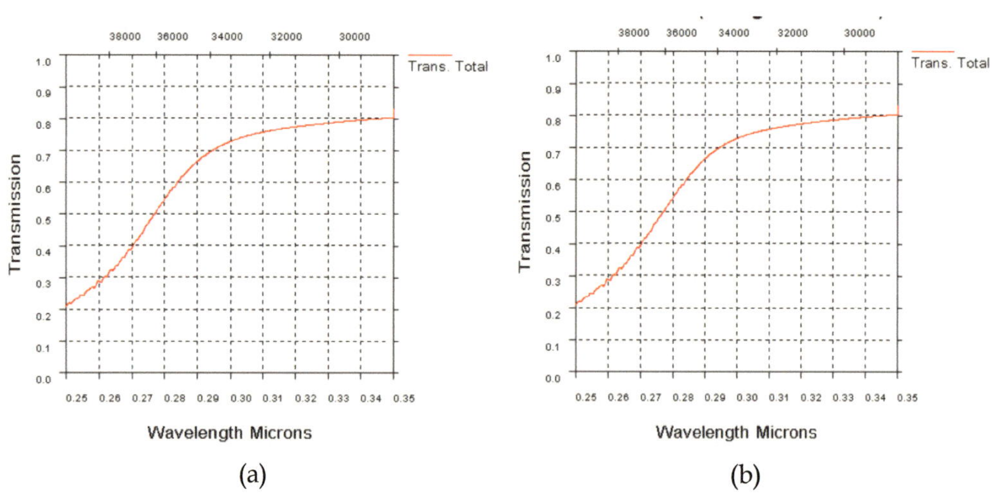

Fig. 19. UV atmospheric transmission over a horizontal path in clear maritime (23 km visibility) conditions, 1 km range (19 a). UV transmission through atmosphere over a horizontal path in desert extinction conditions, 1 km range (19b)

For each of these weather conditions, UV transmission was modeled for 1 km horizontal paths through the atmosphere 5 m above the surface (Figures 1-6). In addition, slanted paths at angles of 15°, 30°, 45°, and 60° from the horizon for a distance of 10 km were simulated in clear maritime conditions (Figures 19-23) to show the relative effects of propagation though different lower atmospheric layers on attenuation. In relation to free-space optical communication networks, a 0° horizontal path would involve ground based communication extending no further than the horizon, while an upward slanted path might involve communication between a ground-based unit or sea vessel and an aircraft [43].

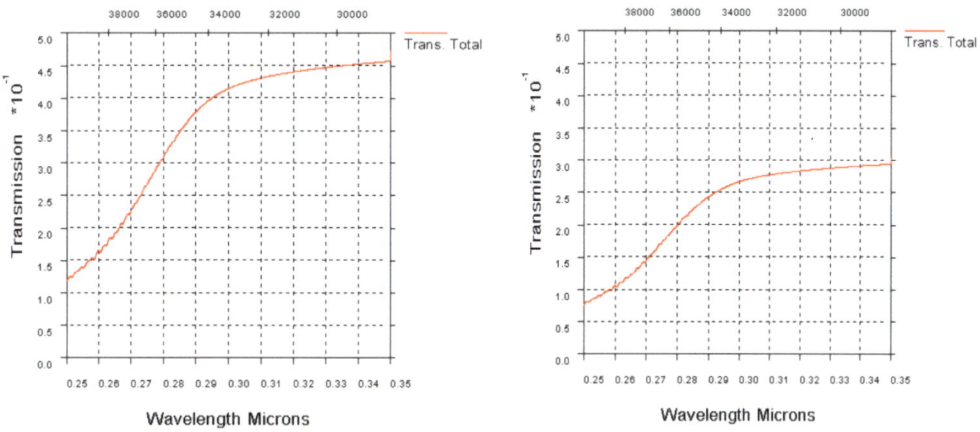

(a) (b)

Fig. 20. UV atmospheric transmission over a horizontal path in drizzle (2 mm/hr) conditions, 1 km range (20a) and UV transmission over a horizontal path in light rain (5 mm/hr) conditions, 1 km range (20b).

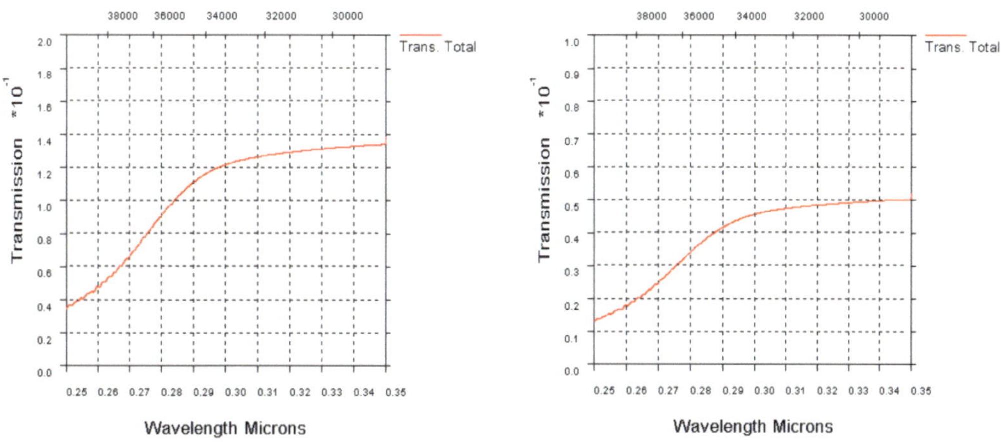

(a) (b)

Fig. 21. UV transmission through atmosphere over a horizontal path in moderate rain (12.5 mm/hr) conditions, 1 km range (21a). UV transmission over a horizontal path in heavy rain (25 mm/hr) conditions, 1 km range (21b)

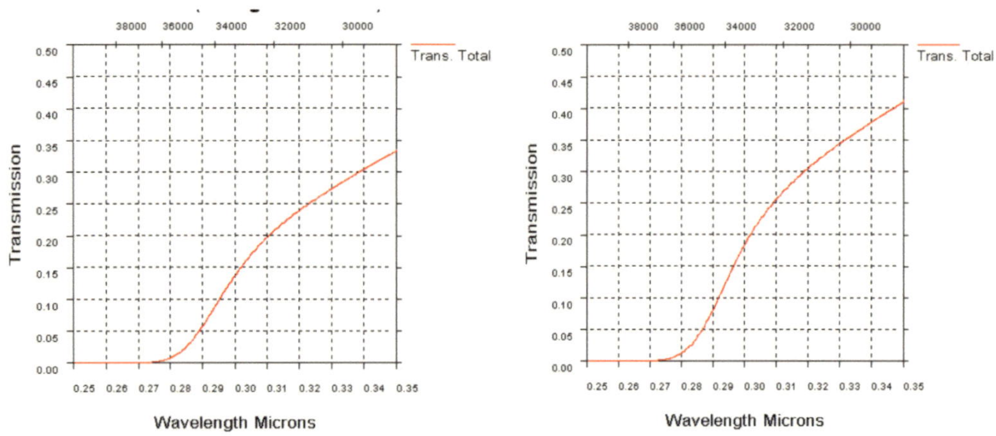

(a) (b)

Fig. 22. UV atmospheric transmission over a slanted path at 15° elevation in clear maritime conditions, 10 km range (22a). UV transmission over a slanted path at 30° elevation in clear maritime conditions, 10 km range (22b)

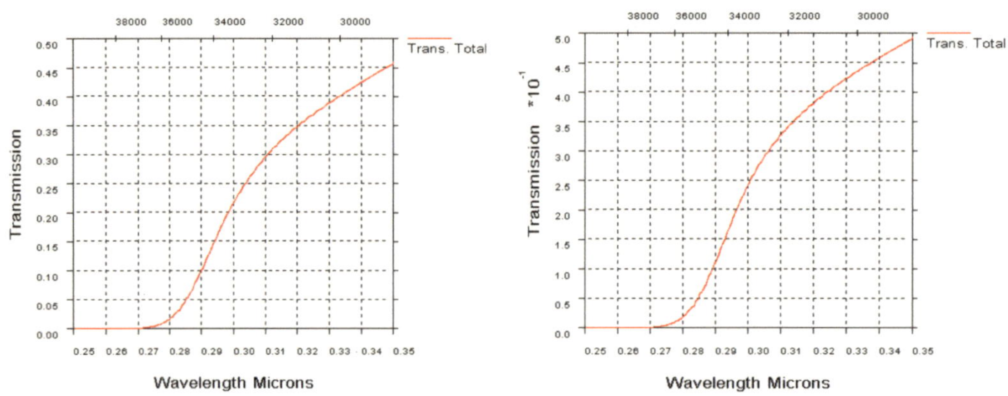

(a) (b)

Fig. 23. UV transmission over a slanted path at 45° elevation in clear maritime conditions, 10 km range (23a). UV atmospheric transmission over a slanted path at 60° elevation in clear maritime conditions, 10 km range (23b)

12. Summary

In this chapter, we have discussed growth, fabrication and characterization of ZnO nanostructures materials and devices on a variety of substrates. ZnO nanostructures offer potential for a variety of optical, electronic and biological devices for nanoscale applications. The Zinc oxide nanostructures are also attractive for energy generation devices and photovoltaic applications. Significant research effort is underway on ZnO nanostructures, their unique properties for application in transparent electronics, ultraviolet (UV) light emitters, piezoelectric devices, chemical sensors and spin electronics.

An assortment of ZnO nanostructures, such as nanorods, nanotubes and nanorings, have been successfully grown via a variety of methods including chemical vapor deposition, thermal evaporation, hydrothermal technique, sol-gel and electro-deposition. These nanostructures have been evaluated for electronic devices, UV detectors, light emitting diodes, lasers, gas sensors, and other biological applications. Further work is underway that will enhance our understanding of ZnO nanostructures and their applications for future system applications.

13. Acknowledgement

The authors would like to acknowledge the contributions of many students at Georgia Tech that have over the years contributed to the exciting work, which has been published by them and presented here in the chapter. We would also like to acknowledge the modeling effort by Dr. John Zeller of NUWC, Newport, RI. The authors would also like to acknowledge the financial support by numerous agencies that have contributed to the ZnO nanostructures and their applications.

14. References

[1] J. R. Choi and Dennis. L. *Polla Journal of Micromechanical Micro-engineering* Volume 3, 60-64, 1993
[2] E. J. Egerton, A. K. Sood, R. Singh, Y. R. Puri, R. F. Davis, J. Pierce, D. C. Look and T. Steiner, *Journal of Electronic Materials* Vol.34, No.6, 2005
[3] Z. L. Wang , A Review Paper, *Journal of Physics: Condensed Matter* 16, 829-858, 2004
[4] X. Wang, J. Song, C. J. Summers, J. H. Ryou, P. Li, R. D. Dupuis, and Z. L. Wang , *J. Phys. Chem. B*, 110 (2006) 7720-7724
[5] X. Wang , J. Song , P. Li , J. H. Ryou , R. D. Dupuis , C. J. Summers and Z. L. Wang , *J. Am. Chem. Soc.*, 127 (2005) 7920-7923
[6] X. Wang, C. J. Summers and Z. L. Wang, *Nano Letters*, 3 (2004) 423-426.
[7] Z. L. Wang and J. Song, *Science*, 312 (2006) 242-246
[8] Z. Fan and J.G. Lu, *Journal of Nanoscience and Nanotechnology*, Volume 5, 1561-1573 (2003)
[9] R.S. Yang, Y. Ding and Z.L. Wang, *Nano Lett.*, 4, 1309 (2004)
[10] A.K. Sood, Y.R. Puri, P. Gao, C. Lao, Z.L. Wang, D.L. Polla, M.B. Soprano, *Proceedings of SPIE*, Volume 6556, 6556IL (2007)
[11] L. Luo, B. Sosnowchik and L.W. Lin, *Applied Physics Letters* 90, 093101 (2007)
[12] W. Lee, M. C. Jeong and J.M. Myoung, *Acta Materialia*, 52, 3949-3957(2004)
[13] J. B. Baxter and E.S. Aydil, *Journal of Electrochemical Society*, 156 (1), H52-H58 (2009)
[14] P. X. Gao and Z. L. Wang, *J. Phys. Chem.* B 106, 12653 (2002)
[15] P. X. Gao, Y. Ding and Z. L. Wang, *Nano Lett.* 3, 1315 (2003)
[16] Z. W. Pan, Z. R. Dai and Z. L. Wang, *Science* 291, 1947(2001)
[17] M. H. Huang, Y. Y. Wu, H. Feick, N. Tran, E. Weber and P. D. Yang, *Adv. Mater.* 13, 113 (2001)
[18] M. S. Arnold, P.H. Avouris, Z. W. Pan and Z. L. Wang, *J. Phys. Chem.* B 107, 659 (2003)
[19] Y. Cui and C. M. Lieber, *Science* 291, 851 (2001)
[20] P. G. Collins, M. S. Arnold and Ph. Avouris, *Science* 292, 706 (2001)
[21] J. Kong, N. Franklin, C. Wu, S. Pan, K. J. Cho and H. Dai, *Science* 287, 622 (2000)
[22] Wang, X.D., Ding, Y. Summers, C.J. and Wang, Z.L., *J. Phys. Chem. B*. 108, 8773 (2004)

[23] X. Bai, E. G. Wang, P. Gao and Z. L. Wang, *NanoLetters*, 3, 1147 (2003)
[24] P. M. Morales and C. M. Lieber, *Science* 279, 208 (1998)
[25] G.F. Zheng, F. Patolsky, Y. Cui, W.U. Wang, and C.M. Lieber, *Nature Biotechnology* 23 (2005) p. 1294
[26] X.D. Bai, P.X. Gao, Z.L. Wang and E.G. Wang, *Appl. Phys. Letts.* 82, p. 4806 (2003)
[27] Z. L. Wang, and J. H. Song, *Science* 312 , p. 242 (2006)
[28] J.H. Song, J. Zhou, and Z.L. Wang, *Nano Lett.* 6 p. 1656 (2006)
[29] M.H. Zhao, Z.L. Wang, and S.X. Mao, *Nano Lett.* 4 (2004) p. 587
[30] Z.L. Wang, X.Y. Kong, and J.M. Zuo, *Phys. Rev. Letts.* 91 (2003) p. 185502
[31] Z.L. Wang and J. H. Song, *Science*, 312 , 242-246 (2006)
[32] P.X. Gao, J.H. Song, J. Liu and Z.L. Wang, *Advanced Materials*, 19, 67-72 (2006)
[33] X.D. Wang, J.H. Song, J. Liu and Z.L. Wang, *Science*, 316, 102-105 (2007)
[34] D. Sridhar, J. Xie, J.K. Abraham, V. K. Varodan and S.H. Choi, *Proceedings of SPIE*, Volume 6172 (2006)
[35] W. J. F. Beck, L. H. Slooft, M. J. Wienk, J. M. Kroon, and R.A. J. Janssen, *Proceedings of SPIE*, Volume 5938 (2005)
[36] A. M. Piero, P. Ravirajan, K. Govender, D.S. Boyle, P.O. O'Brien, D.C. Bradley, J. Nelson and J.R. Durrant, *Proceedings of SPIE* , Volume 5938 (2005)
[37] J. R. Baxter and E.S. Aydil, *Applied Physics Letters*, 86, 053114 (2005)
[38] For Silicon Solar Cells, i.e. Evergreen Solar (Marlboro, MA), Schott Solar (Burlington, MA), and other Silicon Solar Cell Manufacturers). The cell efficiency varies from 14-18 percent for Polysilicon to Single Crystal Silicon Solar Cells.
[39] T. Takamoto, E. Ikeda, H. Kurita and M. Ohmori, *Applied Physics Letters*, 70, 381 (1997)
[40] J. Wu, W. Walukiewicz, K. M. Yu, W. Shan, J. W. Ager, E.E. Haller, H. Lu, W.J. Schaff, W.K. Metzer and S. Kurtz, *Journal of Applied Physics*, Volume 94, 6477 (2003)
[41] G. Ganguly, D.E. Carlson, S. S. Hyedus, D. Ryan , R. Gordon, D. Pang and R. C. Reedy, *Applied Physics Letters* 85, p 479 (2004)
[42] Z. Longyue, D. Songyuan, X. Weiwei and K. Wang, *Plasma Science and Technology*, Volume 8, No2, March 2006
[43] J. Zeller and T. Manzur, *Proceedings of SPIE*, Volume 7833, 783313 (2010)

10

Hybrid Optoelectronic and Photovoltaic Materials based on Silicon Nanocrystals and Conjugated Polymers

Vladimir Svrcek
Research Center for Photovoltaics,
National Institute of Advanced Industrial Science and Technology (AIST),
Central 2, Umezono 1-1-1, Tsukuba, 305-8568
Japan

1. Introduction

Hybrid material, which combines advantages of both organic and inorganic materials, might offer potential for design of novel type of low cost devices with superior performance (Schneider et al., 2000; Liu et al., 2008). The large choices for the organic and inorganic structures offer the possibility to obtain materials with attractive physical and chemical properties. The ***morphology*** of hybrid material at nanoscale might lead to very different ***properties*** from crystalline solids. Particularly molecular structure polymers, conformation and orientation can have a major effect on the macroscopic properties of novel material (Coakley et la.,2003). Nanotubes are a promising subclass of nanomaterials owing unique one-dimensional geometric features that can allow engineering the polymer based material morphology at low cost. Nevertheless, nanotubes fabrication with diameter comparable with exciton diffusion lengths of polymers (~ 15 nm) is still a problem. It has to be noted that among the nanotubes with small diameter (< 5 nm), carbon-based discovered by Iijima (Iijima 1991) were the first to gain recognition in academia (Marte et al., 2001; Harris, 2002; Wang et al., 2009). Novel synthetic strategies for generating nanotubes from inorganic materials have been recently also widely investigated and developed (Tenne et al., 1992; Zhao, et al., 2004). It is believed that fiber/nanotube-polymer based arrays of material have much lower reflectance and enable fabrication of thicker devices with increased absorption compared with thin films.

One of the promising type of polymers used in variety of applications are the conjugated conductive ones (Inzelt, 2008). It has to be noted that in traditional polymers e.g. polyethylenes, the valence electrons are bound in sp^3 hybridized covalent bonds (Inzelt, 2008). Therefore sigma-bond electrons have low mobility and do not contribute to the electrical conductivity. Contrary to that conducting polymers have backbones of contiguous sp^2 hybridized carbon centers. As a result a valence electron on each center resides in a p_z orbital, which is orthogonal to the other three sigma-bonds. Then electrons within the band become more mobile particularly when it is partially emptied. This advantage combined with the mechanical properties (flexibility, toughness, malleability, etc.) make them favorable also for optoelectronic applications as an active material.

On the other hand, quantum dots, sometimes called as nanocrystals as well, are a special class of semiconductor. They range in size from 2-10 nanometers in diameter. As a result the excitons in ***quantum dots*** are confined in all three spatial dimensions (Murray, et al., 2000). Particularly, silicon nanocrystals (Si-ncs) have many advantages over the other nanocrystal materials (Canham, 1990; Hirsmman 1996). For instance, some of these materials contain toxic elements such as lead or cadmium, and others rely on elements such as indium that are in limited supply in nature. Silicon is no toxic and abundant. Newly observed phenomenon in Si-ncs - so called multiple excitons generation - favorites Si-ncs as promising material for photodetectors and solar cells (Beard,et al., 2007; Sukhovatkin et al., 2009). Namely the Si-nc can produce two or three electrons per photon of high-energy sunlight and could lead to a new type of solar cell with more than twice as efficient as nowadays used one. ***Colloidal Si-ncs*** compared to solidly embedded in matrix allow easier processibility and fabrication of device at low cost. [6] Free-standing Si-ncs and conjugated polymers blends shown be a promising optoelectronic and photovoltaic composite material (Švrček et al., 2008a; Lui et al., 2009).

In first part of this chapter we show that electrochemical etching and laser nanosecond laser processing in liquid media is suitable for preparing doped (boron and phosphorus) colloidal and surfactant free Si-ncs. Blends optoelectronic properties consisting of doped Si-ncs and two conjugated polymers (e.g. (poly(3-hexylthiophene) (P3HT) and poly[methoxy-ethylexyloxy-phenylenevinilene] (MEH PPV)) are discussed in details. It is demonstrated that such Si-ncs can be successfully used for fabrication of room temperature photoluminescent and photoconductive blends. The role of selected Si-ncs synthesis techniques on the photoelectric properties of blends is compared. We argue that the luminescence and transport properties of the blends are controlled by Si-ncs properties and could be assigned to quantum confinement of excitons in Si-ncs. We demonstrate that the transport properties of the blend can be tuned by processing conditions. The blends containing Si-ncs produced by the laser ablation clearly evidence superior photovoltaic properties due to the enhanced bulk-heterojunction surface area and improved charge transport.

The morphology of the bulk-heterojunction can be significantly affected by various fabrication parameters during the device formation. In second part of the chapter, in order to achieve an efficient performance of the bulk-heterojunction, both the size distribution and mesoscopic ordering of blend in nantubes is discussed. It is shown that a fiber- and/or vertical 1D-like order of photosensitive bulk-heterojunction gives considerable advantages over the thin film technology, because it provides larger interfacial area for efficient exciton dissociation and straight path for photogenerated carries. As a result, fibers help to avoid circuit shorts and interruption of percolation paths for carriers to their respective electrodes. In this respect, a capillary induced filtering and assembly of blends in nanoporous templates is discussed. We show that the titanium/alumina dioxide (TiO_2/Al_2O_3) nanotubes template could be suitable candidate for vertical order of photosensitive based blends.

2. Experimental methods

Colloidal and surfactant free Si-ncs with quantum confinements effects were prepared by two independent techniques. First by electrochemical etching and second by laser ablation of silicon wafer in water. For this purpose, boron-doped wafer with a resistivity of 0.5–

0.75Ocm (p-type, B concentration of $3x10^{16}cm^3$) and phosphorous-doped wafer with a resistivity of 0.5–2Ocm (n-type, P concentration $2x10^{16}cm^3$) were used. The wafers were electrochemically etched in a mixture of hydrofluoric acid with pure ethanol ($HF:C_2H_5OH$ 1:4). In order to obtain a similar size distribution of Si-ncs, a constant current density 3.2 mA/cm^2 and a constant etching time 60 min were used for the B-doped Si-ncs. In the case of P-doped Si-ncs, we kept a constant current density at $1.6mA/cm^2$. In this process, the etching time was 90 min and a halogen lamp illuminated the P-doped silicon substrate during the electrochemical etching. The B- and P-doped and red luminescent Si-ncs were harvested by mechanical scratching (Švrček et al., 2004). Second technique goes after the synthesis route of blue luminescent Si-ncs based on a water-confined nanosecond laser ablation process [Švrček et al., 2006; Švrček et al., 2009]. Particularly, the Si-ncs are prepared by nanosecond excimer pulsed laser (KrF, 245 nm 20 Hz, 20 ns). Crystalline silicon doped wafers with same characteristics as used for electrochemical etching is used for synthesis of the Si-ncs. The wafers are adhered to the bottom of a glass container and immersed in 10 ml water. The laser beam is focused onto a 1.5 mm diameter spot on the wafer surface by a lens with a focal length 25 cm. The ablation process is continued for 2 hours at room temperature and ambient pressure. Following the ablation process the aqueous solution with produced Si-ncs is left to age in ambient conditions (Švrček et al., 2009a). In order to obtain enough high Si-nc concentration for blend fabrication the process was repeated several times.

Blends consisting of red and blue luminescent Si-ncs and two conjugated polymers (poly(3-hexylthiophene) (P3HT) and poly(methoxy ethylexyloxy phenylenevinilene) (MEH PPV)) were fabricated as follow. A commercially available (ALDRICH) polymer P3HT was dissolved in chlorobenzene (12 mg of polymer in 10 g of chlorobenzene). The Si-ncs powder was mixed with polymer in order to make films. The 300 nm thin films were spun on quartz substrates. The samples were dried at 140 °C for 30 min in vacuum. Furthermore, we have fabricated solar cell based on the Si-ncs prepared by laser ablation and electrochemical etching. Glass substrates with an indium tin oxide (ITO) electrode were used for solar cells fabrications. The PEDOT:PSS is used as electron blocking layer and the hole collector in this structure. An alumina contact film (~100 nm) deposited on the top serves as electron collector. In order to smooth the ITO surface the poly(3,4-ethylenedioxythiophene):poly(styrene sulfonic acid (PEDOT:PSS) was deposited by spin-casting (~ 100 nm). Subsequently, the Si-ncs/P3HT solution was spin-coated on the PEDOT:PSS/ITO substrate. Then, an alumina top electrode of 100 nm was deposited on top of the active layer.by vacum evaporation. The active area of the cell was 4 mm^2.

In order to perform the alignment of the blends two types of nanotube arrays with rather same diameter were used: alumina and titanium dioxide. The nanoporous alumina membranes were fabricated by the two step anodization of 99.99% pure aluminum foil under the constant potential of 40 V in 0.3 M oxalic acid electrolyte at 5°C. The first anodization step lasted 15 h to promote spatial self-organization of nanopores. Then the thick nanoporous layer was selectively dissolved in the mixture of chromic (20 g/l) and phosphoric (66 g/L) acids leaving a highly ordered concave pattern on the surface of aluminum. The ordered concaves acted as a self-assembled mask for the nucleation of nanopores during the second anodization step. In order to perform the filtering of Si-ncs blended with an MEH-PPV polymer on the membrane was glued to an O-ring. The aluminum retained on the back side was selectively dissolved in saturated HgCl2 solution at room temperature. The barrier oxide layer was etched in 5% phosphoric acid with the

formation of a through-hole membrane (Švrček et al., 2010). All samples were carefully cleaned in distilled water and then observed in a scanning electron microscope (SEM). To perform the filtering it was necessary to decrease the viscosity of the blend by adding of 2 g of chlorobenzene into the blend solution before dropping it on the membrane. The O-ring that sealed the edge membrane did not allow the outlet of the solution (Švrček et al., 2010). The formation of Si-ncs/MEH-PPV fibers occurred when the viscosity of the Si-ncs/MEH-PPV blend dropped on the top of nanoporous alumina. The formation of the Si-ncs/MEHPPV nanofibers was confirmed by the dissolution of the alumina membrane in 5% phosphoric acid. After the dissolution, the fibrous structures were observed in SEM. In order to infiltrate the bulk-heterojunction, the TiO_2 nanotubes with rather the same inner diameters (~ 90 nm) were prepared as follows. A 99.99% pure titanium foil was anodized in 0.3 wt% solution of NH4F in ethylene glycol in a two-electrode configuration under constant potentials of 40 V at room temperature (Švrček et al., 2009b). The titanium foil served as anode and platinum mesh as the counter electrode. The transformation of amorphous as grown anodic TiO_2 into crystalline anatase was performed by annealing in air at 450 C.
The photoluminescence (PL) measurements of colloidal solutions were carried out at room temperature using fluorophotometer (Shimadzu Corporation, RF-5300PC) with excitation by Xe lamp at 300 nm for Si-ncs prepared by laser ablation and 400 nm by electrochemical etching, respectively. For the temperature dependence of the PL, a HeCd laser (325 nm) has been used. The samples were placed in the cryostat with a varying temperature from 4 to 300 K. Spectrally resolved photoconductivity was measured under illumination with monochromatic light from a Xe lamp at room temperature and N2/air atmosphere. In all cases the irradiation intensities were calibrated by a standard a-Si solar cell. A small droplet of obtained colloidal solution was deposited onto a copper grid with carbon film for high resolution transmission electron (HR-TEM) and scanning electron microscope (SEM) observations. The SEM images were done with a Hitachi SI 4800 microscope with 30 kV acceleration voltage. The HR-TEM studies were performed on a microscope with 200 kV acceleration voltage (JEOL, JEM 2010).

3. Freestanding room temperature luminescent Si-ncs

Figure 1 (a) shows the room-temperature photoluminescence (PL) spectra intensity as a function of wavelength of n- and p-type doped silicon nanocrystals (Si-ncs) excited at 325 nm prepared by electrochemical etching. We tuned the nanocrystal size distribution, the shape and the position of PL spectra by electrochemical etching conditions (Švrček et al., 2009c) and prepared freestanding Si-ncs with rather similar size distributions. This allowed as better revealing the effects of the dopands (Švrček et al., 2009c). The n-type and p-type doped freestanding Si-ncs show orange PL under at room temperature. For both types the broad PL spectra with a maximum located around 600 nm appeared. In the case of Si-ncs prepared by ns laser ablation in water aging in water is required to obtain visible PL at room temperature (Švrček et al., 2009a). After this procedure our results showed that the PL is stable and does not degrade under continuous exposure to the air or water. Fig 1(b) represents PL typical spectra of doped Si-ncs prepared by an excimer KrF laser ablation in water and aged in water for several months. Visible blue-room temperature PL is observed from both n-type doped crystalline silicon wafer (red line) and image p-type doped (black line) colloids under He:Cd laser excitation at 325 nm. The Si-ncs dispersed in colloidal solution show blue-bands with maxima at around 420 nm. Compared to the Si-ncs made from p-type doped wafer, the PL band for n-type doped is stronger more than 2 times due to the surface passivation and presence of dopants (Lulease, et al., 2002). The origin of the broad band behavior is most likely due to multiple electronic states in Si-ncs. In both preparation methods the large size distribution results in the broadening of the PL spectra. The quantum confinement in Si-ncs with different sizes is responsible for the broad emission.

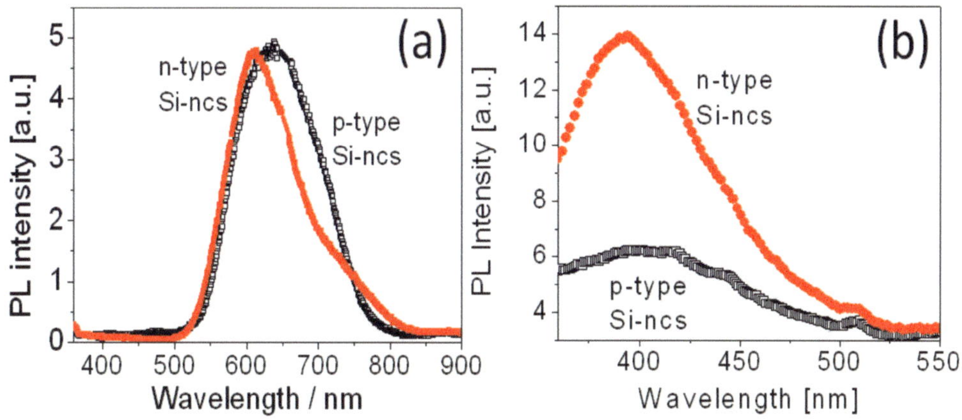

Fig. 1. (a) Room-temperature photoluminescence (PL) spectra intensity as a function of wavelength of n- and p-type doped silicon nanocrystals (Si-ncs) prepared by electrochemical etching. (b) PL spectra of doped Si-ncs prepared by an excimer KrF laser ablation in water and aged in water for 7 months

Temperature-dependent PL analysis gives an inside of the PL origins (Švrček et al., 2009a; Švrček et al., 2009c). Our results show that for Si-ncs prepared by both techniques two bands could be identify by low temperature analysis. The temperature dependence of the PL intensity for a P-doped and for a B-doped Si-ncs prepared by electrochemical etching and laser ablation showed two broad PL bands at various temperatures. For electrochemical etching the short wavelength-band locates at 500–700 nm with a maximum around 590 nm and the long-wavelengths band locates at 650–850nm with a maximum at 740 nm. In the case of laser ablation both bands are blue shifted and the maxima are centered at ~410 nm and ~525 nm, respectively. Short wavelength bands originating from delocalized carriers at the nanocrystal core with the oxide interface do not showed any temperature dependence behavior. Contrary to that, in the long wavelength band, the PL intensities of the shorter wavelengths decrease more slowly due to the state-filling effect, which results in a red shift. Figure 2 represents the shift of PL maxima as a function of temperature for Si-ncs prepared by electrochemical etching and laser ablation in water. It is supposed that the red shift is mainly attributed to the decrease in the band gap energy with temperature, which is typical for semiconductor materials due to the many possible electronic states in Si-ncs (Kovalev, et 1994; Zhuravlev et al., 1998; Garcia et al., 2003). At 4 K, the contribution from quantum confinement is distinguished quite well and one could evaluate real Si-ncs size. It is assumed that the PL spectra with a maximum at 590 nm represent the contribution from excitons confined in nanocrystalites with a corresponding size of 2-3 nm prepared by electrochemical etching (Švrček et al., 2009c) and 1-2 nm by laser ablation (Švrček et al., 2009a).

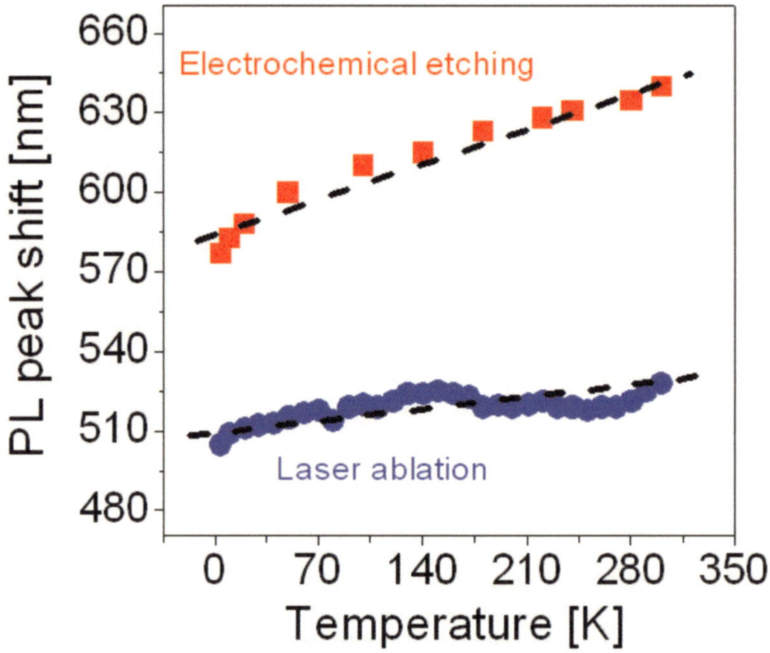

Fig. 2. Shift of PL maxima as a function of temperature for p-type doped Si-ncs prepared by electrochemical etching and laser ablation in water is shown

Photographs in Fig. 3 represent room temperature emission of Si-ncs dispersed on the quartz substrate. Photograph (a) shows the red PL of Si-ncs prepared by electrochemical etching and photograph (b) blue PL of Si-ncs fabricated by laser ablation. The Si-ncs are by HeCd laser at a wavelength of 325 nm excited. The PL emission and intensity in both cases is enough high as observed by naked eye. The wavelength of the PL maximum and that of the emission from the 525 nm band remain stable. During air exposure, oxygen penetrates near the core or on the core interface forming a Si—O—Si layer on the nanocrystal core surface. Subsequently, the Si—Si or Si—O—Si bonds become weaker or even break apart because of a larger induced stress (Ourmazd, et al., 1987). A higher number of Si-O double bonds is likely to be formed that contribute to stabilize the nanocrystal interface by terminating two dangling bonds [Herman et al., 1981). Overall, an improved surface stabilization and passivation for the dangling bonds results in a PL increase for the large wavelength bands. It has been reported that the quantum yield of blue PL colloidal Si-ncs prepared by other techniques decreases during oxidation since a recombination channel for the Si-ncs and the expected PL due to quantum confined excitons is quenched (Jurbergs et al., 2006).

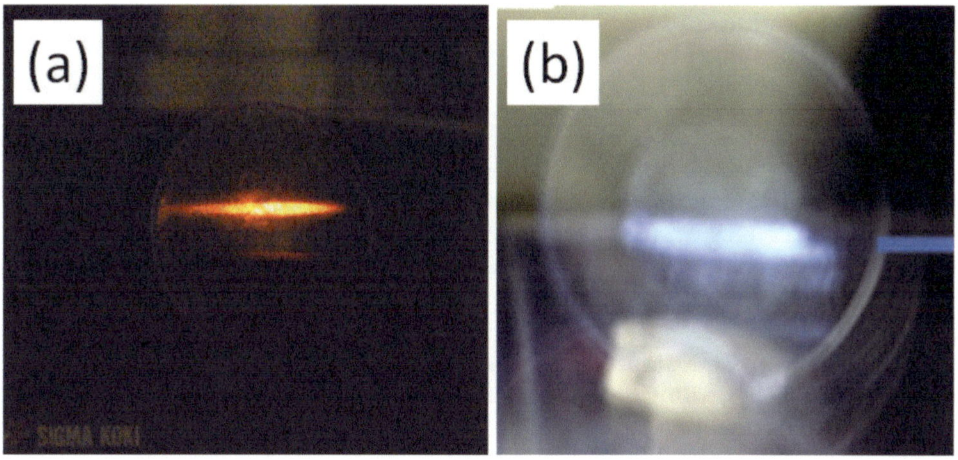

Fig. 3. Photographs of red-orange and blue light emission from Si-ncs dispersed on quartz prepared (a) by electrochemical etching and (b) by laser ablation in water. Si-ncs were excited by a He:Cd laser at 325 nm at room temperature

Systematic studies of high-resolution transmission electron microscopy (HR-TEM) observations showed that such powder contains both single Si-ncs and Si-ncs micrograins (Švrček et al., 2004). Figure 4a shows TEM images of Si-ncs prepared by electrochemical etching. While also the Si-ncs micrograins consist of many nanocrystals (ranging from about 2 to 6 nm) interconnected and kept together by amorphous tissue particles with size several tenths nanometers are observed. Figure 4 (b) represents a typical TEM image of Si-ncs particles prepared by laser ablation processing in water by ns KrF laser. Detailed TEM (APL) revealed that spherical particles with average size of 40 nm containing the Si-ncs with size ~ 1-3 nm. Independently to TEM also Raman spectra, XRD structural analyses of particles prepared by two independent techniques confirm the presence of Si-ncs with the crystalline silicon diamond-like structure (Švrček et al., 2004), indeed, smaller sized Si-ncs were clearly achieved by laser ablation in liquid media (Švrček et al., 2006).

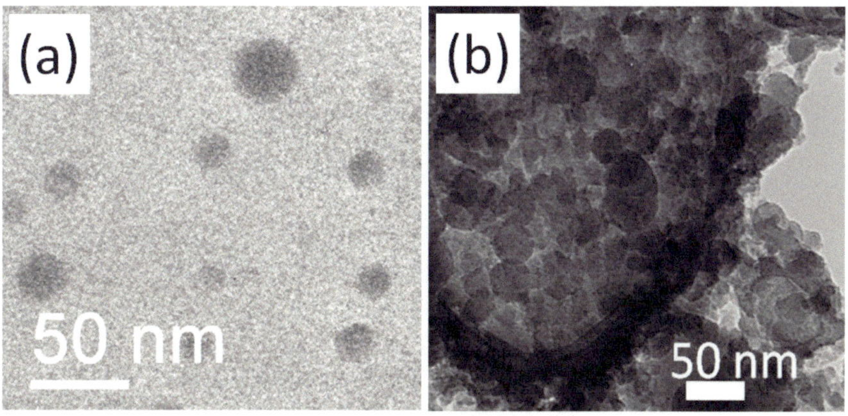

Fig. 4. Transmition electron microscopy (TEM) images of silicon nanocrystals (Si-ncs) prepared (a) by electrochemical etching and (b) by nanosecond excimer KrF laser ablation in water

4. Si-ncs and conjugated polymer s based composites

We have shown recently that freestanding Si-ncs allows the fabrication of hybrid organic-inorganic composites with a high Si-nc with nonlinear optical properties (Švrček et al., 2002; Švrček et al., 2004). In addition, polymers properties could be flexible tuned for example by doping or type of polymer. On the other hand the transparency of silicon dioxide based polymer allows preparing the Si-ncs by nanosecond laser ablation of the Si target in this type of polymer (Švrček et al., 2008b). The confinement of laser generated enhances the formation of Si-nc and the silicon-dioxide-based SOG inhibited Si-nc aggregation compared to the case of laser ablation in water. The silicon-dioxide-based SOG during the solidification process accelerates nanocrystal surface oxidation resulting in visible room temperature PL with a maximum located at ~ 400 nm. The solidification of Si-nc in SOG solution enabled the formation of self-supporting films with well-defined Si-nc concentrations that could be simply varied by laser fluence (Švrček et al., 2008b).
As the conductivity is crucial for optoelectronic application (e.g. LED), however, the SOG polymers shows very low transport properties. Contrary to that conjugated polymers can allow to overcome the problem. Between them (poly[2-methoxy, 5-(2-ethylhexoxy)-1,4-phenylene vinylene]) *MEH-PPV* and P3HT are particularly well studied and already applied for optoelectronics and photovoltaics device development. As reprted elsewhere, a series of commercial organic/inorganic materials were blended into those polymers and used for instance as electron acceptors in the bulk heterojunction structure based devices (Yue et al., 2010).

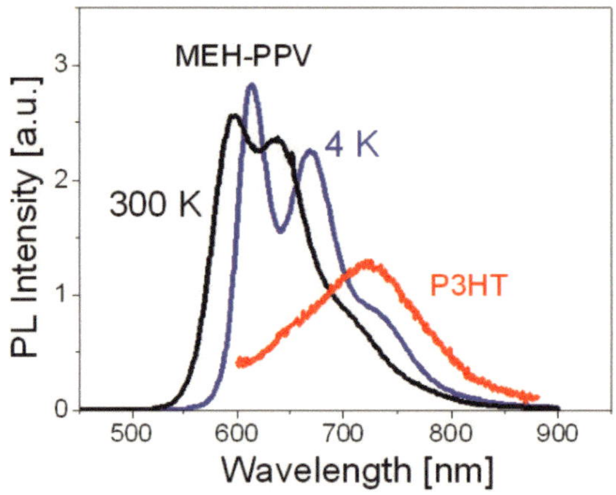

Fig. 5. Photoluminescence (PL) spectra taken at 4 and 300 K of pure MEH-PPV polymer under the same excitation conditions, for comparison. Red line represents typical PL spectrum for P3HT polymer at room temperature

Figure 5 shows typical PL spectra of our used conjugated polymer. In figure 5 blue line show PL spectra taken at 4 and black one at 300 K of pure MEH-PPV polymer, respectively. In this case three characteristic PL peaks, correlating with a zero phonon transition and two phonon replicas are detected. Red line represents the PL spectrum for P3HT polymer at room temperature and is shown for comparison. After Si-ncs introduction within both polymers high absorption of both polymers doesn't allow simple visualization of Si-ncs. The question arisen whether after blending we could see Si-ncs contribution to overall optical properties of such hybrid material. Figure 6 presents' typical PL spectra of the Si-ncs prepared by electrochemical etching and blended in (a) MEH-PPV (b) P3HT conjugated polymers. Figure 6(c) shows the spectra of the Si-ncs prepared by laser ablation in water and blended in P3HT polymer. All spectra are taken under laser c.w. laser excitation (at 325 nm) at 4 and 300 K, respectively. For both polymers based on both doped Si-ncs and prepared by different technique a similar PL emission behavior was observed. Fig 6d represents a summary of the PL peaks position of the blends made of Si-ncs fabricated by electrochemical etching (red circles) and laser ablation (blue squares) as a function of the temperature. Peak position of the pure P3HT polymer film only is shown for comparison (black symbols). We note that the PL behavior of both polymers as a function of the temperature is rather the same and a clear blue shift of ~60 meV is recorded in the same temperature range (4–300 K). Contrary to the pure polymers, the PL spectra are shifted in the opposite direction as a function of temperature. As the temperature was increased in the range from 4 to 300 K the PL intensity and the band shifts varied monotonically. Namely the PL intensities of the shorter wavelengths decreased more slowly due to the state filling effect. Delocalized carriers at the nanocrystal–polymer (P3HT and MEH-PPV) interface also contributed to Si-ncs emission. It is supposed that the red shift is mainly attributable to a decrease in band gap energy with temperature, typical of semiconductor materials (Švrček et al., 2009c).

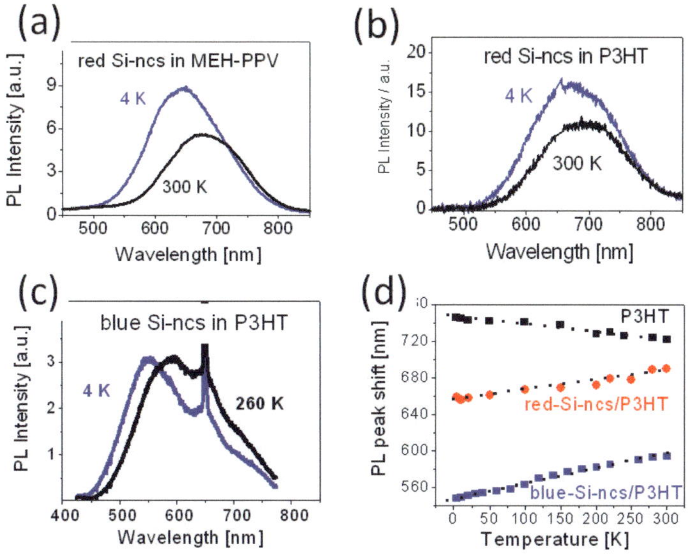

Fig. 6. Typical photoluminescence (PL) spectra of the Si-ncs prepared by electrochemical etching and blended in (a) MEH-PPV (b) P3HT conjugated polymers. (c) The spectra of the Si-ncs prepared by laser ablation in water and blended in P3HT polymer. The spectra are taken under laser cw laser excitation (at 325 nm) at 4 and 300 K, respectively. (d) Summary of the PL peaks position of the blends made of Si-ncs fabricated by electrochemical etching (red circles) and laser ablation (blue squares) as a function of the temperature are shown. Peak position of the pure P3HT polymer film only is shown for comparison (black symbols)

We show that a top-down by electrochemical etching and bottom up by laser ablation are suitable approaches for preparing doped (boron and phosphorus) freestanding and surfactant-free Si-ncs. Then they can be successfully used for the fabrication of room temperature photoluminescent and photosensitive blends based on MEHPPV or P3HT copolymer. We argue that the PL properties of the blends are controlled by the Si-ncs properties and could be assigned to quantum confinement of excitons in nanocrystalites. Generally it is accepted that when a nanocrystal is capped by an organic layer (ligands) the excitation transfer is dominated by the Förster mechanism and attributed to a dipole-dipole coupling process [Förster, 1959; Chang, et al., 2004). In our case, we prepared the Si-ncs surface without any surfactant (i.e. ligands), which limits dipole-dipole coupling. In this case the acceptor-donor pairs are in close distance, the excitation transfer is dominated by the Dexter mechanism, which is a charge exchange process (Dexter, 1953; Greenham, et al., 1996). We assume that in our case the donor and acceptors are in direct contact or at least in van der Waals contact. The Dexter energy transfer rate (R_{ET}) between a donor-acceptor pair at a distance (*d)* can be written as follow

$$R_{\mathrm{ET}}(d) = \frac{2\pi}{\hbar\Psi^2}\int P_{\mathrm{D}}(E)A_{\mathrm{A}}(E)dE \tag{1}$$

where $P_{\mathrm{D}}(E)$ is donor emission and $A_{\mathrm{A}}(E)$ is acceptor absorption. The excitation-transfer energy from the respective polymer to nanocrystal leads to a red shift of the blends PL emission. Even the Forester mechanism cannot be completely ruled out it is assumed that after the Dexter-like energy transfer of the PL band can be attributed mainly to zero-phonon electron–hole recombination in Si-ncs due to a strong enhancement of the quantum confinement effect.

5. Bulk-heterojunction and photovoltaic properties of Si-ncs/P3HT polymer composites

Figure 7(a) shows an energy band diagram of the conjugated polymer and Si-ncs. Also optical measurements suggested that electronic interaction between both types of conjugated polymers occurred. As a result the blending of Si-ncs in both conjugated polymers led to the establishment of a bulk-heterojunction between the Si-ncs and polymer. It is assumed that a fraction of the excitons dissociated at the Si-nc/polymer interface (Švrček et al., 2008b, Švrček et al., 2009c). The HOMO for P3HT and MEH-PPV was 5 eV under vacuum, with the lowest unoccupied molecular orbital (LUMO) at 3 eV. Although the exact work function for Si-ncs is not yet known, it is expected that the value will not be significantly different from the position of the conduction band of bulk silicon (4.1 eV). The optical band gap of the silicon nanocrystallites used is 2 eV in the case of electrochemically etched Si-ncs and 2.9 eV prepared by laser ablation in water, respectively. Thus allow proper bands adjustment and e–h separation in both cases. The different electron affinities and ionization potentials provided a driving force for dissociation when the excitons were generated under irradiation. Compared to MEH-PPV photoconductive response is more important in the case of the lamella like P3HT (Švrček et al., 2009c). This mainly due to a simplify transport of photogenerated carrier in P3HT (Švrček et al., 2009c). In order to determinate the influence of the Si-ncs size on photocurrent generation we have fabricated device based on Si-ncs prepared by laser ablation and electrochemical etching. Fig 7(b) represents schematic structure of Si-ncs/P3HT blend based device. Glass substrate with an ITO electrode was used and an alumina top electrode was deposited on top of Si-ncs/P3HT active layer.

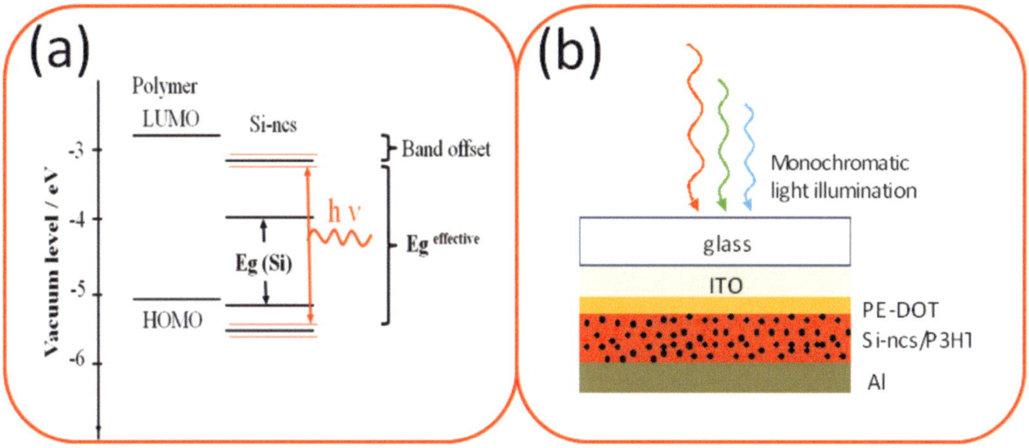

Fig. 7. (a) Energy band diagram of the MEH-PPV polymer and Si-ncs. (b) Schematic structure of Si-ncs/polymer blend based device

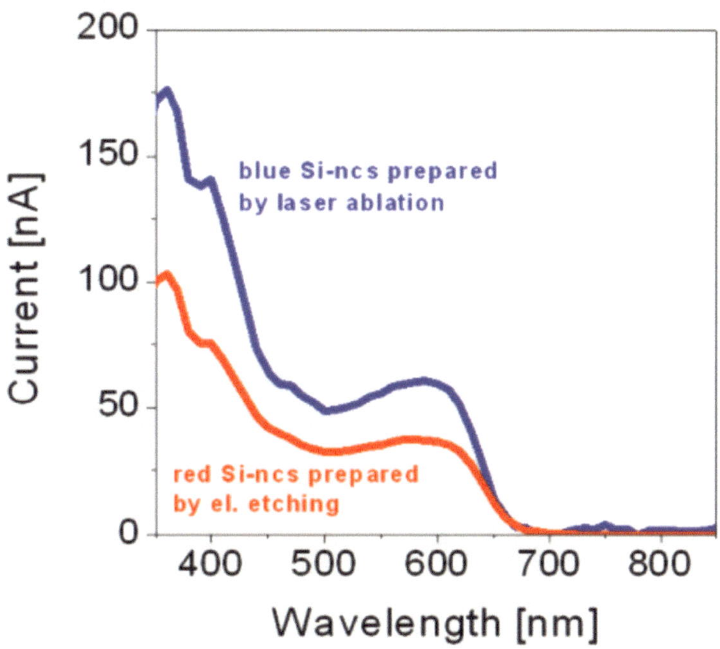

Fig. 8. Photocurrent generation in the Si-ncs/P3HT blends (~38 wt. % concentration of Si-ncs) synthesized by two methods. Red line represents blend with the Si-ncs prepared by electrochemical etching and the blue line by nanosecond laser ablation in water

Figure 8 compares photocurrent generation as a function of the wavelength for blends containing of the Si-ncs blended at ~ 38 wt. % concentration. Red line corresponds to the Si-ncs prepared by electrochemical etching and blue line to Si-ncs synthesized by laser ablation in water. A more important increase in photoconductivity is noticed for the films containing of the blue-luminescent nanocrystals prepared by laser ablation. Similar to disccused above (Švrček et al., 2009c), it is believed that large fraction of the excitons dissociates at Si-nc/polymer interface. Since Si-ncs prepared by laser ablation are smaller in radius (1.5-3 nm) the bulk-heterojunction inteface for exciton dissociation is larger. Then the exciton separation rate is superior to the blend fabricated from the electrochemically etched Si-ncs where the radius is 3-5 nm in average.

6. Filtering and ordering of Si-ncs/co-polymers hybrid composites in nanotubes

It is believed that arrangement of flexible Si-ncs based blends in a form of long nanofibers could resolve multiple problems. It is supposed that the fibrous-like structure will lead to an alignment of Si-nc polymer chains and allow the formation of thicker devices. The nanofiber perpendicular orientation might improve charge collection efficiency through the creation of uninterrupted percolation paths for photogenerated carriers generated by bulk-heterojunction. Noteworthy to say that the nanofiber structure provides a larger interfacial area and has a much lower reflectance compared with the thin films. Several nanotubular architectures have been investigated for potential enhancement of electron percolation pathways in bulk-heterojunction (Coakley et la.,2003). We believe that the nanoporous anodic alumina and TiO_2 are the most feasible large-scale templates that can be used for blend alignment. Furthermore, the size distribution of freestanding Si-ncs can be improved by filtering across the anodic alumina membrane (Švrček et al., 2010). The filtering through the alumina membrane can be used to narrow the size distribution of Si-ncs blended with MEHPPV polymer as well. Figure 9 (a) presents typical top-view SEM images of the nanoporous membrane with an average diameter of 85 nm. The SEM image shows a well-ordered hexagonal packing of the alumina nanopores with a uniform pore size distribution. The high surface energy and O-ring that seal the edges of the membrane kept the liquid on the top of the alumina membrane within the O-ring (Švrček et al., 2010). Then, induced capillary forces from the bottom establish the liquid flow through the opened alumina membrane.

Wettability is another important factor, which contributes to the establishing the flow across one membrane (Švrček et al., 2010). The Young–Laplace equation is commonly used to describe the pressure drop across (Marle, 1981). Once the flow is established, the hydrodynamic flow can be described by the Hagen–Poiseuille equation (Blom, et al., 1996, Švrček et al., 2009c). In this case we deal with the large diameter of pores (Švrček et al., 2010), the flow rate (R) across the alumina nanoporous membrane can be written as

$$R = \frac{\pi d^4}{128\eta}\left(\frac{P}{L}\right) \tag{2}$$

where P is the pressure drop across the pore, d is the pore diameter, η is the solvent viscosity, and L is the membrane thickness. It has to be noted that we achieved the capillary induced wetting of the nanopores when the viscosity of the Si-ncs/MEH-PPV blend dropped on top of nanoporous alumina was high enough to prevent free flow of the blend through the membrane. Compared to TiO_2 advantageous of alumina membrane is that it is easily to be dissolved without dissolution of blend. After dissolution of the membrane in weak 5 wt % phosphoric acid, we could observe the presence of a Si-ncs/MEH-PPV nanofibrous blend. Figure 10 shows typical SEM image of the Si-ncs/MEH-PPV nanofibers after dissolution of the nanoporous alumina membrane. Our observation showed the formation of randomly oriented long fibers only, without the presence of nanotubes. The diameter of the fibers corresponds to the diameter of the alumina nanopores ~80 nm. The wetting phenomena and the presence of Si-ncs in the polymer did not allow formation of nanotubes. However, the nanoporous alumina membrane was quite thick 50 μm and the perpendicularly ordered fibers collapsed. The EDS spectra reveal less oxygen in the blend when after filtering (Švrček et al., 2010). It is supposed that an improvement of both the Si-nc size distribution and the blend morphology limits the oxygen diffusion into the blend. It is well known that oxygen is one of the main factors that lead to the degradation of conjugated polymer-based devices (Scott, et al., 1996]). As this processing can be performed at room temperature it might find application in the future development of new types of optoelectronic and photovoltaic devices.

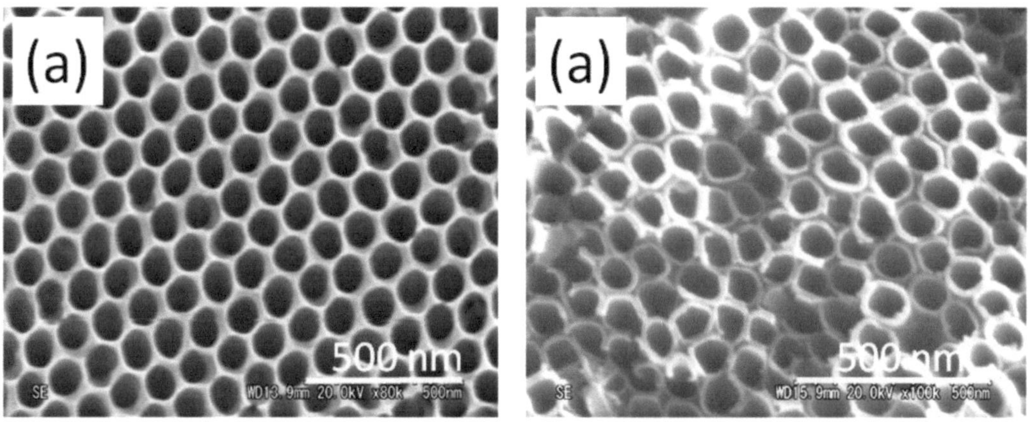

Fig. 9. (a) Top-view SEM images of the nanoporous membrane with average diameter of ~85 nm used for experiments. (b) SEM image of the nanotubular TiO_2 template with diameter of ~90 nm used for vertical order of Si-ncs conjugated polymer based blend

Previous experiments confirmed that the blend can flow through pores with such a diameter. Following of those experiments we have used TiO_2 nanotube template with rather same diameter that the same time serves as an electrode for hybrid device. The morphology of the TiO_2 nanotube with average diameters of 90 nm after thermal treatment is shown in Fig. 9 (b). The transformation of amorphous TiO_2 into crystalline anatase was confirmed by XRD measurements and after the annealing process two most important TiO_2 anatase peaks corresponding to (101) and (200) orientations were clearly recorded (Švrček et al., 2009b). It is observed that infiltration of photosensitive Si-ncs/P3HT blend in the TiO_2 nanotubes results in increase in the photovoltage generation. One can expected that the alignment of Si-ncs/P3HT bulk-heterojunction within TiO_2 nanotubes, which are perpendicular to the contact, will facilitates charge transfer along the nanotubes and reduces losses incurred by charge hopping across the Si-ncs/P3HT blend (Švrček et al., 2009b). Figure 11 (a) shows a

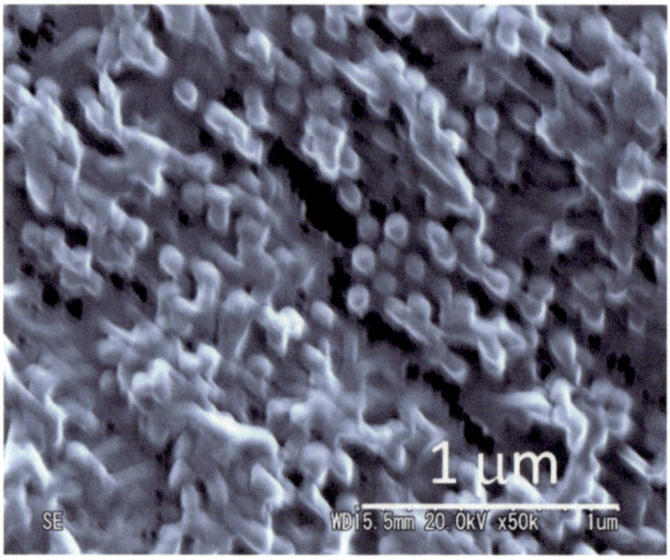

Fig. 10. Scanning electron microscopy image of the Si-ncs/MEH-PPV nanofibers after dissolution of the nanoporous alumina membrane

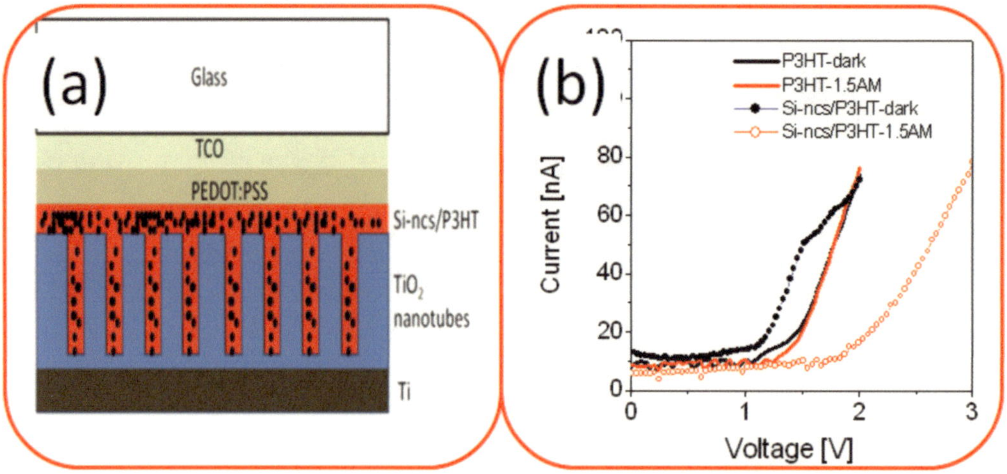

Fig. 11. (a) Schematic sketch of the structure for ordering of Si-ncs/P3HT bulk-heterojunction in TiO_2 nanotube arrays. (b) The current voltage (I-V) characteristics from the pure P3HT polymer and Si-ncs/P3HT blend aligned in within TiO_2 nanotube arrays with the diameters of 90 nm in dark and under illumination AM1.5

schematic sketch of the structure used for alignment of Si-ncs/P3HT bulk heterojunction. The corresponding I-V characteristics of Si-ncs/P3HT heterojunction in vertical configuration in the dark and under illumination at AM1.5 are compared in Fig. 11 (b). The I-V characteristics for P3HT polymer only and Si-nc/P3HT based blend show a p-i-n behavior. However, superior open voltage and ratio between the photo- and dark-conductivity are observed in the case of infiltration of the blend into TiO_2 nanotubes (symbols in Fig. 11 (b)). The photovoltage generation in the bulk-heterojunction most likely results in increase in open circuit voltage. The presence of the Si-ncs in polymer makes the blend to be more semiconductor rather than an insulator; therefore, a high-voltage open-circuit can be partially explained by assuming a Schottky barrier at the Si-ncs/P3HT/PEDOT:PSS interface. The fiber-like geometry provides larger interfacial area, which results in enhanced exciton dissociation. The nanofiber morphology of the blend embedded into nanotubular templates contributed to high open-circuit voltage.

7. Conclusion

We showed that electrochemical etching and nanosecond laser ablation in water are suitable techniques for preparing freestanding and surfactant-free Si-ncs with quantum confinements effects. It is demonstrated that such Si-ncs can be successfully used for the fabrication of room temperature photoluminescent and photosensitive hybrid materials. The materials consisting of Si-ncs with rather different energy band gaps and conjugated polymers (poly(3-hexylthiophene) (P3HT) and poly(methoxy ethylexyloxy phenylenevinilene) (MEH PPV)) with rather similar energy band gaps were fabricated. We argue that the luminescence and transport properties of the blends can be controlled by the Si-ncs properties and could be assigned to quantum confinement of excitons in nanocrystalites. The difference in electron affinity and ionization potential between nanocrystals and polymer leads to bulk-heterojunction formation and excitons desociation. An enhanced photovoltaic effect has been achieved by blending of Si-ncs prepared by laser ablation. The device performances have been improved through enhanced bulk-heterojunction interface area and charge transport. Next, we showed that the nanotube arrays can be efficient for a vertical 1D-like order of photosensitive hybrid material. An arrangement of the Si-ncs/P3HT bulk-heterojunction within ordered Al_2O_3/TiO_2 nanotubes prepared perpendicular to the contact has been shown. It is believed that the Si-ncs non-toxicity, easy integration into well-established silicon technologies and polymer flexibility might bring considerable benefit for hybrid optoelectronic and photovoltaic device development. Needless to mention that the blend flexibility might allow easy and cost effective design new types of nanostructures for enhancement device properties.

8. Acknowledgements

This study was also partially supported by Industrial Technology Research Grant Program from the New Energy and Industrial Technology Development Organization (NEDO) of Japan.

9. References

Canham, L. T. (1990), Silicon Quantum Wire Fabrication by Electrochemical and Chemical Dissolution of Wafer, *Appl. Phys. Lett.* 57, pp.1046-1048.

Beard, M.C.; Knutsen, K.P.; Yu, P.; Luther, J.M.; Song, Q.; Metzger, W.K.; Ellingson, R.J.; Nozik, A.J., (2007), Multiple Exciton Generation in Colloidal Silicon Nanocrystals , *Nano Lett. 7, pp.* 2506-2512.

Blom, P. W. M.; De Jong, M. J. M. ; Vleggaar, J. J. M. (1996), Electron and hole transport in poly(*p*-phenylene vinylene) devices, *Appl. Phys. Lett.*, 68 pp.3308(1-3).

Chang, T. W. F.; Musikhin, S.; Bakueva, L.; Levina, L.; Hines, M.A.; Cyr, P.; Sargent, E.H.; (2004), Efficient excitation transfer from polymer to nanocrystals, *Appl. Phys. Lett.* 84, pp. 4295-4298.

Coakley, K.M.; Liu, Y.X.; McGehee, M.D.; Frindell, K.L.; Stucky, G.D. (2003), Infiltrating semiconducting polymers into self-assembled mesoporous titania films for photovoltaic applications, *Adv. Funct. Mater.* 13, pp. 301-316 .

Dexter, D.L., (1953), A Theory of Sensitized Luminescence in Solids *J. Chem. Phys.* 21, pp. 836-851.

Förster,T., (1959), 10th Spiers Memorial Lecture. Transfer mechanisms of electronic excitation *Discuss. Farraday Soc.* 27, pp. 7-17.

Garcia, C.; Garrido, B.; Pellegrino, P.; Ferre, R.; Moreno, J.A.; Morante, J. R.; Pavesi, L.; Cazzanelli, M., (2003), Size dependence of lifetime and absorption cross section of Si nanocrystals embedded in SiO2, *Appl. Phys. Lett.* 82, pp. 1595 -1598.

Greenham, N. ; Peng, X.; Alivisatos, P. (1996), Charge separation and transport in conjugated-polymer/semiconductor-nanocrystal composites studied by photoluminescence quenching and photoconductivity, *Phys. Rev. B* 54, pp. 17628-17637.

Harris, P.F.J. (2002). *Carbon nanotubes and related structures* (1st ed.). Cambridge University Press. pp. 213–32

Herman. F.; Kasowski, R. V., (1981), Electronic structure of defects at Si/SiO_2 interfaces, *J. Vac. Sci. Technol.* 19, pp. 395-401.

Hirsmman, K. D.; Tsybeskov, L.; Duttagupta, S. P.; Fauchet, P. M. , (1996), Silicon-based visible light-emitting devices integrated into microelectronic circuits, *Nature* 384, pp. 338-341.

Inzelt, G.; (2008). *Conducting Polymers* A New Era in Electrochemistry. *Springer. pp. 265–269.*

Iijima, S. (1991). Helical microtubules of graphitic carbon, *Nature,* 354, pp. 56-58.

Jurbergs, D.; Rogojina, E.; Mangolini, L.; Kortshagen, U., (2006), Silicon Nanocrystals with Ensemble QuantumYields exceeding 60%, *Appl. Phys. Lett.* 88, pp.233116 -233119.

Kovalev, D. I.; Yaroshetzkii, I. D.; Muschik, T.; Petrovakoch, V. F.; Koch P., (1994), Fast and slow visible luminescence bands of oxidized porous Si, *Appl. Phys. Lett.* 64, pp.214 - 217.

Lulease, J.; Fujii, M.; Mimura, A.; Hayashi, S., (2002) *Phys. Rev. Lett.* 89 pp.296805-.296809.

Liu, J.L.; Yan, B.; (2008), Molecular Construction and Photophysical Properties of Luminescent Covalently Bonded Lanthanide Hybrid Materials Obtained by Grafting Organic Ligands Containing 1,2,4-Triazole on Silica by Mercapto Modification, *J. Phys. Chem. C* 112, pp. 14168-14178.

Lui, C., Holman, Z., Kortshagen, U., (2009), Hybrid Solar Cells from P3HT and Silicon Nanocrystals , *Nano Lett.* 9, pp. 449-452.

Marle, C. M. (1981), *Multiphase Flow in Porous Media,* Gulf, Houston, TX .

Martel, R.; Derycke, V.; Lavoie, C.; Appenzeller, J.; Chan, K. K.; Tersoff, J.; Avouris, Ph. (2001), Ambipolar Electrical Transport in Semiconducting Single-Wall Carbon Nanotubes, *Physical Review Letters* 87: pp. 256805-256809.

Murray, C.B.; Kagan, C.R.; Bawendi, M. G. (2000), Synthesis and Characterization of Monodisperse Nanocrystals and Close-Packed Nanocrystal Assemblies, *Annual Review of Materials Research* 30 (1): pp. 545–610.

Ourmazd, A.; Taylor, D. W.; Rentschler, J.A.; Bevk, J. (1987), Si→SiO_2 transformation: Interfacial structure and mechanism," *Phys. Rev. Lett.* 59, pp. 213-216.

Scott, J. C. ; Kaufman, J. H. ; Brock, P. J. ; DiPietro, R. ; Salem, J.; Goitia, J. A.; (1996), Degradation and failure of MEH-PPV light-emitting diodes, *J. Appl. Phys.*, 79, pp. 2745-2752.

Schneider, M.; Muller, K.,; (2000), Hybrid Materials Doped with Covalently Bound Perylene Dyes through the Sol−Gel Process , *Chem. Mater.* 12 , pp. 352-362.

Sukhovatkin, V; Hinds, S.: Brzozowski, L.; Sargent, E.H.; (2009), Colloidal Quantum-Dot Photodetectors Exploiting Multiexciton Generation, *Science* 324 *pp.* 1542-1544.

Švrček, V.; Pelant, I.; Chvojka, T.; Valenta, J.; Dian, J.; Rehspringer, J.-L.; Gilliot, P.; Ohlmann, D.; Crégut, O.; Hönerlage, B., (2002), Photoluminescence properties of sol-gel derived SiO_2 films doped with porous silicon, *Materials Science & Engineering C* 19, pp.233-238.

Švrček, V.; Slaoui, A.; Muller, J.-C. (2004), Ex-situ prepared Si nanocrystals: Their elaboration and characterization in embedded silica glass, *J. Appl. Phys.* 95, pp.3158-3164.

Švrček, V. Sasaki, T.; Shimizu, Y.; Koshizaki, N., (2006), Blue luminescent silicon nanocrystals prepared by ns pulsed laser ablation in water, *Appl. Phys. Lett.*, 89, pp.213113(1-3).

Švrček, V.; Fujiwara, H.; Kondo M.; (2008a), Improved transport and photo-stability of poly[methoxy-ethylexyloxy-phenylenevinilene] polymer thin films by boron doped freestanding silicon nanocrystals, *Appl. Phys. Lett.* 92, p.143301(1-3).

Švrček, V.; Sasaki, T.; Shimizu, Y.; Koshizaki, N.; (2008b), Blue luminescent silicon nanocrystals prepared by ns laser ablation and stabilized in electronically compatible spin on glasses, *J. Appl. Phys.*, 103, pp.023101-023108.

Švrček, V.; Mariotti, D.; Kondo, M.; (2009a), Ambient-stable blue luminescent silicon nanocrystals prepared by nanosecond-pulsed laser ablation in water, *Optics Express*, 17 pp. 520-527.

Švrček, V.; Turkevych, I.; Kondo, M.; (2009b) Photoelectric properties of silicon nanocrystals/P3HT bulk-heterojunction ordered in titanium dioxide nanotube arrays, *Nanoscale Research Letters*, 4 pp.1389-1395.

Švrček, V.; Fujiwara, H.; M. Kondo, M.; (2009c) Luminsecent Properties of doped freestanding silicon nanocrystals embedded in MEH-PPV, *Solar Energy Materials & Solar Cells*, 93, pp. 774-780.

Švrček, V.; Fujiwara, H.; Kondo, M.; (2009c) Top-down silicon nanocrystals and a conjugated polymer-based bulk heterojunction : Optoelectronic and photovoltaic applications, *Acta Materialia* 57 pp.5986-5997.

Švrček. V; Turkevych, K.; Kondo, M.; (2010) Filtering and assembly of Si nanocrystals/conjugated polymers blend with reduced oxygen penetration, *Journal of the Electrochemical Society* 157 (9) pp. K194-K200.

Tenne R, Margulis L, Genut M, Hodes G (1992), Polyhedral and cylindrical structures of tungsten disulphide, *Nature* 360 (6403): pp. 444–446.

Yue, D.; Bai, X.; Zhao,S.; Miao,X.; Li, M.; Dong, J.; Ibrahim,K.; Wang, J.; Zhao,Y.; Yuan, H.; Xing, G.; Sun, B.; (2010), First Endohedral Metallofullerene-Containing Polymer:

Preparation and Characterization of Gd@C_{82}-Polystyrene, *The Journal of Physical Chemistry C 114* (17), pp. 7631-7636.

Zhao, X.; Liu, Y.; Inoue, S.; Suzuki, T.; Jones, R. O.; Andol, Y. (2004), Smallest Carbon Nanotube is 3 Å in Diameter, *Physical Review Letters* 92 (12) pp. 125502-125506.

Zhuravlev, K. S.; Gilinsky, A.M.; Kobitsky, A.Y., (1998), Mechanism of photoluminescence of Si nanocrystals fabricated in a SiO_2 matrix, *Appl. Phys. Lett.* 73, pp.2962-2965.

Wang, X.; Li, Q.; Xie, J.; Jin, Z.; Wang, J.; Li, Y.; Jiang, K.; Fan, S. (2009), Fabrication of Ultralong and Electrically Uniform Single-Walled Carbon Nanotubes on Clean Substrates, *Nano Letters* 9 (9): pp. 3137–3141

11

Synthesis, Self-assembly and Optoelectronic Properties of Monodisperse ZnO Quantum Dots

Ting Mei[1,2] and Yong Hu[3]
[1]*Institute of Optoelectronic Materials and Technology, South China Normal University, Guangzhou 510631,*
[2]*School of Electrical and Electronic Engineering, Nanyang Technological University, 639798,*
[3]*Institute of Physical Chemistry, Zhejiang Normal University, Jinhua, 321004,*
[1,3]*China*
[2]*Singapore*

1. Introduction

Quantum dot (QD), also called artificial atom, is a semiconductor nanocrystal with the size on the order of a few nanometers. By modifying its composition, size and shape, its density of electronic states can be engineered and thus many physical properties can be widely and easily adjusted. Strong quantum confinement of electronic carriers at nanometer scale makes band-gap and luminescence energies size and shape dependent. These dots have now been widely employed as targeted fluorescent labels for biomedical applications.

Zinc oxide (ZnO) is a wide-band-gap semiconductor that presents interesting luminescent properties, which are seen in the recent demonstration of ultraviolet lasing from nanowires. These properties have stimulated the search for new synthetic methodologies for well-controlled ZnO nanostructures. However, many applications require the organization of nanoparticles into two- or three-dimensional (2D or 3D) superlattices. This very efficient organization was obtained by physical methods, whereas chemical approaches toward such organizations would also be of interest since they are easy to perform and allow a facile scale-up procedure.

For the interest of size dependent physical properties, intensive research has been focused on fabricating ZnO QDs with ultra small size that gives strong quantum confinement. By far, various kinds of synthetic approaches have been realized in fabricating such small ZnO QDs, which can be roughly divided into two categories, chemical and physical. Chemical methods involves sol-gel method, hydrothermal growth, thermal decomposition, electrochemical method, while physical methods involves pulsed laser ablation (PLA or PLD), metalorganic chemical vapor deposition (MOCVD), radio frequency (RF) sputtering, flame spray pyrolysis (FSP) method, vapor phase transport (VPT) deposition, etc.. Chemical process takes advantage of the very exothermic reaction of the organometallic precursor bis(cyclohexyl) zinc with water to produce crystalline zinc oxide on one hand, and on the other hand the kinetic control of the decomposition by using long-alkyl-chain amine ligands. However, if the use of amine ligands allows access to ZnO nanoparticles with low size dispersity, these particles are not monodispersed, and consequently no 2D organization was observed. For metal-oxide nanoparticles, a few 2D organizations resulting from particle self-assembly on a surface have been described. ZnO nanoparticles are produced by evaporating the solvent of the reaction solution, a method used in molecular chemistry to form monocrystals. In addition, the synthesis of ZnO nanoparticles is stabilized by combination of amine and acid ligands and their spontaneous organization into 2D ordered superlattices from colloidal solutions.

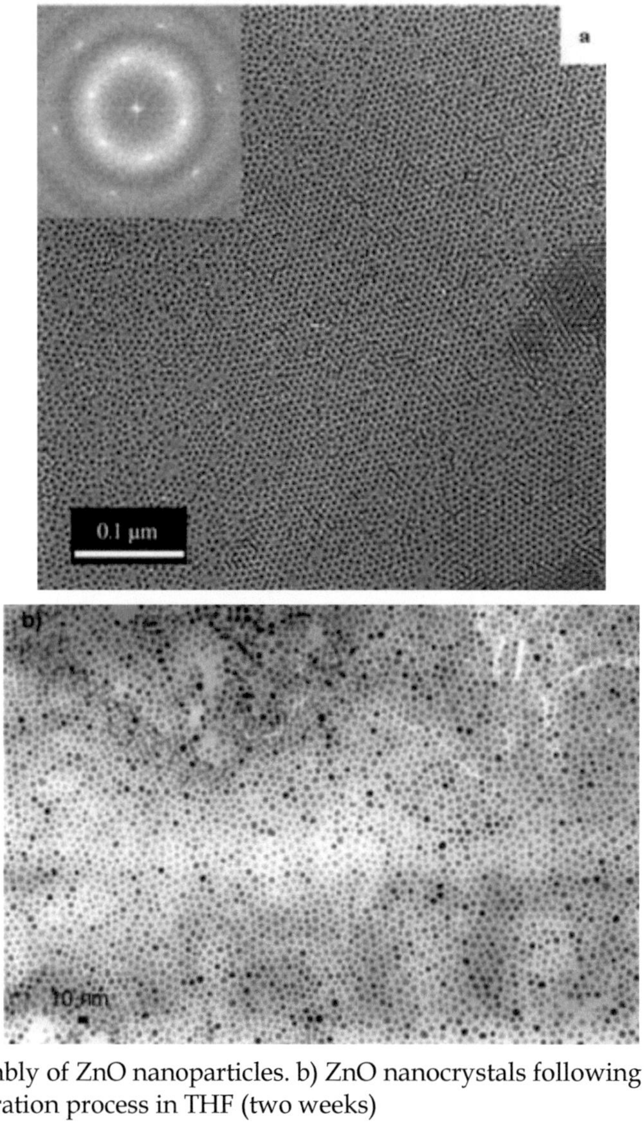

Fig. i. a) 2D assembly of ZnO nanoparticles. b) ZnO nanocrystals following a slow oxidation/evaporation process in THF (two weeks)

If the synthesis is carried out in similar conditions but in the absence of solvent, a liquid fraction is formed even if the reagents are initially solids. For whatever the proportions of the components, the reaction yields isotropic nanoparticles, in contrast to the long nanorods obtained when only longalkylchain amines are present. This probably results in modification of the self-organization of the ligands. The size of the particles obtained under these conditions is independent of the reaction conditions and similar to those resulting from the reaction in THF, but with greater dispersity. This therefore explains why 2D organizations have been previously obtained only with nanoparticles synthesized in THF (see Figure i).

The fundamental study of electronic properties of ZnO QDs is crucial for developing their future applications in nanoelectronics. Electronic transport measurements have been performed on individual ZnO nanowires and nanorods. Recently, the electroluminscence from 2D ZnO nanocrystals have been reported. Intrinsic optical properties of ZnO QDs are being intensively studied for implementing photonic devices. Photoluminescence (PL) spectra of ZnO QDs have also been extensively reported.

In this chapter, we investigate the synthesis and self-assembly of ZnO QDs and attempt to correlate with the optoelectronic properties. The results provide an insight on optoelectronic applications of ZnO QDs.

2. Synthesis of ZnO QDs

In past decades, zero-dimensional (0D) semiconductor nanocrystals, mainly known as QDs or quantum particles, have attracted much attention due to their potential for investigating the dependence of electronic transport or photoelectron emission properties on dimensionality and size reduction (so-called quantum confinement) [1, 2]. Applications of QDs on biology, photovoltaic devices, solar cells, sensors and QD-LEDs also have gained great interest in their relative research fields [3-7]. Much effort has been devoted to fabricating and developing compound semiconductor QDs, such as CdSe, CdTe, GaAs, InAs, InP, ZnS and ZnO [8-14]. A number of synthetic approaches have been used to fabricate those QDs, such as sol-gel [15], hydrothermal growth [16], thermal decomposition [17], pulsed laser ablation [18], flame spray pyrolysis [19], etc. [20, 21].

Zinc oxide (ZnO) is a direct band-gap (E_g = 3.37eV) semiconductor with an exciton binding energy as large as 60 meV at room temperature. The properties of near UV emission and transparent conduction can be exploited in optoelectronic devices such as solar cells, ultraviolet detectors, blue LEDs and LDs [22, 23]. ZnO is a bio-safe material that can be safely used in biological application such as bio-luminescence and bio-sensors [24, 25]. To make use of the size-dependent physical properties, intensive research has been focusing on fabrication of ZnO QDs with ultra small size (<10nm) for better quantum confinement. By far, various kinds of synthetic approaches have been realized in fabricating such small ZnO QDs, which can be roughly divided into chemical methods and physical methods. Chemical methods involves sol-gel method [26-40], hydrothermal growth [41-44], thermal decomposition [45-50], electrochemical method [51-54], while physical methods involves PLA or PLD [55-59], MOCVD [60-63], sputtering [64-66], FSP method [67-70] and VPT deposition [71, 72]. At present, all the methods can be divided into two parts as follow:

3. Chemical methods on synthesis of ZnO QDs

1. Sol-gel method

Sol-gel method, one of the widely used wet chemical methods in synthesizing various nanomaterials [15, 73-75], has been used for fabrication of ZnO QDs [26-40]. Among a number of reports on fabrication of ZnO QDs over the past decades, sol-gel method was the mostly used one due to its simplicity and cheapness. In this method, neither expensive equipments nor harsh conditions are needed. Tokumoto et al and many other research groups have reported synthesis ZnO QDs at low temperature, i.e., less than 80⁰C [27-32]. Meulenkamp has reported a synthetic process with the highest temperature at only room temperature to make as-prepared ZnO QDs at a size of 3-6nm [29].

In typical preparation of ZnO QDs, zinc acetate dehydrate ($Zn(CH_3COO)_2 \cdot 2H_2O$) or purity zinc acetate ($Zn(CH_3COO)_2$) was mostly chosen as Zn precursor, while it is much more flexible to choose other reagents as oxygen source, such as lithium hydroxide (LiOH) [4-8], sodium hydroxide (NaOH) [9-13], Monoethanolamine (MEA) [14,15], Diethylene Glycol (99.5% DEG, ethylenediamine-tetra-acetic acid (EDTA)) [16,17], Triethanolamine (TEA) [18] and water [19,20]. Various alcohols like ethanol [4-8,12,14,], propanol [9-13,19,20] were used as solvents. The procedures of fabricating ZnO QDs are quite simple. Zn precursor and oxygen source were dissolved into alcoholic solvent respectively at a certain temperature with vigorous stirring before cooling down to or below room temperature. Then, these two alcoholic solutions were mixed by adding one solution into the other drop by drop with vigorous stirring at a certain temperature near freezing point. After that, aging or heating was performed to obtain solid ZnO QDs from ZnO sol-gels. Figure 1 shows some TEM images of ZnO QDs prepared by Yadav et al [28]. The size of as-prepared ZnO QDs is arranged from 1nm to 5nm.

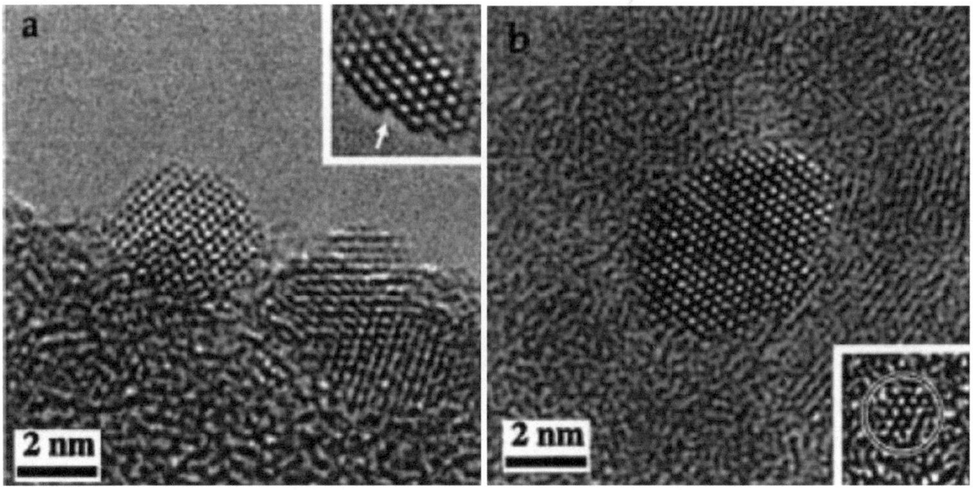

Fig. 1. ZnO QDs synthesized from zinc acetate dehydrate and sodium hydroxide (a) ZnO QDs about 3.7nm in diameter and approximately spherical in shape. The QD shown in the inset demonstrates strong faceting with a surface step consisting of a single atomic layer. (b) A single-crystal ZnO QD that exhibits evidence of faceting and the inset shows a ZnO QD approximately 1.5nm in diameter. (Courtesy of H. K. Yadav et al., [28])

Some detailed studies showed that the growth kinetics of ZnO QDs from colloidal suspensions follows the Lifshitz-Slyozov-Wagner theory for Ostwald ripening and the radius of as-prepared QDs is directly proportional to growth time [31, 32]. It was also shown that the average size of QDs can be tailored by the concentration of zinc precursor [37].

2. Hydrothermal growth

Solvent-thermal growth method is one of the widely used methods in liquid phase. As reactions were most probably occurred in aqueous solution, solvent-thermal growth method was well known as hydrothermal growth method. It is a simple, reliable and effective method in synthesizing nanomaterials with different shapes under low supersaturation and mild conditions [16, 76-78]. ZnO QDs [41-43] as well as other nanostructures, such as nanowires, nanorods, nanotubes and other special shapes [79-83], has been fabricated using hydrothermal growth method. Hu et al [41] has reported fabrication of ZnO QDs with an average size of 3.5nm and good monodispersion (figure 2).

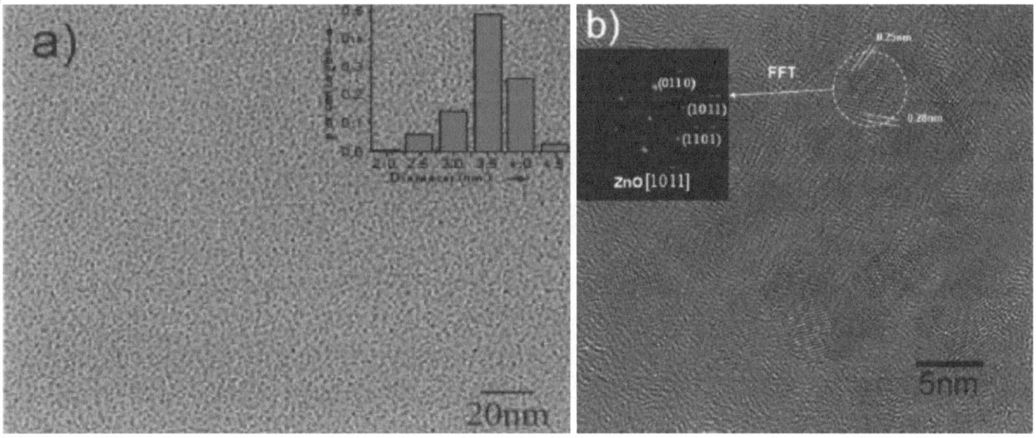

Fig. 2. Characterization of monodisperse ZnO QDs prepared by hydrothermal growth method. (a) Bright field TEM image and histogram of the size distribution (inset). (b) HRTEM image with fast Fourier transform (FFT) of the ZnO QDs. [41]

In a typical synthesis procedure, both inorganic zinc salts, such as $ZnCl_2$ [41, 42], $ZnSO_4$ [43], and organic zinc salts like $Zn(CH_3COO)_2 \cdot 2H_2O$ [44] were used as zinc source, while NaOH [41, 42, 44] and KOH [43] were used as oxygen source. Both alcohol (ethanol [41, 42] and methanol [44]) and deionized water [41-43] could be used as solvent. Zinc salts were firstly dissolved in solvent with ultrasonic vibration and the same treatment was done for the alkali metal hydroxides at room temperature. After mixing those two prepared solutions by stirring, the final solution was transferred to a sealed autoclave and heated at relatively high temperature for hours. After centrifugation of the solution, ZnO QDs was obtained by washing with deionized water and alcohol as well as drying in the air.

Different from sol-gel method, sealed autoclaves were used to get high pressure and higher temperature was needed in hydrothermal growth process. The introduction of high pressure and high temperature accelerated the processes of synthesis. Therefore, less time was needed compare to the long time procedure of sol-gel process.

3. Thermal decomposition

Thermal decomposition method, based on decomposition of metal salts or metal oxides through heating, is theoretically and experimentally simple. To the best of our knowledge, there have been many metal oxides nanoparticles, such as Fe_2O_3, NiO, Co_3O_4 and ZnO [84-87], synthesized via thermal decomposition method. However, there are not many reports on synthesizing ZnO quantum dots with ultra small size [45-50]. Fabrication of ZnO QDs with an average size of 1-10nm by decomposing zinc salts have reported by Cozzoli et al (Figure 3) [47].

The thermal decomposition of ZnO precursors could be taken place either in liquid or air environments. As for synthesis taken place in liquid environment, high temperature resistance solvents were needed, such as n-hexadecylamine (HDA), n-dodecylamine (DDA), tri-n-octylamine (TOA), 1-hexadecanol (HD), 1-octadecene (OD) and purified mineral oil [45-47], and $Zn(CH_3COO)_2 \cdot 2H_2O$ was used as precursors. Before putting Zn source into solvents, it was firstly dissolved in water or oleic acid (OA). The mixture was then heated up to a certain high temperature with stirring and kept for hours. After that, ZnO QDs were obtained by filtering, washing and drying. For synthesis taken place in air environment, instable zinc salts or instable zinc oxides were synthesized in solution or in air first [48-50]. After filtering and washing precipitation formed by mixing zinc source and oxygen source, as-obtained precursors of ZnO QDs were heated in air for hours until decomposition was completed.

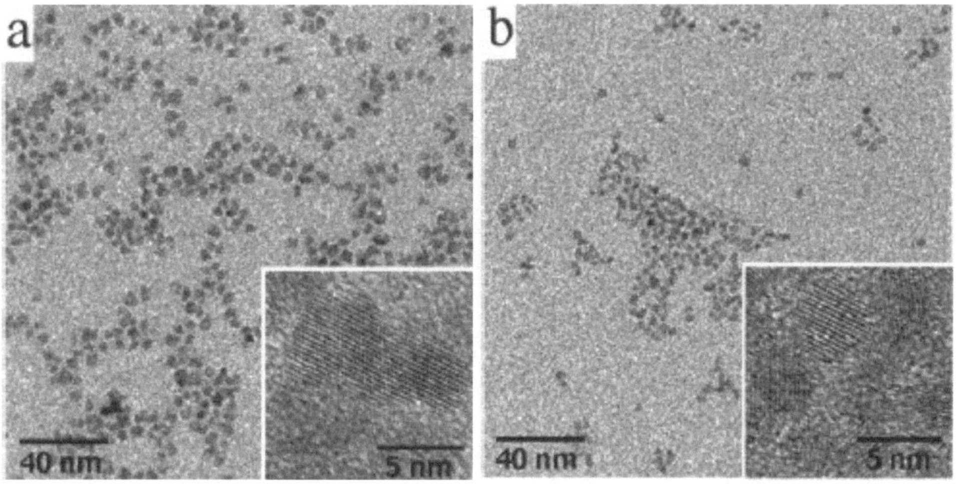

Fig. 3. TEM images of ZnO QDs prepared by thermal decomposition method: ZnO QDs prepared in *n*-hexadecylamine (HDA) with different molar ratio (R) of *tert*-Butylphosphonic acid (TBPA) and zinc acetate ($ZnAc_2$), (a) R=0.12 (b) R=0.52. In the inset, HR-TEM images show ZnO QDs with diameters less than 5nm. (Courtesy of P. D. Cozzoli et al [47])

4. Electrochemical growth method

Electrochemical growth method is a novel and simple method to fabricate ZnO QDs. Only a few reports have been shown up on it. Mahamuni and co-workers [51-54] have reported a lot of their work on synthesizing ZnO QDs using electrochemical growth method. Figure 4 shows TEM images of as-prepared ZnO QDs with average sizes of 6.1nm and 9.3nm [51].

In typical preparation of ZnO QDs, an electrochemical bath consisted of a mixture of acetonitrile, tetrahydrofuran (THF) and tetra-octyl-ammonium-bromide (TOAB). Metal zinc served as a sacrificial anode while platinum served as cathode. Electrolysis was taken for a few hours in constant current mode in oxygen ambient at room temperature. ZnO QDs deposited at the bottom of the electrochemical bath and then, were obtained after centrifugation.

Compared with most of other chemical methods for synthesis of ZnO QDs using OH-ion solution as the reagent, electrochemical growth method can effectively prevent ZnO QDs suffering from OH-ion. Otherwise, it is very easy for ZnO QDs to coat with a $Zn(OH)_2$ layer to form $ZnO/Zn(OH)_2$ core/shell QDs or nanoparticles [88, 89]. The properties of ZnO QDs can be strongly influenced by $Zn(OH)_2$ shell. Electrochemical growth method has provided a novel way to fabricate high quality ZnO QDs.

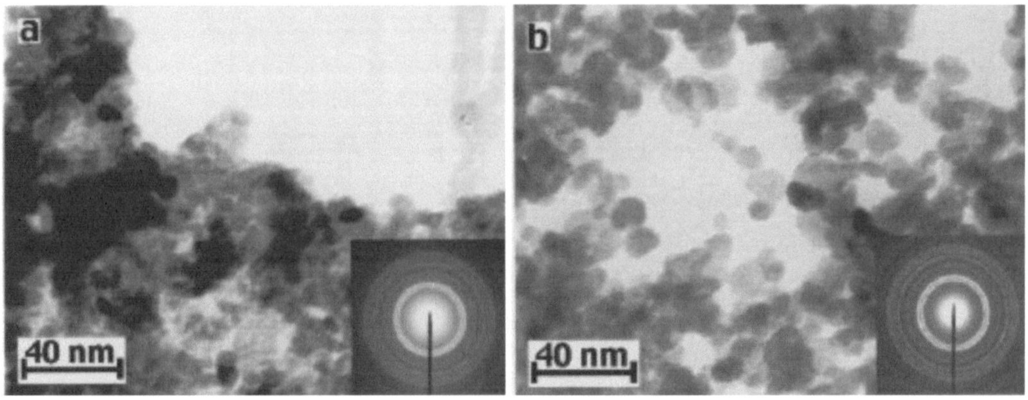

Fig. 4. Bright-field TEM images of: (a) 6.1 ± 2.0nm and (b) 9.3 ± 3.1nm ZnO QDs prepared by electrochemical growth method. Selected area diffraction patterns matching wurtzite ZnO are shown in the insets. (Courtesy of S. Mahamuni et al [51])

4. Physical methods on synthesis of ZnO QDs

1. Pulsed laser ablation (PLA or PLD)

Pulsed Laser Ablation is a novel technique to fabricate nanomaterials, which is based on ablating metals or metal compound materials using a laser with high pulsed power, [90-94]. The whole process can take place in either liquid environment or a vacuum chamber, known as Liquid-phase Pulsed Laser Ablation (LP-PLA or PLA) and Pulsed Laser Deposition (PLD), respectively.

Pulsed laser ablation method, which has been proven to be an effective method for fabrication of nano-sized metal QDs, including Au, Ag, Ti and Ni [90, 95-97], is also a promising method to fabricate QDs of compound semiconductors, such as FeO, ZnO, CdS and TiO_2 [55, 91, 92, 98]. Ajimsha et al [55] prepared ZnO QDs with an average size of 7nm in various liquid media using PLA method (Figure 5).

In preparation of ZnO QDs, third harmonic of Nd:YAG laser (355nm), amplified Ti:Sapphire laser (800nm) and KrF excimer laser (248nm) were usually used. A sintered ZnO bulk or a Zn metal plate was used as the target immersed in different liquid media like deionized water, methanol, and ethanol or in aqueous solutions with different surfactants like sodium dodecyl sulfate (SDS), HCl and NaOH. The target was rotated during the ablation to avoid repeated ablation on the same spot by continuous irradiation of the focused laser beam. ZnO QDs were finally dispersed in the liquid media or solution.

PLD was widely used in fabrication of thin film as many reports have been shown up over the last decade [94, 99,100]. However, it has quite rarely reported fabrication of QDs because of aggregation. Barik et al [58] and Chen et al [59] have realized fabrication of ZnO QDs using PLD method. PLD method is quite similar to PLA method, whereas the only difference is the synthetic environment. Vacuum condition is needed in PLD process. The as-prepared ZnO QDs were finally deposited on the substrate fixed by the target source. Detailed study has demonstrated that the size of as-prepared ZnO QDs was influenced by the deposition time, ambient temperature and laser power.

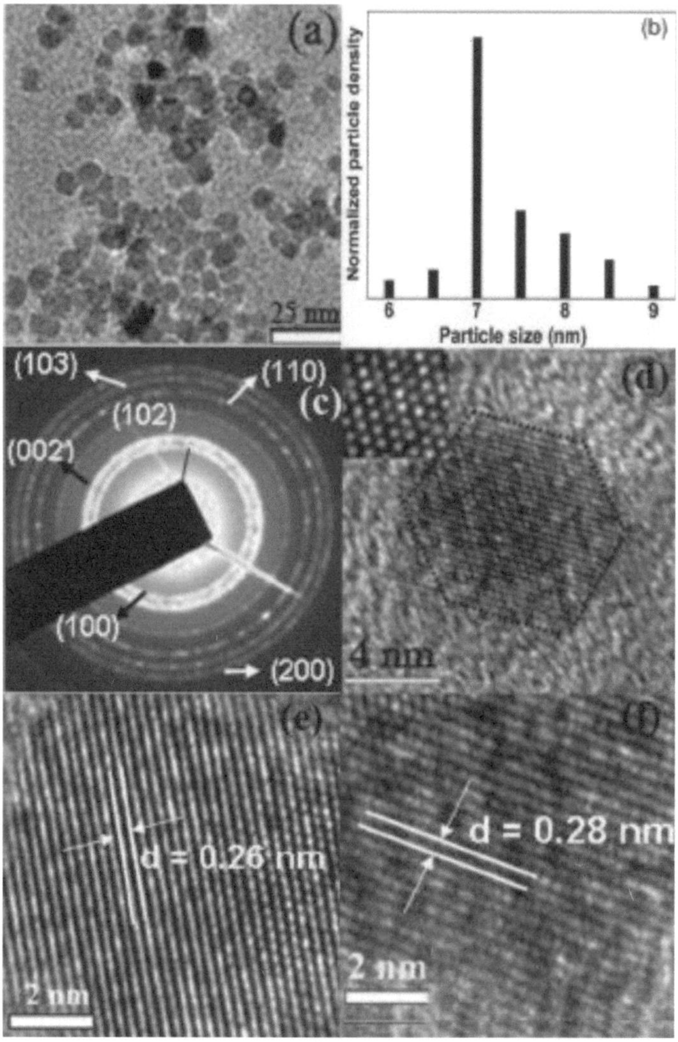

Fig. 5. TEM images and particle-size distribution of ZnO QDs prepared by PLA: (a) TEM image, (b) particle-size distribution, and (c) SAED pattern of ZnO QDs obtained by laser ablation with a fluence of 25 mJ/pulse in water. (d) HR-TEM image for a single QD and (inset) arrangement in the hexagonal close-packed mode. (e and f) HR-TEM showing (002) and (100) planes, respectively. (Courtesy of R. S. Ajimsha et al. [55])

2. Metalorganic chemical vapor deposition (MOCVD)

This method known as metalorganic chemical vapor deposition (MOCVD) is widely used in synthesis of various species of materials with different shapes, especially thin films or multilayer devices [101-103]. As nano-sized ZnO has attracted a lot of attention in the past decade, MOCVD also began to be used in fabricating ZnO nanostructures, such as nanowires, nanorods [104, 105]. However, there are not many reports on synthesis of ZnO QDs using MOCVD [60-63]. Tan et al [60] reported fabrication of ZnO QDs embedded films on Si substrates (Figure 6). Detailed study showed that with increasing substrate temperature, the diameters of QDs were increased with less dispersion.

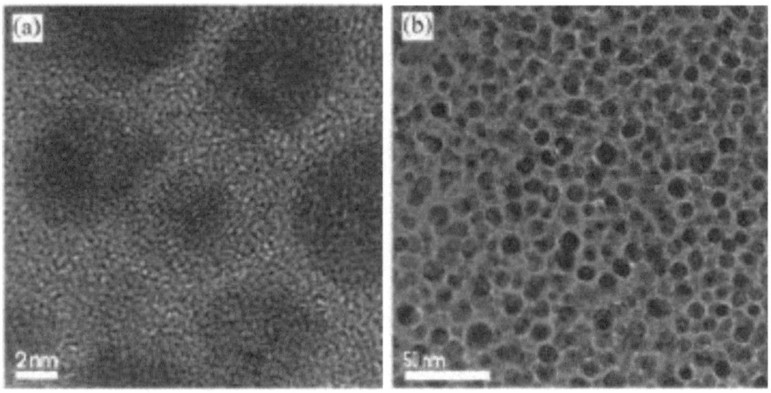

Fig. 6. TEM images of ZnO QDs prepared by MOCVD: (a) high magnification and (b) lower magnification of ZnO QDs embedded film fabricated on Si substrates. (Courtesy of S. T. Tan et al, [60])

In typical preparation of ZnO QDs, diethylzinc (DEZn) was chosen as the source of zinc, while the source of oxygen can be obtained from different gases, such as nitrous oxide (N_2O), nitrogen dioxide (NO_2) and oxygen (O_2). Nitrogen (N_2) was mostly employed as the carrier gas. However, most probably, there is a SiO_2 layer on the surface of the Si substrate due to oxidative environment at high temperature. Therefore, as-prepared ZnO QDs may be embedded in the SiO2 layer or deposited on the SiO2/Si substrate.

Kim et al [62, 63] reported fabrication of ZnO QDs assisted with focused ion-beam. Nano-sized patterns were made on the substrate by focused ion-beam and high aligned ZnO QDs were prepared along the pattern by MOCVD.

3. Radio frequency (RF) sputtering

Radio Frequency (RF) Sputtering is an effective method for preparation of thin films [106-108]. However, reports on preparation of QDs using RF sputtering is quite rare, mainly due to the aggregation of QDs on substrate which leads to form thin films. Recently, ZnO QDs embedded in SiO_2 and SiO_xN_y films were prepared via RF sputtering method by Peng et al [64, 65] and Ma et al [66].

In the preparation of ZnO QDs, high purity ZnO pieces were fixed at the certain distance from the substrate or directly on the substrate during sputtering. Si or Si_3N_4 were used as substrates in order to fabricate different ZnO QDs embedded films. The whole sputtering procedure was carried out in a low ambient pressure with a mixture of Ar and O_2 introduced into the chamber. After deposition, SiO_2 or SiO_xN_y film with ZnO QDs embedded were obtained on the substrates.

It was also reported that ZnO QDs embedded SiO_2 films were prepared by annealing ZnO thin films on SiO_2/Si substrates. The ZnO films were grown on Si substrates at room temperature by RF sputtering [109].

4. Flame spray pyrolysis (FSP) method

Flame spray pyrolysis (FSP) is a promising technique in flame-made technology. This method has been widely used in fabrication of many nano-sized materials including TiO_2, ZrO_2, Al_2O_3, SiO_2, SnO_2, ZnO, CeO_2 QDs [110-116] and multicomponent oxide QDs, such as ZnO/SiO_2, TiO_2/Al_2O_3 and CeO_2/ZrO_2 [117-119]. In the FSP process, the heat generated from fuel combustion was used to decompose the metalorganic precursors. After decomposition, gas-phase metals rapidly react with oxygen to form metal oxide particles.

Over the past decade, Madler et al [68] has successfully fabricated ZnO QDs with diameter about 1.5nm using FSP method. Figure 7 shows HR-TEM images of as-prepared ZnO QDs [68].

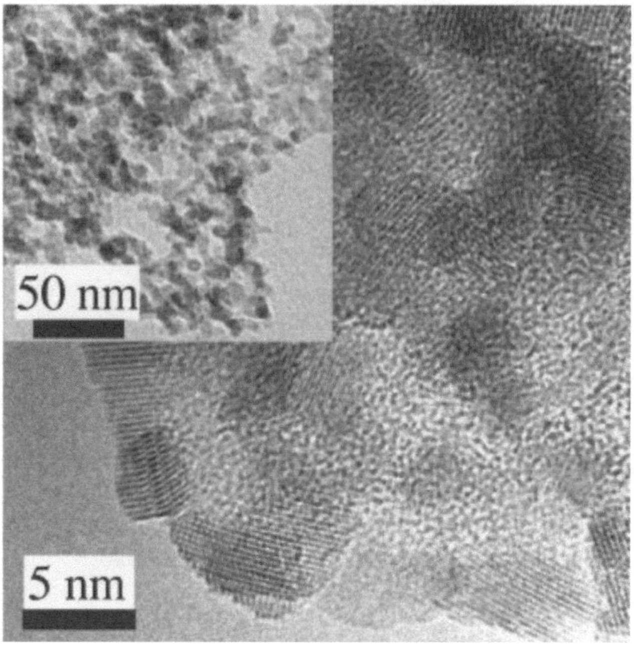

Fig. 7. ZnO QDs prepared by flame spray pyrolysis (FSP) method. (Courtesy of L. Madler et al. [68])

In typical preparation, zinc acrylate was chosen as zinc precursor and dissolved in the solvent, a mixture of acetic acid and methanol. As-obtained transparent solution was fed by a glass syringe into the oxygen-assisted atomizer nozzle, and then sprayed into the high temperature reaction zone which was surrounded by supporting flamelets. With the aid of a vacuum pump, the produced QDs were collected on a glass fiber filter. The highest temperature of the flame in the center was estimated to be above 2000K. ZnO was easily obtained from decomposing metalorganic precursors and reacting with oxygen.

Liewhiran et al [70] fabricated ZnO QDs with a diameter of 10-20nm which presented high performance in ethanol sensors. Against aggregation, reducing the precursor feed rate helps to decrease the diameter of ZnO QDs [68].

5. Vapor phase transport (VPT) deposition

The vapor phase transport (VPT) deposition process which needs very simple equipment has been extremely wide and successful used in fabricating nano-sized compound materials, including CdS, ZnS, ZnO, etc. [120-124]. The only equipment used in VPT process is a horizontal tube furnace with high-temperature capacity. ZnO nanomaterials have obtained great development via introducing this promising technique. The typical structures, such nanowires, nanorods or array of nanorods and nanobelts, have been realized in the past decade [122, 125-128]. 0D ZnO nanomaterials have been rarely reported until recently, Lu et al [71, 72] reported fabrication of ZnO QDs with controllable size using VPT process. Figure 8 shows as-prepared ZnO QDs with a diameter around 7nm [71].

In typical preparation, Znic acetate dihydrate ($Zn(CH_3COO)_2 \cdot 2H_2O$) was used as the zinc source, while two gases, O_2 and Ar, were input as oxygen source and carrier gas. A one-end-sealed quartz tube was used to hold source materials and Si substrates. The whole quartz tube was placed in a sealed horizontal tube furnace. High temperature was needed to decompose metalorganic to vaporize metals and react with oxygen. ZnO QDs were deposited at the low temperature zone where substrates were located.

Detailed study has been experimentally demonstrated that smaller size with better dispersion of ZnO QDs could be achieved by reducing the growth time at the highest temperature (Figure 8).

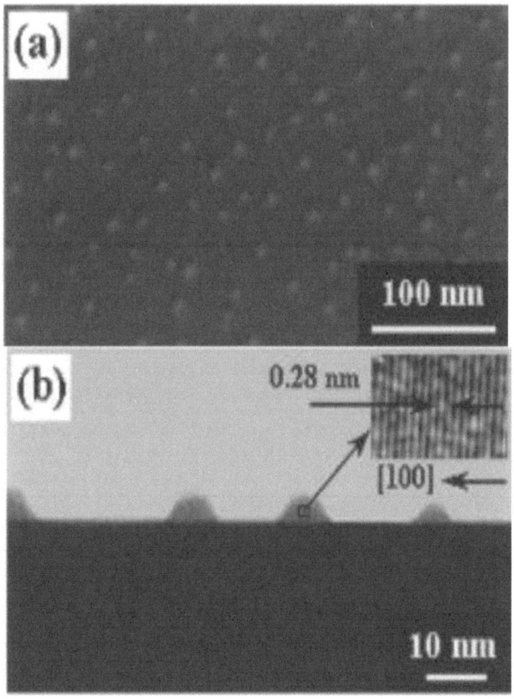

Fig. 8. (a) Plan-view and (b) cross-sectional TEM images of as-prepared ZnO QDs. An inset high-resolution TEM image shows that the ZnO QDs are single crystal. (Courtesy of J. G. Lu et al. [71])

5. Self-assembly of ordered 2D and 3D superlattices of ZnO QDs

ZnO is a luminescent, wide-band-gap semiconductor material that may find applications in a wide range of domains. However, many applications require the organization of the nanoparticles into two- or three-dimensional (2D or 3D) superlattices.[129] Recently, Bruno Chaudret and co-workers reported a study concerning the conditions governing the formation in solution and the dissolution of superlattices of ZnO QDs in the presence of long-alkyl-chain amines, carboxylic acids, or binary amine/carboxylic acid mixtures.[130-131]

The synthesis of ZnO QDs is carried out by slow hydrolysis in air of a THF solution containing the precursor $ZnCy_2$ as well as one equivalent of a long-alkyl-chain amine and half an equivalent of a long-alkyl-chain acid. Thus, when the reaction mixture was exposed to air and left withstanding room temperature, the solvent slowly evaporated, left with white and luminescent products. 2D organizations were spontaneously obtained by depositing one drop of the colloidal solution onto a carbon coated copper grid. Table 1 summarizes the results obtained under various reaction conditions and Figure 9 illustrates some examples of the materials obtained.

Ligand[a] (per mole $ZnCy_2$)	Solvent	Nanoparticle size [nm][b]
HDA/0.5 OcA	THF	3.5 ±1.0
HDA/0.5 OlA	THF	3.1 ±1.1
HDA/0.5 LcA	THF	2.8 ±0.8
DDA/0.5 OcA	THF	3.5 ±0.9
DDA/0.5 OlA	THF	3.5 ±1.7
DDA/0.5 LcA	THF	3.5 ±1.0
OA/0.5 OcA	THF	3.1±0.9
OA/0.5 OlA	THF	3.7±0.9
OA/0.5 LcA	THF	3.2±0.7
OA/0.5 OcA		3.6 ±1.1
OA/0.5 OlA		3.8 ±1.8
OA/0.5 LcA		3.7 ±1.4
2OA/0.5 OlA		3.6 ±1.4
5OA/0.5 OlA		3.4±1.5

Table 1. All procedures were carried out at room temperature, with an incubation time of 17 hours. The particles then were maintained in moist air for 4 days. For each experiment, the concentration of the zinc precursor was 0.042 M. [a]: HDA, DDA, OA, OcA, OlA, and LcA stand for hexadecylamine, dodecylamine, octylamine, octanoic acid, oleic acid, and lauric acid, respectively. [b]: The mean diameter of the obtained nanoparticle was evaluated by fitting of the histogram with a Gaussian curve. The first value corresponds to the centre of the peak whereas the second one corresponds to two times the standard deviation of the Gaussian distribution, or approximately 0.849 times the width of the peak at half height. (Courtesy of M. L. Kahn et al. [130])

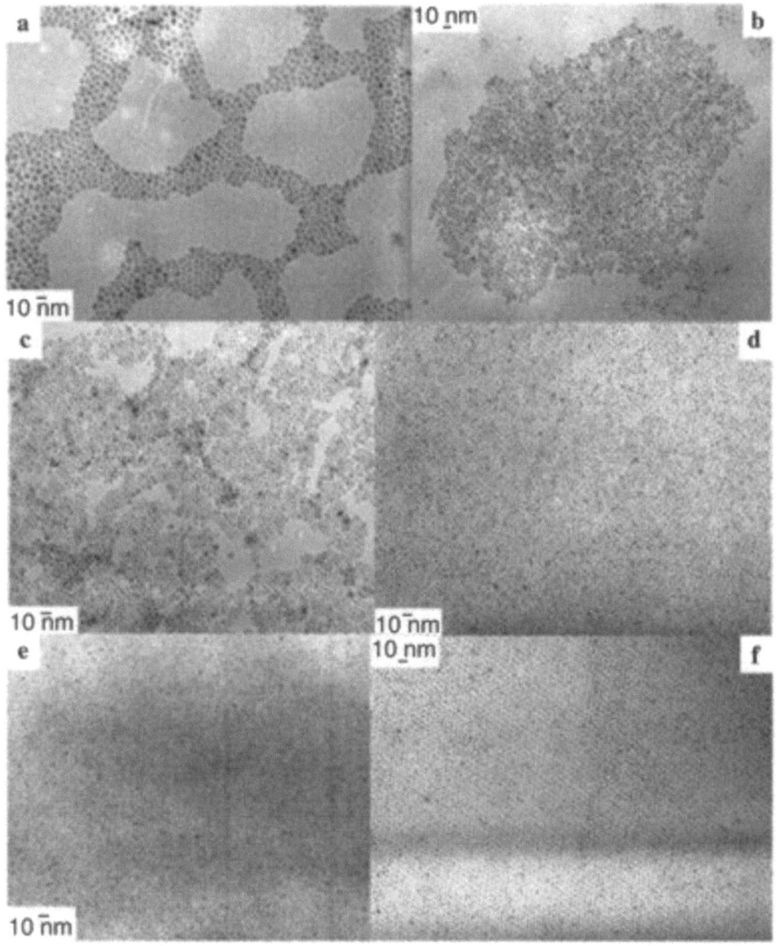

Fig. 9. TEM pictures of ZnO nanoparticles obtained in the presence of: a) OA and OlA (1:1) without other solvent; b) OA and OlA (2:1) without other solvent; c) OA and OlA (10:1) without other solvent; d) OA and LcA (1:1) in THF; e) DDA and OcA (1:1) in THF; f) HDA and OlA (1:1) in THF. (Courtesy of M. L. Kahn et al. [130])

3D superlattices were obtained through slow evaporation of colloidal THF solutions. The superlattices are not monodisperse in size but display uniform hexagonal shapes. Interestingly, this shape may be related to the intrinsic hexagonal crystallographic structure of ZnO, as previously observed for the organization of iron nanocubes.[132, 133] Figure 10 shows SEM image of small 3D superlattices of 3.1 nm ZnO nanoparticles obtained from HDA and OlA. Larger 3D superlattices may be obtained when crystallization is performed until the quasi-complete evaporation of the solvent.

In addition, Chen et al. [134] also reported that highly ordered, free-standing, complex, 2D or 3D nanoparticle superlattice arrays were created by either evaporation-induced self-assembly or precipitation-induced self-assembly from the size-monodispersed ZnO nanoparticles functionalized by carboxylic and alkylthiol ligands recently. The synthesis of the constituent ZnO nanoparticles is both novel and facile and is suggested as a generic means of producing oxide nanomaterials by reaction of metal cations and oxide anions in alcohol solutions at room temperature. The regularity of the nanoparticle products and their chemical functionalization provides the means to achieve superlattice thin films on indium tin oxide (ITO)-coated glass slides (Figure 11).[134]

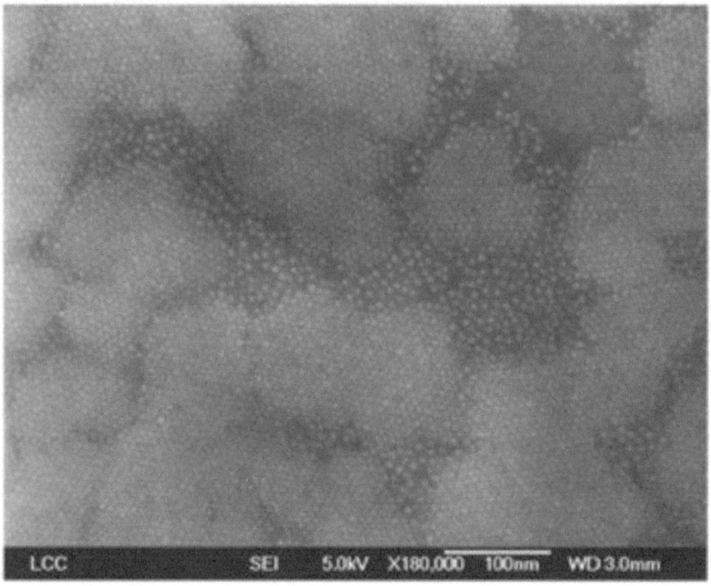

Fig. 10. SEM image of 3D superlattices of ZnO QDs prepared in the presence of HDA and OlA. (Courtesy of M. L. Kahn et al. [130])

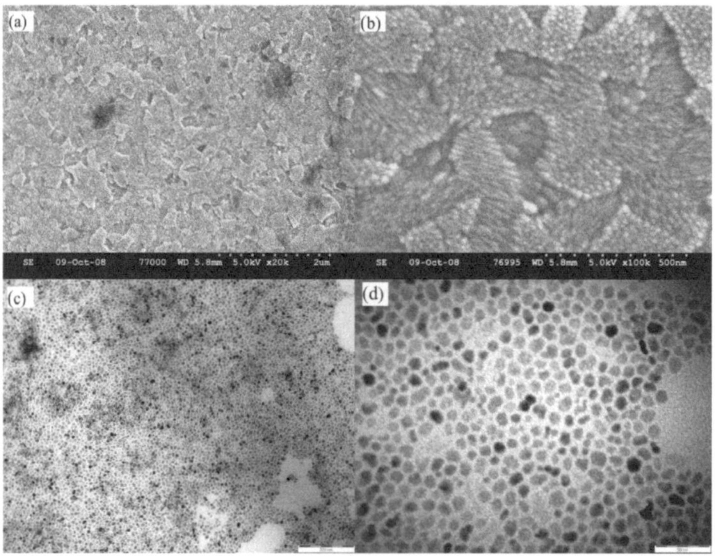

Fig. 11. EISA of OLA-ZnO QD superlattices. The (a) low- and (b) high-magnification SEM images of the layered 3D OLA-ZnO NP superlattice film and the (c) low- and (d) high-magnification TEM images of the 2D OLA-ZnO NP superlattice array self-organized on ITO slides. (Courtesy of L. Chen et al. [134])

The functionalization of these ZnO nanoparticles by long-chain carboxylic acids or alkylthiols improves the size monodispersivity, and semiconduction properties dramatically. The 2D or 3D self-assembly process is driven by the interfacial energy minimization between polar solvents and the hydrophobic moieties of OLA-ZnO NPs where well-organized, ordered nanoparticle superlattice structures are produced by either evaporation-induced self-assembly (EISA) or precipitation-induced self-assembly (PISA).

6. ZnO QDs optoelectronic properties and application

1. The Photoluminescence and Bandgap of ZnO QDs

The photoluminescence (PL) spectrum of the as-prepared ZnO QDs sample reported by Hu et al. [41], was measured at room-temperature with the excitation at a wavelength of 325 nm, as shown in Figure 12. A strong UV light emission peak is clearly observed at 362 nm, corresponding to transition energy of 3.42 eV. A remarkable blue shift of 0.16 eV with respect to the gap energy (3.26 eV) of bandedge emission of bulk ZnO crystals is obtained due to the quantum confinement effect in ZnO QDs. [135] The corresponding band-gap energy is larger than the emission peak energy by the excitonic binding energy. Such an enlarged band gap implies a diameter of 3.5 nm of QDs.[136]

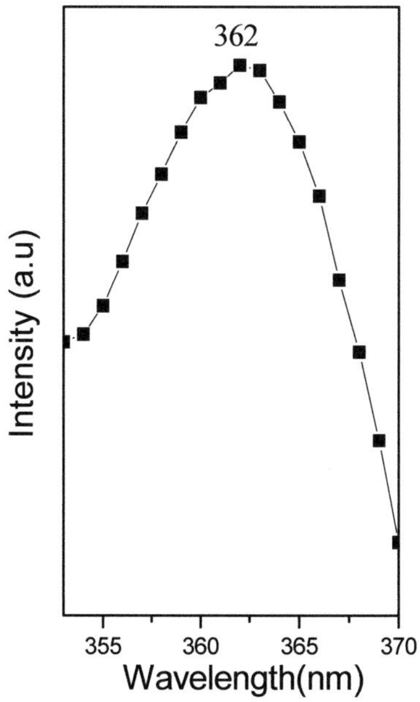

Fig. 12. PL spectrum of ZnO nanocrystals (λ_{ex}=325nm). [41]

The ZnO nanodots were assembled into an ITO/ZnO/SiO_2/Al structure (see Scheme 1) in order to further determine the band alignment of ZnO nanodots on ITO via analysis of I-V characteristics. Figure 13 presents the band diagrams of the ITO/ZnO/SiO_2/Al structure under the two opposite biases. Due to high barriers in both conduction and valence bands at the Al/SiO_2 interface, electron and hole injections from the SiO_2 end were negligible. This is evidenced by the experimental result that the current ratio of ITO/ZnO/SiO_2/Al to ITO/SiO_2/Al is about 100. Figure 14a shows the forward biased I-V characteristics of these two structures. Therefore, we can consider unipolar carrier injection in the ITO/ZnO/SiO_2/Al structure. For sufficiently high bias as illustrated in Figure 13, holes or electrons may tunnel through the ZnO nanodot layer and contribute an appreciable current that follows the Fowler-Nordheim (FN) tunneling equation [137, 138]

$$I_{FN} = C_{FN}E^2 exp(-\frac{4}{3}\frac{(2m^*)^{1/2}}{e\hbar}\frac{\phi^{3/2}}{E}),$$

where I_{FN} is the current, E is the electric field across oxide, m^* is the electron or hole effective mass, e is the electronic charge, $\hbar$ is the reduced Planck's constant, and ϕ is the barrier height. As shown in Figure 14b and c, the FN plots $\ln(I/E^2)$ vs. $1/E$ at high electric field show a good linearity as the sign of the FN tunneling mechanism. The slopes of -2×10^{10} and -3×10^9 can be determined from the linear regions in forward bias and reverse bias, respectively. Apart from the linear regions, other current injection mechanisms may be dominant, e.g. space charge limited current, etc.. Given effective masses m_h^* (=0.59m_0) and m_e^* (=0.19m_0) for hole and electron,[139, 140] the barrier heights are calculated to be 2.44eV and 1.0 eV, indicating the band alignment of the ZnO nanodots on ITO. The barrier heights sum up to be 3.44eV, which agrees to the value of the band gap of ZnO nanodots derived from the PL test very well.

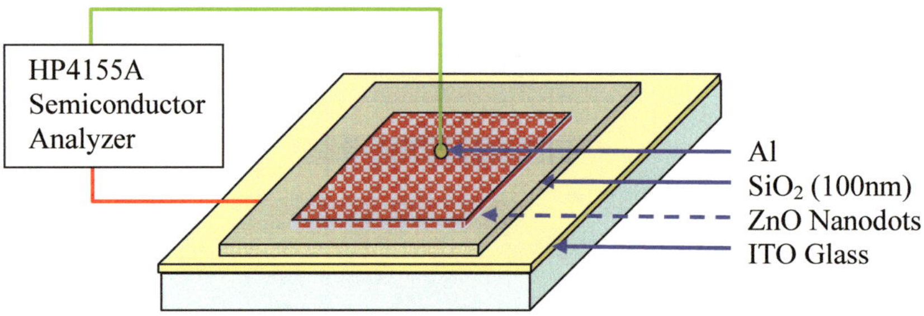

Scheme 1. The fabrication of ITO/ZnO/SiO_2/Al structure

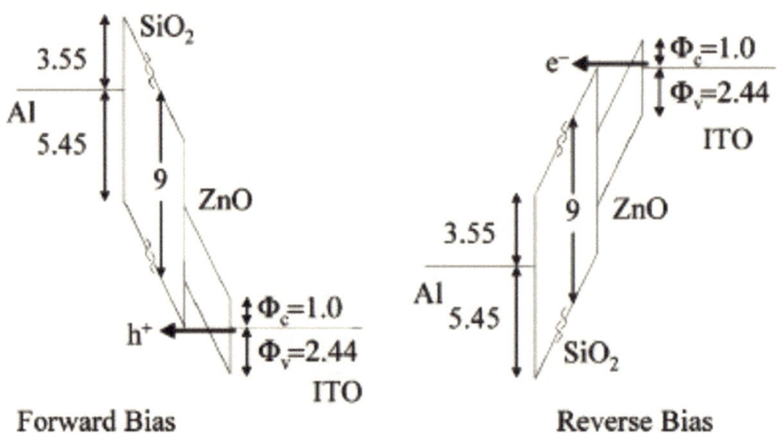

Fig. 13. Energy band diagrams for the ITO/ZnO/SiO_2/Al structure biased under a high field showing hole and electron tunneling, respectively. [41]

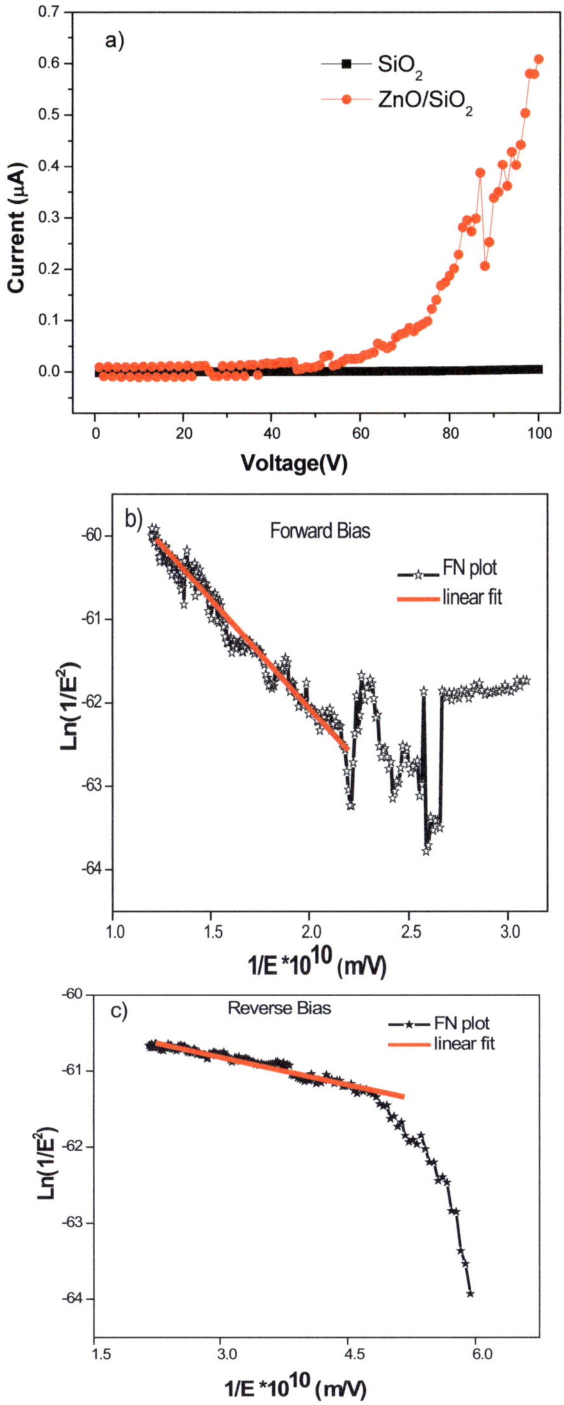

Fig. 14. a) I-V characteristics of the ITO/ZnO/SiO_2/Al and the ITO/SiO_2/Al structures at forward bias; and Fowler-Nordheim (FN) plots for the ITO/ZnO/SiO_2/Al structure at b) forward biases and c) reverse biases. [41]

2. The Electroluminescence of monolayer ZnO nanoparticles

Ultraviolet electroluminescence from ZnO nanoparticle-based devices prepared by the dry-coating technique has been investigated by Lee et al. This section was developed based on their report [141].

With dry-coating process, the structure of the ZnO nanoparticle monolayer (90 nm) in the device can be easily achieved. The method reduces the density of pinhole defects in the ZnO nanoparticles. In this work, they prepared three types of ZnO EL devices formed with a host polymer, ZnO nanoparticle monolayer, and an electron transporting layer. The following host matrix materials dissolving in chloroform with different concentration are chosen respectively: 0.7 wt % poly(fluorine) (PF), 1 wt % poly(*N*-vinylcarbazole) (PVK), and 0.7 wt % poly(3-hexylthiophene) (P3HT). For the electron transporting layer, they employed small-molecular aluminum tris-8-hydroxyquinoline (Alq3). The ZnO nanoparticles were purchased from Aldrich.

The device structure is ITO/host matrix/ZnO nanoparticles (monolayer)/Alq3/Al. The schematic of the device structure is shown in Figure 15a. In order to get the optimized film-forming property, the different host matrixes have the different thickness by different spin-coating condition (PF: 200 nm, PVK: 900 nm, and P3HT: 110 nm). These samples are baked at 170 °C (device I), 170 °C (device II), and 120 °C (device III) for 2 h individually. In addition, in order to reduce the density off pinhole defects in ZnO nanopaeticles, the dry-coating technique is used for development of the ZnO nanoparticels monolayer. During the dry-coating process, the ZnO nanoparticles are adsorbed on the host matrix using ZnO nanoparticle smog during dry-coating process. The ZnO nanoparticle smog is making by homemade nanosmog-making machine, as shown schematically in Figure 15b.

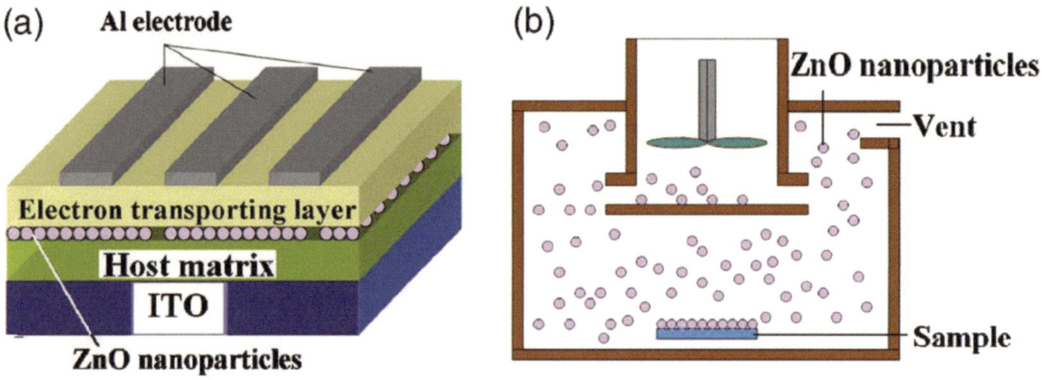

Fig. 15. a) Cross-sectional schematic of the ZnO EL device structure; b) The schematic of the dry-coating machine. (Courtesy of C. Y. Lee et al. [141])

The EL characteristics of the ZnO nanoparticle devices are also measured. Figure 16a shows the EL spectra of the three kinds of the ZnO nanoparticle devices under forward bias of 9 V. For the device with PF (device I, curve a), the EL spectrum shows a strong ZnO band-gap emission peak at 380 nm with the broad background emission from PF. The full width at half maximum of the spectrum is 100 nm. For the device with PVK (device II, curve b), it shows a broadband spectrum from 380 to 700 nm, which are the characteristics of the emissions of the ZnO nanoparticles (380 nm) and PVK (553 nm). The emissions indicate that the radiative recombination occurs in the ZnO nanoparticles and at the host matrix/Alq3 interface. For the device with P3HT (device III, curve c) at the same forward bias of 9 V, it also has a peak around 380 nm contributed from ZnO. To investigate the origin of the EL band, PL spectra of host matrix/ ZnO nanoparticles excited using the 266 nm of a Nd-YAG (yttrium aluminum garnet) laser are also measured at room temperature and shown in Figure 16b. For the device with the higher-energy host matrix (PF/PVK), the PL spectra show the emission band in the ZnO band-gap emission (380 nm) with a weak emission band from the host matrix. For the device with the lowerenergy host matrix (P3HT), the PL spectrum shows the decreased ZnO band-gap emission and the strong host matrix emission, which is due to the Förster energy transfer from the higher-energy ZnO to the lower-energy host matrix. Thus, the ZnO EL device with the lower-energy host matrix causes the poor ZnO band-gap emission.

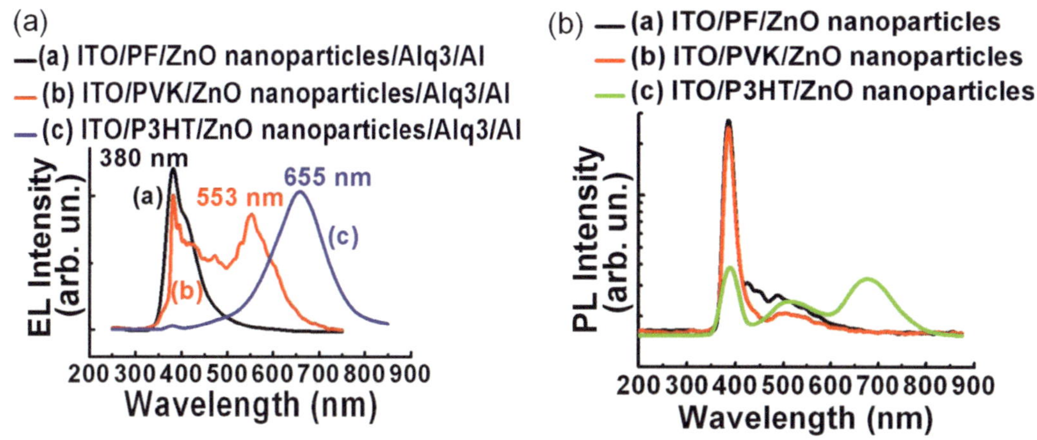

Fig. 16. a) Comparison of EL spectra of ZnO-based nanocomposties with different host matrixes, b) Room temperature PL spectra of three host matrixes (PF, PVK, and P3HT) with ZnO nanoparticles (The intensities are plotted in logarithmic scale, in order to distinguish from the different intensities of the different samples clearly). (Courtesy of C. Y. Lee et al. [141])

7. ZnO QDs for a nanophotonic signal transmission device

Recently, Yatsui et al reported a self-assembly method that aligns nanometer-sized QDs into a straight line. Photonic signals can be transmitted along the QD line by near-field optical effects. This section was developed based on their report [144].

Innovations in optical technology are required for continued development of information processing systems. One potential innovation, the increased integration of photonic devices, requires a reduction in both the size of the devices and the amount of heat they generate. Chains of closely spaced metal nanoparticles that can convert the optical mode into nonradiating surface plasmonic waves have been proposed as a way to meet these requirements.[142, 143] ZnO QDs were bound electrostatically to DNA to form a one-dimensional QD chain. The photoluminescence intensity under parallel polarization excitation along the QDs chain was much greater than under perpendicular polarization excitation, indicating an efficient signal transmission along the QD chain. As optical near-field energy can transmit through the resonant energy level, nanophotonic signal transmission (NST) devices have a number of potential applications, such as wavelength division multiplexing using QDs of different sizes. [144]

To observe the optical properties of the NST, they stretched and straightened the QD-immobilized DNA on the silicon substrate using the molecular combing technique (Figure 17a). [145] First, the silicon substrate was terminated with the silane coupling agent so that the anionic DNA was adsorbed on the cationic silicon substrate. Second, the solution including the DNA and the QDs was dropped onto the cationic silicon substrate. Finally, the glass substrate was slid over the droplet. To check the alignment of DNA-QDs alignment, they obtained an emission image of the cyanine dye attached to the DNA using its 540 nm emission peak under halogen lamp illumination. As shown in the optical image taken with a charge-coupled device camera (Figure 17b), the DNA with QDs stretched in the direction determined by the slide direction of the glass substrate; also, these stretched DNA were found to be isolated.

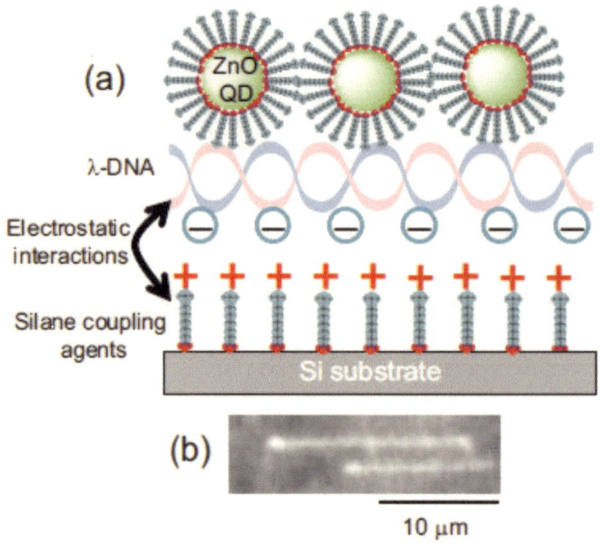

Fig. 17. Schematic of the molecular combing technique: (a) Schematic of the alignment of the DNA with QDs on the cationic silicon substrate. (b) Charge-coupled device image of the stretched λ-DNA. (Courtesy of T. Yatsui et al. [144])

Using the linearly aligned ZnO QDs, they evaluated the PL polarization dependence. A fourthharmonic of a Q-switched Nd:YAG laser (neodymium-doped yttrium aluminum garnet; Nd:$Y_3Al_5O_{12}$ laser, λ=266 nm) with a spot size of approximately 2 mm was used to excite the ZnO QDs at various polarization angles (Figure 18a). From the polarization dependence of the PL at a wavelength of 350 nm (Figure 18b), corresponding to the ground state of 5 nm ZnO QDs, stronger PL emission was obtained by exciting the parallel polarization along the QD chains (E_0) than was obtained under the perpendicular polarization (E_{90}; Figure 18c). Since the decay time of ZnO QDs is more than 20 times longer than the energy transfer time to adjacent QDs,[146] it is possible that the dipoles between adjacent QDs were coupled by an optical near-field interaction, indicating that the signals were transmitted through the QD chain. Furthermore, QD chains have great dipolar strength (see the inset of Figure 18c) that can be realized when the QDs are coherently coupled.[147, 148] If M QDs are coherently coupled and the coherent length along the z-axis is N times greater than that along the x-axis, the equivalent total dipolar strength is given by $Me\times d$ (E_{90}; see Figure 18d) and $Me\times Nd$ (E_0; see Figure 18e), where e is the electrical charge excited in the QD and d is the coherent length along the x-axis, which is equivalent to the width of the QD chain. The resulting emission intensities are $(Me\times d)^2$ and $(Me\times Nd)^2$ for E_{90} and E_0, respectively. Therefore, we obtained N^2 times greater PL intensity with E_0 than with E_{90}.

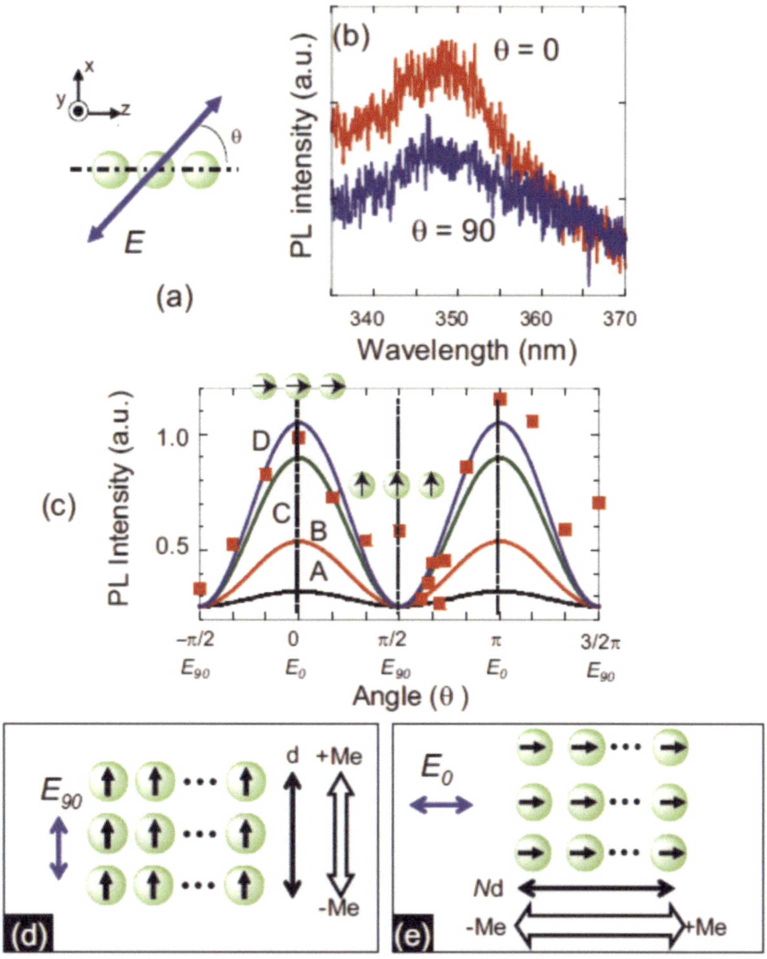

Fig. 18. Polarization dependence of the linearly aligned QD chain: (a) Incident light polarization dependence of PL intensity. θ: Polarization angle with respect to the direction along the QD chains (x-axis), (b) Typical PL spectra obtained at θ =0 and 90. (c) Incident light polarization dependence of the PL intensity obtained at λ=350 nm. Curves A to D correspond to N=3, 6, 9, and 10, respectively. Schematics of the equivalent total dipole strength under (d) E_{90} and (e) E_0. (Courtesy of T. Yatsui et al. [144])

The polarization-dependent PL from the QD chain revealed that the coherent length along the QDs chain was 150 nm, indicating efficient signal transmission through the QD chain.

As optical near-field energy can transmit through the resonant energy level, NST devices have a number of potential applications, such as wavelength division multiplexing using QDs of different sizes.

8. Summary

This chapter describes most of the synthetic methods for synthesis of ZnO QDs. They are frequently used and experimentally demonstrated to be effective on synthesis of ZnO QDs. This chapter also clearly shows that the formation of ZnO superlattices in solution does not result from a kinetic condensation of particles but from a thermodynamic equilibrium. This may be compared to a molecular crystallization process in which an increase in concentration leads to generation of crystals whereas dilution leads to dissolution of them. The system is indeed complex, since several species are involved in the stabilization of the ZnO nanoparticles, and that these species are in thermodynamic equilibrium. Furthermore, we described in detail the origin of UV PL in ZnO QDs, discussing recombination of confined excitons or surface-bound acceptor–exciton complexes, a PL peak with a blue shift of 0.16 eV has been observed in the PL spectrum, showing good size uniformity of nanocrystals. The band gap has also been verified in the analysis of I-V characteristics of an ITO/ZnO/SiO_2/Al structure, high bias, holes or electrons may tunnel through the ZnO nanodot layer and contribute an appreciable current that follows the Fowler-Nordheim (FN) tunneling equation, from which the band alignment of the ZnO nanodots on ITO has been determined. This work demonstrates an efficient synthesis of nanodots and an easy approach to studying physical properties of nanocrystals that will help the material optimization in device application. This chapter also outlines the application of ZnO QDs optoelectronic property, such as EL and NST device. The results described in this chapter are important for the future development of ZnO technology and optoelectronic applications.

9. Acknowledgments

The authors gratefully acknowledge the financial support the Singapore National Research Foundation under CRP Award No. NRF-G-CRP 2007-01, T. Mei acknowledges support from Department of Education of Guangdong Province, China (Grant No. C10131). Y. Hu acknowledges support from the Educational Commission of Zhejiang Province of China (Grant No. Z200909406) and Zhejiang Qianjiang Talent Project (2010R10025).

10. References

[1] M. Haase, H. Weller, A. Henglein. *J. Phys. Chem.* 1998, 92, 482
[2] E. M. Wong, P. C. Searson. *Appl. Phys. Lett.* 1999, 74, 2939
[3] J. M. Klostranec, W. C. W. Chan. *Adv. Mater.* 2006, 18, 1953
[4] S. A. Mcdonald, G. Konstantatos, S. G. Zhang, et al. *Nature Materials.* 2005, 4, 138
[5] A. J. Nozik. *Physica* E. 2002, 14, 115
[6] E. T. Kim, A. Madhukar, Z. M. Ye, et al. *Appl. Phys. Lett.* 2004, 84, 3277
[7] S. Fafard, K. Hinzer, S. Raymond, et al. *Science.* 1996, 274, 1350
[8] C. A. Leatherdale, W. K. Woo, F. V. Mikulec, et al. *J. Phys. Chem.* B. 2002, 106, 7619
[9] S. F. Wuister, I. Swart, F. V. Driel, et al. *Nano Lett.* 2003, 3, 503

[10] S. I. Erlingsson, Y. V. Nazarov, V. I. Fal'ko. *Phys. Rev. B*. 2001, 64, 195306
[11] O. B. Shchekin, D. G. Deppe. *Appl. Phys. Lett*. 2002, 80, 3277
[12] O. I. Micic, S. P. Ahrenkiel, A. J. Nozik. *Appl. Phys. Lett*. 2001, 78, 4022
[13] M. A. Malik, P. O'Brien, N. Revaprasadu. *J. Mater. Chem*. 2001, 11, 2382
[14] Y. L. Wu, C. S. Lim, S. Fu, et al. *Nanotechnology*. 2007, 18, 215604
[15] B. Capoen, A. Martucci, S. Turrell, et al. *J. Molecular Structure*. 2003, 651, 467
[16] W. Y. Mao, J. Guo, W. L. Yang, et al. *Nanotechnology*. 2007, 18, 485611
[17] D. S. Wang, T. Xie, Q. Peng, et al. *Chem. Eur. J*. 2008, 14, 2507
[18] A. Iwabuchi, C. K. Choo, K. Tanaka. *J. Phys. Chem. B*. 2004, 108, 10863
[19] R. M. Laine, J. C. Marchal, H. P. Sun, et al. *Nature Materials*. 2006, 5, 710
[20] B. Ludolph, M. A. Malik, P. O'Brien, et al. *Chem. Commun*. 1998, 1849
[20] O. I. Micic, S. P. Ahrenkiel, D. Bertram, et al. *Appl. Phys. Lett*. 1999, 75, 478
[21] J. G. Cederberg, F. H. Kaatz, R. M. Biefeld. *J. Crystal Growth*. 2004, 261, 197
[22] D. M. Bagnall, Y. F. Chen, Z. Zhu, et al. *Appl. Phys. Lett*. 1997, 70, 2230
[23] D. C. Look. *Materials Science and Engineeriing* B. 2001, 80, 383
[24] S. Krishnanoorthy, T. Bei, E. Zoumakis, et al. *Biosens. Bioelectron*. 2006, 22, 707
[25] L. P. Bauermann, J. Bill, F. Aldinger. *J. Phys. Chem. B*. 2006, 110, 5182
[26] L. Spanhel, M. A. Anderson. *J. Am. Chem. Soc*. 1991, 113, 2826
[27] M. S. Tokumoto, S. H. Pulcinelli, C. V. Santilli, et al. *J. Phys. Chem*. B 2003, 107, 568
[28] H. K. Yadav, K. Sreenivas, V. Gupta, et al. *J. Mater. Res*. 2007, 22, 2404
[29] E. A. Meulenkamp. *J. Phys. Chem*. B 1998, 102, 7764
[30] D. W. Bahnemann, C. Kormann, M.R. Hoffmann. *J. Phys. Chem*. 1987, 91, 3789
[31] N. S. Pesika, Z. Hu, K. J. Stebe, et al. *J. Phys. Chem*. B 2002, 106, 6985
[32] E.M. Wong, J.E. Bonevich, P.C. Searson. *J. Phys. Chem*. B 1998, 102, 7770
[33] L. Guo, S. Yang, C. Yang, et al. *Appl. Phys. Lett*. 2000, 76, 2901
[34] Z. Hu, G. Oskam, P. C. Searson. *J. Colloid Interface Sci*. 2003, 263, 454
[35] S. H. Li, M. S. Toprak, Y. S. Jo, et al. *Adv. Mater*. 2007, 19, 4347
[36] K. F. Lin, H. M. Cheng, H. C. Hsu, et al. *Appl. Phys. Lett*. 2006, 88, 263117
[37] K. F. Lin, H. M. Cheng, H. C. Hsu, et al. *Chemical Physics Letters*. 2005, 409, 208
[38] M. Vafaee, M. S. Ghamsari. *Mater.Lett*. 2007, 61, 3265
[39] R. Viswanatha, P. K. Santra, C. Dasgupta, et al. *Phys. Rev. Lett*. 2007, 98, 255501
[40] Z. Hu, D. J. E. Ramirez, B. E. H. Cervera, et al. *J. Phys. Chem. B* 2005, 109, 11209
[41] Y. Hu, Z. M. Jiang, C. D. Xu, et al. *J. Phys. Chem. C* 2007, 111, 9757
[42] Y. Hu, T. Mei, J. Guo, et al. *Inorg. Chem*. 2007, 46, 11031
[43] K. Sue, K. Kimura, K. Arai. *Mater. Lett*. 2004, 58, 3229
[44] P. M.Aneesh, K. A. Vanaja, M. K.Jayaraj. *Proc. SPIE*. 2007, 6639, 66390J
[45] T. Andelman, Y. Y. Gong, M. Polking, et al. *J. Phys. Chem*. B. 2005, 109, 14314
[46] I. D. Kosobudsky, N. M. Ushakov, G. Y. Yurkov, et al. *Inorg. Mater*. 2005, 41, 1172
[47] P. D. Cozzoli, M. L. Curri, A. Agostiano, et al. *J. Phys. Chem*. B. 2003, 107, 4756
[48] L. L. Yang, J. H. Yang, X. Y. Liu, et al. J. *Alloys Compd*. 2008, Article in press
[49] Y. C. Zhang, X. Wu, X. Y. Hu, et al. *J. Crystal Growth*. 2005, 280, 250
[50] L. M. Shen, L. C. Guo, N. Z. Bao, et al. *Chem. Lett*. 2003, 32, 826
[51] S. Mahamuni, K. Borgohain, B. S. Bendre, et al. *J. Appl. Phys*. 1999, 85, 2861
[52] B. S. Bendre, S. Mahamuni. *J. Mater. Res*. 2004, 19, 737
[53] M. Rajalakshmi, A. K. Arora. *J. Appl. Phys*. 2000, 87, 2445
[54] S. D. Kshirsagar, V. V. Nikesh, S. Mahamuni. *Appl. Phys. Lett*. 2006, 89, 053120

[55] R. S. Ajimsha, G. Anoop, A. Aravind, et al. *Electrochemical. Soild-State Lett.* 2008, 11, K14
[56] A. Said, L. Sajti, S. Giorgio, et al. *J. Phys :Conference Series.* 2007, 59, 259
[57] H. Usui, Y. Shimizu, T. Sasaki, et al. *J. Phys. Chem. B.* 2005, 109, 120
[58] S. Barik, A. K. Srivastava, P. Misra, et al. *Solid State Commun.* 2003, 127, 463
[59] L. Chen, Z. Q. Chen, X. Z. Shang, et al. *Solid State Commun.* 2006, 137, 561
[60] S. T. Tan, X. W. Sun, X. H. Zhang, et al. *J. Crystal. Growth.* 2006, 290, 518
[61] L. M. Yang, Z. Z. Ye, Y. J. Zeng, et al. *Solid State Commun.* 2006, 138, 577
[62] S. W. Kim, M. Ueda, M. Funato, et al. *J. Appl. Phys.* 2005, 97, 104316
[63] S. W. Kim, S. Fujita, S. Fujita, et al. *Appl. Phys. Lett.* 2002, 81, 5036
[64] Y. Y. Peng, T. E. Hsieh, C. H. Hsu. *Appl. Phys. Lett.* 2006, 89, 211909
[65] Y. Y. Peng, T. E. Hsieh, C. H. Hsu. *J. Mater. Res.* 2008, 23, 1155
[66] J. C. Ma, Y. C. Liu, C. S. Xu, et al. *J. Appl. Phys.* 2005, 97, 103509
[67] T. Tani, L. Madler, S. E. Pratsinis. *J. Nanoparticle. Res.* 2002, 4, 337
[68] L. Madler, W. J. Stark, S. E. Pratsinis. *J. Appl. Phys.* 2002, 92, 6537
[69] C. Liewhiran, S. Phanichphant. *Sensors.* 2007, 7, 185
[70] C. Liewhiran, A. Camenzind, A. Teleki, et al. *IEEE Trans. Nanotechnol.*2007, Jan, 672
[71] J. G. Lu, Z. Z. Ye, Y. Z. Zhang, et al. *Appl. Phys. Lett.* 2006, 89, 023122
[72] J. G. Lu, Z. Z. Ye, Y. Z. Zhang, et al. *Appl. Phys. Lett.* 2006, 88, 063110
[73] Y. S. Chang, Y. H. Chang, I. G. Chen, et al. *J. Crystal Growth.* 2002, 243, 319
[74] T. A. Kuriakose, S. N. Kalkura, M. Palanichamy, et al. *J. Crystal Growth.* 2004, 263, 517
[75] G. M. Wu, J. Wang, J. Shen, et al. *Materials Research Bulletin.* 2001, 36, 2127
[76] X. Wang, Y. D. Li. J. *Am. Chem. Soc.* 2002, 124, 2880
[77] M. Mo, J. H. Zeng, X. M. Liu, et al. *Adv. Mater.* 2002, 14, 1658
[78] E. Hosono, S. Fujihara, K. Kakiuchi, et al. *J. Am. Chem. Soc.* 2004,126,7790
[79] J. M. Wang, L. Gao. *Solid State Commun.* 2004, 132, 269
[80] B. Liu, H. C. Zeng. *J. Am. Chem. Soc.* 2003, 125, 4430
[81] Y. Sun, D. J. Riley, M. N. R. Ashfold. *J. Phys. Chem. B.* 2006, 110, 15186
[82] J. B. Liang, J. W. Liu, Q. Xie, et al. *J. Phys. Chem.* B. 2005, 109, 9463
[83] J. C. Ge, B. Tang, L. H. Zhou, et al. *Nanotechnology.* 2006, 17, 1316
[84] M. Stefanescu, O. Stefanescu, M.Stoia, et al. *J. Therm. Anal. Cal.* 2007, 88, 27
[85] X. L. Li, X. X. Zhang, Z. R. Li, et al. *Solid State Commun.* 2006, 137, 581
[86] J. R. A. Sietsma, J. D. Meeldijk, J. P. D. Breejen, et al. Angew. *Chem. Int. Ed.* 2007, 46, 4547
[87] S. Maensiri, P. Laokul, V. Promarak. *J. Crystal Growth.* 2006, 289, 102
[88] H. Zhou, H. Alves, D. M. Hofmann, et al. *Appl. Phys. Lett.* 2002, 80, 210
[89] H. Zhou, H. Alves, D. M. Hofmann, et al. *Phys. Stat. Sol.* 2002, 229, 825
[90] J. P. Sylvestre, S. Poulin, A. V. Kabashin, et al. *J. Phys. Chem.* B. 2004, 108, 16864
[91] P. S. Liu, W. P. Cai, H. B. Zeng. *J. Phys. Chem.* C. 2008, 112, 3261
[92] K. V. Anikin, N.N. Melnik, A. V. Simakin, et al. *Chem. Phys. Lett.* 2002, 366, 357
[93] J. L. Zhao, X. M. Li, J. M. Bian, et al. *J. Crystal Growth.* 2005, 276, 507
[94] D. H. A. Blank, H. Hilgenkamp, A. Brinkman, et al. *Appl. Phys. Lett.* 2001, 79, 394
[95] A. V. Simakin, V. V. Voronov, N. A. Kirichenko, et al. *Appl. Phys. A.* 2004, 79, 1127
[96] G. A. Shafeev, E. Freksz, F. B. Verduraz. *Appl. Phys. A.* 2004, 78, 307
[97] J. Zhang, C. Q. Lan. *Materials Letters.* 2008, 62, 1521
[98] C. H. Liang, Y. Shimizu, T. Sasaki, et al. *J. Mater. Res.* 2004, 19, 1551
[99] J. K. Muth, R. M. Kolbas, A. K. Sharma, et al. *J. Appl. Phys.* 1999, 85, 7884
[100] K. Y. Yun, M. Noda, M. Okuyama. *Appl. Phys. Lett.* 2003, 83, 3981

[101] J. D. Ye, S. L. Gu, S. M. Zhu, et al. *J. Crystal Growth*. 2002, 243, 151
[102] P. A. Williams, J. L. Roberts, C. Jones, et al. *Chem. Vap. Deposistion*. 2002, 8, 163
[103] F. Hamelmann, G. Haindl, J. Schmalhorst, et al. *Thin Solid Films*. 2000, 358, 90
[104] M. C. Jeong, B. Y. Oh, W. Lee, et al. *J. Crystal Growth*. 2004, 268, 149
[105] S.K.Lee, H. J. Choi, P. Pauzauskie, et al. *Phys. Stat. Sol*. 2004, 241, 2775
[106] T. Minemoto, T. Negami, S. Nishiwaki, et al. *Thin Solid Films*. 2000, 372, 173
[107] S. H. Jeong, B. S. Kim, B. T. Lee. *Appl. Phys. Lett*. 2003, 82, 2625
[108] K. Zhang, F.R. Zhu, C. H. A. Huan, et al. *Thin Solid Films*. 2000, 376, 255
[109] K. W. Kim, N. Koguchi, Y. W. Ok, et al. *Appl. Phys. Lett*. 2004, 84, 3810
[110] A. Teleki, S. E. Pratsinis, K. Kalyanasundaram, et al. *Sens. Actuators B*. 2006, 119, 683
[111] R. Mueller, R. Jossen, H. K. Kammler, et al. *AICHE J. 2004*, 50, 3085
[112] A. Killian, T. F. Morse. *Aerosol Sci. Technol*. 2001, 34, 227
[113] R. Mueller, L. Madler, S. E. Pratsinis. *Chemical Engineering Science*. 2003, 58, 1969
[114] T. Sahm, L. Madler, A. Gurlo, et al. *Sens. Actuators B*. 2004, 98, 148
[115] R. Jissen, S. E. Pratsinis, W. J. Stark, et al. *J. Am. Ceram. Soc*. 2005, 88, 1388
[116] L. Madler, W.J. Stark, S. E. Pratsinis. *J. Mater. Res*. 2002, 17, 1356
[117] T. Tani, K. Takatori, S. E. Pratsinis. *J. Am. Ceram. Soc*. 2004, 87, 365
[118] S. Kim, J. J. Gislason, R. W. Morton, et al. *Chem. Mater*. 2004, 16, 2336
[119] H. Schulz, W. J. Stark, M. Maciejewski, et al. *J. Mater. Chem*. 2003, 13, 2979
[120] J. Zhang, F. H. Jiang, L. D. Zhang. *J. Phys. Chem*. B. 2004, 108, 7002
[121] J. Y. Lee, D. S. Kim, J. Park. Chem. Mater. 2007, 19, 4663
[122] C.X. Xu, X. W. Sun, Z. L. Dong, et al. *J. Crystal Growth*. 2004, 270, 498
[123] T. Gao, T. H. Wang. *J. Crystal Growth*. 2006, 290, 660
[124] D. W. Zeng, B. L. Zhu, C.S. Xie, et al. *Materials Science and Engineering* A. 2004, 366, 332
[125] M. H. Huang, Y. Y. Wu, H. Feick, et al. *Adv. Mater*. 2001, 13, 113
[126] X. H. Zhang, S. J. Chua, A. M. Yong. et al. *Appl. Phys. Lett*. 2007, 90, 013107
[127] C. Li, G. J. Fang, J. Li, et al. *J. Phys. Chem. C*. 2008, 112, 990
[128] H. J. Fan, B. Fuhrmann, R. Scholz. Nanotechnology. 2006, 17, S231
[129] S. J. Pearton, D. P. Norton, K. Ip, Y.W. Heo, T. Steiner, Superlattices Microstruct. 2003, 34, 3
[130] M. L. Kahn, M. Monge, E. Snoeck, A. Maisonnat, B. Chaudret, *Small 2005*, 1, 221–224.
[131] C. Pages, Y. Coppel, M. L. Kahn, A. Maisonnat, B. Chaudret, *Chemphyschem 2009*. 10. 2234.
[132] K. Soulantica, A. Maisonnat, M.-C. Fromen, M.-J. Casanove, B. Chaudret, Angew. *Chem. Int. Ed*. 2003, 42, 1945.
[133] F. Dumestre, B. Chaudret, C. Amiens, P. Renaud, P. Fejes, *Science* 2004, 303, 821.
[134] L. Chen, J. Xu, J. D. Holmes, M. A. Morris, J. Phys. Chem. C 2010, 114, 2003.
[135] Wang, L. J., Giles, N. C. *J. Appl. Phys*. 2003, *94*, 973.
[136] Lin, K. F.; Cheng, H. M.; Hsu, H. C.; Lin, L. J.; Hsieh, W. F. *Chem. Phys. Lett*. 2005, *409*, 208.
[137] Zeghbroeck, B. V. *Principles of Semiconductor Devices*, 2004.
[138] Chanana, R. K.; Donald, K. M.; Ventra, M. D.; Pantelides, S. T.; Feldman, L. C.; Chung, G. Y.; Tin, C. C.; Williams, J. R.; Weller, R. A. *Appl. Phys. Lett*. 2000, *77*, 2560.
[139] Sze, S. M. *Physics of Semiconductor Devices*, 1981.
[140] Fan, Z.; Lu, J. G. J. Nanosci. *Nanotechnol*. 2005, 5, 1561.
[141] C. Y. Lee, Y. Y. Hui, W. F. Su. C. F. Lin, *Appl. Phys. Lett*. 2008, 92, 261107.

[142] S. A. Maier, P. G. Kik, H. A. Atwater, S. Meltzer, E. Harel, B. E. Koel, A. A. G. Requicha, *Nature Mater.* 2003, 2, 229._.
[143]W. Nomura, M. Ohtsu, and T. Yatsui, *Appl. Phys. Lett.* 2005, 86, 181108.
[144] T. Yatsui, Y. Ryu, T. Morishima, W. Nomura, T. Kawazoe, T. Yonezawa, M. Washizu, H. Fujita, M. Ohtsu1, *Appl. Phys. Lett.* 2010, 96, 133106.
[145] H. Oana, M. Ueda, and M. Washizu, *Biochem. Biophys. Res. Commun.* 1999, 265, 140.
[146] T. Yatsui, H. Jeong, and M. Ohtsu, *Appl. Phys. B: Lasers Opt.* 2008, 93, 199.
[147]M. Tammer, L. Horsburgh, A. P. Monkman, W. Brown, H. Burrows, *Adv. Funct. Mater.* 2002, 12, 447.
[148] M. Campoy-Quiles, Y. Ishii, H. Sakai, H. Murata, *Appl. Phys. Lett.* 2008, 92, 213305.
[149] B. Gil, A. V. Kavokin, *Appl. Phys. Lett.* 2002, 81, 748.

12

In-Situ Analysis of Optoelectronic Properties of Semiconductor Nanostructures and Defects in Transmission Electron Microscopes

Yutaka Ohno[1], Ichiro Yonenega[1] and Seiji Takeda[2]
[1]*Institute for Materials Research, Tohoku University*
[2]*Osaka University*
Japan

1. Introduction

A wide variety of optoelectronic devices such as photovoltaic (including solar cells and photo detectors), photoemitting (including lasers) and photocatalytic devices have been developed for more than five decades. The physical nature of such devices, especially operated with visible and near-infrared light, is electronic transitions between pairs of energy levels, typically in semiconductors. In addition to the conduction band and the valence band, defect levels, i.e., localized energy levels associated with nanostructures and lattice defects, are responsible for the electronic transitions. Various kinds of nanostructures can be fabricated, spontaneously or artificially, inside and onto semiconductors. Also, lattice defects, such as point defects (including vacancies, interstitials, dopant and impurity atoms) and extended defects (including grain boundaries, stacking faults, dislocations, and point defect clusters), can be introduced, inevitably or accidentally, during crystal growth and device fabrication processes. Therefore, in order to fabricate optoelectronic devices with advanced and ultimate functions, the structural properties of semiconductor nanostructures and defects, as well as their optoelectronic properties such as the possible presence of defect levels, should be understood with a high spatial resolution simultaneously with a high spectral resolution.

Optical measurements such as luminescence and photoabsorption spectroscopy are powerful techniques to determine defect levels. Since the spectral resolution is higher than the order of 10^{-3} eV, this techniques are useful to study the energy levels in the low energy range between the band edges of semiconductors (of the order of 10^{0} eV at most), which dominate the optoelectronic properties of the device products made of the materials. Therefore, when the measurements are performed in a transmission electron microscope (TEM), such as near-field optical measurements in a TEM (Ohno, 2010a), we can examine the optoelectronic properties simultaneously with the structural properties in small regions observed in-situ with the TEM.

With an extremely high spectral resolution in comparison with the other spectroscopic techniques in TEM, such as energy dispersive x-ray spectroscopy (e.g., Terauchi et al, 2010) and electron energy-loss spectroscopy (e.g., Kikkawa et al, 2007) with a resolution less than about 10^{-1} eV, the optical measurements in TEM enable us to assess in detail defect levels in small regions. It is interesting to note that we can examine in-situ the optoelectronic and structural properties of nanostructures and defects *inside* a material, which properties are affected by the surrounding material and are not determined in-situ by the optical measurements in scanning probe microscopes, such as scanning near-field optical microscopy. As the continuing miniaturization and integration of optoelectronic devices, the optical measurements in TEM have been established as an indispensable micro-characterization technique. In this chapter, principles of the optical measurements in TEM, i.e., cathodoluminescence (CL) spectroscopy and transmission electron microscopy under light illumination are briefly summarized, and the resent analysis of some semiconductors for optoelectronic devices are reviewed.

2. Principles of the optical measurements in TEM

2.1 CL spectroscopy in TEM

CL is a phenomenon of light emission induced by electron irradiation. CL light is emitted from a region in which electrons are irradiated, and the optical parameters, such as the photon energy, intensity and polarization, vary depending on the electronic structure in the region. Therefore, CL spectroscopy performed in a TEM enables us to examine the electronic structure simultaneously with the atomic structure in small regions observed in-situ with the TEM, where electrons are irradiated. For example, the structural and compositional variation, as well as the defect concentration and distribution, can be determined. Also, the electronic properties such as defect levels and their carrier capture cross sections, which are associated with the carrier lifetime and diffusion length, can be analyzed. In this subsection, the principles underlying the generation and interpretation of CL signals are summarized. The detailed descriptions of the principles including the pioneer work in 1978 (Petroff et al, 1978) are provided in a review (Yacobi and Holt, 1990).

2.1.1 Spatial resolution of CL measurements

Electrons irradiated into a material can undergo elastic and inelastic scattering. The irradiated material is excited via inelastic electron scattering, and this excitation results in the formation of x-rays, Auger electrons, secondary electrons, electron-hole pairs, and so forth. CL lights are emitted via the recombination of electron-hole pairs. The spatial resolution of CL measurements is, therefore, determined by the distribution of electron-hole pairs.

When electrons are irradiated into a material, each electron changes its direction via an elastic scattering, and it reduces its kinetic energy via an inelastic scattering. As a result of the scattering processes, the original trajectories of the electrons are randomized. For a thin solid material through which most incident electrons can transmit, used as a specimen in a TEM, the shape of the electron penetration range (a so called generation volume) is conical and the radius of a generation volume, which is the maximum at the electron exit surface, is determined (Goldstein, 1979). Electrons and holes are generated inside a generation volume, via some electron-electron interactions. They can diffuse in a material, and the distribution of electron-hole pairs is dominated by the diffusion of minority carriers. The stationary density of minority carriers at a position $\mathbf{r}$, $\Delta n(\mathbf{r})$ obeys the differential equation of continuity, $\mathrm{div}[\mathrm{grad}\ \Delta n(\mathbf{r})]D - \Delta n(\mathbf{r})/\tau(\mathbf{r}) + g(\mathbf{r}) = 0$, in which D is the diffusion constant for minority carriers, τ is the mean recombination lifetime, and g is the generation rate of electron-hole pairs. The distribution of minority carriers has been discussed theoretically (e.g., Everhart & Hoff, 1971, Donolato & Venturi, 1982), and the range in which minority carriers exist is expected to be twice as large as a generation volume at most, for thin materials. Therefore, the spatial resolution of CL measurements can be approximated to the maximum diameter of a generation volume. The typical resolution is the order of 10^2 nm.

2.1.2 CL spectroscopy and analysis

CL lights are emitted via various radiative electronic transitions. One transition is the recombination between an electron in the conduction band and a hole in the valence band (a band-to-band transition), which is typical in direct gap semiconductors at high temperatures. At a temperature T at which kT (k is the Boltzmann factor) is smaller than the binding energy for free excitons, the excitonic level for the excitons is formed just below the conduction band edge and the decay of the excitons (the free exciton transition between the excitonic level and the valence band) results in a CL emission. When a defect exists, it may induce donor and/or acceptor levels. Electrons and holes generated by electron irradiation are trapped at the levels, and CL lights may be emitted via a donor-to-valence band transition, a conduction band-to-acceptor transition, and a donor-to-acceptor transition. They are the typical transitions in indirect gap semiconductors at high temperatures. Similar to the case of free excitons, the excitonic levels bound to a donor and an acceptor are formed at low temperatures and CL lights may be emitted via the decay of the excitons (bound exciton transitions). Also, when an impurity with incomplete inner shells (such as a rare earth ion or a transition metal) exists, its excitation and radiative deexcitation (an inner shell transition) results in a CL emission. Detailed recombination processes are reviewed, for example, by Henderson & Imbusch (Henderson & Imbusch, 1989) for inner shell transitions, and by Yu & Cardona (Yu & Cardona, 2001) for the other transitions.

The photon energy of a CL light emitted via a transition is written as $\Delta E+\delta$, in which ΔE corresponds to the energy difference between the energy levels concerning the transition. δ is a positive value for a donor-to-acceptor transition (Yu & Cardona, 2001), and it is zero for the other transitions. CL energy measurements have been used to assess energy levels in a small volume, which are connected to the structure including defects and composition in the volume.

When the number N of nanostructures or defects of the same kind, with the cross section for radiative recombination of σ, exist in a small volume, the number of photons emitted from the volume per a unit time is proportional to $N\sigma$. CL intensity measurements have been applied to assess the density of point defects acting as radiative recombination center (e.g., Graham et al, 1994, Mitsui et al, 1996) and those acting as non-radiative one (e.g., Ohno et al, 1999, Ohno & Takeda, 1996). Also, the techniques have been used to assess the carrier dynamics such as excited carrier paths towards the lower energy states (e.g., Akiba et al, 2004, Merano et al, 2005, Sonderegger et al, 2006), as well as carrier lifetimes.

Some extended defects such as dislocations (Yamamoto et al, 1984, Mitsui & Yamamoto, 1997), twins (Ohno et al, 2007b), platelets (Ohno, 2005a) and strains induced by a uniaxial stress (Wang et al, 1993, Ohno, 2005b), as well as low-dimensional nanostructures such as nanowires (Ishikawa et al, 2008, Yamamoto et al, 2006), quantum wells (Ohno, 2005a) and superlattices (Ohno & Takeda, 2002, Ohno et al, 2007b), form anisotropic defect levels, and electronic transitions via the levels result in the emission of polarized CL lights. CL polarization analyses have helped us to assess quantitatively the atomistic structures of such nanostructures and defects.

2.1.3 CL imaging and analysis

A CL image (or a CL intensity map) is obtained by detecting the CL intensity as an electron beam scans a raster over a specimen. When there is a microstructure, such as a defect, in which the cross section for radiative or non-radiative recombination differs from that in the surrounding homogeneous material, contrast arises in CL images. For example, the spatial distribution of defects in degraded devices (e.g., Sieber et al, 1993, Cheng et al, 1995, Bonard et al, 1998) have been determined.

CL contrast due to a defect varies depending on the atomistic structure, size and location of the defect. Among them, effects of dislocations on CL contrast have been determined theoretically (e.g., Jakubowicz, 1986, Schreiber & Hildebrandt, 1994, Holt & Napchan, 1994), and the radiative and non-radiative lifetime and diffusion length for minority carriers around a dislocation have been assessed.

2.2 Electron microscopy under light illumination

CL spectroscopy in TEM is a useful technique to determine defect levels of nanostructures and defects acting as radiative recombination center. However, a defect level acting as non-radiative recombination center is, in general, difficult to assess. Also, it is difficult to examine the process of an electronic excitation (i.e., an electronic transition from an energy level to a higher one) via a defect level, since various kinds of electronic excitaitions take place simultaneously during the irradiation of electrons whose energy is much higher than the band gap energy. In order to determine such electronic transitions with a high spatial resolution, transmission electron microscopy under light illumination has been used, since the pioneer work in 1984 (Suzuki et al, 1984). In this subsection, the principles underlying the electronic transitions induced by light illumination are summarized.

2.2.1 Electronic transitions due to light illumination

Unlike electronic transitions due to electron irradiation, induced via an interaction between electrons, electronic transitions due to the illumination of a light, especially in visible and near-infrared region, are mainly induced via an interction of photons with electrons. For a photo-induced electronic transition between two energy levels, the transition rate is high when the photon energy of the illuminating light is close to the energy difference between the levels (in a resonant photo-excitation condition), according to the Fermi's golden rule. Therefore, by choosing a proper photon energy which corresponds to the energy difference between a defect level and an another level, we can selectively determine the defect level. When the technique is applied in TEM, it is useful to determine the nanostructures and defects whose atomistic structures vary under photo-excitation conditions.

Under the illumination of a light with a proper photon energy, electron-hole pairs can be generated. Therefore, the spatial resolution of optical measurements with photo-excitation is approximated to the diameter of the generation volume, similar to CL measurements. Since a light beam almost goes straight inside a thin material, the resolution is approximated to the diameter of the beam. In case of the illumination of a far-field light in TEM with a conventional optical system, such as an optical fiber (Suzuki et al, 1984, Yoshida et al, 2004), optical lens (Ohno & Takeda, 1995), and parabolic mirror (Ohno, 2010a), the resolution is much less than 10^3 nm due to the diffraction limit of light. The spatial resolution beyond the limit, higher than 10^2 nm, is acheived in a TEM with a near-field light probe (Ohno, 2010a).

2.2.2 In-situ photo-excitation in TEM and analysis

Transmission electron microscopy under light illumination has been used to study in-situ photo-induced structural variations such as the dislocation motion related to photo-plasticity (Suzuki et al, 1984) and degradations in photoemitters (Ohno, 2005b, 2010a, Ohno et al, 2009b); the glide of dislocations under light illumination is observabed in-situ. Photocatalytic reactions in catalysts are also observable in-situ, and a photo-induced decomposition (Yoshida et al, 2004), structural transformation (Yoshida et al, 2006), and nucleation of nanostructures (Yoshida et al, 2007) are observed in a photocatalytic system. Such a structural variation is induced via a photo-induced electronic excitation, and depends on the photon energy, intensity and polarization of the illuminating light. A defect level related to the photo-excitation can be estimated with a proper photon energy, since the degree of the structural variation is the maximum when the photon energy is close to the energy difference between the defect level and an another level (Ohno, 2005b, Ohno et al, 2009b, Ohno, 2010a).

Semiconductors under light illumination can emit photoluminescence (PL) light, similar to CL, so as to release some of the photon energy absorbed from the illuminating light. PL spectroscopy has an advantage over CL spectroscopy when various kinds of nanostructures and defects coexist in a small region. A defect level of nanostructures or defects of the same kind can be determined selectively, since the emission efficiency is increased in a resonant photo-excitation condition in which the illuminating light is absorbed via the defect level. Also, Raman spectroscopy, by which the phonon structure of nanostructures and defects can be determined, is available under light illumination, in a resonant photo-excitation condition. PL and Raman spectroscopy with a far-field light is demonstrated in TEM (Ohno & Takeda, 1995, 1996, Ozaki et al, 1998, Ohno et al, 1999, Kohno et al, 2004, Ohno, 2010a).

2.2.3 In-situ photo-excitation systems in TEM

As a convenient system, an optical fiber is installed inside a TEM so that the end of the fiber is located near the area observed with the TEM (Suzuki et al, 1984, Yoshida et al, 2004). Even though the system is easily constructed in any TEMs, the spatial resolution is rather low; almost all areas on a TEM specimen are illuminated. Here a high resolution system for in-situ photo excitation is reviewed.

Figure 1 shows an apparatus for in-situ high-resolution optical measurements with a near-field light probe in a TEM (Ohno, 2010a). A metal tip is located close to the area observed with the TEM. The tip is then illuminated with a far-field light, and an intense near-field light is induced in the immediate vicinity of the tip apex, due to a surface enhanced effect (Inouye & Kawata, 1994).

In detail, a specimen in a TEM is illuminated with a laser light by using a parabolic mirror. The mirror is truncated so that it is inserted inside a small space in the TEM and the specimen is illuminated with an electron beam for TEM observation. A laser light is introduced into the TEM so that the optical pass is parallel to the rotation axis of the parabolic mirror, and the introduced light is focused onto the focal point of the mirror. With a CL technique (Ohno & Takeda, 1995), the focal point is set onto the area observed with the TEM. A specimen generally exhibits CL emission during TEM observation, and the CL light emitted from the observed area is detectable optically. Also, a laser light focused on a specimen is detectable. Each light is observed as a bright spot with an optical zoom microscope. By adjusting the location of the parabolic mirror so that the laser spot coincides to the CL spot, the laser light is precisely focused onto the area observed with the TEM. In order to form a near-field light, a metal tip is located on the observed area with a TEM-STM holder (Figs. 1c and 1d). The tip is set in the side illumination configuration in which the tip axis is about normal to the optical pass of a laser light for inducing a near-field light. With the configuration, an intense near-field light can be induced at the tip apex, due to a *p*-polarization effect (Hayazawa et al, 2002).

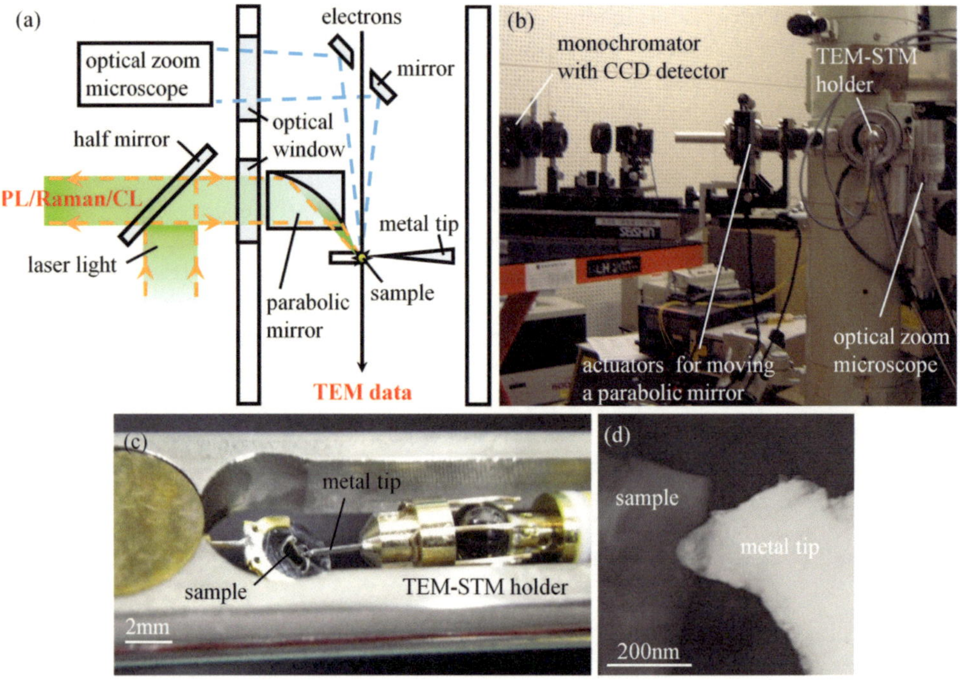

Fig. 1. (a) Schematic and (b) external views of an apparatus for in-situ near-field optical measurements in a TEM. (c) Optical and (d) TEM images of a specimen and a metal tip in a TEM-STM holder

3. Optoelectronic properties of semiconductor nanostructures and defects

3.1 Dislocations in wide gap semiconductors

Extended defects in semiconductors frequently affect the optoelectronic properties. Among them, dislocations are key topics for developing the relevant devices fabricated with wide gap semiconductors, such as GaN and ZnO, since they are inevitably introduced a high density of dislocations. In this subsection, the recent knowledge of the optoelectronic properties of dislocations, mainly revealed by in-situ optical measurements in a TEM, is reviewed. The properties in GaN, which is the pioneer of the commercial photoemitting device operated in the short wavelength range, are first summarized, and those in ZnO, which is attracting keen interest as a promising analog to GaN, are shown.

GaN and related materials have been applicated to short-wavelength optoelectronic devices since the successful fabrication of long-lifetime blue photoemitting diodes operated at room temperature (Nakamura et al, 1994). Due to the lack of a suitable growth substrate, there exist mismatches in lattice constant and in thermal expansion coefficient between a GaN epilayer and its substrate (e.g., sapphire or SiC), and the mismatches induce a high density (10^6-10^{10} cm^{-2}) of dislocations. However, the epilayer has a high efficiency of light emission.

It is well known that most dislocations act as non-radiative recombination center in III-V (e.g., Petroff et al, 1980, Hutchinson et al, 1984, Wang et al, 1992) and II-VI (e.g., Mitsui & Yamamoto, 1997, Mitsui et al, 1996) compound semiconductors, and the dislocations affect the device lifetimes even when the density is as low as 10^4 cm^{-2}. Therefore, the optical properties of dislocations in GaN have been a subject of controversy. Individual dislocations, grown-in (Yamamoto et al, 2003, Nakaji et al, 2005, Albrecht et al, 2008) and introduced intentionally (Albrecht et al, 2002, 2008), are examined in TEM, and the dislocations act as recombination-active centers. Actually, they often exhibit recombination-enhanced glide (Maeda et al, 1999). The dislocations lying on a basal plane (i.e., misfit dislocations) seem to be undissociated with the Burgers vector **b** of $1/3<11\underline{2}0>$ (Albrecht et al, 2002), even though they can be dissociated (Suzuki et al, 1994), presumably due to a high energy barrier for an undissociated dislocation to dissociate (Blumenau et al, 2003). Among the dislocations, 60° dislocations are proposed to act as radiative recombination center that emit a light with photon energy of 2.9 eV at low temperatures, while the other dislocations act as non-radiative recombination center (Albrecht et al, 2002). The screw dislocations have a small recombination activity in comparison with the edge dislocations (Yamamoto et al, 2003), even though their velocity for recombination-enhanced glide is faster in comparison with the edge dislocations (Maeda et al, 1999). Also, threading dislocations, i.e., dislocations lying on pyramidal planes with **b** = $1/3<11\underline{2}0>$ (Albrecht et al, 2008) and those on prismatic planes with **b** = $1/3<11\underline{2}0>$ or $1/3<11\underline{2}1>$ (Yamamoto et al, 2003), act as non-radiative recombination center. Even though the threading dislocations have a smaller recombination activity in comparison with the misfit dislocations (Yamamoto et al, 2003), they reduce overall luminescence intensities (Yonenaga et al, 2006). The diffusion length of minority carriers around a dislocation is estimated (Yamamoto et al, 2003, Nakaji et al, 2005, Ino & Yamamoto, 2008), and it is smaller in comparison with the other semiconductors; e.g., about 200 nm, 150 nm, and 60 nm, respectively, for non-doped, Mg-doped ($1x10^{17}$ cm^{-3}), and Si-doped ($3x10^{18}$ cm^{-3}) GaN, even at low temperatures. The estimated length is smaller than the nearest-neighbor distances of dislocations, and this would be a reason why GaN can tolerate a much higher dislocation density than in the other semiconductors.

Wide gap II-VI compound semiconductors had been expected to be the most promising materials for short-wavelength optoelectronic devices, but most researches were closed after the commercialization of GaN-based devices. However, there is a revival of interest in the materials because of their potential applications in optoelectronic devices employing excitonic effects in the short wavelength range, due to their large exciton binding energy, etc. Among them, ZnO has rapidly emerged as a promising analog to GaN, after the demonstration of the excitonic emission at elevated temperatures up to 550 K (Bagnall et al, 1998). Despite considerable success in optimizing the growth conditions and structural quality such as the fabrication of blue photoemitters (Tsukazaki et al, 2005), ZnO epilayers still contain a number of defects. The most characteristic defect, investigated with a TEM, is a high density (10^9−10^{11} cm^{-2}) of dislocations (e.g., Setiawan et al, 2004). In ZnO, dislocations are introduced with lower stress and are much mobile, in comparison with GaN (Yonenaga et al, 2008). The knowledge of the influence of dislocations is, therefore, required for the practical use of this material.

In order to determine the effects of dislocations on the optoelectronic properties, an arbitrary number (10^9-10^{10} cm^{-2}) of dislocations on a basal plane are introduced in wurtzite ZnO single-crystals, at elevated temperatures (923-1123 K) comparable to the typical temperatures for the fabrication of ZnO-based devices, by compressive deformation (Ohno et al, 2008a). The dislocations seem to be undissociated with **b** = a/3<11$\underline{2}$0>, even though they can be dissociated (Suzuki et al, 1994), similar to wurtzite GaN (Suzuki & Takeuchi, 1999, Albrecht et al, 2002). Most dislocations are of mixed type including 60° dislocations, and screw and edge dislocations also exist. Macroscopic PL measurements reveal that, unlike wurtzite GaN, the introduction of the dislocations do not influence all the emission bands existing in as-grown specimens (Fig. 2a). Also, additional emission bands, with photon energies E_L of 3.100 and 3.345 eV at temperature of 11 K, as well as their LO phonon replicas (with the separation between the nearest-neighbor emission lines of 72 meV), are introduced (Fig. 2b). From the analysis of the thermal quenching processes, the depth of the defect level ΔE associated with the 3.100 eV emission and that with the 3.345 eV one are estimated to be 0.3 eV and 0.05 eV, respectively. Since the PL intensities increase with increasing dislocation density, the emission bands are associated with the dislocations introduced at elevated temperatures (Ohno et al, 2008b).

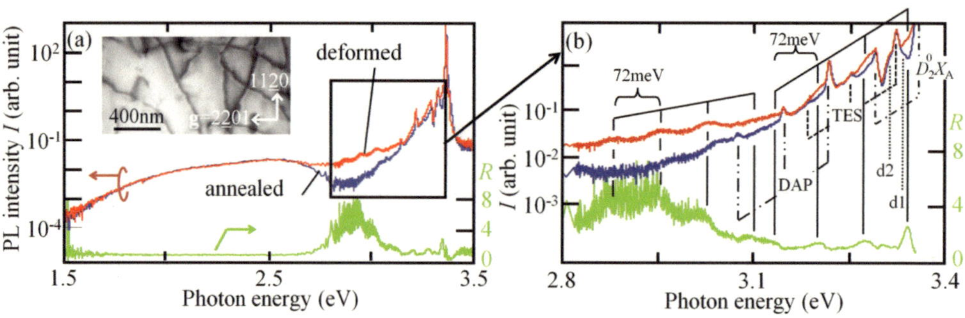

Fig. 2. PL intensity vs photon energy for a specimen deformed at 923 K (the red curve) or annealed at 923 K without stress (the blue curve). The ratio of those PL intensities, $R = I_{\mathrm{deformed}}/I_{\mathrm{annealed}}$, is also shown (the green curve). A part of (a) is magnified in (b). The inset in (a) shows a TEM image of dislocations in the deformed specimen. (Ohno et al, 2008b)

In-situ TEM observation with 80 keV electrons reveals that (Ohno et al, 2009b), screw and edge dislocations glide forming mixed dislocation segments when electrons are irradiated at temperature of 120 K, as well as at room temperature (Fig. 3). Effects of the point defects introduced by the irradiation is negligible, since the electron energy is much smaller than the threshold electron energy for displacement damage in ZnO (310 keV). The glide velocity for a dislocation is independent of irradiation temperature, and it increases with increasing electron flux. Therefore, the dislocations glide via an electron-hole recombination (Maeda et al, 2000), and the screw and edge dislocations act as non-radiative recombination center. The glide velocity for a screw dislocation is more than 10^1 times larger in comparison with an edge dislocation. Those properties are similar to wurtzite GaN (Maeda et al, 1999).

In order to determine the optoelectronic properties of each individual dislocation, the recombination-enhanced glide is determined in detail by TEM observation under the illumination of a light, in the energy range of 2.25-2.92 eV, at temperature of 120K (Ohno et al, 2009b). A screw dislocation glides when a light with photon energy around 2.5 eV is illuminated (Fig. 4). Since the estimated energy is smaller than the band gap energy E_g (about 3.4 eV at 120 K) and is larger than the activation energy of about 1 eV for thermal dislocation glide (Yonenaga et al, 2008), electrons and holes would be recombined at a defect level associated with the dislocation, similar to the case of 90°α partial dislocations in ZnSe (Ohno, 2005b). Namely, the depth of the defect level ΔE is about 0.9 eV (E_g–2.5 eV). Since the recombination-enhanced glide of edge dislocations is not observed, their defect level would locate deeper than 1.15 eV (E_g-2.25 eV) and they might glide under the illumination of a light with the corresponding photon energy.

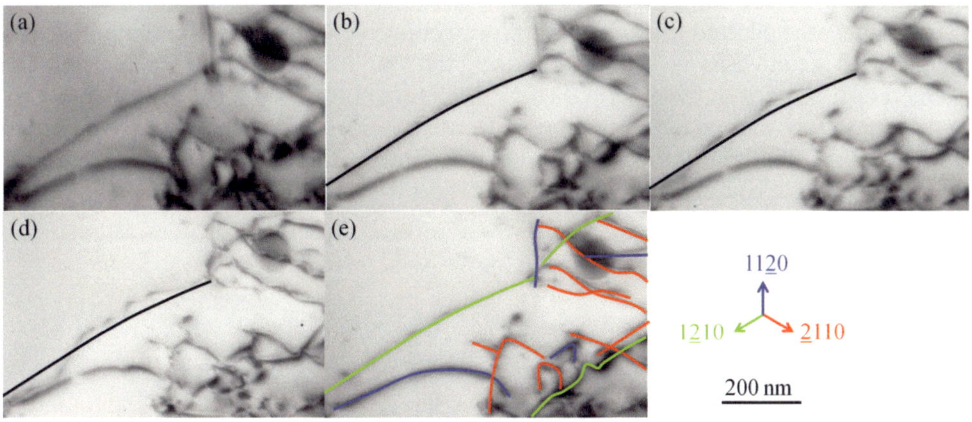

Fig. 3. Dislocation glide due to electron irradiation. The specimen is irradiated with electrons with flux of 5x10^{16} *e* mm^{-2} s^{-1} at room temperature for 3600 s (a to b). It is subsequently irradiated with flux of 5x10^{17} *e* mm^{-2} s^{-1} at room temperature for 900 s (b to c), then irradiated with the same flux at temperature of 120 K for 900 s (c to d). The black curves in b-d indicate the location of a screw dislocation before electron irradiation. (e) A schematic view of the dislocations with **b** = a/3[1$\underline{2}$10] (green curves), a/3[11$\underline{2}$0] (blue curves), and a/3[$\underline{2}$110] (red curves) in (a). (Ohno et al, 2009b). Copyright Wiley-VCH Verlag GmbH & Co. KGaA. Reproduced with permission

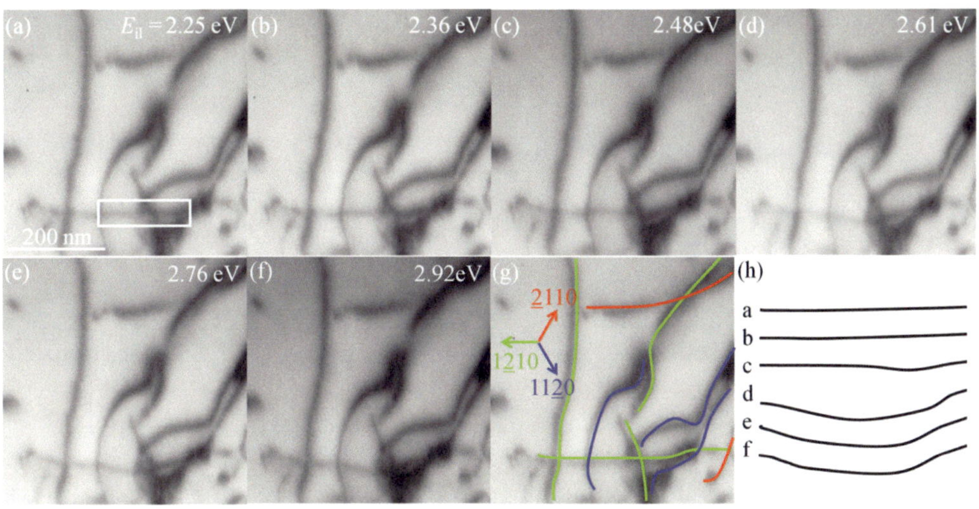

Fig. 4. (a - f) Dislocation glide due to light illumination. Dislocations are illuminated in turn with a light with the photon energy E_{il}. (g) A schematic view of the dislocations with **b** = a/3[1$\underline{2}$10] (green curves), a/3[11$\underline{2}$0] (blue curves), and a/3[$\underline{2}$110] (red curves) in (a). (h) Variation of the line shape of the screw dislocation in the squared area in (a) during light illumination. (Ohno et al, 2009b). Copyright Wiley-VCH Verlag GmbH & Co. KGaA. Reproduced with permission

The emission band peaking at 3.1 eV, due to a defect level of 0.3 eV depth associated with dislocations (Ohno et al, 2008b), is determined by CL spectroscopy (Ohno et al, 2009b). Since a CL light emitted from a dislocation is rather weak, CL lights are obtained from an area in which a number of dislocations exist (the total length of the dislocation lines of about $2x10^3$ nm). The lights are collected after the light illumination with the same illumination sequence as in Fig. 4, and the intensity is unchanged during the illumination (Fig. 5). The intensity is also unchanged during electron irradiation (Fig. 6). Therefore, the emission is not influenced by the conversion of screw and edge dislocations into mixed dislocations, via their recombination-enhanced glide under light illumination or electron irradiation. Those results indicate that the origin of the emission band is neither the screw dislocations nor the edge ones.

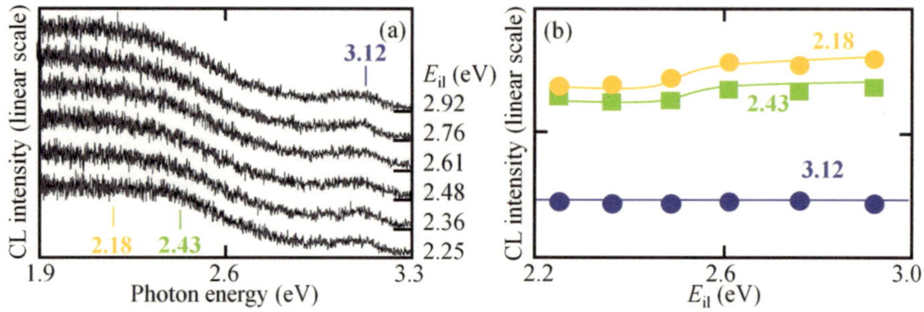

Fig. 5. (a) CL spectra after the illumination of a light with the photon energy E_{il} as in Fig. 4. (b) The CL intensity at 3.12, 2.43, or 2.18 eV vs E_{il}. (Ohno et al, 2009b). Copyright Wiley-VCH Verlag GmbH & Co. KGaA. Reproduced with permission

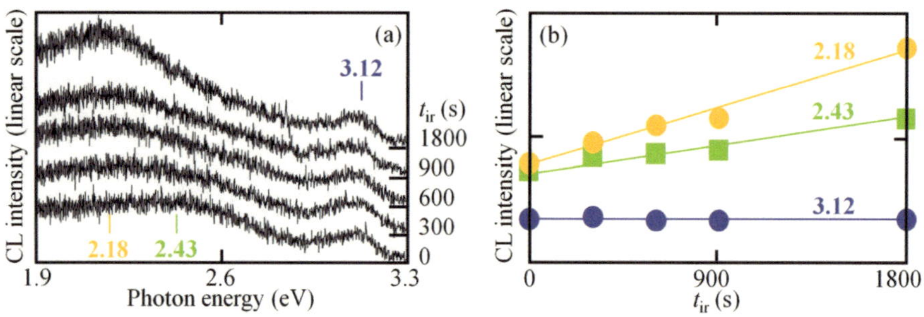

Fig. 6. (a) CL spectra after the irradiation of electrons, with flux of $5x10^{17}$ e mm^{-2} s^{-1} at temperature of 120 K, for the irradiation time t_{ir}. (b) The CL intensity at 3.12, 2.43, or 2.18 eV vs t_{ir}. (Ohno et al, 2009b). Copyright Wiley-VCH Verlag GmbH & Co. KGaA. Reproduced with permission

It is theoretically expected that a 60° dislocation in wurtzite GaN, which is similar to ZnO in the band gap energy, form a defect level 0.3 eV (~$0.09E_g$) above the valence band edge (Blumenau et al, 2003), and the emission due to the corresponding level is proposed (Albrecht et al, 2002). Similar emission bands related to dislocations with photon energy E_L are also observed in a wurtzite II-VI compound of CdS (Negrii & Osipyan, 1978) and another compounds with the zinc blende structure such as ZnSe (Hilpert et al, 2000), ZnTe (Naumov et al, 1993), CdTe (Shreiber et al, 1999) and GaAs (Fravacque et al, 1989), even though the dislocations are dissociated, and it is proposed that the origin of the emission bands is 90° dislocations with a defect level of ΔE ~ $10^{-1}E_g$ (Table 1). Those results suggest that a 60° complete dislocation in ZnO is dissociated into a pair of a 90° partial dislocation and a 30° one, with a narrow separation which is difficult to observe with a conventional TEM technique, and the 90° partial dislocation would induce the 3.1 eV emission.

	structure	E_g [eV]	E_L [eV]	ΔE [eV]	$\Delta E/E_g$	model
GaN	wurtzite	3.5	2.9	0.3	0.086	60° or 90° partials
ZnO	wurtzite	3.44	3.100	0.3	0.087	90° partials with point defects?
ZnSe	zinc blende	2.82	2.61	0.22	0.078	Se(g) 90° partials
CdS	wurtzite	2.58	2.430 2.439 2.447	0.153 0.144 0.136	0.059 0.056 0.053	?
ZnTe	zinc blende	2.39	2.185 2.148	0.21 0.24	0.088 0.100	Te(g) 90° partials?
CdTe	zinc blende	1.61	1.48	0.13	0.081	Te(g) 90° partials
GaAs	zinc blende	1.52	1.13	<0.2	<0.132	As(g) 90° partials

Table 1. Emission bands related to dislocations

Figure 7 summarizes the atomistic structure of dislocations on a basal plane introduced at elevated temperatures, proposed in this ZnO work. 90° partial dislocations in the dissociated 60° dislocations would form a defect level of 0.3 eV depth acting as radiative recombination center, while the screw and edge dislocations form a defect level near the mid gap acting as non-radiative recombination center, similar to wurtzite GaN (Albrecht et al, 2002).

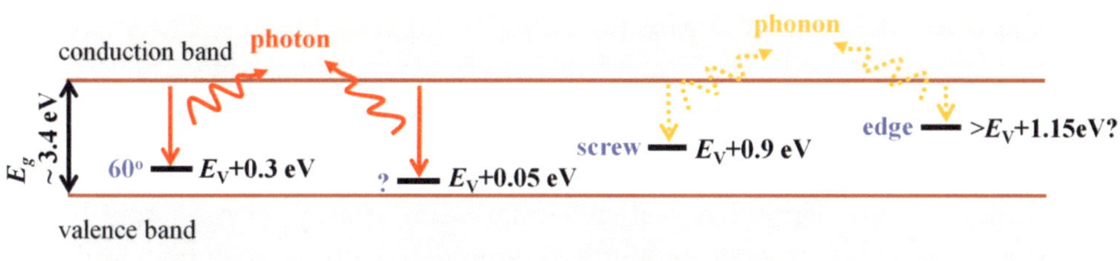

Fig. 7. A schematic view of the defect levels of misfit dislocations in ZnO

Misfit dislocations in ZnO introduced at elevated temperatures do not influence the emissions except the dislocation-related emissions, as in Fig. 2a. On the other hand, almost all emissions in ZnO are suppressed due to the introduction of misfit dislocations at room temperature (Coleman et al, 2006, Takkouk et al, 2005). Since those dislocations are similar in type, the atomistic structure of a dislocation would be modified at elevated temperatures, presumably due to an interaction of the dislocation with point defects. Indeed, a dislocation involving point defects is the candidate for the 3.1 eV emission band (Ohno et al, 2008b). Also, the intensity of the green (2.43 eV) and yellow (2.18 eV) emissions, associated with O- and Zn-vacancies (Zhao et al, 2005), varies when dislocations modify their structure via their glide (Figs. 5 and 6). This characteristic of ZnO may be an advantage over GaN, since all emissions in GaN are suppressed when dislocations are introduced even at elevated temperatures (Yonenaga et al, 2006), as well as at room temperature. (Kucheyev et al, 2000). The interaction of dislocations with point defects is a key to elucidate the influence of dislocations on the optoelectronic properties in ZnO. In order to understand the properties at an atomistic level, it is needed to determine an individual dislocation with a higher spatial resolution, as well as with a high spectral resolution. Near-field optical measurements in a TEM (Fig. 8) will be a unique technique to reveal the subject.

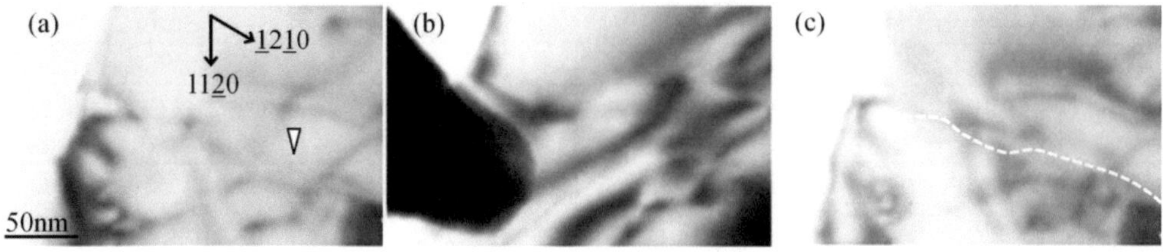

Fig. 8. A preliminary result for in-situ examination of a photo-induced glide of an individual dislocation in ZnO. TEM images of the screw dislocation indicated with the arrowhead in (a); before (a), under (b), and after (c) the illumination of a near-field light with photon energy of 2.41 eV. The broken curve in (c) indicates the dislocation line before the illumination. (Ohno, 2010a)

It is shown that transmission electron microscopy under light illumination is available to determine quantitatively the defect levels acting as non-radiative, as well as radiative, recombination centers, with a high spatial resolution. By means of this technique, combined with another in-situ techniques such as environmental TEMs (e.g., Yoshida et al, 2008, 2009) and in-situ electrifying (e.g., Nogami et al, 2009, Kohno et al, 2009), the dynamic properties of nanostructures and defects under electronic excitation will be determined.

3.2 Quantum nanostructures in middle gap semiconductors

After the successful fabrication of quantum wells based on middle gap III-V compound semiconductors including AlGaAs, around 1970, various kinds of quantum nanostructures have been fabricated. For example, by using the differences in the surface migration of adatoms and using the Stranski-Krastanov growth, quantum wires and quantum dots have been self-organized. Also, by means of modern crystal-growth techniques such as metal-organic chemical-vapor deposition and molecular-beam epitaxy, superlattices based on periodic changes of either material compositions or doping patterns have been grown. Such quantum nanostructures have potentials for enhancing the device performance, and also their electron confinement produces quantum electronic states which provide an important system for fundamental physics.

Recently, superlattices based on periodic changes of crystal directions, i.e., twinning superlattices (Ikonic et al, 1993), have been proposed in III–V (Xiong & Eklund, 2006, Ohno et al, 2007b, Algra et al, 2008, Bao et al, 2008), II-VI (Ikonic et al, 1996), and VI (Hibino et al, 1998, Fissel et al, 2006) semiconductors and in metals (e.g., Kobayashi & Uchihashi, 2010). It is expected that they can offer as much versatility in tailoring the miniband structure as there exists in heterostructure-based superlattices (e.g., Ikonic et al, 1993, Nakamura & Natori, 2006). In this subsection, the optoelectronic properties of twinning nanostructures self-organized in indirect gap $Al_{0.5}Ga_{0.5}As$ epilayers, revealed by CL spectroscopy in a TEM, are reviewed.

An $Al_{0.5}Ga_{0.5}As$ epilayer without extended defects is traditionally grown on a substrate that is flat and electrically neutral; the substrate is oxidized intentionally, to form a flat interface between the oxide and the substrate, and the oxide is then removed by the annealing in an As atmosphere, for fear of the sublimation of As atoms on the substrate. On the other hand, when a substrate is not treated the pretreatments, multiple twin boundaries are induced on the substrate; the boundaries are induced in an epilayer grown on a rough As-deficient surface (Ohno et al, 2008c). Twin boundaries of Σ3 type are formed on (111) and $(\bar{1}\bar{1}1)$ (Fig. 9). No compositional fluctuation is detected around the boundaries, indicating that heterostructures, such as $Al_xGa_{1-x}As/Al_yGa_{1-y}As$ superlattices and impurity agglomerates, are not self-organized around the boundaries.

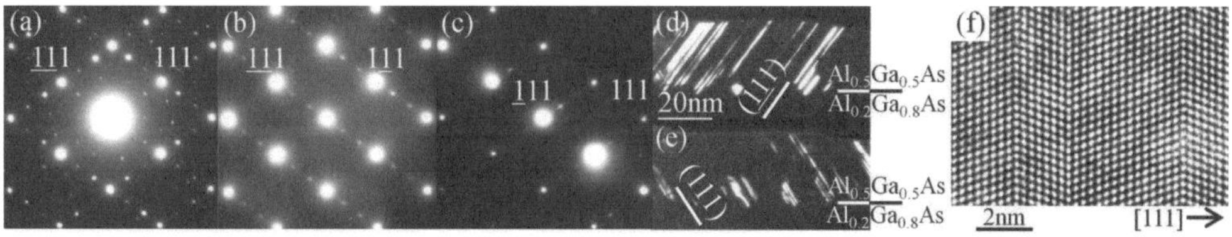

Fig. 9. Diffraction patterns of a twinned $Al_{0.5}Ga_{0.5}As$ epilayer taking along (a) $[\bar{1}10]$, (b) [011], and (c) nearly $[0\bar{1}1]$. TEM images taking with a twin spot at (d) 1/3, 1/3, 5/3 and (e) $\bar{1}/3$, $\bar{1}/3$, 5/3. (f) A high-resolution TEM image of twin boundaries. (Ohno et al, 2007b)

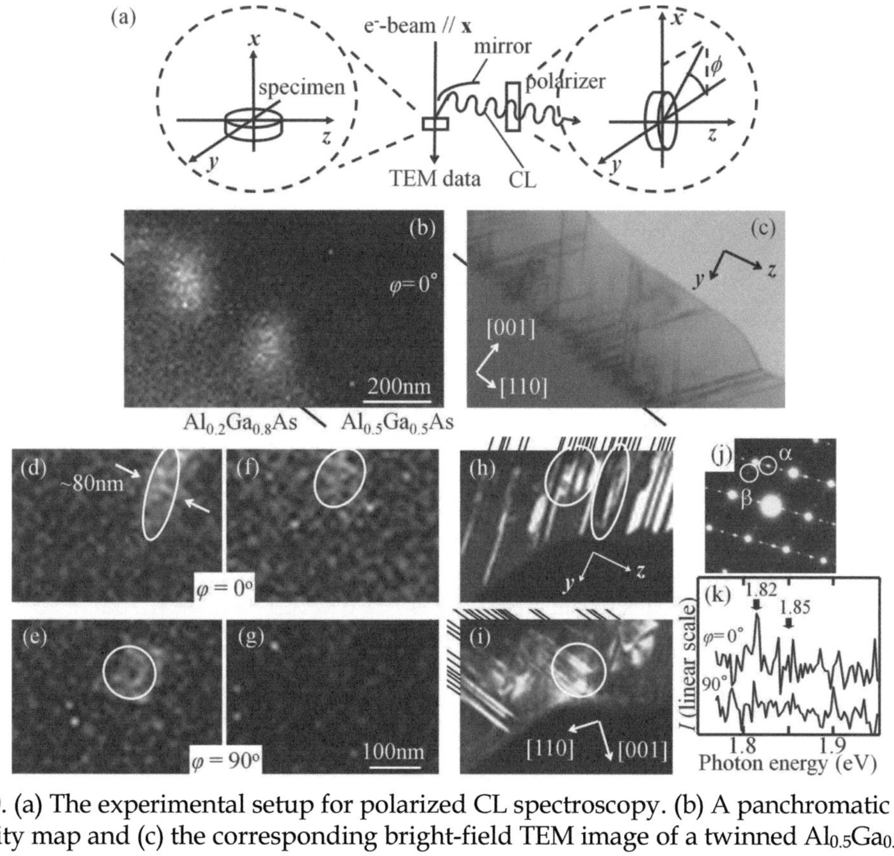

Fig. 10. (a) The experimental setup for polarized CL spectroscopy. (b) A panchromatic CL intensity map and (c) the corresponding bright-field TEM image of a twinned $Al_{0.5}Ga_{0.5}As$ epilayer grown on an $Al_{0.2}Ga_{0.8}As$. (d)-(g) Monochromatic CL intensity maps, and (h)-(i) the corresponding dark-field TEM images in which the locations of boundaries are indicated with the solid lines. The photon energy of the CL light is; (d, e) 1.82 eV or (f, g) 1.85 eV. ϕ = (d, f) 0° or (e, g) 90°. (h) or (i) is, respectively, taken with a twin spot of α or β in (j). (k) Polarized CL spectra for ϕ = 0° and 90° from the square area of Fig. (h). (Ohno et al, 2007b)

CL mapping measurements reveal that (Ohno et al, 2007b), a CL light is emitted from a twinned layer, and the intensity is stronger in comparison with the band-to-band emission in direct-gap $Al_{0.2}Ga_{0.8}As$ (Figs. 10b-10c), even though the layer is an indirect gap semiconductor. A CL light is emitted from an area in which many parallel boundaries are arranged at similar intervals of nanometer length, and no CL light is detected in the other areas (Figs. 10d-10i).

The distribution of the boundaries in a light source is characterized with the intervals between the nearest-neighbor boundaries and the distribution function of boundaries $g(l)$ (Ohno, 2010b). $g(l)$ is defined as the number of boundaries in the range of l to $l+\Delta l$ with the origin at a boundary divided by the number of boundaries N, similar to a distribution function for an extended defect (Ohno et al, 2007a): $g(l) = 1/N\ \Sigma^{N}_{i=1}n(l)/\Delta l$ in which $n(l)$ is the number of boundaries in the range of l to $l+\Delta l$ for the i-th reference boundary, l is the distance from the i-th reference boundary. Figure 11 shows an analysis of a light source, indicated with α in Fig. 11b. There is no translational symmetry in the arrangement of the boundaries, even though the intervals are much the same (Fig. 11c). However, they do not distribute randomly but orderly in a short range. There are three peaks arranged at similar intervals (peaking at the intervals of 2–5, 7–9 and 12–14 nm) in the distribution function for the $(\underline{11}1)$ boundaries (Fig. 11e). Therefore, four or more $(\underline{11}1)$ boundaries are arranged with the period of about 4 nm. Similar results are obtained at the areas from which a monochromatic CL light is emitted.

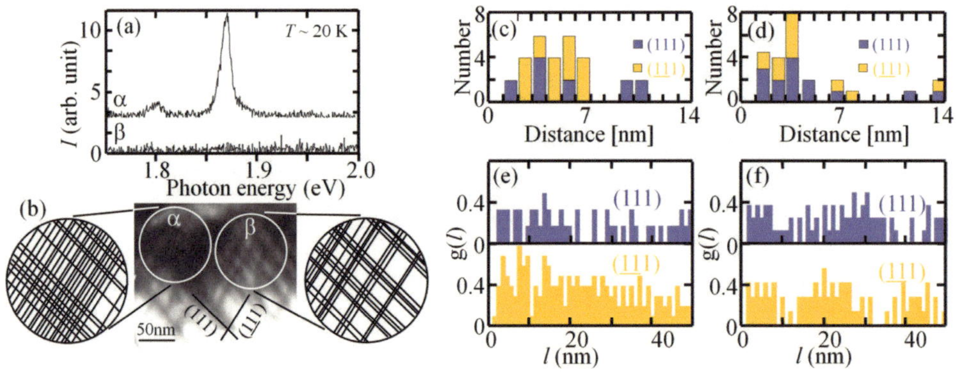

Fig. 11. (a) CL spectra obtained from the encircled areas α and β in (b) [the insets in (b) show the locations of the boundaries in the areas]. (c) or (d), respectively, shows the histogram of the intervals between the nearest-neighbor boundaries in α or β. (e) or (f), respectively, shows the distribution function of boundaries in α or β. (Ohno, 2010b)

The intensity profile of a CL light emitted from a set of parallel boundaries vs the photon energy is fitted with a Lorentz function (e,g., Fig. 11a), and the half-width at half-maximum (5–6 meV) is narrower in comparison with the band-to-band emission in direct-gap $Al_{0.2}Ga_{0.8}As$ (9–10 meV) and with an impurity emission (about 20 meV). The CL light is polarized parallel to the boundaries (Figs. 10d-10i, Fig. 12), and the photon energy increases with decreasing the interval of the boundaries. Those results indicate that a light source is a set of parallel twin boundaries ordered in a short range.

By means of temperature T-dependent CL spectroscopy (Ohno, 2010b), the energy level induced by ordered boundaries is determined (Fig. 13). The intensity of a CL light with the photon energy of E_L, I can be fitted with a function: $I(T) = I_0/[1 + C\exp(\Delta E/kT)]$ where ΔE is the activation energy for the thermal quenching process and I_0 or C is a constant (Holtz et al, 1985). ΔE for a CL light due to ordered boundaries is higher in comparison with bound excitons, and it increases with decreasing E_L. The estimated energy is close to the difference between the band gap energy E_g and E_L, indicating that the light is emitted via an electronic transition between an energy band and a defect level associated with the boundaries; thermal escape of the carriers trapped in the level is responsible for the quenching.

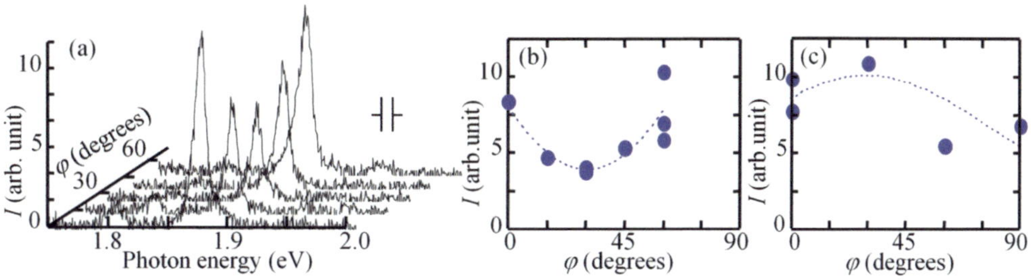

Fig. 12. (a) Polarized CL spectra for different ϕ obtained from α in Fig. 11b. The experimental setup is similar to that in Fig. 10a, and the specimen is set so that the z axis is parallel to [001]. The CL intensity I vs ϕ; the z axis is parallel to (b) [001] or (c) [110] (Ohno et al, 2007b)

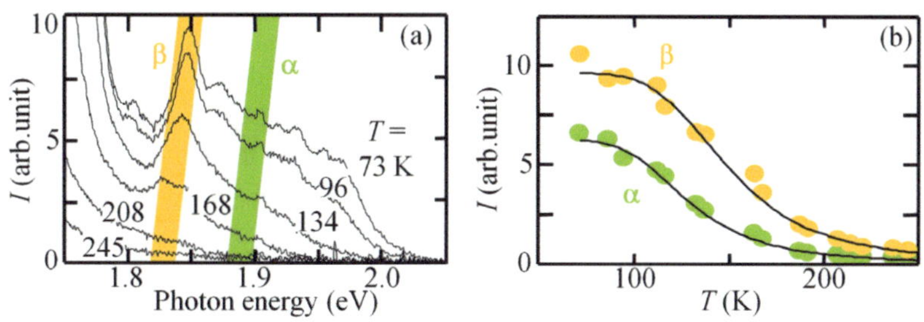

Fig. 13. (a) CL spectra obtained at various temperatures T. (b) T-dependence of the CL intensity for the emission line α and that for β in (a). (Ohno, 2010b)

Electron flux-dependent CL spectroscopy reveals that (Ohno et al, 2007c), the photon energy of a CL light due to a set of ordered boundaries is independent of electron flux. The intensity obeys a power law (Schmidt et al, 1992), and the power α is about 2.0. These results indicate that the CL light is emitted via a band-to-band electronic transition, rather than a donor–acceptor pair recombination ($\alpha < 1$) and an excitonic transition ($1 < \alpha < 2$).

Figure 14 shows the energy diagram based on a twinning superlattice theory (Ikonic et al, 1993). Suppose a twinning superlattice structure, which forms a narrow miniband whose energy is lower than the energy at the conduction band edge, is embedded in the indirect gap $Al_{0.5}Ga_{0.5}As$ crystal. According to the model, an intense monochromatic CL light with photon energy E_L, polarized parallel to the boundaries, is emitted via the direct transition between the miniband and the valence band. The thermal escape of electrons from the miniband into the conduction band in the adjacent $Al_{0.5}Ga_{0.5}As$ crystal is responsible for the quenching of the CL emission, and the activation energy for the quenching ΔE is $E_g - E_L$.

It is shown that CL spectroscopy in a TEM is available for the quantitative analyses of the optoelectronic and structural properties of nanostructures with a high spatial resolution (higher than about 80 nm as in Fig. 10d). The technique will be applied to explore novel nanostructures and defects with useful optoelectronic functions, as well as to assess the functions at an atomistic level, inside semiconductors.

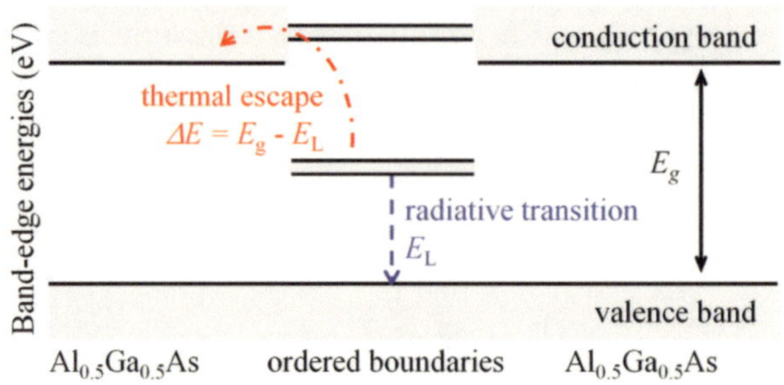

Fig. 14. A schematic view of band-edge energies for an ordered boundaries embedded in an $Al_{0.5}Ga_{0.5}As$ layer in which no twin boundary exists

3.3 Silicon and related materials as optoelectronic materials

Si and related materials are still the most important semiconductor for various kinds of photovoltaic and electronic devices. However, small numbers of the optical assessments have been performed in TEM since Si is an indirect-gap semiconductor. Even though 60° and Lomer-Cottrell dislocations are believed to act as radiative recombination center (Higgs et al, 1992, Sekiguchi & Sumino, 1996), they have not been determined in TEM. Some preliminary optical measurements for Si nanostructures which emit visible light, such as porous-Si (Itoh et al, 1996) and Si nanowires (Ozaki et al, 1998, 2005, Kohno et al, 2004), have been performed in a TEM. With a recent progress in optical measurement systems used in near-infrared region, optoelectronic properties in Si including nanostructures and extended defects, which would be modified by the interactions with point defects (e.g., Inoue et al, 2008, 2009, Ohno et al, 2009a, 2010), will be elucidated at an atomistic level.

4. Concluding remarks

In-situ optical measurements in TEM are powerful and unique techniques to examine the atomistic structures of small regions inside materials, with an extremely high spectral resolution. This in-situ technique in TEM will be applied permanently to discover and assess various novel nanomaterials.

5. Acknowledgements

This work was partially supported by the Ministry of Education, Culture, Sports, Science, and Technology, Japan, Grant-in-Aid for Scientific Research (B) #19310072 (2007–2009) and for Scientific Research on Priority Areas #21016002 (2009-2010), and the Nano-Materials Functionality Creation Research Project in IMR, Tohoku University.

6. References

Akiba, K.; Yamamoto, N.; Grillo, V.; Genseki, A. & Watanabe, Y. (2004). Anomalous temperature and excitation power dependence of cathodoluminescence from InAs quantum dots. *Phys. Rev. B.* Vol. 70, No. 16, pp. 165322/1-9.

Albrecht, M.; Strunk, H. P.; Weyher, J. L.; Grzegory, I.; Porowski, S. & Wosinski, T. (2002). Carrier recombination at single dislocations in GaN measured by cathodoluminescence in a transmission electron microscope. *J. Appl. Phys.* Vol. 92, No. 4, pp. 2000-2005.

Albrecht, M.; Weyher, J. L.; Lucznic, B.; Grzegory, I. & Porowski, S. (2008). Nonradiative recombination at threading dislocations in *n*-type GaN: Studied by cathodeluminescence and defect selective etching. *Appl. Phys. Lett.* Vol. 92, No. 23, pp. 231909/1-3.

Algra, R. E.; Verheijen, M. A.; Borgstrom, M. T.; Feiner, L. F.; Immink, G.; van Enckevort, W. J. P.; Vlieg, E. & Bakkers, E. P. A. M. (2008). Twinning superlattices in indium phosphide nanowires. *Nature.* Vol. 436, No. 7220, pp.369-372.

Bagnall, D. M.; Chen, Y. F.; Zhu, Z.; Yao, T.; Shen, M. Y. & Goto, T. (1998). High temperature excitonic stimulated emission from ZnO epitaxial layers. *Appl. Phys. Lett.* Vol. 73, No. 8, pp. 122077/1-3.

Bao, J.; Bell, D. C.; Capasso, F.; Wagner, J. B.; Matensson, T.; Tragardh, J. & Samuelson, L. (2008). Optical properties of rotationally twinned InP nanowires heterostructures. *Nano Lett.* Vol. 8, No. 3, pp. 836-841.

Blumenau, A. T.; Fall, C. J.; Elsner, R.; Jones, R.; Heggie, M. I. & Frauenheim, T. (2003). A theoretical investigation of dislocations in cubic and hexagonal gallium nitride. *Phys. Stat. Sol. (c).* Vol. 0, No. 6, pp. 1684-1709.

Bonard, J. M.; Ganiere,J. D.; Vanzetti, L.; Paggel, J. J.; Sorba, L. & Franciosi, A. (1998). Combined transmission electron microscopy and cathodoluminescence studies of degradation in electron-beam-pumped $Zn_{1-x}Cd_xSe$/ZnSe blue-green lasers. *J. Appl. Phys.* Vol. 84, No. 8. pp. 1263-1273.

Cheng, Y. M.; Herrick, R. W.; Petroff, P. M.; Hibbsbrenner, M. K. & Morgan, R. A. (1995). Degradation studies of proton-implanted vertical cavity surface emitting Lasers. *Appl. Phys. Lett.* Vol. 67, No. 12. pp. 1648-1650.

Coleman, V. A.; Bradby, J. E.; Jagadish, C. & Phillips, M. R. (2006). Observation of enhanced defect emission and excitonic quenching from spherically indented ZnO. *Appl. Phys. Lett.* Vol. 89, No. 8, pp. 082102/1-3.

Donolato, C. & Venturi, P. (1982). Influence of the generation distribution on the calculated EBIC contrast of line defects. *Phys. Stat. Sol. (a)* Vol. 73, No. 2, pp. 377-387.

Everhart, T. E. & Hoff, P. H. (1971). Determination of kilovolt electron energy dissipation vs penetration distance in solid materials. *J. Appl. Phys.* Vol. 42, No. 13, pp. 5837-5846.

Fissel, A.; Bugiel, E.; Wang, C. R. & Osten, H. J. (2006). Formation of Si twinning-superlattice: First step toward Si polytype growth. *Mater. Sci. Eng. B.* Vol. 134, No. 2-3, pp. 138-141.

Farvacque, J. L.; Vignaud, D.; Depraetere, E.; Sieber, B. & Lefebvre, A. (1989). Electrical and optical properties of dislocations in GaAs. *Inst. Phys. Conf. Ser.* Vol. 104, pp. 141-150.

Goldstein, J. I. (1979). Principles of thin film X-ray microanalysis. In.: *Introduction to Analytical Electron Microscopy*, Ed. Hren, J. J.; Goldstein, J. I. & Joy, D. C. pp. 83-120, Plenum Press, ISBN 0-306-40280-7, New YorK.

Graham, R. J.; Shaapur, F.; Kato, Y. & Stoner, B. R. (1994). Imaging of boron dopant in highly oriented diamond films by cathodoluminescence in a transmission electron microscope. *Appl. Phys. Lett.* Vol. 65, No. 3, pp. 292-294.

Hayazawa, N.; Tarun, A.; Inouye, Y. & Kawata, S. (2002). Near-field enhanced Raman spectroscopy using side illumination optics. *J. Appl. Phys.* Vol. 92, No. 12, pp. 6983-6986.

Henderson, B. & Imbusch, G. F. (1989). *Optical Spectroscopy of Inorganic Solids*, Clarendon Press, ISBN 0-19-851372-0, Oxford. pp. 315-386.

Hibino, H.; Sumitomo, K. & Ogino, T. (1998). Twinned epitaxial layers formed on Si(111) √3x√3-B. *J. Vac. Sci. Technol. A.* Vol. 16, No. 3, pp. 1934-1937.

Higgs, V.; Lightowlers, E. C. & Tajbakhsh, S. (1992). Cathodoluminescence imaging and spectroscopy of dislocations in Si and $Si_{1-x}Ge_x$ alloys. *Appl. Phys. Lett.* Vol. 61, No. 9, pp. 1087-1089.

Hilpert, U.; Schreiber, J.; Worschech, L.; Horing, L.; Ramsteiner, M.; Ossau, W. & Landwehr, G. (2000). Optical characterization of isolated Se(g)-type misfit dislocations and their influence on strain relief in thin ZnSe films. *J. Phys. Condens. Matter.* Vol. 12, No. 49, pp. 10169-10174.

Holt, D. B. & Napchan, E. (1994). Quantitation of SEM EBIC and CL signals using Monte Carlo electron-trajectory simulations. *Scanning.* Vol. 16, No. 2, pp. 78-86.

Holtz, P. O.; Monemar, B. & Lozykowski, H. J. (1985). Optical-properties of Ag-related centers in bulk ZnSe. *Phys. Rev. B.* Vol. 32, No. 2, pp. 986–996.

Hutchinson, H. J. & Myhajlenko, S. (1984). The cathodoluminescence study of recombination behavior at dislocations in indium-phosphide. *Philos. Mag. B.* Vol. 50, No. 4, pp. L49-L53.

Ikonic, Z.; Srivastava, G. P. & Inkson, J. C. (1993). Electronic properties of twin boundaries and twinning superlattices in diamond-type and zinc-blende-type semiconductors. *Phys. Rev. B.* Vol. 48, No. 23, pp. 17181-17193.

Ikonic, Z.; Srivastava, G. P. & Inkson, J. C. (1996). Electronic structure of natural self-organized PbS-Bi_2S_3 twinning superlattices. *Phys. Rev. B.* Vol. 55, No. 15, pp. 9286-9289.

Ino, N. & Yamamoto, N. (2008). Low temperature diffusion length of excitons in gallium nitride measured by cathodoluminescence technique. *Appl. Phys. Lett.* Vol. 93, No. 23, pp. 232103/1-3.

Inoue, K.; Yano, F.; Nishida, A.; Takamizawa, H.; Tsunomura, T.; Nagai, Y. & Hasegawa, M. (2009). Dopant distribution in gate electrode of *n*- and *p*-type metal-oxide-semiconductor field effect transistor by laser-assisted atom probe. *Appl. Phys. Lett.* Vol. 95, No. 4, pp. 043502/1-3.

Inoue, K.; Yano, F.; Nishida, A.; Tsunomura, T.; Toyama, T.; Nagai, Y. & Hasegawa, M. (2008). Three dimensional characterization of dopant distribution in polycrystalline silicon by laser-assisted atom probe. *Appl. Phys. Lett.* Vol. 93, No. 13, pp. 133507/1-3.

Inouye, Y. & Kawata, S. (1994). Near-field scanning optical microscope with a metallic probe tip. *Opt. Lett.* Vol. 19, No. 3, pp. 159-161.

Ishikawa, K.; Yamamoto, N.; Tateno, K. & Watanabe, Y. (2008). Characterization of individual GaAs /AlGaAs self-standing nanowires by cathodoluminescence technique using transmission electron microscope. *Jpn. J. App. Phys.* Vol. 47, No. 8, pp. 6596-6600.

Itoh, M.; Yamamoto, N.; Takemoto, K. & Nittono, O. (1996). Cathodoluminescence imaging of n-type porus silicon. *Jpn. J. Appl. Phys.* Vol. 35, No. 8, pp. 4182-4186.

Jakubowicz, A. (1986). Theory of cathodoluminescence contrast from localized defects in semiconductors. *J. Appl. Phys.* Vol. 59, No. 6, pp. 2205-2209.

Kikkawa, J.; Takeda, S.; Sato, Y. & Terauchi, M. (2007). Enhanced direct interband transitions in silicon nanowires studied by electron energy-loss spectroscopy. *Phys. Rev. B.* Vol. 75, No. 24, pp. 245317/1-5.

Kobayashi, K. & Uchihashi, T. (2010). Electronic states of Ag thin films with a laterally periodic insertion of stacking faults. *Phys. Rev. B.* Vol. 81, No. 15, pp. 155418/1-12.

Kohno, H.; Mori, Y.; Ichikawa, S.; Ohno, Y.; Yonenaga, I. & Takeda, S. (2009). Transformation of a SiC nanowire into a carbon nanotube. *Nanoscale.* Vol. 1, No. 3, pp. 344-346.

Kohno, H.; Yoshida, H.; Ohno, Y.; Ichikawa, S.; Akita, T.; Tanaka, K. & Takeda, S. (2004). Formation of silicon/silicide/oxide nanochains and their properties studied by electron holography. *Thin Solid Films.* Vol. 464, pp. 204-207.

Kucheyev, S. O.; Bradby, J. E.; Williams, J. S.; Jagadish, C.; Toth, M.; Phillips, M. R. & Swain, M. V. (2000). Nanoindentation of epitaxial GaN films. *Appl. Phys. Lett.* Vol. 77, No. 21, pp. 3373-3375.

Maeda, K.; Suzuki, K.; Ichihara, M.; Nishiguchi, S.; Ono, K.; Mera, Y. & Takeuchi, S. (1999). Electronically induced dislocation glide motion in hexagonal GaN single crystals. *Physica B.* Vol. 273, pp. 134–139.

Maeda, K.; Suzuki, K.; Yamashita, Y. & Mera, Y. (2000). Dislocation motion in semiconducting crystals under the influence of electronic perturbations. *J. Phys.: Cond. Matter.* Vol. 12, No. 49, pp. 10079–10091.

Merano, M.; Sonderegger, S.; Crottini, A.; Collin, S.; Renucci, P.; Pelucchi, E.; Malko, A.; Baier, M. H.; Kapon, E.; Deveaud, B. & Ganiere, J. D. (2005). Probing carrier dynamics in nanostructures by picosecond cathodoluminescence. *Nature.* Vol. 438, No. 7067, pp. 479 -482.

Mitsui, T. & Yamamoto, N. (1997). Distribution of polarized-cathodoluminescence around the structural defects in ZnSe/GaAs(001) studied by transmission electron microscopy. *J. Appl. Phys.* Vol. 81, No. 11, pp. 7492-7496.

Mitsui, T.; Yamamoto, N.; Tadokoro, T. & Ohta, S. (1996). Cathodoluminescence image of defects and luminescence centers in ZnS/GaAs(100). *J. Appl. Phys.* Vol. 80, No. 12, pp. 6972-6979.

Nakaji, D.; Grillo, V.; Yamamoto, N. & Mukai, T. (2005). Contrast analysis of dislocation images in TEM–cathodoluminescence technique. *J. Electron Microsc.* Vol. 54, No. 3, pp. 223-230.

Nakamura, J. & Natori, A. (2006). Dielectric discontinuity at structural boundaries in Si. *Appl. Phys. Lett.* Vol. 89, No. 5, pp. 053118/1-3.

Nakamura, S.; Mukai, T. & Senoh, M. (1994). Candela-class high-brightness InGaN/AlGaN double-heterostructure blue-light-emitting diodes. *Appl. Phys. Lett.* Vol. 64, No. 13, pp. 1687-1689.

Naumov, A.; Wolf, K.; Reisinger, T.; Stanzl, H. & Gebhardt, W. (1992). Luminescence due to lattice-mismatch defects in ZnTe layers grown by metalorganic vapor phase epitaxy. *J. Appl. Phys.* Vol. 73, No. 5, pp. 2581-2583.

Negrii, V. D. & Osipyan, Y. A. (1978). Influence of dislocations on radiative recombination processes in cadmium sulfide. *Sov. Phys. Solid State.* Vol. 20, No. 3, pp. 432-436.

Nogami, T.; Ohno, Y.; Ichikawa, S. & Kohno, H. (2009). Converting an insulating silicon nanochain to a conducting carbon nanotube by electric breakdown. *Nanotechnology.* Vol. 20, No. 33, pp. 335602/1-5.

Ohno, Y. (2005a). Polarized light emission from antiphase boundaries acting as slanting quantum wells in GaP/InP short-period superlattices. *Phys. Rev. B.* Vol. 72, No. 12, pp. 121307(R)/1-4.

Ohno, Y. (2005b). Photoinduced stress in a ZnSe/GaAs epilayer containing 90°α partial dislocations. *Appl. Phys. Lett.* Vol. 87, No. 18, pp. 181909/1-3.

Ohno, Y. (2010a). Development of an apparatus for in-situ near-field optical measurements in a transmission electron microscope, to be published.

Ohno, Y. (2010b). In-situ analysis of optoelectronic properties of twin boundaries in AlGaAs by polarized cathodoluminescence spectroscopy in a TEM. *J. Electron Microsc.* Vol. 59, No. S1, pp. S141-S147.

Ohno, Y.; Kawai, Y. & Takeda, S. (1999). Vacancy-migration-mediated disordering in CuPt-ordered (Ga, In)P studied by in situ optical spectroscopy in a transmission electron microscope. *Phys. Rev. B.* Vol. 59, No. 4, pp. 2694-2699.

Ohno, Y.; Koizumi, H.; Taishi, T.; Yonenaga, I.; Fujii, K.; Goto, H. & Yao, T. (2008a). Light emission due to dislocations in wurtzite ZnO bulk single-crystals freshly introduced by plastic deformation. *Appl. Phys. Lett.* Vol. 92, No. 1, pp. 011922/1-3.

Ohno, Y.; Koizumi, H.; Taishi, T.; Yonenaga, I.; Fujii, K.; Goto, H. & Yao, T. (2008b). Optical properties of dislocations in wurtzite ZnO single-crystals introduced at elevated temperatures. *J. Appl. Phys.* Vol. 104, No. 7, pp. 073515/1-6.

Ohno, Y.; Shoda, K.; Taishi, T.; Yonenaga, I. & Takeda, S. (2008c). Formation of multiple nanoscale twin boundaries that emit intense light in indirect-gap AlGaAs epilayers. *Appl. Surf. Sci.* Vol. 254, No. 23, pp. 7633-7637.

Ohno, Y.; Shirakawa, T.; Taishi, T. & Yonenaga, I. (2009a). Interaction of phosphorus with dislocations in heavily phosphorus doped silicon. *Appl. Phys. Lett.* Vol. 95, No. 9, pp. 091915/1-3.

Ohno, Y.; Taishi, T.; Tokumoto, Y. & Yonenaga, I. (2010). Interaction of dopant atoms with stacking faults in silicon crystals. *J. Appl. Phys.* Vol. 108, No. 7, pp. 073514/1-4.

Ohno, Y.; Taishi, T. & Yonenaga, I. (2009b). In-situ analysis of optoelectronic properties of dislocations in ZnO by TEM observations. *Phys. Stat. Sol. (a).* Vol. 206, No. 8, pp. 1904-1911.

Ohno, Y. & Takeda, S. (2002). Analysis of polarization by means of polarized cathodoluminescence spectroscopy in a TEM. *J. Electron Microsc.* Vol. 51, No. 5, pp. 281-290.

Ohno, Y & Takeda, S. (1995). A new apparatus for in-situ photoluminescence spectroscopy in a transmission electron microscope. *Rev. Sci. Instrum.* Vol. 66, No. 10, pp. 4866-4869.

Ohno, Y. & Takeda, S. (1996). Study of electron-irradiation-induced defects in GaP by in-situ optical spectroscopy in a transmission electron microscope. *J. Electron Microsc.* Vol. 45, No. 1, pp. 73-78.

Ohno, Y.; Takeda, S.; Ichihashi, T. & Iijima, S. (2007a). Fabrication of short-range ordered nanoholes on silicon surfaces by electron irradiation. *Jpn. J. Appl. Phys.* Vol. 46, No. 1, pp. 434–439.

Ohno, Y.; Yamamoto, N.; Shoda, S. & Takeda, S. (2007b). Intense monochromatic light emission from multiple nanoscale twin boundaries in indirect-gap AlGaAs epilayers, *Jpn. J. Appl. Phys.* Vol. 46, No. 35, pp. L830-L832.

Ohno, Y.; Yamamoto, N.; Taishi, T.; Yonenaga, I. & Takeda, S. (2007c). Electronic properties of nanoscale multiple twin boundaries in AlGaAs. *Physica B.* Vol. 401, pp. 270-274.

Ozaki, N.; Ohno, Y.; Kikkawa, J. & Takeda, S. (2005). Growth of silicon nanowires on H-terminated Si {111} surface templates studied by transmission electron microscopy. *J. Electron Microsc.* Vol. 54, No. S1, pp. I25-i29.

Ozaki, N.; Ohno, Y. & Takeda, S. (1998). Silicon nanowhiskers grown on a hydrogen-terminated silicon {111} surface. *Appl. Phys. Lett.* Vol. 73, No. 25, pp. 3700-3702.

Petroff, P. M.; Lang, D. V.; Logan, R. A. & Strudel, J. L. (1978). Scanning transmission electron microscopy techniques for simultaneous electronic analysis and observation defects in semiconductors. *Scanning Electron Microsc.* Vol. 1, pp. 325-332.

Petroff, P. M.; Logan, R. A. & Savage, A. (1980). Nonradiative recombination at dislocations in III-V compound semiconductors. *Phys. Rev. Lett.* Vol. 44, No. 4, pp. 287-291.

Schmidt, T.; Lischka, K. & Zulehner, W. (1992). Excitation-power dependence of the near-band-edge photoluminescence of semiconductors. *Phys. Rev. B.* Vol. 45, No. 16, pp. 8989–8994.

Schreiber, J. & Hildebrandt S. (1994). Basic dislocation contrasts in SEM-CL/EBIC on III–V semiconductors . *Mater. Sci. Eng. B.* Vol. 24, No. 1-3, pp. 115-120.

Schreiber, J.; Horing, L.; Uniewski, H.; Hildebrandt, S. & Leipner, H. S. (1999). Recognition and distribution of A(g) and B(g) dislocations in indentation deformation zones on {111} and {110} surfaces of CdTe. *Phys. Status Solidi (a).* Vol. 171, No. 1, pp. 89–97.

Sekiguchi, T. & Sumino, K. (1996). Cathodoluminescence study on dislocation in silicon. *J. Appl. Phys.* Vol. 79, No. 6, pp. 3253-3260.

Setiawan, A.; Vashaei, Z.; Cho, M. W.; Yao, T.; Kato, K.; Sano, M.; Miyamoto, K.; Yonenaga, I. & Ko, H. J. (2004). Characteristics of dislocations in ZnO layers grown by plasma-assisted molecular beam epitaxy under different Zn/O flux ratios. *J. Appl. Phys.* Vol. 96, No. 7, pp. 1785852/1-6.

Sieber, B.; Farvacque, J. L.; Wang, J. & Steeds, J. W. (1993). 1st step of degradation mechanisms in AlGaAs/GaAs laser-like structures. *Mater. Sci. Eng. B.* Vol. 20, No. 1-2, pp. 29-32 (1993).

Sonderegger, S.; Feltin, E.; Merano, M.; Crottini, A.; Carlin, J. F.; Sachot, R.; Deveaud, B.; Grandjean, N. & Ganiere, J. D. (2006). High spatial resolution picosecond cathodoluminescence of InGaN quantum wells. *Appl. Phys. Lett.* Vol. 89, No. 23, pp. 232109/1-3.

Suzuki, K.; Ichihara, M. & Takeuchi, S. (1994). High-resolution electron microscopy of extended defects in wurtzite crystals. *Jpn. J. Appl. Phys.* Vol. 33, No. 2, pp. 1114-1120.

Suzuki, K.; Ichihara, M.; Takeuchi, S.; Nakagawa, K.; Maeda, K. & Iwanaga, H. (1984). In-situ TEM observation of dislocation-motion in II-VI compounds. *Phillos. Mag. A.* Vol. 49, No. 3, pp. 451-461.

Suzuki, K. & Takeuchi, S. (1999). Electron microscopy of dislocations introduced into GaN by plastic deformation. *Phillos. Mag. Lett.* Vol. 79, No. 7, pp. 423-428.

Takkouk, Z.; Brihi, N.; Guergouri, K. & Marfaing, Y. (2005). Cathodoluminescence study of plastically deformed bulk ZnO single crystal. *Physica B.* Vol. 366, No. 1-4, pp. 185-191.

Terauchi, M.; Koike, M.; Fukushima, K. & Kimura, A. (2010). Development of wavelength-dispersive soft X-ray emission spectrometers for transmission electron microscopes - an introduction of valence electron spectroscopy for transmission electron microscopy. *J. Electron Microsc.* Vol. 59, No. 4, pp. 251-261.

Tsukazaki, A.; Ohtomo, A.; Onuma, T.; Ohtani, M.; Makino, T.; Sumiya, M.; Ohtani, K.; Chichibu, S. F.; Fuke, S.; Segawa, Y.; Ohno, H.; Koinuma, H.& Kawasaki, M. (2005). Repeated temperature modulation epitaxy for p-type doping and light-emitting diode based on ZnO. *Nature Mater.* Vol. 4, No. 1, pp. 42-46.

Wang. J. N.; Steeds, J. W. & Hopkinson, M. (1993). Microstructure and cathodoluminescence of MBE-grown (001) $In_xGa_{1-x}P/GaAs$ strained-layer heterostructures. *Semicond. Sci. Technol.* Vol. 8, No. 4, pp. 502-508.

Wang, J. N.; Steeds, J. W. & Woolf, D. A. (1992). The study of misfit dislocations in $In_xGa_{1-x}As/GaAs$ strained quantum-well structures. *Phillos. Mag. A.* Vol. 65, No. 4, pp. 829-839.

Xiong, Q. & Eklund, P. C. (2006). Coherent twinning phenomena: towards twinning superlattices in III-V semiconducting nanowires. *Nano Lett.* Vol. 6, No. 12, pp. 2736-2742.

Yacobi, B. G. & Holt, D. B. (1990). *Cathodoluminescence Microscopy of Inorganic Solids,* Plenum Press, ISBN 0-306-43314-1, New York. pp. 89-120.

Yamamoto, N.; Bhunia, S. & Watanabe, Y. (2006). Polarized cathodoluminescence study of InP nanowires by transmission electron microscopy. *Appl. Phys. Lett.* Vol. 88, No. 15, pp. 153106/1-3.

Yamamoto, N.; Itoh, H.; Grillo, V.; Chichibu, S. F.; Keller, S.; Spence, J. S.; DenBaars, S. P; Mishra, U. K.; Nakamura, S. & Salviati, G. (2003). Cathodoluminescence characterization of dislocations in gallium nitride using a transmission electron microscope. *J. Appl. Phys.* Vol. 94, No. 7, pp. 4315-4319.

Yamamoto, N.; Spence, J. C. H. & Fathy, D. (1984). Cathdoluminescence and polarization studies from individual dislocations in diamond. *Philos. Mag. B.* Vol. 49, No. 6, pp. 609-629.

Yonenaga, I.; Makino, H.; Itoh, S.; Goto, T. & Yao, T. (2006). Photoluminescence properties of GaN with dislocations induced by plastic deformation. *J. Electron. Mater.* Vol. 35, No. 4, pp. 717-721.

Yonenaga, I.; Koizumi, H.; Ohno, Y. & Taishi, T. (2008). High-temperature strength and dislocation mobility in the wide band-gap ZnO: Comparison with various semiconductors. *J. Appl. Phys.* Vol. 103, No. 9, pp. 093502/1-4.

Yoshida, H.; Shimizu, T.; Uchiyama, T.; Kohno, H.; Homma, Y. & Takeda, S. (2009). Atomic-scale analysis on the role of molybdenum in iron-catalyzed carbon nanotube Growth. *Nano Lett.* Vol. 9, No. 11, pp. 3810-3815.

Yoshida, H.; Takeda, S.; Uchiyama, T.; Kohno, H. & Homma, Y. (2008). Atomic-scale in-situ observation of carbon nanotube growth from solid state iron carbide nanoparticles. *Nano Lett.* Vol. 8, No. 7, pp. 2082-2086.

Yoshida, K.; Nanbara, T.; Nozaki, T.; Hirayama, T. & Tanaka, N. (2007). In situ high-resolution transmission electron microscopy of photocatalytic reactions by excited electrons in ionic liquid. *J. Electron Microsc.* Vol. 56, No. 5, pp. 177-180.

Yoshida, K.; Nanbara, T.; Yamasaki, J. & Tanaka, N. (2006). Oxygen release and structural changes in TiO_2 films during photocatalytic oxidation. *J. Appl. Phys.* Vol. 99, No. 8, pp084908/1-8.

Yoshida, K.; Yamasaki, J. & Tanaka, N. (2004). In situ high-resolution transmission electron microscopy observation of photodecomposition process of poly-hydrocarbons on catalytic TiO_2 films. *Appl. Phys. Lett.* Vol. 84, No. 14, pp2542-2544.

Yu, P. Y. & Cardona, M. (2001). *Fundamentals od Semiconductors,* Springer, ISBN 3-540-41323-5, New York. pp. 345-426.

Zhao, Q. X.; Klason, P.; Willander, M.; Zhong, H. M.; Lu, W. E. & Yang. J. H. (2005). Deep-level emissions influenced by O and Zn implantations in ZnO. *Appl. Phys. Lett.* Vol. 87, No. 21, pp. 211912/1-3.

13

Investigating Optoelectronic Properties of the NbN Superconducting Nanowire Single Photon Detector

Zhizhong Yan
University of Waterloo
Canada

1. Introduction

Superconducting optoelectronic devices have been explored as mixers, and detectors to detect single photons, high energy radiations, e.g. X-rays. This category of devices can be made from either high temperature or low temperature superconductive materials. Recent progresses in the semiconducting counterparts have demonstrated competing and surpassing performance over the most of the superconductive optoelectronic devices. However, nanofabrication technique has been widely used in state of the art superconducting optoelectronic devices, such as single photon detectors and hot electron mixers.
One type of Single Photon Detector (SPD) that are sensitive to a wide spectrum of phone energies, namely Superconducting Nanowire Single Photon Detector (SNSPD) based on niobium nitride (NbN) thin films has drawn intensive interests. Such detectors exhibit remarkable performance, such as high counting rate, low dark count, and low jitter time. The fundamental reason is because of the nano size of the superconducting condensate that is comprised of only a few number of charge conducting carriers, i.e. Cooper pairs. The low operating temperature also helps reduce the thermal noise. These advantages result in a superior environment for converting single photon energies into electrically measurable signals. On the other hand, semiconducting materials, such as InGaAs, although technically it is possible to make geometrically and structurally similar device of semiconductor nanowire meander lines, it is impossible to achieve the same photon-to-electrical (quantum optoelectronic) gain, due to the fact that the charge carriers are Fermionic, i.e. the total wave function of them lacks coherence based on the Pauli-Exclusion Principle; superconducting SPDs, instead are comprised of Cooper pairs, which lead to the Bosonic ensemble acting as a quantum optoelectronic detector. Such unique property not only allows superconducting nanowire device for large dynamic range, but also make it sensitive to much lower energy photons, such as telecom wavelength, or even longer wavelength single photons.
Therefore, investigating photon interaction with Cooper pairs in the superconducting nanowire structure will be the key operation to understand the single photon detection, and to devise novel superconducting optoelectronic sensor, in particular to discover its quantum detecting nature. The photon detection can be viewed as one strong nonlinear effect particularly exhibited in those nanowires. In principle, the strong nonlinearity stems from the strong depletion of Cooper pair density. Either bias current, or incident photon flux, or increased temperature can drive the nanowire into deep Cooper pair depletion regime.

One efficient way to probe the Cooper density is to measure the kinetic inductance of the nanowire structure. The inductance measurement requires an incident RF/microwave power and comparing the reflection power for both amplitude and phase. Since the nanowire makes a very sensitive detector, the incident RF/microwave probing power should be as low as possible. In this chapter, we will present a one-port S-parameter measurement that fulfills the above requirement; and apply the developed method to further investigate the superconductive nonlinearity under combinations of temperature, bias current and optical power.

Microwave measurements have been exploited to examine the nonlinear properties of superconducting materials for many decades. The merit of nonlinearity can be studied by measuring the parametric variation of the Quality factor (Q-factor), surface reactance and resistance of superconducting resonators. Such variation always behaves as a function of frequencies and power levels of the input signal, external magnetic field, or ambient temperature. The nonlinearity can even be found by measuring higher orders of harmonic power spectra.

However, studies on dc current or voltage induced nonlinearity have not yet been performed in the above devices, mainly because they can be treated only distributively, due to their relatively larger geometrical sizes with respect to the operation wavelength. Recently, NbN nanowire meanderlines have extensively been used to make the SNSPD. The total length of the meanderline is typically 500 μm, much shorter than the center wavelength of a photon-induced microwave signal. One can treat the SNSPD as a lumped circuit element, terminating at one end of a normal transmission line, typically a 50 Ω Coplanar Waveguide (CPW) made of Au/Ti. Therefore, new technique is required to explore the DC-bias induced nonlinearities, particularly in the kinetic inductance or natural damped resonance frequencies associated with the meanderline.

Since relatively large but nonlinear kinetic inductance is an intrinsic property of the SNSPD, the relationship between the kinetic inductance and bias current might reveal geometrical constrictions that limit the quantum efficiency (QE). The value of the kinetic inductance also determines the rise and fall times of the photo-induced signal in the SNSPDs. The fall time is associated with the device recovery time, a key timing for the operational speed of SNSPDs. In addition, the nonlinearity of the kinetic inductance can be potentially useful to devise and integrate novel superconducting single photon detectors with parametric amplification transmission lines. Therefore, kinetic inductance and its relationship with electrical bias becomes important figure of merit for SNSPDs.

The NbN nanowire is essentially a one-port passive device, one can readily explore combinations of resistive and reactive components, when it is treated as the load of a transmission line by the S-parameter measurement. However, only the kinetic inductance has been reported in the current bias mode. In this chapter, we will employ such measurements not only to explore the kinetic inductance, and other relevant reactive and resistive components, under either dc current or voltage bias; but also to find the resonant frequency and quality factor of the packaged NbN meanderline. Furthermore, the detail of a one-port S-parameter measurement including the calibration at low temperature is presented in this chapter.

In the following sections, we first construct the model of equivalent circuitry; derive mathematical relations to compute values of the circuit elements; apply our experimental results to obtain the kinetic inductance, the capacitance and the series resistance; and then

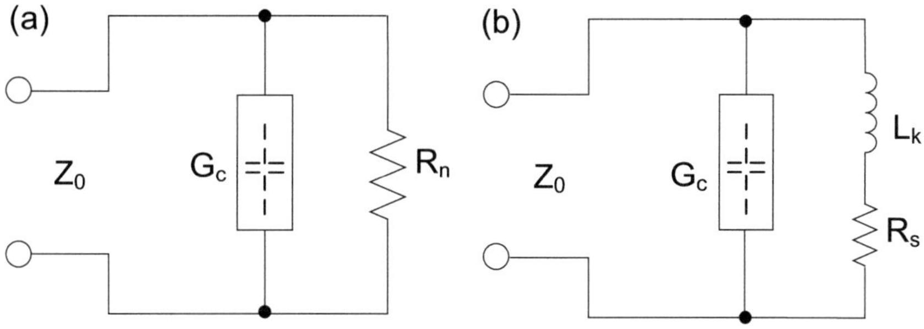

Fig. 1. (a) Room temperature equivalent circuits; (b) Low temperature 4.2K equivalent circuit.

justify the results by the measurement of resonant frequencies and quality factors under various biasing conditions.

More importantly, we would like to highlight that although the methods and experimental procedure in this work are primarily dealing with a NbN superconducting nanowire, these methods and procedures are applicable to other nano size optoelectronic devices, such as high temperature superconducting or even semiconducting optoelectronic devices.

2. Equivalent circuit model for S-parameter measurement

2.1 The construction of the equivalent circuit model

We use the packaged SNSPD identical to what we have reported in (Yan et al., 2009) to undertake all the measurements. However, in the context of current chapter, NbN nanowire meanderlines of the SNSPD are taken as the device under test (DUT). The nanowire meanderline has the same geometrical sizes as what the SNSPD has. For DUT critical temperature T_c, we use an alternative method to find its value. It is estimated to be close to 10.4 K, by the best fitting to the $1-(T/T_c)^4$ like relation based on kinetic inductance-temperature measurement. We will elaborate how to achieve the estimation in later sections. The exact critical current density varies with different DUTs, but all are in the range above $10^6 A/cm^2$.
Since the measurement has to be performed in a packaged device, while the active area of nanowire is relatively small, any distributed effect must be taken into account, including geometrically induced magnetic inductance and capacitance, which are in addition to the kinetic inductance L_k. The former most likely arises from interconnection wires; the latter has been reported either originating from the stray capacitance (Yan et al., 2007; Semenov et al., 2007), or from the dc bias circuitry (Yang et al., 2007). In addition, when the device is dc biased, either with or without the presence of an RF signal, some resistance possibly due to either phase slips or Joule heating may present along the nanowire (Tinkham et al., 2003). We model such lossy contributions by a resistance term of R_s in series with L_k. Therefore, the construction of an equivalent circuit model, which helps to exclude these unavoidable elements, becomes the key to accurately measure L_k, under conditions of both current and voltage bias schemes.

Fig. 1 shows the circuit model at both room and cryogenic temperatures. The procedures to determine values of C, L_k and R_s are deduced from the model, provided that we know the impedance Z_0 of transmission line, and room temperature resistance R_n (usually between 1.5 to 2 MΩ) of the NbN nanowire meanderline. The two-step procedure comprises room and cryogenic temperature S-parameter measurements. Firstly, we perform a room temperature

one port S-parameter measurement to find the value of lumped reactance stemming from geometrical factors. The equivalent circuit is illustrated in Fig. 1 (a). Fig. 2 shows the results of the S-parameter measurement. From the plot, one can find a phase lag whose angle is proportional to the testing frequency. For mathematical simplicity, the phase angle is expressed in wrapped format at $\pm 180°$. The phase angle in fact constantly decreases with each increment of measuring frequencies. Such observation confirms that a capacitance does exist. The value of such capacitance can be well evaluated at frequencies of a few hundred MHz, typically at 100 MHz, to avoid errors arising from magnetic inductances.
Note that two assumptions have been made here:

1. The magnetic inductance L_m has not been taken into account. Fig. 2 displays two resonance frequencies at room temperature measurement. Based on the plot, one can estimate the magnetic inductance L_m to be less than 1 nH, which is not subject to the presence of L_k. In comparison with the kinetic inductance L_k of the nanowire of interests, we estimated L_m is less than 1% of L_k. However, it can become significant when very high RF frequency is applied, typically above 5 GHz. In the frequency range below 300 MHz, where our measurements were made, neglecting L_m should not cause any error in the final results of our measurements.
2. One should be aware that the origin of R_s is not entirely clear yet, but it is definitely different from a stable hotspot resistance, especially when the resistance in the voltage bias scheme, after the critical current point. Quasiparticle excitation is most probably responsible for R_s in this range, where electrically spontaneous relaxations can be observed, and L_k exhibits less than three times L_{k0} (the zero bias L_k) change. Beyond this range, where a stable hotspot has formed, we observe an abrupt L_k variation, associated with a similar style of resistance change. We suspect that the long range coherence of the order parameter has been quenched; a hotspot will thus develop by the process of self Joule-heating.

Based on these assumptions, we conclude that the impedance of the capacitance is dominant (illustrated in Fig. 1 as G_c), while the impedance due to the magnetic inductance is negligible in comparison with the impedance of kinetic inductance L_k, at low frequencies typically below 300 MHz. Secondly, when we cool down the packaged NbN meanderline to the cryogenic temperature, the corresponding circuit model is depicted in Fig. 1 (b). Assuming that the capacitance (G_c) keeps constant, the kinetic inductance L_k in series with R_s can be determined by performing a similar one port S-parameter measurement.

2.2 Scattering parameter analysis

Elements of the circuit model shown in Fig. 1 are determined based on S-Parameter (S_{11}) measurement, which can be done by a Vector Network Analyzer (VNA) in a certain frequency range. The inputs of the following mathematical relations are from S_{11} measured at both room and cryogenic temperatures.
The S_{11} can also be expressed as the reflection coefficient Γ. At each given frequency ω, the magnitude $\rho = |\Gamma|$ and phase angle θ can be written as $\Gamma = \rho e^{i\theta}$. We can define the complex conductance G_c as the termination of the transmission line whose characteristic impedance is 50 Ω

$$G_c = \frac{i(\Gamma - 1)}{\omega(\Gamma + 1) Z_0}. \tag{1}$$

where $Z_0 = 50\Omega$, and the complex $G_c = C - \frac{iG_n}{\omega} = C - \frac{i}{\omega R_n}$. The capacitance and the normal state resistance in parallel can be obtained by the knowledge of S_{11} measurement in the room

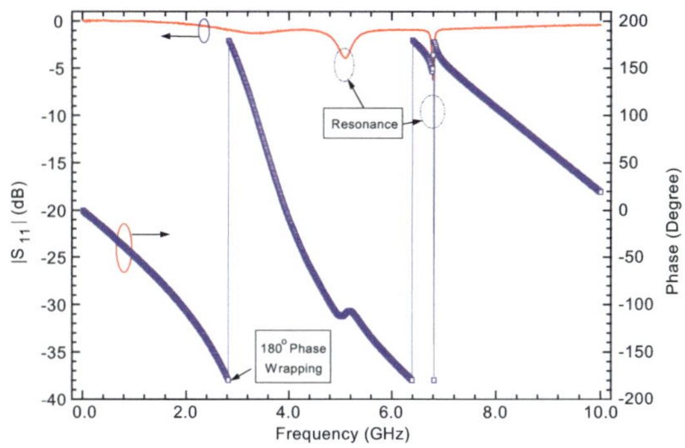

Fig. 2. Room temperature S-parameter measurement, the phase trace (denoted as hollow blue squares) is wrapped at every $\pm 180^\circ$ angle.

temperature as

$$C = \frac{-2\rho \sin\theta}{\omega Z_0 \left(1 + 2\rho\cos\theta + \rho^2\right)}, \tag{2a}$$

$$R_n = \frac{Z_0 \left(1 + 2\rho\cos\theta + \rho^2\right)}{1 - \rho^2}. \tag{2b}$$

The room temperature S-parameter measurement indicates that there exists a capacitor that plays a role in the equivalent circuit model. The contribution of this capacitance leads to the resonance behavior. When the meanderline has been cooled down to its superconducting state, the normal state R_n becomes the combination of a kinetic inductance L_k and a series resistance R_s. The series resistance has very small values when the bias current is lower than the critical current I_c. We introduce a complex resistance Z_{Lk} to represent the circuit components when the NbN nanowire is in the superconducting state. It comprises the contribution from the kinetic inductance L_k and the series resistance R_s. The relation between the measured S-parameter S_{11} or Γ and the capacitance C to obtain Z_{Lk} is expressed as

$$Z_{Lk} = \frac{i\left(\Gamma + 1\right) Z_0}{\omega\left[\left(1 + iZ_0\omega C\right)\Gamma - \left(1 - iZ_0\omega C\right)\right]}. \tag{3}$$

After obtaining Z_{Lk}, the kinetic inductance $L_k = \mathrm{Re}\left(Z_{Lk}\right)$ and series resistance $R_s = -\omega \mathrm{Im}\left(Z_{Lk}\right)$ can be measured by separating the real and imaginary parts of the complex Z_{Lk}. With the increased value of R_s, the resonance frequency deviates from the undamped LC resonance frequency ω_o ($\omega_o = \frac{1}{\sqrt{L_k C}}$) as illustrated in the Fig. 1 (b). The deviation of ω_o degrades to the damped resonant frequency ω_d

$$\omega_d = \omega_o \sqrt{1 - Q^{-2}} \tag{4}$$

where the Quality factor Q can be calculated through the relationship $Q = \frac{1}{R_s}\sqrt{\frac{L_k}{C}}$.

3. Experimental results of kinetic inductance measurement

3.1 Experimental setup and measurement principle procedures

The RF/Microwave characterization system of the packaged NbN meanderline employs an Agilent E8364B PNA Network Analyzer (10 MHz to 50 GHz), a Pasternack bias Tee

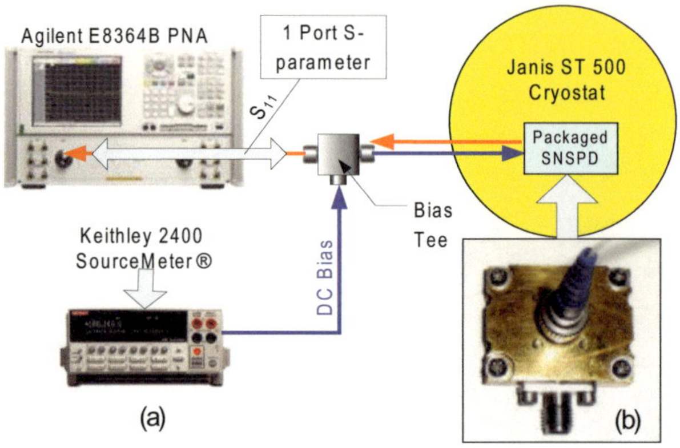

Fig. 3. The experimental setup for the measurement of circuit model elements: (a) The configuration diagram, (b) The photograph of the packaged SNSPD.

10MHz-6GHz, a Keithley 2400 SourceMeter, and a Janis ST-500 cryostat. Their connection is illustrated in Fig. 3 (a). The DUT is packaged in an ETP copper housing equipped with a 40 GHz bandwidth k-connector to maintain the high frequency performance free from the cryogenic temperature stress. The k-connector dielectric layer is made of air/vacuum rather than Teflon in SMA connectors that suffers contraction. In addition, the dielectric constant of Teflon may vary such that it may lead to unwanted variation of the port impedance. The packaged housing is first plated with nickel followed by gold to improve the thermal conduction with the cryostat cold plate. The NbN meanderline is electrically connected by silver epoxy to glue one contact pad to the microwave center pin, and the remaining two pads to the wall of the housing. The final result of the device package is shown in Fig. 3 (b).

The measurement assumes that any extra time delay arising from interconnections within the device packaging is negligible, because of two considerations: (1) The S-parameter S_{11} can be technically measured at only the reference plane, located at input/output port of the packaged NbN meanderline. The reference plane is set by calibration with VNA; (2) The working wavelength (in meters) is much longer than the total electrical wire length (in one hundredths of a meter) including the packaging and the coplanar waveguide on the substrate. Thus, the circuit model in conjunction with its mathematical description has been justified.

It is challenging to do S-parameter measurement under cryogenic temperature and high vacuum because of two limiting factors: (1) We have to limit the amplitude of the ac signal to avoid the device latching when the dc bias is very close to its critical point; (2) The interconnection loss along the transmission line, especially the connector loss, weakens the reflected power. As a result, the ratio of signal to noise becomes extremely low. Since the standard one-port calibration can be performed only in the range of room temperature (approximately 25-30 °C) and pressure (10^3 mbar), the forward tracking error terms in a standard one-port calibration algorithm can not be kept accurate enough, because the portion of the measuring coaxial cable has been cooled far below its calibrating temperature; and has been placed in an ambient of high vacuum condition ($< 10^{-5}$ mbar).

We have observed that the amplitude of Γ in the frequency range experiences about 0.1 dB variation on average. According to our calculation, R_s should be very sensitive to $|\Gamma|$ when the dc bias current is very close to zero, because R_s is expected to be very close to zero. The associated phase angle variation can be approximated in the complex plane by simple geometrical relations and the result indicates the phase angle variation is less than 0.6 degree.

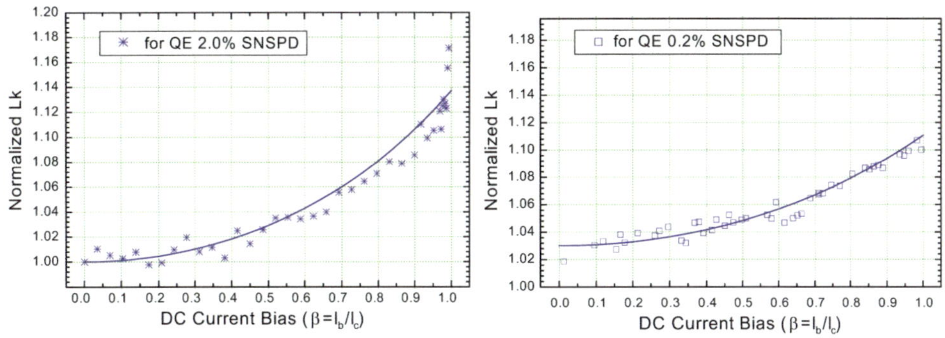

Fig. 4. The measured kinetic inductance for two NbN nanowire meanderlines with different film quality; solid lines are results from theoretical simulations.

Thus, the inductance measurement can achieve a good accuracy, even though the capacitance reduces the phase angle as a function of frequency ω.
In the experiment, we treat the bias tee and cryogenic cable as one lossy transmission line. The calibration is done by setting the reference plane to the end of the cryogenic cable, i.e. the k-connector input of the packaging. Thus, losses from the cryogenic cable, the bias tee and interconnecting components, which are in the ambient temperature, have been compensated by the calibration. The calibration frequency ranges from 10 MHz to 300 MHz. The dc bias source is programmed in either current or voltage mode. The microwave probing power is set at -76 dBm, corresponding to 1 μA peak-to-peak ac current at a 50 Ω coaxial cable. However, considering the cryogenic cable loss, the effective ac current amplitude at the device side falls to about 0.25 μA, which poses much less perturbation to the system during the measurement. Meanwhile, we have observed that when the frequency is higher than about 70 MHz, the S_{11} magnitude variation due to the incident microwave signal becomes negligible. Therefore, the measurement of circuit elements L_k, C and R_s starts with S_{11} frequency points higher than 70 MHz and averages them to the upper frequency limit (in this measurement it is 300 MHz).
Moreover, the choice of calibration power is -30 dBm, whereas the measurement power is set to -76 dBm. This calibration power level is chosen by considering that the minimum noise floor for this VNA preamplifier is -110 dBm. If the calibration power is too low, the forward tracking error term can not be well resolved after the calibration. We can use the data track smooth algorithm as well when we switch the power level down to -76 dBm. However, the smoothing data track can not lead to a higher signal to noise ratio. Instead, we can reduce the *IF bandwidth* to 10 Hz to increase the measurement accuracy. Based on our experiment, further decreasing *IF bandwidth* has no appreciable improvement. In addition, *stepped frequency scanning* is adopted in lieu of the default adaptive scan mode. Each stepped frequency has a 1 ms delay to make sure no residue power affects the adjacent frequencies.

3.2 Kinetic inductance measurement results

Initially, we apply the current bias scheme in the measurements. Fig. 4 shows the measured kinetic inductance for two identical packaged devices, made of NbN nanowire meanderlines. Both measurements of L_k have been normalized to their zero bias kinetic inductance, respectively. The capacitance has been measured in the room temperature via one port S-parameter measurement. The measured capacitance after postprocessing indicates that its value is device dependant. In this work, the capacitance for higher quality film nanowire

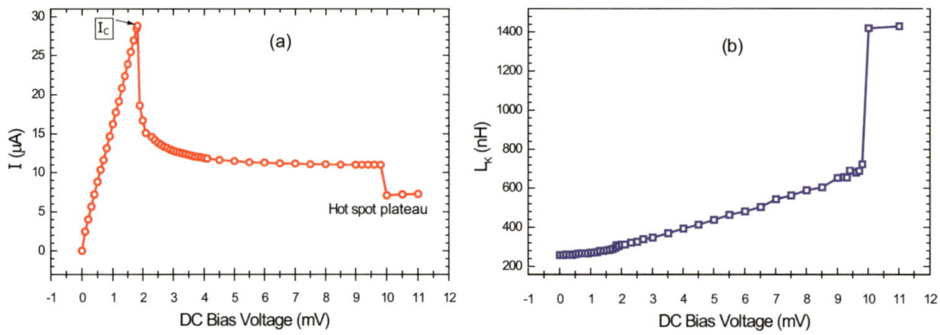

Fig. 5. The linkage of the I-V scan and kinetic inductance variation, (a) the I-V curve under voltage bias scheme; (b) the measured kinetic inductance L_k.

is 1.6 pF; whereas for the lower film quality [1] one, its capacitance is 1.4 pF. On the other hand, the former has zero bias kinetic inductance at about 210 nH; while the latter is 230 nH. In both cases, the nonlinear kinetic inductance displays a trend of higher L_k associated with a higher bias ratio. Detailed discussions on the theoretical background are presented in Appendix A. The relationship of L_k with respect to the bias current ratio is given by (5); and the simulation results are plotted as solid curves in Fig. 4.

However, neither in this work nor in the relevant literatures (Miki et al., 2008; Hadfiled et al., 2007) have shown any abrupt change of L_k when the dc bias current is very close to the so-called critical current I_c, even when we are able to drive the bias current ratio above 99%. Since the kinetic inductance L_k by definition should be inversely proportional to the Cooper pair density of the nanowire condensate, this seems to imply that the I_c may only be a local effect, possibly a resistive spot somewhere along the nanowire. Further increasing the bias current above I_c will cause device permanent latching .

Based on the preceding analysis, we switch our dc bias scheme into voltage bias. Now all of the measurement conditions are kept the same except the bias method. Fig. 5 demonstrates the linkage between the different I-V regimes and the kinetic inductance L_k of the NbN meanderline. The curve shows that even when the bias current decreases after I_c, the kinetic inductance increases constantly with the increasing bias voltage. This suggests that Cooper pairs are proportionally accelerated by applied voltages without apparent resistance barriers along the meanderline path, until the hotspot formation plateau occurs. At that transition point, the kinetic inductance changes very abruptly to 1500 nH at 4.2 K. In order to see the connection of the increasing kinetic inductance L_k with the series resistance R_s, Fig. 6 shows the measured L_k and R_s at each bias voltage point. The figure suggests that when the bias voltage surpasses the point where I_c occurs, both L_k and R_s follow the same trend of a constant increment, due to dissipations caused by the quasiparticle excitation. At lower dc bias, such dissipation is weak, so R_s is negligible. R_s increases smoothly from zero to a quasi stable value close to 120 Ω prior to the formation of an initial hotspot, whose stable value is about 500 Ω. This process can *not* be found in the dc current bias scheme. Instead, when it is switched to the voltage bias, we observe a continuous transition for kinetic inductance L_k and series resistance R_s. They both display an abrupt change at the point where the dc bias current drops below the return current I_{ret}, as shown in Fig. 5(a). The abruptly changed value of R_s is in the range of the hotspot resistance R_h. Very large variations of L_k (more than four times $L_k(\beta = 0)$ have

[1] The higher SNSPD QE, the better film quality. So the better thin film quality one is equivalent to higher QE SNSPD.

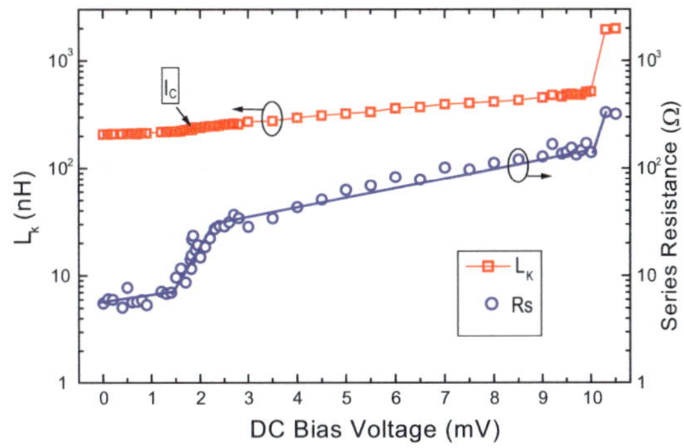

Fig. 6. The measured kinetic inductance and series resistance under the voltage bias scheme for the SNSPD whose QE is 2%.

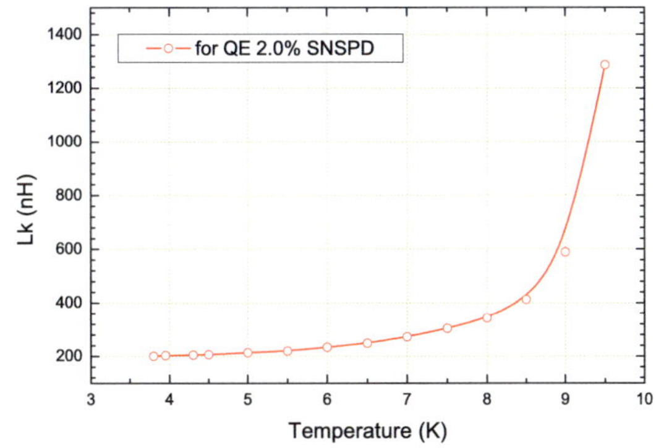

Fig. 7. The measured kinetic inductance by varying the thermal bath temperature in the absence of optical irradiations.

been observed, and show good agreement with the Ginsburg-Landau (G-L) theory prediction (Yan et al., 2007), when R_s becomes comparable with the R_h. Such experimental observation suggests that when the NbN nanowire meanderline is used as SNSPD, its photon-response due to the formation of hotspot strongly correlates to the abrupt change of kinetic inductance. We performed the L_k measurement by changing the DUT temperature. The temperature is monitored and controlled by the Lakeshore 332 two channel temperature monitor/controller. This temperature controller is equipped with two independent channels of heater and thermal sensor interface. The helium flow of the cryostat is also adjusted in order to track the change of temperature and to keep the temperature of the ST-500 cold plate as stable as possible. The temperature has the fluctuation less than 0.1 K. We average our measurements over one particular temperature for multiple times to reduce the uncertainty.
The same one-port S-parameter measurement as what we have performed in the first section is undertaken. L_k is measured as a function of temperature T. The result is depicted in Fig. 7. The open circle in the plot is the experimental measurement; the solid line is the fitting curve using a asymptotic relation of $1 - (T/T_c)_0^{\alpha}$. We take $\alpha_0 = 4$, and $T_c = 10.4K$ for the best fit to the measurements.

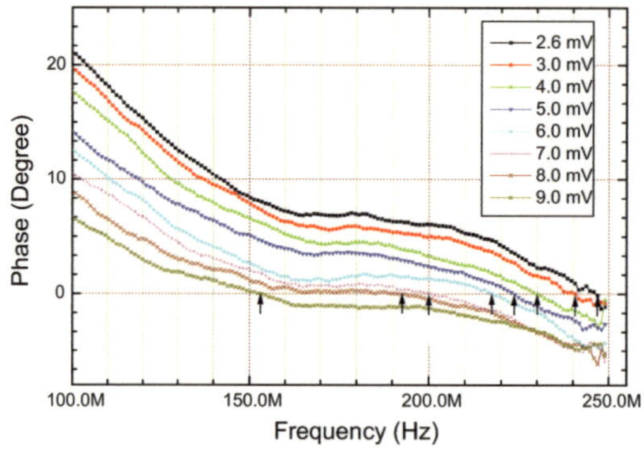

Fig. 8. The measured kinetic inductance and series resistance under the voltage bias scheme for the SNSPD whose QE is 2%, the arrows indicate frequency points of each damped resonance.

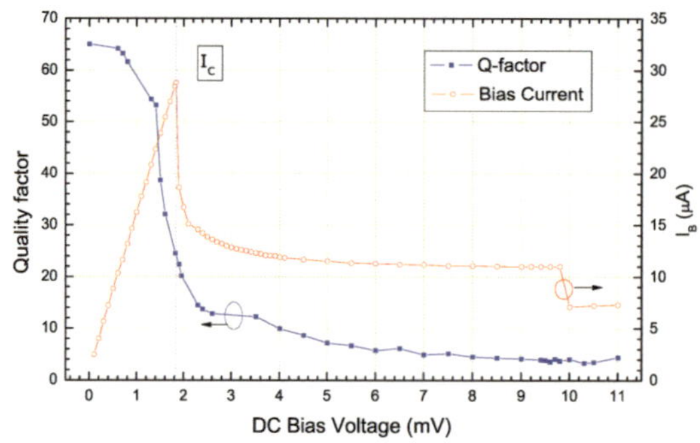

Fig. 9. The quality factor -Q of the equivalent R_s-L_k-C circuit as the resonating circuitry

3.3 The resonating circuit analysis

To further confirm the validity of the R_s-L_k-C circuit model and the related model parameter measurement, one can resort to resonance frequency measurement, i.e. by looking at the frequency points where the phase angle of S_{11} is zero. In this case, the load at the end of the transmission line has no imaginary component. The undamped resonance happens when R_s is zero or negligibly small. What is important for this work is that the damped resonance frequency ω_d deviates from the undamped resonance frequency (ω_o) governed by Eq. (4), due to large values of R_s. Fig. 8 shows the experimental results of the phase with different frequencies of the RF/microwave excitation signals under different bias voltages. The values of L_k and R_s for each branch can be found in Fig. 6.

Fig. 9 displays the extracted quality factor as a function of bias voltage. The plot indicates that the quality factor for this photon detector is not high. Unlike the high-Q resonator detection mechanism, the highest Q of the SNSPD is less than 100 at 4.2K. We think that it is beneficial for the fast resistive state formation. The Q-value is expected to increase with lower temperatures than 4.2 K, but it makes the transition from superconducting to resistive state hard. Therefore, experimentally the QE of SNSPD has not been reported to be further improved when its operating temperature is below 1.8 K.

Table 1 tabulates the calculated undamped resonance frequency (ω_o) and the directly measured damped resonance frequency (ω_d). The corresponding Quality factor Q is obtained

Voltage	L_k	Undamped $f_0 = \frac{1}{2\pi\sqrt{L_k C}}$	Damped Exp. resonance f_d	Q-factor[a]	Q-factor[c]
(mV)	(nH)	(MHz)	(MHz)	Q	Q
9.0	456.00	186.33	153.40	1.75	4.15
8.0	415.16	195.28	192.25	5.56	4.52
7.0	395.50	200.07	200.65	12.50 [b]	4.90
6.0	362.13	209.09	217.45	3.45 [b]	5.75
5.0	323.69	221.15	223.75	6.67 [b]	7.18
4.0	297.35	230.74	230.5	12.5	9.93
3.0	272.66	240.96	240.55	16.67	14.46
2.6	257.50	247.95	246.85	11.11	12.80

a. The quality factor Q is calculated by $Q = \left(\sqrt{1 - \left(\frac{f_d}{f_0}\right)^2}\right)^{-1}$.

b. Absolute value $\left|1 - \left(\frac{f_d}{f_0}\right)^2\right|$

c. The quality factor Q is directly obtained by the definition $Q = \frac{1}{R_s}\sqrt{\frac{L_k}{C}}$

Table 1. Measured resonance frequencies and kinetic inductance.

by both methods of damped resonance frequency, and the Q factor is defined as the function of L_k, R_s and C. The comparison of both approaches shows a good agreement when R_s is relatively large, e.g.$> 100\Omega$.

The resonating R_s-L_k-C circuit does not display any absorption dips at the resonating frequencies, if we focus just on the amplitude of S_{11}, because of two factors: First, the equivalent circuit shown in Fig. 1(b) is neither a parallel nor a serial typical RLC resonating circuit as depicted in (Pozar, 1998). One can convert our equivalent circuitry into a typical parallel R_s-L_k-C resonator circuit by invoking circuit theories, but the converted parallel resistance becomes considerably larger than the 50Ω transmission line impedance, resulting in a pure parallel L_k-C circuitry approximately. Second, at the non-resonating frequencies, the total impedance of our equivalent circuit becomes primarily reactive, so the unitary reflection will be most probably measured.

4. Ginsburg-Landau relation for kinetic inductance and bias current

In the past, many experimental measurements showed that (a). in dc I-V measurement, even at 4.2 K, there is a non-trivial dc current dissipation, i.e resistance, (b) the residual resistance close to 20 K has the highest value compared with that at room temperature, with resistivity of 200 $\mu\Omega$-cm. To seek more insights, we explored extensively in the relevant literatures; and two well-established evidences have been brought to our attention:

1 The fabrication process of NbN thin films involves sputtering pure niobium target in a mixed gas flow of nitrogen and argon (Keskar et al., 1974). The film properties have been studied microscopically in (Shy et al., 1973)-(Tyan & Lue, 1994), in order to systematically show the weaklink properties of NbN thin films. The electrical properties of NbN thin films have been studied in (Kampwirth & Gray, 1981)-(Gray et al., 1985). Based on values of the normal state resistivity, which is in the range of some 200 $\mu\Omega$-cm, granular boundaries inherently existing in NbN thin films were attributed to tunneling weaklinks. Very interestingly, ref. (Kampwirth & Gray, 1981) gives a good approximation

to the critical current density based on quasiparticle tunneling by Ambegaokar *et al.* (Ambegaokar & Baratoff, 1963)-(Ambegaokar & Baratoff, 1963). We have found out that the normal state resistivity of the nanowire is about 200 $\mu\Omega$-cm, which is the range of granular tunneling range at 20 K regime as discussed in (Kampwirth & Gray, 1981)-(Gray et al., 1985).

2 Microwave nonlinear surface resistance Z_s: NbN has been long employed in microwave applications, primarily useful for making transmission lines and resonators. The power handling capability of such devices has been a topic for many decades (Chin et al., 1992)-(Boin & Safa, 1991). The nonlinear dependency of the surface resistance has been attributed to the granular nature of NbN thin films (**?**), (Chin et al., 1992)-(Boin & Safa, 1991), that has been prepared by the RF sputtering technique. The dependency has been modeled on the basis of tangential magnetic field that is parallel with the superconducting surface. The model involves magnetic field penetration depth. This does not seem applicable to our case directly, because the DUT acts only as a lumped load; but the dissipation of microwave power by granular boundary remains the same mechanism to account for the nonlinearity.

Additionally, we also calculated the phase-slip induced I-V relationship at the transition from the superconducting state to the normal state when the temperature increases across the transition temperature T_c. The exponential increase of resistance indicates the pronounced existence of phase-slip centers in the nanowire meander-line. We suspect that these phase slip centers stem from the granular boundaries during the formation of thin film. It is well known about the process of sputtering thin films. Some seeds of metallic grains expand and gradually merge into one continuous film. Thus, we conclude that the superconducting nanowire is patterned on such thin film and its weaklink effect has to be considered in almost every testing, such as dc I-V and RF/microwave measurements.

Thus, we heuristically conclude that the strong-weaklink network (**?**), or inhomogeneities of the order parameter in one dimension superconducting nanowire, is responsible for both nonlinear kinetic inductance and resistance. The amplitude of complex order parameter has been reduced on average, allowing the G-L equation to be valid in a wider temperature range, e.g. at $0.5T_c$.

We have developed a relation of current bias ratio with respect to kinetic inductance L_k for a clean, uniform bulk superconducting 1-D wire, with the absence of magnetic field, in (Yan et al., 2007), and the analytical solution reads

$$L_k(\beta) = \frac{L_{k0}}{1 - 4\cos^2\left(\frac{\pi}{3} - \frac{1}{3}\arctan\left(\frac{\sqrt{1-\beta^2}}{\beta}\right)\right)} \tag{5}$$

where L_{k0} is the zero bias kinetic inductance. It is device dependent, and can be measured via the method in this paper when the setting dc bias to zero. $\beta = {}^{I_b}/_{I_c^{GL}}$ is the ratio of dc bias current I_b to that of bulk critical current I_c^{GL}. Since the order parameter has been reduced due to the strong weaklink network, the (5) is valid for temperature at 4.2 K. However, attention has to be paid to the fact that $I_c^{GL} \gg I_c$, where I_c is defined as the critical current in the context of SNSPD. In terms of thin film quality, or QE of a given SNSPD in the literature, we substitute β with $\beta_b = \gamma\beta$, in which γ is a figure of merit of the thin film quality, defined as $\gamma = \frac{I_c}{I_c^{GL}}$, and $\beta_b = \frac{I_b}{I_c}$.

Parameter	SNSPD # 1[1]	SNSPD # 2[2]
J_c (A/m^2)	6.2×10^{10}	3.7×10^{10}
L_k (nH) [3]	200	230
ρ $(\mu\Omega \cdot cm)$	195	214
λ_L (0) (nm)	383	411
λ_{GL} (0) (nm)	400.7	407.9
λ (0) (nm)	490.6	499
ξ_0 (nm)	6.18	5.91
l_{tr} (nm)	3.97	4.00
κ_{GL}	70.88	73.45

[1]Device QE is 2%
[2]Device QE is 0.2%
[3]Extracted at $T = 4.2$ K with $T_c = 10.4$ K for all calculations

Table 2. Computed superconducting material parameters

Note that technically I_c^{GL} can not be measured directly, but can be deduced through the ratio of $\frac{L_k(\beta_b)}{L_{k0}}$ when $\beta_b \simeq 1$ based on the relationship of (5). Evidently, the highest γ can be close to 0.6 at temperatures of 4.2 K . It would become relatively higher when the temperature is decreased, but we anticipate that γ will saturate at some point due to the possibility that the weaklink will limit I_c from going close to I_c^{GL}. For two of our DUTs, the higher quality one has $\gamma = 0.5$; and $\gamma = 0.4$ for the lower one.
Finally, the effort to account for nonlinear resistance is more complicated, because it involves combinations of material, processing, electrical and thermal physics. The difficulty is added because of the quasiparticles are involved in the presence of RF field. The strong fluctuations of R_s result not only from the quasiparticle generation and recombination process, but also from their tunneling through the weaklinks. We think that detailed discussions are much beyond the scope of this thesis, and we will address this topic in the future works.

5. NbN superconducting thin film properties

We treat superconducting nanowire devices as a nonlinear load of a 50 Ω transmission line. The incident RF/microwave power (ac current) is weakly applied. The reflected power is measured and analyzed as discussed in the preceding sections. Series of experimental results to measure kinetic inductance L_k to study the nonlinearities stemming from dc bias by one port S-parameter measurement. The weak ac current was used (less than 1 μA) as a probe for all measurements.
By knowing the kinetic inductance of the superconducting nanowire via the method described in the preceding section, we are able to obtain some insights to the material properties of the superconducting parameters that is prepared for making NbN SNSPD. After post processing the measured S-parameter, we are able to retrieve London penetration depth, coherence length, and very importantly the first and second critical magnetic field H_{c1} and

H_{c2}. However, the second critical magnetic field intensity can not be simply computed for NbN thin film, especially for the ultra thin film (e.g. 4 nm).
The extraction of values of the penetration depth λ is based on the relations in Ref. (Orlando et al., 1979). The normal state electron specific heat $\gamma_0 = 2.1 \times 10^{-4}\ J \cdot cm^{-3} \cdot K^{-2}$ is from (Johnson et al., 1996). All of the input parameters, including the kinetic inductance L_k in conjunction with the critical current density J_c and the normal state resistivity ρ_n at 20 K are measured parameters. We have also evaluated BCS coherent length ξ_0 and electronic mean free path l_{tr}. Interestingly, we have found that they are comparable. Consequently, all values of different penetration depth are becoming comparable accordingly, which is significantly different from the bulk properties of NbN. This exceptionally long mean free path seems to arise from the two magnitude lower electron gas density at Fermi surface than that of bulk superconductors. Table 2 lists all of input parameters and output results for this evaluation. Oates *et al.* (Oates et al., 1991) reported the penetration depth $\lambda(0)$ obtained from their surface impedance measurement in the microwave range. The comparison shows that our penetration depth value is in good agreement with their results provided that thin film modifies the values of BCS coherent length ξ_0 and mean free path l_{tr}. As a result, their London penetration depth $\lambda_L(0)$ becomes considerably lower than $\lambda(0)$ due to the high ratio of ξ_0 to l_{tr}.

6. DC I-V measurement in the absence of optical power

The dc I-V measurement of the NbN nanowire shows four distinct regimes. The first regime is the superconducting regime and the slope of I-V curve is primarily determined by the cryogenic cable resistance; The second is the switching state (Lindgren et al., 1996), in which spontaneous thermo-electrical relaxation oscillations is dominantly observed; The third regime is the formation and expansion of a stable hotspot, which is the starting point of a self-heating effect. The hotspot grows up with the increment of dc bias to occupy one entire segment of the nanowire. After this point, the fourth regime is entered when one segment of the nanowire becomes a hotwire and starts to heat up the adjacent segments of nanowires.

6.1 Superconducting and switched states

For the first two regimes as demonstrated in Fig. 10, the figure depicts the forward I-V curve obtained via two scan speeds: fast computerized scan and slow manual scan, in which the inset panel displays a typical computerized scan in both forward and backward directions. The manual mode is persistently useful for the small signal S-parameter measurements. In both manual and computerized scan modes, the return current, which manifests the plateau of a hotspot formation, shows the same value. However, the occurrence of this point in the manual scan mode is different from the computerized fast scan mode. For the former mode, it occurs at higher voltage than does the latter, but the latching current (approximately 7 μA) is almost identical. We attribute this to three possible mechanisms: scanning speed, scanning direction and electrical grounding. The manual mode always has slower variation of bias voltage. The voltage relaxation induced by quasiparticle excitations results in a metastable states, which is very susceptible to external perturbations, such as a sudden bias voltage variation, and then stabilizes itself at the latching current.
In addition, the scanning direction depends on whether it starts from a higher bias voltage down to zero (backward), or vice versa (forward). Within the switched states as defined in (Lindgren et al., 1996), the nanowire will exhibit a hysteresis I-V curve due to the thermal healing of the hotspot, i.e. the voltage scans of the forward and backward mode show the hysteresis behavior; and the measured current remains at the level of latching current until

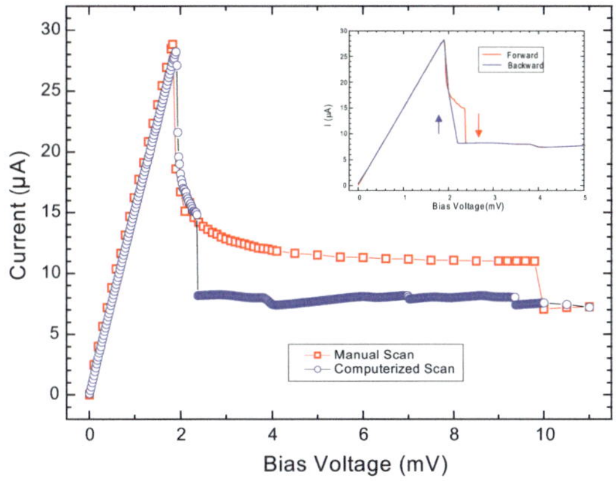

Fig. 10. Experimental voltage and current relationship measured via both manual and computerized scan modes under forward voltage bias scheme; inset panel is a computerized scan curve in both directions.

the bias voltage decreases below the forward latching voltage point. The detail theory has been discussed by Tinkham *et al.* (Tinkham et al., 2003). However, the theory can not account for the different hysteresis loops in two scan modes. We suspect that the electrically shared ground of the analog and digital parts in the source meter is responsible for the difference. In the computerized scan mode, communication between the source meter and computer can induce extra ground voltage fluctuations. Consequently, a smaller hysteresis I-V curve results as demonstrated in the inset panel of Fig. 10.
The dc I-V curve measurement also indicates that when I_b is greater than $\sqrt{2}I_{ret}$, it is no longer a single value function of bias voltage. On the other hand, for any given bias voltage, there exists a single-value current. Lindgren *et al.* used to bias an HTS bridge in the switched state to measure photoresponses (Lindgren et al., 1996). The values of R_s by small signal RF/microwave measurement results in less than 20 Ω, even when I_b is 0.5 μA less than I_c (28.5 μA in this case), which is much lower than the stable resistance of a hotspot by Joule heating.

6.2 Temperature dependent I_c

The critical current I_c can be experimentally determined for a given SNSPD by inspecting the maximum current in Fig. 10. [2] The temperature dependence of I_c can be experimentally found by inspecting the maximum current in the dc I-V measurement at all temperatures. Fig. 11 illustrates the measured zero maximum current for the DUT with higher QE. In the figure, the dot line is a linear fitting curve, whose slope is -5.88 $\mu A/K$; the zero temperature I_c is projected to be 50.74 μA. The red square in the plot is the DUT provider measured I_c at 2 K, which is below the linear fitting trend line. When the temperature ranges from 7.5 K to 9.5 K, approximately a second slope can be observed. Higher 9.5 K, we can only measure submicron ampere current, which is hard to identify it as a maximum current. The linear relation of I_c with the decreased temperature is not consistent with what Ginsburg-Landau theory predicts in the bulk superconductors, from which we expect a relation like $I_c \propto T^{-3/2}$ curve. We

[2] when the bias current I_b is close to I_c, the superconducting nanowire will exhibit a considerable dc resistance primarily due to voltage spikes arising from dark count. Consequently, this will affect the accuracy of the I_c measurement, as well as latching problem in the experiment.

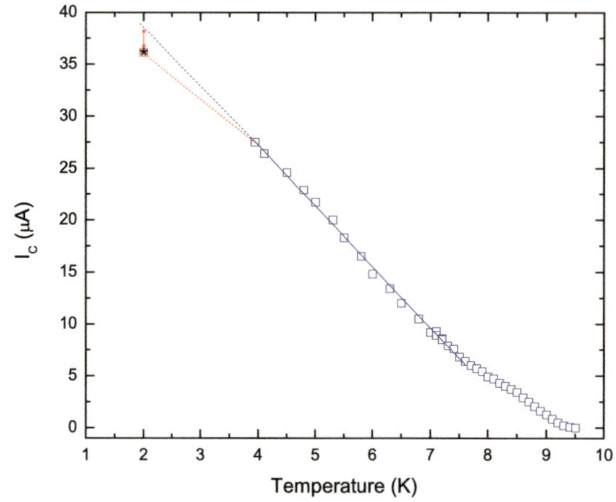

Fig. 11. The zero voltage maximum current, or critical current I_c vs temperature dependence for the SNSPD of higher QE.

attribute the linear relationship of $I_c(T)$ particulary exhibited in the SNSPDs to the NbN thin film, in which the superconducting weaklinks formed during its fabrication process.

The NbN thin film is sputtered on sapphire substrates; and is studied in (Kampwirth & Gray, 1981)-(Gray et al., 1985). The weaklink inherently presents during the ultra thin film sputtering, due to the formation of granular grain boundaries. Kampwirth *et al.* (Kampwirth & Gray, 1981) approximated the critical current based on the quasiparticle tunneling, which is studied by Ambegaokar (Ambegaokar & Baratoff, 1963)-(Ambegaokar & Baratoff, 1963). We can relate I_c to the granular size of the NbN thin film a_0

$$a_0 = \frac{\pi A \Delta(T)}{2 I_c q_e \rho_{20K}} \tanh\left[\frac{\Delta(T)}{2k_B T}\right] \tag{6}$$

where k_B is the Boltzmann constant; q_e is the electron charge; $\Delta(T)$ is the temperature dependent BCS energy gap; A is the cross section area of the nanowire, and ρ_{20K} is the resistivity and will be addressed later in this section. The value of a_0 provides a quantitatively commencement point in the self-heating model for the SNSPD nanowire that will discussed next.

6.3 Modeling of superconducting and switched state

The I-V curve shown in Fig. 5 (a) prior and after the critical current I_c point, is the superconducting and switched state (Lindgren et al., 1996), respectively. For the latter, it is characterized by a spontaneous relaxation oscillation at about 800 KHz counting rate, when we use the method to perform photon/dark count characterization in (Yan et al., 2009).

Hadfield *et al.* (Hadfield et al., 2005) suspected about 5000 Josephson Junction-like behavior exhibited in their SNSPD, which has the identical geometrical size as that of our DUT. It is actually in consistency with what we calculated based on what Kampwirth (Kampwirth & Gray, 1981) estimated about the single granular size of 100 nm. We further developed a lump 5000-stacked Josephson Junction circuit model to simulate to the total current driven by a voltage source. The voltage source spans from 0 to 10 mV. The simulation results is plotted in a solid curve in Fig. 12. The comparison to the experimental result in the figure indicates that the simulation does not display a relatively flat current in the switched state. This is

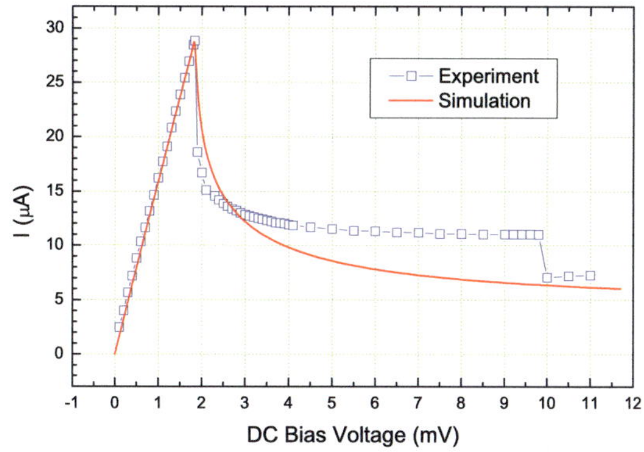

Fig. 12. The simulated dc I-V scan by voltage bias in superconducting and switched states

Name	Return Current	Ambient Temperature	Heat Transfer Coefficient	Resistivity at 20K
Description	I_r (μA)	$T_b(K)$	$\alpha(W/m^2 \cdot K)$	$\rho_{20K}(\Omega \cdot m)$
DUT ♯1	5.66	4.2	2×10^4	2.05×10^{-6}
DUT ♯2	3.96	4.5	2×10^4	2.14×10^{-6}

Table 3. The self-heating hot spot model parameters

primarily due to the fact that the spontaneous voltage oscillation at about 800 KHz has not been considered in the simulation. The strong voltage pulses can act as another resistance, which is beyond our stacked Josephson Junction model.

6.4 Modeling 1-D self-heating

We are not able to observe any relaxation oscillations when the device is latching. We believe that when the bias voltage is higher than the latching voltage point, the nanowire formulates a non-reversible hotspot that can not be thermally healed by the heat bath. Then the nanowire is believed to enter the third regime dominated by self-heating effect.

Carefully examining the I-V curve in Fig. 13 (a) and (b), we notice that there exist a series of zigzag current fluctuations with the increment of bias voltage. We anticipate that the self-heating hotspot expansion along one segment of the nanowire is quantized as resistance staircases. We suspect that those resistive steps are associated with granular nature of the ultra thin NbN thin films.

With the increment of dc voltage bias, a stable hotspot will develop at a certain section of the nanowire. The voltage width of the switched state depends also on the scanning direction and mode. Usually manually increasing the voltage will result in an accurate voltage point of the hotspot formation. Upon its formation, thermo-electrically Joule heating will apply. The first segment of the nanowires gradually exhibits normal state I-V properties as the dc bias voltage increases. The self-heating model by Skocpol *et al.* (Skocpol et al., 1974) is the best to account for this physical process. The current and voltage as the functions of the hotspot size χ_0 are

$$I_b\left(\chi_0\right) = I_r \sqrt{1 + \coth\left(\frac{\chi_0}{\eta_S}\right) \coth\left(\frac{\chi_0}{\eta_S} - \frac{L}{2\eta_n}\right)} \tag{7}$$

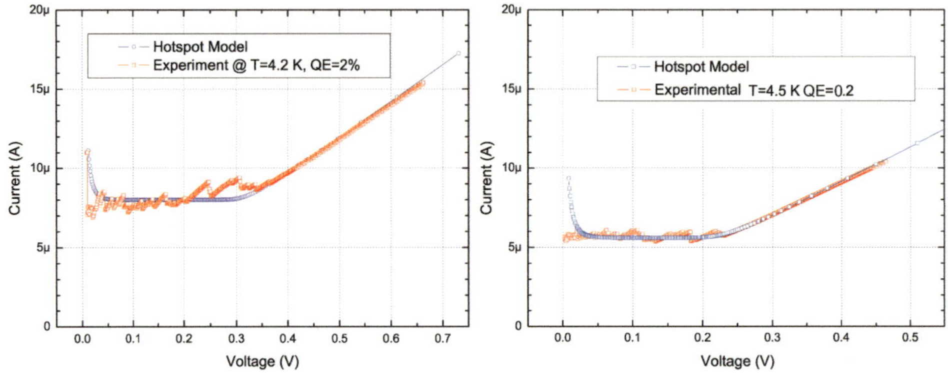

Fig. 13. Experimental and simulation results of the dc voltage and current characterization: (a) Hotspot formation under voltage bias, for DUT I; (b)Hotspot formation under voltage bias, for DUT II.

$$V_b(\chi_0) = I(\chi_0)\,\chi_0\rho_{20K}/Wd \tag{8}$$

in which we treat the thermal healing length η_S to be the same as η_n. The simulation results are displayed in Fig. 13 (a) and (b). The comparison to the experimental results shows a very good agreement between the theory and experiment. The parameters used in simulation are listed in Table 3.

In this model, one of the free parameters is ρ_{20K}, which can be extracted via the dc resistance measured during the sample cooling down. The notation of 20K can be understood as the maximum resistivity before the transition temperature T_c, which usually has the maximum value close to 20 K. The other free parameter, i.e. the return current I_r is determined from the I-V curve scan during the dc characterization measurement. It is the minimum current platform, almost constant current during the bias current. It may reveal the regularity along the segment of nanowire, because essentially this is based on one-dimensional ideal wire whose width is a constant of 120 nm, which can not be maintained to be true during the fabrication, i.e. the lithography. Our estimation based on this model shows that the thermal resistance is 0.048 Kcm^2/W at 4.2 K, which is a very good estimation based on self-heating model hot spots in (Skocpol et al., 1974). Note the range of χ_0 in Eqs. (7) and (8) is $a_0 < \chi_0 < l \sim 10\mu m$, in which a_0 is determined by Eq. 6 and assumed to be completely normal in this regime.

7. Summary

In this chapter, we first present the methods and results achieved from small signal RF/microwave measurement on the NbN nanowires. The equivalent circuit model is obtained via the two-step S-parameter measurement. We have justified the existence of R_s by checking the damped resonance frequency points under different dc voltage bias values (R_s increases when dc bias voltages increases). The theoretical calculation compares well to the experimental measurements when the R_s is large in the circuit model (see Fig. 1 (b)). The Ginsburg-Landau theory is employed to account for the L_k variation as the function of dc bias current ratio β. The analytical relation of L_k to β is obtained from G-L theorem. The weaklink effect is revealed to explain limitation of I_c from reaching the bulk I_c^{GL}. We also anticipate the device quality can be determined by the parametrical variation of the kinetic inductance, which we think it becomes a figure of merit for a given SNSPD in terms of its quantum efficiency. Then we apply the small signal RF/microwave measurement method to investigate the thin film property. The measured results agrees with the expected value reported in the literature. This further confirms our method is a very effective tool to explore the change of Cooper pair density in the nanowires. Lastly, the I-V measurements showed four distinctive regime of a nanowire: i.e. superconducting, switched-relaxation, hotspot growth and hot wire regime, although the last regime is still under investigation, because not sufficient simulation has been carried on.

8. References

Z. Yan; M. K. Akhlaghi; JL. Orgiazzi; & A. H. Majedi (2009) Optoelectronic characterization of a fiber-coupled NbN superconducting nanowire single photon detector *Journal of Modern Optics*, vol. 56, pp. 380-4.

Z. Yan; A. H. Majedi & S. Safavi-Naeini, (2007) Physical Modeling of Hot-Electron Superconducting Single-Photon Detectors, *Applied Superconductivity, IEEE Transactions on*, vol. 17, pp. 3789-3794.

A. D. Semenov, P. Haas, B. Günther, H. -. Hübers, K. Il'in, M. Siegel, A. Kirste, J. Beyer, D. Drung, T. Schurig & A. Smirnov, (2007) An energy-resolving superconducting nanowire photon counter, *Superconductor Science and Technology*, vol. 20, pp. 919-924.

J. K. W. Yang, A. J. Kerman, E. A. Dauler, V. Anant, K. M. Rosfjord & K. K. Berggren, (2007) Modeling the electrical and thermal response of superconducting nanowire single-photon detectors, *IEEE Trans. Appl. Supercond.*, vol. 17, pp. 581-5.

M. Tinkham, J. U. Free, C. N. Lau & N. Markovic, (2003) Hysteretic I-V curves of superconducting nanowires, *Physical Review B*, vol. 68, pp. 134515-1.

S. Miki, M. Fujiwara, M. Sasaki, B. Baek, A. J. Miller, R. H. Hadfield, S. W. Nam & Z. Wang, (2008) Large sensitive-area NbN nanowire superconducting single-photon detectors fabricated on single-crystal MgO substrates, *Appl. Phys. Lett.*, vol. 92, pp. 061116.

R. H. Hadfield, P. A. Dalgarno, J. A. O'Connor, E. Ramsay, R. J. Warburton, E. J. Gansen, B. Baek, M. J. Stevens, R. P. Mirin & S. W. Nam, (2007) Submicrometer photoresponse mapping of nanowire superconducting single-photon detectors, *Appl. Phys. Lett.*, vol. 91, pp. 241108.

D. M. Pozar, (1998) *Microwave Engineering*. New York ; Chichester, England: John Wiley & Sons, ch-6.

K. S. Keskar, T. Yamashita, Y. Onodera, Y. Goto & T. Aso (1974) rf reactively sputtered superconducting NbN_x films, *Japn. J. Appl. Phys.*, vol. 45, pp. 3102-3105.

Y. M. Shy, L. E. Toth & R. Somasundaram (1973) Superconducting properties, electrical resistivities, and structure of NbN thin films *J. Appl. Phys.*, vol. 44, pp. 5539-5545.

J. Tyan and J. T. Lue (1994) Grain boundary scattering in the normal state resistivity of superconducting NbN thin films," *J. Appl. Phys.*, vol. 75, pp. 325-331.

R. T. Kampwirth and K. E. Gray, "NbN materials development for practical superconducting devices," *IEEE Trans. Magn.*, vol. 17, pp. 565-8.

K. E. Gray, R. T. Kampwirth, D. M. Capone II and R. Vaglio (1985) Microscopic investigation of NbN sputtered films *Physica B & C*, vol. 135B+C, pp. 164-7.

V. Ambegaokar and A. Baratoff, (1963) Tunneling between superconductors," *Phys. Rev. Lett.*, vol. 10, pp. 486-489.

V. Ambegaokar and A. Baratoff, (1963) Erratum: tunneling between superconductors," *Phys. Rev. Lett.*, vol. 11, pp. 104.

C. C. Chin, D. E. Oates, G. Dresselhaus & M. S. Dresselhaus (1992) Nonlinear electrodynamics of superconducting NbN and Nb thin films at microwave frequencies *Physical Review B (Condensed Matter)*, vol. 45, pp. 4788-98.

B. Abdo, E. Segev, O. Shtempluck & E. Buks (2006) Nonlinear dynamics in the resonance line shape of NbN superconducting resonators *Phys. Rev. B*, vol. 73, pp. 134513-11.

B. Bonin and H. Safa, (1991) Power dissipation at high fields in granular RF superconductivity *Supercond Sci Technol*, vol. 4, pp. 257-61.

T. P. Orlando, E. J. McNiff Jr., S. Foner & M. R. Beasley, (1979) Critical fields, Pauli paramagnetic limiting, and material parameters of Nb_3Sn and V_3Si *Physical Review B (Condensed Matter)*, vol. 19, pp. 4545-61.

M. W. Johnson, A. M. Herr & A. M. Kadin (1996) Bolometric and nonbolometric infrared photoresponses in ultrathin superconducting NbN films *J. Appl. Phys.*, vol. 79, pp. 7069-7074.

D. E. Oates, A. C. Anderson, C. C. Chin, J. S. Derov, G. Dresselhaus & M. S. Dresselhaus, (1991) Surface-impedance measurements of superconducting NbN films *Phys. Rev. B*, vol. 43, pp. 7655-63.

M. Lindgren, M. Currie, C. A. Williams, T. Y. Hsiang, P. M. Fauchet, R. Sobolewski, S. H. Moffat, R. A. Hughes, J. S. Preston & F. A. Hegmann, (1996) Ultrafast photoresponse in microbridges and pulse propagation in transmission lines made from high-Tc superconducting Y-Ba-Cu-O thin films *IEEE Journal of Selected Topics in Quantum Electronics*, vol. 2, pp. 668-78.

R. H. Hadfield, A. J. Miller, S. W. Nam, R. L. Kautz and R. E. Schwall (2005) Low-frequency phase locking in high-inductance superconducting nanowires *Appl. Phys. Lett.*, vol. 87, pp. 203505-1.

W. J. Skocpol, M. R. Beasley & M. Tinkham, (1974) Self-heating hotspots in superconducting thin-film microbridges *J. Appl. Phys.*, vol. 45, pp. 4054-66.

14

Band Structure and Magneto- Transport Properties in II-VI Nanostructures Semiconductors - Application to Infrared Detector Superlattices

Abdelhakim Nafidi
Laboratory of Condensed Matter Physics and Nanomaterials for Renewable Energy
University Ibn-Zohr,
Agadir
Morocco

1. Introduction

To observe phenomena by the image through invisible radiation to the eye is probably doubtless an old dream of humanity. The way followed to observe the invisible is to study an artificial retina constituted by a matrix of elements sensitive to radiation and to associate it with a system of acquisition and data processing. In brief a camera or a digital camera sensitive to other radiations than the visible light. The detectors are constituted of simple elementary p - n junctions realized in a semiconductor material whose height band gap is adapted to the energy of infrared photons which we wish to detect.

The work of (Essaki & Tsu, 1970) caused a big interest to the study of superlattices made from alternating layers of two semiconductors. The development of molecular beam epitaxy (MBE) was successfully applied to fabricate different quantum wells and superlattices. Among them III-V superlattices [$Ga_{1-x}Al_xAs$-GaAs (Essaki & Tsu, 1970) – (Dingle et al., 1975) - type I], IV-IV [InAs/GaSb (Sakaki et al, 1978) - type II] and later II-IV superlattice [HgTe/CdTe (Bastard, 1982) - type III]. The later is a stable alternative for application in infrared optoelectronic devices than the $Hg_{1-x}Cd_xTe$ alloys. Especially in the region of second atmospheric window (around 10 μm) which is of great interest for communication.

HgTe and CdTe crystallize in zinc –blend lattice respectively. The lattice-matching within 0.3 % yield to a small interdiffusion between HgTe and CdTe layers at low temperature near 200 °C by MBE. HgTe is a zero gap semiconductor (due to the inversion of relative positions of Γ_6 and Γ_8 edges (Tuchendler et al, 1973)) when it is sandwiched between the wide gap semiconductor CdTe (1.6 eV at 4.2 K) layers yield to a small gap HgTe/CdTe superlattice which is the key of an infrared detector.

A number of papers have been published devoted to the band structure of this system (Bastard, 1982) as well as its magnetooptical and transport properties (Nafidi, 2002). The aim of this work is to show the correlation between calculated bands structures and magneto-transport properties in two HgTe/CdTe nanostructures superlattices.

Our calculations of the specters of energy $E(d_2)$, $E(k_z)$ and $E(k_p)$, respectively, in the direction of growth and in plane of the superlattice;were performed in the envelope function formalism. Our two samples are denoted SL1 and SL2.

In the SL1, X-ray diffraction, conductivity, Hall effect, Seebek and Shubnikov-de Haas effects and angular dependence of the transverse magnetoresistance were measured. The profile of the (222) Bragg reflection indicated a modulated structure. At 4.2 K, the sample exhibits p type conductivity with a Hall mobility of 8200 cm^2/Vs. This allowed us to observe the Shubnikov-de Haas effect with p = 1,80 10^{12} cm^{-2}. Using the calculated effective mass (m^*_{HH} =0,297 m_0) of the degenerated heavy holes gas, the Fermi energy (2D) was E_F=14 meV in agreement with 12 meV of thermoelectric power α. In intrinsic regime, $\alpha \sim T^{-3/2}$ and R_H $T^{3/2}$ indicates a gap $E_g = E_1\text{-}HH_1$= 190 meV in agreement with calculated $E_g(\Gamma$, 300 K) =178 meV. The formalism used here predicts that the system is semiconductor for d_1/d_2 = 1,87 and d_2 < 150 Å. Here, d_2=30 Å and E_g (Γ,4.2 K) = 111 meV so this sample is a two-dimensional modulated nanostructure medium-infrared detector semiconductor (7μm<λ<11μm).

In the SL2, the angular dependence of the transverse magnetoresistance follows the two-dimensional (2D) behaviour with Shubnikov-de Haas oscillations. While the hall voltage goes to zero when the field is parallel to the plane. At low temperature, the sample exhibits p type conductivity with a hole mobility of 900 $cm^2/V.s$. A reversal the sign of the weak-field Hall coefficient occurs at 25 K with an electron mobility of 3.10^4 cm^2/Vs. In intrinsic regime, the measured $E_g \approx$ 38 meV agree with calculated $E_g(\Gamma$,300K)=34 meV witch coincide with the Fermi level energy. The formalism used here predicts that the system is semi metallic for d_1/d_2 =4.1 and $d_2 \geq$ 30 Å. In our case d_2 = 44 Å and the gap $E_g(\Gamma$,4.2 K) =3 meV. In spite of it, this sample exhibits a semi metallic p type conduction mechanism, with a quasi-two-dimensional behavior and is a far-infrared detector (50 μm< λ< 450 μm). The investigated thicknesses of the SL2 situate it at the semiconductor to semimetal electronic transition.

In conclusion, we will show that the HgTe/CdTe nano-superlattice is a stable alternative for application in infrared optoelectronic devices than the alloys $Hg_{1-x}Cd_xTe$.

2. Experimental techniques

Our two HgTe/CdTe superlattices samples SL1 and SL2, grown by molecular beam epitaxy (MBE) on a CdTe (111) substrate at 180 °C, had a period $d=d_1+d_2$ (90 layers) of HgTe (d_1=5,6 nm) / CdTe (d_2 =3 nm) and (100 layers) of HgTe (d_1=18 nm) / CdTe (d_2=4.4 nm) respectively. Each sample was cut from an epitaxial wafer with a typical size of 5×1×1 mm^3. The ohmic contacts were obtained by chemical deposition of gold from a solution of tetrachloroauric acid in methanol after a proper masking to form the Hall crossbar. Carriers transport properties were studied in the temperature range 1.5-300K in magnetic field up to 17 Tesla. Conductivity, Hall Effect, Seebek effect, X-ray diffraction and angular dependence of the transverse magnetoresistance were measured. The measurements at weak magnetic fields (up to 1.2 T) were performed into standard cryostat equipment. The measurements of the magnetoresistance were done under a higher magnetic field (up to 17 T), the samples were immersed in a liquid helium bath, in the center of a superconducting coil. Rotating samples with respect to the magnetic field direction allowed one to study the angular dependence of the magnetoresistance.

3. Theory of structural bands

Calculations of the spectra of energy E (k_z) and E(k_p), respectively, in the direction of growth and in plane of the superlattice; were performed in the envelope function formalism (Bastard, 1982) (Nafidi, 2002) with a valence band offset Λ between heavy holes bands edges of HgTe and CdTe of 40 meV determined by the magneto-optical absorption experiments (Guldner, 1984)

The general dispersion relation of the light particle (electron and light hole) subbands of the superlattice is given by the expression (Bastard, 1982) :

$$\cos k_z(d_1+d_2) = \cos(k_1 d_1)\cos(k_2 d_2) - \frac{1}{2}\left[\left(\xi+\frac{1}{\xi}\right)+\frac{k_p^2}{4k_1k_2}\left(r+\frac{1}{r}-2\right)\right]\sin(k_1d_1)\sin(k_2d_2) \quad (1)$$

where the subscripts 1 and 2 refer to HgTe and CdTe respectively. k_z is the superlattice wave vector in the direction parallel to the growth axis z. The two-dimensional wave vector $k_p(k_x,k_y)$ describes the motion of particles perpendicular to k_z. Here,

$$\xi = \frac{k_1}{k_2} r \text{ , and } r = \frac{E - \varepsilon_2}{E + |\varepsilon_1| - \Lambda} \quad (2)$$

E is the energy of the light particle in the superlattice measured from the top of the Γ_8 valence band of bulk CdTe, while εi (i =1 or 2) is the interaction band gaps $E(\Gamma_6)$ - $E(\Gamma_8)$ in the bulk HgTe and CdTe respectively. At given energy, the two–band Kane model (Kane, 1957).gives the wave vector $(k_i^2+k_p^2)$ in each host material:

$$\left.\begin{aligned} &\frac{2}{3} P^2 \hbar^2 (k_1^2 + k_p^2) = (E-\Lambda)(E - \Lambda + |\varepsilon_1|) && \text{for HgTe} \\ &\frac{2}{3} P^2 \hbar^2 (k_2^2 + k_p^2) = E(E - \varepsilon_2) && \text{for CdTe} \end{aligned}\right\} \quad (3)$$

P is the Kane matrix element given by the relation:

$$\frac{2P^2}{3|\varepsilon_1|} = \frac{1}{2m^*} \quad (4)$$

where m* = 0.03 m_0 is the electron cyclotron mass in HgTe (Tuchendler et al, 1973). For a given energy E, a superlattice state exists if the right-hand side of Eq. (1) lies in the range [-1,1]. That implies $-\pi/d \le k_z \le -\pi/d$ in the first Brillion zone.

The heavy hole subbands of the superlattice are given by the same Eq. (1) with :

$$\xi = \frac{k_1}{k_2} r, \textit{ and } r = 1 \quad \text{with} \begin{cases} -\dfrac{\hbar^2(k_1^2+k_p^2)}{2\,m^*_{HH}} = E - \Lambda & \text{for HgTe} \\ -\dfrac{\hbar^2(k_2^2+k_p^2)}{2\,m^*_{HH}} = E & \text{for CdTe} \end{cases} \quad (5)$$

m^*_{HH} =0.3 m_0 (Tuchendler et al, 1973) is the effective heavy hole mass in the host materials. The band structure computation consists of solving Eq. (1) which represents the dispersion relations (i.e. finding the values of energy E which are roots of the Eq. (1) for a given value of the carrier wave vector). Here, we are interested in studying the states of energy of light particles and heavy holes in HgTe/CdTe superlattice as function of k_z when k_p=0 and as function of k_p when k_z=0 and when $k_z=\pi/d$. The solving procedure used for studying E as function k_z in the case where k_p=0 consists of going, with a steep δE, through the studied range of energy E and then finding, for each value of E, the value of k_z which satisfies the dispersion relations. The same procedure is used for studying E as function k_p in the case where k_z=0 and $k_z=\pi/d$. It is noteworthy that, for a given value of E, Eq. (1) may have more than one root in k_p. It appears, from Eq. (3)-(5), that the carrier wave vectors k_1, k_2, and k_p are either real or imaginary (i.e. complex) and then using complex numbers in the calculation seems to be more adequate.

4. The SL1: HgTe(d_1 = 5,6 nm) / CdTe(d_2 = 3 nm)

4.1 Theoretical results and discussions¶

The energy E as a function of d_2, at 4.2 K, in the center Γ of the first Brillouin zone and for our d_1 = 1.87 d_2, is shown in Fig. 1. (a) (El Abidi et al, 2010). The case of our sample (d_2 = 30 Å) is indicated by the vertical broken line. Here the cross-over of E_1 and HH_1 subbands occurs. d_2 controls the superlattice band gap $E_g=E_{E1}-E_{HH1}$. For weak d_2 the sample is semiconductor with a strong coupling between the HgTe wells as in the SL1. At the dot T(d_2 = 150 Å, E= 39 meV) the gap goes to zero with the transition semiconductor- semimetal. When d_2 increases, E_1 and h_1 states drops in the energy gap [0, Λ] and become interface state with energy

$$E_I = \frac{\Lambda \varepsilon_2}{|\varepsilon_1| + \varepsilon_2} = 34\ \text{meV} \tag{6}$$

for infinite d_2 obtained from Eq. (1). Then the superlattice has the tendency to become a layer group of isolated HgTe wells and thus assumes a semimetallic character. The ratio d_1/d_2 governs the width of superlattice subbands (i.e. the electron effective mass). A big d_1/d_2, of 4.09 as in the case of the SL2, moves away the material from the two-dimensional behavior.
In the Fig. 1. (b) we can see that the band gap $E_g(\Gamma)$ increases, presents a maximum at 40 meV and decreases when the valence band offset Λ between heavy hole band edges of HgTe and CdTe increase. For each Λ, $E_g(\Gamma)$ increases with T. Our chosen value of 40 meV is indicated by a vertical dashed line. This offset agrees well with our experimental results contrary to 0 meV used by (Bastard, 1982) and 360 meV given by (Johnson et al, 1988). The later offset give E_g (300 K) =135 meV in the right of Fig. 1 whereas, in intrinsic regime, $\alpha \sim T^{-3/2}$ and $R_H T^{3/2}$ indicates a measured gap $E_g \approx 190$ meV in agreement with calculated $E_g(\Gamma, 300\ K)$ = 178 meV.
Fig. 2 shows that, for each d_2, $E_g(\Gamma)$ decreases when d_1/d_2 increases. For each d_1/d_2, when d_2 increases $E_g(\Gamma)$ decreases, go to zero at the transition dot T and became negative for a semimetal conductivity. In the right of Fig. 2, the cut-off wavelength $|\lambda_c|$ diverge at T with d_2=54 Å, 100 Å, 150 Å,… respectively for d_1/d_2=4.09, 2.67, 1.87, … Thus, the transition goes to high d_2 when d_1/d_2 decreases. In the case of our samples the transition occurs, respectively, at d_2 =150 Å and 54 Å in the SL1 and the SL2.

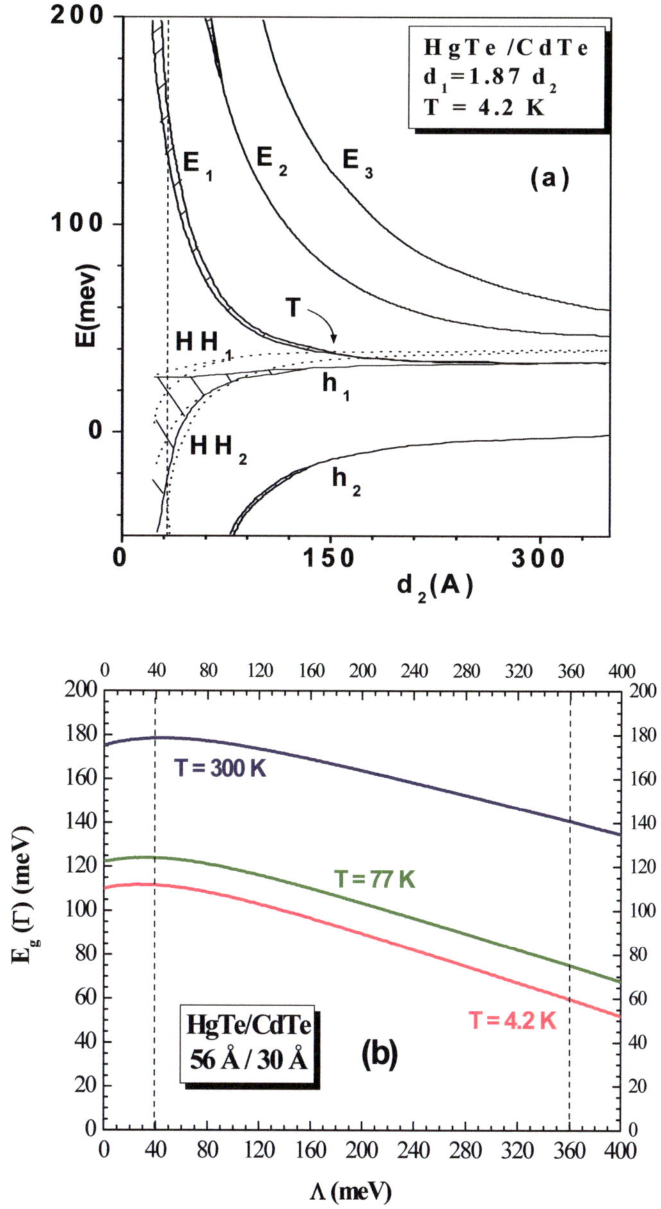

Fig. 1. (a) Energy position and width of the conduction (E_i), heavy-hole (HH_i), and the light-hole (h_1) subbands calculated at 4,2 K in the center Γ of the first Brillouin zone as a function of layer thickness d_2 for HgTe/CdTe superlattice with d_1=1.87 d_2, where d_1 and d_2 are the thicknesses of the HgTe and CdTe layers, respectively. T is the point of the transition semiconductor- semimetal. (b) the band gap E_g (Γ), in the center Γ of the first Brillouin zone, as function of temperature and valence band offset Λ between heavy holes bands edges of HgTe and CdTe for the investigated HgTe/CdTe superlattice

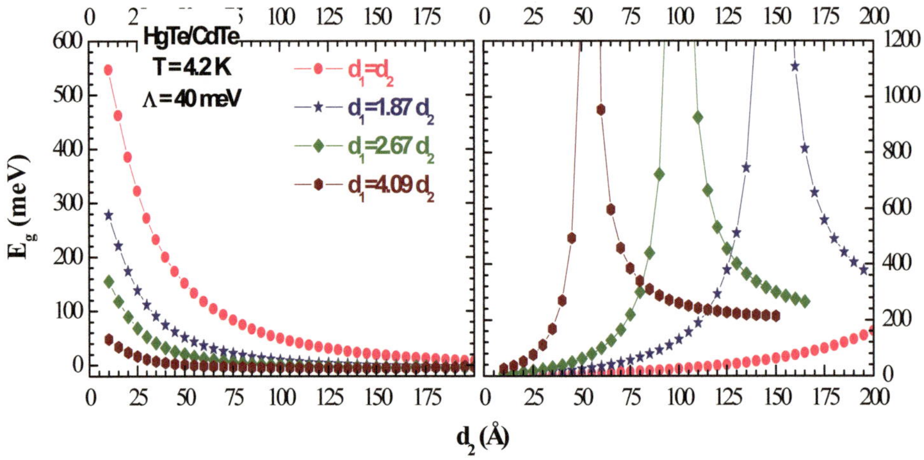

Fig. 2. E_g (Γ) and | λc | as function of d_2 for various d_1/d_2 at 4.2 K

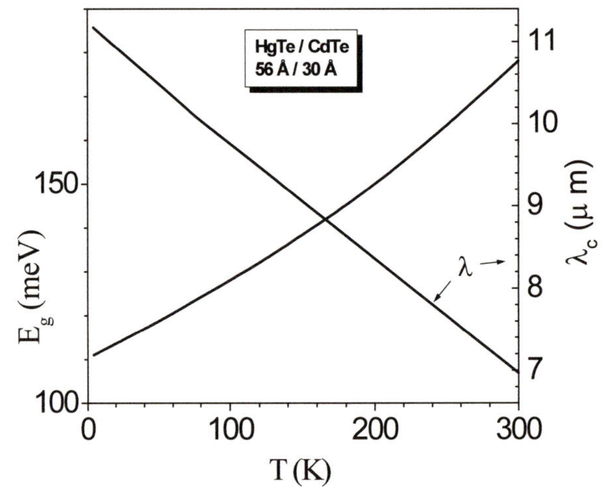

Fig. 3. Temperature dependence of the band gap E_g and the cut-off wavelength λ_c, at the center Γ of the first Brillouin zone

Using the value of the gaps ε_1 and ε_2 at different temperatures between 4.2 K and 300 K (Weiler, 1981) and taking P temperature independent, this is supported by the fact that from Eq. (4) $P \approx \varepsilon_i(T)/m^*(T) \approx$ cte, we get the temperature dependence of the band gap E_g, in the center Γ of the first Brillouin zone in Fig. 3. Note that E_g increases from 111 meV at 4.2 K to 178 meV at 300K. We calculated the detection cut-off wave length by the relation

$$\lambda_c(\mu m) = \frac{1240}{E_g(meV)} \tag{7}$$

In the investigated temperature range 7 μm< λ <11 μm situates our sample as a medium infrared detector.
In Fig. 4 we can see the specters of energy $E(k_z)$ and $E(k_p)$, respectively, in the direction of growth and in plane of the superlattice at 4.2 K. Along k_z the subbands E_1 and h_1 are wide and non-parabolic. Along k_p, E_1 and h_1 increase with k_p whereas HH_n decreases. This yield

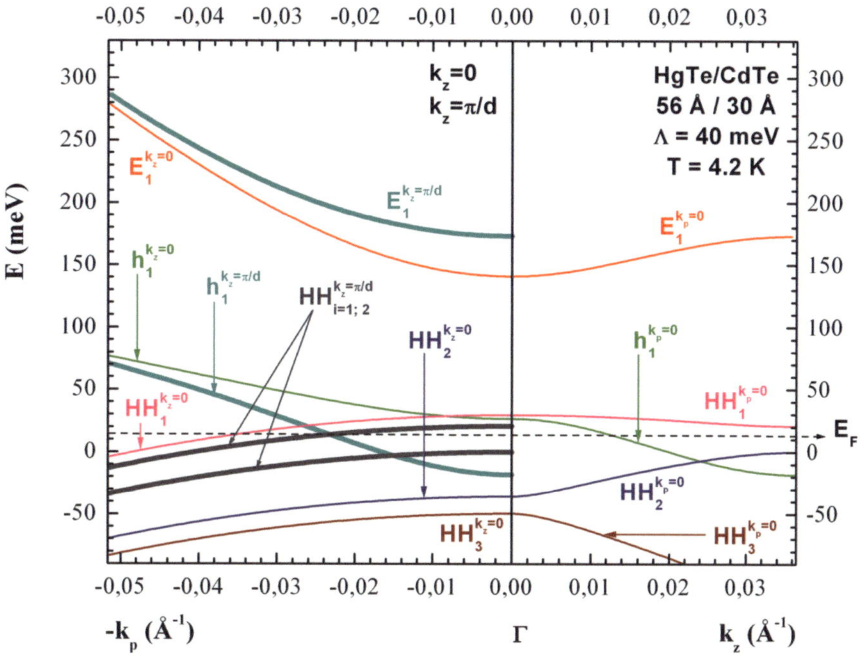

Fig. 4. Calculated bands along the wave vector k_z in the right and in plane $k_p(k_x,k_y)$ in the left, of the HgTe/CdTe superlattice at 4.2 K. E_F is the energy of Fermi level

to an anti-crossing of HH_1 and h_1 at k_p= 0.01 $Å^{-1}$ corresponding to a magnetic field of 4 Teslas according to the Formula (13).

4.2 Experimental results and discussions

In Fig. 2, |λc| as function of d_2, at 4.2 K, shows that there is a multiple choice of d_2 for medium infrared detectors with |λc|≤ 10μm for each d_1/d_2. Our choice of d_2= 3 nm for d_1=1.87 d_2 give a calculated $E_g(\Gamma,300\ K)$ = 178 meV corresponding to E_g (Γ, 300 K) =183 meV for the well used random alloys $Hg_{0.8}Cd_{0.2}Te$ (Hansen et al, 1982).
In Fig. 5(a) we can see that the angular dependence of the transverse magnetoresistance vanishes, when the field is parallel to the plane of the SL, indicating a two dimensional (2D) behavior confirmed by the observation of SDH oscillations in Fig. 7(a).
At low temperatures, the sample exhibits p type conductivity (confirmed by our thermoelectric power measurements in Fig. 6(b)) with a concentration p=1.84 10^{12} cm^{-2} and a Hall mobility μ_p = 8200 cm^2/Vs in Fig. 5(b) (Nafidi et al, 2004). The decrease of R_H (1/T) at 55 K shown by arrow in Fig. 5(c) can be due to coupling between HgTe well (small d_1/d_2 and d_2), to the widening of carrier sub-bands under the influence of the magnetic field

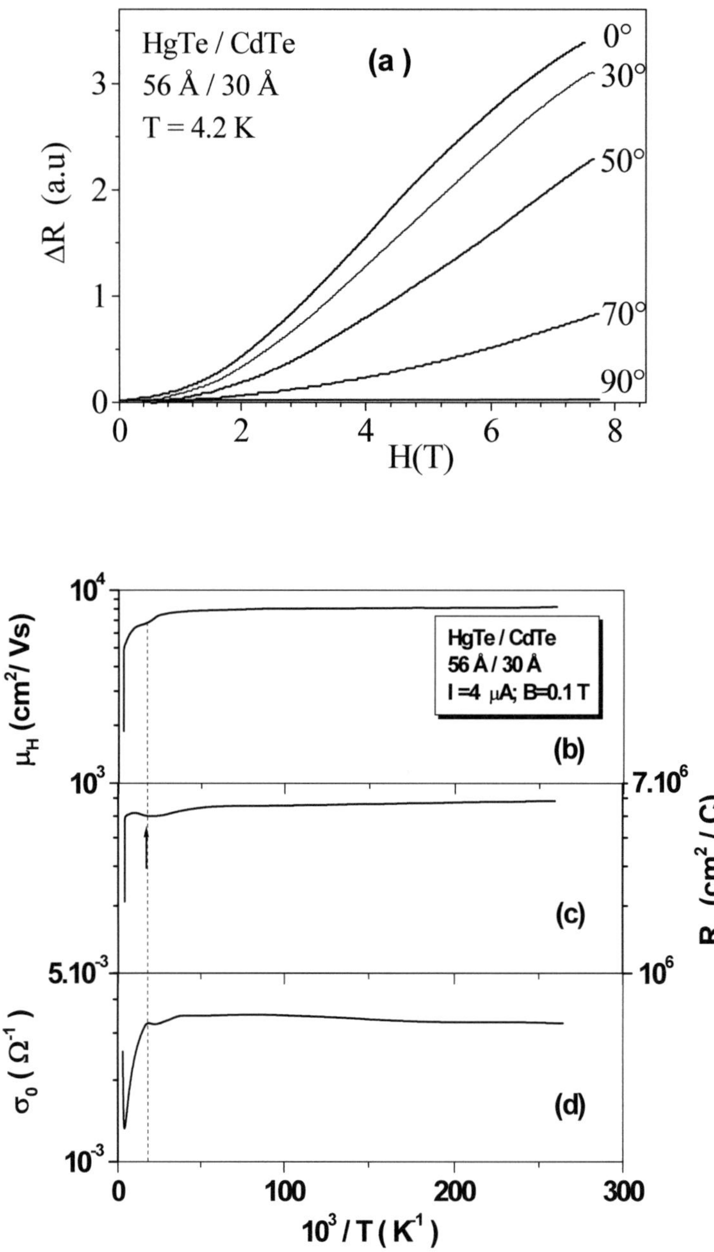

Fig. 5. (a) Variation of magnetoresistance of the sample with various angles between the magnetic field and the normal to the HgTe/CdTe superlattice surface. Temperature dependence of the Hall mobility (b), weak-field Hall coefficient (c) and conductivity (d) (Nafidi et al, 2002) in the investigated HgTe/CdTe superlattice

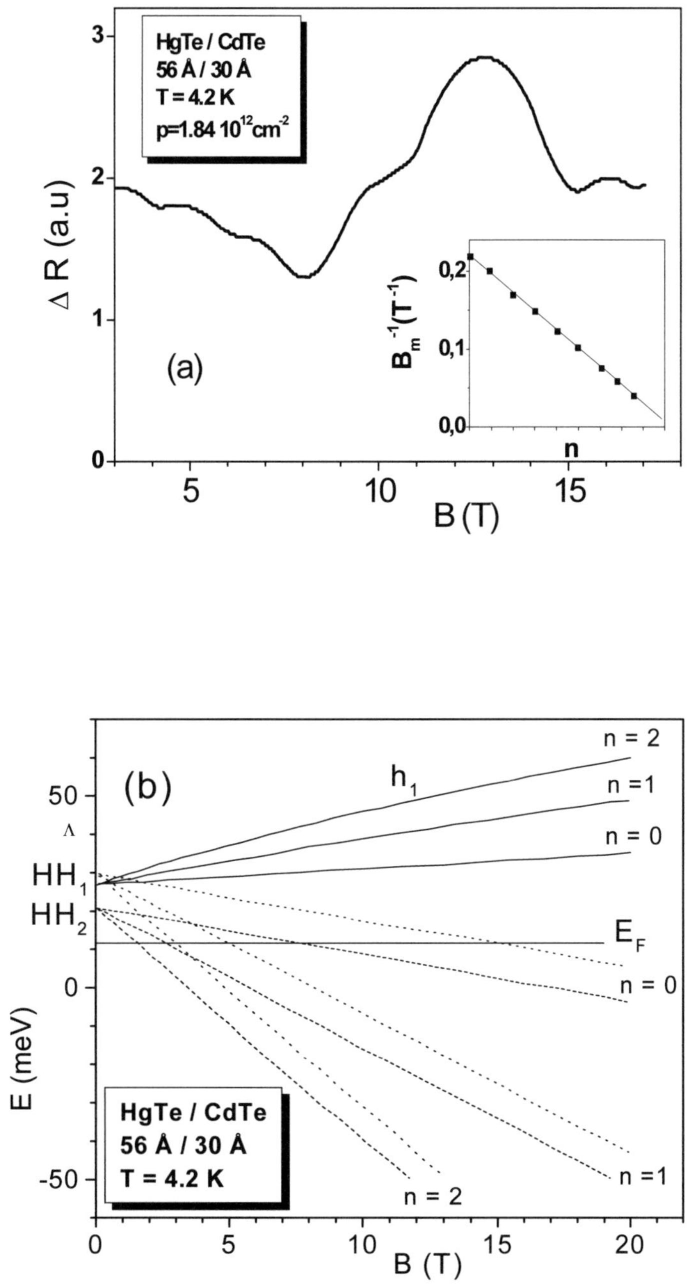

Fig. 6. (a) Variation of transverse magnetoresistance, with magnetic field of HgTe/CdTe superlattice, at 4.2 K (b) calculated Landau levels as a function of applied magnetic field of the HgTe/CdTe SL at 4,2 K. E_F is the Fermi level energy

and/or to the overlap between involved carriers sub-bands (HH_1) and (h_1) at (k_z, k_p) = (0, 0.008 Å^{-1}) and (π/d, 0.023 Å^{-1}) along E(k_p) in Fig. 4. Only the later cross over at E_F can be taking into account. This increases the concentration p of holes (decreases of R_H (1/T) in Fig. 5 (c)) and decreases of conductivity at 55 K (in Fig. 5 (d)) yielding to a decrease of mobility (in Fig. 5 (b)) due to the increase of effective mass from m_{h1} to m_{HH1}. In intrinsic regime (245 K $\leq$ T $\leq$ 300 K), the measure of the slope of the curve R_H $T^{3/2}$ indicates a gap E_g=E_1-E_{HH1}= 190 meV witch agree well with calculated E_g (Γ, 300 K) =178 meV.
This relatively high mobility μ_p = 8200 cm^2/Vs allowed us to observe the Shubnikov-de Haas effect (SDH) until 17 Tesla in Fig. 5. (a) (Nafidi et al, 2006). Its well knows that, for a two dimensional electronic gas, the oscillations of the magnetoresistance is periodic with respect to 1/B. The period Δ(1/B) is related to the concentration p of the holes by the relation:

$$p = \frac{e}{\pi \hbar \Delta\left(\frac{1}{B}\right)} \tag{8}$$

In the insert of Fig. 6(a) we have plot the inverse of the minima's 1/B_m as a function of the entire n following the formula:

$$\frac{1}{B_m} = \Delta\left(\frac{1}{B}\right)\left(n + \frac{1}{2}\right) \tag{9}$$

The linear line slope gives Δ(1/B) =0.027 T^{-1} and p = 1.80 x 10^{12} cm^{-2} (in good agreement with 1.84x10^{12} cm^{-2} of weak field Hall effect from Fig. 5 (c)).
At low temperature, the superlattice heavy holes dominate the conduction in plane (Fig 4). The HH_1 (and h_1) band is parabolic with respect to k_p^2 (from Fig. 4). That permits us to estimate effectives masses $m^*_{HH_1}$ =0.297 m_0, $m^*_{h_1}$ =0.122 m_0 and the Fermi energy (2D) at 4.2 K

$$\left|E_F - E_{HH_1}\right| = \left|\frac{p\pi\hbar^2}{m^*_{HH_1}}\right| = 14 \text{ meV} \tag{10}$$

Around the (222) Bragg reflection in Fig. 7(a), a series of steps are observed corresponding to the oscillating counterparts described by (Arch et al, 1986). The difference with our results comes from the monocromacity and the power of their synchrotron source. Our SL1 is a modulated structure with high quality of interfaces.
The thermoelectric power α measurements shown in Fig. 6(b) (Nafidi et al, 2006) indicate a p-type conductivity, confirmed by Hall effect measurements in Fig. 5(c). At low temperature, α ~$T^{0.8}$ (in the top insert of Fig. 7(b)) is in agreement with Seebeck effect theory deduced from the relaxation time resolution of the Boltzmann equation (Seeger, 2002). For our degenerate holes gas the Seebeck constant is described by the formula:

$$\alpha = \frac{\left[(\pi k_B)^2 T(s+1)\right]}{3eE_F} \tag{11}$$

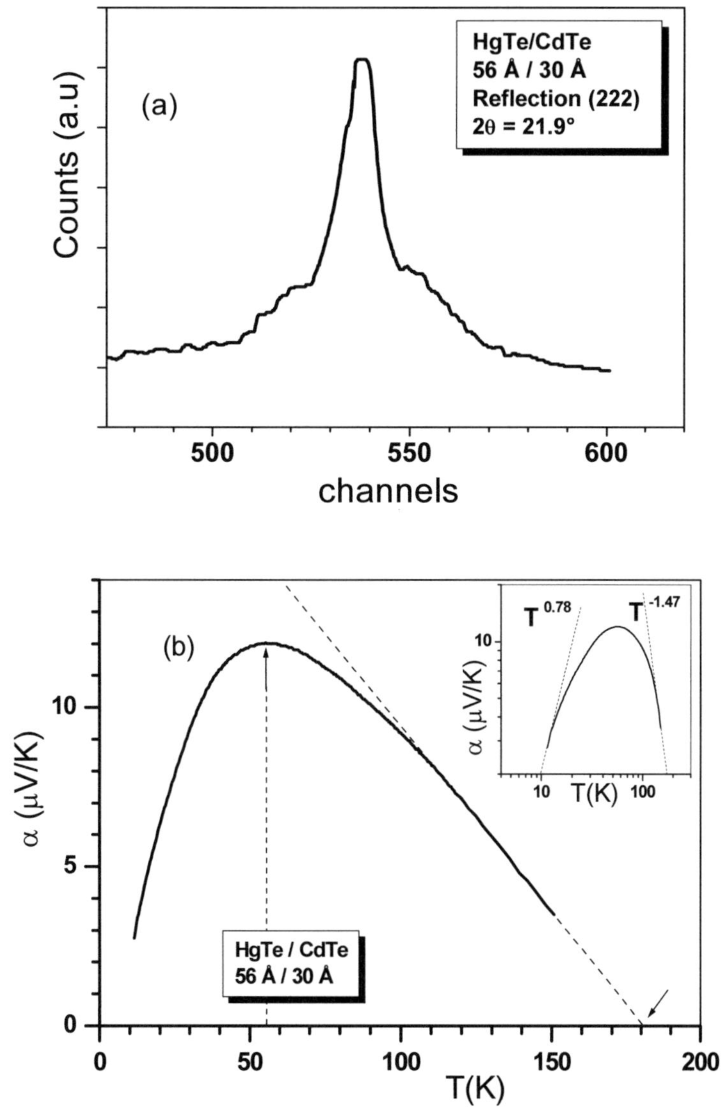

Fig. 7. (a) Room temperature X-ray diffraction profile around the (222) Bragg reflection of the HgTe/CdTe SL. (b) Measured thermoelectric power as a function of temperature of the HgTe/CdTe superlattice

and the collision time $\tau \sim E^{s-(1/2)}$. This permits us to estimate the Fermi energy at E_F = 12 meV (in Fig. 6(b)) in agreement with the calculated $|E_F\text{-}E_{HH1}|$=14 meV with s= 2.06 corresponding to holes diffusion by ionized impurities. It is relevant here to signal that the maximum of α at T=55.2 K correspond to the shift of Hall mobility in Fig. 5(b). Here μ_{HH1}= 2.434 μ_{h1} at E_F. At T=190 K, a reversal of α sign will occur corresponding exactly to the minimum of the conductivity σ_0 in Fig. 5(d). From p = 1.80x10^{12} cm^{-2} of SDH we have deduced the effective masse of the degenerate heavy holes to be

$$m^*_{HH_1} = \frac{\pi p \hbar^2}{E_F} = 0.308 m_0 \quad (12)$$

(in agreement with the theoretical m^*_{HH1}=0.297 m_0). We calculated the energy of the Landau levels (LL) by transposing the quantification rule of the wave vector in the plane of the SL:

$$k_p^2 = \frac{(2n+1)eB}{\hbar} \quad (13)$$

where n are the quantum orders of LL. The crossing of E_F with LL in Fig. 6(b) indicated the same magnetic field positions as those of the observed SDH oscillations minima B_m in Fig. 6(a).

This HgTe/CdTe superlattice is a stable alternative for application in medium infrared optoelectronic devices than the random alloys $Hg_{0.8}Cd_{0.2}Te$ because the small composition x=0.22, with E_g (Γ, 300 K) =183 meV given by the empiric formula for $Hg_{1-x}Cd_xTe$ (Hansen et al, 1982).

$$E_g(x,T) = -0{,}302 + 1{,}93x - 0{,}810x^2 + 0{,}832x^3 + 5{,}035 \times 10^{-4}(1-2x)\,T \quad (14)$$

is difficult to obtain with precision while growing the ternary alloys and the transverse effective masse in superlattice is two orders higher than in the alloy. Thus the tunnel length is small in the superlattice (A. Nafidi et al, 2011).

5. The SL2: HgTe(d_1=18 nm) / CdTe(d_2= 4.4 nm)

5.1 Theoretical results and discussions

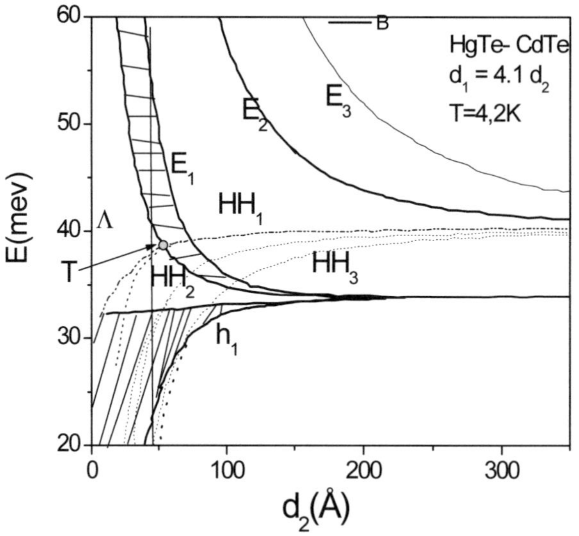

Fig. 8. Energy position and width of the conduction (E_n), heavy-hole (HH_n), and the first light-hole (h_1) subbands calculated at 4.2 K in the center Γ of the first Brillion zone as a function of layer thickness d_2 for HgTe/CdTe superlattice with d_1 = 4.1 d_2. T is the point of the transition semiconductor- semimetal

The energy E as a function of d_2, at 4.2 K, in the center Γ of the first Brillion zone and for d_1 = 4.1 d_2, is shown in Fig. 8. The case of our sample (d_2 = 44Å) is indicated by the vertical solid line. Here the cross-over of E_1 and HH_1 subbands occurs. d_2 controls the superlattice band gap Eg = E_1-HH_1. For weak d_2 the sample is semiconductor with a strong coupling between the HgTe wells. At the point T(d_2 = 53.7 Å, E= 38.7 meV) the gap goes to zero with the transition semiconductor- semimetal. When d_2 increases, E_1 and h_1 states drops in the energy gap [0, Λ] and become interface state with energy E_I=34 meV for infinite d_2. Then the superlattice has the tendency to become a layer group of isolated HgTe wells and thus assumes a semimetallic character. The ratio d_1/d_2 governs the width of superlattice subbands (i.e. the electron effective mass). A big d_1/d_2, as in our case, moves away the material from the two-dimensional behavior.

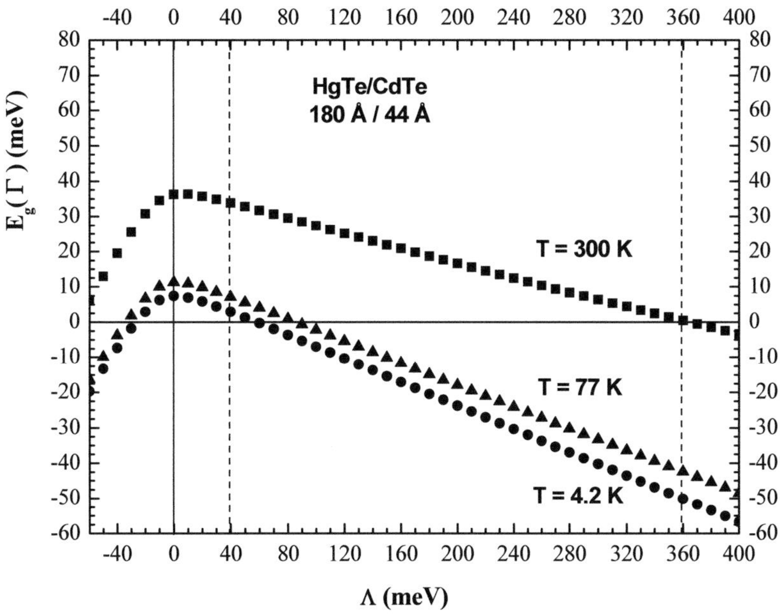

Fig. 9. The band gap E_g (Γ), at the center Γ of the first Brillion zone, as function of temperature and valence band offset Λ between heavy holes bands edges of HgTe and CdTe for the investigated HgTe/CdTe superlattice

In Fig. 9 we can see that the band gap E_g(Γ) increases, presents a maximum at 0 meV and decreases when the valence band offset Λ between heavy hole band edges of HgTe and CdTe increase. For each Λ, E_g(Γ) increases with T. Our chosen value of 40 meV is indicated by a vertical dashed line. This offset agrees well with our experimental results contrary to 0 meV used by (Bastard, 1982) and 360 meV given by (Johnson et al, 1988). The later offset give a zero gap whereas, in intrinsic regime, our measured E_g = 38 meV agree with calculated E_g(Γ, 300 K) = 34 meV.

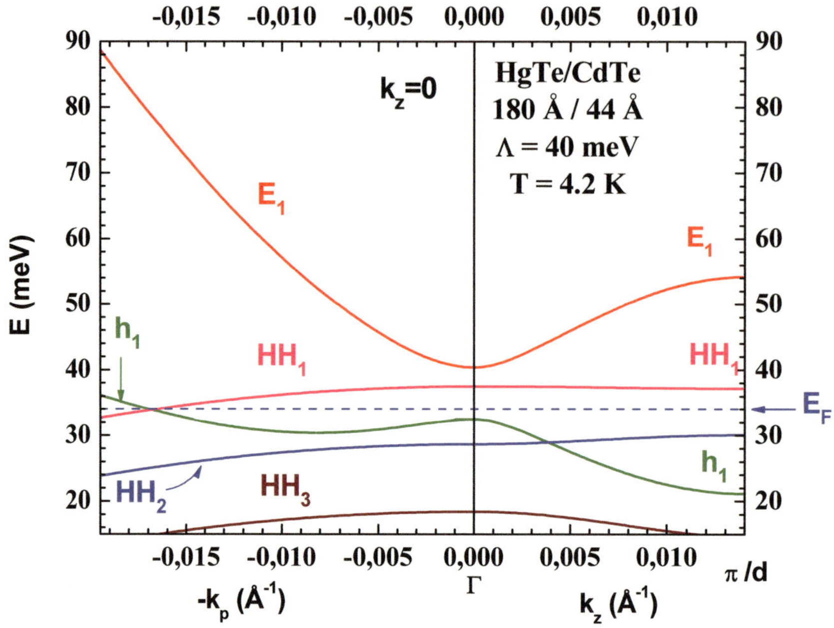

Fig. 10. Calculated bands along the wave vector k_z (a) and in plane $k_p(k_x,k_y)$ (b), of the HgTe/CdTe superlattice at 4.2 K. E_F is the energy of Fermi level

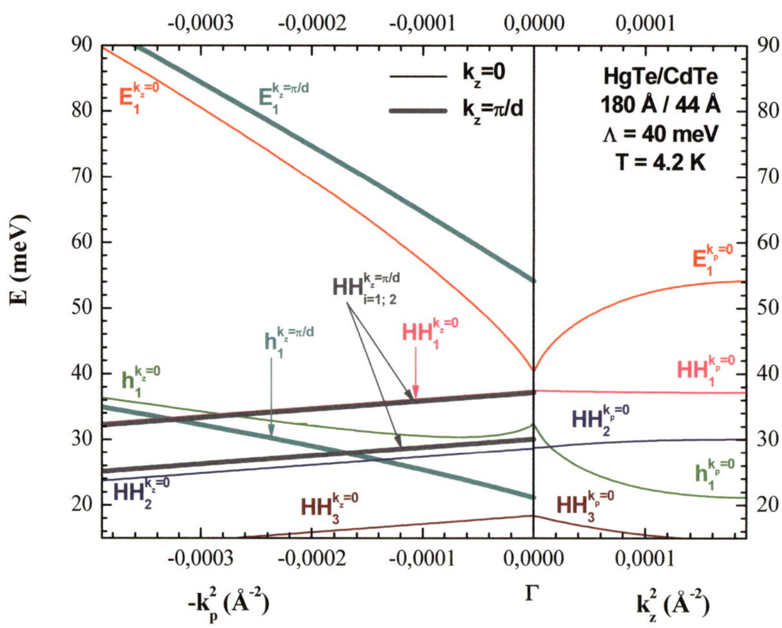

Fig. 11. Calculated bands as a function of k_z^2 and k_p^2 of the HgTe/CdTe superlattice at 4.2 K

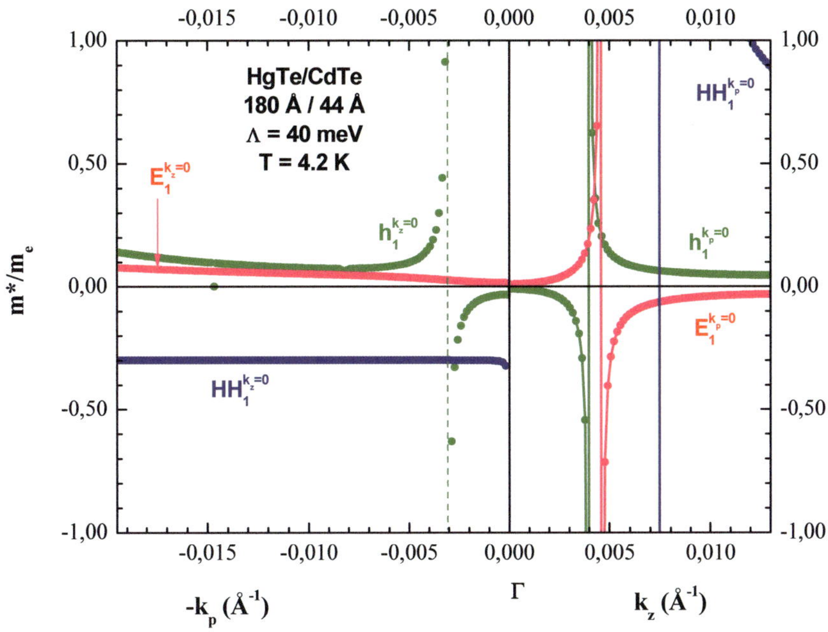

Fig. 12. Calculated relative effective mass bands along the wave vector k_z and in plane k_p of the HgTe/CdTe superlattice at 4.2 K

Fig. 10. show the spectra of energy $E(k_z)$ and $E(k_p)$, respectively, in the direction of growth and in plane of the superlattice at 4.2 K. Along $E(k_p)$, E_1 and h_1 increase with k_P whereas HH_n decreases. This yields to a anti-crossing of HH_1 and h_1 at k_p=0.0168 $Å^{-1}$. The gap is $E_g(\Gamma$,4.2 K) = 3 meV. Note that E_F(4.2 K) =34 meV= E_I and then the conduction is assumed by heavy and light holes. As seen in Fig. 11., along k_z and k_p the light particles (electrons and light holes) subbands are not parabolic whereas the heavy hole subbands are parabolic.

For an anisotropic medium, such as the HgTe/CdTe superlattices, the effective mass is a tensor and its elements along μ and ν directions are given by the following expression [Kittel, 2001]:

$$\left(\frac{1}{m^*}\right)_{\mu\nu} = \frac{1}{\hbar^2}\frac{\partial^2 E_{k_{\mu\nu}}}{\partial k_\mu \partial k_\nu} \quad (15)$$

By carrying out second derivative of the energy E_1, h_1 and HH_1 along k_z and k_p in Fig. 10 we calculated the effective mass bands in Fig. 12. Along k_p, the effective mass of heavy holes m^*_{HH1}= -0.300 m_0 and the effective mass of electrons m^*_{E1} increases from 0.011 m_0 to 0.080 m_0 whereas the effective mass of the light holes h_1 deceases from 0.140 m_0 by half to a minimum of 0.074 m_0. After it increases and diverges at k_p = 0.003 $Å^{-1}$ assuming a electronic conduction. After it increases to -0.034 m^0 at the center Γ of the first Brillion zone assuming a light hole conduction.

Using the same procedure as for the SL1 we get the temperature dependence of the band gap E_g, in the center Γ of the first Brillouin zone in Fig. 13. Note that E_g increases from 3.6

meV at 4.2 K to 34 meV at 300 K. In the investigated temperature range, the detection cut-off wave length verify : $40\mu m \leq \lambda_c \leq 500\mu m$. This situates our sample as a far infrared detector.

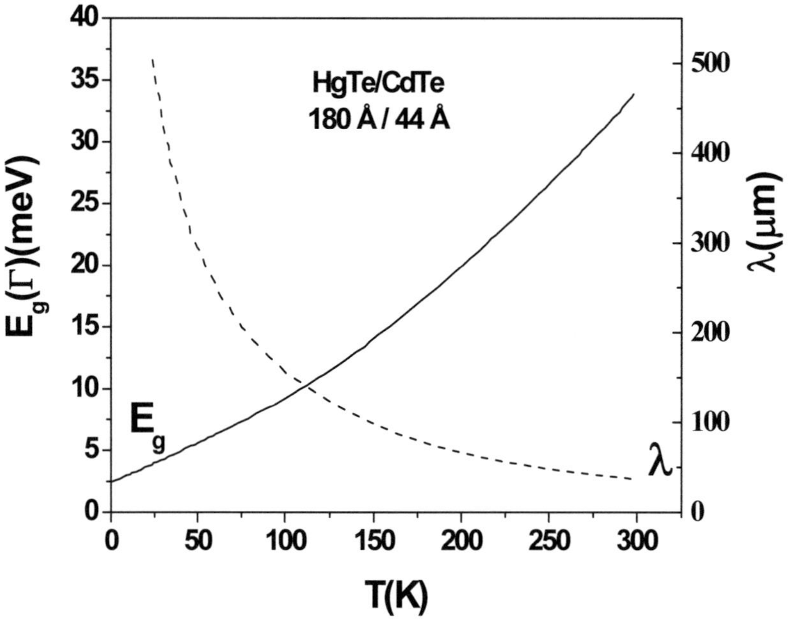

Fig. 13. Temperature dependence of the band gap E_g and cut-off wavelength, at the center Γ of the first Brillion zone, in the investigated HgTe/CdTe superlattice

5.2 Experimental results and discussions

The transverse magnetoresistance ρ/ρ_0, in Fig. 14, follows the two-dimensional (2D) dependence with manifestation of the beginning of the Shubnikov-de Haas oscillations (with a weak $[(\Delta\ \rho/\rho_0)max = 0.2]$). However, over the entire investigated magnetic field range, a non-vanishing magnetoresistance is observed when the field is parallel to the plane. This can be due to the inter-diffusion between HgTe well (big d_1/d_2 and small d_2) and/or to the widening of carriers subbands under the influence of the magnetic field along $E(k_p)$. While the Hall voltage V_H goes to zero at this configuration in Fig. 15. This suggests quasi two-dimensional conductivity behavior (between three-dimensional (3D) and two-dimensional (2D).

At low temperature, the sample exhibits p type conductivity with a hole mobility $\mu_p \approx 900$ $cm^2/V.s$. As seen in Fig. 16 (b), a reversal of the sign of the weak-field Hall constant R_H occurs at 25 K. It may be attributed to trapping of carrier charges in intrinsic state E_I. Such a reversal of the sign of the Hall voltage may be inferred by the existence of at least two types of carriers, which suggests a semimetallic character of the conduction mechanism.

Fallowing the relaxation time resolution of the Boltzmann equation (Seeger b, 2002) (Kittel, 2001) the Hall constant at weak magnetic field in Fig. 16 (b) is described by the well knows formula:

$$(R_H)_w = \frac{1}{|e|}\frac{p - nb^2}{(p+nb)^2} \quad \text{with} \quad b = \frac{\mu_n}{\mu_p} \tag{16}$$

At low temperature in the saturation regime:

$$(R_H)_{sat} = \frac{1}{|e|p} \tag{17}$$

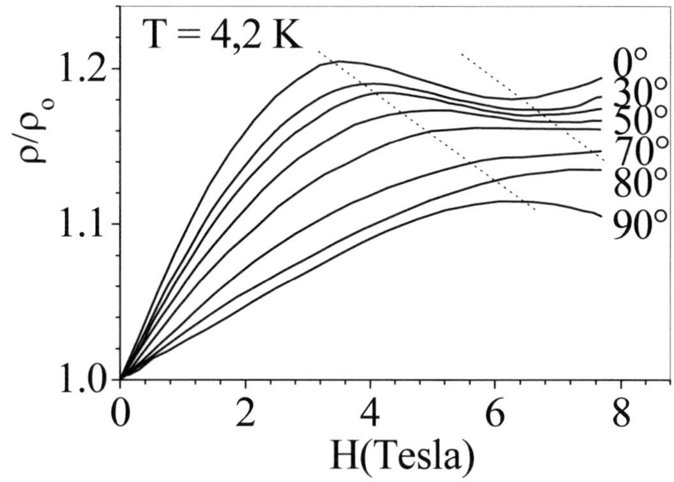

Fig. 14. Transverse magnetoresistance of the sample, at various angles between the magnetic field and the normal to the HgTe/CdTe superlattice surface, at 4.2 K

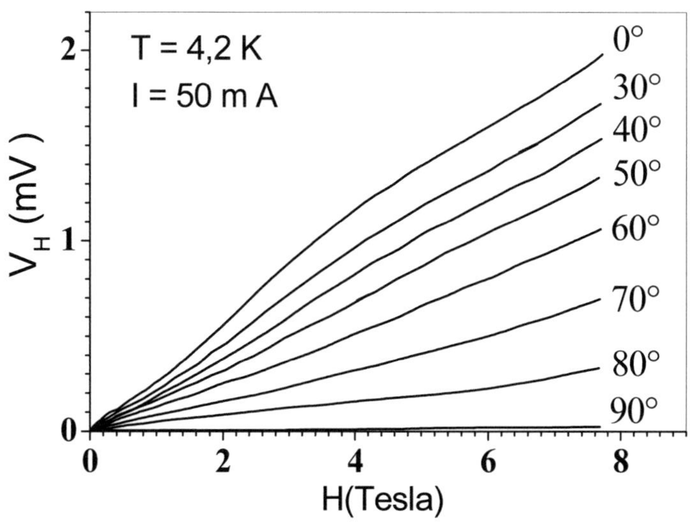

Fig. 15. Field dependence of Hall voltage, at various angles between the magnetic field and the normal to the HgTe/CdTe superlattice surface, at 4.2 K

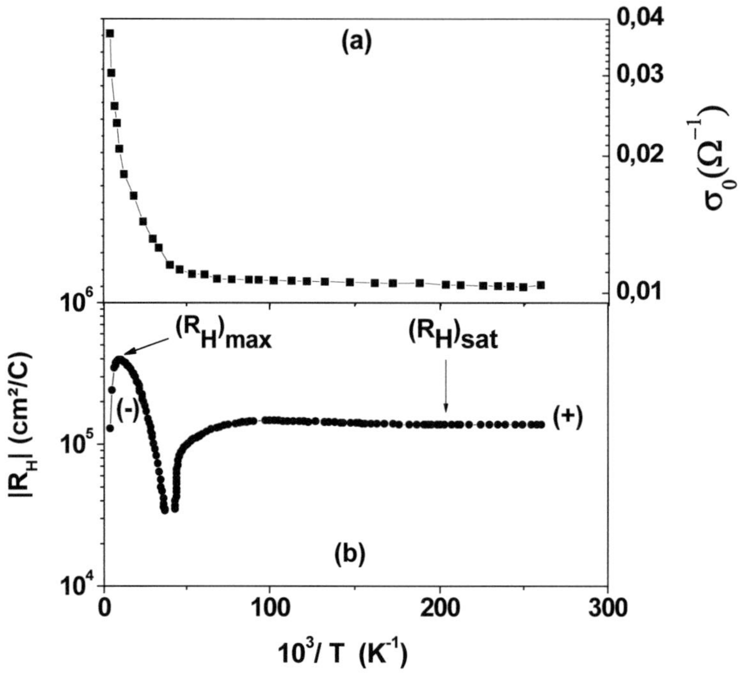

Fig. 16. Temperature dependence of the conductivity (a) and weak-field Hall coefficient (b), in the investigated HgTe/CdTe superlattice

Near the intrinsic regime, the maximum of the Hall constant is given by:

$$(R_H)_{max} = \frac{(b-1)^2}{|e|4bp} \tag{18}$$

This results in the ratio:

$$\frac{4(R_H)_{max}}{(R_H)_{sat}} = \frac{(b-1)^2}{b} \approx b = 32 \tag{19}$$

Eq. (15) implies for $R_H = 0$:

$$b^2 = 1024 = \frac{p}{n} \tag{20}$$

This implies an electron mobility of $\mu_n = 3\times10^4$ cm²/V.s. Such a low value of the electron mobility can be connected, from the one side, with the superlattice electron effective mass which is much higher than that of the bulk material (Nafidi et al, 2011) and, from the other side, with different types of imperfections of the investigated structures, including strong compensation.

From Eq. (3) we have:

$$\frac{2}{3}P^2\hbar^2k_F^2 = E_F(E_F - E_g) \tag{21}$$

We extract the Fermi level energy:

$$E_F = \frac{E_g}{2} \pm \sqrt{\left(\frac{E_g}{2}\right)^2 + \frac{2}{3}P^2\hbar^2k_F^2} \tag{22}$$

Where the Fermi wave vector is $k_F(3D) = \left(3\pi^2 p\right)^{1/3}$ and $k_F(2D) = \left(2\pi p\right)^{1/2}$, respectively for three-dimensional and two-dimensional motion of carriers charges. Here we have chosen the minus sign because the sample is p-type at low temperature. Fig. 10 shows that the band gap and the 3D Fermi level energy increase with temperature whereas 2D Fermi level energy and HH_i bands energy remain constant. In all cases the conductivity is assumed by light and heavy holes. So we have quasi-two-dimensional and semimetallic conductivity. In table1 bellow we summarized the electronic transport parameters in the investigated two superlattices.

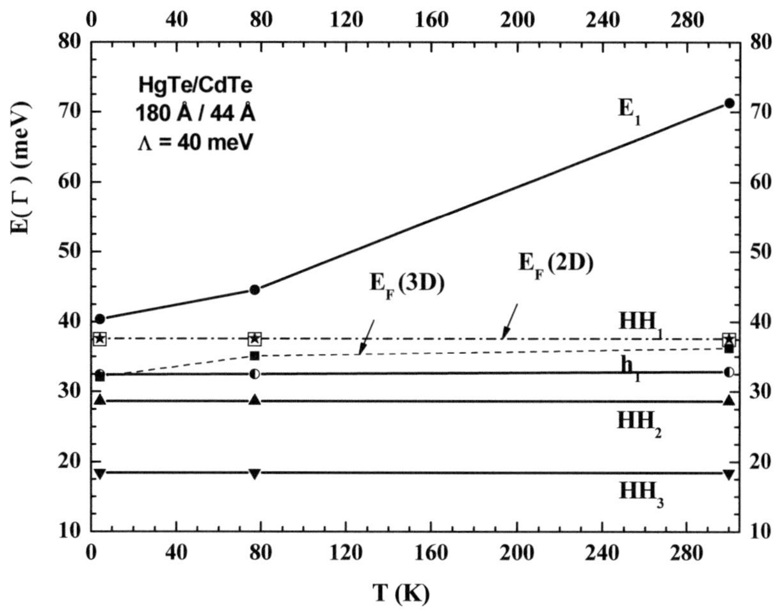

Fig. 17. Temperature dependence of the energy for the three-dimensional (3D) and two-dimensional (2D) Fermi levels in the investigated HgTe/CdTe superlattice

	E_g (meV) Theory 4.2 K	λ_c (µm) Theory 4.2 K	E_F(meV) Exper. 4.2 K	p (cm^{-2}) Hall ef. 4.2K	μ_p (cm^2/Vs) Hall ef. 4.2K
SL1 (p-type)	111	11 (MIR)	14 (2D)	1.84×10^{12}	8200
SL2 (p-type)	3.6	464 (FIR)	32 (3D) 38 (2D)	4.51×10^{13}	900

Table 1. Electronic transport parameters in the investigated two HgTe/CdTe superlattices

7. Conclusions

The fundamental main ideas of this work are:

- HgTe is a zero gap semiconductor (or semimetal) when it is sandwiched between the wide gap semiconductor CdTe (1.6 eV at 4.2 K) layers yield to a narrow gap HgTe/CdTe superlattice which is the key of an infrared detector.
- Before growing our two superlattices, we calculated the bands structures $E(d_2)$ and the gap for each ratio thickness d_1/d_2. After we choose the SL1 in the semiconductor conductivity zone and the SL2 at the transition semiconductor-semimetal

We reported here remarkable correlations between calculated bands structures and magneto-transport properties in two HgTe/CdTe nanostructures superlattices SLi (i=1 and 2). Our calculations of the specters of energy $E(d_2)$, $E(k_z)$ and $E(k_p)$, respectively, in the direction of growth and in plane of the superlattice; were performed in the envelope function formalism.

In the SL1, the formalism predicts that the system is semiconductor, for our HgTe to CdTe thickness ratio $d_1/d_2 = 1.87$, when $d_2 < 14$ nm. In our case, d_2=3 nm and E_g (Γ, 4.2 K) = 111 meV. In spite of it, the sample exhibits the features typical for the semiconductor type p conduction mechanism. In the used temperature range, this simple is a medium-infrared detector, narrow gap and two-dimensional p-type semiconductor.

In the SL2, the formalism predicts that the system is semimetallic, for our HgTe to CdTe thickness ratio d_1/d_2 =4.1 for $d_2 \geq 3$ nm. In our case d_2 = 4.4 nm and the gap E_g(Γ,4.2 K) =3 meV corresponding to thermal energy necessary to change the sign of $R_H(T)$. In intrinsic regime, the measurements indicates $E_g \approx$38 meV in good agreement with calculated E_g (Γ, 300 K) =34 meV = E_F (4.2 K)=E_I. In spite of it, the sample exhibits the features typical for the semimetallic conduction mechanism, which agree well with the overlap between carrier subbands along $E(k_p)$ with a quasi-two-dimensional behavior and is a far-infrared detector.

Our HgTe/CdTe superlattices SL1 and SL2 are a stable alternative for application in infrared optoelectronic devices than the alloys $Hg_{0.8}Cd_{0.2}Te$ and $Hg_{0.66}Cd_{0.34}Te$ respectively. The theoretical and magnetotransport parameters are in good agreement. Measurements performed by us on others' samples indicate an improvement of quality of the material manifested by higher mobility.

8. Acknowledgements

This work is supported by research budget of University Ibn Zohr in Agadir. We thank Li Rukang from the School of Chemistry, University of Birmingham, for verifying our X-ray diffraction Rietveld refinements.

9. References

Arch, D. K.; Faurie, J. P.; Staudenmann, J.L.; Hibbs-Brenner, M. & Chow, P. (1986). Interdiffusion in HgTe–CdTe superlattices. *J. Vac. Sci. Technol. A*, Vol, 4, (1986) (2101-2106), DOI: 10.1116/1.574035.

Bastard, G. (1981). Superlattice band structure in the envelope-function approximation. *Phys. Rev. B*, Vol,24, (November15, 1981) (5693–5697), DOI: 10.1103/PhysRevB.24.5693.

Bastard, G. (1982). Theoretical investigations of superlattice band structure in the envelope-function approximation. *Phys. Rev. B*, Vol, 25, (June 15, 1982) (7584–7597), DOI: 10.1103/PhysRevB.24.5693.

Cava, R. J. *Science* 247, pp. 656-662 (1990).

Dingle, R.; Gossard, A. C.; & Wiegmann, W. (1975). Direct Observation of Superlattice Formation in a Semiconductor Heterostructure. *Phys. Rev. Lett.*, Vol, 34, (May 26, 1975) (1327–1330), DOI:10.1103/PhysRevLett.34.1327.

El Abidi, A.; Nafidi, A.; Chaib, H.; El Kaaouachi, A.; Braigue, M.; Morghi, R.; EL Yakoubi, E.Y. & M. d'Astuto, (2010). Application of the transition semiconductor semimetal in modulated nanostructures for communication as infrared optoelectronic device. *Physica B: Physics of Condensed Matter*, Vol, 405, (1 February 2010) (936-940), DOI:10.1016/j.physb.2009.10.019 .

Esaki, L. & Tsu, R. (1970). Superlattice and negative differential conductivity in Semiconductors. *IBM J. Res. Development*, Vol., 14, (Jan. 1970) (61-65), ISSN: 0018-8646.

Guldner, Y.; Bastard, G.; Vieren, J.P.; Voos, M.; Faurie, J.P. & Million, A. (1984). Magneto-optics in a II–VI superlattice: HgTe-CdTe. *Surface Science*, Vol., 142, (1 July 1984) (593-597), DOI:10.1016/0039-6028(84)90367-4

Hansen, G. L.; Schmit, J. L. & Casselman, T. N., (1982).Energy gap versus alloy composition and temperature in $Hg_{1-x}Cd_xTe$. *J. Appl. Phys*, Vol., 53, (10 June 1982) (7099-7101), DOI:10.1063/1.330018.

Johnson, N. F.; Hui, P. M. & Ehrenreich, H. (1988). Valence-Band-Offset Controversy in HgTe/CdTe Superlattices: A Possible Resolution. *Phys. Rev. Lett.*, Vol, 61, (24 October 1988) (1993–1995), DOI: 10.1103/PhysRevLett.61.1993

Kane, E. (1957). Band structure of indium antimonide. *Journal of Physics and Chemistry of Solids*, Vol. 1, Issue 4,(January 1957) (249-261). DOI: 10.1016/0022-3697(57)90013-6

Kittel, C. (2001). *Introduction to solid stat physics*, 3d edition, John Wiley and Sons, Inc, New York, 333.

Nafidi, Ab.; El Kaaouachi, A.; Sahsah, H.; Nafidi, Ah. (2002). Band structure and magneto-transport in HgTe/CdTe superlattice. *Book of the International Conference on Theoretical Physics (HT 2002)*, pp. 274-275, ISBN 3-7643-2433-3, , 22-27 July, Paris, France Birkhäuser Verlag, Germany.

Nafidi, A.; El Kaaouachi, A.; Nafidi, Ah.; Faurie, J.P.; Million, A.; Piaguet, J. (2002). Some Transport Properties of HgTe/CdTe Superlattices. *Physica status solidi (b)*, Vol. 229, (January 2002) (573-576), DOI: 10.1002/1521.

Nafidi, Ab.; EL Abidi, A.; El Kaaouachi, A.; & Nafidi, Ah. (2004). Electronic Band Structure and New Magneto-transport Properties in p-type Semiconductor Medium-infrared HgTe / CdTe Superlattice. *27th International Conference on the Physics of Semiconductors - ICPS-27, AIP Conference Proceedings*, Vol. 772, pp. 1001-1002, DOI:10.1063/1.1994448, July 26-30, 2004, in Flagstaff, Arizona, USA.

Nafidi, A.; El Abidi, A. & El Kaaouachi, A. (2006). Seebeck and Shubnikov-de Haas Effects in a Two-Dimensional p-type HgTe/CdTe Superlattice. *24th International Conference on Low Temperature Physics - LT24 , AIP Conference Proceedings* Vol. 850,pp. 1359-1360, ISBN: 0735403473, 10 - 17 Aug 2005, Orlando, FL, USA.

Nafidi A., Bouallal A., El Kaaouachi B., A. and Chaib, H., *IEEE Transactions on Applied Superconductivity*. Vol 17 , NO 02, pp 2969-2972, (2007).

Nafidi A., Bouallal B., El Kaaouachi A., Chaib H. and Sahsah H., Advance in Cryogenic Engineering, *Transactions of the International Cryogenic Materials conference-ICMC*, American Institute of Physics, AIP CP986, Vol.54, pp.551-558, (2008).

Nafidi, A., et al, (2011) to be published

Sakaki, H.; Chang, L. L.; Sai-Halasz, G. A.; Chang, C. A. & Esaki, L. (1978). Two-dimensional electronic structure in InAs-GaSb superlattices. *Solid State Communications*, Vol. 26, (June 1978) (589-592), DOI:10.1016/0038-1098(78)90770-6

Seeger, K. a (2002). *Semiconductor physics: an introduction,* Springer, ISBN: 9783540438137, Edition Number: 8, chap.4, p.121.

Seeger, K. b (2002). *Semiconductor physics: an introduction,* Springer, ISBN: 9783540438137, Edition Number: 8, chap.6, p.159.

Tokura, Y. Takagi, H and Uchida, S. *Nature* 337, pp. 345-347 (1989).

Tuchendler, J.; Grynberg, M.; Couder, Y.; Thomé, H. & Le Toullec R. (1973). Submillimeter Cyclotron Resonance and Related Phenomena in HgTe. *Phys. Rev. B,* Vol., 8, (15 October 1973) (3884–3894), DOI: 10.1103/PhysRevB.8.3884.

Weiler, M. H. (1981). Magnetooptical Properties of Hg1-xCdxTe Alloys, In: *semiconductors and Semimetals,* Vol., 16, Willardson, R. K. & Beer, A. C. (Ed), (119-191) (Academic, New York), ISBN: 9780127521169

Part 3

Optoelectronic Measurements in Spatial Domain

15

3D Body & Medical Scanners' Technologies: Methodology and Spatial Discriminations

Julio C. Rodríguez-Quiñonez[1], Oleg Sergiyenko[1], Vera Tyrsa[2], Luís C. Básaca-Preciado[1], Moisés Rivas-Lopez[1], Daniel Hernández-Balbuena[1] and Mario Peña-Cabrera[3]
[1]Autonomous University of Baja California, Mexicali-Ensenada,
[2]Polytechnic University of Baja California, Mexicali,
[3]Research Institute of Applied Mathematics and Systems (IIMAS - UNAM)
Mexico

1. Introduction

Medical practitioners have traditionally measured the body's size and shape by hand to assess health status and guide treatment. Now, 3D body-surface scanners are transforming the ability to accurately measure a person's body size, shape, and skin-surface area (Treleaven & Wells, 2007) (Boehnen & Flynn, 2005). In recent years, technological advances have enabled diagnostic studies to expose more detailed information about the body's internal constitution. MRI, CT, ultrasound and X-rays have revolutionized the capability to study physiology and anatomy in vivo and to assist in the diagnosis and monitoring of a multitude of disease states. External measurements of the body are more than necessary. Medical professionals commonly use size and shape to production of prostheses, assess nutritional condition, developmental normality, to analyze the requirements of drug, radiotherapy, and chemotherapy dosages. With the capability to visualize significant structures in great detail, 3D image methods are a valuable resource for the analysis and surgical treatment of many pathologies.

Taxonomy of Healthcare 3D Scanning applications				
Application	**Epidemiology**	**Diagnosis**	**Treatment**	**Monitoring**
Size	Anthropometric surveys	Growth defects	Scoliosis	Fitness and diet
Shape	Screening	Abdominal shape	Prosthetics	Obesity
Surface area		Lung volume	Drug dosage	Diabetes
Volume	Eczema		Burns	
Head Visualization		Melanomas	Eating disorders	
Chest Visualization			Facial reconstruction	
Hole Body Visualization			Cosmetic surgery	

Table 1. Taxonomy of Healthcare 3D Scanning applications

1.1 Scanning technologies

Three-dimensional body scanners employ several technologies including 2D video silhouette images white light phase measurement, laser-based scanning, and radio-wave linear arrays. Researchers typically developed 3D scanners for measurement (geometry) or visualization (texture), using photogrammetry, lasers, or millimeter wave (Treleaven & Wells, 2007).

Taxonomy of 3D Body Scanners		
Technique	**Measurement**	**Visualization**
Millimeter Wave	Radio Waves	
Photogrammetry	Structured light Moire fringe contouring Phase - measuring profilometry	Close-range photogrammetry Digital surface photogrammetry
Laser	Laser Scanners Laser range Scanners	

Table 2. Taxonomy of 3D Body Scanners

In the following section it will be described the diverse measurement techniques (see table 2) used in medical and body scanners. Listing applications, scanners types and common application areas, as well of how they operate.

2. Millimeter wave

Millimeter wave based scanners, send a safe, lower radio wave toward a person's fully clothed body; most of the systems irradiate the body with extremely low-powered millimeter waves a class of non-ionizing radiation (see Figure 1) not harmful to humans. The amount of radiation emitted in the millimeter-wave range is 10^8 times smaller than the amount emitted in the infrared range. However, current millimeter-wave receivers have at least 10^5 times better noise performance than infrared detectors and the temperature contrast recovers the remaining 10^3. This makes millimeter-wave imagine comparable in performance with current infrared systems.

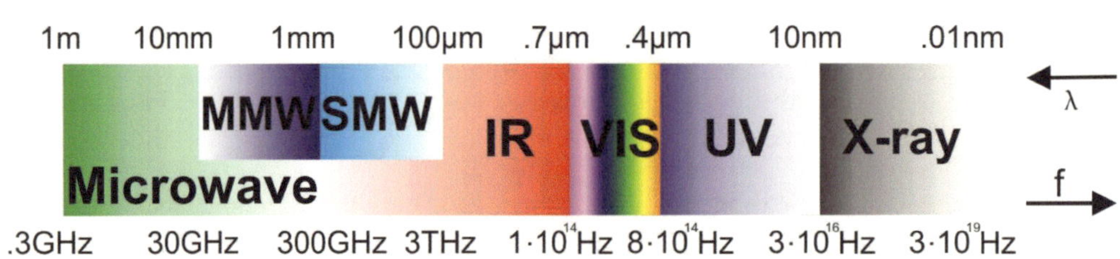

Fig. 1. Electromagnetic spectrum showing the different spectral bands between the microwaves and the X-rays

Millimeter (MMW) and Submillimeter (SMW) waves fill the gap between the IR and the microwaves (see Figure 1). Specifically, millimeter waves lie in the band of 30-300 GHz (10-1 mm) and the SMW regime lies in the range of 0.3-3 THz (1-0.1 mm). MMW and SMW radiation can penetrate through many commonly used nonpolar dielectric materials such as paper, plastics, wood, leather, hair and even dry walls with little attenuation (Howald et al., 2007) (Liu et al., 2007). Clothing is highly transparent to the MMW radiation and partially transparent to the SMW radiation (Bjarnason et al., 2004). Consequently, natural applications of MMW and SMW imaging include security screening, nondestructive inspection, and medical and biometrics imaging. Low visibility navigation is another application of MMW imaging

Is also true that MMW and SMW open the possibility to locate threats on the body and analyze their shape, which is far beyond the reach of conventional metal detection portals. A recently demonstrated proof-of-concept sensor developed by QinetiQ provides video-frame sequences with near-CIF resolution (320 x 240 pixels) and can image through clothing, plastics and fabrics. The combination of image data and through-clothes imaging offers potential for automatic covert detection of weapons concealed on human bodies via image processing techniques (Haworth et al., 2006). Other potential areas of application are mentioned below.

Medical: provide measurements of individuals who are not mobile and may be difficult to measure for prosthetic devices.

Ergonomic: provide measurements and images for manufacturing better office chairs, form-fitting car and aviation seats, cockpits, and custom sports equipment.

Fitness: provide personal measurements and weight scale for health and fitness monitoring.

2.1 3D Body millimeter wave scanner: Intellifit system

The vertical wand in the Intellifit system (see Figure 2) contains 196 small antennas that send and receive low-power radio waves. In the 10 seconds it takes for the wand to rotate around a clothed person, the radio waves send and receive low-power signals. The signals don't "see" the person's clothing, but reflect off the skin, which is basically water (Treleaven & Wells, 2007). The technology used with the Intellifit System is safer than using a cell phone. The millimeter waves are a form of non-ionizing radiation, which are similar to cell phone signals but less than 1/350th of the power of those signals, and they do not penetrate the skin. When the wand's rotation is complete, Intellifit has recorded over 200,000 points in space, basically x, y, and z coordinates. Intellifit software then electronically measures the "point-cloud", producing a file of dozens of body measurements; the raw data is then discarded.

Fig. 2. Intellifit System, cloth industry application and point cloud representation of the system

Although the system is functional to obtain a silhouette of the body, object detection as a security system and as a tool in the cloth design industry, the problem of this system is the inaccurate measurements that are closed to 1cm, which makes the system not appropriate for medical applications.

3. Photogrammetry

Photogrammetry is the process of obtaining quantitative three-dimensional information about the geometry of an object or surface through the use of photographs (Leifer, 2003). Photogrammetric theories have on a long history of developments for over a century. Intensive research has been conducted for the last 20 years for the automation of information extraction from digital images, based on image analysis methods (Emmanuel, 1999). In order for a successful three-dimensional measurement to be made, targeting points, each of which is visible in two or more photographs, are required. These targets can be unique, well-defined features that already exist on the surface of the object, artificial marks or features attached to the object, or a combination of both types. The accuracy of the reconstruction is directly linked to the number and location of the targets, as well as number of photographs and camera positions chosen. Intricate objects generally require more targets and photographs for a successful reconstruction than do flat or near-flat surfaces. (Leifer, 2003). The latest shift in photogrammetry has been the passage to fully digital technologies. In particular, low cost digital cameras with high pixel counts (> 6 mega-pixels image sensors), powerful personal computers and photogrammetric software are driving a lot of new applications for this technology. (Beraldin, 2004). As shown in Table 2, the measurement photogrammetry techniques can by refer as show below.

3.1 Structured-light systems

One of the simplest systems consists of a projector that emits a stripe (plane) of light and a camera placed at an angle with respect to the projector as shown in Figure 3. At each point

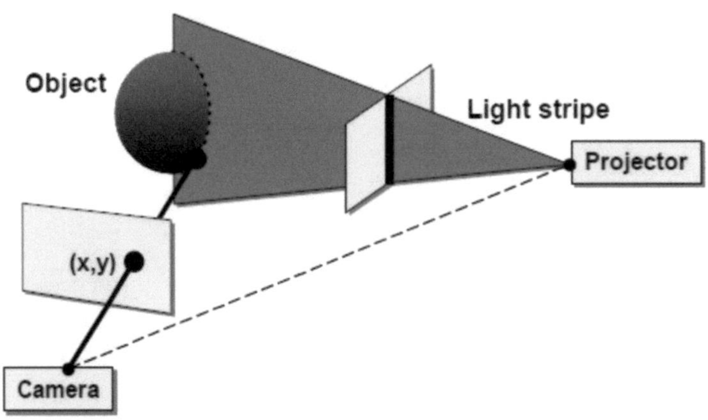

Fig. 3. Schematic layout of a single-camera, single-stripe-source triangulation system

in time, the camera obtains 3D positions for points along a 2D contour traced out on the object by the plane of light. In order to obtain a full range image, it is necessary either to sweep the stripe along the surface (as is done by many commercial single-stripe laser range scanners) or to project multiple stripes. Although projecting multiple stripes leads to faster data acquisition, such a system must have some method of determining which stripe is which (Rusinkiewicz et al., 2002). There are three major ways of doing this: assuming surface continuity so that adjacent projected stripes are adjacent in the camera image, differentiating the stripes based on color, and coding the stripes by varying their illumination over time. The first approach (assuming continuity) allows depth to be determined from a single frame but fails if the surface contains discontinuities. Using color allows more complicated surfaces but fails if the surface is textured. Temporal stripe coding is robust to moderate surface texture but takes several frames to compute depth and, depending on the design, may fail if the object moves (Rusinkiewicz et al., 2002).

3.1.1 Body and medical 3D structured light scanner: Formetric 3D/4D

The system Formetric 3D/4D is based on structured light projection. The scanning system consists of four main components: electro-mechanical elevating column for height adjustment, projector, camera and software. The projection unit emits a white light grid onto the dorsal surface of the patient standing in a defined way toward the projection device, which then obtains measuring data on the dorsal profile by means of a video-optic device from another direction (Hierholzer & Drerup, 1995). Rasterstereography excels by its precision (methodic error < 0.1 mm) and allows a radiation-free representation of the profile. For angular data, the reproducibility of an individual rasterstereographic shot is indicated with 2.8°. The measuring speed of 0.04 seconds can be considered as quick, and the total dorsal surface is registered simultaneously (Lippold et al., 2007). An automatic recognition of anatomical structures by means of the connected software provides the basis for a reconstruction of the three-dimensional profile of the dorsal surface. Figure 4 shows the Formetric 3D/4D Scanning System. By means of mathematical algorithms, a two-dimensional median sagittal or frontal-posterior dorsal profile is generated (Lippold et al., 2007). The gained information is of use for analysis and diagnosis.

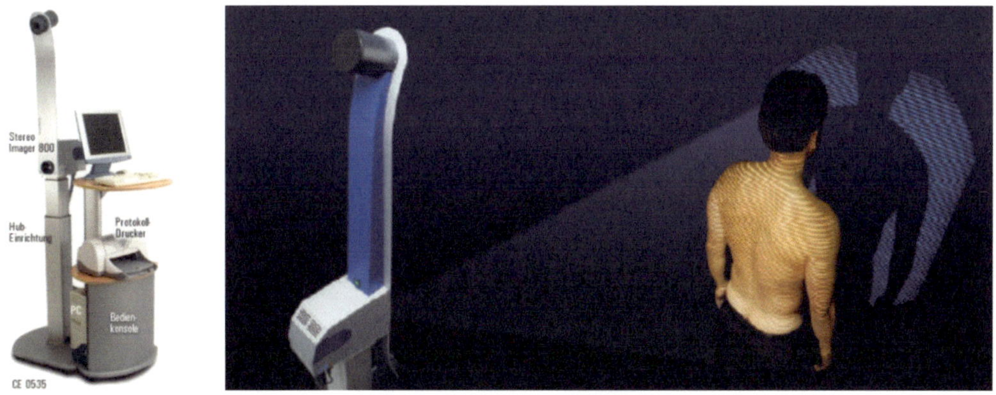

Fig. 4. Formetric 3D/4D Scanning System

However, one of the disadvantages of this procedure is when a 360° view of an object is required; it is unable to use simultaneously multiple systems around the object because of interference between multiple light projections. It can give inaccurate data. Although, multiple systems use in sequence will increment the scanning time.

3.2 Moiré fringe countering

In optics moiré refers to a beat pattern produced between two gratings of approximately equal spacing. It can be seen in everyday things such as the overlapping of two window screens, the rescreening of a half-tone picture, or with a striped shirt seen on television (Creath & Wyant, 1992). The moiré effect is obtained as a pattern of clearly visible fringes when two or more structures (for example grids or diffraction gratings) with periodic geometry are superimposed. It has also been verified that the obtained fringes are a measure of the correlation between both structures. Additionally, it has been shown that the moiré effect can be obtained when other types of structures are superimposed, such as random and quasi-periodic ones or fractals. Fringe projection entails projecting a fringe pattern or grating over an object and viewing it from a different direction. It is a convenient technique for contouring objects that are too coarse to be measured with standard interferometry. A simple approach for contouring is to project interference fringes or a grating onto an object and then view it from a different direction (Calva et al., 2009). The first use of fringe projection for determining surface topography was presented by Rowe and Welford in 1967. Fringe projection is related to optical triangulation using a single point of light and light sectioning where a single line is projected onto an object and viewed in a different direction to determine the surface contour Moiré and fringe projection interferometry complement conventional holographic interferometry, especially for testing optics to be used at long wavelengths. Although two-wavelength holography (TWH) can be used to contour surfaces at any longer-than-visible wavelength, visible interferometry environmental conditions are required. Moiré and fringe projection interferometry can contour surfaces at any wavelength longer than 10-100 μm with reduced environmental requirements and no intermediate photographic recording setup (Creath & Wyant, 1992). However doesn't exist commercial scanners who take advantage of the combine technique of moiré fringe.

3.3 Phase Measuring Profilometry (PMP)

A well-known non-contact 3D measurement technique has been extensively developed to meet the demands of various applications. In such system (see Figure 5), generally, periodic

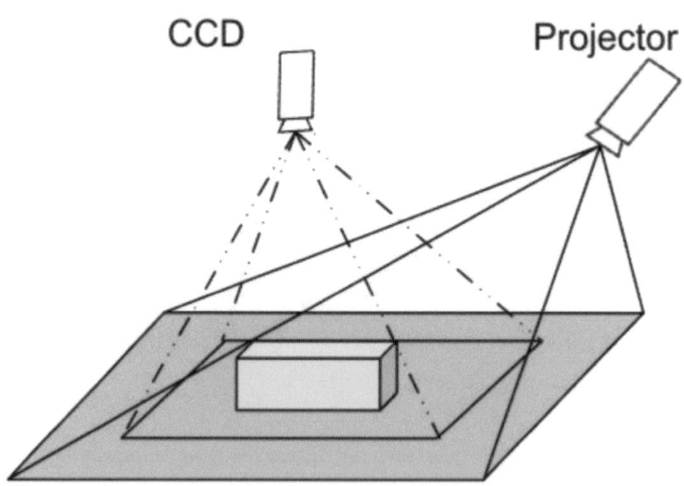

Fig. 5. The Phase Measuring Profilometry system

fringe patterns are projected on the objects surface, and the distorted patterns caused by the depth variation of the surface are recorded. The phase distributions of the distorted fringe patterns are recovered by phase-shifting technique or the method based on Fourier transformation analysis and then the depth map of the object surface is further reconstructed. Currently, light pattern is designed and generated by computer and Digital Light Projector (DLP) is popularly used to project the periodic sinusoidal fringe patterns on object surfaces. It is more flexible and accurate than conventional approaches in which grating is used for generating the sinusoidal fringe images. However, some problems still exist in PMP using DLP. One of them is that the inherent gamma nonlinearity of the DLP and CCD camera affects the output. As a result, the actual obtained fringe waveform is nonsinusoidal (Di & Naiguang 2008).

3.3.1 White light scanners by 3D3 solutions

The scanning system (see figure 6) consists of three main components: Projector (2200 Lumens to 2700 Lumens, 1024 + resolution), two 5MP high-speed HD machine vision cameras and a PC with FlexScan3D image capture software. The scanner use a projector to emit a white light pattern on to the surface of an object, two simple video cameras placed at different position scan the object and the software by triangulation of patterns renders the model in three dimensions. The first step in the scan procedure is the camera calibration using a pattern board, which the software needs to interpret the position of both cameras. When the pattern is projected the cameras provide the information to the software and render the image. The system needs a minimal 4 scans for a 360° view and is Recommended 8 scans for a full 360° view, the working range is 0.4 meters to 5 meters, and the scan speed is 1 to 6 seconds depending on scanner configuration. The common applications are: scanning faces for cosmetic surgery and burn treatments (in table 1 are presented medical applications for 3D scanners), bracing products (Knees, elbows and ankles), dental scanning replaces the need to create physical dental molds for patients.

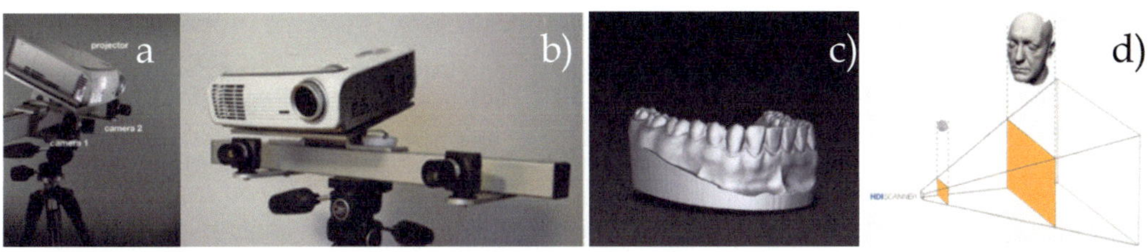

Fig. 6. a) Right view of 3D3 scanning system b) Front View of scanning system c) Dental scanning d) Field of view and face scanning

However this system only generates a 3D image and does not give as an output dimension measurements.

4. Laser scanning

Most of the contemporary non-contact 3D measurement devices are based on laser range scanning. The simplest devices, and also the least reliable, are based on the triangulation method. Laser triangulation is an active stereoscopic technique where the distance of the object is computed by means of a directional light source and a video camera. A laser beam is deflected from a mirror onto a scanning object. The object scatters the light, which is then

collected by a video camera located at a known triangulation distance from the laser (Azernikov & Fischer, 2008). Using trigonometry, the 3D spatial (XYZ) coordinates of a surface point are calculated. The charged couple device (CCD) camera's 2D array captures the surface profile image and digitizes all data points along the laser. The disadvantage of this method is that a single camera collects only a small percentage of the reflected energy. The amount of collected energy can be drastically increased by trapping the whole reflection conus. This improvement significantly increases the precision and reliability of the measurements. The measurement quality usually depends on surface reflection properties and lighting conditions. The surface reflection properties are dictated by a number of factors: a) angle of the laser ray hitting, b) surface material, and c) roughness. Owing to these factors, with some systems the measured object must be coated before scanning. More advanced systems provide automatic adaptation of the laser parameters for different surface reflection properties (Azernikov & Fischer, 2008).
There are a number of laser scanning systems on the market specifically engineered to scan manufactured parts smaller (10" L x 10" W x 16" H) than the human body. These systems are smaller than the typical laser body scanners mentioned below and employ a different scanning mechanism. The industrial units may pass a single laser stripe over the part or object multiple times at different orientations or rotate the part on a turntable. The smaller systems often have increased accuracy and resolution in their measurements when compared to their larger counterparts because of their reduced size and different scanning mechanisms. (Lerch et al., 2007)

4.1 Spatial discrimination

Given the nature of light there are discriminations to be performed in laser scanning systems, for example even in the best emitting conditions (single mode), the laser light does not maintain collimation with distance (e.g. check the beam divergence on scanner specifications sheets). In fact, the smaller the laser beam, the larger is the divergence produced by diffraction. For most laser scanning imaging device, the 3D sampling properties can be estimated using the Gaussian beam (see Figure 7) propagation formula and the Rayleigh criterion. This is computed at a particular operating distance, wavelength and desired spot size within the volume. Figure 4 illustrates that constraint (λ = 0.633 µm) (Beraldin, 2004).

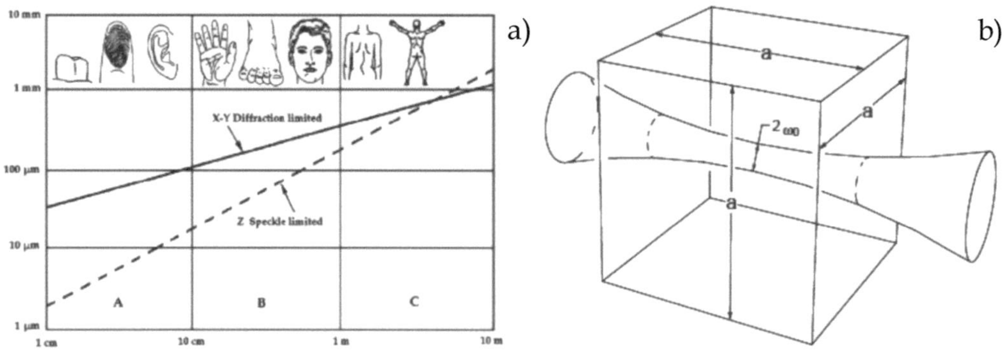

Fig. 7. a) Physical limits of 3D laser scanners as a function of volume measured. Solid line: X-Y spatial resolution limited by diffraction, Dashed line: Z uncertainty for triangulation-based systems limited by speckle. b) Gaussian Beam (Beraldin, 2004)

4.2 Body and medical 3D laser scanners

Of the diverse current methods for body scanning, laser scanners are used to graphically represent the silhouette and perform accurate measurements. The following systems are appropriate to perform the representation task but they have disadvantages which can decrease its measurement precision.

4.2.1 Vitus Smart 3D laser scanner

The scanning system developed by Human Solutions consists of two main components: the scanning assembly or booth and a PC with image reconstruction software. The scanning assembly is 4′ wide by 4′ deep by 10′ high. (See figure 8) with a structural frame to keep the device stationary; curtains are hung from the frame to minimize outside light. Located in each of the four corners is a vertical column containing the essential scanning equipment: a low energy laser, and two charge coupled device (CCD) cameras, all of which ride together in an elevator assembly that travels up and down in the vertical column. When the system is calibrated correctly, the four elevator assemblies travel down the columns in unison, sweeping the scanning zone with a horizontal plane of laser light.
The laser light illuminates the contours of an object standing within the scanning zone and the CCD cameras record discrete points on these contours at each horizontal plane. The entire scan takes approximately 12 seconds (Lerch et al., 2007).

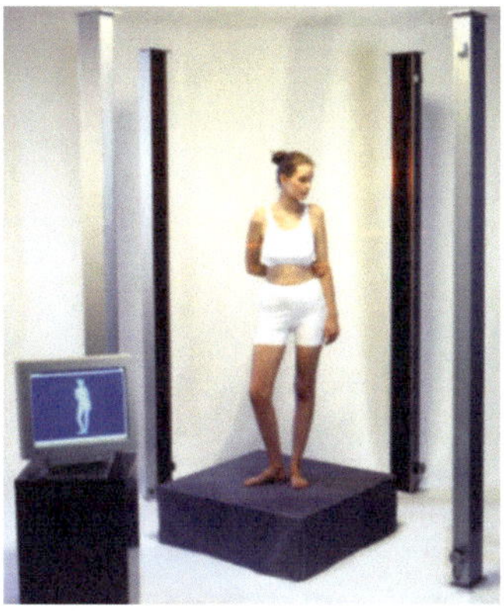

Fig. 8. Vitus 3D Laser Scanning

A computer attached to the scanner contains the user interface, data acquisition/reconstruction, and data analysis software, while interfacing with the motor controller. The computer software acquires data from the A/D converter and triangulates the discrete points for all of the horizontal planes, creating a point cloud representation of the object scanned. This process takes approximately 2 minutes to complete. After the data acquisition/reconstruction program is completed, a 3D image of the object is displayed on the computer screen. The point cloud data can be exported into proprietary and standard file formats (obj. dxf sdl. ascii) which can be imported into various computer aided design (CAD), finite element analysis (FEA). and rapid prototyping software packages (Lerch et al., 2007).
The elevated production costs of hardware components for the Vitus 3D Laser Scanning could be considered as a disadvantage. Moreover, precision electric motors should be used

for the displacement of the scanner units. Lastly, the whole scanner system must be calibrated so that the geometrical disposition of all the elements can be accurately determined. Any error in calibration will result in inaccurate measurements because there is no gap uncertainty in the calibration.

4.2.2 Konica Minolta 910

The Vivid 910 scanner (see figure 9) from Konica Minolta consists on a single camera and laser stripe, and acquires 3D data using triangulation. According to Konica the scanning process is comfortable, although subjects can see a quick flash of red when the laser stripe crosses the pupil. The laser is eye safe so the subject's eyes can remain open during scanning. The scan takes approximately 2.5 seconds and the subject must remain motionless during that time or a poor scan will result. The Vivid 910 managed to be accurate with a repeatability of 0.003 mm. (Boehnen & Flynn, 2005). There are three different zoom lenses available and an automatic focus system that allows scanning at a wide variety of distances from the camera (there is a tradeoff between image resolution and standoff). It is somewhat sensitive to lighting conditions and is necessary to operate on indoors environments (Boehnen & Flynn, 2005).

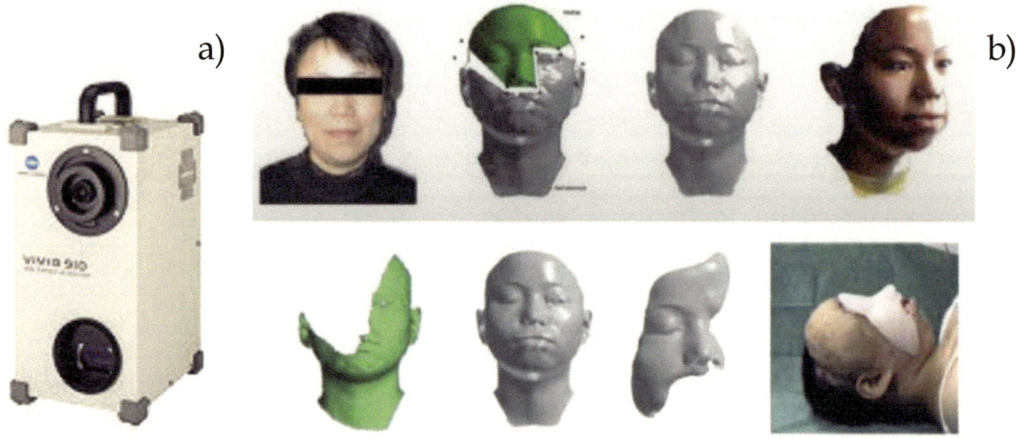

Fig. 9. a) Vivid 910 b) Rough procedures to create the missing part for visualization using Vivid scanner

4.2.3 3D Dynamic Triangulation scanner

The scanning system consists of four main components: electro-mechanical inclining angle system, laser beam projector, photodetector and software. A laser beam is projected onto the body and is detected by a photodetector which sets the angle of incidence. The system has a rotating system that allows inclining the angle for a complete scan. The system reduces measurement error because doesn't have independent elements to coordinate like Vitus Smart. The precision is 0.04 mm and allows a radiation-free representation of the profile.

The laser and the collimator are installed in own laser positioning system (PL) see figure 10. PL has its step drive, which on a command from the onboard computer can turn PL in a horizontal plane at each for one angle pitch (Rivas et al., 2008). On the other end of the bar is located a scanning aperture (SA) (Sergiyenko et al, 2009). Bi is the angle detected and Ci is the output angle of the laser. The system works in the next way. By the command from the computer the bar is installed so that the SA rotation axis becomes perpendicular to plane XOY of reference system. PL puts the laser with the collimator, for example, in an extreme right position. The axis of the collimator (with the help of PV-step drive) then takes extreme top position (above the horizon). The laser and the SA are switched on. SA is rotated by the electromotor EM. At each SA turn a laser ray should hit an obstacle, is reflected diffusely by it (point Sij) and returns to mirror in SA. At the moment when three objects - the point of reflection Sij, the perpendicular to mirror and the vertical axis of SA - takes their common plane, perpendicular to plane XOY while SA is rotating, an optical signal, having travelled a path "Sij - mirror M - objective O - optical channel OC - photoreceiver PR ". It makes an electrical stop signal. A start signal is previously formed by SA by means of a zero-sensor (installed on a bar b axis) (Rivas et al., 2008).

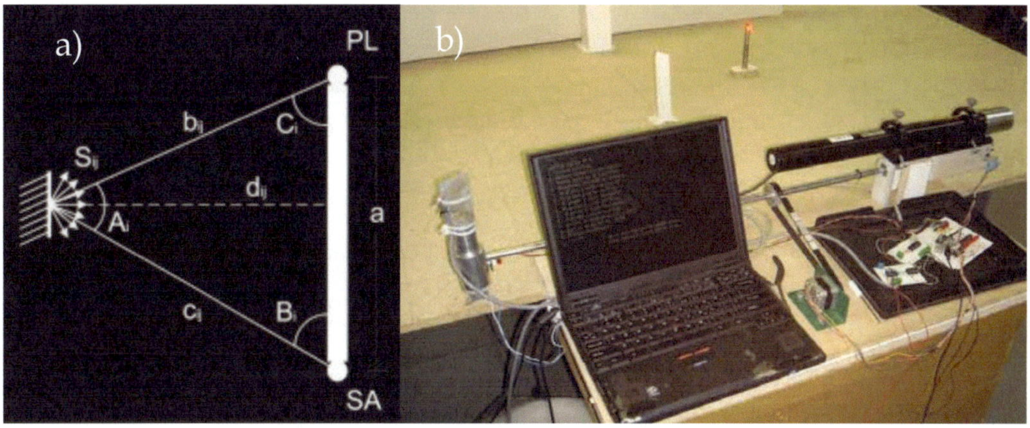

Fig. 10. a) Triangulation scheme, b) Dynamic triangulation scanner

The principle of this system is promising, although it has multiples disadvantages when the system is actually developed and running. The usage of the timing belts for the angular rotation of the system is one of them. Moreover, the system must undergo a thoroughly calibration to guarantee that the mirror rotates parallel to the system, and the receptor motor is not sufficient to guarantee constant rotational speed. Lastly, there are some components that vibrate and generate unwanted noise.

4.2.4 3D Rotational Body Scanner

The Rotational Body Scanner uses the principals of Dynamic Triangulation Scanner. (Basaca & Rodriguez, 2010). Increases its precision, decreases the mechanic noise sources and makes the addition of a stationary rotation system independent of timing belts (Rivas et al.,2008). The system receptor (see Figure 11) consist of 5 main components A) 45 degree rotational mirror, whose principal function is to direct the laser light beam towards the lenses (targets). B) Targets, whose function is to concentrate the light beam onto photodetector. C) DC Motor, which rotates the mirror. D) Photodetector, it captures the light beam located within the frequency range of the laser. E) Flat Bearing, allows the rotation in the angular axis of the system.

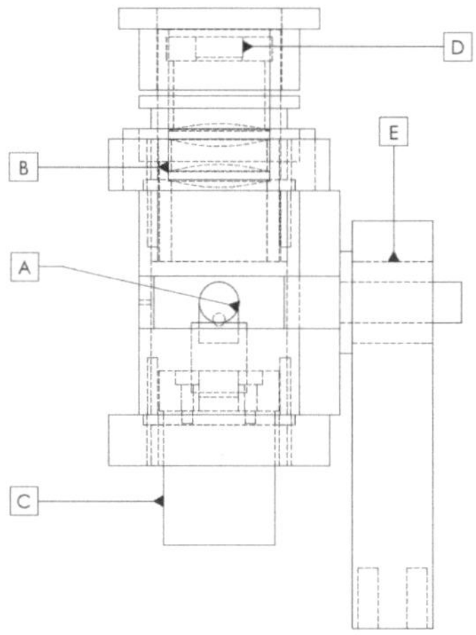

Fig. 11. System receptor

The system projector has 5 main components (see figure 12), which are the following: 1) Step Motor of angular rotation, whose main function is to control the rotation of the entire system. 2) Step motor for the mirror rotation, which controls the mirror rotation. 3) System's rotation gear, increases the precision of the system since it gives a 10:1 ratio gear-motor. 4) Mirror's rotation gear increases the precision of the system giving a 10:1 ratio gear. 5) Mirror, reflects the laser light beam towards the scanning body.

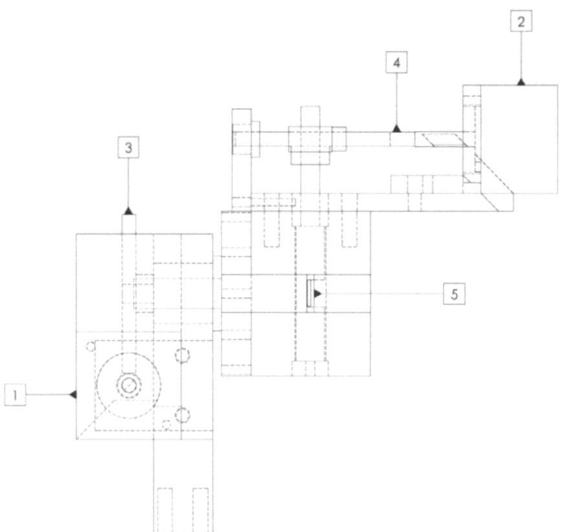

Fig. 12. System projector

The laser light projector emits the light at different angles towards the body. And at the same time the receptor rotates until it detects the light deflected by the body. When the mirror of the receptor deflects the scattered light towards the target and concentrates the light towards the photodetector, an electronic pulse is emitted which indicates the point has been detected. A relationship between the rotation time and detection time shows the angle in which the receptor detects the point. Since the projector rotation is controlled by the user, the angle of the projector is known at all times. The relationship between the 2 angles and the known distance between the projector and receptor gives each of the captured coordinates.

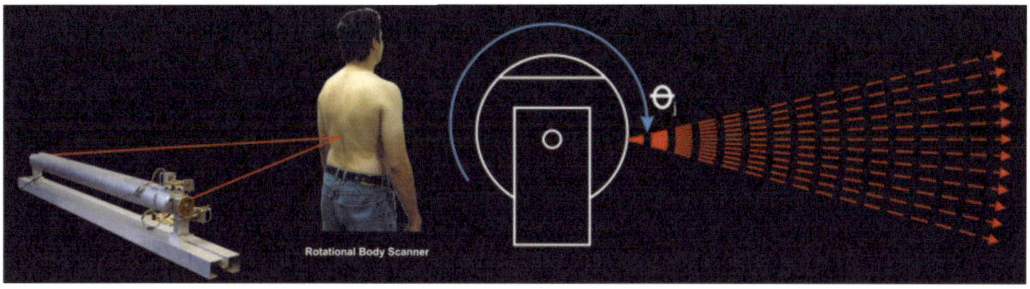

Fig. 13. 3D Rotational Body Scanner

As shown in figure 13, the projector and receptor are separated by a bar that gives the exact distance of 1 meter between them, and located in the bar is the laser light source. Within the bar the laser also gets aligned and locked avoiding measurement errors. The triangulation principle used is well known, and some of the advantages given by this system is the angular rotational mechanism (see figure 13) which allows the rotation with no chains, an increment in resolution of 10 times by using gears that gives 1 rotation for each 10 rotations that gives the step motor, inaccuracy caused by friction are decreased by using polytetraflourtethylene flat bearings which has the lowest friction coefficient of all materials, and the fabrication cost is economic.

4.3 Traceable 3D laser imaging metrology

The statement of uncertainty is based on comparisons with standards traceable to the national units (SI units) as requested by ISO 9000-9004. For example, manufacturers of theodolites and CMM manufacturers use specific standards to assess their measuring instruments. A guideline called VDI/VDE 2634 has been prepared in Germany for close range optical 3D vision systems. It contains acceptance testing and monitoring procedures useful for practical purposes for evaluating the accuracy of optical 3D measuring systems based on area scanning - bundle of rays. These systems work according to the principle of triangulation, e.g. fringe projection, Moiré techniques and photogrammetric/scanning systems based on area scanning (Beraldin et al., 2007). According to National Institute of Standards and Technology (NIST) in the Proceedings of the LADAR Calibration Facility Workshop, Gaithersburg, June 12 - 13, 2003 the steps to perform a 3D laser scanning calibration could be the following.

Calibration of the direction component: Using theodolite-type scanners, the direction affecting instrumental errors of the laser-scanner could be calibrated by procedures known from theodolites These are:

1. Vertical axis wobble, which acts as a lever effect, if the scanner does not correct this influence by inclination sensors.
2. Eccentricity of scan center.
3. Collimation axis error.
4. Horizontal axis error.

However no internationally recognized standard or certification method exists; the evaluation of the accuracy, resolution, repeatability or measurement uncertainty of a 3D imaging system still remains the responsibility of the user.

5. Conclusions

Not all scanning methods are as accurate as the diverse applications demands. None of the systems is superior in every area of applications.

The MillimeterWave based systems are sufficient for object detection but underdeveloped to be used in the medical environment where accuracy is needed. The main disadvantage of these systems is that their accuracy and contrast are sacrificed to be able to perform real time scanning.

The diverse techniques used in Photogrammety are appropriate to perform the modeling representation of the scanned objects, although not all techniques have the capability to perform measurements, such as the White Light Scanner by 3D3 Solutions mentioned above. This is one of the main reasons why the laser scanner based systems are preferred when measurements and surface areas are needed to be known, due to their attributes such as accuracy and efficiency.

If one of the system requirements to be met is that the 3D Model can be digitally rotated to offer its view in different angles, multiple laser scanner based systems can be used simultaneously. The speed of the laser scanning will be proportional to the number of systems used, since the simultaneously measurements of the multiple systems do not interfere between them. This laser scanning system attribute differs with the Photogrammetry based systems since they cannot perform the scan operation simultaneously due to the light projections interference, such as Formetric 3D/4D, which makes the speed ratio inversely proportional.

The 3D Rotational Body Scanner increases by 10 times its resolution in comparison with the former 3D Triangulation method. This is possible by using gears that gives 1 rotation per each 10 that gives the step motor. The increase in accuracy given by this improved method can be potentially used in other applications, for example, the scan of small parts of the human body, such as fingers and teeth.

Moreover, the 3D Rotational Body Scanner decreases significantly the mechanical sources of noise, and guarantee less calibration since is a more stable than the former 3D Dynamic Triangulation scanner.

The combination of the photogrammetry method and the 3D dynamic triangulation method could be an interesting area of opportunity. The image modeling phase could be obtained through the photogrammetry techniques and the accuracy and dimensional measurements could be complemented by the improved 3D Rotational Body Scanner system, although this is yet to be explored.

6. References

Azernikov, S.; Fischer, A. (2008). Emerging non-contact 3D measurement technologies for shape retrieval and processing, *Virtual and Physical Prototyping*, Vol. 3, No. 2 (June 2008) pp 85-91, ISSN: 1745-2759.

Básaca, Luis C.; Rodríguez, Julio C.; Sergiyenko, Oleg;. Tyrsa, Vera V; Hernández, Wilmar;. Nieto Hipólito, Juan I; Starostenko, Oleg. 3D Laser Scanning Vision System for

Autonomous Robot Navigation, *Proceedings of IEEE (ISIE-2010)*, Bari, Italy, July 4-7, 2010, pp.1773-1779, ISBN 978-1-4244-6390-9.

Beraldin, J.A.; Rioux, M.; Cournoyer, L.; Blais, F.; Picard, M.; Pekelsky, J (2007). Traceable 3D Imaging Metrology, *Proc. SPIE 6491*, ISBN: 9780819466044, California, USA, January 2007, SPIE, San Jose.

Beraldin, J.A. (2004) Integration of Laser Scanning and Close-range Photogrammetry the Last Decade and Beyond, *ISPRS Journal of Photogrammetry and Remote Sensing*, Volume XXXV, No. B5 (July 2004) pp. 972-983, ISSN: 1682-1750.

Bjarnason, J. E.; Chan, T. L. J.; Lee, A. W. M.; Celis, M. A.; Brown, E. R. (2004) Millimeter-wave, terahertz, and mid-infrared transmission through common clothing, *Applied Physics Letters*, Vol. 85, No. 4, (June 2004) pp. 519 -521, ISSN: 0003-6951.

Boehnen, C.; Flynn, P. (2005) Accuracy of 3D Scanning Technologies in a Face Scanning Scenario, *Proceedings of 3-D Digital Imaging and Modelling*, pp 310-317, ISBN: 0-7695-2327-7, Ontario Canada, June 2005, IEEE, Ottawa.

Calva, D.; Calva, S.; Landa, A.; Rudolph, H.; Lehman1, M. (2009). Face recognition system using fringe projection and moiré: characterization with fractal parameters, *IJCSNS*, Vol.9, No.7, (July 2009) pp 78 – 84, ISSN: 1738-7906.

Creath, K.; Wyant, J. C. (1992). Moiré and Fringe Projection Techniques, *Optical Shop Testing*, Editor: Daniel Malacara pp. 653 – 685, John Wiley & Sons, ISBN: 0-471-52232-5, New York.

Di, W.; Naiguang, L (2008). A pre-processing method for the fringe image in phase measuring profilometry, *Proc. SPIE*, Vol. 6623, No. 66231A, (March 2008) pp 1 – 8, ISSN: 0277-786X.

Emmanuel P. Baltsavias (1999). A comparison between photogrammetry and laser scanning. *ISPRS Journal of Photogrammetry and Remote Sensing*, Vol. 54, No. 2-3, (March 1999) pp. 83 – 94, ISSN: 0924-2716.

Haworth, C.D.; De Saint-Pern, Y.; Clark, D.; Trucco, E.; Petillot, Y.R. (2006). Detection and Tracking of Multiple Metallic Objects in Millimetre-Wave Images, *International Journal of Computer Vision*, Vol. 71, No. 2, (June 2006) pp 183-196, ISSN: 0920-5691

Hierholzer, E. Drerup, B. (1995). High-resolution rasterstereography. *Three-Dimensional Analysis of Spinal Deformities*, Editors: Amico, D' M. Merolli, A. Santambrogio, G.C. pp. 435 – 439, The Netherlands: IOS Press, ISBN: 90-5199-181-9, Amsterdam.

Howald, R.L.; Clark, G.; Hubert, J. Ammar, D. (2007), Millimeter Waves: The Evolving Scene, *Technologies for Homeland Security IEEE Conference on HST*, pp.234-239, ISBN: 1-4244-1053-5, Massachusetts USA, June 2007, IEEE, Woburn.

Leifer, J. (2003) A close-range photogrammetry laboratory activity for mechanical engineering undergraduates, *Frontiers in Education, 2003. FIE 2003. 33rd Annual*, Vol 2, No F2E, (November 2003) pp 7 – 12, ISSN: 0190-5848.

Lerch, T.; MacGillivary, M.; Domina, T. (2007). 3D Laser Scanning: A Model of multidisciplinary research, *Journal of Textil and Apparel, Technology and Management*, Vol. 5, No. 4 (October 2007) pp 1-21, ISSN: 1533-0915.

Lippold, C.; Danesh, G.; Hoppe, G.; Drerup, B.; Hackenberg, L. (2007). Trunk Inclination, Pelvic Tilt and Pelvic Rotation in Relation to the Craniofacial Morphology in Adults, *Angle Orthodontist*, Vol. 77, No 1, (January 2007) pp 29 – 35, ISSN: 0003-3219.

Liu, H.B.; Zhong, H.; Karpowicz, N.; Chen, Y. (2007) Terahertz Spectroscopy and Imaging for Defence and Security Applications, *Proceedings of the IEEE,* Vol. 95, No. 8, (October 2007) pp.1514-1527, ISSN: 0018-9219.

Rivas Lopez, M.; Sergiyenko, O. & Tyrsa, V. (2008). Machine vision: approaches and limitations, In: *Computer vision,* Xiong Zhihui, (Ed.), pp. 395-428. Editorial: IN-TECH, ISBN 978-953-7619-21-3, Vienna, Austria.

Rusinkiewicz, S.; Hall-Holt, O.; Levoy, M. (2002). Real-time 3D model acquisition. *ACM Transactions on Graphics (TOG),* Vol. 21, No. 3, (July 2002) pp. 438 - 446, ISSN: 0730-0301.

Sergiyenko, O.; Hernandez, W.; Tyrsa, V.; Devia Cruz, L.; Starostenko, O. & Pena-Cabrera, M. (2009). Remote Sensor for Spatial Measurements by Using Optical Scanning, *Sensors,* 9(7), July, 2009, MDPI, Basel, Switzerland, pp. 5477-5492. ISSN 1424-8220.

Treleaven, B.; Wells, J. (2007). 3D Body Scanning and Healthcare Applications. *Computer,* Vol. 40, No. 7, (August 2007) pp. 28 - 34, ISSN : 0018-9162.

16

Research and Development of the Passive Optoelectronic Rangefinder

Vladimir Cech[1] and Jiri Jevicky[2]
[1]*Oprox, a.s., Brno*
[2]*University of Defence, Brno*
Czech Republic

1. Introduction

1.1 Basic specification of the problem

The topographical coordinates of an object of interest (the target), which is represented by one contractual point $T = (E, N, H)_T$, need to be determined indirectly in many cases that occur in practice, because an access to respectively the target and the target point T is disabled due to miscellaneous reasons at a given time. Hereafter we will confine to methods that make use of specialized technical equipment (rangefinders) to determine coordinates of the target point T - Fig. 1.

The point $P_{RF} = (E, N, H)_{RF}$ represents a contractual position of the rangefinder in the topographical coordinate system, D_T is the target slant range measured by means of the rangefinder. This value D_T represents the estimate of the real slant range of the target D_{T0} that is equal contractually to the distance of points P_{RF} and T. The angle ε_T is the measured estimate of the elevation of the target ε_{T0} and the angle a_T is the measured estimate of the target azimuth a_{T0}. The coordinates (D, ε, a) are relative spherical coordinates towards the contractual position of the rangefinder which is represented by the point P_{RF}.

The rangefinder is a device that, from the view of Johnson's criterion for optical systems classification, functions to locate the target (target coordinates $(E, N, H)_T$) and usually it also functions to determine motional parameters of the target that are primarily represented by the instantaneous target velocity vector $\mathbf{v}_T$ - Fig. 2.

Typical measured ranges interval for ground targets is from 200 to 4000 m and for aerial or naval targets from 200 to 10000 m or more.

1.2 Passive optoelectronic rangefinder (POERF)

The passive optoelectronic rangefinder (POERF, Fig. 1, 8) is a measurement device as well as a mechatronic system that measures geographic coordinates of objects (targets) selected by an operator in real time (in online mode). In the case of a moving object, it also automatically evaluates its velocity vector $\mathbf{v}_T$ and simultaneously extrapolates its trajectory - Fig. 2.

Active rangefinders for measurement of longer distances of objects (targets), e.g. pulsed laser rangefinders (LRF), emit radiant energy, which conflicts with hygienic restrictions in many applications and sometimes with given radiant pollutions limitations, too. In security and military applications there is a serious defect that the target can detect its irradiation. The use of POERF eliminates mentioned defects in full.

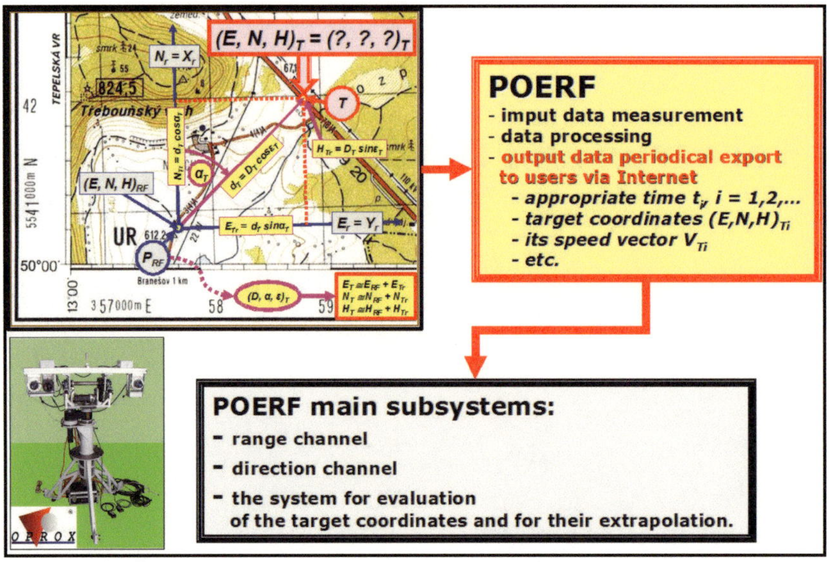

Fig. 1. Input/Output characteristics of POERF (the demonstration model 2009)

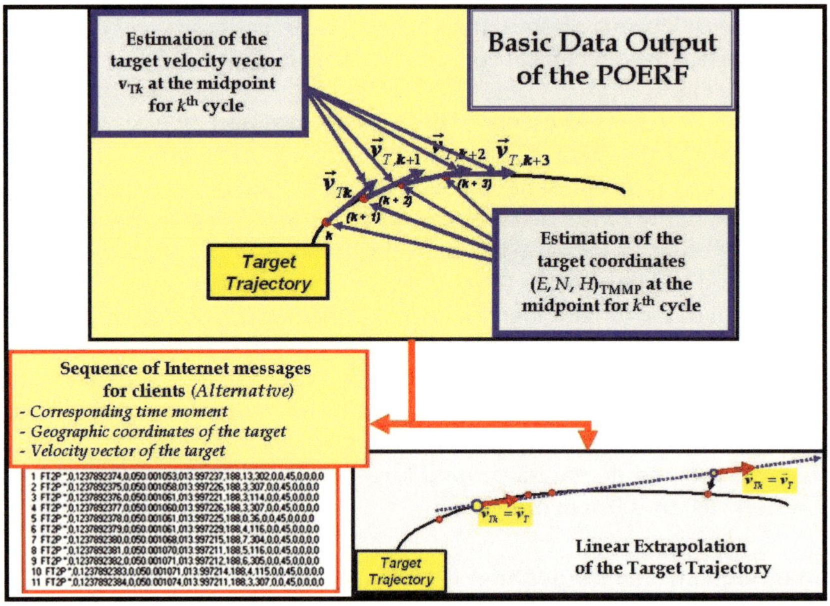

Fig. 2. Principle of measurement of the target trajectory and the data export to users (clients)

The POERF measurement principle is based on the evaluation of information from stereo-pair images obtained by the sighting (master) camera and the metering (slave) one (see the subsection 4.1 and the Figure 9). Their angles of view are relatively small and therefore a spotting camera with zoom is placed alongside the sighting camera – Fig. 1, 9. This spotting camera is exploited by an operator for targets spotting. After operator's steering the cameras towards a target, the shots from the sighting camera serve to evaluate angle measured errors and to track the target automatically (see the section 3).

The POERF is able to work in two modes: online and offline (processing of images saved in memory - e.g. on the hard disc). The offline mode enables to measure the distance of fleeting targets groups in time lag to approx. 30 seconds. The active rangefinders are not able to work in a similar mode (see the section 2).
In general, the POERF continues to measure the UTM coordinates (Fig. 1, 2) of moving target with rate from 10 to 30 measurements per second and extrapolates its trajectory. All required information is sent to external users (clients) via the Internet in near-real-time whereas the communications protocol and the repetitive period (for example 1 s - Fig. 2) are preconcerted. The coordinates can be transformed to the coordinate system WGS 84 and sent to other systems - in accordance with the client's requirement.
Presumed users of the future system POERF are the police, security agencies (ISS - Integrated Security Systems, etc.) and armed forces (NATO NEC - the NATO Network Enabled Capability, etc.).

1.3 The state of POERF research and development, used methods and tools, results

1.3.1 Demonstration model of the POERF

A demonstration model of the POERF (Fig.1, 8) was presented to the opponent committee of the Ministry of Industry and Trade of the Czech Republic within the final opponent proceeding in March 2009. The committee stated that POERF is fully functional and recommended continuing in its further research and development. This chapter will give basic information about the research and development of this POERF demonstration model.
The working range of measured distances is circa from 50 m to 1000 m at the demonstration model (see the subsection 4.1).

1.3.2 Simulation programs Test POERF, Test POERF RAW and the Catalogue of targets

In this chapter the basic possibilities of simulation program Test POERF (see the section 5) are presented. This program serves to simulate functions of the range channel core of the POERF. It allows verifying the quality of algorithms for a target slant range finding from taken stereo pair images of the target and its surroundings. These images are generated as a virtual reality by a special images generator in the program - Fig. 12.
Next, we present consequential simulation software package Test POERF RAW which works with taken images of a real scene (see the section 5, too). The package presently consists of three separate programs: the editing program RAWedi, the main simulation program RAWdis and the viewer RAWpro. The editor RAWedi allows editing of stereo pair images of individual targets and supports the creation of the Catalogue of targets. The simulation program RAWdis serves for testing algorithms for estimation of horizontal stereoscopic disparity (stereo correspondence algorithms) which are convenient for the use in POERF. Simulation experiments can also help to solve problems in the development process of the software for a future POERF prototype.
In publications that deal with problems of stereoscopic disparity determination there is constantly emphasized the deficiency of quality stereoscopic pairs of varied object images, which are indispensable to testing the functionality and quality of various algorithms under real conditions. Considering the POERF specifics, we have decided to create own database of horizontal stereo pair images of targets with accurately known geographic coordinates - shortly the "Catalogue of Targets" (see the section 5 and the Figures 16, 17).

The stationary "targets" (73 objects) were chosen, so that on the one hand they cover slant ranges from c. 100 m to c. 4000 m and on the other hand their appearance and placement should be convenient for unique determination of their stereoscopic disparity - Fig. 17. The number of successive stereo pair images of every target is minimally 512, which is precondition for statistical processing of simulation experiments results.

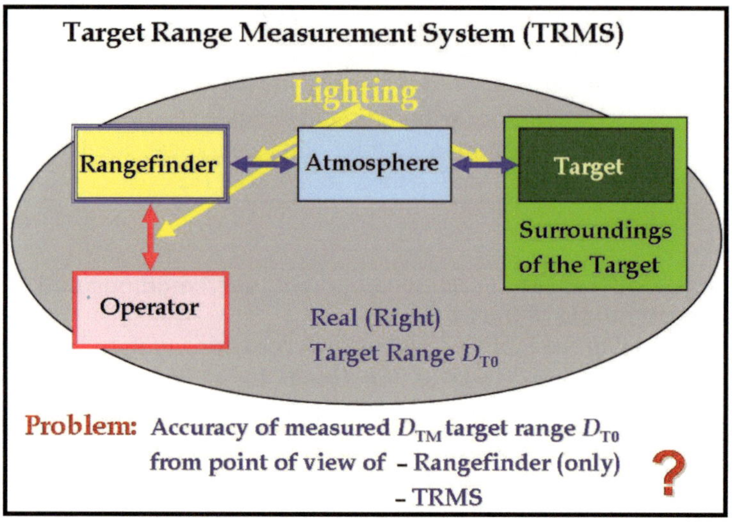

Fig. 3. Target Range Measurement System - TRMS

1.4 What has been done by other researchers

The principle of passive optoelectronic rangefinder has been known minimally since the 50's or 60's of the 20th century (see the subsection 3.1). The development was conditioned primarily by progress in the areas of digital cameras and in miniature computers with ability to work in field conditions (target temperature limit from -40 to +50 °C, dusty environment, etc.) and to realize the image processing in the real-time (frame rate minimally from 5 to 10 frames per second, ideally from 25 to 50 fps).

Our development started initially on a department of Military Academy in Brno (since 2004 University of Defence) in the year 2001 in cooperation with the firm OPROX, a.s., Brno. The centre of the work was gradually transferred into OPROX that has practically been the pivotal solver since 2006 (see the subsection 3.2).

The patents of POERF components have been published since the end of 1950's but there are no relevant publications dealing with the appropriate research and development results. We have not found out that similar device development is being carried out somewhere else. We have found only one agency information that a POERF was developed in Iran (www.ariairan.com, date: 20.7.2008). The problem itself consists particularly in users' unshakable faith in limitless possibilities of laser rangefinders and probably in the industrial/trade/national security directions (see the section 2).

Similar principle is applied to focusing system of some cameras as well as mobile robots navigation/odometry systems. Measured distance range is within order one up to tens of meters, therefore the hardware and software concepts in these systems are different from concepts in the POERF system. Sufficient literature sources cover these problems.

1.5 Future research

At present we have started the new period (2009 - 2012), in which we intend to fully handle the measurement of the target coordinates (for stationary and moving target) inclusive of the target trajectory extrapolation by POERF that can be set *on a moving platform*.

This work is supported by the Ministry of Industry and Trade of the Czech Republic - project code FR - TI 1/195: "Research and development of technologies for intelligent optical tracking systems". Also this chapter has originated under the support of financial means from this project.

2. Target Range Measurement System and the problem of fleeting targets

The accuracy of the target range measurement depends not only on properties of the rangefinder itself, but also on the whole system composed of the rangefinder, the atmosphere, a target, a target's surroundings, an operator and lighting - Fig. 3. Dependability and accuracy of the range measurement is characterized especially by the use of

- the probability of successful measurement of "whatever" range p_M (estimated by the relative frequency),
- the (sample) mean of measured range D_{T0} (resp. D_{Taver}),
- the (sample) standard deviation of measured range σ_D (resp. s_{TM}),
- the (sample) relative standard deviation of measured range $\sigma_{DR} = \sigma_D / D_{T0}$ (resp. $s_{DR} = s_{TM}/D_{Taver}$) and
- the probability p_D of the right (real) target range measurement, i.e. a range from the interval $\langle D_{T0} - \Delta D, D_{T0} + \Delta D \rangle$, where ΔD is chosen in compliance with the concrete situation, e.g. 10 m or 50 m.

Instead of the (sample) standard deviations σ (resp. s), corresponding probable errors E (resp. e) are often used. It is valid for normal distribution

$$E \approx 0.6745 \cdot \sigma \,. \tag{1}$$

The value of the relative probable error E is usually required less than 2 to 4% in a requisite interval of ranges under good conditions - daylight and meteorological visibility s_M (or MOR - meteorological optical range) over 10 km. This error is regarded as the error of appropriate Target Range Measurement System (TRMS), because the same error for measurement by the means of customary stadia methods (en.wikipedia.org/ ... /stadiametric_ rangefinding - targeting reticle) is usually 7 to 15% (in dependence on the operator training and tiredness; it is valid under nocturnal conditions, too).

In the case of pulsed laser rangefinders (LRF), the value ΔD = 5, 10 or 15 m is frequently adduced as the indicator of their accuracy and, due to *advertising reasons*, it evokes the notion, that the probability p_D is almost 100% for the appropriate range interval and that is valid also for LRF maximal working range, e.g. 8 or even 20 km, and that it is the characteristic of the whole TRMS. We will explain shortly, what the reality is.

The precondition for range measurement by means of LRF (it is valid similarly for all active rangefinders - also radars, sonars) is the target irradiance by emitted laser beam - Fig. 4. The contractual target point T always lies on the beam axis. The usual divergence 2ω of LRF beam is from 0.5 to 1 mrad and for eyesafe LRF (ELRF) is lesser - circa to 0.3 mrad. In the case of fleeting target (the target is appearing surprisingly on shot time periods), it is extremely difficult - or quite impossible - to aim at such target accurately enough and to realize the measurement. In the frequent case of relatively small target (e.g. a distant one), a very small part of the beam cross-section area falls to the target and the rest falls on the target surroundings - Fig. 4, 5. So, an estimate of surroundings range D_{N0} is usually measured, but the system is not able to distinguish it. This range is then presented as the estimate of the target range D_{T0}. It is a gross error of measurement. LRFs are equipped with a certain cleverness that allows helping in the gross error detection. Operator's experience is its fundamental. Nevertheless, these systems fail practically in the case of fleeting targets.

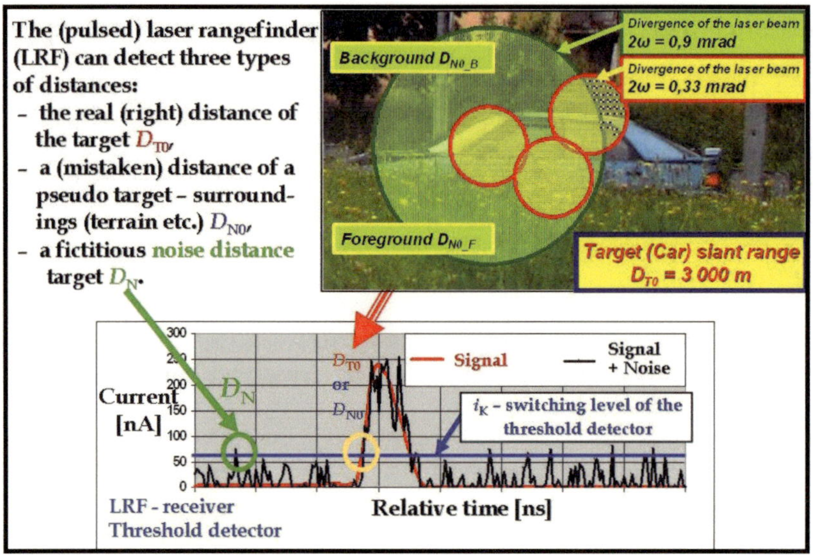

Fig. 4. Principle of influence of the laser beam divergence on the occurrence of gross errors in the target range measurement; more closely in (Apotheloz et al., 1981; Cech et al., 2006)

As clarified above, the aiming accuracy is decreasing in the cases of a fleeting target and an increase of tiredness and nervousness of the operator. The aiming accuracy will be characterized by the standard deviations in elevation σ_φ and in traverse (line) σ_ψ. We will assume a circular dispersion and hence $\sigma_A = \sigma_\varphi = \sigma_\psi$ is the (circular) standard deviation of ELRF. The example in the Figure 5 is from (Cech & Jevicky, 2005). It follows evidently, that the probability p_D of the right target range measurement depends significantly on the meteorological visibility and on the aiming accuracy.

The decrease of p_D under increasing range corresponds with the increase of the relative standard deviation σ_{DR}, and it is substantially greater than 5 or 10 m, as it can be incorrectly deduced from advertising materials.

However, it generally holds that the use of ELRF with the divergence of laser beam $2\omega < 0.5$ mrad requires the utilization of systems for aiming and tracking the target with extreme accuracy of the level $\sigma_\varphi \approx \sigma_\psi \approx \sigma_A \leq (0.1 \text{ to } 0.2)$ mrad.

Mentioned problems can be overcome by the use of POERF, which is able to work in both modes - online and offline. It is sufficient for measuring the target range that the target is displayed in fields of view of both cameras (sighting and metering), whereas their angles of view are in compliance with the system determination from 1.5° to 6° and therefore relative large aiming errors are acceptable.

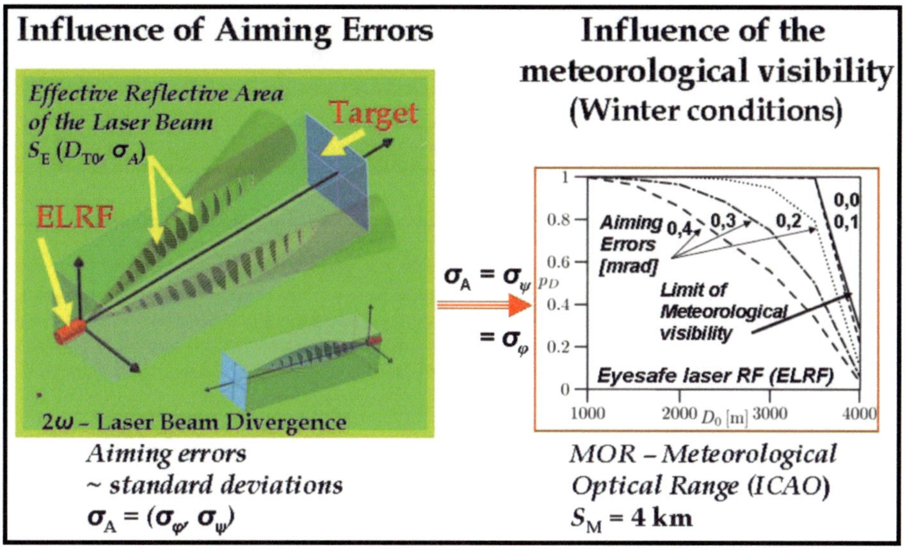

Fig. 5. Simulation experiment outputs - example (Cech & Jevicky, 2005): target 2.3 × 2.3 m and reflectance $\rho(\lambda) = 0.1$ for $\lambda = 1.54$ µm; $2\omega = 0.33$ mrad; $\Delta D = 5$ m

3. Short overview of the optical rangefinders evolution

3.1 General development of optical rangefinders

We will only deal with a subset of optical rangefinders (see en.wikipedia.org/… /Range_imaging), especially those ones which are based on measuring of parameters of the telemetric triangle lying in the triangulation plane and on consequential computation of estimate of the target slant range D_T. It is a special task solved within the frame of photogrammetry - more details in (Kraus, 2000), (Hanzl & Sukup).

These rangefinders are usually divided into three main groups: with the base in the ground space, with the base in the device (inner base) and with the base in the target.

Henceforth, we will not deal with rangefinders with the base in the target - see more details in (en.wikipedia.org/…/stadimeter).

The oldest system is an optical range-finding system with the base in ground space - Fig. 6. Ever since antiquity two "theodolites" placed at ends of the base have been used. It is possible to use only one theodolite which is transferred between ends of the base. A short history of theodolite development can be found in (Wallis, 2005).

Special theodolites (photogrammetric tracking theodolites) were progressively developed for measuring immediate positions of moving targets. They can be divided into two groups: without and with continuous recording of measurement results. Theodolites without continuous recording of measurement results were used for measuring positions of ships (en.wikipedia.org/…/Base_end_station), balloons and airplanes (Curti, 1945).

Theodolites with continuous recording of measurement results were used since 1930s for measuring positions of balloons (e.g. Askania Recording Balloon Theodolite - pibal theodolite), airplanes (en.wikipedia.org/…/Askania; e.g. Askania Cinetheodolite - kine-theodolite), (Curti, 1945) and projectiles (Hännert, 1928; Curti, 1945). The basis of these kine-theodolites was a special movie-picture camera. In connection with measuring positions of flying projectiles the term ballistic photogrammetry is used. Besides theodolites with photographic registration, the ballistic cameras have been used for measuring positions of flying projectiles since 1900s (Hännert, 1928; Curti, 1945), e.g. Wild BC2 Ballistic Camera (since 1938), whose basis is a still camera modified for multiple repeated exposition of the projectile position on the same photographic plate.

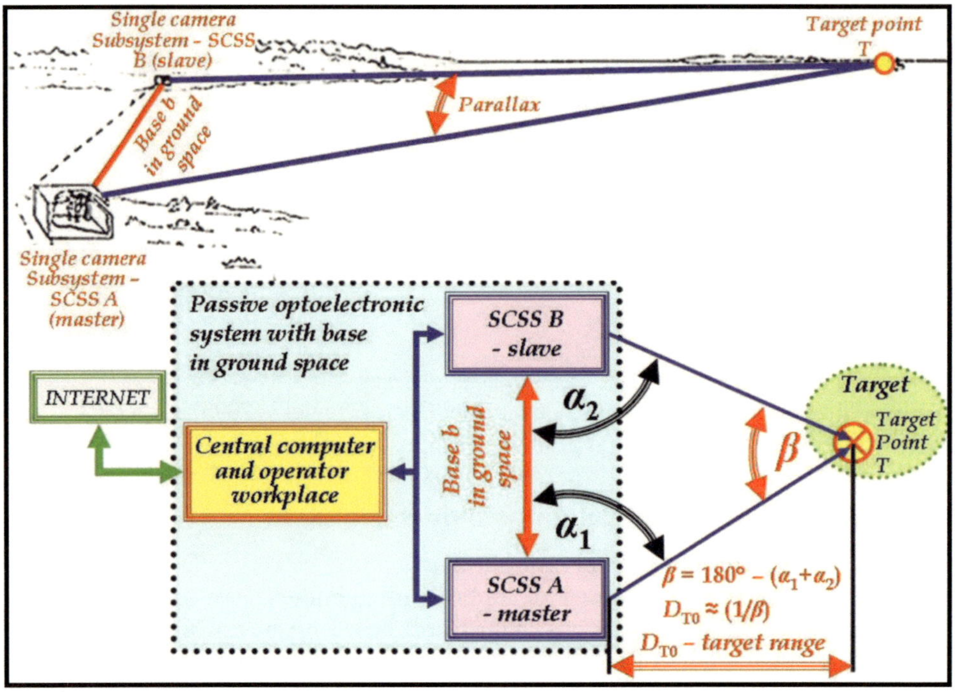

Fig. 6. Principle of the Optical Rangefinding System with the base in ground space

The next developmental step since 1940s (reference resources are not for disposal) could have been the usage of theodolites with cameras with video camera tube (pickup tube) (en.wikipedia.org/ .../Video_camera), which have already made possible the picture watching on CRT monitor. Since 1956 there has been a possibility to record the picture on a video tape recorder - VTR (en.wikipedia.org/.../Video_recorder).

Our task (see the subsection 1.5) is the development of a single camera subsystem - Fig. 6 - with the usage of digital camera and Tit and Pan Device (System, Assembly).

Optical rangefinders with the base in the device are divided into coincidence and stereoscopic rangefinders. The production of both types started already in 1890s. The first coincidence rangefinders were made by Scottish firm Barr and Stroud (Russall, 2001). The first stereoscopic rangefinders were made by German firm Zeiss. Theory, projection and adjustment are published in (Keprt, 1966). The construction principles and utilization of these rangefinders can be found in (Composite authors, 1958; Curti, 1945). One of the first constructions of POERF is described in (Gebel, 1966). It is a modification of a coincidence rangefinder with the utilization of one piece of a special pick-up transducer tube (U.S. Patent 2 969 477, author Gebel, R. K. H.). U.S. Patent presupposing utilization of two televisions sensors (Gilligan, 1990), which is a modification of stereoscopic rangefinder, adverts to older patents, whereas the oldest patent is U.S. Patent 2 786 096 Television rangefinder (Palmer, march 1957). Subsequent patent applications of POERF presuppose the usage of linear array CCD sensors, for instance the application No. PCT/AU1990/000423 Passive-Optoelectronic Rangefinding. Patent applications presupposing the use of digital matrix sensors (CCD or CMOS) have not been found till now.

The first commercially offered CCD sensor (100 × 100 pixels) was produced by the firm Fairchild Imaging in the year 1973. The first really digital cameras did not originate until the half of 1980s. The serial cameras with resolution e.g. 640 × 480 pixels were not offered until the half of 1990s.

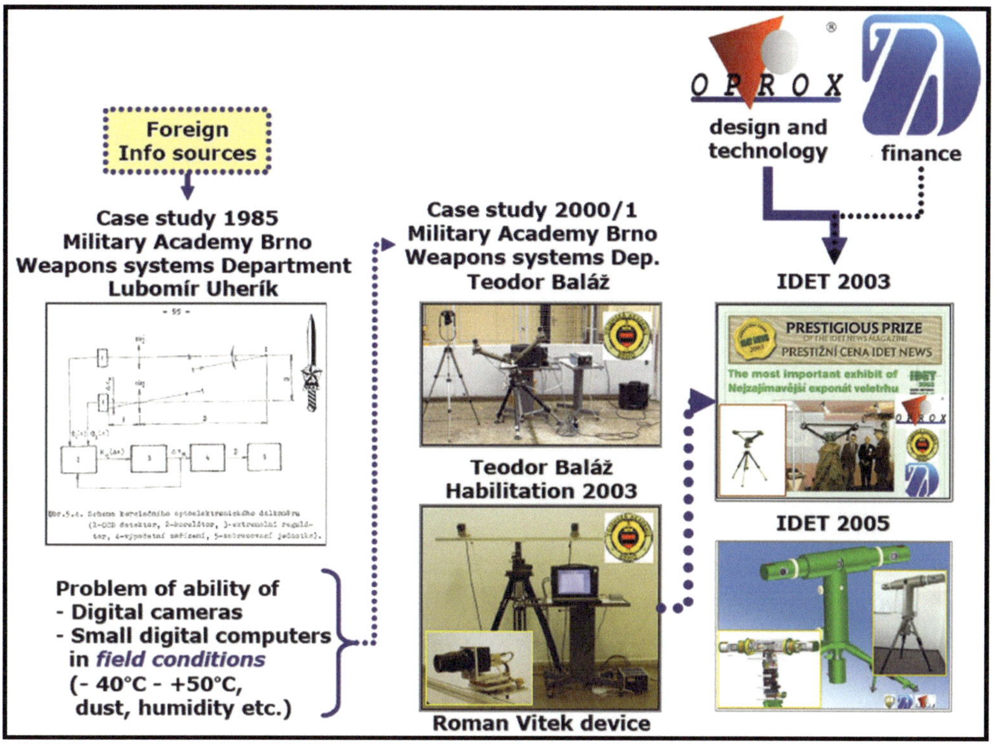

Fig. 7. Brief summary about the research and development of POERF in the Czech Republic

If we observe the development of sufficiently small and efficient computers whose construction is resistant to environmental influences, we find out that they appeared in the market as lately as the second half of 1990s.

According to the article (Jarvis, 1983), one of the oldest algorithms for stereoscopic disparity finding from which the estimate of the target range is computed - the cross-correlation algorithm - was already published e.g. in (Levine et al., 1973). The fundamental classification and comparison of algorithms for finding of stereoscopic disparity can be found for instance in (Scharstein & Szelisky, 2002). The date of this publication corresponds to the period when the basic hardware means (cameras and computers) have begun to satisfy requirements for the construction of components for fully digital POERF.

3.2 POERF development in the Czech Republic

The development of the passive optoelectronic rangefinder has proceeded in the Department of Weapon Systems of the Military Academy in Brno (since the year 2005 the

Department of Weapons and Ammunition of the University of Defence) and in firms cooperating with the department, especially in the firm Oprox, a.s.
Based on the study of foreign sources, the fundamental properties of POERF were analyzed in the study (Uherik et al, 1985) - Fig. 7. The research and development of POERF started as late as the year 2001 after accomplishment of the objective properties.
The development can be divided into three periods as it is shown in the Figures 7 and 8.

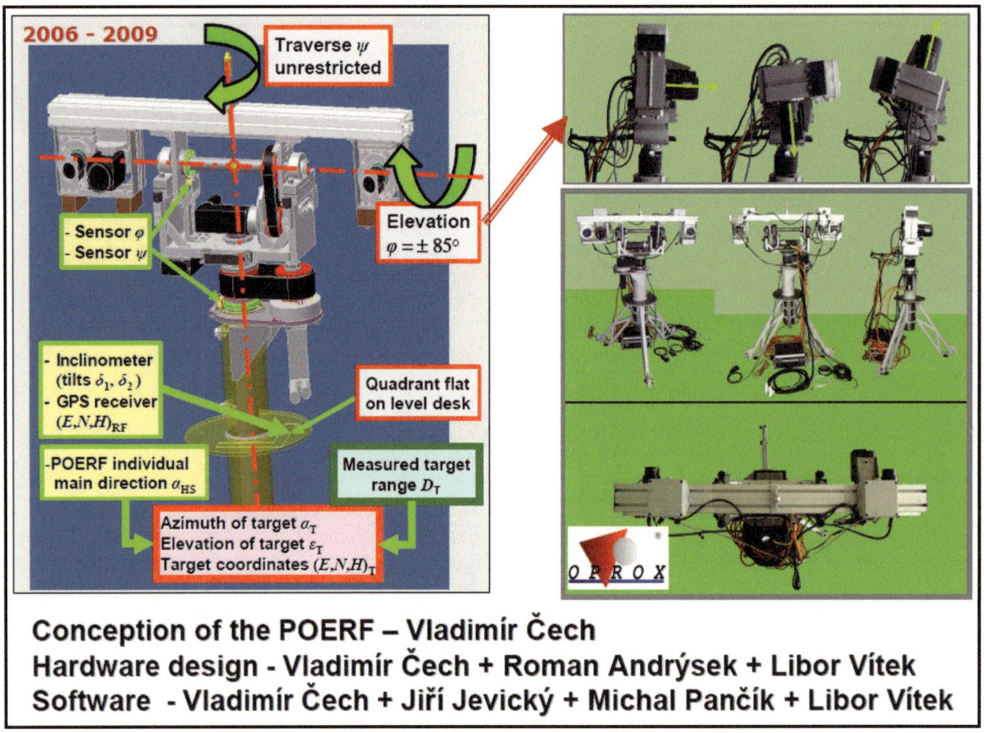

Fig. 8. The third demonstration model of POERF from the year 2009; the project manager was Jozef Skvarek

In the first period (2001 to 2003), the basic principles were verified (Balaz, 2003). The first demonstration model of POERF was awarded in the 7th International Exhibition of Defence and Security Technologies and Special Information Systems in Brno (IDET 2003). The development was supported during a certain period by the firm Z.L.D., s.r.o., Praha.
In the second period (2003 to 2006), the technology of the range measurement of a stationary target was handled (Skvarek, 2004). The second demonstration model was developed and introduced at the exhibition IDET 2005. Authors of this chapter have joined in the research and development of POERF in the year 2003.
In the third period (2006 to 2009), the measurement of coordinates of a moving target and its trajectory extrapolation (Cech et al., 2009a) belonged among main extensions of POERF functionality. Starting this period, the crux of the work was transferred into the firm Oprox, a.s. The third demonstration model was the final result of the research and development in this period - Fig. 8.
At present we have started the fourth period (see the subsection 1.5).

4. POERF – demonstration model 2009

From the system view, the POERF as a mechatronic system is composed of three main subsystems, Fig. 1, 8 (Cech & Jevicky & Pancik, 2009d):
- the range channel,
- the direction channel and
- the system for evaluation of the target coordinates and for their extrapolation.

The task of the range channel is on the one hand automatic recognition and tracking of the target which has been selected by the operator in semiautomatic regime and continuous measuring of its slant range D_T (c. 10 measurements per second at present, which is identical to cameras frame rate) and on the other hand the evaluation of angle measured errors (e_φ, e_ψ) that are transferred to input of the direction channel - Fig. 12.
The direction channel - its core consists respectively of two servomechanisms and of special Pan and Tilt System (Device, Assembly) - ensures continuous tracking of the target in the automatic and semiautomatic regime and measuring of angle coordinates of the target (the elevation φ and the traverse ψ) - Fig. 1, 8. The elevation range is c. ±85° and the traverse potential range is not limited - Fig. 8. The real range of the traverse is limited to c. ±165° by two terminal sensors due to safeguard protection of cables - Fig. 8. The optical sensors SIGNUM™ RESM 20 μm by the firm RENISHAW® are used for the detection of elevation and traverse. The spherical coordinates of the target (D_T, φ, ψ) are transformed into the UTM coordinates by the system for evaluation of the target coordinates and their extrapolation - Fig. 1, 2, 15. Withal, the knowledge of the POERF geographic coordinates $(E, N, H)_{RF}$ and the POERF individual main direction a_{HS} (Fig. 8) is utilized. In the case of moving target, required extrapolative parameters are consecutively evaluated (coordinates of the measurement midpoint, corresponding time moment and the velocity vector of the target). The extrapolative parameters (UTM coordinates of the target are transformed into geographic coordinates WGS 84) are sent periodically to a user in near-real-time (at the present with the period 1 second, i.e. the data "obsolescence" is c. 0.5 seconds) - Fig. 2, 15.
POERF must be adjusted so that the traverse axis is vertical. Due to this, the setscrews are situated in the bottom ends of the support legs - Fig. 8. The main tool for the adjustment is the level or the quadrant which can be placed on the quadrant flats on the level desk. Two inclinometers placed perpendicularly to each other will be used in the future - Fig. 8.
The demonstration model 2009 works only in the online mode.

4.1 Range channel

As mentioned above, the main task of the range channel is on the one hand automatic tracking of the target which has been selected by the operator in semiautomatic regime and continuous measuring of its slant range D_T and on the other hand the evaluation of angle measured errors that are transferred to the input of the direction channel.
The core of hardware consists of three digital cameras fixed through adjustable suspensions to the cameras beam - Fig. 1, 8, 9. The camera of type Basler A101p (image size 2/3"; C Mount; monochromatic CCD sensor SONY IXL085AL with 256 brightness levels, the number of columns is $n = 1\,300$, $c = 1, 2, \ldots, n$; the number of rows is $m = 1\,030$, $r = 1, 2, \ldots, m$; square pixels $\rho(c) = \rho(r) = \rho = 6.7$ μm) was chosen for the sighting and metering cameras. The type IQ 753 by the firm IQinvision (image size 1/2"; CS Mount; the number of columns is $n = 2048$, the number of rows is $m = 1536$, square pixels $\rho(c)$

$= \rho(r) = \rho = 3.1$ µm, exploited 256 monochromatic brightness levels) was used as the spotting camera.

The algorithm for computation of estimate of a slant range D_T is based on solution of the telemetric triangle that lies in the triangulation plane - Fig. 9, 10. The input data are ordinal numbers c_{T1}, c_{T2} of columns of matrix sensors in which images T'_1, T'_2 of the target point T are projected. In particular, it is sufficient to determine their difference Δc_T (horizontal stereoscopic disparity) that is proportional to the appropriate parallactic angle β. Therefore algorithms for computation of estimate of the difference Δc_T are crucial (the correspondence problem algorithms). We work with algorithms for an estimate of Δc_T, which involve the definition of 2D model of the target image (shortly "target model"). We use a rectangular target model for the present (Marik et al., 2003). This size (in pixels: rows × columns = $(2m_M + 1) \times (2n_M + 1)$) is adjustable (the default setting is 51 × 51 pixels) - Fig. 12. Apex of the main aiming mark lies always in the centre of the target model - Fig. 12. The contour of the target model is not displayed in the image from the sighting camera - Fig. 13.

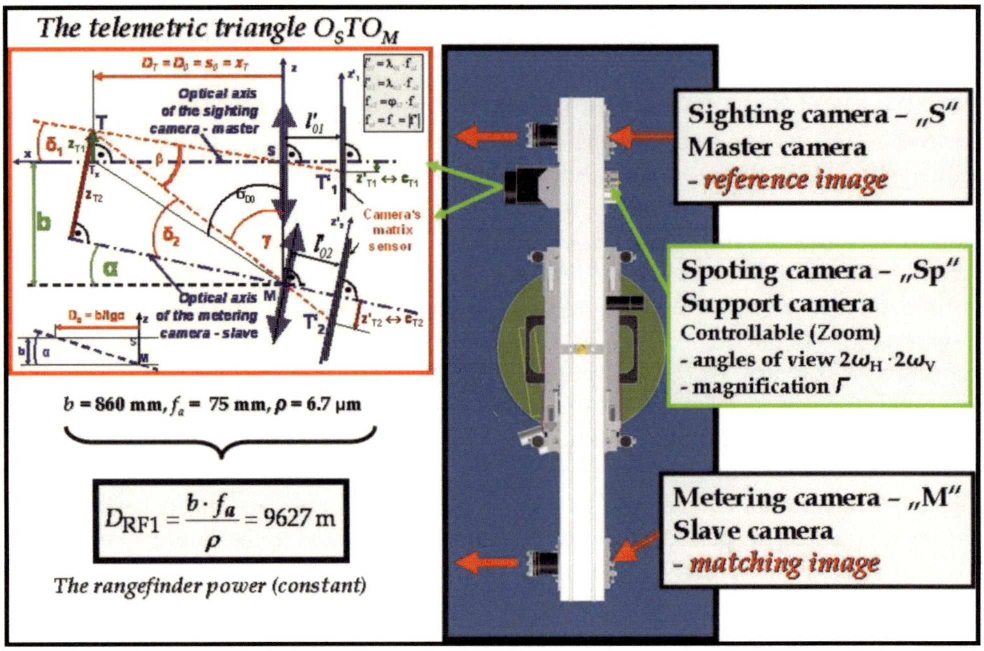

Fig. 9. Basic arrangement of the range channel hardware

The positive value $d_T = C_{ORF} - \Delta c_T$ is usually regarded as the disparity. The sign convention is elected so that $\Delta c_T \geq 0$ is valid for $D_T \geq D_a$ - Fig. 9, 10, where $D_a = b/\tan a_\Sigma$. The size of the convergence angle a_Σ (resp. a) - Fig. 9, 10 - is chosen with respect to the requirement that the measurement of the given minimal range D_{Tmin} of the target should be ensured. In our case D_a = c. 50 m. The columns $c_{20} \approx c_{10} \approx 1300/2 = 650$ determine the horizontal position of the principal points of autocollimation/projection. If the target is in infinity (the Sun, the Moon, stars), then its disparity is just $\Delta c_T = C_{ORF}$. The rated value $C_{ORF} = 190.317$ pixels - Fig. 10.

The rangefinder power (constant) D_{RF1} is the basic characteristics of potential POERF accuracy - Fig. 10, 11. With increasing value of the power, the accuracy of measurement increases too. The power of POERF demonstration model is $D_{RF1} = 9627$ m - Fig. 9, 10. The size of D_{RF1} depends on the width of rows of pixels $\rho(c)$, on the absolute value of the image focal length f_a and on the size of the base b.

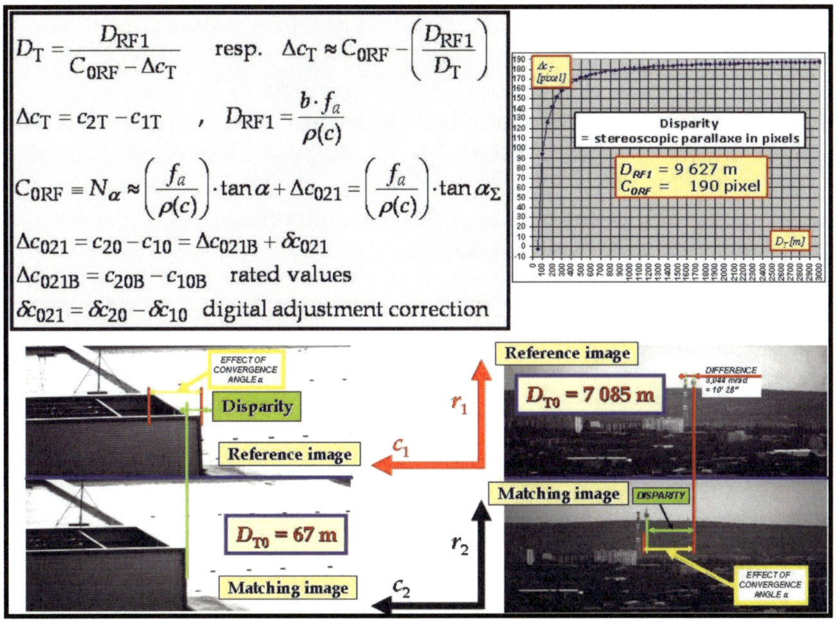

Fig. 10. The main relations for computation of the estimate of the target slant range D_T from the estimate of the horizontal disparity Δc_T (the coordinates $(r, c)_i$, $i = 1, 2$ are coordinates of the digital matrix sensors of the sighting and metering cameras)

The choice of the size $\rho(c)$ (resp. ρ) is a compromise between the effort to achieve the maximum potential well depth, which is increasing with the size of ρ, and the minimal image size of the sensor, whereas many other demands on the camera parameters must be reflected. The choice of size of the (absolute value) image focal length f_a results from the requirement that the sorted type of the target (e.g. passenger vehicle) must be identifiable in the requisite maximum spotting range $D_{T\_spot\_max}$ of the rangefinder ($D_{T\_spot\_max} \geq D_{Tmax}$ - the maximum working range). In accordance with Johnson criterion (50% successfulness of the target identification under excellent meteorological visibility $s_M \geq 10$ km), the target has to be displayed minimally on 16 times 16 pixels (Holst, 2000), (Balaz et al., 1999). In practice, the resolution of the target image should be minimally 32 times 32 pixels (Cech et al., 2009).
The real maximum spotting range of the sorted target type $D_{T\_spot\_max}$ depends simultaneously on the up-to-date horizontal meteorological visibility s_M. The final choice of the value f_a is influenced by the demands imposed on the lens. It affects chiefly the size of angles of view and these angles determine potential possibilities of POERF in the offline mode (see the section 2). For example, it is valid for the horizontal angle of view

$$2\omega_H = 2 \cdot \arctan\left(\frac{n \cdot \rho(c)}{2 \cdot f_a}\right). \tag{2}$$

It is evident from this relation that it is advantageous to use the camera with sensor with a large number of columns n. The lenses PENTAX B7518E (1″ format Auto-Iris DC, C Mount;

f_a = 75 mm, the minimal aperture ratio a_{min} = 1.8) have been chosen for the sighting and metering cameras. Their horizontal angle of view is 6.65° and vertical 5.27°. The lens PENTAX H15ZAME-P with the zoom 1 to 15× (1/2″ format Auto-Iris DC; C Mount; 15× Motorized zoom - DC, f_a = 8 to 120 mm, minimal aperture ratio a_{min} = 1.6 (resp. 2.4)) has been chosen for the spotting camera.

The last parameter that influences the size D_{RF1} is the length of the base b. Its size is selected with respect to the demand for accomplishment of requisite size $D_{T0} = D_{Tmax}$ - the maximum working range, in which the relative size of the probable error E_{DR} of the range measurement attains the given size, e.g. 3% - Fig. 11. The size of the base b depends simultaneously on the size of the standard deviation $\sigma(c)$ (resp. σ_C) of determination of the disparity Δc_T corresponding to the range D_{Tmax}.

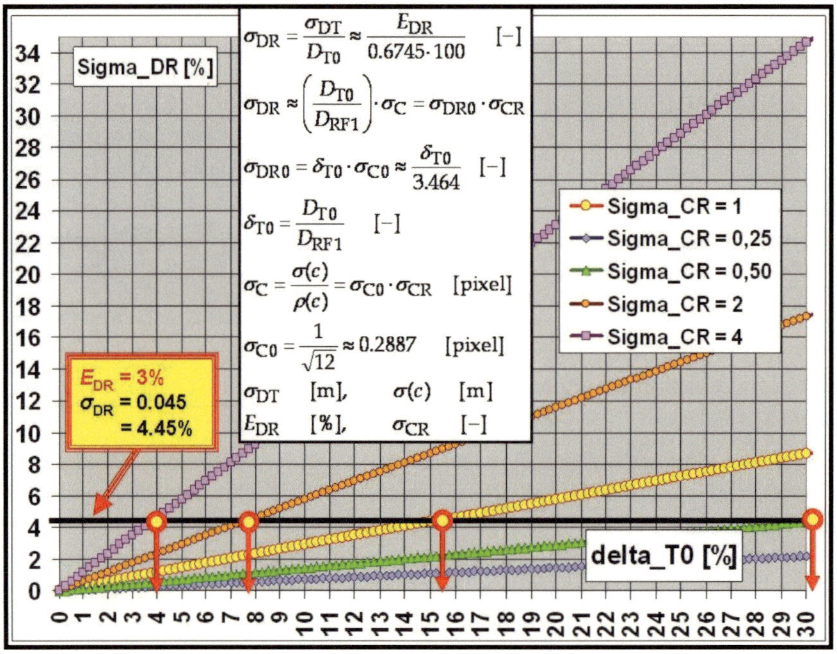

Fig. 11. The main relations for estimate of POERF accuracy

As its basic (standardizing) value σ_{C0} can be elected the standard deviation that originates always in finding the *integer* value of the disparity Δc_T as an unrecoverable discretization (quantizing) noise with the uniform distribution on the interval of the length just one pixel. Then it is $\sigma_{C0} \approx 0.2887$ - Fig. 11. Instead of values σ_C, their relative values σ_{CR} can be used as well. The value σ_C (resp. σ_{CR}) is the quality indicator for appropriate hardware and software of the POERF, especially for algorithms for estimates of sizes of the disparity Δc_T under given conditions (meteorological visibility, atmospheric turbulence, exposure time, aperture ratio, motion blur, etc.). If the value of σ_{CR} increases twice, then it is necessary to elongate the base b also twice with a view to preserve the requisite value D_{Tmax}. Whence it follows that the quality of hardware and software immediately influences the POERF sizes that are directly proportional to the size of base b. The used base is 860 mm long.

The actual values of constants D_{RF1} and C_{0RF} are determined during manufacturing and consecutively during operational adjustments. The adjustment is realized under utilization of several targets whose coordinates are known for high accuracy. The appropriate measurements are processed statistically with the use of the linear regress model (the component part of POERF software - Fig. 14). For example, it was determined for targets 1 to 33 from the Catalogue of Targets (see the section 5, the Figures 16, 17) and for integer estimates of the disparities that $D_{RF1} = 9\,215.5$ m and $C_{ORF} = 195.767$ pixels (the correlation coefficient $r = -\,0.999\,725$).

The starting situation in the process of a target searching and tracking can be characterized as follows. The operator has only common information that a potentially interesting object (the future target) could be in a given area. In the first period, the operator (sometimes with the help of other persons) usually searches an odd object in the area under interest with his eyes only or with the use of tools, e.g. field-glasses, and also with the help of POERF that works in the regime "searching" in which the angles of view of the spotting camera are sufficiently large (ideally c. 40° to 50°).

As soon as the target is identified and localized, the first period is closed and the second period starts. The operator creates the first estimate of model of the target on the monitor from the image provided by sighting camera (master) and passes on the computer. Sizes of the first estimate of the target model must be sufficiently large - under aiming errors that correspond to the actual situation and that are characterized by standard deviations in the elevation σ_φ and in the traverse σ_ψ - because the operator needs to place the real target into the area of model of the target reliably - Fig. 12. Whenever he thinks that he attains it in the process of sighting and tracking of the target, he pushes the appropriate button (Fig. 13) and thus he passes the target model to the use in algorithms of automatic tracking of the target and measuring of its range.

In the third period, the target position and its range are evaluated automatically. The operator tries to reduce sizes of the target model (the POERF demonstration model 2009 does not enable it) and to place it again on the target. In the case of success, he pushes the appropriate button and the system starts the exploitation of a new target model. The whole process is supported by automatic stabilization of positions of optical axes of cameras and eventually also by additional stabilization of the image on operator's monitor (it is not implemented in the demonstration model 2009). The operator can terminate this process as soon as the target model includes pixels with only a part of image of the real target. Complications are caused by objects which are situated in front of the target and are badly visible, e.g. branches of bushes and trees, the grass, but also raised dust. The operator consequently monitors automatically proceeding process. He enters into it in the case of disappearance of the target behind a barrier for a longer time. In the case that information about extrapolated future position of the target is exploited, a short disappearance of the target can be compensated by the automatic system (not implemented in the demonstration model 2009). The level of algorithm ability to learn will determine if the operator's intervention is needful in the case of the target turning to markedly other position towards the POERF.

The program for automatic tracking of a target is based on the utilization of procedures from the library Open CV, specifically on a modification of Lucas Kanade algorithm (Bouguet, 1999). If the target disappears momentarily behind a barrier, then the algorithm collapses. The operator must intervene as it is explained above.

The program starts its functions as soon as the operator pushes the button "Start Measurement" or "Start Tracking". The algorithm then finds the nearest corner (of an object) to the apex of the main aiming mark in the shot from the sighting camera. This point is considered the image T'_1 of the target point T but only due to needs of the target tracking - Fig. 9. Consequently, just two last consecutive shots from the sighting camera are processed. With utilization of Lucas - Kanade Feature Tracker algorithm (Bouguet, 1999) for evaluation of the optical flow, the position of the corner - the point T'_1 - is always estimated in the consequent shot with a subpixel accuracy. The algorithm is robust and that is why it can cope with a gradual spatial slew of the target. The algorithm simultaneously highlights in the image on the monitor the points, which have been identified as appurtenant to the moving target, so that the operator has in his hands the screening control over the system activity. In the case of problems, it is necessary to use 2D model of the target as mentioned above. The point T'_1 is at the same time considered the aiming point T_{AP} - Fig. 12, and so the control deviations (e_φ, e_ψ) are evaluated (as measured errors of angles) for the direction channel control - Fig. 12.

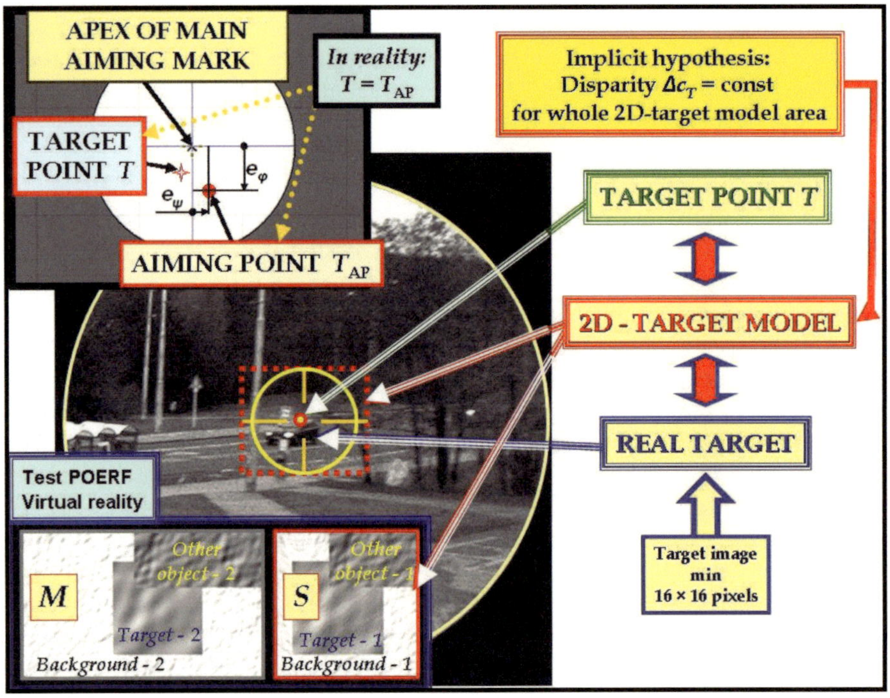

Fig. 12. Relation among the image of the real target in the sighting camera, the target point T and 2D model of the target

The maximum computing speed is required primarily, in order that about 30 range measurements per second are necessary in our applications (POERF). Therefore, we prefer simple (and hence very fast) algorithms. Random errors of measurements are compensated during statistical treatment of measurement results (extrapolation process).

The matching cost function $S(k)$ is used in the meantime (in general it is pixel-based matching costs function) - the sum of squared intensity differences SSD (or mean-squared error MSE) (Scharstein & Szelisky, 2002). The computation of matching cost function $S(k)$ proceeds in two steps.

Firstly, its global minimum with one-pixel accuracy is calculated (the tabulation over all admissible horizontal shifts of the 2D target model on the matching image). Simultaneously, the constriction for the choice of the global minimum - known as the Range Gate (Cech et al., 2009) - is applied.

In the second step, the global minimum is searched with sub-pixel accuracy while using the polynomial approximation (the interpolation and the least-squares method can be alternatively used) in the neighbourhood of the integer point of the global minimum, which has been found in the first step.

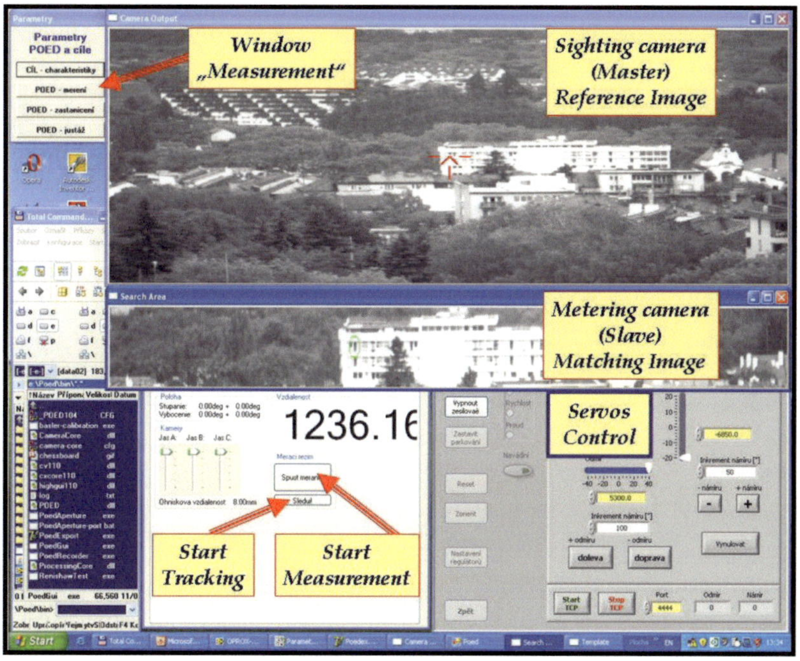

Fig. 13. Five basic windows for the operator during the target spotting, aiming and in the beginning of the measurement or tracking

While using above-mentioned algorithms, it is always presumed that the same disparity Δc_T = const is for all pixels of 2D target model - Fig. 12. This precondition is equivalent to the hypothesis that these pixels depict immediate neighbourhood of the target point T representing the target and this neighbourhood appertain to the target surface (more accurately all that is concerned the image T'_1 of this point and its neighbourhood). These algorithms belong to the group referred to as local, fixed window based methods.

Adduced precondition can be frequently satisfied by a suitable choice of size and location of the target model (i.e. by the aim of a convenient part of the target). The choice is performed iteratively by the operator for the real POERF.

Usual shapes of a target surface (e.g. balconies on a building facade, etc.) have only a little influence on the above-mentioned precondition violation, because the range difference generated by them is usually less than 1 to 2 percent of the "average" target slant range D_T evaluated over the target surface represented by the 2D target model.

Operations over the set of pixels of the 2D target model that generate the matching cost function $S(k)$ and the procedure of its minimum searching can be counted as a definition of special moving average, and - as a consequence - the whole process appears as a low-frequency filtration.

It holds generally that if respectively the meteorological visibility is low and the atmosphere turbulence is strong, then it is necessary to choose a larger size of the 2D target model, i.e. to filter out high spatial frequencies loaded by the largest errors and to work with lower spatial frequencies.

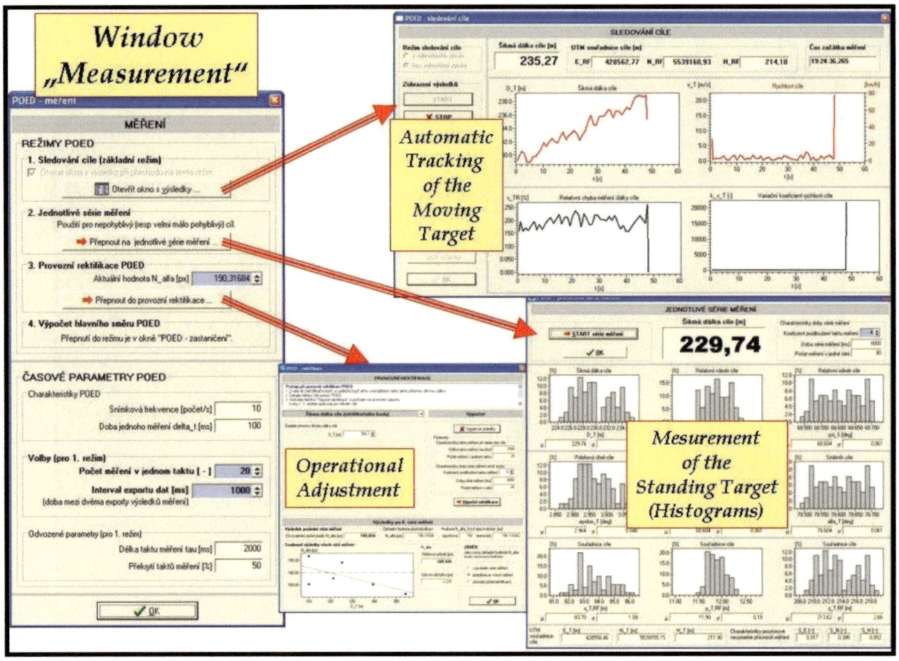

Fig. 14. Auxiliary windows for monitoring of respectively the measurement and the adjustment

In many cases it is inevitable that some pixels of the 2D target model record a rear object or a front object instead of the target - Fig. 12. Simulation experiments with the program Test POERF showed (Cech & Jevicky, 2007) that farther objects have minimal adverse impact on the accuracy of the range measurement, contrary to nearer objects that induce considerably large errors in the measurement of the target range. From the problem merits, these errors are random blunders. Their greatness depends on the mutual position of the front object and the target - Fig. 12. This finding has been also verified in computational experiments by the help of the program RAWdis (Cech & Jevicky, 2010b). It is a specific particular case of more common problem that is known as occluded areas; the specific case is the result of the depth discontinuity (Zitnick & Kanade, 2000).

It is evident from the above that the choice of the position and the size of 2D target model is not a trivial operation and it is convenient to entrust a man with this activity. The operator introduces a priori and a posteriori information into the measurement process of respectively the disparity and the range of a target and this information can be only hardly (or not at all) obtained by the use of fully automatic algorithm.

Algorithms commonly published for the stereo correspondence problem solving are altogether fully automatic - they use the information included in the given stereo pair images, eventually in several consecutive pairs (optical flow estimation). Therefore, it is possible to get inspired by these algorithms, but it is impossible to adopt them uncritically.

In conclusion it is necessary to state that these automatic algorithms are determined for solving the dense or sparse stereo-problems, whereas the algorithms for POERF estimate the disparity of the only point - the target point *T*, but under complicated and dynamically varying conditions in the near-real-time.

4.2 Direction channel

The purpose of the direction channel is already mentioned above.

The core of the direction channel (Cech et al., 2009a) comprises two independent servomechanisms for the elevation φ and the traverse ψ - Fig. 8. Identical servomotors and servo-amplifiers by the firm TGdrives, s.r.o., Brno were used there. AC permanent magnet synchronous motors (PMSM) TGH2-0050 (24 VDC) have a rated torque 0.49 Nm and a rated speed 3000 rpm. The servo-amplifiers are of the type TGA-24-9/20. Furthermore, cycloidal gearheads TWINSPIN TS - 60 from the firm Spinea, s.r.o., Presov with the reduction ratio respectively 47 (elevation) and 73 (traverse) were used. Reduction ratio of the belt drive is respectively 1.31 and 1.06.

The properties of the range channel and of the direction one are bound by the relation (Cech et al., 2009)

$$\theta = \Delta\omega \cdot \Delta t_{\mathrm{E\_lag}} \leq \theta_{\max} = \frac{\delta c_{\max} \cdot \rho(c)}{f_a}, \tag{3}$$

where

$\theta_{\max}$ is the maximum permissible measurement error of the parallactic angle β - see the Figure 9,

$\delta c_{\max}$ is the same error expressed by pixels, e.g. $\delta c_{\max} = 0.1$ (resp. 0.05) pixel,

$\Delta t_{\mathrm{E\_lag}}$ is the absolute value of the time difference between starting the exposition in the sighting camera and in the metering one,

$\Delta\omega = | v_{\mathrm{Tp}}/D_{\mathrm{T}} - \omega_{\mathrm{S}} |$ is the absolute value of the error of the immediate angular velocity in the elevation/ traverse,

$\mathbf{v}_{\mathrm{Tp}}$ is the appropriate vector component of the relative velocity of the target in the plane, which is perpendicular to the radius vector of the target (i.e. perpendicular to the vector determined by the points P_{RF} and T),

ω_{S} is the appropriate immediate angular velocity in the elevation/ traverse, which is generated by the servo-drives.

It is evident from the relation (3) that the primary attention should be paid to the exact time synchronization of expositions of the sighting camera and the metering one (resp. to the synchronous sampling of the all relevant data), and that the increase of demands on the precision of servo-drives is less important. It is interesting to retrace solutions of the problem in former times (en.wikipedia/…/Base_end_station).

4.3 Target trajectory prediction

Algorithms used in the demonstration model POERF have been evolved by authors of this chapter (Cech & Jevicky, 2009b). Consequently, they have created appropriate software. Firstly, they developed a tuning and test simulation program and secondly, they have programmed procedures for the library POED.DLL (these procedures are exploited by the control program of POERF).

Trigonometric calculations relate to points P_{RF} (coordinates $(E, N, H)_{RF}$) and T (coordinates $(E, N, H)_{T}$), but rangefinder and target are spatial objects with nonzero sizes. It arises the fundamental problem, how and where to set unique contractual point on the rangefinder, and analogously, how and where to set (preferably uniquely) contractual point on the area of target image in the sight.

Chosen position of the point T in the target image determines simultaneously its position in the space. This point T is conventionally described as the "target point", i.e. reference point that represents the target at given moment due to needs of measurement of the target position.

Rangefinder construction can require aiming by the sight not into the target point T, but into so-called "aiming point" T_{AP}. Its position must be chosen in accordance with instructions for the work with rangefinder. In our case, the aiming point T_{AP} is identical with the target point T - Fig. 12.

As a result of aiming errors (Fig. 12), the position of the apex of the main aiming mark in the rangefinder sight (that represents the position of sensitive axis of the rangefinder in the space) does not coincide with the position of the aiming point T_{AP} image in the sight at the moment of range measurement. It is usually the source of additional errors in measurement of the target position in the space because the range to the point T' is measured (and it is possible that this point lies off the target), but the range is interpreted as range to the target point T. In this case, a gross error appears in the target range measurement.

By reason of simple derivation of seeking dependencies, it is necessary to introduce several coordinate systems. Detailed analysis of this problem was already presented in (Cech et al., 2009a).

The measurement point (j-th point of measurement $T_j = T(t_j)$ denotes position of the target point T at the moment t_j that characterizes contractually the moment of taking the stereo-pair images, from which the target slant range D_{Tj} is evaluated.

Data record (j-th record) - means a process beginning by preparation for taking the stereo-pair images (time $t_{STARTj} = t_{Sj}$) and ending (time $t_{STOPj} = t_{Kj}$) by completion of export of evaluated estimate of the target coordinates (generally $(E, N, H)_{Tj}$), that are contractually related to the "measurement moment", i.e. in the time t_{STOPj}, the target coordinates are given to the next use for all system. The length of record continuance is $T_{Zj} = t_{STOPj} - t_{STARTj}$.

Observing period is the time interval between two consecutive records (exports of data - the target coordinates) $t_{OPj} = t_{Kj} - t_{Kj-1}$. This period is usually constant, $t_{OPj} = t_{OP} = \text{const}$.

On the basis of information from publications and supposed accuracy of the test device POERF, the linear hypothesis about target motion was selected (presumption of uniform straight-line motion of the target with constant speed) as the most robust hypothesis from applicable ones. This hypothesis, in the case of immovable target, degenerates automatically into hypothesis of stationary target. Measured data are smoothed by linear regress model. Application of Kalman filter is problematic enough, especially due to low frequency of the target slant range measurement. This frequency is c. 10 to 100 times lower than it is usual in radiolocation. Needful organization of all processes follows from adduced preconditions - Fig. 15.

Total N_k data records - measurements ($j = j_{mink}, \dots, j_{maxk}$) are evaluated together in the k-th cycle. In our model there is $N_k = \text{const}$ for $k = 2, 3, \dots$ and $N_1 = N_2 \cdot P_{1,0}$. Linear regress model is applied on data from these records.

One measurement period Δt_{MES1}, as the interval between two successive measurement points, is estimated from the rate frame [fps]. Measurement cycles overlap $P_{k,k-1} = 1 -$

(N_{SHk}/N_k) is in functional relation with the interval of data export T_{SHk}. The overlap of measurement cycles denotes what relative number of records (measurements) is shared by two successive cycles.

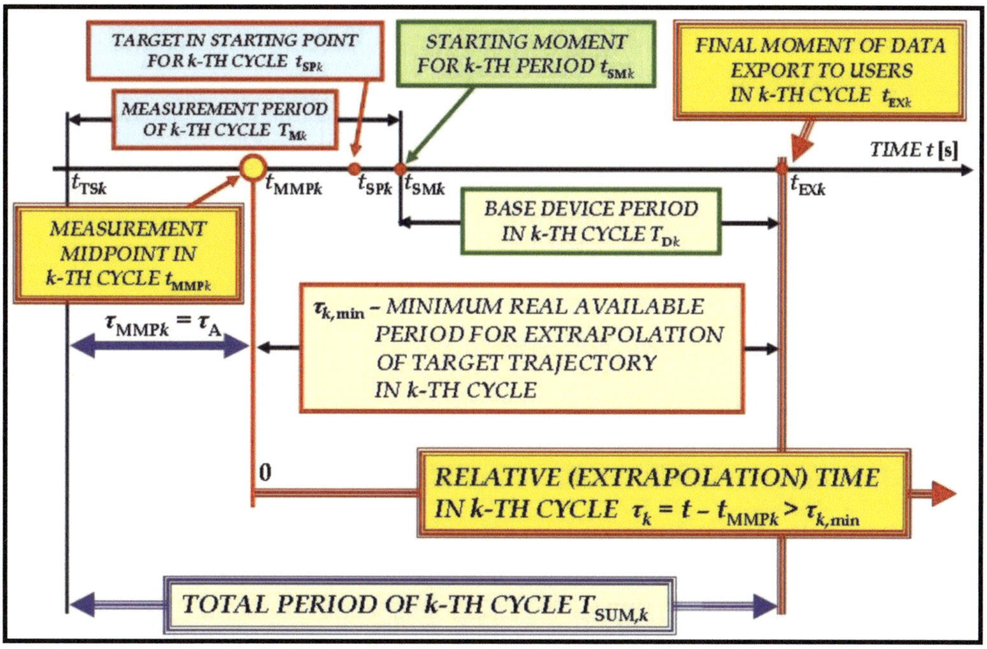

Fig. 15. Fundamental relations among time data needful for the target trajectory extrapolation

Two terms refer to the last record used in the k-th cycle. At the moment of taking the last stereo-pair images, the target point T lies at the starting point for the k-th cycle. The moment of export ending of the last record is denoted as the starting moment for the k-th cycle - Fig. 15. From the starting moment, all needful data for regress model processing is fully at disposal and can be evaluated.

Appropriate calculations and data export to users proceed during base device period in the k-th cycle T_{Dk} - Fig. 15. From the view of the user, the (total) device period in the k-th cycle $T_{DSk} = T_{Dk} + T_{DUk}$ consists of the base device period T_{Dk} and the user device period T_{DUk}, in which the user assumes data, executes preparatory operations and calculations, and only then he acquires extrapolated coordinates of the target for the time t. As it is evident from the Fig. 15, the time t must satisfy the condition of feasibility of extrapolation calculation in the k-th cycle $t > (t_{MMPk} + T_{DSk})$.

We have introduced the term measurement midpoint in the k-th cycle - Fig. 15. It is a point in the space, in which the target point T lies at the contractually selected moment t_{MMPk}. Linear regress model allows the estimate of coordinates $(E, N, H)_{TMMPk}$ of the target point and the estimate of the vector $\mathbf{v}_{Tk}$ of the target speed in this point (or at the time t_{MMPk} respectively).

Input to linear regress model is created by coordinates in coordinate system of the base $(x, y, z)_{TBj}$ and corresponding times t_j, $j = j_{mink}, \dots ,j_{maxk}$. For notation simplification, we will use these denotations: t_i, $(x, y, z)_i$, $i = 1, 2, \dots ,N_k$, so $i = 1$ corresponds to $j = j_{mink}$, etc.

Furthermore, we will introduce common denotation q_i for x_i or y_i or z_i. For all three coordinates, it is valid the same linear regress model

$$\hat{q} = \hat{q}_0 + v_q \cdot \tau, \qquad \tau = (t - t_1) \geq 0 \ , \tag{4}$$

where $(\hat{q}_0, v_q)$ are unknown parameters of linear regress model; the coefficient v_q has sense of coordinates of the speed vector (v_{TBx}, v_{TBy}, v_{TBz}).
The time for measurement midpoint is chosen (contractually - Fig. 15) as follows

$$t_{MMP,k} = t_{TS,k} + \tau_{MMP,k} \ ,$$

$$\text{where} \quad \tau_{MMP,k} = \tau_q + \Delta\tau_q \ , \quad \Delta\tau_q = t_{j\min,k} - t_{TS,k} \ , \quad \tau_q = \frac{1}{N_k}\sum_{i=1}^{N_k} \tau_i \ , \quad \tau_i = t_i - t_1 . \tag{5}$$

Estimates of coordinates of measurement midpoint for the k-th cycle are then (q_{MMPk} corresponds respectively to x_{MMPBk} and y_{MMPBk} and z_{MMPBk})

$$q_{MMP_k} = \hat{q}_0 + v_q \cdot \tau_q \ . \tag{6}$$

5. Simulation programs and Catalogue of targets

The principal purposes and characteristics of the simulation programs Test POERF and Test POERF RAW - including the Catalogue of targets - have been already introduced in the subsection 1.3.2.
The third version of the program Test POERF is described in (Cech& Jevicky, 2009c). Together four results, which have been obtained during simulations and which influence radically the solution of hardware and software of the passive optoelectronic rangefinder, are discussed here. (The four main results from the hitherto simulation experiments are presented inside the foregoing text.)
The Test POERF simulation program is an open development environment being continuously supplemented with further functions. We intend to upgrade radically the program in order to simulate the process of the moving target range measurement.
As mentioned before, the software package Test POERF RAW works with records from real scenes and consists of three separate programs: the editing program RAWedi, the main simulation program RAWdis and the viewer RAWpro.
The program RAWedi (Cech & Jevicky, 2010a) serves primarily to create horizontal stereo pair images of targets from shots that have captured wider area of a scene (a "standardization" of horizontal stereo pair images of targets and their nearest surroundings or the target image cut outs). These stereo pairs form a database part of the Catalogue of Targets. Simultaneously it allows editing stereo pair images for other purposes. The program is an analogy of the part of older program Test POERF, which is denoted as a generator of stereo pair images.
We have selected image formats REC (a special variant of RAW format) and BMP for images of the Catalogue of Targets (Cech & Jevicky, 2010a). The catalogue is a live system to which images of additional targets can be appended. For the present, we work with a database that was created from July to September 2009. The initial set has 76 stationary targets (buildings) and several other records with moving objects, especially vehicles. Meanwhile, we are dealing with stationary objects.

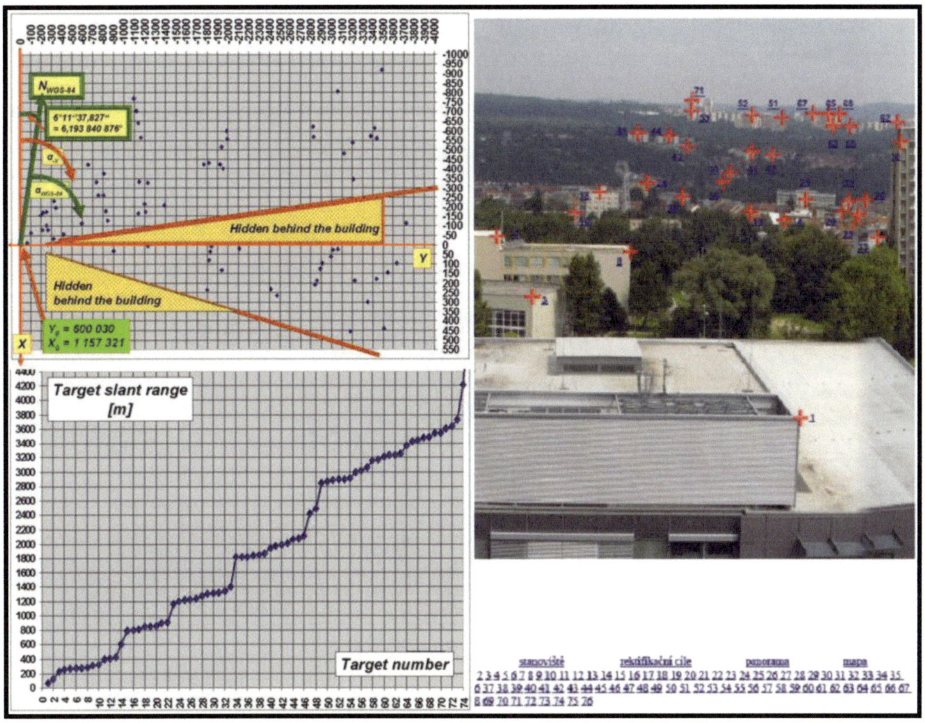

Fig. 16. Stationary targets registered in the Catalogue of Targets

Records of horizontal stereo pair shots of scenes were elected so that every target lies near to the centre of cameras field of view (suppression of possible distortion of objectives). Because of achievement of the record rate c. 10 image-pairs per second, the vertical binning was used, i.e. every two rows were aggregated into one row in all images (a special loss compression for RAW data); this compression technique does not influence the size of horizontal stereoscopic disparity.

Coordinates (X, Y) of every target and the POERF standpoint in the coordinate system of unified trigonometric cadastral network (S-JTSK) were determined with the use of Geographic information system (http://nahlizenidokn.cuzk.cz). Super-elevation angles of targets were measured by a theodolite - Fig. 16.

The program RAWdis (Cech & Jevicky, 2010b) that corresponds to the core of program Test POERF serves, as mentioned above, to determine the horizontal stereoscopic disparity of stereopairs from Catalogue of Targets (Fig. 17) and to estimate consequently the target slant range. We suppose that the simulation program RAWdis will be further developed and supplemented by new functions. The paper (Cech & Jevicky, 2010b) contains program outputs of the program version from May 2010. The problem of influences of front objects on the accuracy of the range measurement is also deeper discussed there (see the subsection 4.1 and the Figure 12).

Furthermore, in the paper there is a short problem specification of the influence of the spatial noise, whose source is partly the recording system (i.e. cameras, lenses and the basic digital image processing) and partly properties of the optical signal transmission channel - atmosphere (Roggeman & Welsh, 1996). The key role in the accuracy of the range measurement plays the atmospheric turbulence - (Rehor, 2004), (Cech et al., 2009a).

Adduced influences take effect in increasing the value of standard deviation σ_{CR} (see the subsection 4.1 and the Figure 11).

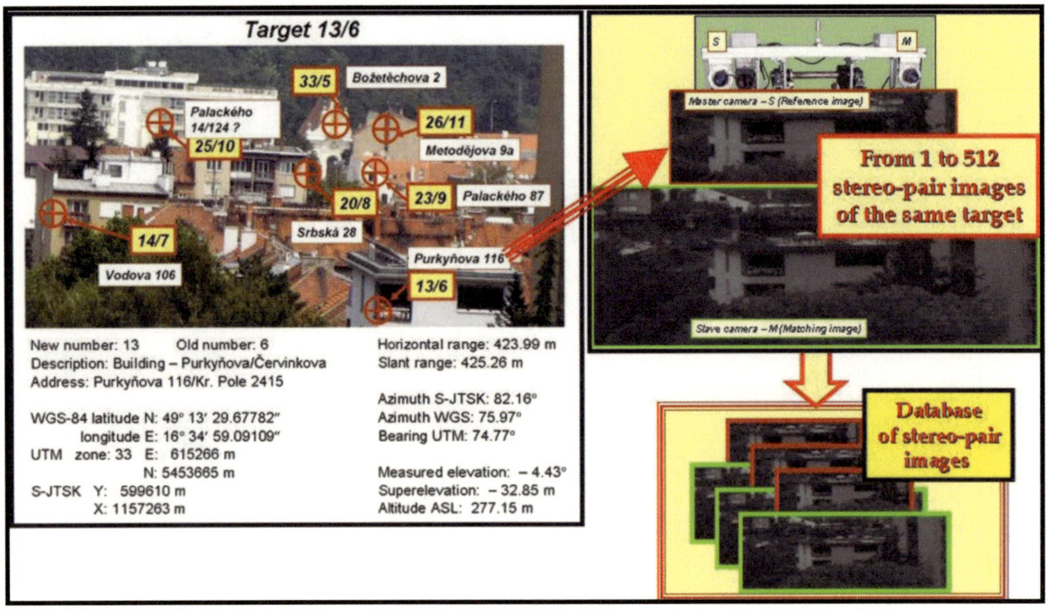

Fig. 17. The sample of "Catalogue card" for the target No. 13 from the Catalogue of Targets (it corresponds to a set stereo-pair shots in the Database of images)

6. Conclusion

As it results from subsections 1.5 a 3.2, at the present we are working on the consequential project of the research and development of the passive optoelectronic rangefinder (the fourth period of the system development). In accordance with the project plan, we endeavor to solve progressively - on a qualitatively higher level - the problems mentioned in this chapter.

7. References

Apotheloz, R. et al. (1981) *Oerlikon Taschenbuch*. Werkzeugmaschinenfabrik Oerlikon - Bühерle A.G., Zürich, 706 p.

Balaz, T.; Bucholcer, J.; Golian, M. & Racek, F. (1999) Probabilities of Target Discrimination. In *Proceedings of the 4th Conference on Weapon Systems*, Brno, May 1999, p. 203–207.

Balaz, T. (2003) *Utilization possibilities of the passive optoelectronic rangefinder in current tank fire control systems*. (In Czech.) Habilitation dissertation. Military Academy in Brno, 2003. 139 p.

Bouguet, J.-Y. (1999) *Pyramidal Implementation of the Lucas Kanade Feature Tracker*. Description of the algorithm. Intel Corp., Microprocessor Research Labs. Open CV Documents. From http://sourceforge.net/projects/opencvlibrary.

Cech, V. & Jevicky, J. (2005) The Influence of Laser Rangefinder Parameters on the Hit Probability in Direct Tank Fire. In *Proceedings of 22nd International Symposium on*

Ballistics, Vancouver BC, Canada, 2005. DEStech Publications, vol. 1, p. 144-151. ISBN 1-932078-52-5.

Cech, V.; Balaz, T.; Jevicky, J. & Racek, F. (2006) Analysis of Influence of Radiant Flux Distribution in the Laser Beam on the Probability of Right Target Range Measurement II. In *Proceedings of 40th International Conference Modeling and Simulation of Systems - MOSIS´06*. Prerov, 2006, p. 153-160. ISBN 80-86840-21-2.

Cech, V. & Jevicky, J. (2007) The Development of Algorithms for Range Channel of the Passive Optoelectronic Rangefinder. In *Book of Extended Abstracts of Engineering Mechanics 2007* and Proceedings on CD-ROM, Svratka, 2007, p. 27-28, ISBN 978-80-87012-06-2.

Cech, V. et al. (2009a) *Research of high-end technologies and methods for recognition of moving objects, for determining of objects movement parameters and for automatic tracking systems of moving objects*. (In Czech.) Final research report of industrial research project of MIT CR FT - TA3/103. Oprox, a.s., Brno, 2009, 175 p., appendices 404 p.

Cech, V. & Jevicky, J. (2009b) Algorithms of the Target Trajectory Extrapolation. In *Proceedings of 43rd Spring International Conference Modeling and Simulation of Systems - MOSIS´09*. MARQ, Roznov pod Radhostem, 2009, p. 144-153, ISBN 978-80-86840-45-1.

Cech, V. & Jevicky, J. (2009c) Simulation of Target Range Measurement Process by Passive Optoelectronic Rangefinder. In *Proceedings of the First International Conference on Computational Intelligence, Modelling and Simulation*, Brno 2009, IEEE Computer Society 2009, p. 181-186, ISBN 978-0-7695-3795-5.

Cech, V.; Jevicky, J. & Pancik, M. (2009d) Demonstration Model of Passive Optoelectronic Rangefinder. In *Recent Advances in Mechatronics 2008-2009*. Editors Brezina, T. and Jablonski, R., Springer-Verlag Berlin Heidelberg 2009, p. 79-84, ISBN 978-3-642-05021-3.

Cech, V. & Jevicky, J. (2010a) Advanced Simulation of Range Channel of the Passive Optoelectronic Rangefinder. In *Proceedings of 44th Spring International Conference Modeling and Simulation of Systems - MOSIS´ X*. MARQ, Hradec nad Moravici, 2010, p. 101-108, ISBN 978-80-86840-51-2.

Cech, V. & Jevicky, J. (2010b) Development of the Simulation Software Package Test POERF RAW. *Proceedings of the 7th EUROSIM Congress on Modelling and Simulation*. Prague, Czech Republic, September 6-10, 2010. Vol. 1: Book of Abstracts, p. 1. ISBN 978-80-01-04588-6. Vol. 2. Full Papers (CD), 10 pages, ISBN 978-80-01-04589-3.

Composite authors (1958) *Naval Ordnance and Gunnery*. Volume 2, Fire Control, Chapter 16: Radar and Optics. Prepared by the Department of Ordnance and Gunnery United States Naval Academy. Edited and produced by the Bureau of Naval Personnel, NavPers 10798-A. U. S. Government Printing Office, Washington 25, D. C. From http://editionwww.eugeneleeslover.com/USNAVY/CHAPTER-16-F.html.

Curti, P. (1945) *Einführung in die Äussere Ballistik*. Frauenfeld (Schweiz), Verlag Huber and Co. Aktiengesellschaft 1945, p. 408.

Gebel, R. K. H. (1966) Optical Radar and Passive Opto-electronic Rangefinding. *The Ohio Journal of Science*, 66(5), pp. 496-507, ISSN 0030-0950.

Gilligan, L. H. (1990) *Passive Range Finding Apparatus Utilizing Television Sensors*. US Patent No. 4 969 735, 13.11. 1990.

Hänert, L. (1928) Geschütz und Schuss. *Eine Einführung in die Geschützmechanik und Ballistik.* Berlin, Verlag von Julius Springer 1928, 360 ps.

Hanzl, V. & Sukup, K. (2001) *Photogrammetry I.* (In Czech.) University textbook. Brno University of Technology, FME. Academic press CERM, Brno 2001, 94 pages.

Holst, G.C. (2000) *Electro-optical Imaging System Performance* (Second edition). JCD Publishing, Winter Park, FL, USA and SPIE Optical Engineering Press, Bellingham, WA, USA, 2000, 438 p.

Jarvis, R. A. (1983) A Perspective on Range Finding Techniques for Computer Vision. *IEEE Transaction on Pattern Analysis and Machine Intelligence,* vol. PAMI-5, No. 2. March 1983, pp. 122-139, ISSN 0162-8828.

Keprt, E. (1966) *Theory of optical apparatus IV: Theory and Construction of Optical Rangefinders.* (In Czech.) University textbook. Palacky University, Olomouc, Publication No. 1212 - 5331, 220 p.

Kraus, K. (2000) *Photogrammetry.* Volume 1 - Fundamentals and Standard Processes. Dümmler, Köln, 397 p.

Levine, M. D. et al. (1973) Computer determination of depth maps. *Computer Graphics Image Processing,* vol. 2, No. 4, pp. 134-150, Elsevier Science, ISSN 0146-664X.

Rehor, Z. (2004) *Influences of the atmosphere on measuring by the passive optoelectronic system.* (In Czech.) Research report of industrial research project of MIT CR FD - K3/099 Research and development of technology and technical devices for both passive optoelectronic tracking and objects measuring. 1st chapter, 55 p., Oprox, a.s., Brno.

Roggeman, M. C. & Welsh, B. (1996) *Imaging Through Turbulence.* CRC Press.

Russell, I. (2001) Technical Transfer in the British Optical Industry 1888-1914: The Case of Barr and Stroud. *Scottish Industrial History,* 21, pp. 15-33.

Scharstein, D. & Szelisky, R. (2002) A Taxonomy and Evaluation of Dense Two-Frame Stereo Correspondence Algorithms. *International Journal of Computer Vision,* 47 (1-3), pp. 7-42, ISSN 0920-5691.

Skvarek, J. et al. (2004) *Research and development of technology and technical devices for both passive optoelectronic tracking and objects measuring.* (In Czech.) Abbreviated final research report of industrial research project of MIT CR FD - K3/99. Oprox, a.s., Brno. Epitome 50 p.

Uherik, L. et al. (1985) *Small arms aiming.* (In Czech.) Study for ZVS-VVU Brno, government research project SRAZ A/02. VAAZ, Brno, 103 p.

Wallis, D. A. (2005) History of Angle Measurement. Proceedings of *From Pharaohs to Geoinformatics.* FIG Working Week 2005 and GSDI-8, Cairo, Egypt, April 2005, 17 p.

Zitnick, C. L. & Kanade, T. (2000) A Cooperative Algorithms for Stereo Matching and Occlusion Detection. *IEEE Transactions on Pattern Analysis and Machine Intelligence,* vol. 22, No. 7, July 2000, pp. 675–684, ISSN 0162-8828.

17

Methods and Devices of Processing Signals of Optoelectronic Position Transducers

Zbigniew Szcześniak and Adam Szcześniak
University of Technology in Kielce
Poland

1. Introduction

1.1 Precise optoelectronic position transducers

In digital position measurements two opposite ideas of taking measurements can be observed. They both tend towards obtaining the most precise results. The first idea is based on the precise construction of the glass scale and construction of the optoelectronic transducer by using simple circuits of electronic processing. The second idea is based opposite, that is on precise circuits of electronic processing and simple construction of the glass scale and construction of the optoelectronic transducer.
The position transducers can be classified as quantizing (dynamic) and coding (static). The quantization transducers produce an electric impulse during shift of length or angle unit in a way that on the output of the transducer the number of impulses proportional to measured value is obtained. The coding transducers are used to make a conversion from linear or angular displacement to a number code. About measurement accuracy of optoelectronic transducers decides first of all value of discretisation error which depends on number of modulation elements, fall to number of linear or angular shift unit. Among the new optoelectronic transducers, the width of gaps modulating beam of light reaches value of 0,05..0,004 mm. It enables to construct the transducer of resolution 10000 grains per rotation at disc diameter varies from 25 to 320 mm (discretisation error 0,01%) (Heidenhein, 2003) (Holejko, 1981).
In cases where specific work conditions exist (high dustiness, high temperature) the optoelectronic transducers cannot be used. In such cases special very precise magnetization scales are used (Heidenhein, 2003) (thin layers range micrometers - elements of different residual magnetism placed alternately)

2. Optoelectronic quantizing transducers

Presently the most popular are the optoelectronic position transducers which consist of a disc (or a strip used for linear displacement measurements) which has on its circumference a sequence of transparent and non-transparent fields. They respectively let in or absorb the light. (fig.1) The optoelectronic transducers dedicated for length measurement are equipped with a glass scale. Its length is equal to the length of the measured position. In the case of measuring the angle, the optoelectronic transducers are equipped with rotating discs. (Ching-Fen & Mao-Hong, 2005) (Feel-Soon & Sung-Jun, 2005) The glass scale models a beam

of light which drops on the photoelement in such a way that the appropriate number of impulses appear at its end. These are proportional to the linear or angular movement.
The linear and angular transducers (Heidenhein, 2003) (Holejko, 1981) have the glass scale and the reticle made with a period of the reticle 10μm or 20μm. (fig.1)

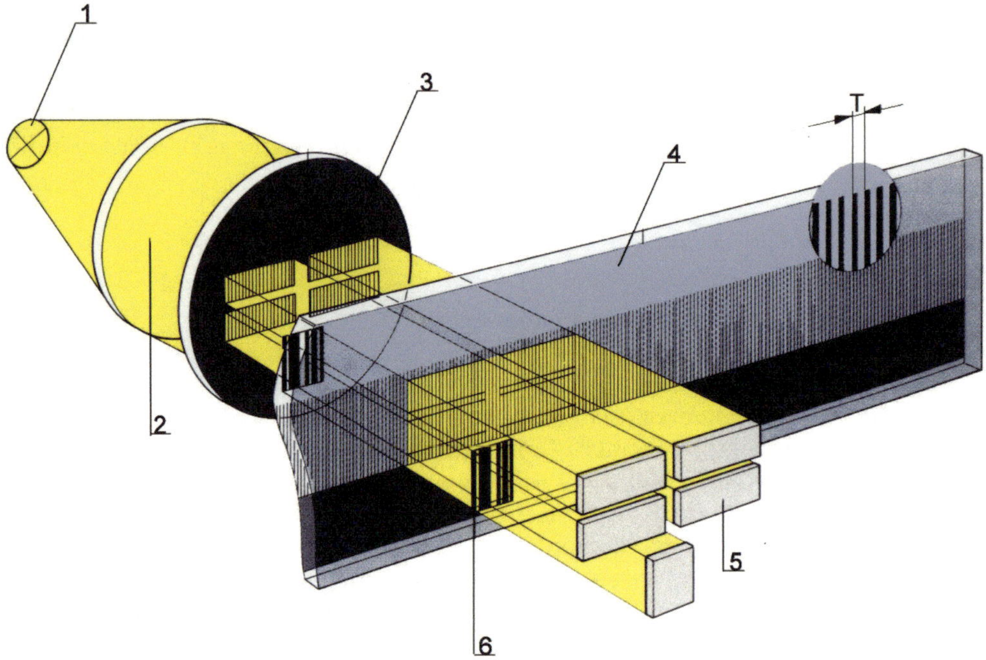

Fig. 1. The principle of photoelectric measurement with the glass scale
1. source of light, 2. condenser lens, 3. scanning reticle, 4. glass scale, 5. solar cells
6. reference mark

The second path (of the glass scale) transfers one or more reference marks. The scanning unit consists of: a source of light, a condenser lens used to obtain parallel beam of light for the scanning reticle, the glass scale and solar cells. When the glass scale is moved relatively to the scanning unit, the lines of the scanning reticle converge alternately with the lines or intervals of the glass scale. As a result periodic fluctuations of light intensity are converted to electrical signals by solar cells.
The transducers which use the steel scale work with reflected light. The steel scale consists of high-reflecting gold lines and light dissipating or absorbing intervals and it is used as a measuring reference. When the scale is displaced, the solar cells of the scanning unit produce periodic signals similar to those produced from the glass scale.
There are also used the linear transducers which use diffraction and interference of light waves. The measuring reference reflects the light-phase reticle with length of pitch approximately 0.2μm, and a scanning unit is the corresponding diffraction grating on transparent glass. They reach high accuracy and enable very precise measuring steps.
In measuring devices (Heidenhein, 2003) (Holejko, 1981) the most important role play division carriers on which the scale, mostly in a form of the cross-hatched reticle, is manufactured. These divisions are produced precisely according to elaborated technology

and are accurate reference of the measuring devices. Divisions on the glass or glass ceramics are made by covering an extremely thin layer of Chromium where accuracy in range of division structure is in range of micrometers and more precise. The divisions on steel carriers consist of highly reflective gold dashes and metal intervals between them. Apart from high precision of laying dashes one from another they have also a considerably high sharpness of edges and they are resistant to mechanical and chemical factors. Apart from this they are sensitive to loads made by vibrations or impacts. They also demonstrate the precisely defined thermal response. Changes of air pressure or humidity do not have any effect on accuracy. Length of the scale - measuring range for the scale made on steal carriers (steel tapes) is up to 30m., whereas for the glass scale is up to 3m..

In the measuring incremental transducers it is not practical to displace an object of high distance due to repeating an algorithm of determining the starting point of the processed object. This problem has been solved by proposing both in the linear and circular scales length-coding reference marks. The scale consists of the reticle and a parallel path of reference marks. By changing, by defined rule, distance between the neighboring reference marks absolute position of each reference mark is coded.

In the linear transducers which have the length-coding reference marks, absolute position value can be once again determined by displacement over the two following reference marks that is after displacement of 20μm. The manufactured angular transducers with the length-coding reference marks have 36 reference marks on the discs with 900 and 18000 lines, while the scales with 36000 lines have 72 reference marks. Rotations which do not cross adequately 20° or 10° are adequate to establish another absolute position.

3. Optoelectronic coding transducers

The main element of the coding transducer (fig.2) is the rotating coding mask. Each position of the coding mask (that is each possible to distinguish by given transducer value of measured quantity) corresponds to the determined combination of transparent and nontransparent fields for stream of light. This combination presenting 0 or 1 digits of binary code is read by photoelectric reading complex which consists of a source of light and photo elements. For each transducer coding path fall separate light-sensitive element in a way that on the output of the optoelectronic transducer static code is formed. This depends on the the actual position of the modulating disc. This code can be used directly as a value of measured quantity. If necessary the decoder enabling the matching code of the transducer disc to the code of the reading device can be used. About accuracy conversion decides first of all digitization error depended on number of bites in the output signal of the transducer. In fig.2 the code transducer was shown, in practice the transducers of 100..1000 grain resolution working in binary-decimal code (8..12 -bit) or 128..1024 grain resolution working in binary code (7..10 -bit). If higher precision is required the transducers with several coding discs coupled to each other by non-backlash gears are used. In this way in the transducer with two gears (1:100) received 10^{6} resolutions. The basic advantage of the coding transducers is the generation of parallel static binary code or binary-decimal code which is the value of the measured quantity. If there is a need parallel code can be easily converted to serial code. Conversion can be achieved by subsequent reading of information from particular coding strips. The drawback of both specified codes is the sensibility to making gross read errors. It comes from strict assembly tolerances of complex: the glass scale (disc) - a reading device. Elimination of this drawback can be achieved by special construction means (Heidenhein, 2003) (Holejko, 1981) (i.e. reader in shape of a reversed letter V) or by applying instead of binary code different codes in which during the change of measured value of 1 grain, and the change of a state occur only on one code path (i.e. Grey code). Absolute measuring devices provide actual information of the position. There are also built converters which additionally produce incremental signals. (fig.2)

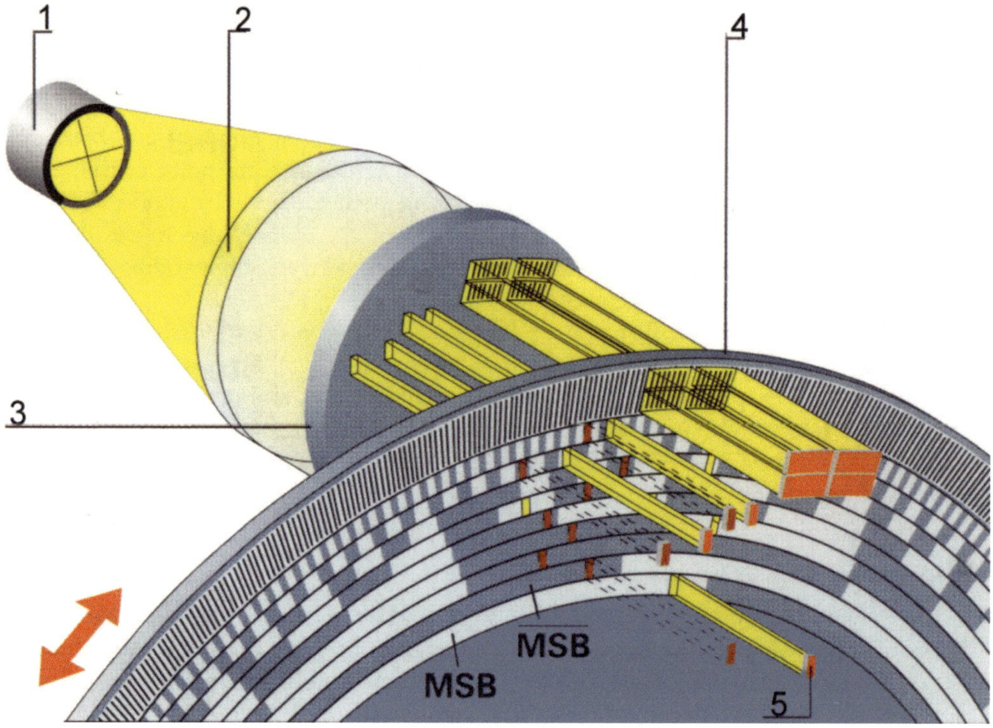

Fig. 2. The construction diagram of the code transducer for measuring angle source of light, 2- condenser lens, 3- scanning reticle, 4- glass scale, 5- solar cells

4. Method of shaping measuring sinusoidal signals of the optoelectronic transducer

On the glass scale 2, four rectangular fields with scanning reticle 1 are placed. They are assigned to individual photovoltaic cells. (fig.3)

The parallel light beam produced by the source of light and the lens is incident through the above mentioned grids on the scale of the scanning reticle and from there on the photovoltaic cell. The grids have identical widths of light, transparent and opaque fields, which is a condition for photovoltaic cells to generate sinusoidal voltage signals (Szcześniak, 2005)d (Szcześniak, 2005)g. In order to obtain sufficient signal value, it is advantageous that the measurement field should consist of N periods T of the displacement division. Then, for a specified field lit with a radiation beam of intensity E_o, at a given efficiency of the optical system, we will obtain an effective beam φ getting to the photooptic system U_a,U_b,U_c,U_d. We can form the beam incident on photoelements by placing a scanning reticle with appropriate relative displacement in relation to the scale index grid 2 of the transducer. The relative displacement for the photoelectric system U_a amounts to 0°. The displacement of the photoelectric system U_b is 1/4 of the period T in relation to the photoelectric system U_a. Its position is defined by the dependence $N_2T + 1/4T$. On the other hand, the system U_c is shifted by 1/2 of the period T with respect to the system U_a and of the same period system U_d is shifted to U_b. The position of this photoelectric system is given by the dependence $N_1T + 1/2T$, $N_4T + 1/2T$. Exemplary positions of the reading fields are presented in the Fig.4. in which N, N1, N2, N3, N4 denote the number of periods T.

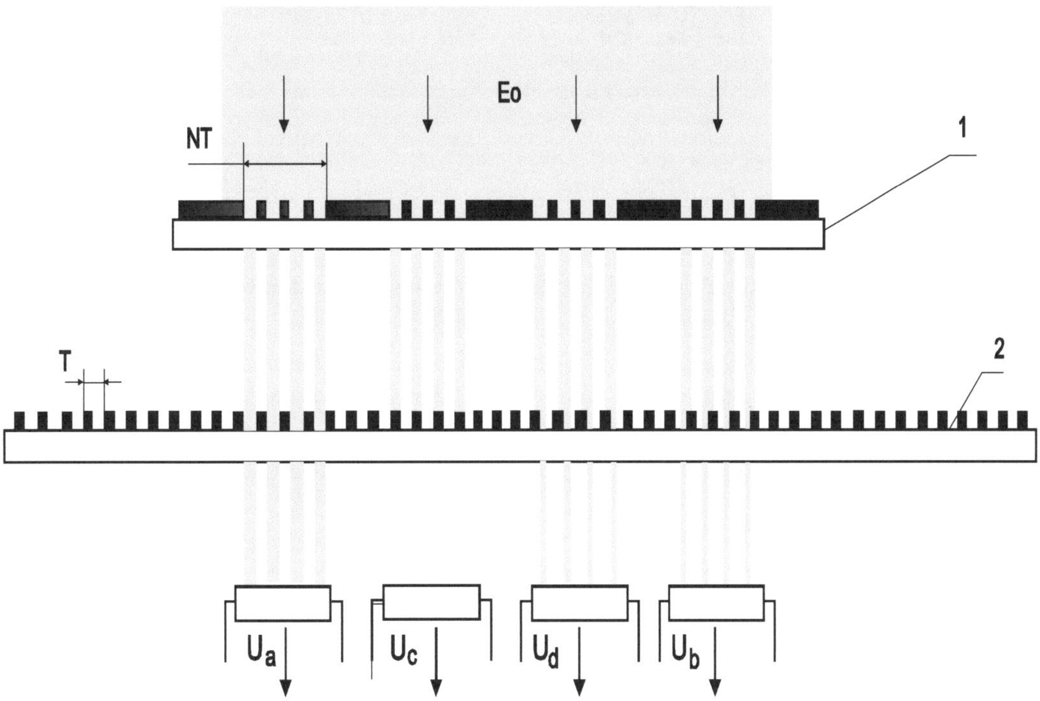

Fig. 3. Method of signal generation in photoelectric position transducer for three photoelectric systems

We can form the beam incident on photoelements by placing a scanning reticle with appropriate relative displacement in relation to the scale index grid 2 of the transducer. The relative displacement for the photoelectric system U_a amounts to 0°. The displacement of the photoelectric system U_b is 1/4 of the period T in relation to the photoelectric system U_a. Its position is defined by the dependence $N_2T + 1/4T$. On the other hand, the system U_c is shifted by 1/2 of the period T with respect to the system U_a and of the same period system U_d is shifted to U_b. The position of this photoelectric system is given by the dependence $N_1T + 1/2T$, $N_4T + 1/2T$. Exemplary positions of the reading fields are presented in the Fig.4. in which N, N1, N2, N3, N4 denote the number of periods T.

When the scanning distribution grating is in motion, photovoltaic cells generate sinusoidal voltage signals :

$$U_a = A_1\sin\varphi + A_o,\ U_c = A_1\sin\varphi + A_o,$$
$$U_b = A_1\cos\varphi + A_o,\ U_d = A_1\cos\varphi + A_o, \qquad (1)$$

where: $\varphi = 2\pi y/T$ in addition: A1- signal amplitude, Ao- signal constant component, y - displacement, T - signal period.

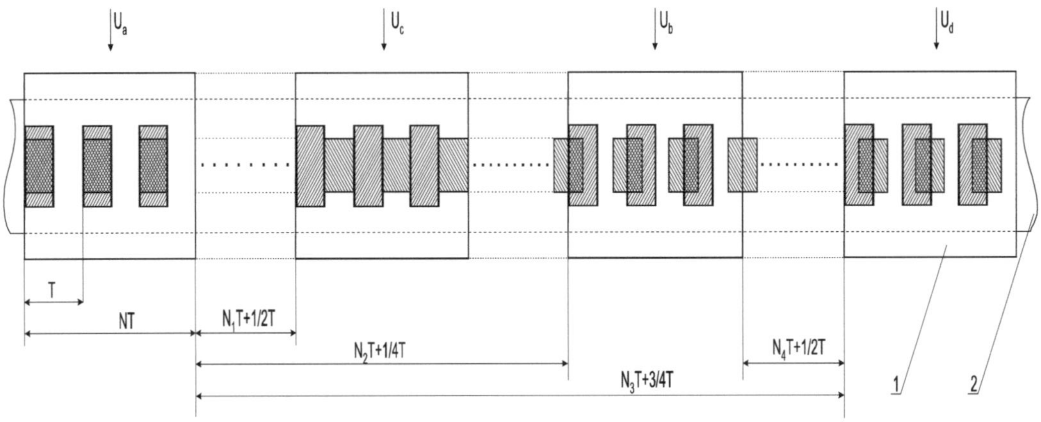

Fig. 4. Diagram of the reading fields of the measurement scale for three photoelectric systems

Subtracting sinusoidal signals U_a from U_c and U_b from U_d (signals of 180° phase shift), we get rid of the DC component and receive sinusoidal voltage signal that is symmetrical in relation to zero and shifted one from another by ¼ of period.

$$U_1 = A\sin\varphi \qquad U_2 = A\cos\varphi \tag{2}$$

The principle of signal processing is presented in Fig.5. The signal period equals the grid period of the scale T. The method presented in the paper for the compensation of the signal DC component takes into account its dynamic changes when the transducer operates under changeable environmental conditions (temperature, humidity, dustiness, etc.).

Generated signals sin, cos are used to determine the motion direction. They can also be processed further in order to increase the accuracy of position measurement (Ching-Fen & Mao-Hong, 2005) (Feel-Soon & Sung-Jun, 2005) (Szcześniak & Szcześniak, VIII.2009) (Szcześniak, 2005)e.

Signals that do not contain a DC component make it possible to generate phase signals and then multiply the frequency of the signals in comparison to input signals. The generated signals are rectangular and have pulse-duty factor 0.5. Due to the lack of constant component in the signal U1, U2 it is possible to directly receive from them rectangular signals Uw1, Uw2 of pulse-duty factor 0.5. Constant and stable pulse-duty factor ensures obtaining uniform measurement step α presented in fig.5. The measuring step was defined as the distance between two subsequent slopes of rectangular signals U_{w1}, U_{w2} (Szcześniak, 2006)b. The number of pulses generated from the slopes of the signals and counted in the course of measurement gauge shift is the measure of position. Rectangular signals shifted in relation to each another by ¼ of the period and of pulse-duty factor 0.5 provide the basis for the determination of the photoelectric transducer motion direction. The method of shaping measuring signals with compensation of DC constant thus minimizing number of devices of measuring optoelectronic transducer to three presented in literature (Szcześniak, 2006)b.

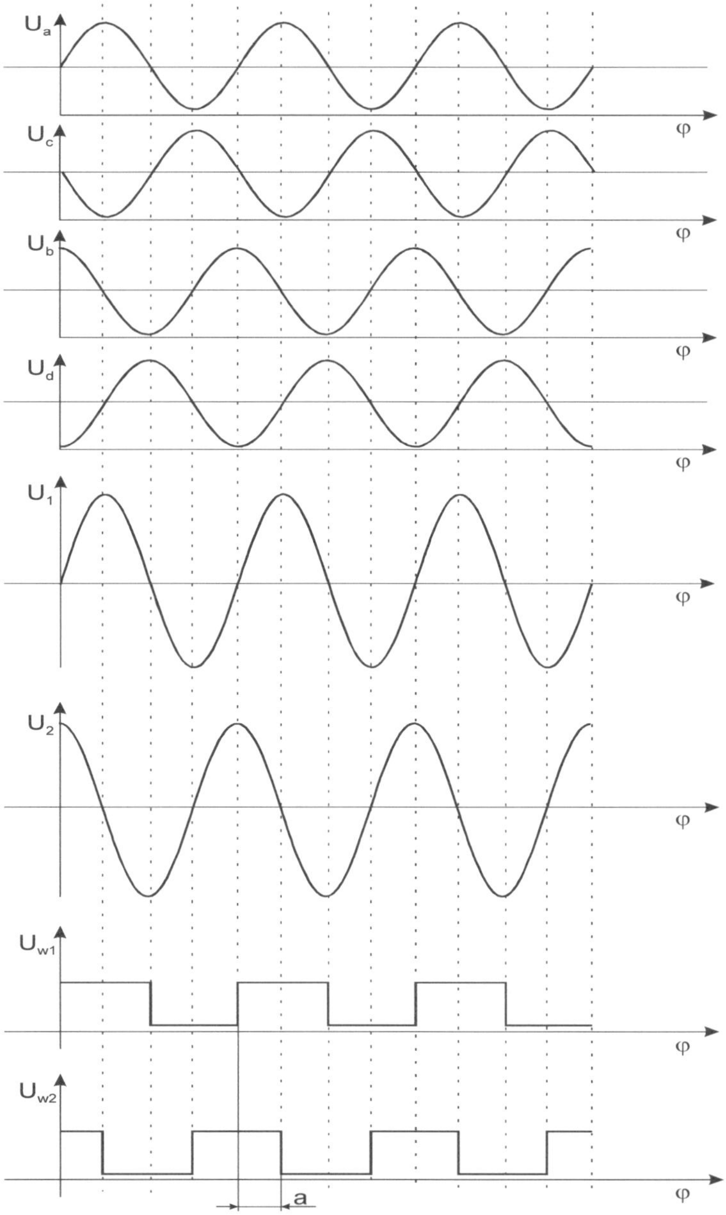

Fig. 5. Principle of signal generation in the optoelectronic quantization transducer for position measurement

5. Methods of processing electric signals of the optoelectronic position transducers to increase the accuracy of measurement and the motion direction determining

The development of specialized analog and digital systems allows for processing fast alternating signals. This means that in practice there are no limits to signal frequency on

account of measuring electronic systems. Much more significant are mechanical limits on account of the dynamic of measured object and its range.
The quantizing transducers (incremental) produce electrical impulse during shift of one unit of length or angle in a way that on the output of the transducer the number of impulses proportional to measured value is obtained (Szcześniak & Szcześniak, IV.2009).
About measurement accuracy of the optoelectronic transducers decides first of all value of discretisation error which is depended on number of modulation elements fall to number of linear or angular shift unit. The input signal of the optoelectronic transducer in a form of variable in time linear or angular velocity is integrated in time resulting in constant increasing of position *y(t)* presented in fig.6. Uniformly spaced sequences of light transparent and nontransparent fields create a measuring scale which conducts rhe operation of quantization horizontally continuous in position time *y(t)*. Next, quantized signal of increasing position $y_{kw}(t)$ is differentiated in the circuit of optoelectronic transducer. Each another increase of position *y* of unit *T* (quantum) cause receiving impulse of output signal U_{wy} of frequency proportional to velocity moving mechanism. An example of time signal functions is presented in fig.6.

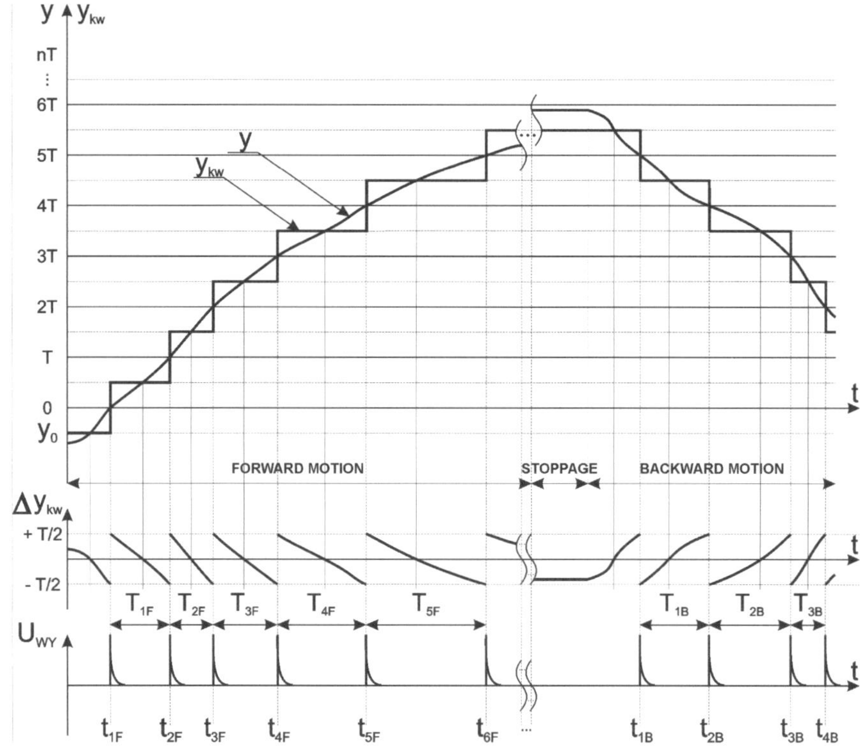

Fig. 6. Time functions of signals in the optoelectronic position transducer

Position increase from initial value y_0 with velocity proportional to value of input signal (velocity moving mechanism) and is quantized horizontally with step equals *T*. The position is presented for three phases of motion: forward motion, stoppage and backward motion. Error of quantization Δy_{kw} consist in a range of ±T/2. In particular moment of time t_{1F}, t_{2F}, t_{3F} position y reaches following values T, 2T, 3T... and on the input of the sensor we receive

impulses spaced one from another in time of T_{1F}, T_{2F}, T_{3F}... creating output signal of motion frequency in the forward direction. For the backward motion direction, impulses are generated in moments t_{1F}, t_{2F}, t_{3F} etc. In the presented figure, motion impulse is generated based on an equal step of quantization T, but in different moments depending on velocity of motion so the method of distinguishing the motion direction of the optoelectronic transducer was used. This method produces one impulse per one period of the glass scale.
Below is presented the electronic method of increasing accuracy of the optoelectronic transducer by applying electronic interpolation (multiplication of signals` frequencies) and digitalization (conversion to digital form) and electronic methods of distinguishing motion direction and measurement of position with possibility of additional increase of optoelectronic transducer accuracy working with an object.

6. Method for the multiplication of the optoelectronic position transducer measurement signal frequency

Optoelectronic transducer output signals are composed of two sinusoidal signals, phase shifted with respect to each other by ¼ of the period. The signal period is equal to the period of the transducer scale grid (measurement bar). The scanning signals from transducers are first amplified and then interpolated. The interpolation method proposed by the authors relies on a network of resistors. It generates phase-shifted signals from two sinusoidal scanning signals by means of their vector summation in accordance with the description below. Voltage signals from the sensor

$$x = A\sin\varphi\, ;\ y = A\cos\varphi \tag{3}$$

where: $\varphi = 2\Pi y/T$ in addition A- signal amplitude; y- displacement; T- signal period.
In method of interpolation from signals x, y signals x1, x2, x3 etc. are created. Their phase is shifted to x signal by angle α, 2α, 3α etc. Initial signals can be written as:

$$\begin{aligned} x &= A\sin\varphi \\ x1 &= x\cos a + y\sin a = A\sin(\varphi + a) \\ x2 &= x\cos 2a + y\sin 2a = A\sin(\varphi + 2a) \end{aligned} \tag{4}$$

In the relationship above, the resistances R_1,R_2 are defined as :

$$R_1 = \frac{R}{|\sin\alpha|} \quad \text{and} \quad R_2 = \frac{R}{|\cos\alpha|} \tag{5}$$

where: α - angle of the assumed phase shift, R – adopted signal reference resistance
Generated voltage on the resistance divider is equal to:

$$U_d = A(\sin\varphi R_1 + \cos\varphi R_2) \tag{6}$$

Taking into account the above dependences and after transformation we receive

$$U_d = A_1 \sin(\varphi + \alpha) \tag{7}$$

in addition A_1- the amplitude of a signal generated with the assumed phase shift α

For the sake of the method description clarity, it is assumed that the input signals are of constant frequencies (periods). This case represents the transducer collaboration with the object moving at constant velocity.

At the first stage of the method, signals are amplified and reversed producing four signals: sin, cos, -sin, -cos, afterwards two appropriate of them get interpolated. This creates twenty sinusoidal functions shift one from another by 18°. The principle of choosing input signals to dividers for individual parts of period presents fig.7. It is essential for the signals selected for interpolation to always have opposite signs in the period quarter that is of interest to us. For quarter I (0° - 90°) sin and –cos signals are used, for quarter II (90°-180°) they are sin and cos, for quarter III (180°-270°) the signals are -sin and cos, whereas for quarter IV (270°-360°), they are -sin and -cos. The signals get to twenty resistance dividers, the values of which are selected in such a way so that twenty sinusoidal waveforms are generated at their outputs. The waveforms are shifted with respect to each other by an assumed angle, which in the method presented amounts to 18°. For so assumed phase shift, for angles 0°, 90°, 180°,270° interpolation does not occur and the signals -sin, cos, sin, -cos are selected for further processing.

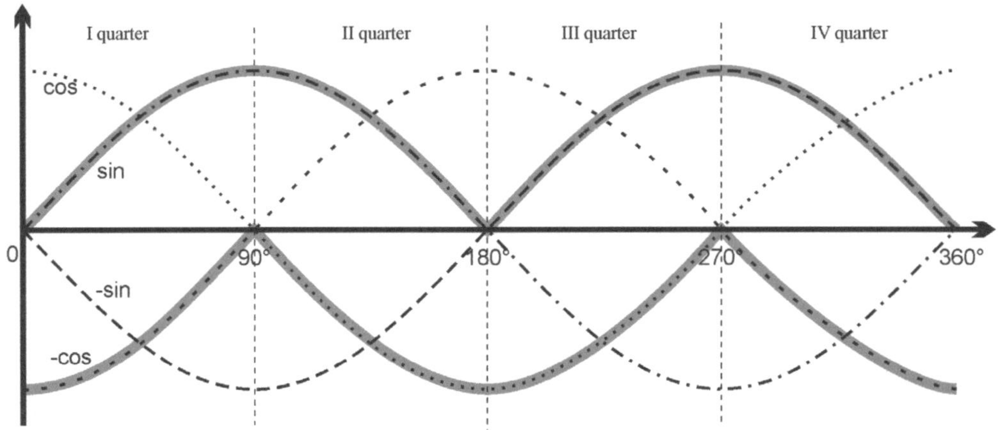

Fig. 7. Selection of signals to interpolation for following parts of period of input signal

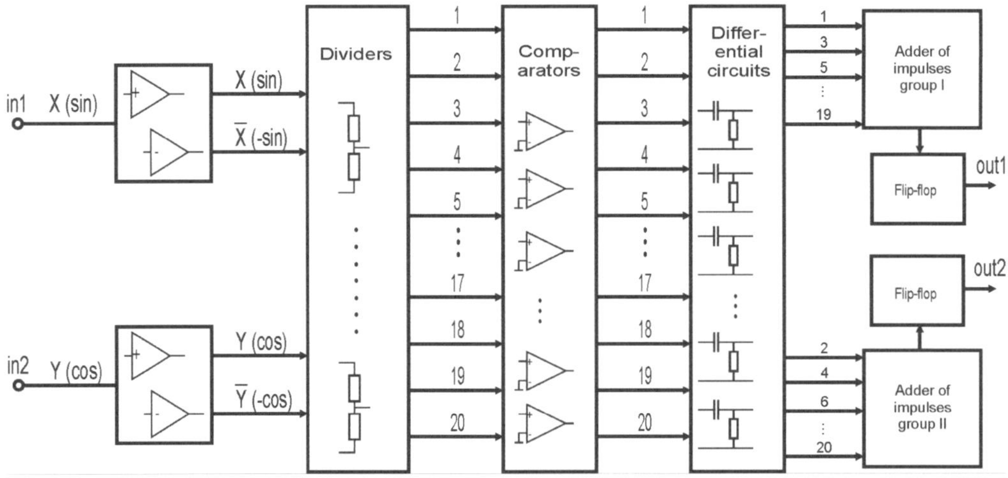

Fig. 8. The block diagram of the system of interpolation and digitalization of transducer signals

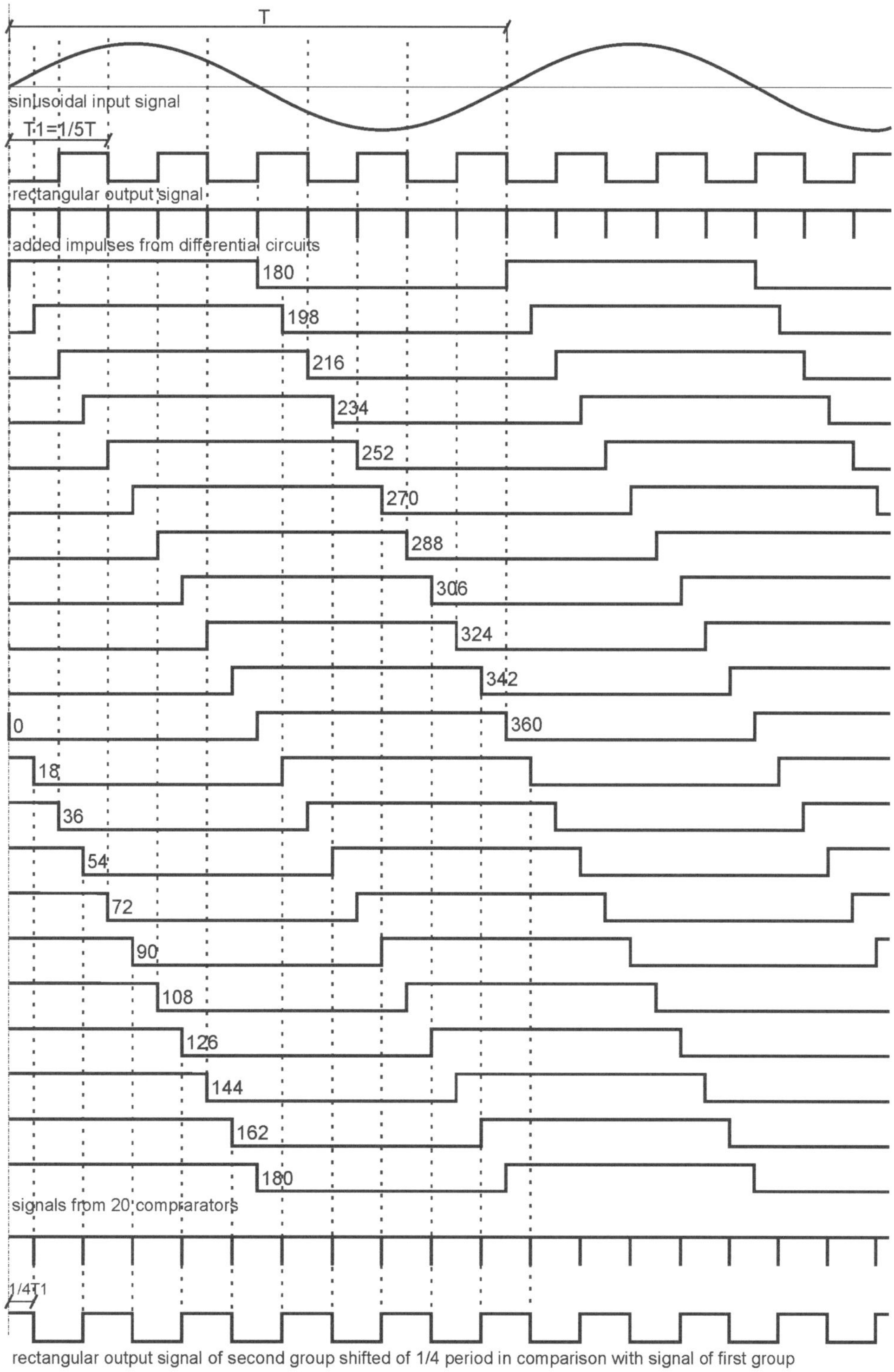

Fig. 9. Time functions of digital part of signals after interpolation

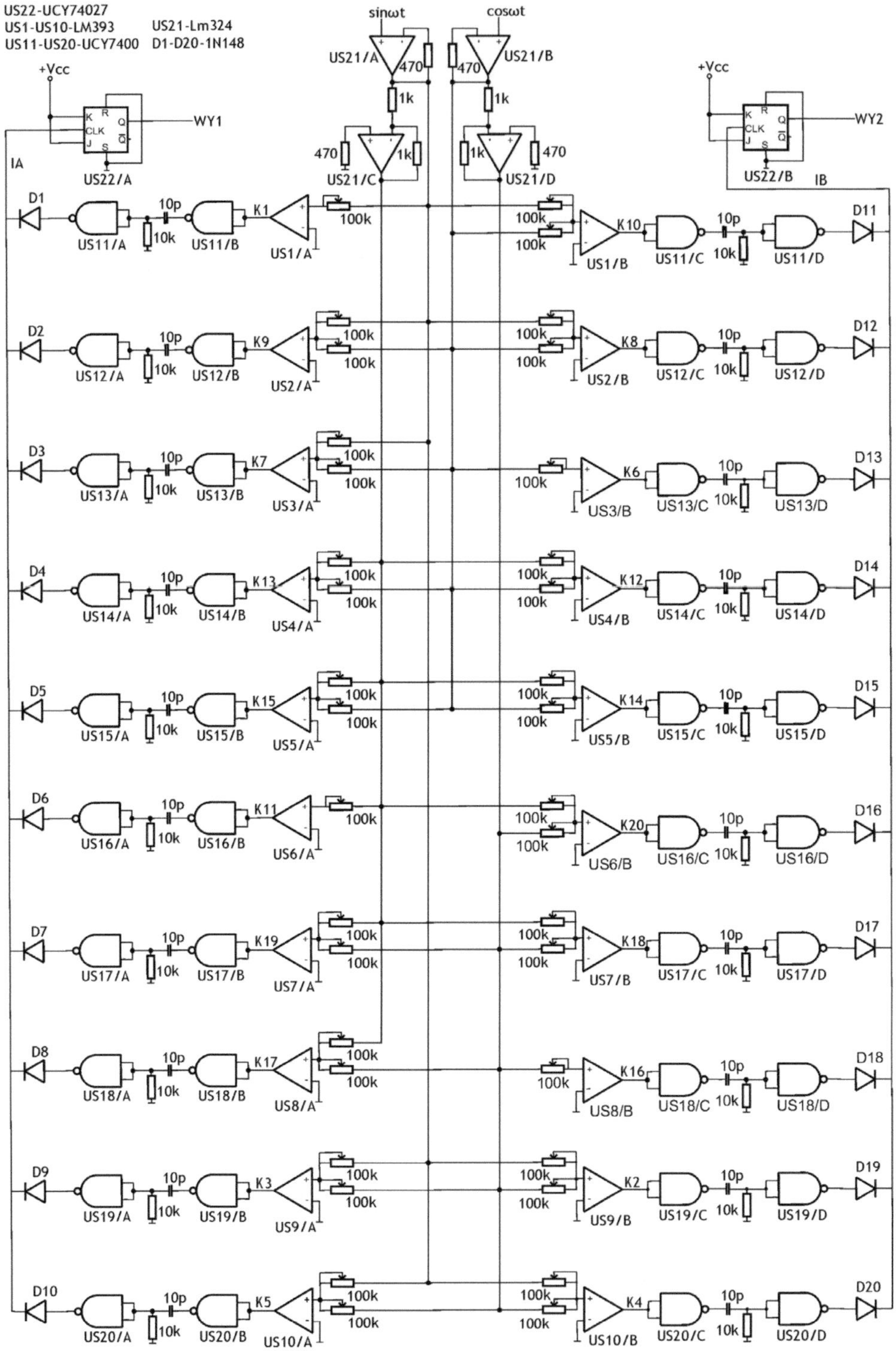

Fig. 10. Schematic diagram of interpolation the circuit and digitalization of transducer signals

Angle α [°]	0	18	36	54	72	90	108	126	144	162
sinα	0,000	0,309	0,588	0,809	0,951	1,000	0,951	0,809	0,588	0,309
cosα	1,000	0,951	0,809	0,588	0,309	0,000	-0,309	-0,588	-0,809	-0,951
R1 [kΩ]	_	64,72	34,02	24,72	21,02	20,00	21,02	24,72	34,02	64,72
R2 [kΩ]	20,00	21,02	24,72	34,02	64,72	_	64,72	34,02	24,72	21,02
Angle α [°]	180	198	216	234	252	270	288	306	324	342
sinα	0,000	-0,309	-0,588	-0,809	-0,951	-1,000	-0,951	-0,809	-0,588	-0,309
cosα	-1,000	-0,951	-0,809	-0,588	-0,309	0,000	0,309	0,588	0,809	0,951
R1 [kΩ]	_	64,72	34,02	24,72	21,02	20,00	21,02	24,72	34,02	64,72
R2 [kΩ]	20,00	21,02	24,72	34,02	64,72	_	64,72	34,02	24,72	21,02

Table 1. Selection of resistor values for dividers from fig. 10

The generated signals are fed to one of twenty comparators which change their shapes making them rectangular signals. In addition, the change in state takes place at the site of the input signal passing zero. In this way twenty rectangular signals shifted with respect to each other by 1/20 period (18°) are generated. The signals are divided into two groups in such a manner that in each group each successive signal was shifted by 1/10 period with respect to the previous one. Two groups of signals are formed, each containing ten signals shifted by 1/10 period within the group and by 1/20 period with respect to the other group (Fig.9, Fig.10).
The next stage consists in such differentiation of waveforms so that each run would yield only one pulse in the period, i.e. one group will produce ten pulses dividing one period of input reference signal into ten equal parts, on the assumption that the frequency of input signals is constant. Pulses from two groups are summed to form two pulse waveforms of the frequency that is five times higher than the input signal (Fig.8, Fig.10). They operate as clock signals for JK triggers arranged in such as manner that each pulse of input clock signal causes the change of state to the opposite one at the trigger output. As a result, two symmetrical rectangular waveforms are created of the input frequency that is 5 times higher, i.e. there is one input sinusoidal signal period to 5 periods of output rectangular signal. The signals are shifted with respect to each other by 1/4 period and the displacement direction depends on the direction of input signal shift, which is dependent on the transducer motion direction (Pizoń et al., 1993) (Szcześniak, 2005)c (Szcześniak, 2005)f.

7. Method of optoelectronic transducer motion direction discrimination based on logical functions of the transducer signals

In the previous chapter, the method of increasing the frequency of the measuring signal in relation to the frequency resulting from quantization of position by the glass scale was presented. The output signal is shaped as two rectangular signals shifted one from another by ¼ period. These signals are used in processing systems which distinguish the motion direction of a drive coupled with the transducer. Appropriate shaping of these signals enables additional increase of processing accuracy of the transducer. (Pizoń et al., 1993) (Szcześniak & Szcześniak, 2009)a (Szcześniak & Szcześniak, 2009)b (Szcześniak, 2006) a. The method of optoelectronic transducer motion direction discrimination of the optoelectronic transducer is illustrated by time functions in fig.11. Circuit structure of optoelectronic transducer motion direction discrimination based on logical functions of transducer signals presents fig.12. During the transducer operation the electric rectangular signals A, G, shifted by quarter period are obtained on their two outputs (Fig.11, Fig.12). The task of the presented electronic system is counting the adequately generated motion impulses and distinguishing their sequence. The method is based on motion pulses generation on the base of the sum of the signals A+G and on the base of their product AG, what is presented in Fig. 1. Making the sum of those pulses

$$WY = \overline{A + G} + AG \tag{8}$$

it is possible to generate, during the period T, 1 or 2 pulses, which are counted in the reverse order.

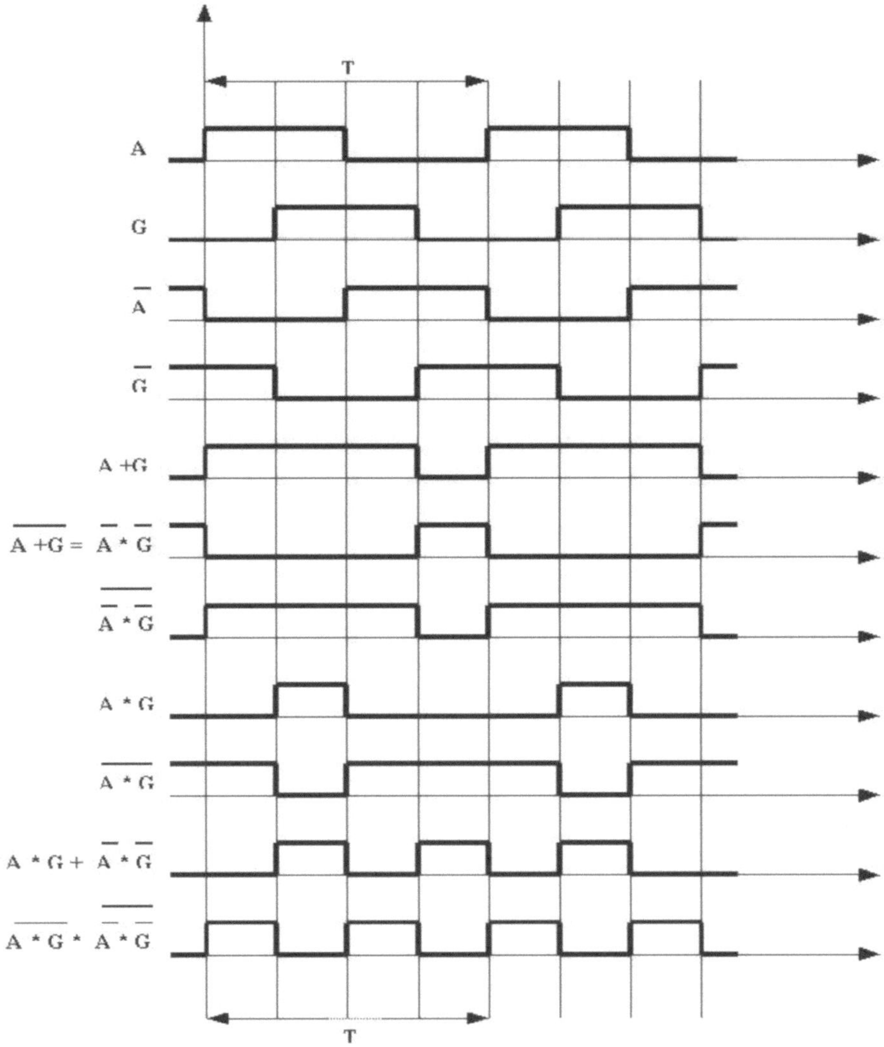

Fig. 11. Method of optoelectronic transducer motion direction discrimination based on logical functions of the transducer signals

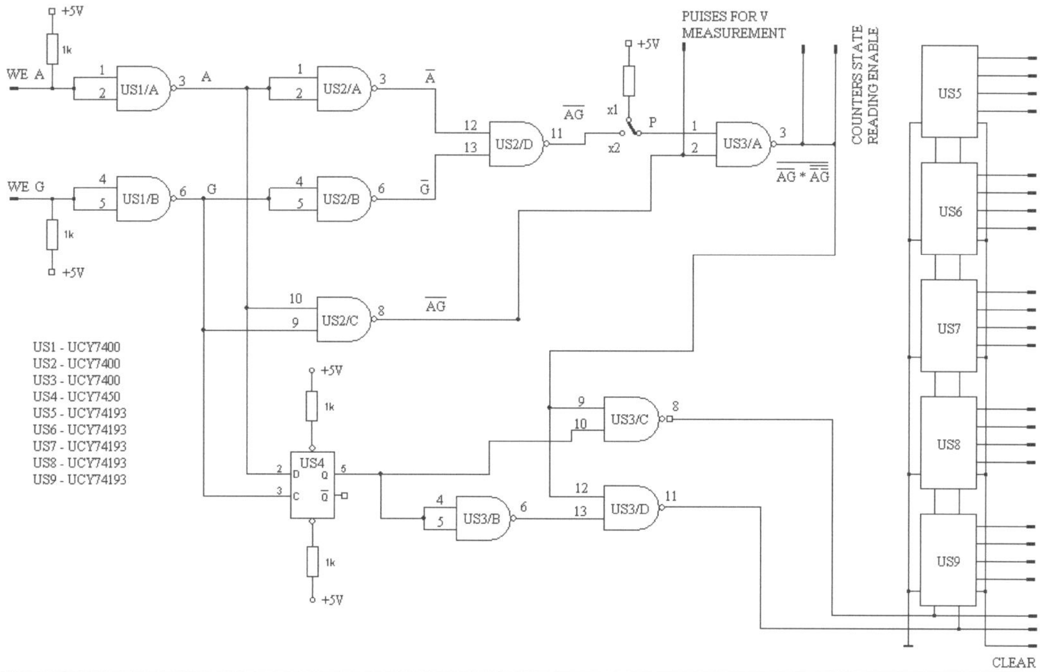

Fig. 12. System of optoelectronic transducer motion direction discrimination based on logical functions of the transducer signals

The counting order is chosen dependently on the shift between the A and G signals, what is set by the trigger. The G signal is delivered to the clock input and the A signal is delivered to the D input of the trigger. Change of the G signal from "0" to "1", when A="1", sets the Q trigger to "1". Counting to the right is then enabled. In the case when change of the G signal appears at A="0", i.e. when the G signal is leading to the A signal (change of the motion direction, the Q trigger changes its state and counting to the left is enabled. The system enables the choice of measurement accuracy x1 or x2, dependently on the setting of the P switch.

8. Method of optoelectronic transducer motion direction discrimination based on logical functions of the transducer signals and motion pulses generated in the RC circuits

In this method (Fig.13, Fig.14.), by summing the suitable products of $A,\bar{A},G,\bar{G}$ signals and the signals arising from generation of pulses of duration τ (from their rising edge) it is possible to count the impulses in the reverse counter, dependently on the transducer motion direction. When moving to the right the WY2 output of the NOR gate is set to "1", and on the WY1 output of the NOR gate a series of impulses repeated with the period of one transducer channel pulses is obtained:

$$WY1 = \overline{\bar{A}\cdot I\bar{G} + A\cdot IG + G\cdot I\bar{A} + \bar{G}\cdot IA} \qquad (9)$$

where: IA,IG - impulses from the A,G signal edge; $\overline{IA},\overline{IG}$ - impulses from the $\overline{A},\overline{G}$ signal edge.
When moving to the left the WY1 output of the NOR gate is set to "1", and on the WY2 output of the NOR gate a series of pulses, similar to the ones when moving to the right are obtained:

$$WY2 = \overline{\overline{A} \cdot \overline{IG} + A \cdot IG + G \cdot \overline{IA} + \overline{G} \cdot IA} \tag{10}$$

Those impulses are subtracted from the reverse counter state when moving to the right.

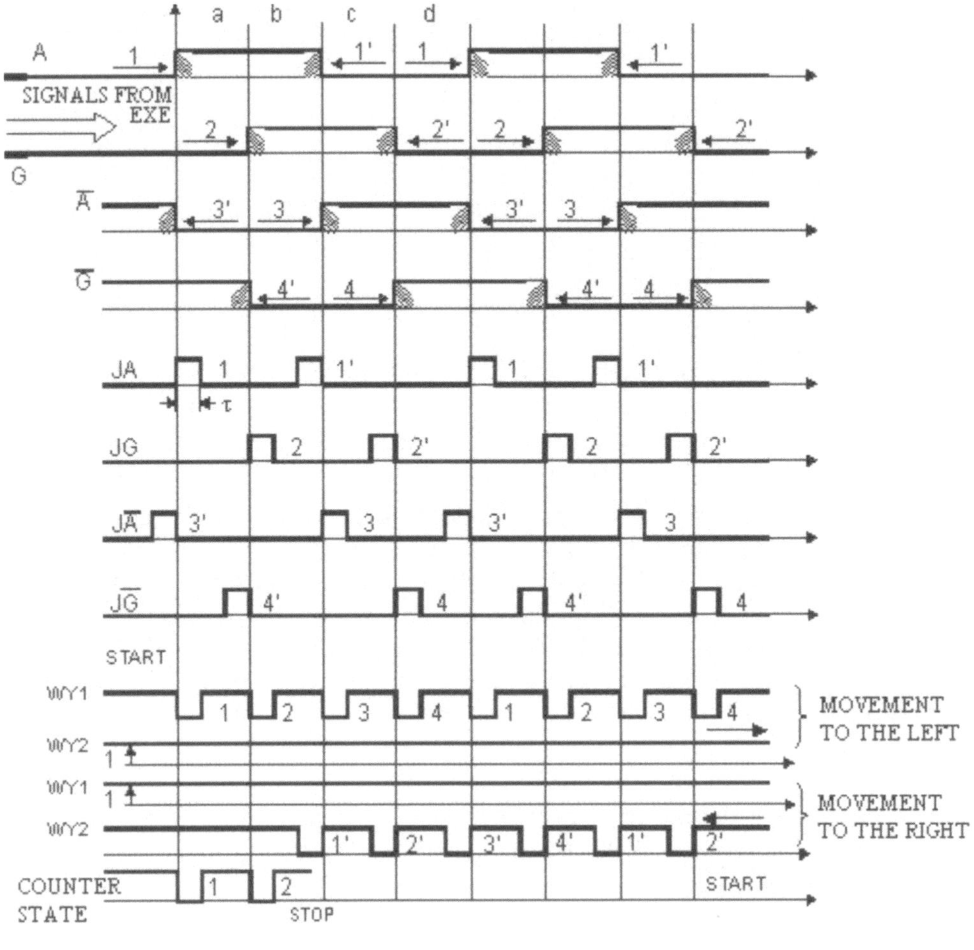

Fig. 13. Method of optoelectronic transducer motion direction discrimination based on logical functions of the transducer signals and motion pulses generated in the RC circuits

Change of the movement direction can appear in four different states a, b, c and d, as depicted in Fig.13. In every one of these states the system must respond identically, so the largest error of movement direction discrimination is equal to $\frac{1}{4}T$ (T - period of pulses in one transducer channel). It should be underlined that the system has the possibility of controlling the number of pulses counted for one period of the input signal (1, 2, 4), dependently on the settings of P1, P2, P3 switches (Fig. 14.). In this way there is a possibility of setting the accuracy of system positioning on 1, 2 or 4 times.

9. Method of optoelectronic transducer motion direction discrimination based on logical functions of the transducer signals and motion pulses generated in the trigger circuits.

Fig. 15 presents the system of direction discrimination and the way of output impulse forming in „+"IA and „-"IG channels. There are NAND gates with Schmitt triggers at the system input. Their task is to form steep slopes of input signals and to narrow the switching zone from 1 to 0 inversely. This is important in the case of possible object vibrations and of moving direction change. The pulse counted in the given direction must be counted again, in the opposite movement direction, after having changed the moving direction.

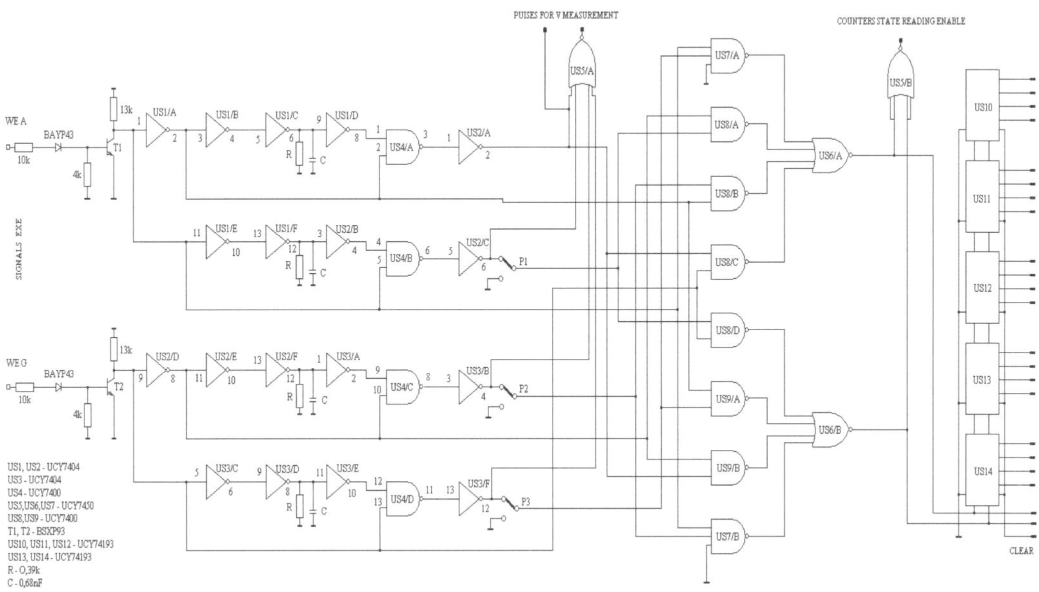

Fig. 14. System of optoelectronic transducer motion direction discrimination based on logical functions of the transducer signals and motion pulses generated in the RC circuits

The Schmitt gates formed output impulses at A and G points (Fig.15) have the polarity of rotary-pulse transducer output pulses because of the negations performed by the photoinsulator (optoelectronic coupler).

The system of JK flip-flops and gates perform basic functions of movement direction discrimination. The JK flip-flops are synchronised with the CLK clock pulses the combinational circuit performs the following function:

$$„+\text{"IA} = (C\,\overline{E}\,F)\ \text{v}\ (\overline{C}\,E\,\overline{F}) \tag{11}$$

$$„-\text{"IG} = (\overline{CE}\,F)\ \text{v}\ (CE\,\overline{F}) \tag{12}$$

The rule of output impulse forming has been presented in Fig.16. Pulses for the positive direction are obtained at A = „1" and the falling slope of the G signal and at A = „0" and the rising slope of the G signal. Impulses for the direction taken as negative are obtained at A = „1" and the rising slope of the G signal and at A = „0" and the falling slope of the G signal.
The system allows the formation of two output impulses in the given direction during one period of the input signal. The pulses are then counted in the impulse counters. The reverse counter state is the object position measure.

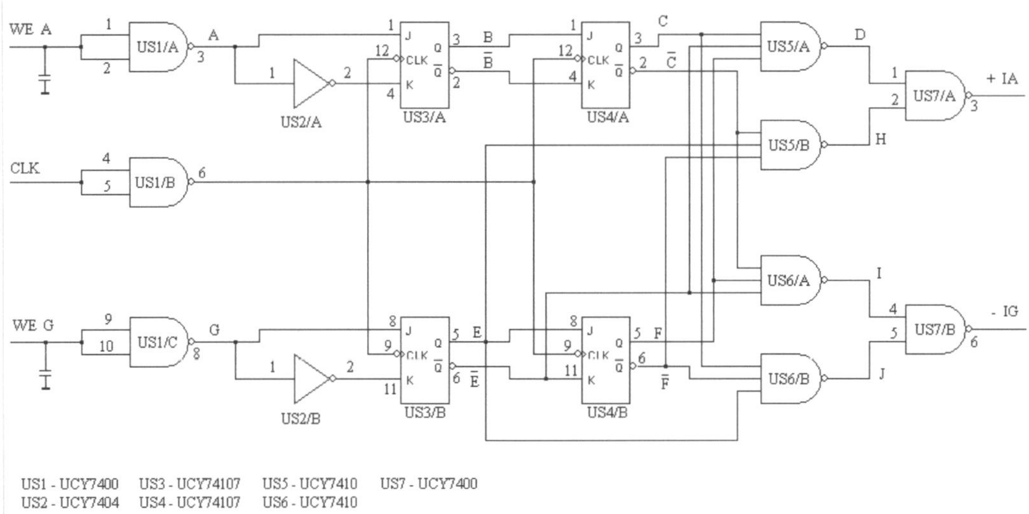

Fig. 15. System of optoelectronic transducer motion direction discrimination based on logical functions of the transducer signals and motion pulses generated in the trigger circuits

10. Application of the optoelectronic transducer to measuring velocity

The angular and linear velocity is the physical value of which measurement is used in almost all system solutions of drive controlling. Angular velocity measurement is made by analog or impulse sensors. Analog methods characterize little precision. A much higher precision of measurement can be reached by using digital methods. During measurement of displacement it is sufficient to use system counting following impulses where each impulse corresponds to specific displacement (Szcześniak, 2005)a (Szcześniak, 2005)b. During measurement of velocity additional timing needs to be assured what with considerable amount of impulses determining displacement gives in result velocity.
Presently the best solution of an impulse counting system, storing needed data, measuring time and doing calculations is a personal computer equipped with appropriate measuring-controlling interfaces (Szcześniak & Szcześniak, IV.2009) (Szcześniak & Szcześniak, VIII.2009) .
In the case of using the incremental (impulse) optoelectronic transducers velocity is determined on the grounds of:

- time measurement between two following impulses generated by the photo-optic disc connected with the drive.

- number of impulses of the transducer counted in particular period of time

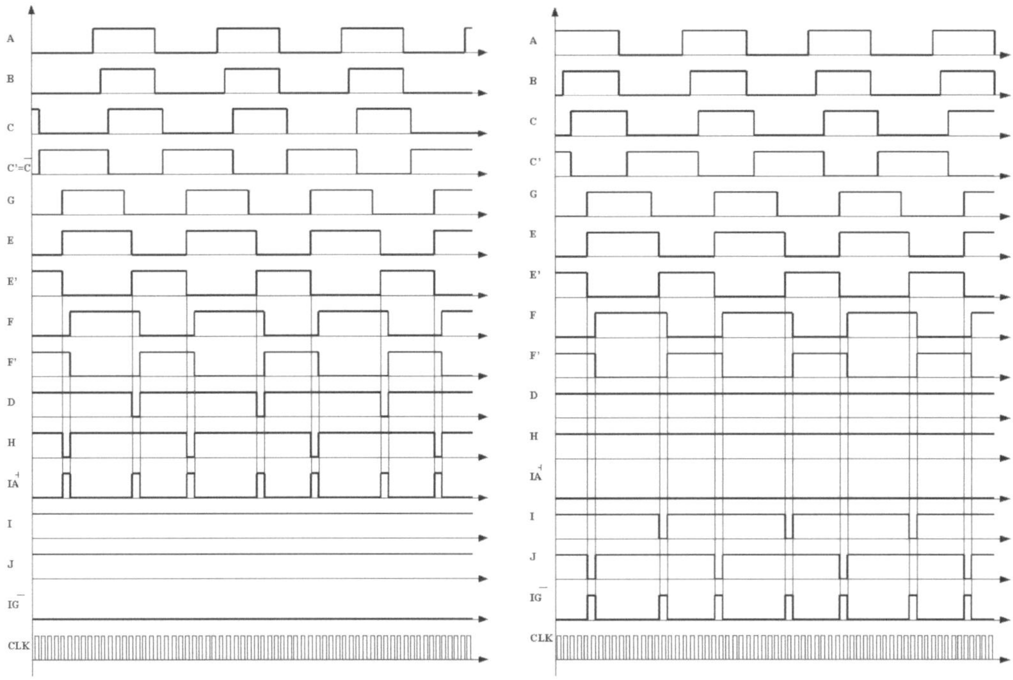

Fig. 16. Method of optoelectronic transducer motion direction discrimination based on logical functions of the transducer signals and motion pulses generated in the trigger circuits for movement to the right and to the left

In this chapter digital methods for measuring angular or linear velocity of drive were presented. Methods of measuring small and large frequencies with using the optoelectronic transducer were compared. A measuring algorithm and measuring system structure was discussed.

A measurement method of measuring high velocities is the most frequently realized by using the impulse transducers. This method is based on counting transducer pulses within the limited time T_w. The time is then sampling time. If the counter counts n impulses within the T_w time, and the path quantum is equal to ΔL or$\Delta\gamma$ and is known from construction of the transducer and co-operating systems construction, then angular speed is determined as:

$$v = \frac{N \cdot \Delta L}{T_W} \tag{13}$$

whereas for angular velocity:

$$\omega = \frac{N \cdot \Delta\gamma}{T_w} \tag{14}$$

These equations show the essence of the measurement. The idea of working of measurement system is shown in fig.17 and fig.18. The transducer coupled with moving element sends sequence of impulses (signal UI). The model generator sends impulses G_w of model frequency f_w. A frequency divider DF sets time of measurement T_w. Time of measurement is set basing two following signals from the frequency divider. Impulses are counted by the counter in time T_w and from this, it is easy to determine velocity using presented previous equations.

The above measurement method is not very precise for small speeds, because of the small number of pulses counted within T_w time. In this situation the method based on measurement of the time between consecutive transducer impulses, corresponding to the specified displacement, should be applied fig.17, fig.18.

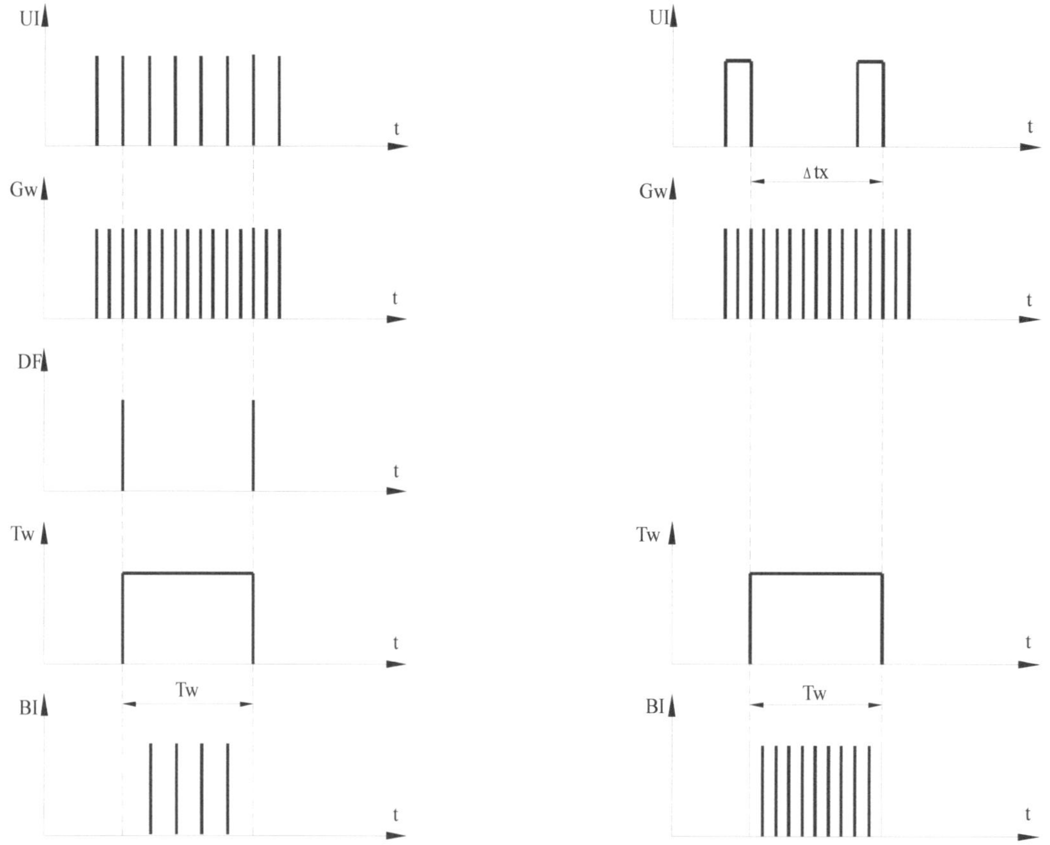

Fig. 17. Diagrams presenting way of measuring

In this method the transducer generates impulses of unknown duration Δt_x .Duration is measured between the following transducer impulses. A standard frequency generator sends standard impulses BI of known period $T_w = \frac{1}{f_w}$. The counter counts the impulses within the Δt_x time. Then linear speed is determined by the equation:

$$v = \frac{\Delta l_w}{\Delta t_x} = \frac{\Delta l_w}{N \cdot T_w} = \frac{\Delta l_w \cdot f_w}{N} \tag{15}$$

whereas angular speed:

$$\omega = \frac{\Delta\gamma_W}{\Delta t_x} = \frac{\Delta\gamma_W}{N \cdot T_W} = \frac{\Delta\gamma_W \cdot f_W}{N} \quad (16)$$

where:

Δl_W $\Delta\gamma_W$ -known displacement, Δt_x - unknown displacement time, $T_W = \frac{1}{f_W}$ - standard generator pulse period, N - number of pulses counted within the Δt_x time.

Measured velocity in method of high velocities is the number of transducer impulses counted in constant measuring time and is proportional to the measured velocity. In the second method velocity is inversely proportional to the number of impulses from the model generator (constant frequency) in fixed time dependant on the appearance of the following impulses from the transducer.

The diagram of the system to measure angular velocity using the incremental optoelectronic transducer, which idea of measurement consists in counting the number of impulses in the determined period of time is presented in fig.18.

The input transducer signals A and G (electrical signals) shifted one from another by 90°, are inserted into the system forming impulses (UI) and to the system distinguishing motion direction (UW_k), which was realized on the D flip-flop. In the (UI) system there is made multiplication of frequency of input signals and appropriate forming of impulses in order to increase resolution of velocity measurement (multiplication of frequency impulses) (Szcześniak, 2006)b.

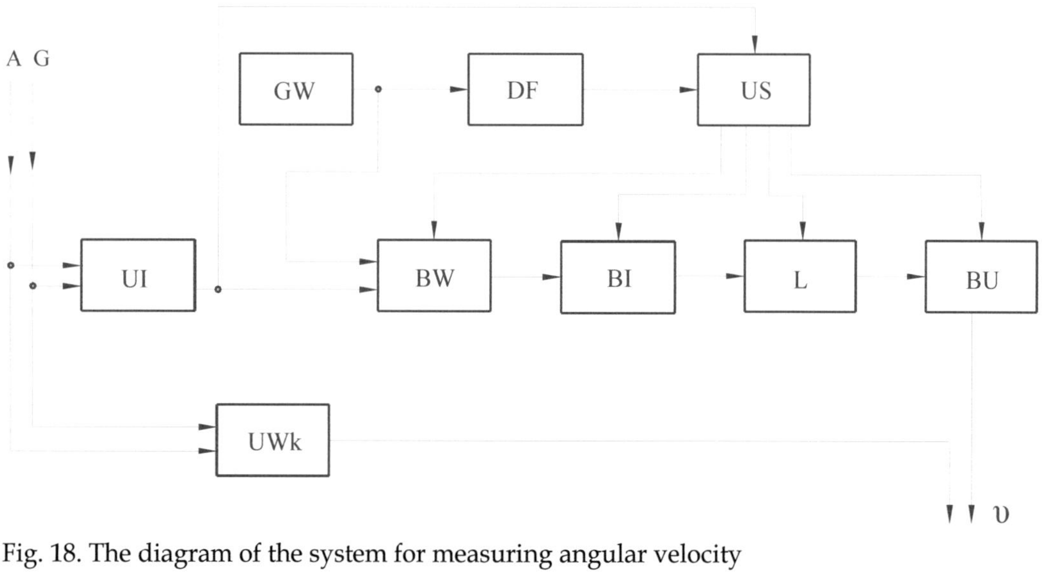

Fig. 18. The diagram of the system for measuring angular velocity

The generator stabilized by quartz (GW) and set of the frequency dividers (DF) produce clock signal. The impulse signal of increased frequency (UI) is multiplied with the input signal from the frequency divider and next is put on the input of the counters (LI). To control the counters and the buffers (BU) output signal from the frequency divider and shaped in the control system (US) was used. Counting impulses is done during high state. Resetting the counters and putting their value to the buffers happens during low state of clock signal. In case of measuring small frequencies, the signal from the generator (choice the gate WS) is multiplied by clock signal created in the control system (US) from the two following transducer signals (UI).

With problem of choosing right measuring method of high or small velocities can help us relative error of measurement, assuming allowed relative error of measuring velocity maximum number of impulses from the transducer is determined. This number is set for determining which method should be used. Simultaneously it determines number of counted impulses from transducer determining time of counting impulses of model frequency. Model frequency (f_w) is determined from border measurement of method that is for a determined time of counting impulses (T_w) and border number (N) of impulses from the transducer (the method of high velocities). Above can be presented as:

$$N = f_W T_W \tag{17}$$

Selection of the method is determined according to the number of appearing transducer impulses in measuring time which is compared with the set number of impulses. The set number of impulses is determined according to desired accuracy of measurement of quant velocity and its range. The set number of impulses is determined according to the dynamic of a measured object (Szcześniak, 2005)d.

11. Summary

The presented ideas have important meaning to the economy consisting in creating new methods, algorithms and devices in order to increase of quality and reliability of displacement controlling. During designing measurement systems it is needed to take into consideration such things as:

1. The production of the precise optoelectronic position transducers requires advanced technologies, which involves high manufacturing costs. There has been a general tendency towards applying the optoelectronic quantization transducers of simpler design and lower accuracy which could be enhanced by electronic means.
2. In methods of processing high alternating signals of the optoelectronic transducer especially when high precision is needed, dynamic compensation of DC-component of their signals is necessary.
3. The electronic method of interpolation (multiplication of signals frequency) and digitalisation (conversion to digital from) presented in the paper makes it possible to enhance accuracy of the photoelectric transducer. The method of 5-time frequency increase is universal and can be applied to another scale of transducer signal frequency multiplication.
4. Stated that the systems of the motion direction of the transducer which shape one, two or four impulses per period can be constructed and they are additionally enable to increase accuracy of measuring displacement. Application of the presented method of interpolation and digitalization and counting impulses enables to increase accuracy of the transducer twenty times.
5. Presented methods enable correct readout of position and velocity measurement of an object independently from the motion direction and also eliminating incorrect readout

in case of vibrations of an object. Upper frequency limit which the transducer can create or the counting system count, determine criteria of selection minimal value of time T_w called measuring or average time.

6. The use of a system of high value of impulse frequency increase has significant impact on increasing accuracy of position and velocity measurement. In dependence of velocity measuring range of drive, method of measuring high or small velocities should be used. Also relative error of measurement determines using the particular method is assumed.

12. Acknowledgements

This chapter is partly funded by the European Union as part of the European Social Fund entitled "Developmental Program of Didactic Potential of the University of Technology in Kielce - education for success". Operation Program Human Capital, Agreement number UDA-POKL.04.01.01-00-175/08-00.

13. References

Ching-Fen, K. & Mao-Hong, L. (2005). Optical encoder based on the fractional Talbot effect, *Optics Communications*, 250, pp. 16–23, Elsevier

Feel-Soon K. & Sung-Jun P. (2005). A linear encoder using a chiaroscuro and its extension to switched reluctance motor drive, *Energy Conversion and Management*, 46, pp. 1119–1128, Elsevier

Heidenhain catalog (2003)

Holejko K. (1981). Precyzyjne elektroniczne pomiary odległości i kątów, WNT, Warszawa.

Pizoń A., Sikora K., Stefański T., Szcześniak Zb. (1993). Microprocessor - based electrohydraulic position system, *Prace VII Międzynarodowej Konferencji nt. „ Pneu - Hidro"*, Miskolc, pp.347 -352

Pizoń A., Sikora K., Stefański T., Szcześniak Zb. (1993). Minimierung der Regelabweichung in einen elektrohydraulischen nichtlinearen mikrorechnergesteuerten Synchronsystem, *Prace IX Międzynarodowej Konferencji nt „Hydraulik und pneumatic"*, Drezno pp. 427 - 436

Szcześniak A. & Szcześniak Zb. (2009). Methods and analysis of processing signals of incremental optoelectronic transducer, *Review of Scientific Instruments*, Vol. 80, pp. 1-6

Szcześniak A., Szcześniak Zb. (IV.2009). Mikroprocesorowe przetwarzanie sygnałów optoelektronicznego przetwornika położenia, *Przegląd Elektrotechniczny (Electrical Review)*, ISSN 0033-2097, vol. 85, pp.153-158

Szcześniak A., Szcześniak Zb. (VI.2009)a. Signals of optoelectronic transducer processed in flip-flop circuits, *8-th European Conference of Research and Scientific Workers TRANSCOM 2009*, University of Žilina, Slovak Republic. pp.81-85

Szcześniak A., Szcześniak Zb. (VI.2009)b. Phase signals in the optoelectronic position transducer system. *Logistyka*, pp 1-7

Szcześniak A., Szcześniak Zb. (VIII.2009) Mikroprocesorowy symulator sygnałów optoelektronicznego przetwornika położenia. *Przegląd Elektrotechniczny (Electrical Review)*, ISSN 0033-2097, vol. 85 NR 8/2009 pp.119-122

Szcześniak Zb. (2005)a. Struktura pomiarowo - informacyjna mikroprocesorowego układu pomiaru wysokości odkuwki w procesie kucia swobodnego, *Zagadnienia Eksploatacji Maszyn*, Polska Akademia Nauk, Vol. 143, No. 3 pp. 101-109

Szcześniak Zb. (2005)b. Cyfrowe metody pomiaru prędkości z zastosowaniem optoelektronicznego przetwornika położenia, *Elektronika,* No. 2-3, Warszawa 2005, pp. 73-74.

Szcześniak Zb. (2005)c. Metoda interpolacji sygnałów elektrycznych optoelektronicznych przetworników w pomiarach położenia, *Elektronika,* No. 1, Warszawa 2005, pp. 74-76

Szcześniak Zb. (2005)d. Analysis of the position and velocity measurement line for the hydraulic press crossbar with the use of photoelectric transducer, *Advances of Manufacturing Science and Technology, Polisch Academy of Sciences* Vol. 29, No 3, pp. 75-84

Szcześniak Zb. (2005)e. Method of processing accuracy enhancement of the photoelectric position transducer signals Metrology and Measurement Systems, *Polisch Academy of Sciences* Vol. 12, No 3, pp. 315-322

Szcześniak Zb. (2005)f. Elektroniczny moduł dziesięciokrotnego zwiększania dokładności przetwarzania sygnałów optoelektronicznego przetwornika położenia, *Polska Akademia Nauk, Elektronika i Telekomunikacja,* Vol. 51, No. 3 pp. 439 - 447

Szcześniak Zb. (2005)g. Metoda kompensacji składowej stałej sygnałów pomiarowych optoelektronicznego przetwornika położenia, *Pomiary Automatyka Kontrola,* No. 6, Warszawa, pp. 12, 13

Szcześniak Zb. (2006)a. Methods of converting of electric signals from photoelectric position transducer for discrimination of its motion direction and for increasing its measurement accuracy, *Polisch Academy of Sciences, Electronics and Telecommunications,* vol. 52, No. 1, pp. 23-30

Szcześniak Zb. (2006)b. Photoelectric position transducer with the optimum compensation of the dynamic constant component of its signals, *Electronics and Telecommunications, Polisch Academy of Sciences,* Vol. 52, No. 2, pp.185-192

18

Optoelectronic Measurements in Science and Innovative Industrial Technologies

Vladimir G. Meledin
Kutateladze Institute of Thermophysics,
Siberian Branch, Russian Academy of Sciences
Institute of Optic-Electronic Information Technologies,
Public corporation, Novosibirsk
Russia

1. Introduction

Optoelectronic measurements makes it possible to find real information about real world quantities (M.Planck) with advantages of the absence of contact, performing operations in a parallel optical form and with modern element base of optoelectronics and photonics. Experiment and practice (according to N.Wiener) play a key role in obtaining adequate information. Adequate experimental information in science allows one to create adequate theoretical models and verify numerical computer calculations and codes. Adequate information, along with resources, has become a decisive industrial factor practically in all fields. To obtain such information, one often needs industrial organization of physical experiments with new data processing methods and modern measurement base of "innovative technologies" in informatics, lasers, and photonics.
Investigations of fine effects are based on the creation and development of methods to extract data from noise-like signals. Bandwidth narrowing and synchronous detection imply a deterministic effect on the object, processing of responses, and organization of information exchange. The interaction with the object is performed at all stages. This enables to extract a maximum possible amount of information. Complete use of the information implies, methodologically, the use of a multidisciplinary approach, the creation of an infrastructure for fast performance of experiments, and a highly qualified optimistic team.
The material is presented as a sequence of sections based on a unified methodology and illustrating, step-by-step, the efficiency and fruitfulness of the fundamental principles and scientific ideas proposed in informatics of optoelectronic measurements illustrated by important practical results. Most significant and prominent results were selected for this presentation. These large-scale and productive results are successfully used in various fields of science and technology.
Laser diagnostics of multiphase flows, information monitoring in hydropower engineering and hydraulic turbine machine building, metallurgy, and railway transport, glass industry and medicine, semiconductor electronics and nanotechnology, atomic and oil and gas industries - this is a list of fields where the fundamental scientific ideas being proposed are successfully used. These results can help to solve, in many cases, the problem of import

product replacement and provide a considerable increase in the effectiveness and safety of numerous new industrial technologies.

2. Semiconductor laser Doppler anemometry of multiphase flows

High-precision nondisturbing optoelectronic measurements of the kinematics and structure of multiphase flows in the regime of back light scattering are needed in experimental mechanics, physics, chemistry, biology, medicine, ecology, and engineering. Velocity information is obtained when measuring the Doppler shift of the frequency of laser radiation scattered by small suspended particles and phase inhomogeneities moving in the flow. Nanotechnologies, experimental hydro-, gas-, and plasma-dynamics of multiphase systems, safety, resources, and ecology of power engineering, aviation, rocket, and nuclear engineering, shipbuilding, ground-based transport, machine-tool building, and instrument making--this list of fields for which such systems are vitally important is far from complete. Recently, of special importance are code verification in Computational Fluid Dynamics (CFD) and computer models of flows in science and innovative industrial technologies. Such laser information systems, which are highly competitive with the best foreign analogs, simple and safe, are also necessary for educational purposes (Meledin, 2009).

This high level of achievements is based on a series of papers, published in the 80s and 90s, on the creation of systems based on gas lasers and development of the first injection laser emitters for optical instrument making - jointly with Ioffe Physico-Technical Institute, RAS, and under the guidance of Zh.I. Alferov, Nobel laureate for physics (Arbuzov et al., 1992). This produced the world's first Doppler meters of velocity and linear sizes of hot rolled metal for metallurgy on semiconductor lasers (the Altair series presented below). This significant scientific progress resulted in creating in 2003 an information-measurement system called LAD-0x within the framework of a program on import product replacement, Siberian Branch of the Russian Academy of Sciences. This took place three years before the appearance of the world's best analog, DANTEC (Meledin, 2008).

Semiconductor lasers (with an efficiency of 20-60 %) made it possible to create information-measurement systems with fundamentally new functional capabilities. These systems are efficient and small in comparison to systems based on gas lasers (with an efficiency of 0.03 %). However, the quality, space, and time coherence of optical radiation and applicability in metrology of semiconductor lasers make them inferior to gas ones (Belousov & Meledin, 1985; Belousov et al., 1988).

In order to replace highly coherent gas lasers by semiconductor ones in information-measurement interferometric systems, some theoretical and experimental investigations have been performed. Theoretical studies on some peculiarities of formation and recording of optoelectronic signals under the photonic constraint and investigations on stability of anamorphotic schemes have been carried out. Also, new methods for parametric stabilization of wavefronts and photoelectronic transformations of optical signals with a dynamic energy range up to 10^{16} have been created (Rahmanov et al., 2006).

An important line of investigation was transition from flows with artificial light-scattering particles used in some foreign methods to processing of signals of natural suspensions with unusual characteristics of light scattering. Only in this case it has become possible to create onboard underwater systems to study the fine microstructure of hydrophysical ocean fields.

A nontrivial scientific task is to create Doppler information-measurement semiconductor laser systems operating with back light scattering from random suspensions in flows. To solve this problem, a series of multidisciplinary and fundamental investigations had to be carried out: in informatics, signal processing methods, quantum optics and optoacoustics, radio physics, one-photon reception, data collection and processing systems, etc (Dubnistchev et al., 1987; Belousov et al., 1988, 1996, a, 1990, a).

In the field of processing and transforming optoelectronic signals from random suspensions, some methods have been created to handle ergodic processes with operations separated in time and having maximum possible field energy concentration in the measurement volume and maximum signal-to-noise ratio (Meledin et al., 1999; Dubnistchev et al., 2000, 2003).

Information-measurement semiconductor Doppler laser systems for diagnostics of multiphase flows (LAD-0x system), designed for noncontact measurement and visualization of two and three components of the velocity vector of gas-liquid multiphase, turbid flows, and concentration of light-diffusing particles, have been created. Also, a variant of the information-measurement scheme of a three-component anemometer based on a combination of two two-component systems has been developed. Each two-component system has all elements of optical and electronic subsystems and operates with an external computer via a standard network channel. Owing to the TCP protocol, the computer can be placed in an arbitrary place (for instance, in another city), and some physical experiments can be performed in a remote mode, in closed and dangerous zones, and at test sites, and can be used for distance education. In these systems, Russian-made algorithms and components are used, which have no foreign analogs.

The LAD-0x information-measurement semiconductor Doppler laser systems are used in the following Russian scientific and educational organizations: the Institute of Hydrodynamics (diagnostics of vortices and internal waves), the Institute of Thermophysics SB RAS, the Institute for Safety of Nuclear Power Engineering RAS (heat and mass exchange, safety of nuclear power engineering, nanotechnologies), Perm State University (investigations of convection), Tomsk State University (investigations of forest fires), Research Center of Power Engineering Problems, and Kazan Scientific Center, RAS. These systems are requested by many of the Russian organizations (Anikin et al., 2004; Meledin et al., 2000, 2001, 2003).

The information-measurement semiconductor Doppler laser systems created for diagnostics of multiphase flows meet the modern requirements and forecasts on the expected development of fundamental and applied science in the nearest future.

At the VII Moscow International Salon of Innovations and Investments in 2007, these developments were rewarded with Golden Medals and First-Rank Diplomas. At the VIII Moscow International Salon in 2008, they were rewarded with diplomas and Silver Medals, received the titles of laureates of the VII Contest of Russian Innovations (nomination: "The best 2008 Industrial Innovation of Russia," and were rewarded with diplomas and small Golden Medals of the Sibpolitekh-2008 International Industrial Exhibition.

3. Hydropower engineering and hydraulic turbine machine building. Optoelectronic for optimization of operation modes and safety of hydropower plants information

3.1 Dynamic monitoring of the geometry of loaded rotors of super-high-power generating units of the Neporozhny Sayano-Shushenskaya HPP

Dynamic monitoring of unique super-high-power generating units requires solving some serious scientific problems. The dynamic shape of a rotating rotor is a very important parameter determining the safety of operation of the aggregate on the whole. The purpose

of this work was to create an information system based on methods and technology of monitoring the geometry of loaded rotors of super-high-power generating units of the Sayano-Shushenskaya HPP to increase the efficiency of their operation and safety.
The power of one hydrogenerator is the same as that of the power-generating unit of a nuclear power plant (735 MW), the stator diameter is 17 m, the rotor diameter is 14 m, the rotor weight is 750 tons, and the linear velocity at the edge of a rotating rotor is 100 m/s. The consequences of a failure with the energy accumulated in the rotor are comparable to those of a failure at a nuclear plant.
Selective mechanical control of the rotor and stator shapes is performed by the operators when the aggregate is unloaded and completely stopped. This information is used to predict changes in the rotor shape while the aggregate is running. It is evident that the deformation of the rotating rotor shape and the dangerous decrease in the gap between the rotor and stator cannot be reliably estimated.
The well-known control methods used when the generator is stopped cannot be employed when the aggregate operates under normal loading. The major reason for this is as follows: there are considerable electromagnetic fields and no free space for recording equipment near the rotor. Information monitoring of the rotor geometry must be continuous, noncontact, and remote. Under conditions of considerable electromagnetic noise, vibrations, temperature differences, and dust, the monitoring must not affect the operation of the existing systems.
From the above scientific investigations, a method for information dynamic monitoring of the rotor shape of a loaded super-high-power energy aggregate at the Sayano-Shushenskaya HPP, based on a semiconductor lidar with coherent-optical feedback and hardware-software processing of optoelectronic signals, was proposed and implemented for the first time. The lidar was placed in the hot air chamber of the stator and adjusted relative to the rotor rotation axis. A sounding beam was directed to the rotor surface through a narrow extended channel of air cooling and ventilation in a 10x20x500 mm stator body.
Scanning of the rotor surface shape, which moved at 100 m/s, by the semiconductor lidar was made in the transverse direction. After software-hardware processing of the lidar signal, the frequency, being directly proportional to the distance to the rotor surface, was determined. Then, phase averaging of the data was made. Using the averaged data, a rotor surface profile was constructed and tolerance control was made.
The figures show some rotor poles and a standard diagram of deviation (in millimeters) from a constant radius of 6700 mm. This information about the rotor surface shape of an operational hydroelectric generator of the most powerful hydropower plant in Russia was obtained for the first time. The tests were performed on operational hydroelectric generator no. 4 of the Sayano-Shushenskaya HPP (type: RO 230-v-677). These full-scale experiments and tests have confirmed that the basic scientific principles on which this development, which is relatively simple to implement, is based are applicable and efficient. The measurement error of the experiments did not exceed 1% (Kulikov et al., 2010).
The results of this series of works were rewarded with a First-Rank Diploma and Golden Medal of the VII Moscow International Salon of Innovations and Investments (2008).

3.2 Hydropower engineering: laser diagnostics of flows downstream of hydroturbines to increase efficiency and safety

The Power Machines - LMZ (Saint-Petersburg) is the largest turbine construction plant in Russia. This open stock company produces power turbines. In the past 90 odd years of

development and creation, more than 2300 turbines with a total power of about 300 million kW, more than 670 hydroturbines with a total power of about 62 million kW, about 1000 hydroturbine regulators, and 60 pre-turbine gates have been produced. The Power Machines - LMZ is a leader in domestic and world hydroturbine construction. Power Machines - LMZ turbines operate at large hydropower plants, such as Volga, Bratsk, Krasnoyarsk,
Ust Ilimsk, Sayano-Shushenskaya (Russia), Maika, Site 1, Jenpeg (Canada), Sobradinho (Brazil), Uites (Mexico), Hoabin (Vietnam), and many others. With its hydraulic turbines operating in 30 countries of Europe, Asia, Africa, and America, the LMZ gained substantial experience in hydroturbine construction. Power Machines - LMZ plans are to supply hydroturbines, regulators, and gates for electric power plants in India, Argentina, Iraq, Kazakhstan, Latvia, and Russia.
The important scientific problem of obtaining information about the real dynamic characteristics of a multiphase cavitating flow at the outlet of a hydroturbine has been solved specially for the Power Machines - LMZ. In 2004 through 2005, new information-measurement methods and 2D laser systems, LAD-05 and LAD-06, and a 3D laser information-measurement system, LAD-056, were created and introduced. These systems are designed for bench testing, optimizing, and certifying various turbine models, verifying CFD calculation codes, and increasing the efficiency of operation and safety of domestic power engineering. Some foreign projects (DANTEC and others) were also considered, but the above systems were chosen owing to their scientific and technical parameters (Meledin et al., 2006, a,b,c).
New Doppler laser information-measurement systems were created on the basis of the fundamental investigations described above. These are used to determine the kinematic parameters of twisted hydrodynamic flows and the influence of parameters of large-scale helical vortex structures on the transfer processes in twisted flows downstream of the hydroturbine wheel. Such systems were installed on a large cavitation stand of the Silovye mashiny-LMZ.
First full-scale experiments for all operational modes of the Frensis turbine have been performed. A 2 % total estimate for the error of the flow rate of a two-phase cavitating flow has been obtained. The three-dimensional velocity fields of a multi-phase flow at the hydroturbine outlet have been constructed (Sadbakov et al., 2004).
The results of comprehensive testing of information-measurement Doppler laser systems for diagnostics of gas-liquid flows downstream of hydroturbines have verified their technical characteristics. The systems have been found to be suitable for exploitation, bench tests, optimization, subsequent certification of various turbine models, and verification of the CFD calculation code. These systems have provided an increase in the efficiency of operation and safety of domestic hydropower engineering.

4. Metallurgy. Laser monitoring of velocities and linear sizes of hot rolled metal body

Obtaining adequate information about the velocities and linear sizes of a hot rolled metal body in metallurgy is a complicated multidisciplinary scientific problem. The light-diffusing surface of a hot rolled metal body (with a temperature of up to 1200°C) is a set of random roughnesses. The resulting optical signal is formed by the superposition of light waves scattered with random phases on a set of centers confined simultaneously within a sounding optical field. Multiple interference increases the phase noise, forms a multi-fractal small-contrast signal, and decreases the accuracy of measurement. Permanent oxidation, which affects the light scattering and disturbs the Doppler spectra, takes place on the hot metal body surface.

Being linearly displaced, the rolled metal body participates in a multiple motion: there are vibrations, turns, inclinations, and bends forming optoelectronic signals with a very large dynamic energy range. The convective flows near a hot rolled metal body cause considerable refraction, which modulates light both in space and in phase, and contain a moving suspended aerosol. There is a lot of mud and water, in particular, in the form of moving films. Measurement of sizes is also a difficult scientific problem of boundary recognition and stable interpolation of signal losses. And, finally, since huge masses and energies are involved in metallurgic processes, high reliability and stability are needed for the entire operation of information-measurement complexes.

On the basis of the fundamental results obtained in our investigations, a number of information laser control systems have been created. These are used to sound moving rolled metal body surfaces, record the Doppler frequency shift of scattered coherent light, which is proportional to the velocities of the objects, accumulate information, perform statistical software-hardware processing, and form standard signals for metrological support, monitoring, and control of the technological processes.

The information systems made it possible to obtain, in real time, adequate information (not available previously) about the velocity and geometry of rolled metal bodies with an error of less than

0.1 %. Also, they provided automatic monitoring and increased stability of the technologies used for unstable and worn materials and fuels in variable environmental conditions. Since the obtained information was adequate and accurate, a considerable economy and an increase in the products of higher quality have been obtained.

The first Russian Qazar-M information-measurement optoelectronic system that was successfully used at the Novosibirsk Metallurgy Plant up to 2006, was developed and introduced in 1986. In period from 1985 through 1987, first laser noncontact Doppler meters of velocity and rolled metal length of the "Altair" series were developed and tested in conditions of real metal production on semiconductor lasers developed at the Ioffe Physico-Technical Institute (by Zh.Alferov and D.Garbuzov). At their creation, these systems were the world's first developments.

In 1986, the Institute of Economics and Industrial Production Organization, Siberian Branch of the USSR Academy of Sciences, made an estimate of the economic effect obtained by introducing the information-measurement laser complexes into ferrous metallurgy. The estimate is of certain interest. This is as follows: the economy in the metal is 1,422,744 tons/year and the total economic efficiency exceeds 105,000,000 rubles (in 1986 prices). The effect was very impressive, and in 1986 the USSR Ministry of Instrument Making, Automatization, and Control Systems and the USSR Ministry of Ferrous Industry adopted a program on the creation and introduction of Doppler laser measurement systems for metallurgical purposes in the industry.

To measure the velocity and rolled metal length on the basis of *He-Ne* and semiconductor lasers, Qazar, Altair, and LI-803 optoelectronic Doppler information-measurement systems have been developed. In the Qazar and LI-803 meters, *He-Ne* lasers are used as sources of coherent radiation. A number of integrated research-and-development works have been performed, and small lots have been produced. In 1993, LI-803m information-measurement systems were created and tested at large metallurgy enterprises, namely, at Nizhnetagilsk and West Siberian metallurgical plants. These systems had an ultra-long-range measurement distance of 3 to 5 m (the maximal distance of the existing foreign meters was 2 m) and a maximal measurement zone depth of more than 200 mm. The major technical characteristics of these information-measurement systems were higher than those of existing foreign analogs and were better suited for real conditions of domestic metal production (Meledin, 1995).

A number of methods for controlling and testing the accuracy characteristics of some information systems oriented to industrial technologies have been validated both theoretically and experimentally. An automated software-hardware stand complex to test and adjust information systems and to assess long-term stability and reproducibility of their metrological characteristics has been created (Belousov et al., 1990 - b, 1996 - b).

Accurate verification of the information obtained in this way is rather easy. A section of 10.00 m length measured with an error of 1 cm can be considered as a reference object of 0.1 % accuracy. As such an object moves at any regime in the measurement zone of the information-measurement complex, the latter, after integration of the measured velocity, must indicate a length of 10.00. The complex is tested in a "black box" regime as a whole, from the basic scientific ideas to the peculiarities of practical realization, because an error in any method, algorithm, or element may make accurate operation of the entire system impossible.

Multiple industrial tests of these systems have been carried out. Specifically, the Qazar-3M meter has been tested at the West Siberian Metallurgical Plant, on a continuous-supply mill 850/730/580 after a roughing train. The temperature of rolled 150x150x80,000 blocks (St3 PS steel) was about 1000°C. The error of measuring the linear sizes of hot rolled metal bodies by these systems in real industrial conditions was 25 times less than that of the standard production system for the inspection of rolled body lengths. The tests have shown that the Doppler laser measurement systems have considerable advantages in accuracy and reliability. It has been proved experimentally that information monitoring of the rolled metal production technology in real time and high quality of adjustment of the successive series of rolling mills in production lines are feasible.

Comprehensive tests of these information-measurement systems were made in 1994 at the Nizhnetagilsk Metallurgical Plant in a rolling shop of broad-flanged beams. This shop is the last in the entire technological chain of the complex, and this new advanced technology provided maximum output.

It was found experimentally that the average deviation of the rolled metal length from the real value measured by the system is less than 0.05 % and the maximal deviation does not exceed 0.099 %. These errors had a compensatory bias component produced by the errors of the setup in tests of the system optoelectronic sensor at a rolling mill. Since that time, the system of information laser control of hot rolled metal, LI-803m, has replaced the standard IDSh measurement complex, since its functional characteristics, accuracy, and reliability are by an order of magnitude higher.

Information-measurement systems to control the transverse size of hot rolled metal based on the original scientific idea of measuring the size of the object by estimating its integral radiance in the near infrared region have been developed. It has been shown, both theoretically and experimentally, that the estimates obtained above are stable to the distortions of visible object boundaries produced by refraction, aerosol diffraction, etc. First accurate measurements of the transverse sizes of hot rolled metal with an error less than

0.2 % at a distance of 5 to 15 m from the dangerous zone have been obtained (Bakakin et al., 1997).

The successful scientific idea has made it possible to create a small inexpensive information-measurement complex. The efficiency of this complex was proved in tests performed at the Nizhnetagilsk Metallurgical Plant. Information about the distributions of sizes of hot rolls and the roll temperature was obtained for the first time. The following data were also obtained for the first time: the flange height variation of hot beams, temperature variation along the roll, and spatial boundaries of "tongues".

A second series of tests of the system was made at the West Siberian Metallurgical Plant in 1997. Information about the spatial distributions of transverse sizes of hot rolled metal in a breakdown mill after the finishing train was obtained for the first time.

Since 1994, the LI-803m information-measurement complexes have been used in the production record system of a broad-flanged beam shop at the Nizhnetagilsk Metallurgical Plant. The continuous reliable operation of the system provided accurate measurements of the velocity and length of hot and cold rolled metal with a total error not exceeding 0.1 % and a daily volume of controlled products of 3,000,000 US dollars (in the prices of 1994). In the process of introducing the information system, it was found experimentally that the economic effect reached 100,000 US dollars per day and the pay-back period was not more than two days.

For the first time, information-measurement systems and laser software-hardware complexes for monitoring and controlling technological information on hot metallurgical rolled metal with characteristics even better than those of the foreign analogs have been created. These measurement complexes have been introduced and successfully used at large ferrous metallurgical enterprises of Russia: at the Nizhnetagilsk Metallurgical Plant (from 1993 up to now), the West Siberian Metallurgical Plant (from 1992 through 2001), and the Novosibirsk Metallurgical Plant (from 1986 up to now).

5. Transport. Information monitoring of the railway wheel geometry

Railway wheels and wheel pairs are very complicated geometrical objects. In production, they are subject to various operations: heating, pressing, rolling, calibrating, turning, and thermal treatment. In operation, they are intensively worn out and periodically turned to restore the profiles of the roll surface.

Information monitoring of the geometry is required at all stages of production control and life cycle of railway wheels. The number of major parameters to be controlled reaches 14. Traditional contact control methods are inefficient and selective, which inevitably increases the percentage of defective articles and accidents.

A joint work was initiated with the producers of railway wheels at the Nizhnetagilsk Metallurgical Plant in 1994. A new method to control the geometry of a rotating hot wheel by scanning the projection of the linear velocity vector at a point of measurement of the instantaneous radius was proposed and implemented. The velocity was effectively measured by a Doppler laser information-measurement system at a distance of several meters.

It was shown that this non-traditional use of anemometers to control the geometrical parameters of axially symmetric industrial products effectively solves the problems of measurement and control of the shapes of bodies with complicated geometries, in particular, when manufacturing hot railway wheels. The measurement error depended on the radius and the rate of radius change (the degree of asymmetry) of the object. It was about ± 0.1 % for an eccentricity of about 20 %.

The method was tested experimentally from 1994 through 1995. The measurement error was less than 0.1 % at a linear wheel surface velocity of up to 10 m/s, a surface temperature of up to 1100°C, and sounding angles $\beta \leq 30°$. Unfortunately, for some reasons in the period from 1995 through 1997, the work on the project was stopped. Nevertheless, a basis for the future fast development of dynamic measurement of the railway wheel geometry was created (Meledin, 1995).

The year 2001 marked a second stage in the development of this theme. A Laboratory on Optoelectronic Information Technologies, organized by the Technological Design Institute of Scientific Instrument Engineering jointly with the Institute of Thermophysics, SB RAS, started works on creating an Automated diagnostic complex to measure the wheel pairs of railroad carriages at approaching a railroad station (Center for Introduction of New Equipment and Technologies of the Russian Ministry of Railways, contract 1733-02/UCV/UPS-17/10 of 2001/08/22), and also joint works with the West-Siberian Railroad on Express diagnostics of moving wheel pairs of railroad carriages (called a Complex and an Express, respectively, under the guidance of Prof. V. Meledin).

The best foreign analog, the Argus system that was developed in 2000, had a length of 18 m, a weight of 11 tons, a power consumption of 30 kW, and operated at air temperatures of -10... +40°C (not suitable for Russia). The system cost was 3 million dollars, and no license was on sale. It is evident that to rapidly create a system for information monitoring of the geometry of railway wheels at reasonable costs, fundamentally new scientific ideas and a nonstandard approach to solving this complicated scientific and technical problem should be used.

At the start of these works, a method for obtaining information about wheel geometry based on laser anemometry was tested. Results of the testing were good, but the method was not simple enough to implement. After some scientific investigations, another method based on dynamic laser triangulation was used. The method had a relatively simple calibration in field conditions at considerable temperature differences and easy organization of additional measurement channels.

A rolling wheel pair of a moving train participates in numerous kinds of motion: translational motion along the rails and rotational motion. As the wheels move, the rails sag resiliently in the vertical direction with an amplitude of up to several centimeters, and the wheel pairs make small transverse damped oscillations. We proposed a new method for information monitoring of the geometry of railway wheels. In this method, the instantaneous coordinates of the ends of four crossed radius-vectors directed to the moving wheel pairs with self-scanning surfaces are recorded synchronously. They are modeled by beams of laser distance meters. The synchronously recorded data are accumulated, and the wheel shape is reconstructed with the use of dynamic models. Tolerance control of the reconstructed 3D virtual images of the wheel pairs is made.

The data processing was made by rejecting the unreliable observations, forming continuous sequences, interpolating the signal dropouts, and matched filtering, which improved considerably the signal-to-noise ratio. The calculations were made in quasi-real time (as far as it was possible for the standard operating system used in calculations).

The above-mentioned information-measurement complex, which used a method of dynamic 3D size control in a virtual system of coordinates, had sets of laser optoelectronic meters on the outer and inner sides of the wheel pair. The meters had special signal processors and communication channels with a calculator and a network server. Special software based on those methods and algorithms was created for the complex.

A peculiarity of the information-measurement complex was an untied base, which decreased the upper frequency of the microseisms, and a passive aerodynamic shield of the optical surfaces. The shield had a special design taking into account the active transport of atmospheric aerosols by the turbulent vortices produced by a moving train.

Because of some unscheduled actions of the Russian Ministry of Communications, the system was made very quickly (about six months since the contract signing; two months from the start of funding to the handing-over of an operational specimen at a time when the Trans-Siberian Railway was specially blocked off for this purpose).

The operational specimen of the Complex system (named after the corresponding project code) provided reliable monitoring of the geometry of railway wheels. This noncontact and remote monitoring can be used for a moving train under real weather conditions and in the presence of dust, mud, and other contaminants. As a result of these works, first laser measurement complexes have been created, which were tested in real conditions (at the Agryz station of the Gor'ky Railroad, from November through December, 2001). The algorithms were proved to be robust under various weather conditions (-20°C; rain, snow, and hoarfrost) and various train velocities (10… 25 km/hour).

Taking into account these successful results, in 2002, the Department of Railroad Cars provided additional funding of this theme. In February 2002, the Complex was given to a special organization headed by S.Plotnikov. Now, a simplified variant of the complex (without dynamic models and virtual systems of coordinates, only local measurements are made) is widely used on the Russian railroads.

At present, more than 38 large Russian railroad stations have been provided with a measurement system of the Complex type (Meledin, 2008, 2009).

6. Molecular electronics. Measurement complex for investigation and optimization of new-generation nonvolatile memory (from 2010 through 2012)

The electronics sector for memory chip production is on the verge of new serious changes. In the last 30 years, the physical principles on which modern memory elements are based have been unchanged. Major efforts of the electronic industry have been focused on the problem of miniaturization of electronic devices. At present, the possibilities of miniaturization have practically been exhausted. Since the need for cheap and versatile memory microchips is increasing, electronic companies carry out intensive investigations of new physical phenomena and materials to create a new generation of memory devices. Molecular and polymer materials are the first among the active and functional materials. These materials allow generation and real development of molecular electronics.

Searching for new conceptual and technological ideas, many electronic companies cooperate with universities and research institutes all over the world. Korean and American companies, which are major producers of memory microchips, have especially active cooperation. Spansion, a joint American-Japanese company, a world leader in the production of flash memory microchips (of the NOR type), is a daughter enterprise of the American AMD company and the Japanese Fujitsu company. The Spansion supports investigations on the creation of a new-type nonvolatile memory in the Siberian Branch of RAS.

The new generation of nonvolatile memory microchips under development has a resistive type of nonvolatile memory based on changing the conductivity of unit nanoelements. The

nanoelement of memory has a sandwich structure, in which the active material located between two electrodes changes the conductivity under the action of pulses of an electric field. This type of memory makes it possible to create passive memory arrays. This decreases considerably the number of transistors required for addressing unit elements.
Resistive memory types are most adequate for the logic of construction of electronic circuits. The new memory type allows one to create multilayer structures. This provides a considerable increase in the information density without changing the design lithographic norms and a minimum in the energy consumption.
Testing, investigating, and optimizing this type of memory cells call for the creation of a fundamentally new type of information-measurement complex. At the start of these works, there were no analogs of such a complex both in research and industry. There are special and rather contradictory requirements on the characteristics of an information-measurement complex to investigate and optimize the nonvolatile memory of a new generation.
When pulses of programming or deleting information (pulse duration: 10 ns, front: from 2 to 3 ns) act on a memory cell, simultaneous data reading (memory cell resistance measuring) takes place. Memory cell resistance may vary in the range from several hundreds of ohms to several hundreds of megohms. At a time constant of 3 ns and a resistance of 150 kiloohms the effective capacity of the system, including the cell, conductors, and electronic circuit, must not exceed 0.02 picofarads.
The creation of electronic devices with such parameters is a complicated scientific and technical problem.
Data recording, reading, and deleting in a memory cell are determined by the size and sequence of the corresponding electric pulses. Since the characteristics of individual resistive elements are widely spread, a mode switching control based on fast feedbacks in current and voltage is performed.
The complicated scientific and technical problem of organizing fast feedback (from 3 to 4 ns) between the resistance measured and the pulse duration of programming or deleting has been solved.
Sufficient complex phase stability and operating speed for recording and exchanging information with a nanoelement in a molecular cell have been provided.
The program interface is implemented via a COM-object. This allowed planning and performing experiments with any program packages supporting COM-automatization (OrignPro, Matlab or Excel).
Stability and long-term cycling operation have been achieved in this information-measurement electrometric complex for the memory element, which is based on modern concepts of molecular electronics. Operation in a multi-bit mode with a typical time constant of 3 ns at a cell resistance of 150 kiloohm has been ensured (Glavny et al., 2007).

7. Technical vision systems and multi-dimensional signal processing. Information monitoring in the glass industry

The purpose of this work was to create an optoelectronic computer method for obtaining information about the mass of a free falling object using its image, in other words, a computer implementation of the well-known method of weighting judged "by eye."
In a classical method to measure the volume V of a complex-shaped body, the body is divided along the vertical axis y by parallel sections of area $S(y)$ and height Δy. The total volume of the body is equal to the sum of N volumes of the thin truncated solids.

Reconstructing the boundary shape requires a series of measurements of transverse sizes of sections at all heights y by rotating the body about the vertical axis. If optical methods are used for measurements, the plane projections of the body at different angles must be recorded. A similar approach is used in tomography.

There exist certain classes of objects with topologically similar parallel sections where the volume can be measured by recording a single plane projection. The topological similarity means that any parallel section has an area S associated with a special transverse size ($2r$) by a coefficient ξ that does not change at any continuous scale transformation. In this case, the condition $S(y) = \xi \cdot r^2(y), \forall y \in [a,b], \xi = const$, is satisfied for any coordinate y.

Here, r is the distance from the vertical axis to the plane section boundary and ξ is a topological calibration coefficient determined by the shape of the section boundary.

When the geometry of sections is determined optically, the volume V of a body with topologically similar parallel sections is calculated by taking into account the topological and geometrical calibrations as $V = \lim_{N\to\infty} \sum_{n=0}^{N} S(y_n)\Delta y_n K_n(\beta) \approx \xi \cdot \sum_{n=0}^{N} r^2(y_n)\Delta y_n K_n(\beta)$, where $r(y)$ is the projection of the transverse size of a body of revolution onto the observation surface as a function of the coordinate y along the vertical axis. The geometrical calibration coefficient K depending on the vertical coordinate y and parameter β is introduced to compensate for the optical hardware distortions of the results of measurements of S and y. The coordinates a of the lower edges and the coordinates b of the upper edges of the body are also determined along the vertical axis.

A particular case of objects with topologically similar parallel sections is glass gobs formed in the production of glass bottles. The shapes of gobs solidifying in flight are close to those of bodies of revolution, and for the latter the condition of topological similarity of plane sections is satisfied. To estimate the volume of such an object, it is sufficient to measure the sizes of projections of sections $r(y)$ onto the recording plane, given the topological ξ and geometrical K calibration coefficients. The measurement error is due to inaccurate determination of transverse size projections at various coordinates $r(y)$ and violation of topological similarity. An important property of the objects with topologically similar parallel sections is insensitivity of volume estimation by plane images to possible vertical axial deformations.

The spatial coordinates $r(y)$, a, and b are determined by a program method of analyzing bitmap images. Spatial coordinates x and y in the image plane are selected and tied to an absolute system of coordinates. The image of a glass gob (A) is transformed into a video signal by a TV camera by line-wise scanning. A scanning coordinate-sensitive photodetector (B) "cuts" the droplet projection into lines. In this case, each line corresponds to one parallel section of the object.

This method was implemented in an information monitoring system for the glass industry. Now it is used as a scientific basis for engineering by other organizations. The system was created for obtaining noncontact real-time information about the masses of freely falling hot gobs of fluid glass in the production of glass works (bottles). The system operation is preliminary optical processing of the images of freely falling hot glass gobs, introducing the video signal into a computer, real-time analyzing the video flow, extracting the geometrical information, calculating the volume and mass of gobs, and statistical processing of the results.

The inner volume of a bottle is determined by the mass of the glass gob (the outer bottle sizes are determined by the geometry of the mould). In dosing fluid glass, the glass parameters are near the crystallization point and greatly affect the stability of operation of the gob former. The number of rejects reaches 30 to 45 %. The mass of a glass gob can be measured only "by eye," as the droplet falls from the gob former to the mould of a suck-and-blow automatic machine. The glass is hot; there are intense convective and aerodynamic stratified flows and strong refraction. The air has considerable amounts of aerosol particles produced by the black-lead lubrication of the gob former scissors. In two-three hours, the unprotected optical devices become completely dirty. The systems used for measurement should not hamper the operation of the available technological equipment and should be located at a distance of no less than 3 to 10 meters from the objects measured. The required accuracy is 0.5 %.

In the system, information undergoes some successive transformations, and each of them introduces its own errors. Temperature is a major perturbing factor. This parameter affects the quantum yield of the photomatrix, the levels of its dark current, the leakage currents of all semiconductor elements, and the accuracy of binding to the levels of black and white in the processing chain of a TV signal. It has been shown that for obtaining an accurate and reproducible estimate of the spatial boundaries of a self-luminous object with an error of 0.2 %, the video camera should be thermally stabilized with an error not exceeding 0.1°C.

A second important factor is luminosity of the object. The temperature of hot glass gobs is about 1000°C, and they have high luminosity. The temperature and luminosity of their surface can change due to intensive cooling by a cold air flow. A method to stabilize the threshold discrimination level of a video signal has been developed. This method performs statistical estimation of the image, levels of black and white, and provides accurate spatial discrimination of the boundaries when the luminosity of the object changes.

A method for passive aerodynamic protection of the optical surfaces of an information-measurement system was proposed. With this method, no atmospheric aerosol can get into the boundary layer, the optical surfaces are not dirty, and the time of undisturbed operation of the system is increased from 3 to 4000 hours.

The image of a falling gob of molten glass is formed by a receiver based on a video matrix. A video signal comes to the computer of a workstation for subsequent real-time processing. The workstation software receives video flow from a TV camera, detects the object in the field of vision, binds the object to the frame boundaries, and performs spatial filtration, spectral selection of the image, noise elimination, and amplitude discrimination by the Heaviside step function. After this, the gob mass is calculated. The calculated mass values are kept in a local database of the workstation. The monitor screen shows the gob mass variation in real time.

If the gob mass is greater than some given limits, the program generates warning and control signals. With these signals, an operator (or an actuating mechanism) introduces corresponding corrections to the technological process. Three watchdog timers controlling the major program modules maintain the operation of the workstation software.

A real-time information monitoring system for noncontact measurement of the masses of freely falling hot glass gobs is used by the Ekran (Novosibirsk). This low-cost system is used to fabricate glass bottles. The system provides noncontact technological control of the masses of hot glass gobs and generates warning signals when the mass of gobs exceeds the tolerance zone boundaries. The mass measurement error is between 0.5 and 1 %. The use of monitoring based on this method in the glass industry has decreased the number of rejects by a factor of 3 to 5 (Meledin et al., 1999, 2000, 2001, 2002).

8. Medicine. Optoelectronic complexes for endoscopy

Endoscopy is one of the most important trends in the development of medical science and practice. Endoscopic operations are widely used for diagnostics and treatment of various human diseases. Karl Storz endoscopes developed in Germany and used in 140 countries are a world standard. Their quality meets the most stringent requirements of medical diagnostics, and the cost reaches a considerable amount of 2000 to 5000 US dollars. Since the optical devices in existing endoscopes are complicated and expensive, they cannot be employed on a mass scale and easily used in scientific investigations and experiments.

The purpose of this work was to create optoelectronic endoscopic complexes of a new generation, which permit operation with disposable tools in medicine, scientific investigations and experiments, and technical introscopy.

A standard endoscope has several high-quality image-transmitting rod lenses with complicated centering (H. Hopkin's scheme) and a fiber highlighter. Because of high temperature, humidity, autoclaving, and chemical treatment, frequent sterilization of the endoscope reduces its lifetime and worsens image quality. The propagation of HIV and acute viral hepatitis makes the problem of creating disposable sterile endoscopes especially important. The performance of hundreds of operations on different patients with the same endoscope without reliable sterilization creates a moral and ethical problem for the doctors and a serious risk for the patients. The limited reliability of modern endoscopes due to high sensitivity of lens centering to mechanical deformations is also a problem.

Our endoscope has been created as a means of optical diagnostics of multiphase flows in closed cavities. A new optical scheme for the endoscope has been created. Its advantages are a high resolution, high reliability, small cost (20 to 100 US dollars), simple construction at good image quality, and high sterility owing to the use of disposable tools.

One can see from the comparative table that the parameters, except for the number of lenses, depth resolution, and admissible deformation, do not differ considerably. A major advantage of this endoscope is that it has a small number of lenses, since the complexity of an endoscope is determined by the complexity of its optics. Advantages of the endoscope are also simplicity, easy assembly and disassembly, and large admissible deformations. The endoscope allows the use of disposable elements of plastic or traditional optics. A number of endoscopes with these advantages have been created.

A scientific problem is obtaining information about the frequency-contrast characteristic and distortion of the optical and optoelectronic schemes of endoscopes under considerable uncompensated aberrations. At the start of this work, there were no suitable methods. A method for obtaining information about the parameters of endoscopes has been proposed and validated. It is based on the reconstruction of the test signal by using a distorted image of the test object, calculation of the frequency-contrast characteristic as the ratio of cross spectrum of the input optical signal and the reconstructed test one to spectrum of the reconstructed test signal, and determining the spatial distribution of distortion curves. The method has been implemented on the above-mentioned optoelectronic stand for obtaining information about the characteristics of endoscopes.

This work was initiated, supported, and awarded by the technology park of the city of Katzrin, Golan (Israel). An international company, Optiscope Technologies, Ltd., has been created. For these endoscopes, a number of patents, among them two American patents, have been taken out. Now further development of the technology of precision plastic casting of a disposable endoscope is being performed by a Carl Zeiss daughter enterprise in Germany (Meledin, 2008).

9. Machine building. Optoelectronic diagnostics of coordinates and shapes of 3d objects

In power engineering and machine building, there is a wide class of products. Specifically, these are: low-rigid large products with complex shapes, such as blades of hydroturbines, steam and gas turbines, and compressors, box and shell casings, airscrews of flying vehicles, screw propellers of ships, impellers, covering discs, etc. The cost of large blanks is high, and requirements on the quality of products are very rigid. Therefore, the problem of high-precision control of coordinates and shapes of 3D objects at all stages of mechanical treatment is very important.

The purpose of this work was to create new methods and information systems as refined and easy to manufacture as possible for obtaining information about the full geometry of 3D objects with sizes from several millimeters to tens of meters.

A new method for stable processing of sets of spatially modulated images with a stepwise shift has been proposed. It minimizes the error in determining the phase with noise of constant mathematical expectation and variance. In this case, the phase difference estimation of sets of spatially modulated fields is independent of the background intensity distribution.

The method is successfully used in the form of a program code. A generalized algorithm to obtain a solution to a system of transcendental equations at arbitrary phase shifts has been constructed. The complexity of the algorithm with respect to the number of interference patterns with different phase shifts has been estimated. The method is insensitive to additive noise with constant mathematical expectation and variance. An estimate of the root-mean-square deviation of the phase obtained and the reconstruction accuracy of the relief surface of the object measured has been obtained.

The method is used in an information-measurement system to obtain information about the geometry of 3D objects. The object under investigation is sequentially illuminated by a structured light, which is a system of parallel bright halftone bands a constant spatial frequency and linearly varying shift of the initial phase. Phase triangulation and the method of phase steps are used to reconstruct the shape of the surface investigated. The method is insensitive to additive random noise and the illumination source and receiver (video camera) characteristics.

The considerable advantages of the method are used in an information-measurement system. In this system, an office projector for exposing objects with typical sizes of tens of meters is used as an illuminator and spatial modulator (Dvojnishnikov et al., 2010).

The information-measurement system created on the basis of the method of stable processing of sets of spatially modulated images with a stepwise shift has modern characteristics, and some parameters of the system surpass those of existing systems. Some peculiarities of the system are as follows: high accuracy of measurements, which is independent of the reflective properties of the surface, automatic calibration and verification of data, dynamic real-time synthesis of images of the surface measured, automatic comparison of measured data with model data, adaptive generation of the structure of lighting of the object investigated to identify its problem zones, and simultaneous parallel measurement of coordinates at more than 600,000 points (Meledin, 2010).

10. Optoelectronic diagnostics in atomic power engineering

10.1 Three-dimensional velocity fields of heat carrier flow in fuel assembly of power reactor

To raise competitive capacity of Russian atomic plants and substantiate safety of the active zones, new methods should be created for obtaining experimental information about the heat-exchange and hydraulic characteristics of water-moderated water-cooled power reactors. The safety can be increased by validating the reserves prior to heat exchange crises determined by the construction of active zones with different spacer grids and flow turbulizers. At the start of this work, there was no detailed comparative information about the fields and turbulent velocity pulsations in assemblies with different spacer grids for geometrical and operational conditions typical for water-moderated reactors.

Since the thermal-hydraulic processes taking place in an assembly are very complicated, full theoretical calculation of such parameters is not possible. The 1D methods used at present provide only conservative forecasts, and the 2D and 3D methods need data on the turbulent characteristics of flows and corresponding information obtained experimentally. The absence of reliable and detailed experimental information hinders not only effective design of the active zone elements of reactor plants, but also development and verification of the corresponding calculation codes.

The purpose of the work was to create a new method for obtaining information about the local and integral hydrodynamic characteristics of heat carrier flows in models of fuel assemblies with various spacer grids.

A new combined method for obtaining information about the detailed local and integral hydrodynamic characteristics of heat carrier flows in a model of fuel assembly with spacer grids has been proposed. The method is based on semiconductor laser Doppler anemometry, video anemometry, and immersion of the optically transparent reactor model into an immersion liquid with a refraction coefficient equivalent to a model one.

If a phase object - a transparent body - is immersed into a liquid with a refraction coefficient equal to that of the object material, the object will be invisible. However, the object boundaries, which determine the hydrodynamics and heat exchange, will remain. The structure of the flows will also remain. There will be no refraction and scattering of light by the boundaries of the object. Thus, it will be possible to obtain reliable information about the processes screened by the object boundaries by optical methods (notice that this idea goes back to H.G.Well's novel "The Invisible Man"). For the first time, it is possible to obtain information from the previously inaccessible space zones by noncontact nondestructive methods.

The method has been tested experimentally with a model problem on the propagation of the heat carrier flow induced by a rotating central rod and disc in a system with multiple rods imitating fuel elements of a power reactor. Semiconductor laser Doppler anemometry and computer video anemometry are used in this method to obtain information about the heat carrier flow at a horizontal section of the reactor model. The method algorithms use spatial transformations that can obtain information at any spatial section of the fuel assembly.

For the first time, 3D vector velocity fields and fields of velocity modules of the heat carrier flows excited by a rotating central rod or a disc have been obtained, and 2D and 3D characteristics of the flows have been determined. The efficiency of turbulizing and mixing grids of the models of power reactor fuel assemblies has been estimated. For the first time, experimental information about 2D and 3D kinematic characteristics of heat carrier flows, which can be used to verify calculation codes and fuel assembly models of power reactors, has been obtained (Bolshov et al., 2006; Meledin, 2006).

10.2 Geometry of spacer grid of power reactor fuel assembly

A method to obtain information about the geometry, shading zone, and spatial structure has been proposed. It is used to obtain, by program methods, spatial information from scanned images of spacer and intermixing grids of fuel assemblies of power reactors.

It has been shown that this method can be implemented to create a noncontact information system to determine the shading areas of spacer and mixing grids with an absolute measurement error of about 2 μm by a standard modern scanner and specially developed software.

An algorithm to process scanned images has been developed. It consists of the following sequence of operations: obtaining of the initial scanned image of the grid; normalization and adaptive threshold discrimination; morphological filtration; segmentation; adaptive classification by threshold estimation of the logarithm of the area of segmented elements; recognition of channels, interchannel spaces and joints; recording into the database; construction of a non-dimensionalized drawing and formation of statistics of the objects.

With the help of the method, the following information is chosen and displayed in the drawing:

1. Channels and their geometrical centers (circles and equivalent diameters).
2. Interchannel spaces and their geometrical centers (red circles).
3. Welds and their geometrical centers.
4. Grid of geometrical centers of channels (lines and sizes).
5. Grid of geometrical centers of interchannel gaps (lines and sizes).
6. Grid of geometrical centers of welded joints (lines and sizes).

This method is simple and inexpensive. It can serve as a basis for a wide range of applications in atomic and other industries where precision size control in a range limited only by the sizes of the scanner used is required (Meledin, 2008, 2009).

11. Computer visualization: dynamic structure of boiling jet

Here, the capabilities of information diagnostics based on image processing and computer visualization of processes in atomic power engineering are demonstrated.

The problem of catastrophic breakdown of a high-pressure pipeline with a moving overheated heat carrier was investigated experimentally (the pipeline diameter is 40 mm, the initial pressure is 8.3 MPa, and the temperature is 235°C). In the process of catastrophic breakdown, due to a fast pressure drop, a steam explosion takes place and a shock wave front, which can destroy the neighboring pipelines, is formed.

The purpose of this work was to extract important information from an image sequence constructed by a high-speed camera. This was done to study in detail the steam explosion mechanisms, shock wave front formation, and to take some precautions.

The method is based on the fact that there is a small difference in the light scattering by aerosols at different wavelengths to different solid angles. The vortex structures of the flow act like separators, selecting droplets in their sizes and light scattering properties. Therefore, it is possible to reliably identify and visualize the space zones of initial vortex structures. This can be done by identifying even small colour gradients in the images and enhancing them by morphological methods. All the image processing operations are made by computer programs. Therefore, the method can be classified as a computer visualization method (Pribaturin et al., 2006).

12. Optoelectronic for physical experiments

12.1 Computer video anemometry of floating bubbles

In recent years, two-phase bubble flows have been intensively studied, since they are widely used in heat and power engineering, metallurgy, pharmaceutics, and chemical industry. By digital processing of images from a high-speed camera, one can obtain versatile information about the motion of bubble media. The purpose of this work was to create a method of computer video anemometry to obtain reliable 2D and 3D characteristics of bubble flows and to test the method on simple processes.

The motion of gas or vapor inclusions in a liquid space is typical for two-phase bubble media. These are the processes of bubble floating in a stationary liquid space, gas- and vapor-liquid flows in pipes and channels of various geometries, boiling of a liquid, breakdown of gas or vapor formations, etc. The liquid and the bubbles have different optical properties. Owing to this fact it is possible to recognize the spatial boundaries of bubbles by computer image processing.

Three-dimensional images can be obtained by placing, with the help of auxiliary mirrors, images of lateral projections in free zones of the field of vision of the camera, recognizing and successive processing the data of orthogonal projections. With a time sequence of information blocks of the same object, one can determine the dynamics of its shape change and obtain quantitative data about its velocity.

A method of computer video anemometry has been developed. It consists of filming the motion of the medium by a digital high-speed camera, recognizing objects of interest in the images, and digital data processing. The speed of filming is determined by the camera, the velocity of the object under study, and the quality of images. The speed can reach 2000 frames per second.

The method allows not only to obtain simple information about a linear velocity, but also to construct 2D and 3D trajectories of motion. These data are very important for physics of two-phase flows. The average velocities and pulsating 2D and 3D velocity vector components were measured. The streamlines and parameters of rotation were reconstructed. Some measurements of turbulent characteristics were made. The typical pulsation frequencies of the shapes and volumes of bubbles have been determined. In fact, the computer anemometry method developed for bubble flows is a dynamic analog of tomography. As in tomography, one can select any section and then decompose it in detail to obtain quantitative characteristics of velocity profiles and gas content.

Some algorithms for computer processing of a video series with modern image processing methods have been developed. Spectral and adaptive spatial filtration, morphological analysis and segmentation, as well as nonlinear field transformations, were used. The noise was decreased by filtering of spectrally averaged frames by a dynamically formed adaptive spatial filter. A dynamic energy range of light signals was provided by compensating the spatial brightness inhomogeneity of sources by choosing relevant brightness levels in the vicinity of bubbles by adaptive spatial loop filters. The image sharpness is increased by inverse convolution with a given point blurring function by the Lucy-Richardson algorithm. Then, optimal threshold discrimination, transition to binary images, distant transformations, morphological analysis, and image segmentation were performed. The use of nonpermutable nonlinear transformations increases considerably the threshold sensitivity of the method. As a result, dynamic fixation of the weak-contrast and small bubbles in a field with nonuniform illumination is provided.

While processing a sequence of images, each bubble to be recognized is assigned a mark. Full information about a bubble was written into the base. Information about the bubble center coordinates, area, linear sizes of the approximating ellipsoid, eccentricity, inclination angle of the main axis, perimeter of the bubble, and shape of its boundaries was recorded. This database made it possible not only to reconstruct the trajectories of motion of any bubble contained in the initial frame and to construct the gas content matrix, but also perform 2D and 3D visualization of the bubble motion. It should be noted that in some cases, by analyzing the database information it is possible to look at certain phenomena from a new viewpoint. For instance, with some virtual reality methods (simultaneous rotation of the object and the camera in a 3D experimental data space), for the first time we have experimentally demonstrated that the gas content has vertical stratification near the bubble generation source. In this case, the spatial discrete levels could be seen only at a certain small angle (Bolshov et al., 2006).

12.2 Gravitational-capillary waves and hydrodynamic quasicrystals

The fundamental problem of exciting gravitational-capillary waves on the surface of a liquid layer by an electric field was first considered in the 60s. Another attempt was made at the Institute of Automation and Electrometry of the Siberian Branch of the USSR Academy of Sciences in the early 80s; the attempt was, however, not very successful. Experimental investigations of such phenomena were hindered by the absence of adequate information systems capable of accumulating statistics and making formalized descriptions of various experimentally obtained wave structures to verify theoretical models. At the start of this work in 1999, no full linear theory of wave formation on the surface of a viscous liquid at any thickness of the dielectric layer and any wave numbers was constructed, and no satisfactory explanation of the contradictory experimental data obtained earlier was available.

Based on these investigations, a method to extract information from shadow pattern images (shadowgraphs) of gravitational-capillary waves by decomposing the 2D Fourier spectrum of the shadowgraphs into a set of spatial wave vectors and sorting the vectors with respect to the vibrational modes being excited was proposed. Wavelet decomposition of the spatial structures reconstructed was used to investigate the nonlinear regimes.

The phenomenon was investigated experimentally and representative statistics was collected on a special unit based on this method. The appearance, evolution, and transformation of ordered wave structures, change in the wave regimes, and transition to spatial chaos by varying the control parameters were investigated.

In the experimental investigations with this method, it has been shown that there is good agreement with a linear theory. It has also been shown that in the nonlinear case in such electrohydrodynamic systems there appear organized dynamic structures called ``hydrodynamic quasicrystals." The figure shows an example of such structures. It should be noted that this pattern is dynamic and the shapes of such ``crystals" periodically change in time. There appear and develop some ``dislocations'' in the crystals, which cause a periodic change in the structural order.

Phase stroboscopy of the wave surface of a liquid dielectric was used for computer visualization on the basis of the above-proposed method. This was done to obtain shadowgraphs and a detailed spectral pattern of the wave structures of various types of quasicrystal.

The figure shows the result of wavelet decomposition of a central transverse gravitational-capillary wave in a strong oscillating electric field. This procedure made it possible to identify information about the spatial-frequency modulation. The shadowgraph profile along a coordinate, 2D spatial Fourier spectrum, and wavelet spectrum are shown. In the wavelet spectrum, one can clearly see spatial nonlinear frequency modulation of a large-scale wave component. It should be noted that it is very difficult to reconstruct this spatial frequency modulation by other existing methods.

For the first time, good agreement between the experimental data and the above-mentioned linear theory has been obtained with this method (Meledin et al., 2005, a,b).

13. Oil production. Multiphase borehole flowmeter survey

Oil-producing companies must provide a survey of their boreholes in accordance with Russia State Standard P8.615-2005 introduced on March 1, 2005 and valid since March 1, 2006. However, "today... is no measurement equipment with parameters of accuracy meeting the requirements of the new national standard. With available expensive foreign systems, the oil production may become economically unprofitable" (*TechSovet* 4(35), April 1, 2006).

Scientifically, the problem of measuring the flow rate of multiphase flows has also not been solved. The main reason is that there is a great variety of the existing hydrodynamic structures and regimes. The structures of flows of a mixture of oil, gas, water, and solid particles vary in a wide range from homogenized to stratified, with different velocities of individual phases. The properties of a multiphase mixture depend on the temperature, pressure, density, air salinity, etc.

Practically all methods of flow rate measurement are optimized for some combination of hydrodynamic regimes and, in explicit or implicit form, use their characteristic physical properties. Since the number of combinations of such properties is large, the number of existing methods is also large. Unfortunately, however, each of them has a rather narrow "ecological niche."

We relate the complicated scientific problem of multiphase flowmeter survey to informatics. The capabilities of modern computer processing systems make it possible to store in databases numerous virtual images of real objects and rapidly solve problems of multidimensional optimization, search, and comparison. The basic scientific idea of our method is not to decrease the variety of existing hydrodynamic regimes by simplifying information flows and signals, but to increase it as far as possible. The variety of existing regimes provides the formation of wideband multidimensional data best suited to calculate correlations with an etalon. A shortcoming of the approach is that either a preliminary training of the system or a pre-constructed etalon should be available.

These ideas were used as the basis for a new method to obtain information about the phase flow rate of a gas-liquid flow. In this new method, acoustic ultrasound Doppler speckle interferometry for the flow regions with singularities is used and the multidimensional signals obtained are compared with a basic etalon by using elements of artificial intelligence with genetic program algorithms.

There are many methods for measuring the oil watering, but almost all of them are sensitive to the presence of salts in water due to formation of ionic conduction. By increasing the operating frequencies to 1 GHz and more the problem is simplified, but not solved. The resulting error is considerable, about tens of percent.

Its bias component is compensated by making calibrations for each borehole, which must be regularly controlled. Therefore, isotopic methods, which are insensitive to fluid ionic conduction, are most widely used now. However, they are very sensitive to the presence of extremely small concentrations of H-ray contrast salts.

A new method, insensitive to ion-salt conduction, was proposed to obtain information about oil watering based on phase diagnostics of superposition of radio frequency waves with a frequency of 2.4 GHz in a resonator of special shape. Phase measurements of superposition of radio frequency waves demonstrated insensitivity to salt in the fluid in the range of concentrations up to 500 g/l.

The creation of experimental stands is an important part of works on methods for obtaining information about multiphase flows. They make it possible to verify the performance of these methods in a wide range of hydrodynamic regimes. A fully automated experimental stand was created. It provides the formation of a multiphase flow with flow rates of liquid up to 120 l/min and of gas up to 400 l/min with contour diameters of 25, 60, and 100 mm. This stand was used to calibrate methods and flow meters and to teach the program systems with elements of artificial intelligence integrated into the flow meters.

In December 2007, this scientific basis was used to create a prototype of a new information-measurement complex for multiphase borehole flow meter survey and optimization of oil production. The parameters of the system met State Standard P8.615-2005. Some industrial tests of the complex were made at the Kalchinsky oilfield of the Tyumenneftegaz company. This experimental specimen for works in real field conditions was created in half a year.

In the industrial tests at the Kalchinsky oilfield, it was found that the experimental information-measurement complex for multiphase borehole flow meter survey at boreholes 300/7 and 301/7 successfully operated uninterruptedly during five days at environmental temperatures from -36°C to -10°C, providing real-time measurements of the borehole fluid flow rate with a measurement error of 3% (Meledin, 2008, 2009).

14. Conclusion

Some basic principles of semiconductor laser Doppler anemometry and location of multiphase flows with natural light diffusers have been proposed. They form a fundamental scientific basis for a wide class of information optoelectronic measurement systems.

First high-power semiconductor stabilized laser emitters for Doppler anemometry and scientific instrument making (jointly with the Ioffe Physico-Technical Institute) have been created.

Methods to form, receive, and perform software-hardware processing of complex optoelectronic and acoustooptical signals and images under the photon constraint have been developed. Also, methods for signal processing, in particular, on the basis of wavelets and multifractal formalism, and information methods for dynamic monitoring of 3D geometry of objects and automatic control of information systems have been created.

Information-measurement systems and laser software-hardware complexes for technological information control of hot rolled metal in metallurgy have been created. (The daily economical effect at the largest enterprise of this industry - the Nizhnetagilsk Metallurgical Plant - was up to 100,000 US dollars; such complexes have been continuously operated since 1994.)
Information-measurement systems for semiconductor laser diagnostics of multiphase flows in hydraulic turbine machine building have been created. They increased the safety of domestic hydropower engineering (Power Machines - LMZ, 2003-2008).
The problem of replacing the imported information-measurement diagnostic semiconductor laser Doppler systems with adaptive time selection and visualization of the velocity vector for investigations of gas-liquid multiphase flows by similar domestic systems has been solved. These systems were created three years before the best foreign analogs and were used at many domestic scientific enterprises and higher educational establishments, from 2003 through 2007.

Russia
1. SU 1116837.-1983
2. SU 1261450.-1984
3. SU 1374939.-1986
4. SU 1433195.-1986
5. SU 1345120.-1987
6. RU 1302865.-1993
7. RU 1431498.-1993
8. RU 1525579.-1993
9. RU 1569715.-1993
10. RU 1610438.-1993
11. RU 1638625.-1993
12. RU 1789932.-1993
13. RU 2006867.-1994
14. RU 2029307.-1995
15. RU 2044267.-1995
16. SU 1832942.-1996
17. RU 2098755.-1997
18. RU 2104479.-1998
19. RU 2108585.-1998
20. RU 2006118624.-2006
21. RU 2334195.-2008

USA
22. US2003233029 A1.-2003.
23. US2004133071 A1.-2004.
24. US7018330 B2.-2006.
25. US7160248 B2.-2007.

European patent
26. EP1519675 A1.-2005.
27. EP1696787 A1.-2006.

Eurasian patent
28. WO03103482 A1.-2003.
29. WO2005053519 A1.-2005.

Australia
30. AU2003224417 A1.-2003.
31. AU2004294838 A1.-2005.
32. AU2004294838 A2.-2005.
33. AU2003224417B B2.-2007.

Brazil
34. BR0311583 A.-2005.
35. BRPI0417224 A.-2007.

Canada
36. CA2488341 A1.-2003.
37. CA2548377 A1.-2005.

China
38. CN1665439 A.-2005.
39. CN1905830 A.-2007.
40. CN100361620C C.-2008.

Mexico
41. MXPA04012215 A.-2005.
42. MXPA06006287 A.-2006.

New Zealand
43. NZ537021 A.-2006.

Taiwan
44. TW258355B B.-2006.

Japan
45. JP2005528182T T.-2005.

Table 1. List of the patent documentation

Software-hardware information-measurement complexes based on systems for technical vision and processing of multidimensional signals for science and new industrial technologies have been created (Table 1). These are:

- first laser information-measurement complex to control the geometry of railway wheels of moving trains and increase the safety of railroad traffic (the Complex system, 2001, widely used on Russian railroads);
- system for information monitoring of masses of freely falling hot glass gobs by using their dynamic images (the Ekran, 2000);
- optical-electronic complexes of a new generation for medicine (disposable endoscopes of the Optiscope Technologies Ltd.; works were performed by a Carl Zeiss branch, from 2000 through 2007);
- information system to monitor the geometry of loaded rotors of super-high-power plants based on laser lidars (the Sayano-Shushenskaya HPP, 2006);
- information-measurement complex for investigation and optimization of nonvolatile memory of a new generation, to be introduced in 2010-2012, which is based on switching the conductivity of unit nanoelements of sandwich structure (the Fujitsu and the AMD, from 2004 through 2006);
- information-measurement and hardware-software complexes of multiphase flow meter, State Standard P 8.615-2005, for borehole flowmeter survey and optimal oil production control (the NPK VT, TNK-BP, Tyumen, 2008).

These fundamental scientific results have made a considerable contribution to the development of physical diagnostics, informatics of optoelectronic measurements, and creation of many new information monitoring technologies to increase efficiency and safety, which are successfully used in science and various fields of technology.

15. References

Anikin, Yu. ; Meledin, V. ; Naumov, I. ; Okulov, V. & Sadbakov, O. (2004). Investigation of Pulsation Characteristics of Twisted Flow in Cubic Container. *Tepl. Aeromekh.*, Vol. 11, No. 4, 2004, pp. 571-576.

Arbuzov, V. ; Belousov, P. ; Dubnistchev, Yu. ; Meledin, V. & Pavlov, V. (1992). Systems for Optical Diagnostics of Dynamics and Phase Structure of Flows. *Sib. Fis.-Tekh. Zh.*, No. 2, 1992, pp. 4-11.

Bakakin, G. ; Belousov, P. ; Dubnistchev, Yu. ; Meledin V. & Pohalchuk, Yu. (1997). Linear Size Measurement by Photometry Optoelectronics. *Opt. Instr. Data Proc.*, No. 4, 1997, pp. 1-6.

Belousov, P. & Meledin, V. (1985). A Simple Precision System for Remote Adjustment of Lasers. *Avtom.*, No. 5, 1985, pp. 105-106.

Belousov, P. ; Dubnistchev, Yu. & Meledin, V. (1988). Optical Velocity Meter on the Basis of Semiconductor Laser. *Kvant. Elektron.*, Vol. 15, No. 3, 1988, pp. 633-635.

Belousov, P. ; Dubnistchev, Yu. ; Meledin, V. & Pavlov, V. (1988). Laser Doppler Anemometer with Time Selection of Orthogonal Components of Velocity Vector, *Avtom.*, No. 2, 1988, pp. 43-49.

Belousov, P. ; Dubnistchev, Yu. ; Meledin, V. & Pavlov, V. (1990). Laser-Doppler Anemometer with Adaptive Temporal Selection of the Velocity Vector. *Opt. Appl.*, Vol. 20, No. 3, 1990, pp. 187-197. - a.

Belousov, P. ; Dubnistchev, Yu. & Meledin, V. (1990). Optical Velocimeters for Moving Surfaces Using Gas and Semiconductor Lasers. *Opt. Laser Techn.*, No. 6, 1990, pp. 335-339. - b.

Belousov, P. ; Dubnistchev, Yu. & Meledin, V. (1996). Optic Methods of Flows Studying by Selecting of the Spatial-Temporal Structure of the Scattered Light, *Proceedings of SPIE. Laser Optics*, pp. 147-153,Vol. 2773, 1996. - a.

Belousov, P. ; Dubnistchev, Yu. & Meledin, V. (1996). Optical Method for Measuring the Size of Objects of Complex Shape. *J. Opt. Techn.*, Vol. 63, No. 10, 1996, pp 794-796. - b.

Bolshov, L. ; Chudanov, V. ; Strizhov, V. ; Alekseenko, S. ; Meledin, V. & Pribaturin, N. (2006). Best Estimate Methodology for Modeling Bubble Flows, *Proc. ICONE 14: 14th Int. Conf. on Nuclear Engineering*, Miami, Florida, USA, 2006.

Dubnistchev, Yu. ; Meledin, V. ; Zhuravel, F. & Pavlov, V. (1987). Laser Doppler Anemometry with Selection of Optical Signal Coherent Component. *Opt. Appl.*, Vol. 17, No. 2, 1987, pp. 71-80.

Dubnistchev, Yu. ; Meledin, V.; Naumov, I. & Sotnikov, V. (2000). Laser Diagnostic of the Low-Velocity Swirled Flows. *Optoelectr., Instr. Data Process.*, No. 5, 2000, pp. 30-39.

Dubnistchev, Yu. ; Meledin, V. ; Pavlov, V. & Yavorskii, N. (2003). Investigation of Energy Separation in Vortex Tube of Square Section. *Tepl. Aeromekh.*, No. 10, 2003.

Dvojnishnikov, S. ; Kulikov, D. ; Meledin, V. (2010). Optoelectronic method of noncontact restoration of three-dimensional objects surface contour of the complex shape. *Metrology*, No. 4. 2010, pp. 15-27.

Glavny, V. ; Bakakin, G. ; Meledin, V. ; Sadbakov, O. & Rakhmanov, V.V. (2006). A Preprocessing Signal Module for Thermophysical Experiments. *Prib. Tekh. Eksp.*, No. 2, 2007, pp. 166-168.

Kulikov, D. ; Anikin, J.; Dvojnishnikov, S. ; Meledin, V. (2010). Lazer technology for definition of geometry of turbine rotor under load. *Power stations,* No. 7, 2010, pp. 39-43.

Meledin, V. (1995). *Laser Devices for the Speed and Length Control of Rolling in Metallurgy.* Novosibirsk: Vortex Technologies, 1995, P/No. 9500001.

Meledin, V. ; Bakakin, G. & Naumov, I. (1999). A Wide-Aperture Precision Photodetector, *Instr. Exp. Techn.*, Vol. 42, No. 1, 1999, pp. 90-93.

Meledin, V. ; Bakakin, G. ; Naumov, I. & Sotnikov V. (1999). Laser Measuring Complex for the Investigation of the Hydroimpulsive Devices, *Proceedings 8th Int. Conf., Laser Anemometry. Advanced and Applications, (Eds.: A.Cenedese, D.Pietrogiacomi),* pp. 45-54, Rome, Italy, September 6-8, 1999.

Meledin, V. ; Bakakin, G. ; Naumov, I. ; Pavlov, V. & Sotnikov V. (2000). Measurement of Volumes of Skew Fields with Topologically Similar Parallel Cuts on Optical Images on the Plane of Registration, *Proceedings of SPIE "Applications of Digital Image Processing,"* Vol. 4115, pp. 669-679, 2000.

Meledin, V. ; Naumov, I. & Okulov, V. (2000). Influence of the Dissymmetry in Rotation Fluids on the Vortex Core Precession and Spectral Characteristics of the Flows, *Proceedings 5th World Conf. on Experimental Heat Transfer, Fluid Mechanics and Thermodynamics,* pp. 2441-2446, Celata, G.P., et al., Eds., 2001, vol. 3,

Meledin, V. ; Naumov, I. & Sotnikov, V. (2001). Machine Vision System for Noncontact Weighing, *Proc. ISA 2001/IMEKO Special Millennium Sessions*, Houston, Texas, USA, 2001, vol. 415, CD ISA, 2001.

Meledin, V.; Bakakin, G. ; Naumov, I. ; Pavlov, V. & Sotnikov, V. (2002). Real Time Machine Vision System for Noncontact Measurements of the Masses of Free Falling Hot Glass Drops, *Current Research on Holography and Interferometric Methods for Measurement of Object Properties: 2000-2002, Sel. Pap. SPIE*, vol. 5134, pp. 139-149.

Meledin, V. ; Naumov, I.V. & Pavlov, V. (2002). An Experimental Investigation of the Flow Produced in a Rectangular Container by Rotating Disk Using LDA, In : *Optical Methods for Data Processing in Heat and Fluid Flow, Chap. 3*, Greated, C., Cosgrove, J., and Buick, J.M., Eds., pp. 25-37, Suffolk, UK: Professional Engineering Publishing, 2002.

Meledin, V. ; Okulov, V. & Naumov, I. (2003). Experimental Investigation of Twisted Flow in Cubic Container. *Techn. Phis.*, Vol. 48, No. 10, 2003, pp. 1249-1254.

Meledin, V. ; Pavlov, V. ; Tsvelodub, O. & Yavorskii, N. (2005). *Poverkhnostnye volny v zhidkom dielektrike* (Surface Waves in Liquid Dielectric), Novosibirsk: Inst. of Thermophysics, SB RAS, 2005, ISBN 5-89017-029-5. - a.

Meledin, V. ; Pavlov, V. ; Tsvelodub, O. & Yavorskii, N. (2005). Gravity-Capillary Waves on the Surface of a Liquid Dielectric. *Dokl. Phys.*, Vol. 50, No. 8, 2005, pp. 426-430. - b.

Meledin, V. (2006). Optoelectronic Computer Science in Energy Saving Technologies of Industrial Corporations. *Energy Saving Technologies in Scientific and Technical Development for Industrial Corporations*, Dortmund: DU Press, 2006, pp. 35-92.

Meledin, V. ; Anikin, Yu. ; Bakakin, G. ; Glavniy, V. ; Dvoinishnikov, S. & Kulikov, D. (2006). Laser Doppler Semiconductor Anemometry of Vortex Flow behind the Vane Wheel Rotor of the Water Turbine. *Proc. SPIE Optical Methods of Flow Investigation*, Vol. 6262, pp. 123-133, 2006. – a.

Meledin, V. ; Anikin, Yu. ; Bakakin, G. ; Glavniy, V. ; Dvoinishnikov, S. & Kulikov, D. (2006). Laser Doppler Diagnostic of Flow in Draft Tube behind Hydroturbine Runner, *Proc. 11th Int. Symp. on Unsteady Aerodynamics, Aeroacoustics and Aeroelasticity of Turbomachnes*, pp. 446-457, 2006, Moscow, CIAM, 2006. - b.

Meledin, V. ; Naumov, I. ; Kuznetsov, I. ; Morkin, O. & Mostovsky, N. (2006). Applying of Specialized Optical Laser and Video Systems for Study of Three-Dimensional Flows in Hydraulic Turbines, *Proc. ''HYDRO-2006, Maximizing the Benefits of Hydropower''* (Porto Carras, Greece, 2006), Aqua Media Int., pp. 1-8. - c.

Meledin, V. (2008). *Informatics of Optoelectronic Measurements*, Novosibirsk: Inst. of Thermophysics, SB RAS, Russia.

Meledin, V. (2009). Informatics of Optoelectronic Measurements: Science and Innovative Industrial Technologies. *Journal of Engineering Thermophysics*, Vol.18, No. 2, 2009, pp. 99-128.

Meledin, V. (2010). 3D-diagnostic: it is simple, exact, accessible. *First-hand Science*, No. 2, 2010, pp. 23-24.

Pribaturin, N. ; Meledin, V. ; Bykov, M. ; Bezrukov, Y. & Onshin, V. (2006). Experimental Investigation of Boiling Liquid Jet from a Pipeline Breaking, *Proc. Int. Conf. ''Nuclear Energy for New Europe,''* Portoroz, Slovenia, 2007.

Rakhmanov, V. ; Bakakin, G. ; Glavny, V. ; Meledin, V. & Naumov, I.V., (2006). A Controllable High-Voltage Stabilized Power Source of Photoelectric Multiplier. *Instruments and Experimental Techniques,* Vol.49, No. 5, 2006, pp.676-678.

Sadbakov, O. ; Meledin, V. ; Okulov, V. ; Naumov, I. ; Anikin, Yu. & Mostovskii, N. (2004). Laser Doppler Diagnostics of Flow Structure Downstream of Hydroturbine Blade Wheel at Optimal and Forced Loads. *Tepl. Aeromekh.,* Vol. 11(4), 2004, pp. 577-582.

Part 4

Optoelectronic Measurements in Frequency Domain

19

Optoelectronic Oscillators

Patrice Salzenstein
Centre National de la Recherche Scientifique (CNRS)
Franche Comté Electronique Mécanique Thermique Sciences et Technologies (FEMTO-ST) Institute, Besançon
France

1. Introduction

Optoelectronic oscillator was invented in 1994 by Yao and Maleki, two researchers of the NASA Jet Propulsion Laboratory [1]. The aim of this oscillator was first to generate microwave signal with particulary low phase noise. The best results is -163 dBc/Hz at 10 kHz from a 10 GHz carrier. This system was initially developed for next generation radar to replace microwave generators. Then new applications appear for time and frequency, telecommunication, and navigation technology. Few years ago were published first Optoelectronic Oscillators (OEO) with fiber loop [2,3], affordable for telecommunication systems with adjustable frequency chosen with band filter value. But optical fiber are still bulky because of their several km packaged lenghts and bring difficulties with temperature control. However with a 4 km optical delay line in OEO, a 10 GHz oscillator prototype exhibits a frequency flicker of 3.7×10^{-12} (Allan deviation) and a phase noise lower than −140 $dB.rad^2/Hz$ at 10 kHz off the carrier [4]. The choice of integrating a mini-resonator is a way to reach problems related to regulation of temperature and to work in limited volume, necessary condition for building transportable sources. Optical fiber delay line is replaced by a whispering gallery mode (WGM) optical mini-resonator in simple topology of OEO. Optical signal can propagate by total internal reflection by WGM inside the crystal resonator. One can then achieve a long equivalent delay line into the few millimeter diameter optical mini-disk resonator. High quality factor were demonstrated [5]. In this chapter are presented main principle of OEO. The interest to build such an oscillator is that the expected microwave frequency that modulate the optic carrier can be increased without

---

[1]: X. S. Yao and L. Maleki, „High frequency optical subcarrier generator," Electronics Letters, 30(18), 1525 (1994)

[2]: A. Neyer, E. Voges, "High frequency electro optic oscillator using an integrated interferometer," Appl. Phys. Lett. 40(1), 6-8 (1982)

[3]: X. S. Yao, L. Maleki, "Optoelectronic microwave oscillator," J. Opt. Soc. Am. B 13(8), 1725-1735 (1996)

[4]: K. Volyanskiy, J. Cussey, H. Tavernier, P. Salzenstein, G. Sauvage, L. Larger, and E. Rubiola, "Applications of the optical fiber to the generation and measurement of low-phase-noise microwave signals," J. Opt. Soc. Am. B 25(12), 2140-2150 (2008)

[5]: I. S. Grudinin, V. S. Ilchenko, L. Maleki, "Ultrahigh optical Q factors of crystalline resonators in the linear regime," Phys. Rev. A 74, 063806(9) (2006)

loosing stability. Main limitation are then in the ability to find stable enough components such as high speed photo detectors.

2. How works an OEO

An OEO is an oscillator typically delivering a microwave signal. Purity of microwave signal is achieved thanks to a delay line inserted into the loop. For example, a 4 km delay corresponds to 20 µs time for optical energy to stay in the line. It is equivalent to a quality factor $Q=2\pi FT$ where F is the microwave frequency and T the delay induced by the delay line. The continuous light energy comping from a laser is converted to microwave signal. The loop of the oscillator consists in an optic and an electric part as systematized on the following figure. Light from the laser goes through a modulator. The modulation microwave signal comes from the output of the microwave amplifier after crossing a -10dB directional coupler. Resonant element can be an optic fiber equivalent to a delay line. It can also be an optical mini-resonator coupled to the optical fiber at the output of the phase modulator. The microwave signal is amplified after the photodiode. OEO can have optic out put with the modulated optical signal and microwave output through a directional coupler. The oscillator consists of an amplifier of gain G and a feedback transfer function $\beta(f)$ in a closed loop. The gain G compensates for the losses, while $\beta(f)$ selects the oscillation frequency. Barkhausen condition gives $G.\beta(f) = 1$.

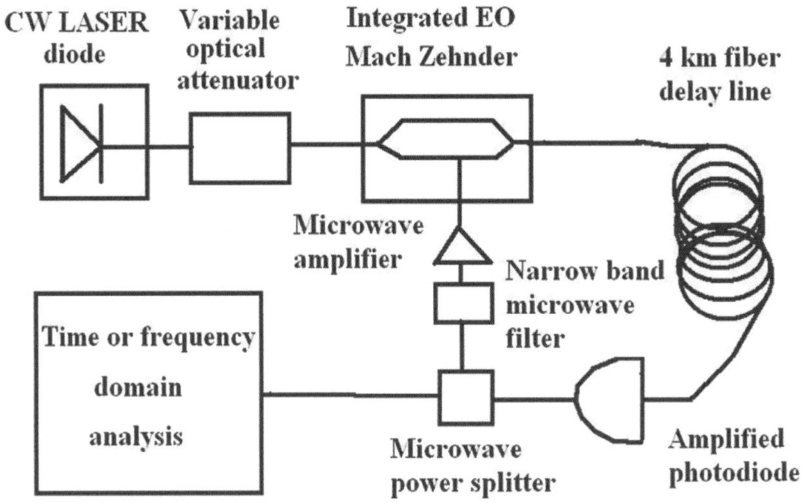

Fig. 1. Typical architecture of a fiber delay line OEO realized at FEMTO-ST institute

The optical fiber is a good choice for several reasons explained in this paragraph. A long delay can be achieved, of 100 µs and more, thanks to the low loss (0.2 dB/km at 1.55 µm and 0.35 dB/km at 1.31 µm). The frequency range is wide, at least of 40 GHz, still limited by the optoelectronic components. The background noise is low, close to the limit imposed by the shot noise and by the thermal noise at the detector output. The thermal sensitivity of the delay (6.85×10^{-6}/K) is a factor of 10 lower than the sapphire dielectric cavity at room temperature. This resonator is considered the best ultra stable microwave reference. In oscillators and phase-noise measurements the microwave frequency is the inverse of the

delay. This means that the oscillator or the instrument can be tuned in steps of 10^{-5}–10^{-6} of the carrier frequency without degrading stability and spectral purity with frequency synthesis. Finer-tuning is possible at a minimum cost in terms of stability and spectral purity.
On the following figure is represented a typical topology of an OEO with a 4 km fiber delay line.

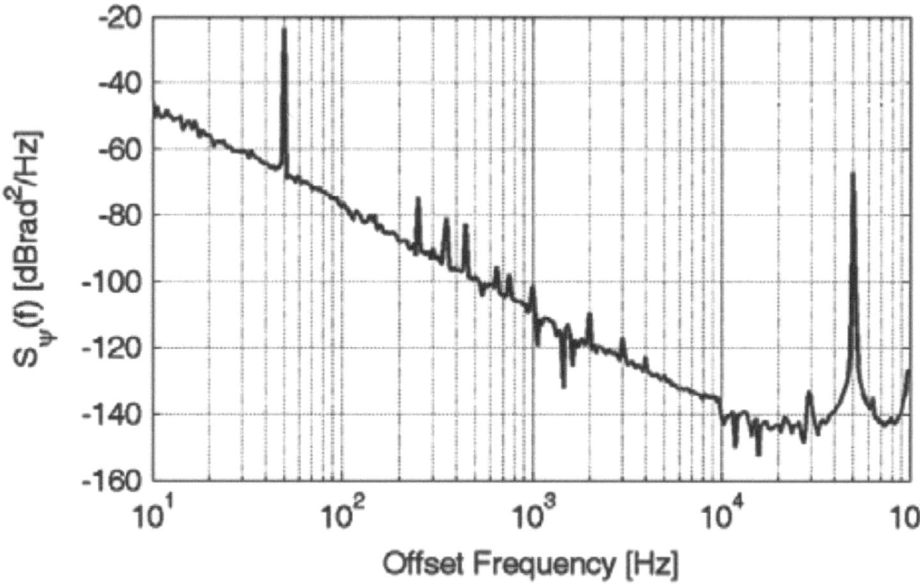

Fig. 2. Phase noise of an OEO realized at FEMTO-ST with a 4 km fiber delay line with a -145 dB.rad^2/Hz noise floor at 10 kHz from a 10 GHz carrier

2. Examples of other topologies

With optical fiber several modes are in competitions. The use of two different loops enable elimination of parasitic peaks. For illustration, a simple topology with two loops is represented on figure 3. We design two optical ways detected by two different photo-detectors. We schematically present on figure 4 how one loop can filter the signal.
A new approach for the generation of ultralow jitter optical pulses using optoelectronic microwave oscillators was proposed. Short pulses are obtained through time-lens soliton-assisted compression of sinusoidally modulated pre-pulses, which are self-started from a conventional single-loop optoelectronic oscillator. The inherent ultra-low phase noise of optoelectronic oscillators is converted into ultra-low timing jitter for the generated pulses. Generation of 4.1 ps pulses along with a microwave whose phase noise is -140 dBc/Hz at 10 kHz from the 10 GHz carrier, with 2.7 fs jitter in the 1-10 kHz frequency band was demonstrated [6]. Figure 5 represents such topology with compression of impulsion.

---

[6]: Y. K. Chembo, A. Hmima, P. A. Lacourt, L. Larger and J. M. Dudley, „Generation of Ultralow Jitter Optical Pulses Using Optoelectronic Oscillators With Time-Lens Soliton-Assisted Compression," J. of Lightwave Technology, 27(22), 5160 – 5167 (2009)

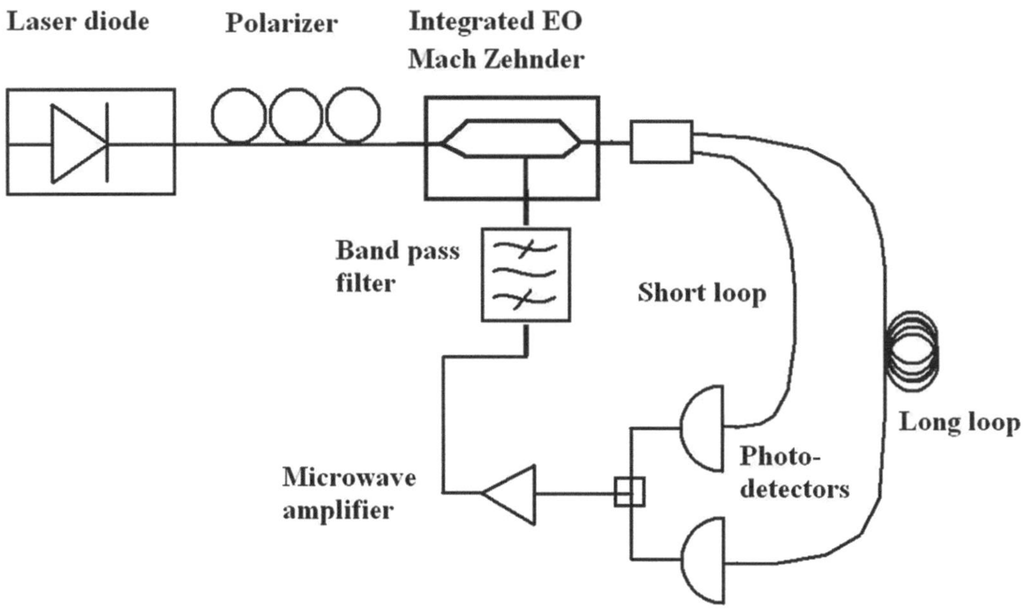

Fig. 3. Double loop topology of OEO

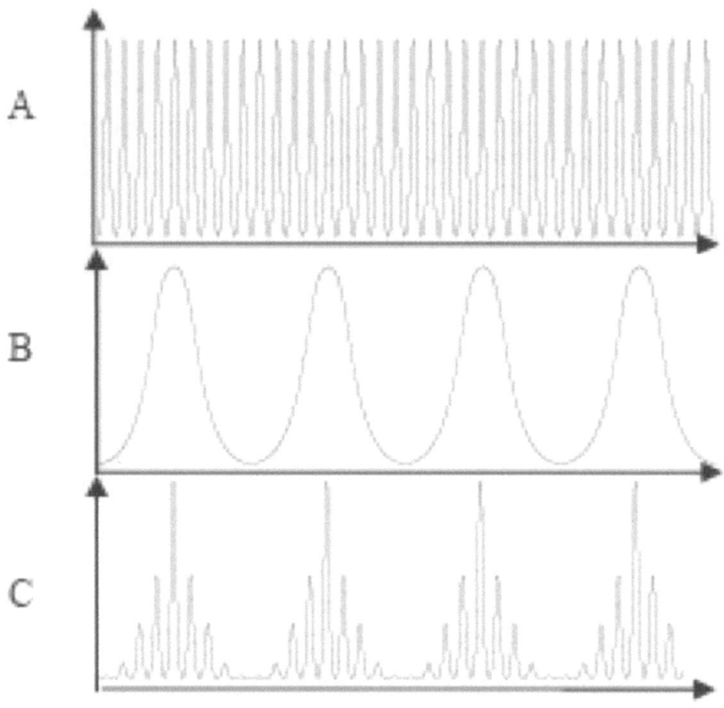

Fig. 4. A and B respectively represent long and short loops, one can see that peaks have been filtered in the sum, C signal. Amplitude is represented versus frequency

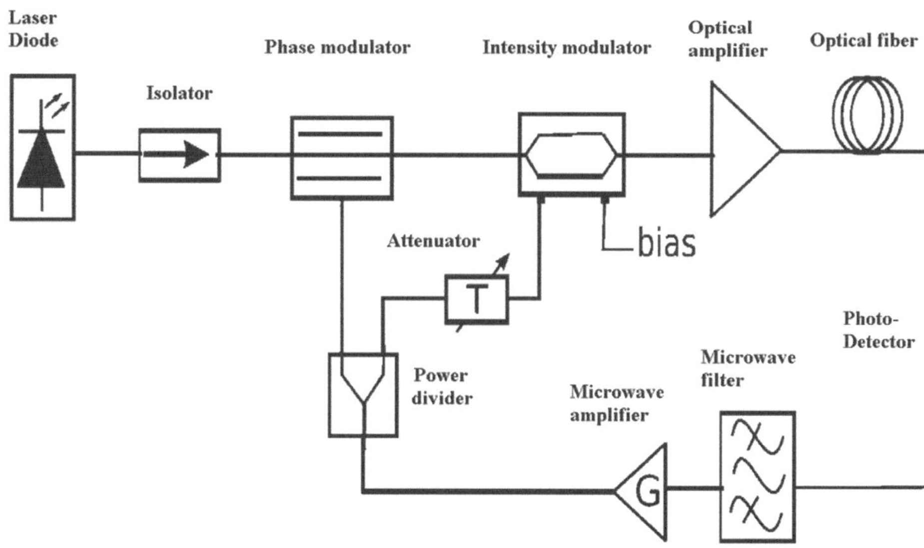

Fig. 5. Topology with compression of impulsion

3. Non linear approach for modelling the OEO

To study the behaviour of the system, a non linear approached was developed. It is based on complex equation for the delay. A Neimark-Sacker bifurcation was demonstrated and shown as a limitation for the performance of the system [7,8]. The possibility of multi-mode propagation according to the starting conditions of the oscillators was also demonstrated [9]. They were experimentally confirmed with a remarkable precision. These results established for the first time theoretical base of the spectral stability of the OEOs : noise floor and characterization of peaks.

4. Exploring the choice of compact resonators

It is interesting to integer a compact resonator and forget a too long and temperature sensitive optical delay line. With its tetragonal crystal and a good behaviour with risk of water pollution, CaF_2 is a good candidate but it has a bad reaction to mechanical shocks. Resonators with MgF2 and fused silica are still interesting for their properties [10,11]. MgF_2 can present low an inversion point around 80°C. Temperature variation of refractive index

[7]: Y. K. Chembo, L. Larger, H. Tavernier, R. Bendoula, E. Rubiola and P. Colet, „Dynamic instabilities of microwaves generated with optoelectronic oscillators," Optics Letters, 32(17), 2571 (2007)

[8]: Y. K. Chembo, L. Larger and P. Colet, „Non linear dynamics and spectral stability of optoelectronic microwave oscillators," IEEE J. Quantum Electron., 44(9), 858 (2008)

[9]: Y. K. Chembo, L. Larger, R. Bendoula and P. Colet, „Effect of gain and banwidth on the multimode behaviour of optoelectronic microwave oscillators," Optics Express, 16(12), 9067 (2008)

[10]: P. Salzenstein, H. Tavernier, K. Volyanskiy, N. N. T. Kim, L. Larger, E. Rubiola, "Optical Mini-disk resonator integrated into a compact optoelectronic oscillator," Acta Phys. Pol. A 116(4), 661-663 (2009)

[11]: H. Tavernier, P. Salzenstein, K. Volyanskiy, Y. K. Chembo and L. Larger, „Magnesium Fluoride Whispering Gallery Mode Disk-Resonators for Microwave Photonics Applications," IEEE Photonics Technology Letters, 22(22), 1629-1631 (2010)

dn_e/dT of MgF_2 is around zero in the range of 80°C (positive at lower temperatures and negative at upper temperatures). It is particularly helpful to achieve stable oscillators as a precise control of the temperature is a quasi-necessary condition to reach high stabilities. Hardness of MgF_2 and CaF_2 is in the range of 6 Mohs and they have both good answer to mechanical shocks that makes less difficult fabrication of mini-disk with these materials. There crystal class is very different as MgF_2 crystal is tetragonal and fused silica is non crystalline, and if MgF_2 is not sensitive to water pollution, it is necessary to have special treatments to minimize H_2O inclusion in fused silica [12]. These two material are relatively easy to manipulate without damaging the surface. A special equipment must be developed for manufacturing resonator. A dedicated polishing machine affords small eccentricity and a precision adapter. System is hold on air bearing support to mechanically prevent influence of vibration on the surface of the external tore surface of the mini-disk resonator. Process is started from an initial crystal optical windows of about 6 mm diameter and 500 µm thickness for X-band applications. The coupling zone has to be reduced to less than 50 µm, that's why two 20 degrees angle bevels can be performed on the disk to form a sharp edge. We need a very good optical quality with very low and regular roughness all around the torical surface of the disk periphery. Powders with decreasing grain size are used. One can also achieve spheric resonators from electric flash in a fibber to perform spheric profile. Some methods exist to choose similar diameters for microspheres, based on the choice of similar diameters in the used powder. Advantage of spheres is to be free with polishing process, one disadvantage consist in the dispersion in the periphery, and difficulty to evacuate temperature.

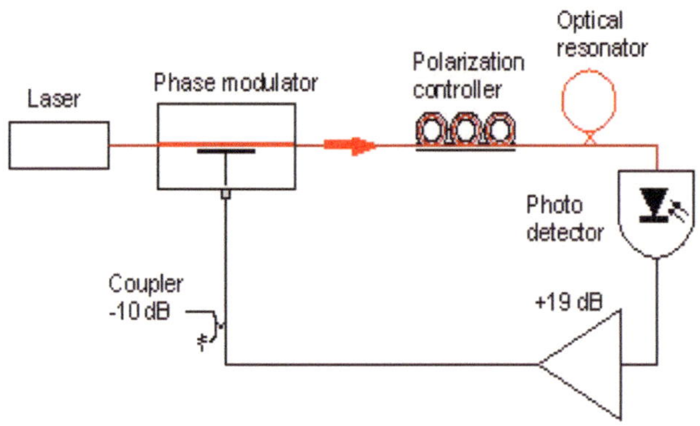

Fig. 6. Typical architecture of an optical resonator based OEO realized at FEMTO-ST institute

In order to introduce into the loop the fabricated resonator, it has to be coupled to the optical light coming from a fiber. Best way to couple is certainly to to use a cut optical fibre through a prism. But a good reproducible way in a lab is to use a tapered fibber glued on a holder. Holder alloy and geometry match the thermal expansion of the glass.

For measuring the resonance [13], one uses the signal from a tunable laser diode . Fast digital real time oscilloscope permits the analysis of the very sharp phenomena at peak resonance.

12: V. G. Plotnichenko, V. O. Sokolov, and E. M. Dianov, "Hydroxyl Groups in High-Purity Silica Glass," Inorganic Materials, 36(4), 404-410 (2000)

13: H. Tavernier, N. N. T. Kim, P. Feron, R. Bendoula, P. Salzenstein, E. Rubiola, L. Larger, "Optical disk resonators with micro-wave free spectral range for optoelectronic oscillator," Proc. of the 22nd European Time and Frequency Forum - Toulouse, France, paper FPE-0179 (2008)

It is necessary to use a high speed resolution oscilloscope for the analysis of very short phenomenas. Oscilloscope is inserted after the photodiode that detects optical signal coming from the mini-disk resonator coupled to the fibre glued on the holder. Resonance peak detection is in single mode excitation. Small taper size selects a thin excitation area. Resonance measurement set-up is in open loop. Although wavelength span is too small to scan a full free spectral range (FSR) and scan rate is 50 Hz, it let be possible Q factor measure with the self homodyne methodology [14]. Optical resonator is then inserted in the loop of the OEO as schematically represented on figure 6. Inside the optical resonator, light propagates with Whispering Gallery Modes (WGM) and the difference of optical index between the optical cavity and air permits a quasi total reflection of the signal inside the resonator, even if it depends on the roughness of the surface that also causes losses. OEO consists in a classic architecture. Phase modulator is optically driven by the laser. The optical mini-resonator is coupled to the optical fibre at the output of the phase modulator. Microwave signal is then amplified after being detected in the photodiode. Modulation microwave signal of the light comes from the output of the microwave amplifier through a directional coupler.

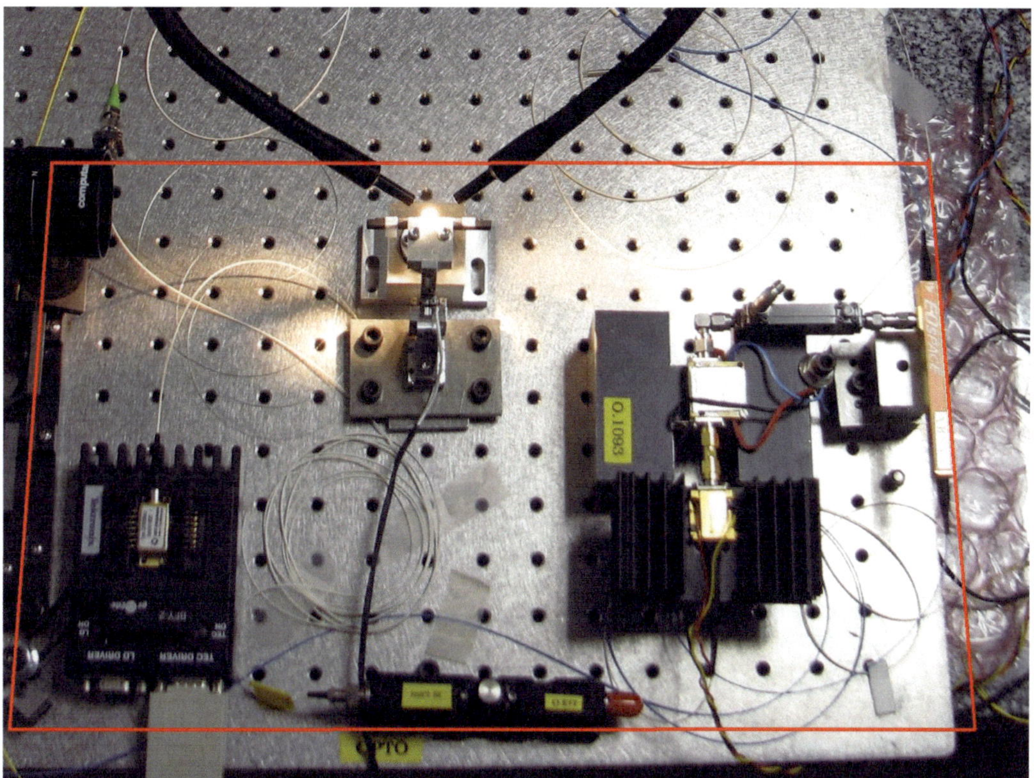

Fig. 7. Photography of an optical resonator based OEO realized at FEMTO-ST institute. The rectangle represents A3 format (297x420mm^2)

[14]: J. Poirson, F. Bretenaker, M. Vallet, A. Le Floch, " Analytical and experimental study of ringing effects in a Fabry-Perot cavity. Application to the measurement of high finesses," J. Opt. Soc. Am. B 14(11), 2811-2817 (1997)

Figure 7 show an OEO with a compact mini-disk optical resonator. We clearly see the positioning system which combines the tapered fiber and the mini-resonator. Resonator coupled to the optic fiber can be seen at the top of the picture. It is under light in order to focus the camera (left top edge of the picture) on the coupling zone to watch on the screen of a camera-connected computer how closed is the fiber from the resonator. The nano-positioning system provides enough space for the moves in a 12x12x12 mm^3 typically volume and is controlled by a joystick including three different speeds to approach the resonator and the selected tapered fiber.

Optical mini-disk resonator helps to increase reduction of the dimension of OEO. Structure could be optimized by the use of a several meters long fiber loop in addition to the optical mini-disk. It could be interesting to work at higher frequencies than in X-band. Working at upper frequencies (20 GHz, 60 GHz or more) could be helpful to achieve a better frequency stability close to the carrier, even if it is to early to think that OEO could replace stable quartz oscillators. By the way OEO with fiber delay line are still promising for such applications. Optical resonators can be good candidate for several connected applications like filtering the frequency, generation of frequency, non linear functions like optical modulator at higher frequencies, use of combs of modes etc. Optical resonators based OEO can also be improved by stabilization of Laser signal and control of the polarization.

5. Measuring performances of an OEO

To measure phase noise of a unique OEO at Fourier frequencies between 10 Hz and 100 kHz from the carrier using dedicated optoelectronic phase noise measurement bench [[15]] because it cannot be locked on another if there is not the same exact frequency. State-of-the art OEO in terms of phase noise are presently manufactured in the USA [[16]]. Fused Silica micro-sphere resonators were already previously fabricated [[17]] and integrated into OEO. Recently, fused silica compact mini-disk optical resonators were also integrated into an OEO and it was demonstrated a upper phase noise floor [[18]]. The used fused silica resonator had a quality factor in the range of 10^8. In order to generate microwave signal in X-band (8.2-12.4 GHz), diameter is in the range of 5 mm and quality of the surface less than few nanometres. Performance in terms of phase noise is certainly lower for an OEO based on compact optical resonator, but a large reduction of the noise should be possible with well optimized coupling conditions and thermal and mechanical environment of the resonator perfecty controlled. Stabilization of the laser on the resonance can be improved by the use of a Pound driver.

---

[15]: P. Salzenstein, J. Cussey, X. Jouvenceau, H. Tavernier, L. Larger, E. Rubiola, G. Sauvage, "Realization of a Phase Noise Measurement Bench Using Cross Correlation and Double Optical Delay Line," Acta Physica Polonica A, 112(5), 1107-1111 (2007)

[16]: V. S. Ilchenko, A. A. Savchenkov, A. B. Matsko, D. Seidel, L. Maleki, "Crystalline resonators add properties to photonic devices," 17 February 2010, SPIE Newsroom. DOI: 10.1117/2.1201002.002536 (2010)

[17]: V. S. Ilchenko, X. S. Yao, and L.e Maleki, "High-Q microsphere cavity for laser stabilization and optoelectronic microwave oscillator," Proc. SPIE, 3611, 190 (1999)

[18]: K. Volyanskiy, P. Salzenstein, H. Tavernier, M. Pogurmirskiy, Y. K. Chembo and L. Larger, „Compact Optoelectronic Microwave Oscillators using Ultra-High Q Whispering Gallery Mode Disk-Resonators and Phase Modulation," Optics Express, 18(21), 22358-22363 (2010)

6. Conclusion

OEO is a particularly interesting system to be studied for fundamental physics with its complex system with delay, but also for its applications. Several aspect already have been explored and its performances help in understanding this system. It should probably play a major rule in the future especially for new generation navigation system applications. Several fields are still open in research, especially by considering new complex architectures regarding existing architectures. One contribution should help with a different approach than usual, i. e. stochastic and non-linear dynamic systems.

7. Acknowledgement

The author acknowledges Dr. Yanne K. Chembo (CNRS researcher, FEMTO-ST, Besançon) and Hervé Tavernier (PhD student, FEMTO-ST, Besançon) for helpful discussions.

8. References

Y. K. Chembo, L. Larger, H. Tavernier, R. Bendoula, E. Rubiola and P. Colet, Dynamic instabilities of microwaves generated with optoelectronic oscillators, *Optics Letters*, 32(17), 2571 (2007)

Y. K. Chembo, L. Larger and P. Colet, Non linear dynamics and spectral stability of optoelectronic microwave oscillators, *IEEE J. Quantum Electron.*, 44(9), 858 (2008)

Y. K. Chembo, L. Larger, R. Bendoula and P. Colet, Effect of gain and banwidth on the multimode behaviour of optoelectronic microwave oscillators, *Optics Express*, 16(12), 9067 (2008)

Y. K. Chembo, A. Hmima, P. A. Lacourt, L. Larger and J. M. Dudley, Generation of Ultralow Jitter Optical Pulses Using Optoelectronic Oscillators With Time-Lens Soliton-Assisted Compression, *J. of Lightwave Technology*, 27(22), 5160 - 5167 (2009)

I. S. Grudinin, V. S. Ilchenko, L. Maleki, Ultrahigh optical Q factors of crystalline resonators in the linear regime, *Phys. Rev. A* 74, 063806(9) (2006)

V. S. Ilchenko, A. A. Savchenkov, A. B. Matsko, D. Seidel, L. Maleki, Crystalline resonators add properties to photonic devices, *17 February 2010, SPIE Newsroom*. DOI: 10.1117/2.1201002.002536 (2010)

V. S. Ilchenko, X. S. Yao, and L.e Maleki, High-Q microsphere cavity for laser stabilization and optoelectronic microwave oscillator, *Proc. SPIE*, 3611, 190 (1999)

A. Neyer, E. Voges, High frequency electro optic oscillator using an integrated interferometer, *Appl. Phys. Lett.* 40(1), 6-8 (1982)

J. Poirson, F. Bretenaker, M. Vallet, A. Le Floch, Analytical and experimental study of ringing effects in a Fabry–Perot cavity. Application to the measurement of high finesses, *J. Opt. Soc. Am. B* 14(11), 2811-2817 (1997)

V. G. Plotnichenko, V. O. Sokolov, and E. M. Dianov, Hydroxyl Groups in High-Purity Silica Glass, *Inorganic Materials*, 36(4), 404-410 (2000)

P. Salzenstein, J. Cussey, X. Jouvenceau, H. Tavernier, L. Larger, E. Rubiola, G. Sauvage, Realization of a Phase Noise Measurement Bench Using Cross Correlation and Double Optical Delay Line, *Acta Physica Polonica A*, 112(5), 1107-1111 (2007)

P. Salzenstein, H. Tavernier, K. Volyanskiy, N. N. T. Kim, L. Larger, E. Rubiola, Optical Mini-disk resonator integrated into a compact optoelectronic oscillator, *Acta Phys. Pol. A* 116(4), 661-663 (2009)

H. Tavernier, N. N. T. Kim, P. Feron, R. Bendoula, P. Salzenstein, E. Rubiola, L. Larger, Optical disk resonators with micro-wave free spectral range for optoelectronic oscillator, *Proc. of the 22nd European Time and Frequency Forum* - Toulouse, France, paper FPE-0179 (2008)

H. Tavernier, P. Salzenstein, K. Volyanskiy, Y. K. Chembo and L. Larger, Magnesium Fluoride Whispering Gallery Mode Disk-Resonators for Microwave Photonics Applications, *IEEE Photonics Technology Letters*, 22(22), 1629-1631 (2010)

K. Volyanskiy, J. Cussey, H. Tavernier, P. Salzenstein, G. Sauvage, L. Larger, and E. Rubiola, Applications of the optical fiber to the generation and measurement of low-phase-noise microwave signals, *J. Opt. Soc. Am.* B 25(12), 2140-2150 (2008)

K. Volyanskiy, P. Salzenstein, H. Tavernier, M. Pogurmirskiy, Y. K. Chembo and L. Larger, Compact Optoelectronic Microwave Oscillators using Ultra-High Q Whispering Gallery Mode Disk-Resonators and Phase Modulation, *Optics Express*, 18(21), 22358-22363 (2010)

X. S. Yao and L. Maleki, High frequency optical subcarrier generator, *Electronics Letters*, 30(18), 1525 (1994)

X. S. Yao, L. Maleki, Optoelectronic microwave oscillator, *J. Opt. Soc. Am.* B 13(8), 1725-1735 (1996)

20

Statistical Tools and Optoelectronic Measuring Instruments

Ionel Sabin and Ionel Ioana
Universitatea "Politehnica" Timişoara
Romania

1. Introduction

In the frame of European research projects, several air quality measuring campaigns in cross roads, streets, parks as well in a non ecological waste deposit were realized. The analyzed signals, representing CO, NO_2, O_3, SO_2 and HC concentrations, were measured with several optoelectronic instruments. Two of the utilized optoelectronic devices are shortly presented at the beginning of the chapter.

Due to their random character, pollutant concentrations signals can be analysed using statistical processing methods. The main statistical functions and parameters taken into account within this chapter are histograms, correlation coefficients, correlation and covariance functions (Ionel et al., 2009). Actually, statistical tools are usually utilized in analysing ecological data (Zuur et al., 2007) but, as far we know, it is not common to imply statistics in a comparative analysis of optoelectronic devices (Ionel et al., 2007).

Specific pre-processing procedures must be used for signal conditioning. Thus, „ideal" low-pass filtering based on fast Fourier transform can be implemented for the rejection of measurement noise and artefacts from the pollutant concentration signals. On the other hand, „ideal" high-pass filtering allows the extraction of the variable component of the pollutant level signals. In order to avoid redundant measurements, one can use interpolation for increasing the number of samples, especially in the case of slowly varying meteorological parameters.

Computer experiments with real pollutant concentration signals lead to some practical recommendations concerning acquisition parameters like data size and sampling frequency. The most important practical rules are as follow: assure the temporal length of the measured signal, assure the necessary resolution on the time axis, and make interactive verifications of the acquisition parameters during the measuring campaign. Guidance on MATLAB software for calculating statistical functions and parameters are provided.

As a particular application, the correlative comparison of two carbon monoxide (CO) measuring instruments is presented. The point source device and the open path optical remote sensing instrument do actually not measure the same quantity but a statistical comparison of the two instruments is still possible. The correlative analysis leads to the expected conclusion that the open path instrument is more suitable for monitoring the pollution level in a large area than the classical point source device.

2. Pollutant concentrations measured with optoelectronic instruments

2.1 The optoelectronic measuring instruments

One of the utilized instruments was the specialized HORIBA APMA-350E CO monitor, which furnishes the local pollution level. Fig. 1 presents a bloc diagram of this instrument working on the classical Non-Disperse Infrared (NDIR) method. The APMA-350E represents a generation of ambient CO monitors designed to eliminate routine calibration cycles and to provide long-term stable measurements and unattended continuous operation. It features a newly developed cross-flow modulation (CFM) technique which results in remarkable improved zero drift performance and sensitivity. The cross-flow modulated analyzer incorporates the basic design features of the conventional NDIR analyzer.

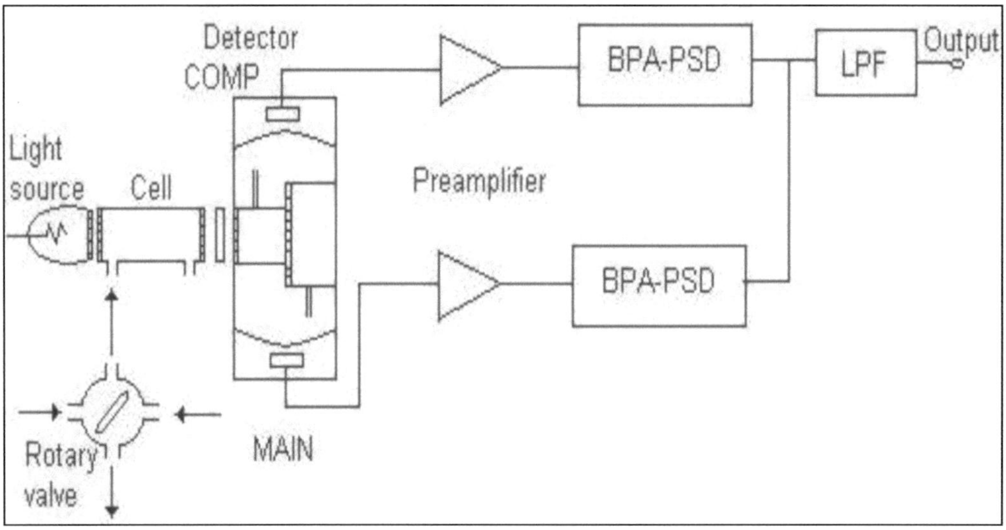

Fig. 1. Bloc diagram of the HORIBA APMA-350E CO monitor

The essential new element in this design, according to Fig. 1, is a rotary valve that alternately directs the sample gas and a reference gas to the one cell of the analyzer. By this method, the distinction between the sample and the reference optical path is eliminated and each path alternately functions as a reference and a sample path. The requirement for an optical chopper to modulate the detector output is thereby eliminated. In the cross-flow analyzer design, sensitivity is inherently increased because the amount of IR (infrared) energy absorbed and translated into the output signal is theoretically doubled for any concentration at the given modulation frequency. In addition, the signal-to-noise ratio is significantly better because the optical chopper which tends to introduce noise in the conventional NDIR instrument is removed in this CFM design. In the CFM scheme, gas flow rates and cell configuration can be selecting providing very smooth modulation. To minimize interference, dual detector system employing a compensating detector located behind the main detector is adopted in this instrument. The two detectors are charged in such a way that response to the interference gas in the second detector is compared to that of the measured gas. The signal from this detector is amplified and subtracted from the main detector signal, in the electronic part of the analyzer.

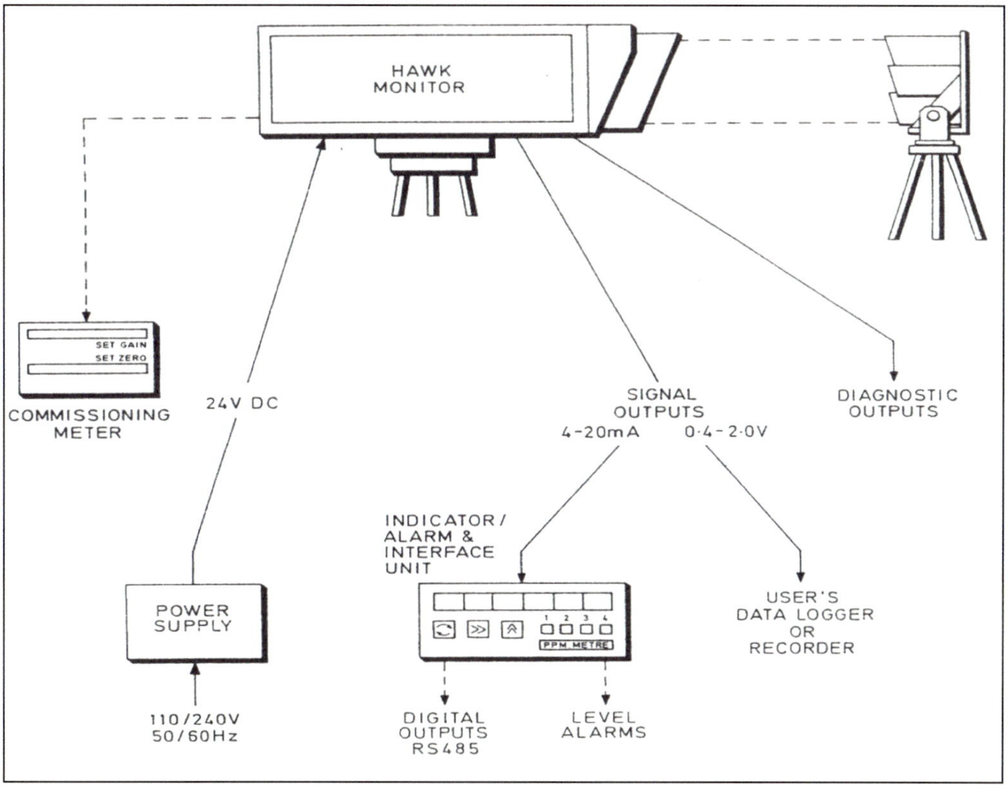

Fig. 2. A schematic diagram of the IR DOAS HAWK instrument

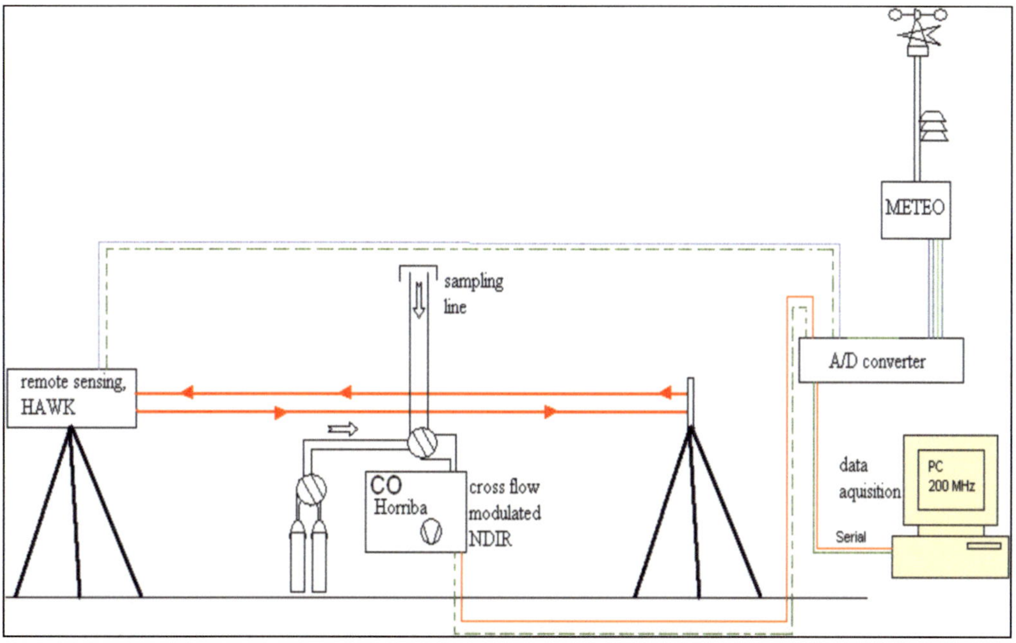

Fig. 3. Measuring setup with SIEMENS-HAWK and HORIBA-APMA 350E instruments

The second utilized instrument was an IR HAWK system from Siemens Environmental Systems, with the schematic diagram presented in Fig. 2. This instrument is an IR DOAS (Differential Optical Absorption Spectroscopy) apparatus, which can be configured to detect several species of pollutants including carbon monoxide. The beam path can be up to 400m and detection is typically better than 50ppb. The HAWK system works by measuring the absorption of infrared radiation passing along the instrument beam path by the gas to be measured. The system consists of a monitor, which contains the source and the detector unit, and a reflector. The total path length is therefore twice the distance between monitor and reflector. The source emits over a range of wavelengths and the beam is modulated after generation. The beam is reflected back to the monitor where it is filtered at a wavelength specific to gas of interest. The filtered beam is focused onto a detector which compares filtered and unfiltered reflected light in order to measure the concentration of the target gas. Open path techniques have an advantage over the point source detectors: the sample volume is mach greater, the non-uniformity of the sample is eliminated and a more representative value of the concentration to be measured is obtained. Under field conditions, the degree of mixing is affected by the local environment, primarily, wind and thermal gradients.
Fig. 3 shows a typical relative setup for the HAWK and HORIBA analysers. One should observe also the meteorological mast, which continuously sent data (15 minutes mean values) to the general data acquisition system.

2.2 Measured and conditioned signals

The CO-concentration signals were measured with both HAWK open path monitor and the HORIBA point monitor at a sampling period of 6 seconds. This corresponds to a sampling frequency $f_S = 600$ cycles/hour and a maximal (Nyquist) frequency of $f_{\max} = 300$ cycles/hour. Each signal contains 4950 samples expressed in $[\,\mathrm{mg/m^3N}\,]$ as represented in Fig. 4. The total registration length covers 8 hours and 15 minutes. Obviously, both signals have a non-stationary random character.

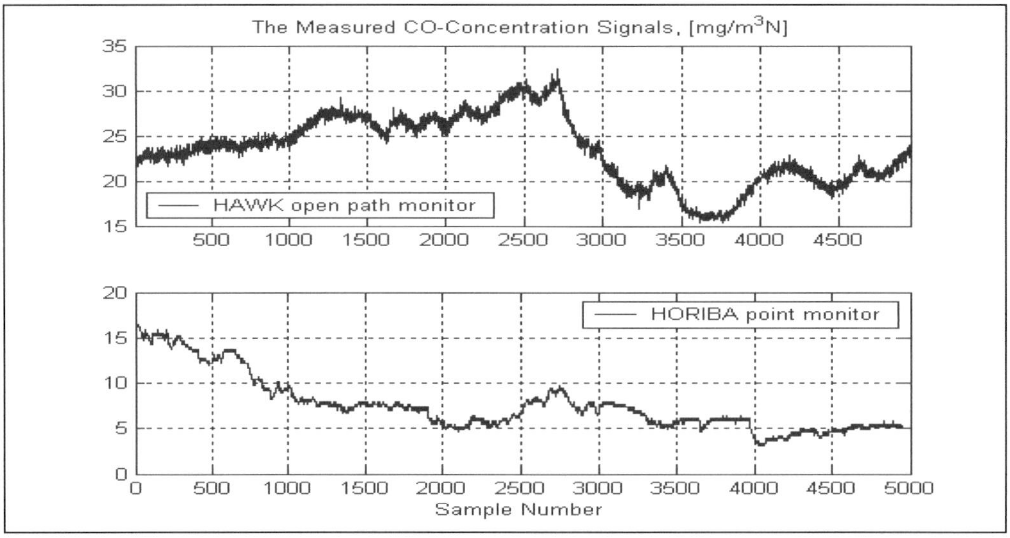

Fig. 4. CO-concentration signals measured with the HAWK open path and the HORIBA point monitor, respectively

During the same time interval of more than 8 hours, some meteorological parameters were also measured. Two of them, namely temperature, in [°C] , and wind velocity, in [m/s], are represented in Fig. 5. The direction of the wind were measured and utilized in the determination of the wind component parallel to the optical axis of the HAWK open path monitor as well the component perpendicular to this axis. The two wind components, in [m/s], are also represented in Fig. 5.

The sampling period for the meteorological parameters was 15 minutes, so that each of the meteorological parameters is determined through 33 values. The corresponding sampling frequency of 2 cycles/hour is a good choice for a slowly varying quantity like temperature, but can become critical in a turbulent environment with rapid changing wind direction or intensity. One can appreciate that, in our case, the sampling frequency was great enough for temperature and for wind velocity as well.

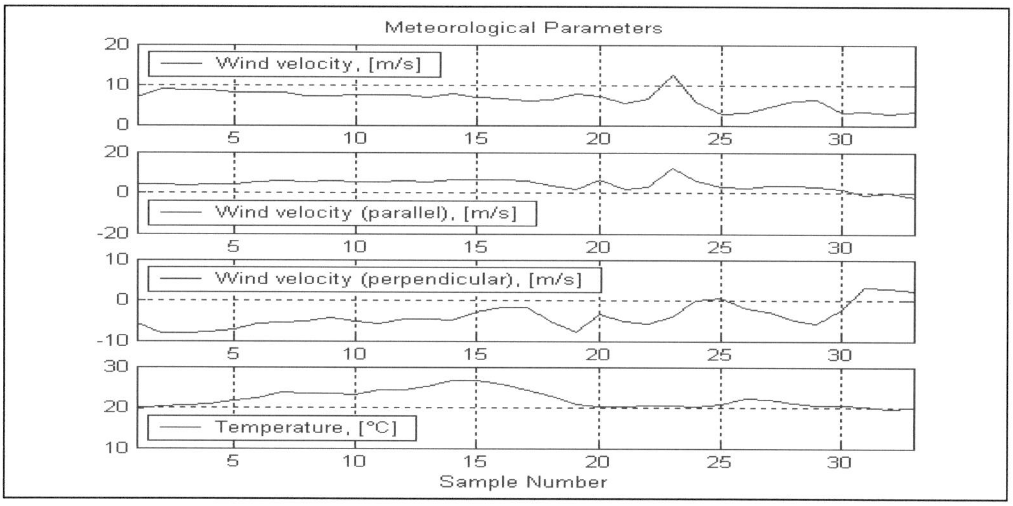

Fig. 5. Meteorological parameters: wind velocity and temperature

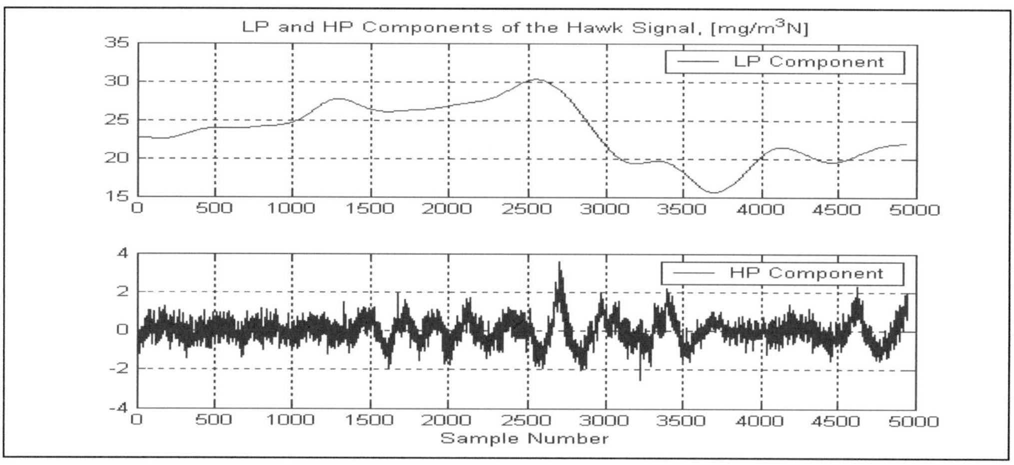

Fig. 6. LP and HP components of the CO-concentration signal measured with the HAWK instrument

Certainly, in a routine measurement choosing very different sampling frequencies for CO-concentrations and meteorological factors is not justified. In our case, the high sampling frequency of the CO-concentrations allows a characterization of the noise associate with these measurements. On the other side, choosing a higher sampling rate for the slowly varying meteorological parameters could be equivalent with storing a large amount of redundant data.

In order to separate the stabile local mean value of the CO-concentration from the measurement noise, an ideal low-pass (LP) and high-pass (HP) filtering of the signals were performed. The two components of the signal measured with the HAWK instrument are represented in Fig. 6. By „ideal" filtering we mean the infinitely sharp frequency characteristic at the cut-off frequency so that the sum of the HP and LP components gives exactly the values of the original unfiltered signal.

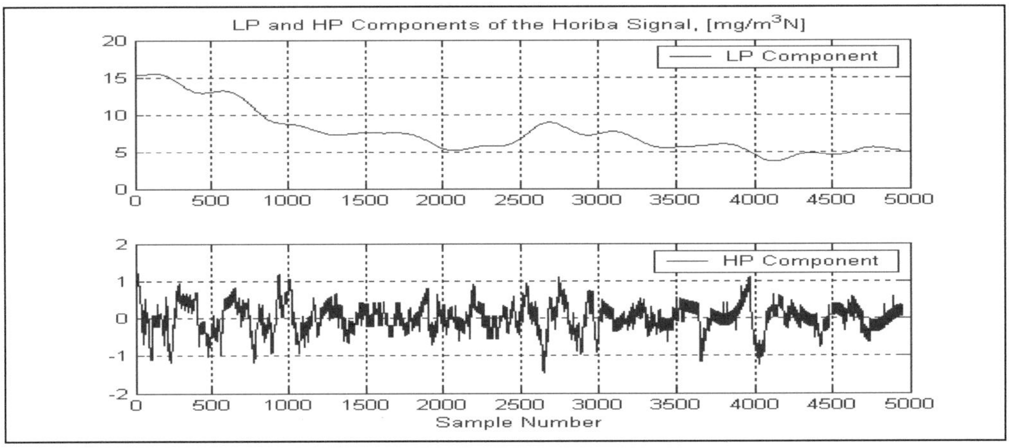

Fig. 7. LP and HP components of the CO-concentration signal measured with the HORIBA instrument

Practically, the ideal filtering was implemented through a direct fast Fourier transform (FFT) followed by a windowing the obtained spectrum with the desired filter characteristic and finally a reverse FFT to obtain the time domain representation of the LP or HP component of the CO-concentration signal.

The cut-off frequency of the LP filter representing also the corner frequency of the HP filter, was empirically chosen to be 2 cycles/hour. So the LP component contains frequencies between 0 and 2 cycles/hour while the HP component covers the range from 2 cycles/hour to 300 cycles/hour. Similar observations are valid for the LP and HP components of the CO-concentration measured with the HORIBA instrument. These are represented in Fig. 7.

Other signal processing techniques, like wavelets or short-time spectra can be used to separate the HP and LP components of the Pollution level signals. However ideal filtering based on Fourier transformation is simpler and very efficient.

The power of the measurement noise, calculated as the mean value of the HP components of the CO-concentration proves to be greater in the case of the HAWK instrument, than for the HORIBA device: $P_{HAWK} = 0.4151$ in comparison with $P_{HORIBA} = 0.1455$. This relation can be observed also in the graphical representation of the unfiltered signals, in Fig. 4. Experiments show that measurement noises may introduce an up to 5% error in the determination of the correlation coefficients. This effect could be neglected. However, the separation of LP and HP components of the CO-concentration signals can be a very useful signal conditioning step which allows a simple elimination of possible artifacts in the measured pollution levels.

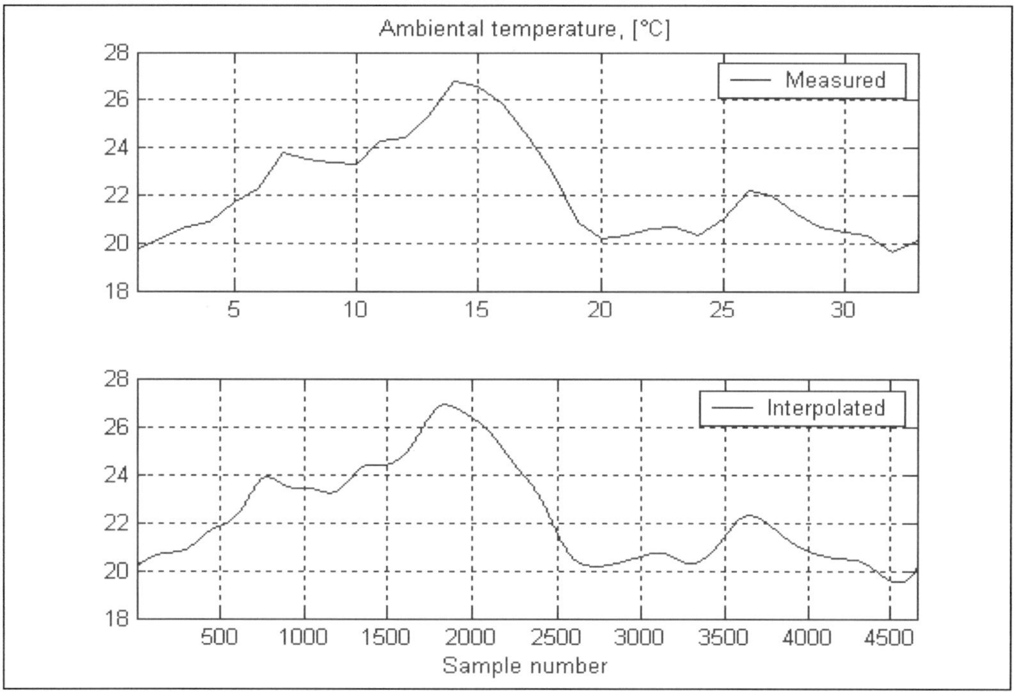

Fig. 8. Ambient temperature: measured and interpolated

Another pre-processing step is concerning the measured meteorological parameters. In order to calculate the correlation coefficients between CO-concentration signals on one side, and the meteorological parameters on the other, we must have the same number of samples in every signal. Therefore, the temperature and wind velocity signals were interpolated using a cubic spline procedure; the number of samples was increased from 33 to 4950. Due to the great interpolating errors at the beginning and the end of the signals, the first 150 and the last 150 samples of the interpolated signals were rejected. For example, Fig. 8 presents the ambient temperature signal in both forms, measured and interpolated. The first and the last 150 samples from the LP components of the CO-concentration signals were also eliminated. Finally, all signals, representing the pollution level, as well as the meteorological factors have the same length: 4650 samples.

3. Histograms and related parameters

The histogram is a representation of the signal amplitudes according to their values, regardless of the time variable: on the pollutant concentration axis a number of equally spaced containers are defined and the histogram returns the number of signal samples in each container (Montgomery & Montgomery, 2006; Navidi, 2008; Peck et al., 2007). This function can be easily implemented using the MATLAB software (Hoffmann & Quint, 2007; Martinez, & Martinez, 2002). Namely, *N=hist(Y)* puts the elements of *Y* into ten equally spaced containers and returns the number of elements in each container while *N=hist(Y,M)*, where *M* is a scalar, uses *M* bins.

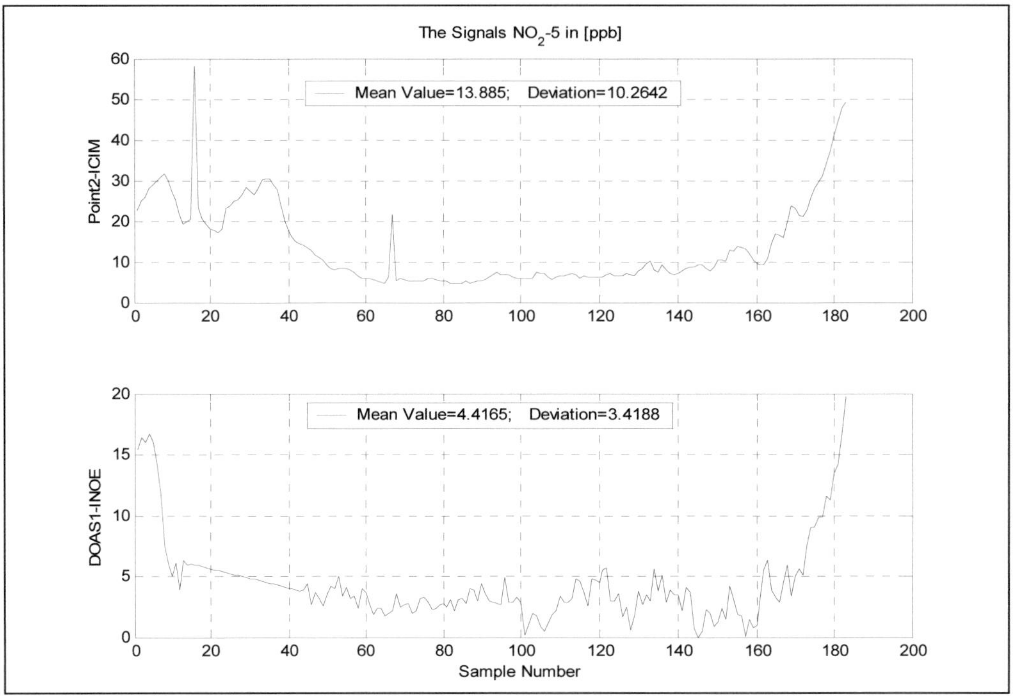

Fig. 9. Two NO_2 signals measured simultaneously

The histogram can be a useful tool for a preliminary characterisation of the pollutant concentration signals, complementary to there usual temporal representation. For example, Fig. 9 presents two NO_2 concentration signals, each containing 183 measured samples. The corresponding histograms of the same signals can be seen in Fig. 10. The realisations represented in Fig. 9 contain the full information about the measured concentrations. Particularly, one can observe the similarity of there evolution. The signal shown in the upper panel has two artefacts which must be eliminated during the pre-processing step. However, the soft limitation of this signal appears with greater clarity in the corresponding histogram from Fig.10. At the same time, the quasi normal (Gaussian) character of the signal represented in the lower panel is more evident in the corresponding histogram then in the temporal representation. The good continuity of both histograms suggests a sufficient size of the measured signals.

The normalized histogram (with area equal to one) is an approximate representation for the probability density function of the analysed signal. Some statistical parameters related to the histogram (through the probability density function) can be utilized for a quantitative or qualitative characterisation of the pollutant concentration signals. Thus, the mean value (Martinez, & Martinez, 2002; Montgomery & Montgomery, 2006; Papoulis, 1991; Therrien, 2007)

$$\bar{x} = \frac{1}{N}\sum_{i=1}^{N} x_i \tag{1}$$

and the standard deviation

$$\sigma = \sqrt{\frac{1}{N}\sum_{i=1}^{N}\left(x_i - \overline{x}\right)^2} \tag{2}$$

usually reported together, as in Fig. 9 and Fig. 10, offer a quantitative description for the location of the pollutant concentration values. Two other statistical parameters can be utilized for a rather qualitative characterisation of the data set in comparison with a normal distribution. Namely, the skewness (the third central moment divided by the cube of the standard deviation)

$$s = \frac{1}{N-1}\frac{\sum_{i=1}^{N}\left(x_i - \overline{x}\right)^3}{\sigma^3} \tag{3}$$

describes the lack of symmetry of the histogram. One can mention that for any symmetrical distribution (e.g. the normal distribution), the skewness is zero.

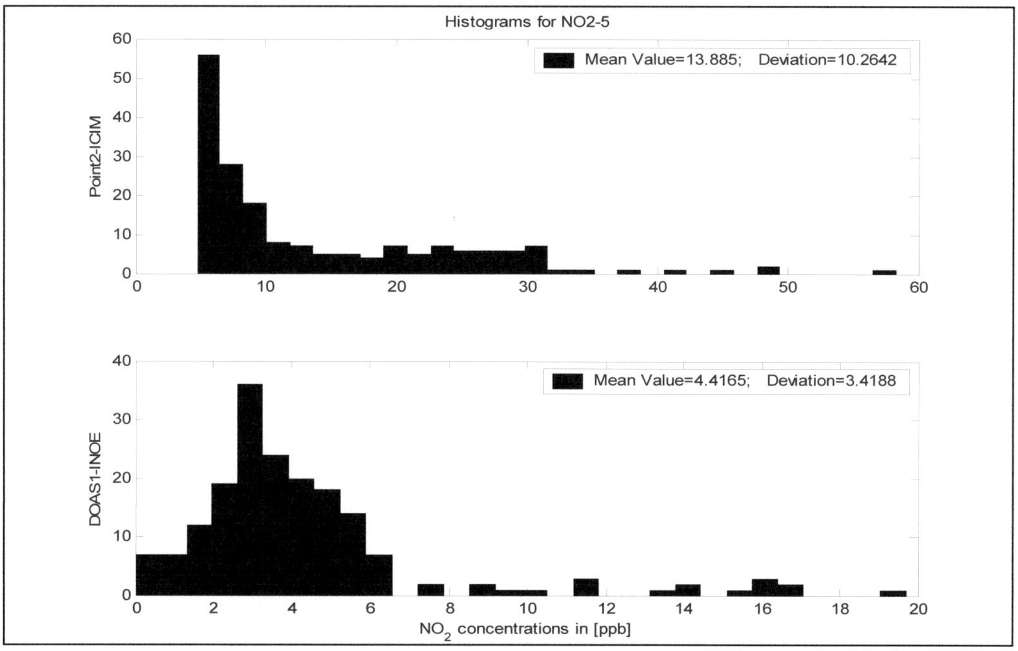

Fig. 10. Histograms of two NO_2 signals

The kurtosis (the fourth central moment divided by fourth power of the standard deviation)

$$k = \frac{1}{N-1}\frac{\sum_{i=1}^{N}\left(x_i - \overline{x}\right)^4}{\sigma^4} \tag{4}$$

offers a characterization of whether the histogram is peaked or flat relative to a Gaussian distribution. The kurtosis of a normal distribution equals three. Pollutant concentration signals with high kurtosis have a sharp peak near the mean value while signals with low

kurtosis have a rather flat top near the mean value. Obviously, in (1), (2), (3) and (4), x_i is the current sample from the N samples of the pollutant concentration signal.
The MATLAB functions *mean(X)*, *std(X)*, *skewness(X)* and *kurtosis(X)* return the sample mean, sample standard deviation, sample skewness and sample kurtosis, respectively (Hoffmann & Quint, 2007; Martinez, & Martinez, 2002). For matrices, the results of this functions are row vectors containing the sample mean, standard deviation, skewness respectively kurtosis of each column. The commands *mean(X,DIM)* and *std(X,FLAG,DIM)* takes the mean respectively the standard deviation along the dimension *DIM* of X. If *FLAG=0*, *std* normalizes by (*N-1*), otherwise *std* normalizes by *N*.

4. The correlation coefficient

4.1 Definitions and practical rules for acquisition parameters

Taking into account the usual lack of stationarity of pollutant concentration signals, the correlation coefficient is, probably, the most useful statistical tool revealing possible influences between such data sets. Particularly, the (Pearson) correlation coefficient (Papoulis, 1991; Therrien, 2007; Zuur et al., 2007) is defined by the following relation:

$$r = \frac{\frac{1}{N}\left[\sum_{i=1}^{N} x_i y_i - \sum_{i=1}^{N} x_i \cdot \sum_{i=1}^{N} y_i\right]}{\sigma_x \sigma_y} \tag{5}$$

where $\left[x_i\right]$ and $\left[y_i\right]$, $i = 1,2,...,N$ are two simultaneously measured pollutant concentration signals, with standard deviations σ_x and σ_y, respectively. One can prove that $|r| \le 1$. The variability of the two data sets can be interpreted according to the range including the particular value of the correlation coefficient. Thus, $-1 \le r \le -0.33$ denotes negative correlation, while $0.33 \le r \le 1$ signifies that the two signals are positive correlated. If $-0.33 \le r \le 0.33$, the two pollutant concentration signals are said to be uncorrelated (independent). Obviously, the defined domains for the correlation coefficient are, to some extent, arbitrary. More than this, in our interpretation of the estimated value of r, one must take care of the confidence interval of this parameter (Martinez, & Martinez, 2002; Papoulis, 1991; Shen & Lu, 2006). Practically, not only the value of the correlation coefficient but also his confidence interval must be in one of the three ranges in order to conclude that the analyzed signals are positive correlated, negative correlated or independent.
The MATLAB *corrcoef* function calculates the correlation coefficients. Thus, in the command *R=corrcoef(X)*, *R* is the matrix of correlation coefficients for the array *X* (each row of *X* is an observation and each column of *X* is a variable). This function has also the possibility to calculate the lower and upper bounds for a 95% confidence interval for each correlation coefficient (Hoffmann & Quint, 2007; Martinez, & Martinez, 2002).
Two acquisition parameters of the pollutant concentration signals are important for a reliable determination of the correlation coefficient: the sampling frequency and the number of samples in the measured data sets. In order to establish some practical rules for the acquisition phase of future pollutant concentration measuring campaigns, several signal processing experiments were carried out. Thus, Fig. 11 presents the evolution of the correlation coefficient calculated for a moving segment of data, containing only 30 samples measured with a sampling period of five minutes. (It is meaningless for this analysis that the two signal pairs were designated as Point1UPT-DOASINOE and Point1UPT-POINT3INOE,

respectively). However, the huge variation of the correlation coefficients shows that the size of the data is totally insufficient, at least for this sampling frequency.

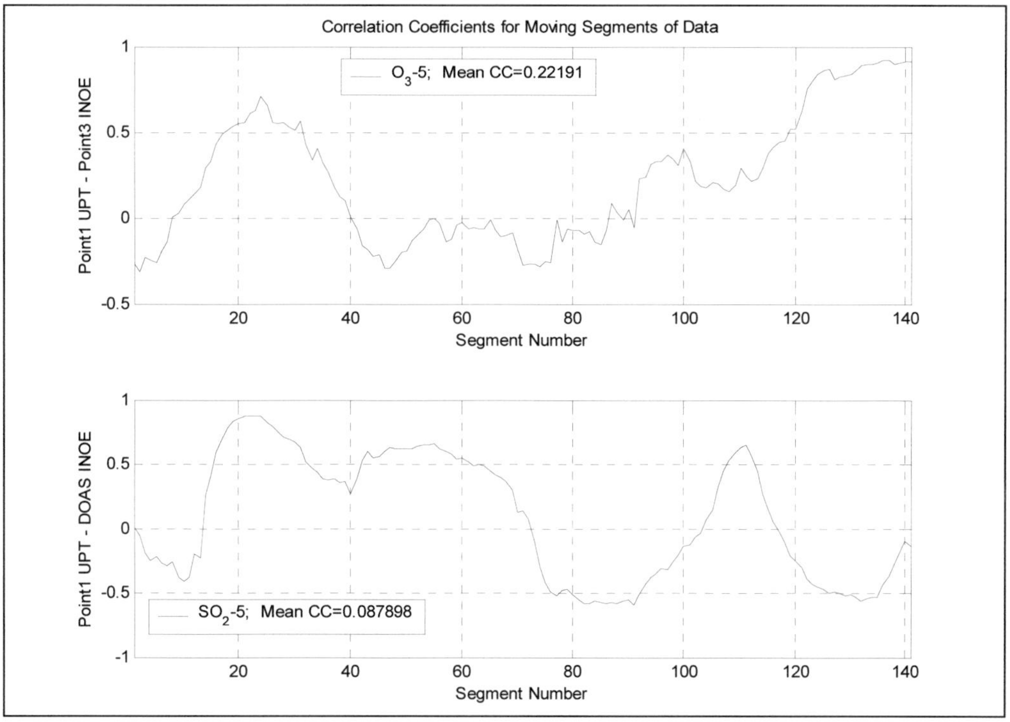

Fig. 11. Correlation coefficients for moving segments of data

Fig.12 shows the dependence of the correlation coefficient on size of the measured pollutant concentration signals. The correlation coefficient shown in the upper panel has a transient behaviour. The final value of the correlation coefficient proves that the two signals are strongly positive correlated. Just 60 measured samples of the signals could be sufficient for a good determination of the correlation coefficient for these signals, achieved at a sampling period of five minutes. Fig. 12 shows also the oscillatory behaviour of the correlation coefficient between Point1UPT and DOASINOE signals measured with a sampling period of five minutes. The low values of the correlation coefficient, proves that the two pollutant concentration signals are practically independent. Taking the confidence interval of the analysed data sets into account, the relative high values of r for segment length between 100 and 140 are not high enough to draw the conclusion that the two signals are positive correlated. The final decay of the correlation coefficient for segment length greater than 140 definitively proves the independence of the two concentration signals.

A practical conclusion can be drawn from this experiment: the calculation of a correlation coefficient with transient behaviour is practically completed as soon as the value of r reaches the quasi-steady state; if the evolution of the correlation coefficient is characterized by small amplitude variations, an optimal segment length is difficult to establish; however, since the duration of a pollutant concentrations measuring campaign is measured in days while the calculation of the correlation coefficient takes a few seconds, the interactive signal processing, during the measuring campaign is highly recommended.

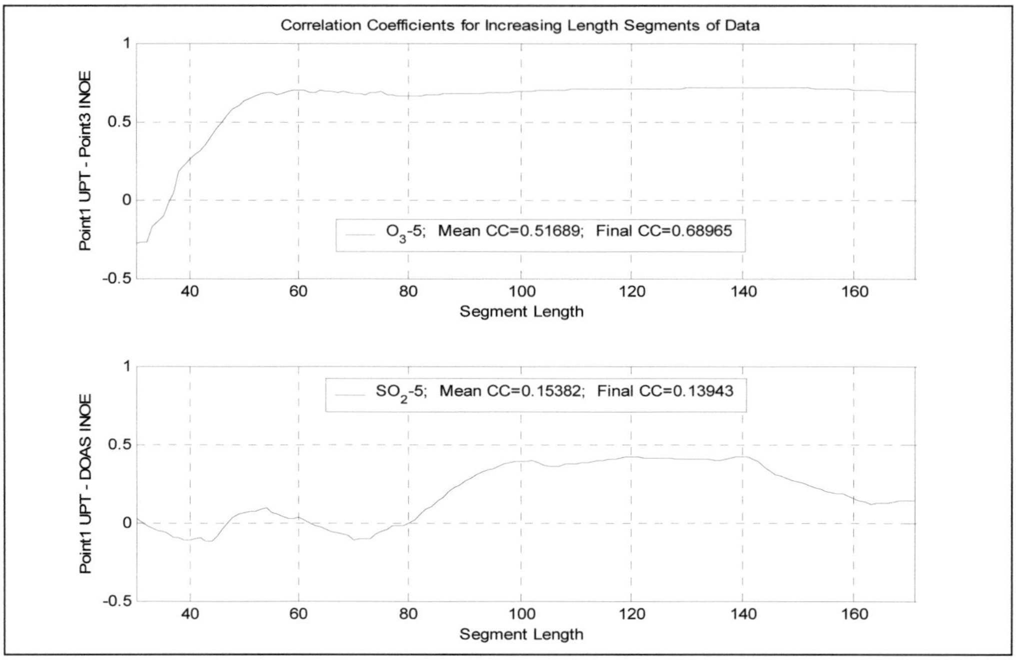

Fig. 12. The dependence of the correlation coefficient on the length of data segments

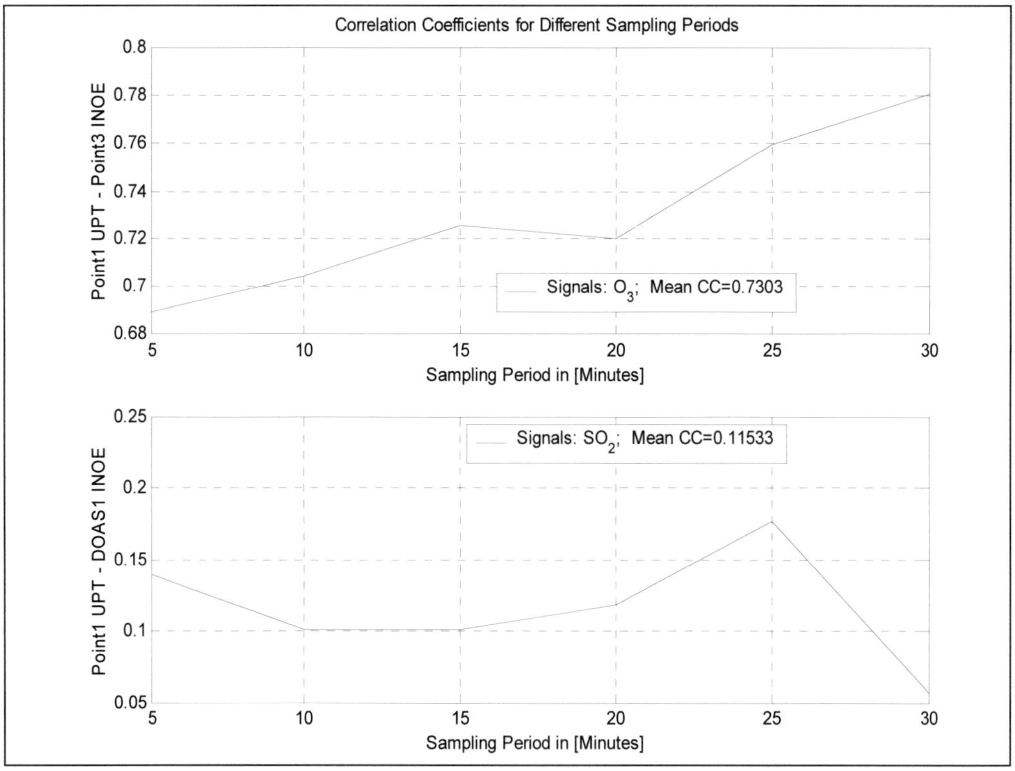

Fig. 13. Correlation coefficients calculated for different sampling periods

The dependence of the correlation coefficient on the sampling period is illustrated in Fig.13. The correlation coefficients are presented for two signal pairs achieved at a sampling period of 5, 10, 15, 20, 25 and 30 minutes, respectively. Corresponding to the increasing sampling period, the number of samples utilized in the calculation of the correlation coefficient was 172, 86, 57, 42, 34 and 28, respectively, so that the temporal signal length is maintained approximately constant (about 850 minutes). Comparing to the whole range [-1 to +1] of possible values for the correlation coefficient, the variations of the results shown in Fig.13 are small. This proves a high positive correlation between Point1UPT and Point3INOE signals (correlation coefficient about 0.73) and statistical independence between Point1UPT and DOAS1INOE signals (correlation coefficient about 0.11).
The practical conclusion of this experiment is that a reduced number of samples in the data sets can be compensated by a corresponding higher sampling period. This conclusion differs from the usual statement found in books on statistics, where the confidence interval of the correlation coefficient is determined by the number of sample in the analyzed signals. Our interpretation of this fact is that pollutant concentration signals are not artificial random signals generated by a certain mathematical rule, but physical signals where the generating mechanism can change in time. Consequently, one must give the signal enough time to show his features, and this can be done assuring a minimal temporal signal length, i.e. a minimal value of the product (Segment Size) · (Sampling Period). Certainly, the condition of a minimal number of measured samples (about 30 samples) must be also fulfilled.

4.2 Correlation coefficients Interpretation: a case study

The sample Pearson product-moment correlation coefficient, *r*, was computed for the following signals: LP component of the CO-concentration signal measured with the HAWK instrument (HAW, shown in Fig. 6), LP component of the CO-concentration signal measured with the HORIBA device (HOR, shown in Fig. 7), the wind velocity (W, shown in Fig. 5), the wind velocity component parallel with the optical axis of the HAWK instrument (Wpar, shown in Fig. 5), the wind velocity component perpendicular to the optical axis of the HAWK instrument (Wper, shown in Fig. 5), and the ambient temperature (T, shown in Fig. 8). Table 1 presents the approximate values (with only two decimals) of the determined correlation coefficients.

HAW	HOR	W	Wpar	Wper	T
1.00	0.23	0.45	0.33	-0.40	0.62
0.23	1.00	0.56	0.11	-0.68	-0.09
0.45	0.56	1.00	0.75	-0.73	0.23
0.33	0.11	0.75	1.00	-0.35	0.49
-0.40	-0.68	-0.73	-0.35	1.00	0.15
0.62	-0.09	0.23	0.49	-0.15	1.00

Table 1. Approximate correlation coefficients

In order to facilitate the interpretation, a diagram of the correlation coefficients is represented in Fig. 14. One can see a positive but small correlation coefficient between the two CO-concentration signals ($r \cong 0.23$). Related to the maximal possible value ($r \cong 1.00$), the actual correlation coefficient reflects on the fact that the two instruments do actually not measure the same quantity. The poor correlation proves the fundamental difference between a local pollution level (measured with HORIBA instrument) and spatial averaged CO-concentrations (from the HAWK system), not only as absolute values but also as variations tendencies.
The averaged signal measured with the open path remote sensing instrument is strongly positive correlated with the temperature ($r \cong 0.62$), while this meteorological parameter has practically no influence on the local pollution level ($r \cong -0.09$). This can be explained by the intensification of the activities in the non ecological waste deposit by daytime, fact clearly sensed by the open path instrument but not by the Horiba device which measures the pollution in a certain point, at soil level. The CO-concentrations measured with both instruments are moderate positive correlated to the magnitude of the wind velocity ($r \cong 0.45$ and $r \cong 0.56$ with the HAWK and HORIBA signals respectively). However, the component of the wind velocity perpendicular to the optical axes is negative correlated with both measured pollution levels. Due to the geometrical arrangement of the measuring systems this wind component tends to clean the air. As a global conclusion, the HAWK CO-concentration signal is better correlated with the meteorological parameters than the HORIBA signal.

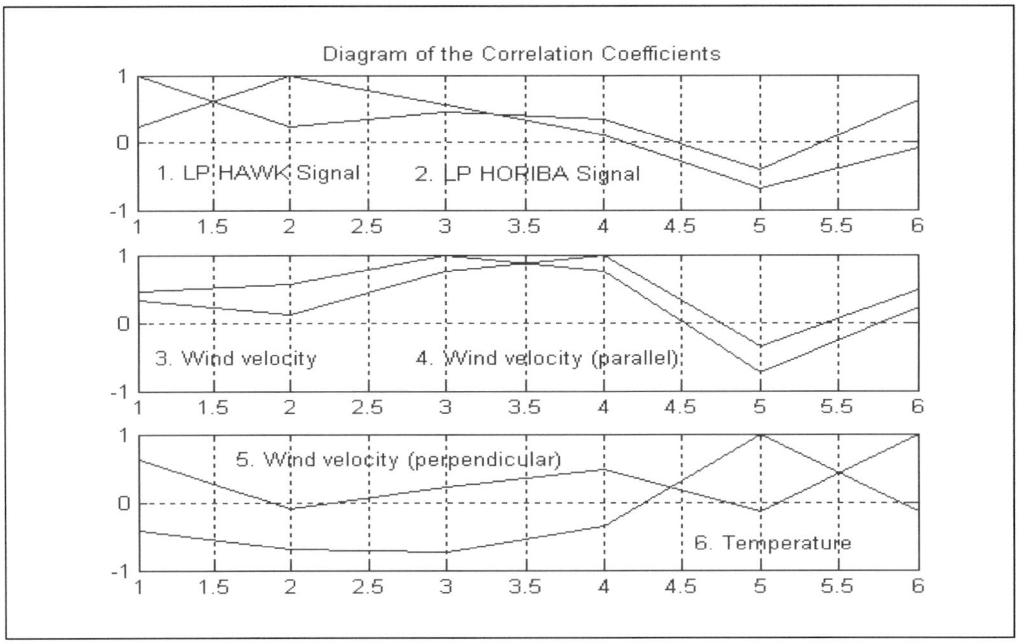

Fig. 14. Graphical representations of correlation coefficients

The 95% confidence intervals of the correlation coefficients were computed using a procedure based on the Fisher transformation (Shen & Lu, 2006). Exact values of the correlation coefficients between HAWK CO-concentration and the other five signals are given in Table 2, together with their confidence intervals. Due to the great number of samples in each signal, the confidence intervals are narrow around the calculated correlation coefficients. Fig.15 gives a graphical image of the narrowness of these intervals.

HAW -95%	HAW	HAW +95%
1.0000	1.0000	1.0000
0.2039	0.2313	0.2583
0.4260	0.4492	0.4719
0.3085	0.3343	0.3596
-0.4281	-0.4043	-0.3800
0.6006	0.6187	0.6361

Table 2. Exact values of the correlation coefficients and their confidence intervals

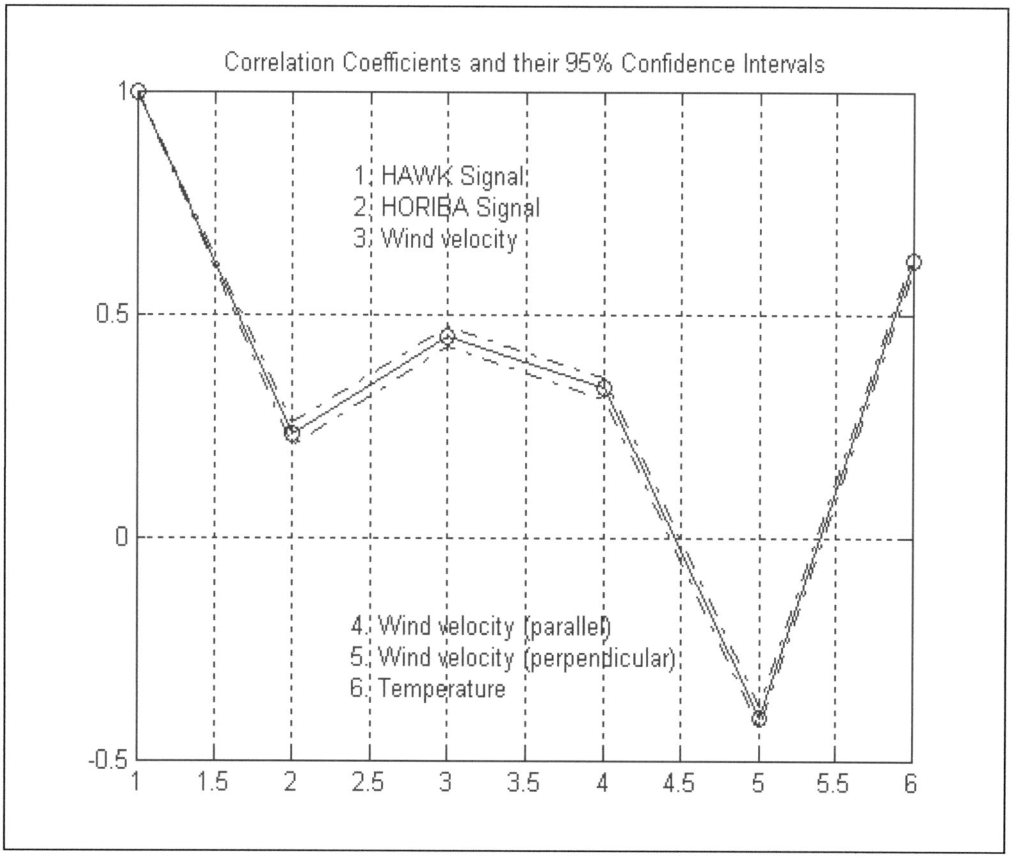

Fig. 15. Correlation coefficients of the HAWK signal (-o) and their 95 % confidence intervals (- ·)

5. Correlation and covariance functions

The correlation and covariance functions can put into evidence power and temporal relations between pollutant concentration signals. The sample crosscorrelation function is defined by the formula (Therrien, 1992):

$$R_{xy}[l] = \begin{cases} \dfrac{1}{N-l} \cdot \sum\limits_{i=0}^{N-1-l} x_{i+l} y_i ; \ for \ 0 \le l < N \\ \dfrac{1}{N-|l|} \cdot \sum\limits_{i=0}^{N-1-|l|} x_i y_{i+|l|} ; \ for - N < l < 0 \end{cases} \tag{6}$$

can be calculated if the measured pollutant concentration signals are stationary, but this is rarely the case. However, for $[y_i] \equiv [x_i]$ we obtain the autocorrelation $R_{xx}[l]$ of the sequence $[x_i]$. Fig. 16 shows the autocorrelation functions for Point1UPT signal (in the upper panel) and Point3INOE (in the lower panel). The value of the autocorrelation at lag zero, $R_{xx}[0]$, gives the power of the corresponding signal: approximately 1024 ppb^2 for Point1UPT and 1078 ppb^2 for Point3INOE signals. After some experience, one can appreciate the spectral content of the signal from the shape of the autocorrelation function.

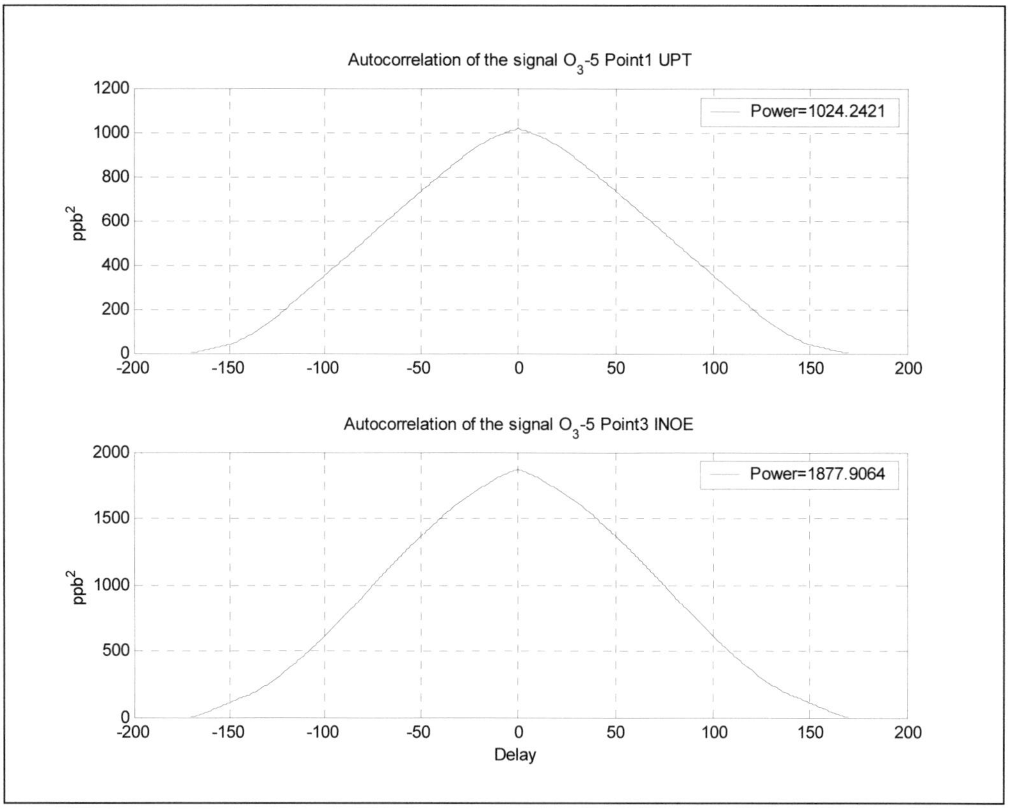

Fig. 16. Typical autocorrelation functions

The sample auto covariance, $C_{xx}[l]$, is the sample autocorrelation of the centred sequence $[x_i]$ (the sequence after removing the mean) while the sample cross covariance, $C_{xy}[l]$, is the cross correlation of the centred sequences $[x_i]$ and $[y_i]$. Thus, Fig. 17 presents the auto covariance functions for the same signals, namely Point1UPT and Point3INOE pollutant concentrations. The auto covariance at zero lag, gives the power of the variable part of the analysed signal: approximately 74 ppb^2 for Point1UPT and 131 ppb^2 for Point3INOE signals. The shapes of the auto covariance functions show that Point1UPT signal has a highly random character while in the Point3INOE signal the spectral components are concentrated in a small range near 0.016 mHz.

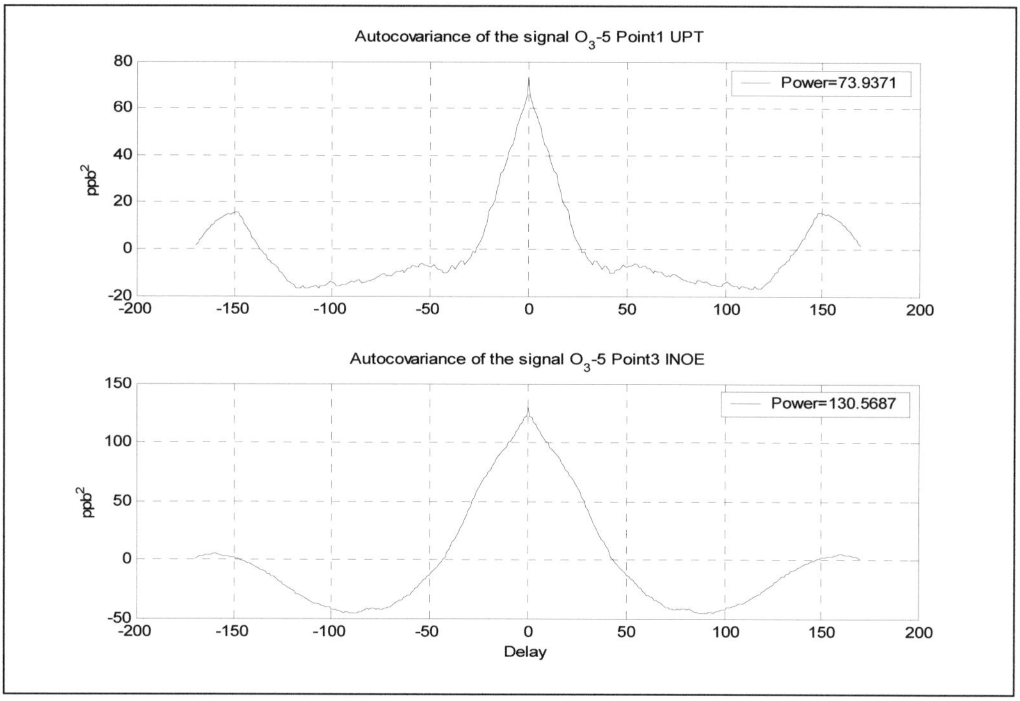

Fig. 17. Typical auto covariance functions

The cross-covariance functions offer an interesting possibility to determine delays between pollutant concentrations measured simultaneously. The position of the peak in the covariance function around zero gives the temporal delay between two signals. Depending on the wind direction and intensity the pollutants can be transported from one place to the other in the experimental area. However such temporal relations can be put into evidence only if the resolution on the time axis is sharp enough to allow the proper localization of the crosscovariance peak. This condition was not fulfilled during the related measuring campaign: the sampling period should be in the range of seconds while the signals were achieved with sampling periods of 5 to 30 minutes. The consequence is that the peak of the autocorrelation function appears in the origin (zero lag) or they have a flat maximum around zero, as shown in Fig.18.

Summarizing, the autocorrelation and auto-covariance functions are useful tools in establishing power relations between pollutant concentration signals measured with optoelectronic instruments. For reliable results, a sufficient temporal length of measured data sets must be assured, so that the signals can manifest there features. Temporal relations i.e. delays between certain signals measured simultaneously can be revealed using the crosscovariance function. But, for this purpose a supplementary condition must be fulfilled: a small sampling period must assure a good resolution on the time axis.

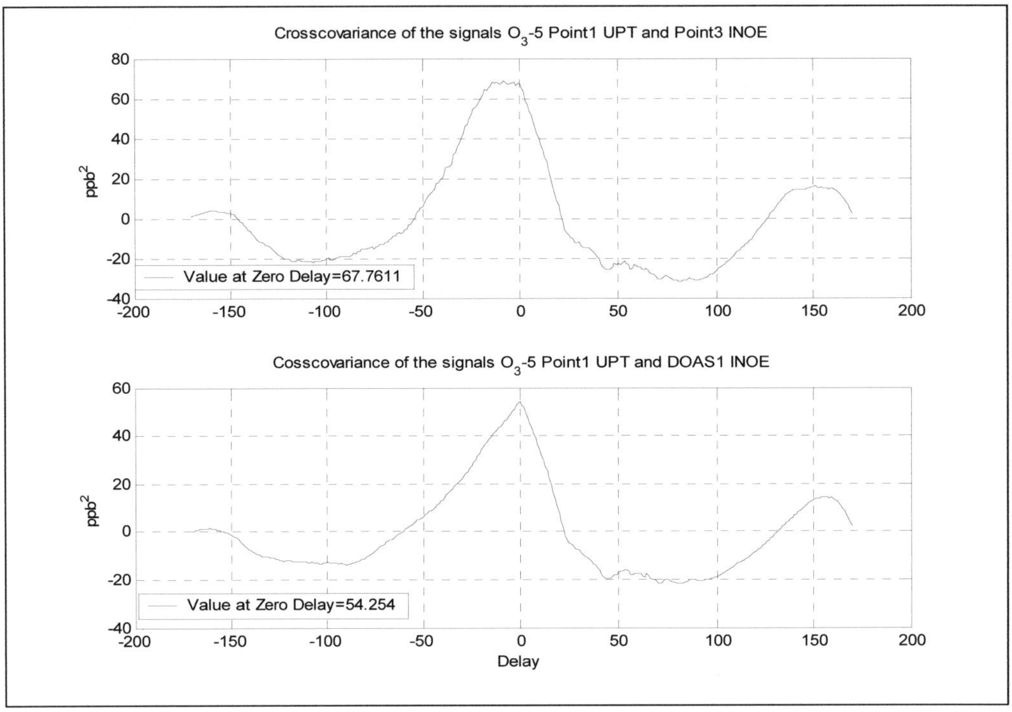

Fig. 18. Experimental cross covariance functions

The MATLAB *xcorr* function produces estimates of the cross-correlation function. For example, the command *C=xcorr(A,B)*, where *A* and *B* are length *M* vectors, returns the length 2**M*-1 crosscorrelation sequence *C*. Particularly, *C=xcorr(A)*, where *A* is a vector, returns the autocorrelation sequence. One can limit the range of lags in the (auto/cross) correlation function to (*-Maxlag, Maxlag*), using the command form *xcorr(...,Maxlag)*. Similarly, the MATLAB function *xcov* produces estimates of the (auto/cross) covariance function (actually, correlation functions of sequences with their means removed).

6. Conclusion

Due to the random character of the pollutant concentrations measured with optoelectronic instruments, statistical signal processing methods are recommended. Histograms, correlation coefficients, (auto/cross) correlation, (auto/cross) covariance functions or statistical parameters like mean, standard deviation, skewness and kurtosis can be useful tools in analyzing such signals. However, according to the purpose of the measuring campaign, the experiment must be carefully designed in order to obtain reliable results. This chapter reveals some practical rules for setting acquisition parameters like data (segment) size and sampling frequency. As a first rule, for reliable correlation coefficient determination one must assure a sufficient temporal length of the concentration signals, i.e. the product between segment size and sampling period must be large enough in order to obtain stable values of the correlation coefficients. This rule is also valid for the calculation of autocorrelation or auto covariance functions. However, if we are interested to use cross correlation or cross covariance function to reveal delays between pollutant concentration and/or meteorological signals, another rule must also be taken into account: assure the necessary resolution on the time axis i.e. the sampling period must be small enough in comparison with expected delays. In any case, interactive verification and setting of the acquisition parameters during the measuring campaign, according to the purpose of every particular experimental research, are recommended.

Signal conditioning procedures must be implemented before determining the statistical functions and parameters. Thus, ideal filtering based on fast Fourier transform is an useful pre-processing step allowing a simple rejection of measurement noise and possible artefacts of the pollution level signals. Interpolation can be used to increase the number of samples of the slowly varying meteorological parameters, avoiding redundant measurements.

The purpose of one measuring campaign was the correlative comparison of two CO-concentration optoelectronic measuring instruments, working on different principles. Within this research the correlation coefficient proved to be the most useful tool in analyzing dependencies between pollution levels and the meteorological factors. The open path remote sensing instrument measures spatial averaged values which show better correlation to the meteorological parameters. Thus, the open path instrument is better suited for monitoring the pollution level in a large area than the classical NDIR device.

7. References

Hoffmann, J. & Quint, F. (2007). *Signalverarbeitung mit MATLAB® und SIMULINK®. Anwendungsorientierte Simulationen*, Oldenbourg Verlag, ISBN 978-3-486-58427-1, München

Ionel, I.; Ionel, S. & Nicolae, D. (2007). Correlative comparison of two optoelectronic carbon monoxide measuring instruments. *Journal for Optoelectronics and Advanced Materials*, Vol. 9, No. 11, pp. 3541-3545 ISSN

Ionel, I.; Ionel, S. & Lie, I. (2009). Statistical Tools in the Analysis of Pollutant Concentrations Measured with Optoelectronic Instruments, *Proceedings of the 11th WSEAS International Conference on Sustainability in Science and Engineering (SSE '09)*, Vol. II, pp.293-298, Timişoara, Romania, May, 2009, Published by WSEAS Press

Martinez, L. W. & Martinez, R. A. (2002). *Computational Statistics Handbook with MATLAB®*, Chapman & Hall/CRC, ISBN 1-58488-229-8, Boca Raton

Montgomery, C. D. & Runger, C. G. (2006). *Applied Statistics and Probability for Engineers*, 4th Edition, John Wiley & Sons, Inc., ISBN 0-471-74589-8, New York

Navidi, W. (2010). *Statistics for Engineers and Scientists*, McGraw-Hill, Inc., 3rd Edition, ISBN-13: 978-0071222051, New York

Papoulis, A. (1991). *Probability, Random Variables, and Stochastic Processes*, McGraw-Hill, Inc., 3rd Edition, ISBN 0-07-100870-5, New York

Peck, R.; Olsen, C. & Devore, J. (2008). *Introduction to Statistics and Data Analysis*, Duxbury Press, 3rd Edition, ISBN-13: 978-0-495-11873-2, Pacific Grove, CA

Shen, D.; Lu, Z. (2006). Computation of Correlation Coefficient and Its Confidence Interval in SAS, www2.sas.com/proceedings/sugi31/170-31.pdf

Therrien, C. W. (1992). *Discrete Random Signals and Statistical Signal Processing*, Prentice-Hall International, Inc., ISBN 0-13-217985-7, Englewood Cliffs

Zuur, A. F.; Ieno, E. N. & Smith, G. M. (2007). *Analyzing Ecological Data*, Springer Science + Business Media, ISBN-13: 978-0-387-45967-7, New York

Part 5

Physical Modeling and Simulations of Optoelectronic Devices

21

Advanced Numerical Simulation of Organic Light-emitting Devices

Beat Ruhstaller[1], Evelyne Knapp[2], Benjamin Perucco[3], Nils Reinke[4], Daniele Rezzonico[5] and Felix Müller[6]
[1,2,3,4] *Zurich University of Applied Sciences, Institute of Computational Physics, 8401 Winterthur*
[5,6] *Fluxim AG, 8835 Feusisberg*
Switzerland

1. Introduction

Organic light-emitting devices (OLEDs) are novel and efficient light sources that consist of a sequence of layers that fulfill distinct electronic and optical tasks. Given the variety of organic semiconductor materials available, the improvement of the device performance is a tedious and demanding task that often involves numerous experimental optimizations of layer materials, thicknesses and sequence. Comprehensive numerical device models have recently been developed that master both the physical complexity of the underlying optoelectronic processes as well as the numerically challenging system of equations. In this chapter we introduce such a comprehensive optical and electronic device model that is able to describe the device performance of OLEDs. We first introduce the key device model equations for charge and exciton transport as well as the dipole emission model for describing the out-coupling of light. Then we present a series of simulation results that are of practical interest when studying and optimizing OLEDs. These include the calculation of current-voltage curves, current transients signals, time-of-flight current transients and impedance spectroscopy data. We show that the physical model can be combined with a nonlinear least-square fitting algorithm for extracting transport parameters from measurements. In terms of optical device characteristics, quantitative outcoupling mode contributions and angular characteristics are presented as well as results of an emission zone extraction method.

Electrical characterization of devices and materials is essential and helps to elucidate the underlying, physical models of charge carrier transport in disordered, organic semiconductors. Besides the commonly used current-voltage curves, dark-injection, electroluminescence and time-of-flight transient measurements as well as impedance spectroscopy offer other ways to validate models for organic LEDs and extract model parameters. By means of a one-dimensional numerical OLED model we are able to simulate these different measurement techniques. Here, we present numerical methods in the physical and numerical framework of reference (Knapp et al, 2010) and solve directly for the steady- and transient state. Further, we conduct a numerical small signal-analysis for OLEDs. The underlying model solves the drift-diffusion equations in a coupled manner for disordered, organic semiconductors. The disordered nature of organic semiconductors affects the density of state, the mobility model, the Einstein diffusion relation as well as charge injection. These novel physical model ingredients constitute a second generation OLED model and are implemented in the simulator SETFOS (Fluxim AG, 2010). It is expected that the second generation OLED model will impact the way OLED characteristics and performance are quantitatively described.

2. Description of the device model

2.1 Charge drift-diffusion model

To describe the main features of charge transport in organic LEDs four processes have to be considered as illustrated in the schematic energy level diagram in Fig. 1. In a first step, charge carriers have to be injected into the organic material (1), secondly they will be transported (2) until they recombine to an exciton (3). Then the excitons decay radiatively or non-radiatively (4). In the following we will first look at the transport process (2). For the description of

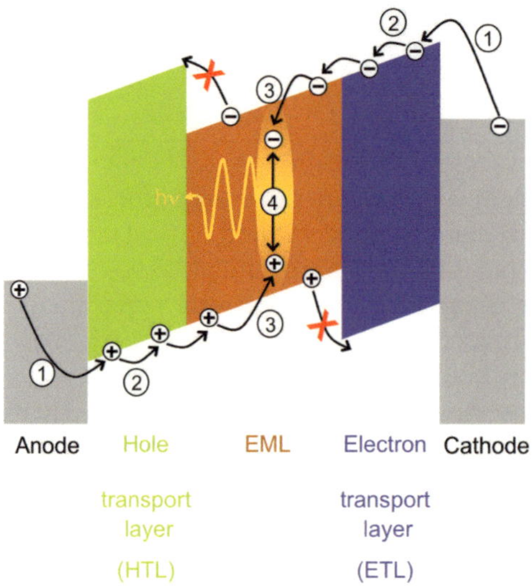

Fig. 1. Main processes in OLED operation: 1) Injection, 2) Transport, 3) Formation of exctions, 4) Radiative decay

charge transport in OLEDs the general semiconductor drift-diffusion equations for electrons and holes are valid. In Poisson's equation

$$\nabla \cdot (\epsilon \nabla \psi) = e(n + n_t - p - p_t), \tag{1}$$

the electrical potential ψ is related to the mobile electron and hole densities n and p and the trapped electron and hole densities n_t and p_t where e is the elementary charge and ϵ the product of the vacuum permittivity ϵ_0 and the relative permittivity ϵ_r of the organic material. The current equations for electrons and holes read

$$\begin{aligned} J_n &= -en\mu_n \nabla \psi + eD_n \nabla n, \\ J_p &= -ep\mu_p \nabla \psi - eD_p \nabla p \end{aligned} \tag{2}$$

where $\mu_{n,p}$ denotes the mobility and $D_{n,p}$ the diffusion coefficient for electrons and holes. Only mobile charges contribute to the current. The conservation of charges leads to the continuity equations for electrons and holes

$$\begin{aligned} \frac{\partial n}{\partial t} &= \frac{1}{e}\nabla J_n - R(n,p) - \frac{\partial n_t}{\partial t}, \\ \frac{\partial p}{\partial t} &= -\frac{1}{e}\nabla J_p - R(n,p) - \frac{\partial p_t}{\partial t} \end{aligned} \tag{3}$$

where R denotes the bimolecular recombination rate given by Langevin and t the time (Langevin, 1903). These equations take charge migration and recombination into account. The trapped electron (n_t) and hole (p_t) charge carriers obey the rate equations for an energetically sharp trap levels as shown in Fig. 2

$$\begin{aligned} \frac{\partial n_t}{\partial t} &= r_c n(N_t - n_t) - r_e n_t, \\ \frac{\partial p_t}{\partial t} &= r_c p(N_t - p_t) - r_e p_t. \end{aligned} \tag{4}$$

where r_e denotes the escape rate, r_c the capture rate and N_t the trap density. Note, that more general trap distributions can be introduced that are described by an exponential or Gaussian density of trap states (Fluxim AG, 2010; Knapp et al, 2010).

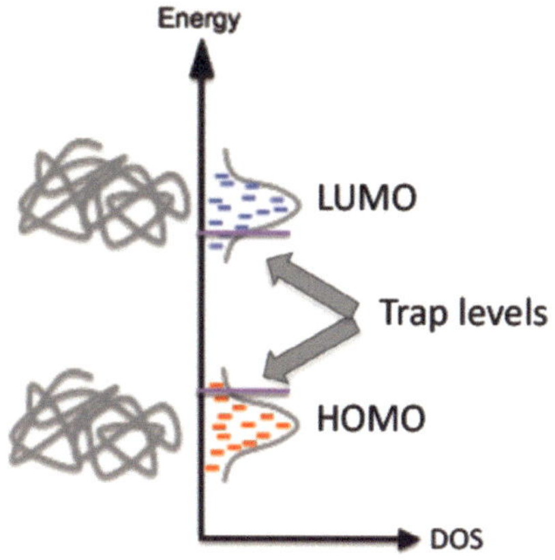

Fig. 2. Gaussian distributions of density of states and trap levels for trapped charges.

As opposed to inorganic semiconductors the density of states for organic semiconductors is described by a Gaussian shape since transport is assumed to occur via a hopping process between uncorrelated sites. Thus, polymers and small molecules have broadened energy levels of their highest occupied molecular orbital (HOMO) and lowest unoccupied molecular orbital (LUMO) as shown in Fig. 2 and are described in the following way

$$N_{Gauss}(E) = \frac{N_0}{\sqrt{2\pi\sigma^2}} \exp\left[-\left(\frac{E - E_0}{\sqrt{2}\sigma}\right)^2\right] \tag{5}$$

where N_0 denotes the site density, σ the width of Gaussian and E_0 the reference energy level. In the extended Gaussian disorder model the Gaussian density of states affects charge diffusion. Tessler pointed out that the use of the generalized instead of the classical Einstein relation is appropriate (Tessler et al., 2002). The generalized Einstein diffusion coefficient is now determined by

$$D = \frac{kT}{q}\mu(T,p,F)g_3(p,T), \tag{6}$$

where the enhancement function g_3 reads

$$g_3(p,T) = \frac{1}{kT}\frac{p}{\frac{\partial p}{\partial E_f}}. \tag{7}$$

Using the expression

$$p(E_f) = \int_{-\infty}^{\infty} DOS(E)f(E,E_f)dE \tag{8}$$

for the density and inserting the Fermi-Dirac distribution and the Gaussian DOS we obtain

$$\frac{D}{\mu} = \frac{kT}{q}\frac{\int_{-\infty}^{\infty} N_{Gauss}(E)\frac{1}{1+\exp\left(\frac{E-E_f}{kT}\right)}dE}{\int_{-\infty}^{\infty}\frac{N_{Gauss}(E)\exp\left(\frac{E-E_f}{kT}\right)}{[1+\exp\left(\frac{E-E_f}{kT}\right)]^2}dE}. \tag{9}$$

We will now turn to the charge mobility model. A mobility model that has been applied for quite some time now is the Poole-Frenkel mobility which is field-dependent and reads

$$\mu = \mu_0 \exp\left(\gamma\sqrt{E}\right), \tag{10}$$

where μ_0 is the zero-field mobility and γ is the field-dependence parameter. However, it has been shown by Bässler with the aid of Monte Carlo simulations that the energetic disorder in organic semiconductors influences the charge mobility (Pautmeier et al., 1990). Experiments have shown that the mobility in hole-only devices can differ up to three orders of magnitude between OLED and OFET device configurations with the same organic semiconductor. An explanation for this difference is a strong dependence of the mobility on the charge density (Tanase et al., 2003). Vissenberg and Matters developed a mobility model that considers such a density-dependent effect (Vissenberg & Matters, 1998). Using a 3D master equation approach to simulate the hopping transport in disordered semiconductors a dependence on the temperature, field and density was determined. Pasveer's model is therefore dependent on the temperature, field as well as the density and accounts for the disorder in the material (Pasveer et al, 2005). The extended Gaussian disorder model (EGDM) is an extension of the Pasveer model by additionally considering diffusion effects. In the EGDM the mobility may be expressed as a product of a density-dependent and field-dependent factor according to van Mensfoort as (Mensfoort et al., 2008a)

$$\mu(T,p,F) = \mu_0(T)g_1(p,T)g_2(F,T), \tag{11}$$

with the enhancement functions $g_1(p,T)$ and $g_2(F,T)$ accounting for the influence of the charge density, the electric field and the energetic disorder. The functions $g_1(p,T)$ and $g_2(F,T)$ are nonlinear and strongly increase with higher disorder parameter (Mensfoort et al., 2008b). In the remainder of this text, we will use either the traditional mobility model (Equation 10) or the advanced density-dependent model (Equation 11).

2.2 Exciton generation, transport and decay

This section describes the generation, transport and decay of excitons in organic semiconductor layers of OLEDs and becomes relevant in bipolar devices. While electron-hole recombination may be locally confined, the resulting excitons can migrate with a characteristic diffusion constant before decaying radiatively or transferring their energy to other exciton species in the same layer. Thus, the rate equation for exciton species S_i contains a generation, a diffusion, transfer and (non-)radiative terms:

$$\frac{dS_i(z)}{dt} = G_i R(z) + \vec{\nabla} \vec{J}_{s_i}(z) - (k_{rad_i}(z) + k_{nonrad_i}) \cdot S_i(z) - k_{annihilation_i} \cdot S_i(z)^2 + + \sum_{j=1}^{n_{exc}} (k_{ji} \cdot S_j(z) - k_{ij} \cdot S_i(z)) \quad (12)$$

In Eq. 12 G_i is the generation efficiency prefactor, $k_{rad_i}(z)$ the position-dependent radiative decay rate, k_{nonrad_i} the non-radiative decay rate and $k_{annihilation_i}$ an annihilation rate. For a comprehensive fully-coupled opto-electronic OLED model, the radiative decay rate is calculated with the dipole emission model (see section 2.3 below) and ref. (Fluxim AG, 2010). In a classical view the generation efficiency for singlet excitons is $G_i = 0.25$ while it would be $G_i = 0.75$ for triplet excitons. The exciton energy transfer term $-k_{ij} \cdot S_i$ shall be used to model energy transfer from exciton S_i to exciton species S_j. Such exciton energy transfer rates can be taken from the literature. Common energy transfer mechanisms are Förster and Dexter transfer (Dexter, 1953; Förster, 1948). Förster transfer is a long-range dipole-dipole interaction while Dexter transfer is a short-range electron-exchange interaction. Both mechanisms rely on the spectral overlap of the emission spectrum of the donating exciton species with the absorption spectrum of the accepting exciton species. Exciton transport happens by diffusion and is therefore mathematically described by

$$J_{s_i} = D_s \nabla S_i \quad (13)$$

where D_s denotes the exciton diffusion constant.

2.3 Optical dipole emission model

In this section we present a model for the dipole emission describing the radiative decay of excitons in planar multi-layer systems. This model accounts for the influence of layer interfaces on the radiative decay rate of fluorescent molecules and is based on a classical theory for the emission of an oscillating electrical dipole. The basic concept of this theory was derived by Sommerfeld to analyze electromagnetic radiaton close to a conductive surface and extended by Chance et al. (Chance et al, 1978) to model fluorescence emission in optical cavities.

The emission characteristic of a single molecule is similar to that of a Hertzian dipole. However, in an infinite medium consisting of isotropically oriented molecules, the emission from the multitude of molecules is homogeneous in all directions. If, on the contrary, the molecules are embedded in a thin layer, interference effects of electromagnetic waves reflected at layer interfaces dominate the emission characteristics.

The dipole emission model allows to quantitatively analyze the individual energy loss channels of excited molecules. These energy loss channels include radiative coupling to the far-field, modes guided inside the multilayer structure as well as absorption losses by the surrounding media. The model is based on the following assumptions:

(1) Each medium of the multilayer system is assumed to be isotropic and only characterized by its dielectric permittivity ϵ.

(2) The medium embedding the emissive dipole is considered to be loss-free (i.e. real dielectric permittivity ϵ)

(3) The interface planes are infinitely extended.

(4) The wavelength λ of the radiation as well as the distance between the emitter and the individual interfaces is large with respect to the spatial extensions of the dipole (point dipole approximation).

(5) The dipole is at least at $\lambda/50$ distance away from the regions with a complex permittivity ϵ.

(6) The materials of the multilayer stack are considered to be non-magnetic.

Despite these restrictions, this model is valid for a wide range of physical situations. The model considers the molecules as driven damped harmonic dipole oscillators. These oscillators are damped by the emission process and driven by electric field components reflected at dielectric interfaces. An illustration of an emitting dipole near an interface is given in Fig. 3. The dynamics of the oscillating dipole moment $\vec{p}$ is described by the differential

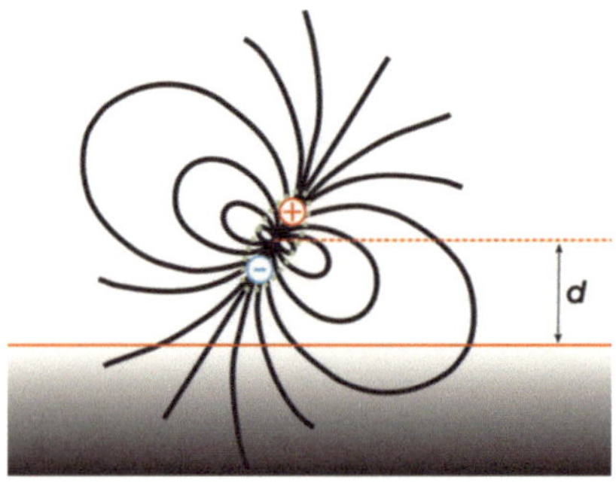

Fig. 3. Illustration of an emissive dipole near a reflecting interface at distance d.

equation

$$\frac{d^2}{dt^2}\vec{p} + b_0\frac{d\vec{p}}{dt} + \omega^2\vec{p} = \frac{e^2}{m}\vec{E}_R(\omega), \tag{14}$$

with the oscillator frequency ω, the elementary charge e, the effective mass m of the dipole, and $\vec{E}_R(\omega)$ the electric field at the dipole position. The radiative power of a dipole located inside an isotropic, infinite medium with a refractive index of n_e is

$$q_0 b_0 = \frac{|p|^2 \omega k_0^3}{12\pi\epsilon_0} n_e,$$

where q_0 is the intrinsic quantum efficiency of the dipole and $k_0 = 2\pi/\lambda_0$. The intrinsic power of a dipole consists of a radiative and a non-radiative component.

The dynamics of a dipole located in an optical multilayer system is affected by the presence of the reflected field E_R and from its radiative character as

$$b = qb + (1-q)b, \tag{15}$$

where q is the modified quantum efficiency. The second term in Equation 15 refers to the *non-radiative* processes within the optical structure. These are not affected by the optical environment. Therefore we summarize the impact of the optical environment on the radiative part and write

$$b = q_0 b_0 \cdot F + (1 - q_0) b_0 \tag{16}$$

where F describes the influence of the electromagnetic field components on the oscillating dipole. F can be expressed by means of a generalized parametrization of the *emission* angle $u = \sin \theta_e$ as

$$F = \int_0^\infty f(u) du. \tag{17}$$

The total lifetime of a dipole in the optical system is consequently related to its intrinsic lifetime τ_0 by

$$\frac{\tau_0}{\tau} = \frac{b}{b_0} = (1 - q_0) + q_0 \cdot F. \tag{18}$$

Please note, that the above equations are modified for distinct dipole orientations and polarizations of the emitted light.

2.3.1 Mode Analysis

As seen above, the total dissipated power from the dipole is proportional to F and therefore, a stepwise integration of Equation 17 based on the index limits of total internal reflection allows to determine the fractional radiated power that can be assigned to out-coupled, guided (in the thin films), and evanescent modes. The integration limits are given by:

$$\begin{aligned} u &= 0 \ldots n_t/n_e && \text{radiative modes} \\ &= n_t/n_e \ldots 1 && \text{guided modes} \\ &= 1 \ldots \infty && \text{evanescent modes,} \end{aligned} \tag{19}$$

n_t and n_e being the refractive indices of the top and the emissive medium, respectively. The multilayer surrounding of the emitting dipole does not only affect its lifetime $\tau = \tau_0 \cdot b_0/b$, but also its radiative quantum efficiency. From Equation 15 and 16, the modified quantum efficiency q is therefore

$$q = \frac{q_0 \cdot F}{b/b_0}. \tag{20}$$

3. Simulation results

In this section, we present a number of simulation results that i) highlight the impact of energetic disorder on steady-state and transient currents as well as impedance spectra (subsections 3.1 to 3.3), ii) exemplify the fully-coupled electronic-optical simulation of multilayer OLEDs (subsection 3.4) and iii) demonstrate the use of time-of-flight transients under the influence of traps (subsection 3.5).

3.1 Steady-State Current-Voltage Curves

In the following we present the simulation results of a symmetric hole-only device under three different operating conditions. Thereby we consider the disordered nature of organic semiconductors that affects the density of states, the mobility model, the Einstein relation as well as charge injection. In this device the highest occupied molecular orbital (HOMO)

is aligned with the workfunctions of the anode and cathode. Therefore, we obtain a well-injecting anode and cathode. The high barrier for injection of electrons from the cathode to the lowest unoccupied molecular orbital (LUMO) level essentially leads to a hole-only device.
The high charge densities at the anode and cathode ($p(0)$ and $p(L)$) are set to half of the density of chargeable sites N_0, i.e. the Gaussian DOS is half-filled at either side due to the energy level alignment at the electrodes.

Parameter	Value	Units
N_0	$8.5 \cdot 10^{26}$	m^{-3}
σ	0.15	eV
μ_0	10^{-10}	$m^2V^{-1}s^{-1}$
V_{bi}	0	V
$p(0)$	$0.5N_0$	m^{-3}
$p(L)$	$0.5N_0$	m^{-3}

Table 1. Simulation parameter set of the hole-only device for the three operating conditions.

First, we look at the best known electrical characterization method: the current-voltage curve. Therefore the steady-state for different voltages is calculated and we obtain current-voltage characteristics as shown in Fig. 4. We compare the steady-state results for the EGDM and the constant mobility case with the analytical solution which is given by

$$J(V) = \frac{9}{8}\varepsilon\varepsilon_0\mu_0\frac{V^2}{L^3} \tag{21}$$

where L denotes the length of the device, V the applied voltage and μ_0 the mobility (Mott & Gurney, 1938). All of the analytical solutions in the following are obtained by neglecting the diffusion and only taking the drift in Equation 2 into account. We observe the effect of diffusion especially at low voltages where the current is increased. In the regime of high bias the analytical solution and the constant mobility case merge whereas the EGDM solution moves away due to the field- and density-dependence of the EGDM and the slope of the EGDM curve is increased. Generally, the current density of the EGDM is higher due to the enhancement functions $g_1(p,T)$, $g_2(F,T)$ and $g_3(p,T)$. The EGDM has recently been successfully applied to current-voltage curves by van Mensfoort et al. (Mensfoort et al., 2008a). A comprehensive analysis of the effects of the EGDM and its components for the steady-state is conducted in reference (Mensfoort et al., 2008b).

3.2 Dark-injection transients

We now turn to the analysis of the time-dependent response to a step voltage. We perform transient simulations for the same device as above shown in Fig. 5. The dark-injection transients are simulated from 2 to 12 V in steps of 2 V. For the initial rise of the transient response the analytical solution is described by as

$$J(t) = \frac{2L\varepsilon\varepsilon_0}{\mu}\frac{1}{(2\tau_{SCF} - t)^2}, \tag{22}$$

where the space-charge-free transit time is defined as $\tau_{SCF} = \frac{L^2}{\mu V}$ (Helfrich & Mark, 1962). In the SCLC at the transit time $\tau_t = 0.786\tau_{SCF}$ the maximum current is obtained and reaches

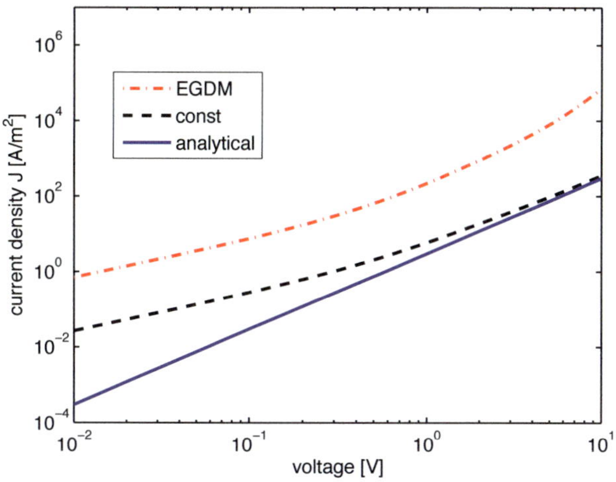

Fig. 4. Current-voltage curve for a hole-only device for a constant mobility and diffusion coefficient, the EGDM and the analytic solution (Mott-Gurney).

a value of $J(\tau_t) = 1.21J_c$ where J_c stands for the stationary current from Mott-Gurney's law (Equation 21). The value of $J(\tau_t) = 1.21J_c$ is marked in Fig. 5 with a star. Note that the slope of the line connecting these stars is minus two which can be explained with Equation 21. The relatively low charge mobility of organic semiconductors leads to space-charge limited transport. In this regime the transient response exhibits a cusp that relates to the transit time and thus the charge mobility. This has been the motivation for carrying out dark-injection transient experiments.
We notice that the peak position for the constant mobility case coincides well with the analytical solution, especially at high bias where the drift current is the dominant part of the total current. At low voltage the agreement becomes worse due to the diffusion current.

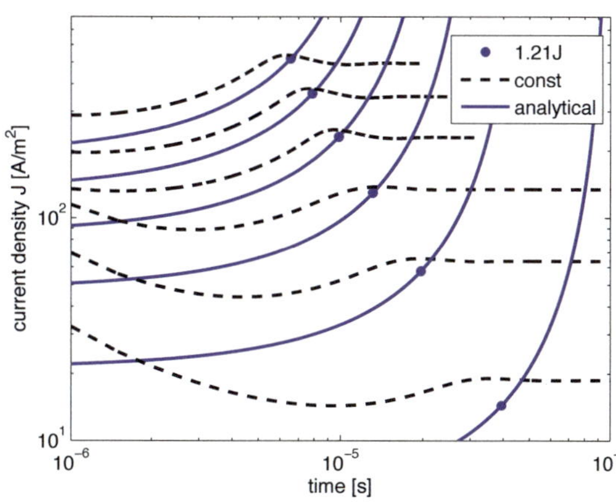

Fig. 5. Dark-injection transients for a hole-only device for a constant mobility and diffusion coefficient, the analytical solution and the generalized Einstein solution. The transients were simulated from 2 to 12 V.

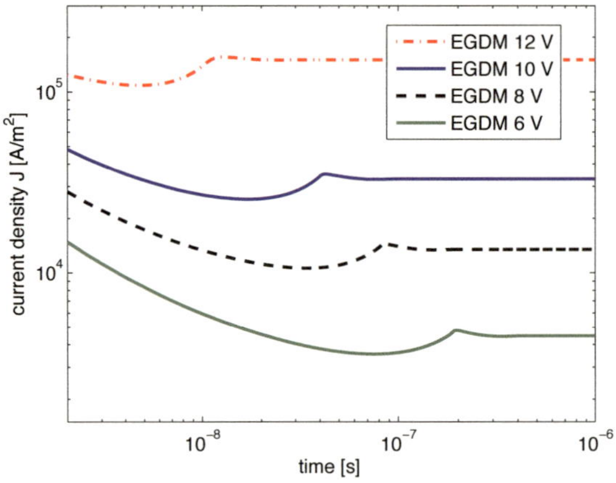

Fig. 6. Dark-injection transients for a hole-only device for the EGDM at different voltages.

If we also consider the enhancement functions for the EGDM as shown in Fig. 6, the peak positions shift to shorter times due to the enhanced mobility and the current density is vastly increased. In the EGDM case no formulas are available that would relate the peak position in the current with EGDM mobility parameters. Therefore, we have to rely on numerical fitting of the model parameters.

3.3 Impedance Analysis

As a third step, we investigate the response of the device to a voltage modulation. We obtain therefrom the impedance $Z(\omega)$ which is related to the admittance $Y(\omega)$ as

$$Y(\omega) = \frac{1}{Z(\omega)}. \tag{23}$$

The admittance is decomposed into an equivalent parallel conductance and capacitance as follows

$$Y(\omega) = G(\omega) + i\omega C(\omega). \tag{24}$$

We analyze the normalized capacitance at 8V in Fig. 7. We varied the energetic disorder parameter σ which is the main EGDM model parameter. We compare our numerical simulation results with the analytical solution for the drift-only case. For the small signal analysis the analytical solution for the impedance $Z(\omega)$ is given by

$$Z(\omega) = \frac{6}{g_0(i\omega\tau_0)^3}[1 - i\omega\tau_0 + \frac{1}{2}(i\omega\tau_0)^2 - \exp(-i\omega\tau_0)] \tag{25}$$

where $\tau_0 = \frac{4}{3}\frac{L^3}{\mu V}$ and $g_0 = \frac{3}{\tau_0}\frac{\epsilon}{L}$ (Shao & Wright, 1961). We also calculated the solution for the constant mobility case. The drift-only model provides a quite accurate solution for the constant mobility case at 8 V. At lower voltages however, the drift-only model differs significantly from the constant mobility model. We observe that the oscillations at high frequencies are damped out when diffusion is present. Also due to diffusion, the capacitance at low frequencies is increased from the typical value $\frac{3}{4}C_{geom}$ of the drift-only solution, even more in case of the EGDM. In the EGDM the peak position is additionally shifted to higher frequencies.

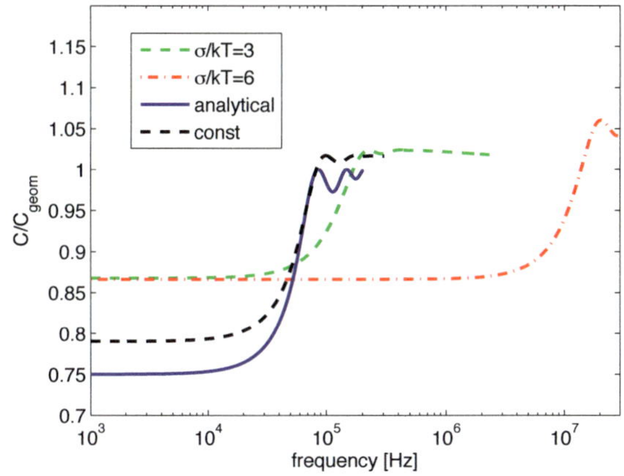

Fig. 7. Frequency dependence of the normalized capacitance at 6 V in a hole-only device for a constant mobility and diffusion coefficient, the EGDM and the analytical solution.

3.4 Fully-coupled electrical-optical OLED simulation

State-of-the art OLEDs are composed of several functional layers whose thickness obeys some optical or electronic criteria and which may be further complicated by dye doping, charge doping etc. The efficient harvesting of excitation energy by use of fluorescent and phosphorescent dye dopants remains a challenge. Here we attempt to give a simulation example of dual emitters in two layers EML1 and EML2 of a hypothetical 4-layer device structure. The rate equation for excitons (see Equation 12) is used here to model the energy transfer among different exciton species in a multilayer small-molecule OLED as depicted in Fig. 8.

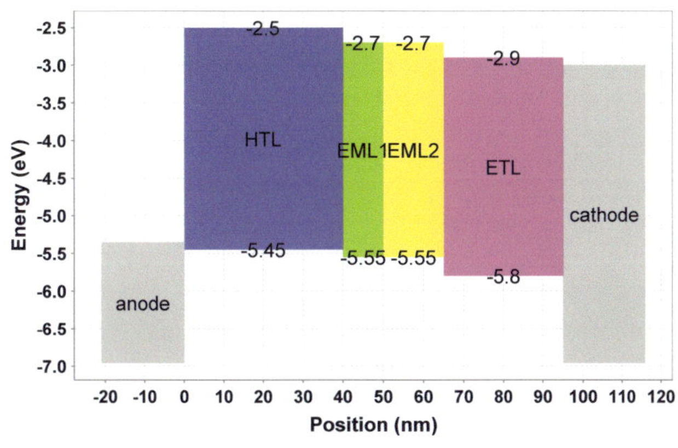

Fig. 8. Energy level diagram of a 4-layer OLED with two emissive layers EML1 and EML2 with two exciton species each.

The middle two layers were assumed to have emissive excitons in the host and guest material. The mobilities were assumed as illustrated in Fig. 10. The resulting charge distribution is shown in Fig. 9. For the chosen parameter values, the charges accumulate at either side of the EML2/ETL interface, thus also leading to a high electric field at that interface, see Fig. 9. While only the excitons of the host materials EML1 and EML2 are excited by recombination events, the guest dye dopants receive their energy from the host by transfer. For this energy

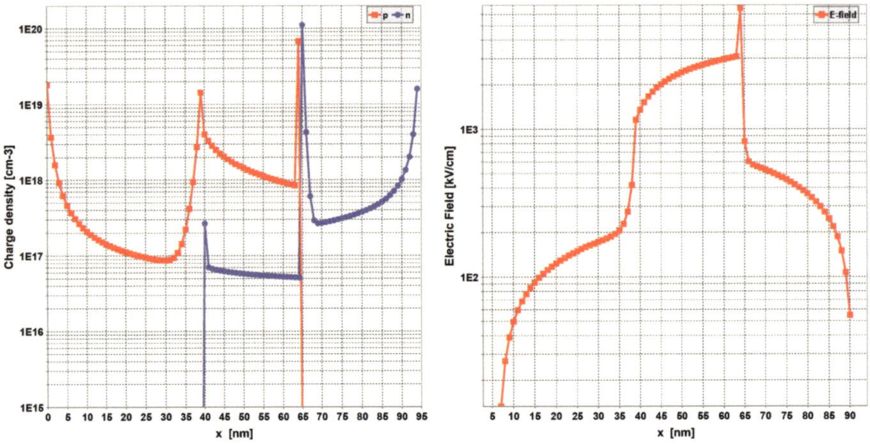

Fig. 9. Simulated profiles of charge carrier densities (left) and electric fields (right) for the hypothetical 4-layer OLED example of Fig. 8.

transfer we employ the transfer rate k_{ij} in Equation 12) and find that the guest excitons equilibriate in a different spatial distribution, due to a different parametrization, see Fig. 10 (right). The excitons are only modeled in the middle two layers.

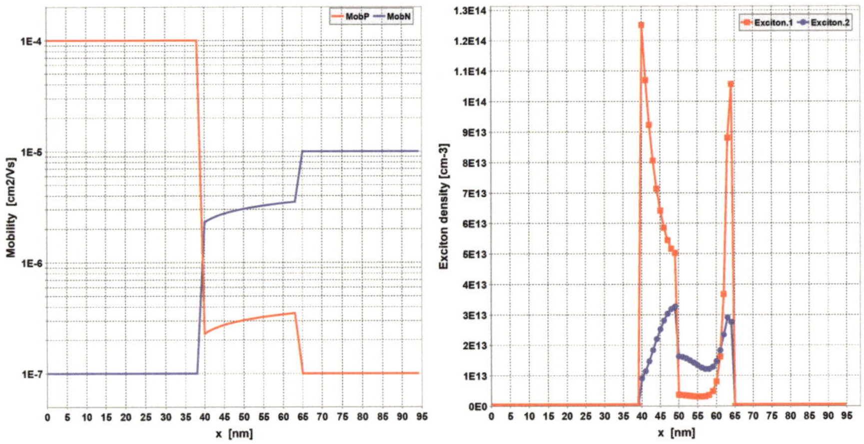

Fig. 10. Simulated profiles of mobilities (left) and Exciton.1 and Exciton.2 in the middle of the OLED structure (right).

If one is only interested in the steady-state, then the governing transport equations 3 are solved with an iterative steady-state solver (Fluxim AG, 2010; Knapp et al, 2010). However, if one is interested in the dynamics of charge transport, transient solution algorithms must be employed. Experimentally, the transient electroluminescence and transient current signals have revealed the dynamics of charge migration into the device (Ruhstaller et al., 2003; 2001). In Fig. 11 one can observe a monotonic increase of the recombination current, which testifies that the charges have first to penetrate the device before they can recombine. For the given hypothetical device, the charges are all consumed in the device, i.e. the recombination current is as large as the device current and therefore the current balance is equal to unity.

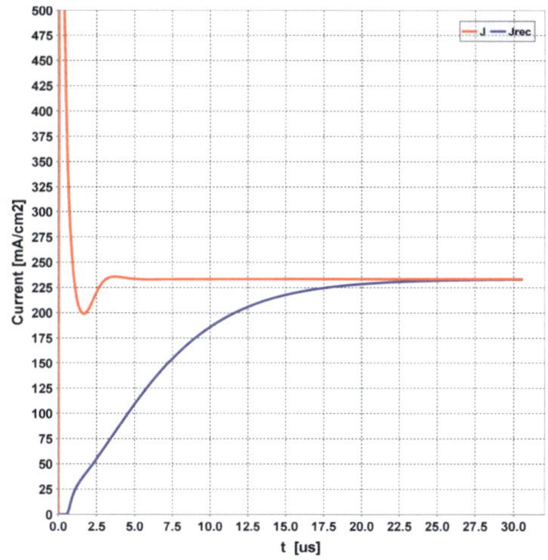

Fig. 11. Simulated device current and recombination current during turn-on.

This example simulation can be completed with a light-outcoupling calculation which takes the exciton distribution as dipole distribution. In this step one has to select which excitons are assumed to decay radiatively and thus contribute to light emission. For instance, we consider the emissive guest excitons in the emission layer EML1 and perform a calculation of the dissipated power versus the wavelength and the normalized in-plane wave vector u in Fig. 12 (left). The various emission modes can easily be identified qualitatively by looking at this plot. The guided modes are visible in the u range between 0.6 and 1. This power seems to exceed the power in the propagating outcoupled mode ($u < 0.6$) while the evanescent modes are displayed at values of u larger than 1. In addition in the left plot of Fig. 12 we report an analysis of emissive modes that takes care of the integration of the dissipated power of each mode in the relevant u ranges. The critical in-plane wavevector for light outcoupling into air is around 0.6 and the corresponding relative mode contribution is not more than 10%, in this example. Outcoupling efficiency numbers around 20% are typical values for OLEDs without any outcoupling enhancement techniques.

3.5 Numerical analysis of time-of-flight measurements

An exhaustive study of transport mechanisms in organic compounds shall comprehend the characterization of the charge dynamics. Time-of-flight (TOF) experiments are among the most common techniques to measure the mobility of the charges. Nevertheless, only a thorough analysis of the measured transient curves can give insight into the behavior of the charges, including the impact of trapped states. Current transients in TOF measurements are generally classified into dispersive and non-dispersive ones, the former being characterized by a monotonic decay of the measured current, while the latter show an initial plateau followed by a clear and sudden drop in the current.

In the following, we will investigate TOF transients by numerical simulations for distinct kinds of samples in order to allocate the specific characteristics of electron trap states. Here, we reproduce the TOF experiment presented by Malliaras et al. for Alq_3 devices that underwent different compound purification steps (Malliaras et al., 2001). The samples of our virtual experiment consisted of 8-μm Alq_3 films sandwiched between a transparent ITO layer and an aluminium (Al) electrode. The samples were reverse biased at a voltage of -1000 V, the

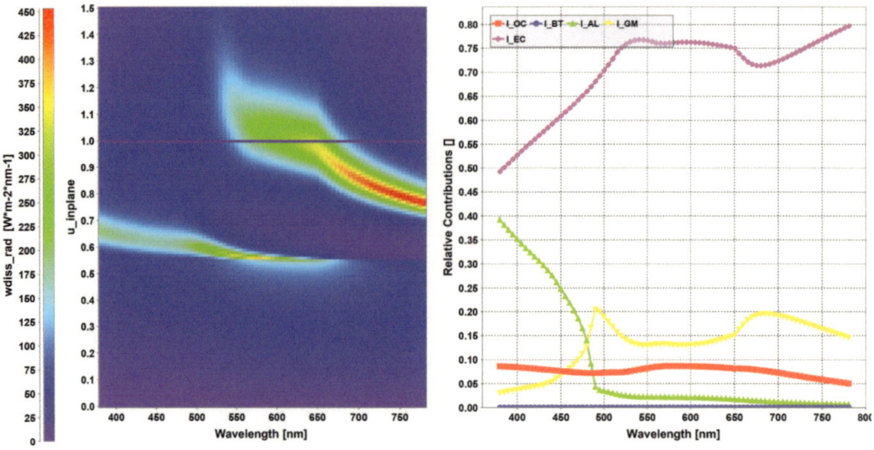

Fig. 12. Dipole emission calculation for the guest excitons in EML1. The dissipated power (left) and the mode analysis (right) are shown.

ITO being the negative electrode. A 10 ps laser pulse at 337 nm wavelength with energy of 1μJ per square meter was used to generate a sheet of free carriers at the ITO-Alq_3-interface. The light intensity was kept sufficiently low to avoid space charge effects. Assuming HOMO and LUMO levels for the organic compound to be 5.6 and 3.0 eV, respectively, the two electrodes were acting as non-injecting layers because of their workfunctions of 5.0 eV for ITO and 4.1 eV for Al. For completeness, the (vanishing) injection was considered as thermionic and modeled after Scott and Malliaras (Scott & Malliaras, 1999).

Despite the energetic disorder of the transport states in organic compounds, the analysis was performed assuming Poole-Frenkel, field-dependent mobility, see Equation 10. The study of the impact of the energetic disorder on the current transient is beyond the goal of this text and it is known that the disorder does not affect the dispersive characteristics of TOF-related current signals (Bässler, 1993). Trapped charge carriers on sharp energy levels were described by Staudigel et al. (Staudigel et al, 1999) by a rate equation for trapped electrons n_t that reads

$$\frac{dn_t(x)}{dt} = \mu(E + \frac{kT}{ed_m})\sigma_t[n(N_t - n_t) - n_t N_0 e^{-\frac{E_{t_n}}{kT}}]. \tag{26}$$

In Equation (26), N_0 is the total number of available sites per unit volume, N_t is the trap concentration, E is the (local) electric field while E_{t_n} is the trap energy depth. Further, $d_m = N_t^{-1/3}$ is the mean trap distance and $\sigma_t = 5d_m^2$ is the effective capture cross section while n is the concentration of free carriers. Note, that Equation 26 is connected to Equation 4 by the relation $r_c = \mu E \sigma_t$ and $r_e = cN_0 e^{-\frac{E_{t_n}}{kT}}$. The electrical parameters for the modeled compound are summarized in Table 2. All calculations were performed with SETFOS (Fluxim AG, 2010). In Figure 13 we present the results for calculations considering no traps, shallow and deep traps.

When no traps are considered, a non-dispersive charge behavior is clearly recognized in the current transient. Contrarily, when deep traps (typically $E_{t_n} > 0.3$ eV at room temperature) are taken into account, the transient characteristics is evidently dispersive since no definite current drop is recognizable when the free electrons reach the aluminium electrode. A more careful analysis is needed in case of shallow traps ($E_{t_n} < 0.25$ eV at room temperature). Due to the relatively low energy barrier, trapped electrons are quickly released into the conduction

	n	p
μ_0	2.9e-9 cm^2/Vs	1e-8 cm^2/Vs
γ	7.3e-3 cm^0.5/V^0.5	0
LUMO/HOMO	3.0 eV	5.6 eV
E_{t_n}	–/0.212/0.33 eV	–

Table 2. Assumed electrical parameters for Alq_3 to reproduce the measurements presented in reference (Malliaras et al., 2001). Three different electron trapping depths, E_{t_n} were considered.

band. The resulting current transient characteristics shows an evident drop as the charges reach the aluminium electrode. Such "quasi-non-dispersive" behavior is often confused with trap-free characteristics (Malliaras et al., 2001), whereas trap states are still present but do not significantly damp the dynamics of the charges.

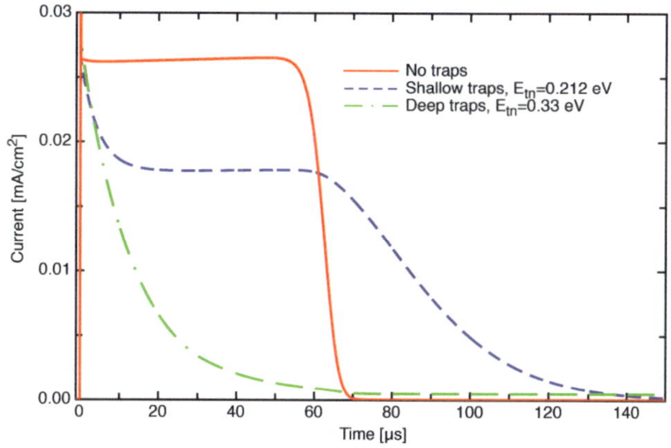

Fig. 13. Current transient characteristics of TOF measurements for samples with no traps (red solid line), shallow traps (blue dashed line) and deep traps (green dot-dashed line). As expected, trap-free compounds show a non-dispersive charge transport and charges in materials with deep traps exhibit a dispersive behavior. Weakly trapped charges on the contrary would indicate a non-dispersive behavior which is often mistakenly confused with the absence of traps (blue dashed line).

The behavior of the electrons in presence of deep or shallow traps can be analyzed looking at the charge density profiles for different times. Immediately after the light pulse that generates the charges a package of free electrons moves away from the ITO layer toward the positively charged aluminium electrode.

In Figure 14, left, we show the profiles of the charges in presence of deep traps, 10 μs after the light pulse. We recognize a symmetric, free electron packet partially overlapping the profile of the trapped electrons. When the free charges reach the aluminium electrode after 60 μs, their number is strongly reduced due to the many trapped states occupied during the flight from one electrode to the other (Figure 14, right). For the trapped electrons, it is energetically unfavorable to release back into a conductive state. They need many hundreds of microseconds to be de-trapped.

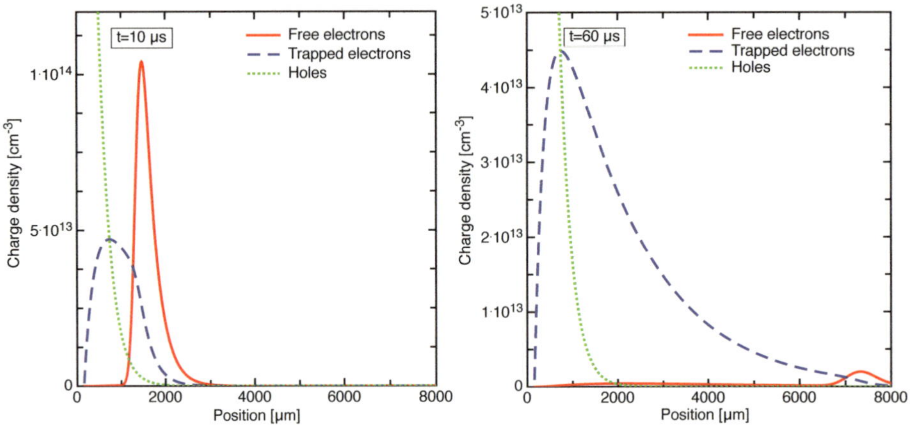

Fig. 14. *Deep traps.* Left: At $t = 10$ μs, the sharp packet of free electrons (red solid line) leaves the ITO electrode at position 0 μm and moves toward the aluminium electrode placed at position 8000 μm. A certain amount of charges remains trapped close to the ITO electrode (blue dashed line). Right: After 60 μs, those few free carriers that have not been trapped reach the aluminium electrode. Notice that trapped charges were "left behind" by the traveling free charge packet and did not move from their original position. Few trapped charges that begin to release back into the conduction band are recognizable through the slight increase of the free charges concentration around 2000 μm.

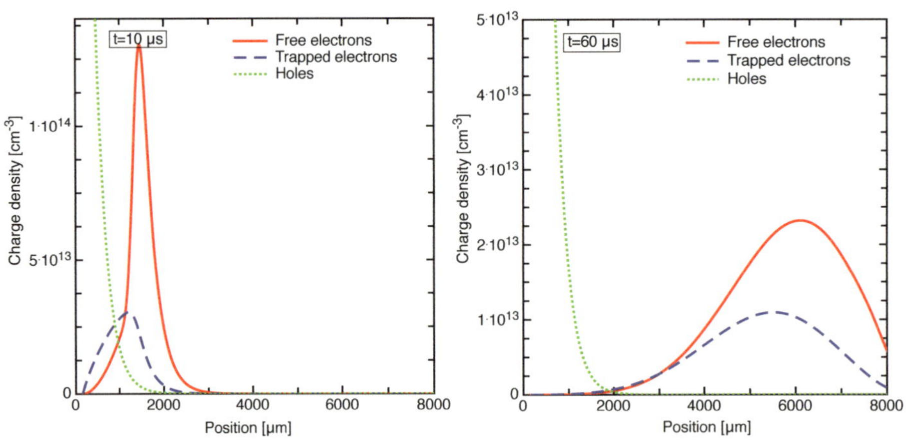

Fig. 15. *Shallow traps.* Left: Already at $t = 10$ μs, the free charge packet (red solid line) appears asymmetric due to the release of the weakly trapped charges (blue dashed line) around 1000 μm. Right: After 60 μs, a significant amount of free charges reaches the aluminium electrode. The fast kinetics of the trapped states makes them release back into the conduction band fast enough to follow the movement of the free electrons.

In Figure 15, left, we show the profiles of the charges in presence of shallow traps, 10 μs after the light pulse. The profile of the free electrons appears asymmetric because of the on-going release process of the temporarily trapped electrons. The kinetics of the trapped charges is so fast that the trapped packet follows the free electron packet in its travel toward the positive electrode. In Figure 15, right, we show the charge profiles at the moment they reach the aluminium electrode after 60 μs: the broad free electron packet overlaps the trapped charge packet so that both types appear to be extracted about at the same time. This explains the non-dispersive-like characteristics of the current transient in case of trapped states that are less than 25 meV below the LUMO level (blue dashed curve in Figure 13).
Despite the lack of precise information on the experimental procedure, the calculated current transients reproduce both qualitatively and quantitatively the measured curves presented by Malliaras et al.(Malliaras et al., 2001). Through our comprehensive numerical investigation, we could show that a qualitative analysis of the shape of the current transient is not enough to characterize trap-free samples. As a matter of facts, weakly trapped charges are able to follow the "flight" of free electrons through the device. Underestimating the presence of traps might lead a.o. to wrong statements on the lifetime of the device.

4. Parameter extraction

The following section introduces some aspects of parameter extraction in organic light-emitting devices (OLEDs). Two main applications are discussed. The first application is the extraction of the emission profile in a light-emitting layer of an OLED by a numerical method. The emission profile describes mathematically how the excitons are distributed inside the emissive layer. The numerical method presented is a linear least-square fit algorithm. The theory of numerical algorithms to extract the emission profile and some applications are also presented in more detail in Perucco et al. (Perucco et al, 2010). The second application is the extraction of EGDM parameters from multiple measured current-voltage curves by a nonlinear least-square algorithm.

4.1 Extraction of emission profiles in OLEDs

The objective of this section is to present and test a numerical fitting algorithm for the extraction of the emission profile and intrinsic source spectrum. The fitting algorithm is evaluated by adequate examples and validated on the basis of consistency checks. This is achieved by an optical model, where a transfer-matrix theory approach for multi-layer systems is used in combination with a dipole emission model. The optical model is implemented in the semiconducting emissive thin film optics simulator (SETFOS) (Fluxim AG, 2010). With SETFOS, the emission spectrum of an OLED based on an assumed emission profile and a known source spectrum is generated. The fitting method is then applied to the calculated emission spectra in order to estimate the emission profile and source spectrum. The comparison between the obtained and assumed emission profile and source spectrum is an indication of how successfully the inverse problem can be solved. Sections 4.1.1 deals with the mathematical derivation of this numerical fitting algorithm. In Section 4.1.2, the applications or consistency checks are presented.

4.1.1 Theory

The theoretical background of the fitting method is introduced in this section. The method is linear in terms how the measured emission spectrum is related to the unknown emission profile. For simplicity the mathematical formulation for the least-square problem is derived for only one emitter. The emitter is characterized by the emission profile. For the moment, it is also assumed that the emission spectrum is measured in normal direction and therefore the light is unpolarized. In Section 4.1.1.1, this approach is extended further to multiple emitters described by several emission profiles and emission spectra measured for several angles θ. The extracted emission profile $P_e(\delta_j)$ is discretized at N relative positions $\delta_j = d_j/L$ in the light-emitting layer, where d_j is expressed as an absolute position and L is the width of the layer. The emission spectrum is divided into M wavelengths λ_i ($i = 1...M$). The fitted emission spectrum $I_f(\lambda_i)$ can be written as

$$I_f(\lambda_i) = \sum_{j=1}^{N} I_c(\lambda_i, \delta_j) \cdot P_e(\delta_j), \tag{27}$$

where $I_c(\lambda_i, \delta_j)$ is the emission intensity for the wavelength λ_i and assuming a discrete emission profile (dirac function) at the relative position δ_j in the layer. The emission intensity is given by

$$I_c(\lambda_i, \delta_j) = I(\lambda_i, \delta_j) \cdot S(\lambda_i), \tag{28}$$

where $I(\lambda_i, \delta_j)$ is the emission intensity for emissive dipoles with spectrally constant intensity. $S(\lambda_i)$ is the source spectrum. Between the measured emission spectrum $I_m(\lambda_i)$ and fitted emission spectrum $I_f(\lambda_i)$, a residuum can be defined and written as

$$r_1(\lambda_i) = I_f(\lambda_i) - I_m(\lambda_i). \tag{29}$$

Equation 29 can be interpreted as a linear least-square problem, written as a system of linear equations

$$r_1(\lambda_i) = \sum_{j=1}^{N} I_c(\lambda_i, \delta_j) \cdot P_e(\delta_j) - I_m(\lambda_i). \tag{30}$$

The system of equations is normally overdetermined (i.e. $M > N$) and thus is ill-posed. In matrix notation, the problem can be formulated as $r_1 = A \cdot x_1 - b_1$. The matrix A has the following structure

$$A = \begin{pmatrix} I_c(\lambda_1, \delta_1) & I_c(\lambda_1, \delta_2) & ... & I_c(\lambda_1, \delta_N) \\ I_c(\lambda_2, \delta_1) & I_c(\lambda_2, \delta_2) & ... & I_c(\lambda_2, \delta_N) \\ ... & ... & ... & ... \\ I_c(\lambda_M, \delta_1) & I_c(\lambda_M, \delta_2) & ... & I_c(\lambda_M, \delta_N) \end{pmatrix}, \tag{31}$$

b_1 is a vector containing the measured emission spectrum $I_m(\lambda_i)$ and the vector x_1 corresponds to the a priori unknown emission profile $P_e(\delta_j)$. The term linear refers to the linear combination between the matrix A and the vector x_1 of unknown weights. In every column of the matrix A, an emission spectrum is calculated for a dirac shaped emission profile at the position δ_j. The emission profile $P_e(\delta_j)$ at the relative position δ_j is the weight of the corresponding spectrum, respectively the column. The mathematical task is to minimize the length of the vector $\| r_1 \|$.

4.1.1.1 Extracting multiple emission profiles

The most general case of the emission spectrum is determined by the emission profile of multiple emitters $P_e(\delta_j^k)$ and emission angles θ_l. Given is the emission spectrum measured at O different angles ($l = 1...O$) and the OLED consists of Q different emitters ($k = 1...Q$) in the same or in separate layers. The relation stated in Equation 27, combined with the definition of the residuum in Equation 30, can be extended to

$$r_2^{s,p}(\lambda_i, \theta_l) = \sum_{j=1}^{N} I_c^{s,p}(\lambda_i, \delta_j^k, \theta_l) \cdot P_e(\delta_j^k) - I_m^{s,p}(\lambda_i, \theta_l). \quad (32)$$

$I_c^{s,p}(\lambda_i, \delta_j^k, \theta_l)$ stands for the s-polarized or p-polarized emission intensity at the wavelength λ_i. We assume a dirac shaped emission profile at the relative position δ_j^k for emitter k and an emission angle of θ_l. $P_e(\delta_j^k)$ is the emission profile at relative position δ_j for emitter k. Equation 32 represents a system of linear equations $r_2^{s,p} = A^{s,p} \cdot x_2 - b_2^{s,p}$, where the matrix $A^{s,p}$ contains the s-polarized and p-polarized emission spectra, the vector x_2 contains the information of several emission profiles and the vector $b_2^{s,p}$ represents the measured emission spectrum. The mathematical task is again to minimize the length of the vector $\| r_2^{s,p} \|$.

4.1.1.2 Extracting the intrinsic source spectrum

In the case of a single emitter, van Mensfoort et al. (Mensfoort et al., 2010) presented a method to extract the source spectrum of the light-emitting material. The source spectrum can be obtained by replacing the emission intensity $I_c^{s,p}(\lambda_i, \delta_j^k, \theta_l)$ by the emission intensity for emissive dipoles with spectrally constant intensity $I^{s,p}(\lambda_i, \delta_j^k, \theta_l)$ in Equation 32. This method is employed and evaluated in Section 4.1.2.2.

4.1.2 Applications

In this section, the reliability and limitation of the linear fitting method is addressed after it was mathematically deduced and described in Section 4.1.1. A given intrinsic source spectrum from a light-emitting material is assumed, together with an emission profile, stating where the dipoles are located in the device. The effects of quenching are disregarded in the presented applications below. First, quenching would likely limit the amount of dipoles close to the electrodes as the lifetime is very short. And secondly, light emitted from the dipoles is also captured in evanescent modes and therefore, does not couple out into air. Finally, the emission spectrum is generated by an optical dipole model described by Novotny (Novotny, 1997) and implemented in the simulator SETFOS (Fluxim AG, 2010). The calculated emission spectrum is used to solve the least-square problem in Equation 32. This allows the extraction of both, source spectrum and emission profile. The comparison of the extracted and assumed emission profile reveals the reliability of the presented algorithm. Throughout this text, an open cavity is used for the consistency checks. But the method here may also be applied to cavity and small-molecule based OLEDs. The OLED investigated here has a broad light-emitting polymer (LEP) of 100 nm. Further, the light-emitting layer is embedded between a 80 nm thick PEDOT:PSS anode and an aluminum cathode of 100 nm. The device is depicted in Figure 16. With respect to an experimental setup, the diameter of the semi-sphere glass lens is at least an order of magnitude larger than the diameter of the OLED. In order to achieve an absolute quantity of the emission intensity and emission profile, the assumed current density in all considered consistency checks is 10 mA/cm^2.

4.1.2.1 Extraction of the emission profile from angularly resolved emission spectra

As an introductory example, this section shows the application to angularly resolved emission intensity spectra. It compares the extracted emission profile from these spectra to an emission

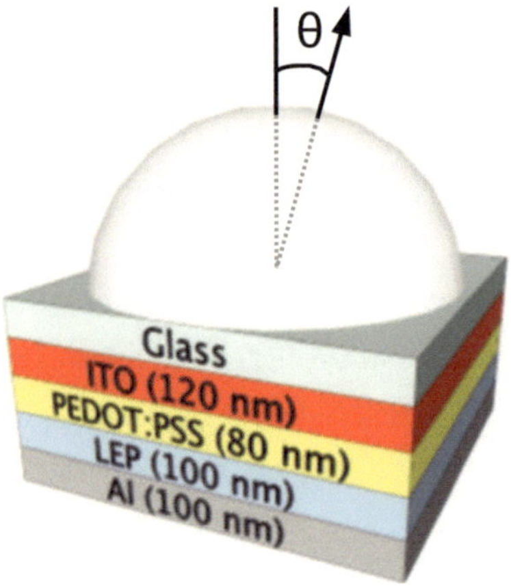

Fig. 16. OLED used to perform the parameter extraction tests with the semi-sphere glass lens. θ stands for the observation angle.

profile extracted from an emission intensity spectrum measured at normal angle. The assumed emission profile is Gaussian shaped, where the peak is set to 0.3 expressed in terms of a relative position in the emission layer. The width of the Gaussian shape is 20 *nm*.

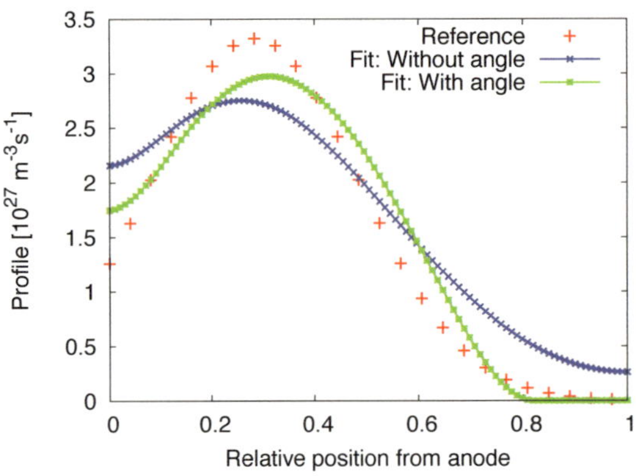

Fig. 17. Comparison between the assumed and extracted emission profiles. The emission profiles were extracted using angularly resolved emission spectra and an emission spectra measured at normal angle.

The comparison between the extracted and assumed emission profiles in Figure 17 shows an improvement of the extracted emission profile when angularly resolved emission intensity spectra are used. The fitted emission intensity spectra match visually perfectly the emission spectra serving as a measurement, as seen from Figure 18.

4.1.2.2 Source spectrum extraction

This section demonstrates the ability of the least-square algorithm to extract the intrinsic source spectrum of a light-emitting material. The same assumptions regarding the parameters of the emission profile are made as in Section 4.1.2.1. The extracted emission profile and source spectrum can be found on the left, respectively on the right in Figure 19.

It can be seen from Figure 19 that the source spectrum can be extracted very accurately. The emission profile is also well extracted and even the peak position is reproduced well.

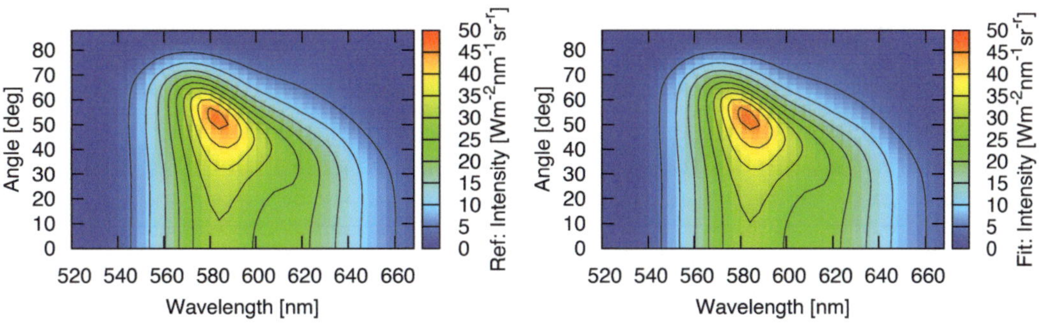

Fig. 18. Left: Angularly resolved emission intensity spectra serving as a measurement. Right: Fitted emission intensity spectra by the linear least-square algorithm.

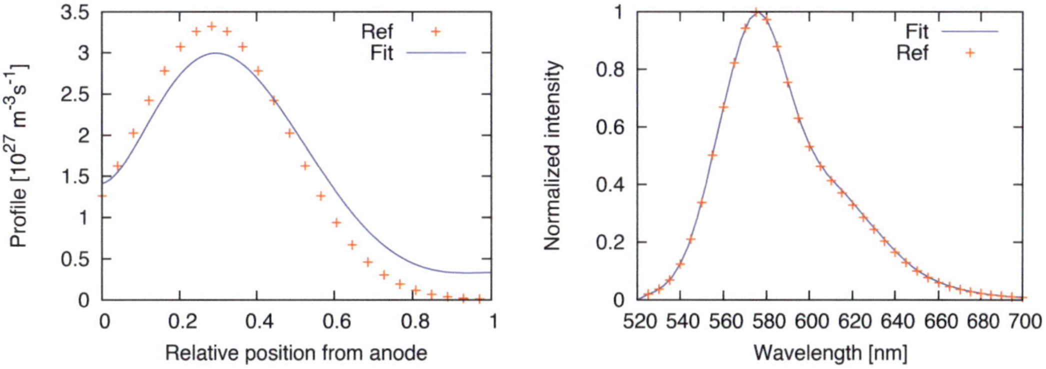

Fig. 19. Left: Comparison between the assumed and extracted emission profile. Right: Relation between the assumed and extracted intrinsic source spectrum by the method discussed in Section 4.1.1.2.

Illustrated in Figure 20 is the comparison between the assumed and fitted emission spectra, which are visually also in perfect agreement.

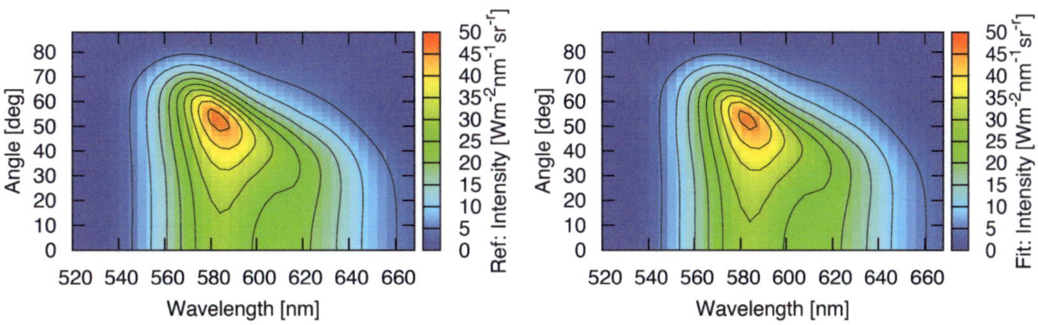

Fig. 20. Left: Angularly resolved emission intensity spectra serving as a measurement to extract the intrinsic source spectrum from. Right: Fitted emission intensity spectra by the linear least-square algorithm.

4.1.2.3 Extracting multiple emission profiles

Equation 32 explains how multiple emission profiles can be extracted from a measured emission intensity spectrum. This section illustrates the application of the method to a multi-emitter OLED. In this example, two emission profiles are extracted. The first assumed emission profile is Gaussian shaped with a peak at 0.7 and a width of 40 nm. The second assumed emission profile is also gaussian shaped, where the peak is at 0.3 and the width is 20 nm. Figure 21 shows the extracted and assumed emission profiles, as well as the reference and fitted emission intensity spectra.

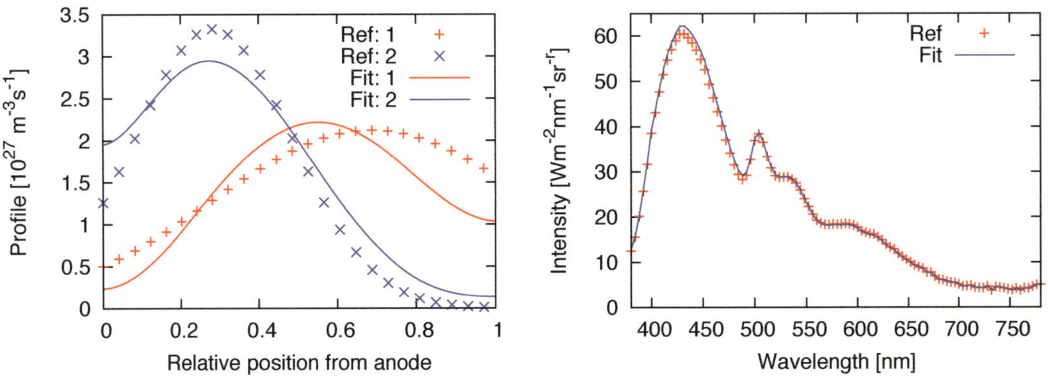

Fig. 21. Left: Showing the differences between the emission profiles from a multi-emitter OLED. The dotted curves represent the assumed emission profiles, whereas the lines stand for the extracted emission profiles. Right: Comparison between the measured and fitted emission intensity spectra.

Despite the fact that the assumed and fitted emission spectra match very well, some differences in the emission profiles are visible. Nonetheless, the general trend is explained by the extraction. For the second emission profile, both peak and width can be reproduced more or less. For the first emission profile, the flat emission profile can be explained as well.

4.2 Extraction of transport parameters from current-voltage curves

The following section deals with the application of a nonlinear least-square fitting algorithm to extract EGDM parameters from measured current-voltage curves. The nonlinear fitting algorithm, as well as the EDGM model is implemented in SETFOS. SETFOS is used to generate three hypothetical measured current-voltage curves at temperatures 320 K, 300 K and 280 K. All three current-voltage curves are simultaneously fitted for extracting the parameters. The device considered is a single-layer, hole-only device where the electrical layer has a thickness of 121.5 nm and the build-in voltage is 1.9 V. The energy diagram of the simulation device is depicted schematically in Figure 22. The following parameters are of interest: the mobility μ_p, the width of the Gaussian DOS σ_p, the density of chargeable sites N_0 and the workfunction at the cathode Φ_c. Meanwhile, the workfunction at the anode is held constant. The parameters represent real EGDM parameters as discussed in van Mensfoort et al. (Mensfoort et al., 2008b). The following parameters are assumed: $\mu_p = 1 \cdot 10^{-7}\, m^2/Vs, \sigma_p = 0.13\, eV, N_0 = 6 \cdot 10^{26}\, 1/m^3$ and $\Phi_c = 3.2\, eV$. The mobility μ_p is related to Equation 11 in the following way: $\mu_0(T) = \mu_p \exp\left(-0.39(\sigma/(kT))^2\right)$. The left hand side of Figure 23 shows that the nonlinear least-square algorithm is capable of extracting all four EGDM parameters as the current-voltage curves at the same temperature match each other visually perfectly.

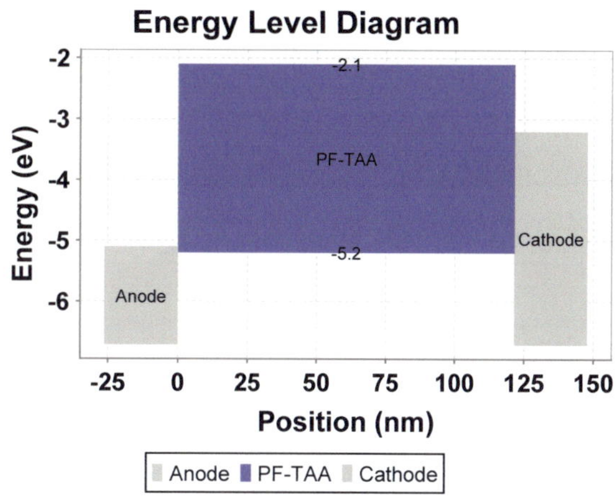

Fig. 22. Energy diagram for the simulated device. The workfunction at the anode side is held constant at 5.1 eV, while the workfunction at the cathode side Φ_c is being optimized. The HOMO and LUMO levels are 5.2 eV, respectively 2.1 eV.

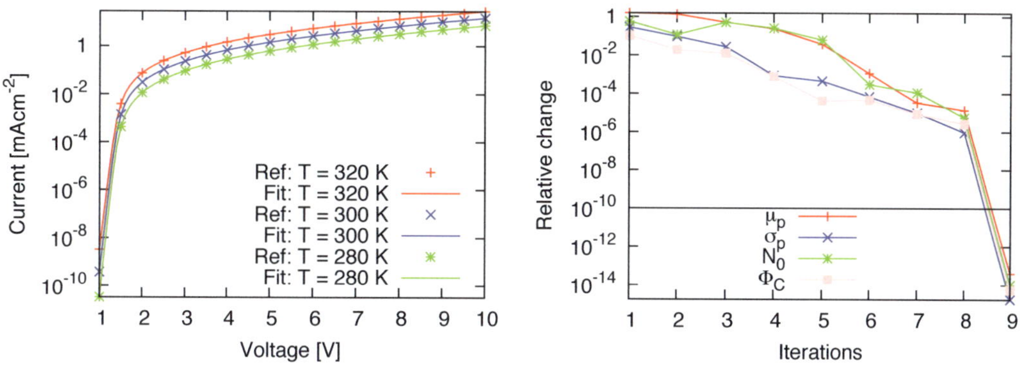

Fig. 23. Left: Comparison between the measured current-voltage curves (points) and the fitted current-voltage curves (lines). The extraction of parameters is done using all three measured curves at different temperatures at the same time in order to get more accurate results. Right: Showing the relative change in the parameters as a function of iterations. The black line represents the relative change which must be achieved as a termination criterion.

The right hand side of Figure 23 illustrates the relative error of the EGDM parameters as a function of the optimization iteration. The relative change is defined as $(p_{i+1} - p_i)/p_{i+1}$, where p_i is the value of parameter p at iteration i. The black line shows the relative change, which must be achieved in order to terminate the optimization process. All four parameters obtained after 9 iterations have a smaller relative change than 10^{-10}. Depending on the parameter, it can be seen from Figure 23 that the relative change in the parameter is initially between 10^{-1} and 1. The assumed, extracted and initial parameters for the fit can be seen in Table 3.

Table 3 shows that the nonlinear least-square algorithm is able to accurately extract the assumed EGDM parameters. A potential limitation of the nonlinear least-square method is that the error landscape should be smooth and the initial parameters should not be too far off from the assumed values, otherwise the method might converge into another local minimum.

Parameter	Assumed	Extracted	Initial
μ_p $[10^{-7}\ m^2/Vs]$	1.400	1.398	3.000
σ_p $[eV]$	0.130	0.130	0.100
N_0 $[10^{26}\ 1/m^3]$	6.000	5.984	1.000
Φ_c $[eV]$	3.200	3.200	3.200

Table 3. Assumed and extracted EGDM mobility model parameters for the test, as well as initial parameters for the nonlinear least-square algorithm.

For such situations and to cross-check the fit result, potentially more time-consuming global optimization routines may be used (Fluxim AG, 2010).
In a post-processing step, the correlation coefficients for the four model parameters can be calculated at the best fit point. These coefficients corresponding to the fit result listed in Table 3 are shown in Table 4.

	μ_p	σ_p	N_0	Φ_c
μ_p	1.00	0.94	0.39	-0.95
σ_p		1.00	0.07	-0.84
N_0			1.00	-0.55
Φ_c				1.00

Table 4. Correlation coefficients for the EGDM parameters at the best fit point in case of a multiple current-voltage curve fit as seen in Figure 23: mobility μ_p, width of Gaussian DOS σ_p, density of chargeable sites N_0 and workfunction cathode Φ_c

A correlation of 0 indicates no correlation between the two parameters, meaning the parameters are independent from each other. Whereas a correlation of -1 or 1 means perfect correlation. A slight change in the same direction on the fitted curve can be achieved by either parameter. As an example we consider the high correlation of -0.95 between the hole mobility μ_p and the workfunction ϕ_c. This means that the current can be increased by either increasing the mobility μ_p or lowering the workfunction ϕ_c. Such correlations do physically make sense, however they can complicate parameter fitting. Overall, our electrical parameter extraction example demonstrates, that rather complex physical models can be fitted to current-voltage curves thus making an efficient numerical analysis of measured data suitable.

5. Conclusions

We have given an overview on state-of-the-art electronic and optical device models for OLEDs. We demonstrated that numerical models are able to simulate different kinds of measurement setups for OLED characterization. We investigate the influence of the EGDM and the Gaussian density of states on the current-voltage curves, the dark-injection transients and the frequency response for a polymer device. All three characterization curves change significantly, if the disorder model ingredients are taken into account. The simulation results were successfully compared with analytical solutions, where applicable. We also show that the physical models can be combined with a nonlinear least-square fitting algorithm for extracting emission profiles and charge transport parameters from (hypothetical) measurements. In terms of optical device characteristics, quantitative outcoupling mode contributions and angular characteristics are presented as well as results of an emission zone extraction method. Simulations of electronic and optical processees in a doped, multilayer OLED are also presented, thus highlighting, that comprehensive OLED device simulations are becoming feasible.

6. Acknowledgement

We acknowledge financial support from the European Community's Seventh Research Framework program for the AEVIOM.eu project under grant agreement no. 213708.

7. References

Bässler H., "*Charge Transport in Disordered Organic Photoconductors a Monte Carlo Simulation Study,*" Phys. Status Solidy B **175**, 1 (1993).

Blom P. W. M., de Jong M. J. M., and Vleggaar J. J. M., "*Electron and hole transport in poly(p-phenylene vinylene) devices,*" Appl. Phys. Lett. 68, 3308 (1996).

Chance R. R., Prock A., Silbey R., "*Molecular Fluorescence and Energy Transfer Near Interfaces,*" Advances in Chemical Physics, Vol. 37, John Wiley and Sons, pp. 16 (1978).

Crone B. K, Davids P. S, Campbell I. H, and Smith D. L., "*Device model investigation of bilayer organic light emitting diodes,*" J. Appl. Phys. 87, 1974 (2000).

Dexter D. L., "*Theory of Sensitized Luminescence in Solids,*" J. Chem. Phys. 21 (5), 836 (1953).

Fluxim AG, "*Semiconducting thin film optics simulator SETFOS,*" Website: http://www.fluxim.com.

Förster T., "*Zwischenmolekulare Energiewanderung und Fluoreszenz,*" Ann. Phys. 6 (2), 55-75 (1948).

Helfrich W. and Mark P., "*Raumladungsbeschränkte Ströme in Anthrazen als Mittel zur Bestimmung der Beweglichkeit von Defektelektronen,*" Z. Phys. **166**, 370 (1962).

Knapp E., Häusermann R., Schwarzenbach H. U., Ruhstaller B., "*Numerical simulation of charge transport in disordered organic semiconductor devices,*" J. Appl. Phys. 108, 054504 (2010).

Langevin P., Ann. Chim. Phys. VII **28**, 433 (1903).

Malliaras G. G., Shen Y., Dunlap D. H., Murata H. and Kafafi Z. H., "*Nondispersive electron transport transport in Alq_3,*", Appl. Phys. Lett. **79**, 16 (2001).

Mott N. P. and Gurney R. W., "*Electronic Processes in Ionic Crystals,*" Oxford Univ. Press, London (1938).

Novotny L., "*Allowed and forbidden light in near-field optics. I. A single dipolar light source,*" J. Opt. Soc. Am. A 14, 91 (1997).

Pasveer W. F., Cottaar J., Tanase C., Coehoorn R., Bobbert P. A., Blom P. W. M. et al., "*Unified description of charge-carrier mobilities in disordered semiconducting polymers,*" Phys. Rev. Lett. **94**, 206601 (2005).

Pautmeier L., Richert R., and Bässler H., Synth. Metals 37, 271 (1990).

Perucco B., Reinke N. A., Müller F., Rezzonico D., and Ruhstaller B., "*The influence of the optical environment on the shape of the emission profile and methods of its determination,*" Proc. SPIE 7722, 77220F (2010).

Perucco B., Reinke N. A., Rezzonico D., Moos M., and Ruhstaller B., "*Analysis of the emission profile in organic light-emitting devices,*" Optics Express, Vol. 18, Issue S2, pp. A246-A260 (2010).

Preezant Y., Roichman Y., and Tessler N., "*Amorphous organic devices - degenerate semiconductors,*" J. Phys. Cond. Matt. **14**, 9913 (2002).

Roichman Y. and Tessler N., "*Generalized Einstein relation for disordered semiconductors: implications for device performance,*" Appl. Phys. Lett. **80**, 1948 (2002).

Ruhstaller B. et al., "*Simulating Electronic and Optical Processes in Multilayer Organic Light-Emitting Devices,*" IEEE J. Sel. Top. Quantum Electron. 9, 3 (2003).

Ruhstaller B. et al., "*Transient and steady-state behavior of space charges in multilayer organic light-emitting diodes,*" J. Appl. Phys. 89 (2001).

Scharfetter D. L. and Gummel H. K., "*Large-signal analysis of a silicon read diode oscillator,*" IEEE Trans. Electr. Dev. 16, 64 (1969).

Scott J. C. and Malliaras G. G, "*Charge injection and recombination at the metalâĂŞorganic interface,*" Chem. Phys. Lett. 299, 115 (1999).

Shao J., Wright G. T., "*Characteristics of the space-charge-limited dieletric diode at very high frequencies,*" Solid-State Electron. 3, 291 (1961).

Staudigel J. et al., "*A quantitative numerical model of multilayer vapor-deposited organic light emitting diodes,*" J. of Appl. Phys. **86**, 7 (1999).

Tanase C., Meijer E. J., Blom P. W. M., and de Leeuw D., Phys. Rev. B 91, 4 (2003).

van Mensfoort S. L. M. and Coehoorn R., "*Effect of Gaussian disorder on the voltage dependence of the current density in sandwich-type devices based on organic semiconductors,*" Phys. Rev. B 78, 085207 (2008).

van Mensfoort S. L. M., Vulto S. I. E, Janssen R. A. J., and Coehoorn R., "*Hole transport in polyfluorene-based sandwich-type devices: Quantitative analysis of the role of energetic disorder,*" Phys. Rev. B, 78, 085208 (2008).

van Mensfoort S. L. M., Carvelli M., Megens M., Wehenkel D., Bartyzel M., Greiner H., Janssen R. A. J., and Coehoorn R., "*Measuring the emission profile in organic light-emitting diodes with nanometre resolution,*" Nature Photonics 4, 329-335 (2010).

Vissenberg M. C. J. M. and Matters M., Phys. Rev. B 57, 12964 (1998).

22

Design and Simulation of Time-Pulse Coded Optoelectronic Neural Elements and Devices

Vladimir G. Krasilenko, Aleksandr I. Nikolskyy and Alexander A. Lazarev
Vinnitsa Social Economy Institute of University "Ukraine",
Vinnitsa National Technical University
Ukraine

1. Introduction

Creation of intellectual systems is impossible without creation of proper tools, proper micro- and macroelement bases (Krasilenko et al., 1995, a; Krasilenko & Magas, 1999). So as any concrete technical decision is a compromise between many contradictory requirements taking into account solvable tasks, the creation of universal, effective, neural module as a macroelement base for neurobiologic (Krasilenko et al., 2002, a; Krasilenko et al. , 2001, a) is, probably, the future task. At the same time, searches for ways of creation of multifunction devices which work on new mathematical, logical, hybrid, neural, neural-fuzzy bases and others like that, especially on the basis of the biologically explained conceptions, will always be actual (Krasilenko et al., 2002, a; Levin, 1990, a; Shimbirev, 1990, a), because just life and time choose the most optimal solutions from all offered and existent.
The strategic direction of solution of various scientific problems, including the problem of creation of artificial intelligence (AI) systems, human brain simulators, robotics systems, monitoring and control systems, decision-making systems, as well as systems based on artificial neural networks, etc., becomes fast-acting and parallel processing of large arrays (2D) of data using non-conventional computational systems, matrix logics (multi-valued, signed-digit, fuzzy logics, continuous, neural-fuzzy and others) and corresponding mathematical apparatus (Krasilenko et al., 2001, a; Masahiko & Yatagai, 1997; Berger & Collings, 1997; Krasilenko et al., 1997,a; Leondes, 1998; Krasilenko, 1988,b). For numerous perspective realizations of optical learning neural networks (NN) with 2D structure (Masahiko & Yatagai, 1997) of recurrent optical NN (Berger & Collings, 1997) of the continuous logic equivalency models (CLEM) NN (Krasilenko et al. , 2001, a; Krasilenko et al., 1997,a; Leondes, 1998), the elements of matrix logic are required. Optic and optoelectronic technologies, methods and principles as well as corresponding element base provide attractive alternative for 2D data processing. These technologies and methods successfully solve problems of parallelism, input-output and interconnections. Signed-digit number system (SDNS) are considered to be most promising and widely used nowadays, among this systems we can find high-order systems, including the quaternion SDNS system (Krasilenko et al., 2001, c). The architectures based on the system embrace such features as high density of recording, they are less complicated and they contain few system components, stages of addition, etc. Only a few of them (Huang & Itoh & Yatagai, 1999) can

be used for processing of 2D data and perform wide range of needed arithmetic and logic operations (Guilfoyle & McCallum, 1996).
There exist a number of generalizations of continuous logic (CL) (Levin , 1990, a), (Shimbirev, 1990, a; Volgin, 1986). Among them we should note sequential logic, hybrid logic (HL), continuous discrete logic (CDL), controlled continuous logic (CCL) or predicate selective algebra (Volgin, 1986), fuzzy logic (Zadeh, 1965). In hybrid logic the threshold and dethreshold operations are basic, they realize transformations over continuous and discrete variables (Shimbirev, 1990, a). Hybrid logic (HL) is basic mathematical body for analysis and synthesis of hybrid (analog-digital) devices and systems. Functions max and min can be realized sequentially performing threshold and dethreshold operators, i.e. by means of hybrid logic (HL). Generalization of scalar two-valued logic on matrix case has led to intensive development of binary images algebra (BIA) (Huang et.al., 1989) and 2D Boolean elements for optic and optoelectronic processors (Huang & Itoh & Yatagai, 1999; Awwal & Iftekharuddin, 1999; Krasilenko et al., 2001, c; Guilfoyle & McCallum, 1996; Krasilenko et al. , 1988,a; Krasilenko et.al., 1991,a; Krasilenko & Dubthak, 1992). We would like to draw the attention to the fact, that both natural neurons and their more complex physical and mathematical models suggest discrete-analog and purely analog means for information processing with different level of accuracy, with the possibility of rearrangement of chosen coding system. This, in its turn, requires corresponding image neuron circuit engineering with programmable logic operations, with transition from analog to discrete processing, to storing etc. Thus, the search of means aimed at construction of elements, especially universal (at least quasi-universal or multifunctional) with programmable tuning, able to perform not only operations of two-valued logic, but other matrix (multi-valued, continuous, neuro-fuzzy, etc.) logic operations is very actual problem(Guilfoyle & McCallum, 1996).
Neuromodule, as it has been shown in works (Krasilenko et al., 1995, a; Krasilenko et al. , 2001, a; Krasilenko et.al , 1995,c) can serve such macroelement basis. Functioning of neural elements of such neuromodule must be described by future more generalized neurobiological (Krasilenko et al., 1995, a; Krasilenko et al. , 2001, a) or the most generalized at present time logics: continuous, hybrid, neural, fuzzy, neural-fuzzy and other logics (Krasilenko et al., 1995,a; Krasilenko et al., 2002, a; Krasilenko et al. , 2001, a; Levin , 1990, a; Shimbirev, 1990, a; Krasilenko et.al , 1995, c).
One of perspective directions of researches on creation of macrobase neural modules with intelligence and quasiuniversality, and which operate on afore-mentioned new mathematical bases, is the use of conception of time-pulse coded signals and the offered approach to creation of different elements and structures for effective and highly productive 1D and 2D data processing (Krasilenko et.al., 2004, a; Krasilenko et.al, 2004, b; Krasilenko et.al., 2005, a). Prospects of new optical and optoelectronic technologies (Masahiko & Yatagai, 1997; Guilfoyle & McCallum, 1996; Baukens et.al., 1997) are limited to narrow special processors. Therefore, to our mind, the central problem of new untraditional parallel architecture, picture logic structures and elements creation, neurocomputers, is the problem of logical universal and intelligence vector-matrix functional base (Krasilenko et al., 1995, a; Krasilenko & Magas 1999; Krasilenko et.al., 2004, a; Krasilenko et.al., 2005, a; Krasilenko et.al , 1995, c; Krasilenko et.al., 2005, b).
One of promising directions of research in this sphere is the application of time-pulse-coding architectures (TPCA), that were considered in works (Krasilenko et.al., 1991,a;

Krasilenko & Dubthak, 1992; Krasilenko, 1991,b) these architectures were generalized in (Krasilenko et al., 1995, a). The time-pulse representation of matrix continuous-logic variables by two-level optic signals not only permits to increase functional possibilities (up to universality), noiseproofnes, stability and decrease requirements regarding alignment and optical system, but also simplify control circuits and adjustment circuits on required functional performed, operation, and keep untouched the whole methodological basis of such universal elements construction irrespective of word length of a code and type of logic. It is shown, that it is possible to create more sophisticated problem-oriented processors, in which the specific time-pulse coded operands are used and to use the elements of only two-valued logic, which will realize functions of different logics, continuous etc. The two-stage structure is used for 2D-arrays processing: at the first stage - elementwise processing; at the second stage - overall processing. At the same time in works (Krasilenko et al., 1995, a; Krasilenko et.al , 1995, c) little attention is paid to such non-traditional basic components of TPCA as width-pulse, phase-pulse, photoconverters, photocurrents comparators. Considerable results, to our mind, for creation of ANN are the development of «equivalence» models and a proper «equivalence» paradigm (Krasilenko & Nikolskyy et.al., 2000, a; Krasilenko & Nikolskyy et.al., 2001, d). Computer simulation results of «equivalence» neural networks models (ENNM) have showed their efficiency and substantial advantages: increase of capacity to (2.5 - 4.0)N; where N is neurons amount, the possibility of spatially invariant recognition of very correlated 2D- and 1D-patterns (Krasilenko et.al., 2008, a; Krasilenko et.al., 2009, a; Krasilenko, 2001, e; Krasilenko et.al., 2002,b). Considerable foundation for development of the proper hardware tools and design of the newest macroelement base was expansion of scalar neurobiologic (NBL) and creation of mathematics of matrix neurobiologic with its base operations of the normalized equivalences (nonequivalence) of two vectors, «equivalence» of spatial functions of 2D-patterns, operations of autoequivalence of nonlinear transformations (Krasilenko et.al., 2009, a; Krasilenko, 2001, e; Krasilenko et.al., 2002,b) and whole family of new «equivalence» operations on the basis of use summarizing operations of continuous fuzzy and neuro-fuzzy logic: fuzzy negation, t-norm and s-norm (Krasilenko et al., 2002, a).

There are tens of different physical models of neurons (Bardachenko et.al., 2004; Krasilenko et.al, 2002, d). But majority of them realized by simplified old mathematic models. Models which have equivalency paradigm properties are described in works (Krasilenko et al., 2002, a; Krasilenko et.al., 2005, a; Krasilenko et.al., 2002, e). It is possible to design such neuron macroelement base on the basis of optical and optoelectronic technologies. The last allows to work out the problems of coupling, density et al. But such optoelectronic models for the equivalence paradigm, which are described in works (Krasilenko et.al., 2005, a; Krasilenko et.al., 2002, e; Krasilenko et.al.,2009, b) have some disadvantages. Realization of normalized equivalence or nonequivalence are considered very little, except paper (Krasilenko et. al., 2009, c). But synthesis and design of hardware representations of normalized equivalence or nonequivalence of two vectors (or even matrices) and their modifications (Krasilenko et al., 2002, a) is the most actual task. In paper (Krasilenko et.al., 2009,c) an approach to realize complement dual neuron elements - neuron equivalentor/nonequivalentor was offered. They are one of variants of more general neuron model. We used conception of pulse-time encoding, as in works (Krasilenko et.al., 2004, a; Krasilenko et.al, 2004, b) . Advantages of such biologically motivated conception are pulse-time (PWM or PPM) principle of signals and variables and operands encoding. Signals at the output of such neurons can be both digital and analog or hybrid, and with two complement outputs. Then they realize principle of dualism that gives a number of advantages. Therefore, the purpose of this work is subsequent research and perfection of such complement time-pulse coded optoelectronic neural elements, which are named by us as the equivalentor-nonequivalentor neuron. This conceptual approach is used for two-valued, multivalued, neural and neuro-fuzzy continuous logics. Our second position consists in the following: the micromodel of neuron that is the model of heteroassociative memory (HAM) can realize any operation of neurobiologic as well as operations of discrete multivalued logic. For every input combination (situation, state) of vector of Boolean variables there are an output variable. Such model can operate in learning mode (tuning to necessary operations of memorizing of input and proper output variables) and in direct mode which consists of two phases: phase of recognition of input state and phase of forming of output signal. The traditional models of neurons that described in work (Rudenko & Bodjanskyy, 2005) have an input vector and a vector of weighing coefficients as a rule. In our view, the model of neuron must have not only one vector of weighing coefficients but a few. Set of such vectors which are memorized by a neuron will allow comparing of their closeness (remoteness), that is equivalence (nonequivalence), for recognition of a situation at inputs.

2. Motivation and short theoretical ground

2.1 Basic neurological operations of normalized equivalence

For solution of recognition problems a number of mathematics theories and methods is involved; the development of these theories and methods is connected with the advent of expert and intelligence systems: theory of fuzzy sets and theory of possibilities (Zadeh, 1965), theory of neural networks (Freeman James, 1992), and other mathematics methods (Madani, 1998). More purposefully the apparatus of algebraic and fuzzy sets of continuous logic (CL); its derivatives and generalizations, including hybrid, fuzzy predicate logic, neural logic etc. (Krasilenko et al., 1995, a; Levin , 1990, a; Shimbirev, 1990, a; Zhiging Wen, 1996) have become to be applied. Among non-parallel single-stage algorithms of multialternative recognition, an important role is played by numerous versions of algorithms, based on distances calculations (Evclidus, Mahalonobise, Hamming etc) (Shirman, 1996). However, widely spread method, used in static recognition, cluster analysis and other spheres, namely the method of classification by minimal space, has a drawback, due to non-sufficient fast action of necessary space between images calculation, vectors of large dimensions represent these images. Besides, quality of recognition, in particular, the number of stored and correctly recognized references in neuroassociative memory, depends greatly on chosen matrix, type, and space. In neural networks models and recognition algorithms in hidden layers, minimal spaces (while teaching) (Leondes, 1998; Freeman James, 1992) and criteria of maximum convergency are used as intermediate criteria, in some new equivalence models for recognition of strongly corrected images (Krasilenko et al. , 2001, a; Krasilenko et al., 1997,a). The interest to these new directions, neurofuzzy models, logic-algebraic apparatus, common neurobionic principles can be explained by the possibility to understand with their help the principles of human brain functioning. As neurophysiological studies show, frequency-dynamic model of neuron (FDMN), would be the most adequate (nowadays), and neurobiologic (NBL) as the most generalized (as compared with other known logics, such as hybrid, continuous, threshold etc.) could be used to describe the functioning of FDMN (Krasilenko et al., 1995, a;

Krasilenko et.al, 1995, c; Antamonov, 1974; Posin, 1970). Realizations of such FDMN are already known (Krasilenko et.al , 1995, c; Posin, 1970; Levshin, 1987) but they are rarely used, that is connected with their electronic realizations. Only new optic and optoelectronics technologies and their corresponding realization of neuromodules (arrays) with a great number of FDMN and simplified parallel one-cycle inputs-outputs, efficient realizations of interconnections by means of holograms or other optic methods are the most promising direction, permitting to decrease considerably the size and weight parameters, at the same time increasing the dimensions of arrays, performance and speed of data processing (Krasilenko et al., 1995, a; Krasilenko et al. , 2001, a). Basic operation of NBL, in authors opinion (Krasilenko et.al., 2000,b) or the set of operations may be not one, even functionally complete operation but also optimal, basic set of operations, which is necessary for most efficient algorithmic (and sequential in time and parallel simultaneously in space) execution of needed transformations over information signals. We realize that on this way it is quite impossible immediately to find out really final decision for all cases of infinite, complex and dynamic existence of nervous system of living beings. But consideration of such neurobiological models and their basic fundamental operations even for scalar NBL is of gnoseological aspect, it permits still at the stages of conceptual approaches and structural-functional and mathematical-model design rely on vitality and a ability to compete of the suggested solutions. In many neural models (with the exception (Krasilenko et al., 1997, a)) used for optical realization of different associative devices and neural nets, only carrier sets ${}^2C_b = \{-1;1\}^N$ and ${}^2C_u = \{0,1\}^N$ are used in bipolar (b) and unipolar (u) coding respectively. Models, proposed in (Krasilenko et al., 1997,a; Krasilenko et.al., 1996; Krasilenko et.al., 1997,b) and called equivalental, are more general and good for representation of bipolar and unipolar signals, including multilevel signals. The connections, especially braking, are described more natural there. In such models, the basic operation is a standard equivalency of vectors. The models are suitable for different methods of weighing. Considering their prospects, in this work we will show how to build associative (auto-associative and hetero-associative) memory of correlated 2-D images, including multilevel (grey scale) images, based on matrix-tensor models. Functioning of most adequate for present day, frequency-dynamic models of neurons is described by means of mathematic apparatus of continuous logic (CL), hybrid logic, neural-fuzzy logic and other logics, but in the long run, by means of more generalized logic - neurobiologic (NBL) (Shimbirev, 1990, a; Krasilenko et al. , 2001, a; Antamonov, 1974), which is the generalization of all known logics. It is impossible to find partial universal solution for all cases of such complex, versatile, infinite and dynamic existence of nervous system of living organisms. But gnoseological study of all known modules and taking into account the results of neurophysiological research, providing neuroelements with adequate properties and principles of information processing, still at the stages of conceptual approaches and structural functional synthesis allows to hope that the suggested technical solutions will be successfully realized and competitive. Functional and system requirements to neuromodules are considered in a paper (Freeman James, 1992) taking into account the list of neurons properties, some of their optoelectronic realizations, performing one of the basic operations of limited difference (LD) ($a \dot{-} b$) have been suggested. The connections among the limited difference (LD) and the basic operations are shown in works (Zadeh, 1965; Freeman James, 1992; Shimbirev, 1990, a; Krasilenko et al., 1997,a): max(a,b), min(a,b), complement, functionally complete system of functions CL{$\wedge$, $\vee$, $-$}; operations of limited sum ($\dot{\cup}$) and limited product ($\dot{\times}$) of neuro-fuzzy logic (Shimbirev, 1990, a), operations of equivalence ($\overset{\vee}{\sim}$), ($\overset{+}{\sim}$), and non-equivalence ($\overset{\vee}{\nsim}$), ($\overset{+}{\nsim}$) of NBL (Krasilenko et al. , 2001, a; Antamonov, 1974) from the whole set of such operations being generalizations of XNOR and XOR of two-valued logic, permitting logically compare continuous (analog) and multilevel normalized unipolar and bipolar signals including scalar, vector and matrix. In the same works (Krasilenko et al. , 2001,a; Antamonov, 1974), the properties of these operations of equivalence (non-equivalence) are shown, that is why we don't concentrate our attention on this point, but we will use them, in case of necessity. Taking into consideration the above-described approach, consisting in universality, let us recollect some known facts regarding the number of functions. The number of Boolean functions of n variables in algebra of two-valued logic (ATL), which is also Boolean algebra, equals 2^{2^n}. In this ATL there are $N_2 = 2^n$ atoms, which are minterms. Functions of n variables k-valued logic (k>2) are reflections $A^n \rightarrow A$, where A={0, 1, ... k-1}, and the number of functions equals $N_k = k^{k^n}$. Algebra, formed by set $^{\wedge}C_u = [0,1]$ or $^{\wedge}C_b = [-1,1]$ is called CL algebra (ACL), and the number of CL functions, as reflections $C_u^n \rightarrow C_u$ depending on the CL algebra can be infinite or finite (the set of reflections is always infinite). CL functions are called only those functions of the set $N_{\wedge}$, which are realized by formulas. The number $N_{\wedge}$ of CL functions in the most developed CL algebra - quasi-Boolean Cleenee algebra ($\Delta = (C_u, \wedge, \vee, -)$), in which any function on any set of arguments takes the value of one of the arguments or its negation, is finite. In this case the number $N_{\wedge}(n)$ of functions of n arguments increases with increase of n very rapidly (Shimbirev, 1990, a): $Ч_{\wedge}(0) = 2$; $N_{\wedge}(1) = 6$; $Ч_{\wedge}(2) = 84$; $N_{\wedge}(3) = 43918$. It should be noted that among fuzzy logics, considered in (Shimbirev, 1990, a) and associated with three variations of fuzzy algebra, only one of them coincides with Cleenee algebra, and the notion of fuzzy functions coincides with the notion of CL-function. That is why the sign of equality not always can be put between terms "continuous" and "fuzzy logic". Thus, our conception consists in creations of time-pulse coding elementary formal neurons, able, by means of their adjustment, (teaching, but not teaching the whole neural networks) to perform any required reflections $C_u^n \rightarrow C$ let even with a certain error. This reflection, in many cases, can be created by means of simple memorizing while teaching and often due to limited taught samples be only approximate, but with sufficient criterial evaluation. For the first time in the work (Krasilenko et.al., 1995, e), for constructing of models of neuronal associative memory (NAM) base binary operations of NBL "equivalence" ($\sim$) and "non-equivalence" ($\nsim$) were used of such types:

$$a \overset{\vee}{\sim} b = \min\left\{\max\left(a,\bar{b}\right),\max\left(\bar{a},b\right)\right\} = \left(a \vee \bar{b}\right) \wedge \left(\bar{a} \vee b\right); \tag{1}$$

$$a \overset{\vee}{\nsim} b = \max\left\{\min\left(a,\bar{b}\right),\min\left(\bar{a},b\right)\right\} = \left(a \wedge \bar{b}\right) \vee \left(\bar{a} \wedge b\right); \tag{2}$$

$$a \overset{+}{\sim} b = 1 - \left|a - b\right|; \tag{3}$$

$$a \overset{+}{\nsim} b = \left|a - b\right|, \text{ where } a,b \in C_u = \left[0,1\right],\ \bar{a} = 1 - a, \bar{b} = 1 - b. \tag{4}$$

In works (Krasilenko et al., 1997, a; Krasilenko et.al., 1997,b) new equivalental operations of NBL were introduced:

$$a \dot{\sim} b = a \cdot b + \bar{a} \cdot \bar{b}\ ; \tag{5}$$

$$a \dot{\not\sim} b = a \cdot \bar{b} + \bar{a} \cdot b\ , \tag{6}$$

with are more convenient in use, properties of there operations were determined, their connection with function of scalar with metric distances, the connection was demonstrated between "equivalental algebra" (EA) and other algebra's of continuous and multivalued logics. The variants of these equivalence operations on a carrier set $^\wedge C_u = [0,1]$ are shown on fig. 1 (a, b, c) respectively for: $eq^1 = a \overset{\vee}{\sim} b = \max\{\min(a,b), \min(\bar{a},\bar{b})\}$; $eq^2 = a \dot{\sim} b = a \cdot b + \bar{a} \cdot \bar{b}$; $eq^3 = a \overset{+}{-} b = 1 - |a - b|$.

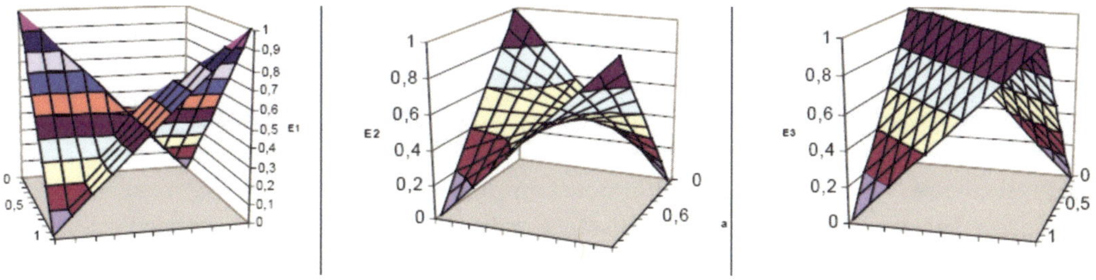

$eq^1 = a \overset{\vee}{\sim} b$ $eq^2 = a \dot{\sim} b$ $eq^3 = a \overset{+}{-} b$

Fig. 1. Operations of equivalence

Their negations are first, second and third nonequivalence respectively. In general case for scalar variables $a,b \in {}^\wedge C = [A,B]$ - continuous line segment, signals themselves and their functions and segment $^\wedge C$ can be brought to segments $[-D,D]$ (or [-1,1]) in bipolar coding and $[0,D]$ (or [0,1]) in unipolar coding. Further, a carrier set $^\wedge C = [0,1]$ and its variables will be considered. Besides, for easier transformations we can limit ourselves at first to operation of equivalence (nonequivalence) of the second type, namely ($\dot{\sim}$) and ($\dot{\not\sim}$) respectively. The terms of equivalental uni-dimension (1-D) and bi-dimension (2-D) functions were introduced: $E(\xi) = f(\vec{a}, \vec{b}_\xi) = \frac{1}{N}\sum_{i=1}^{N}(a_i \sim b_{i-\xi})$, $\tilde{E}(\xi,\eta) = f(A,B) = A \tilde{*} B = \sum_{n=1}^{N}\sum_{m=1}^{M}(a_{n,m} \sim b_{\xi+n,\eta+m})$, terms of systemic equivalental function $\Phi_\sim$ and systemic non- equivalental function $\Phi_{\not\sim} = \bar{\Phi}_\sim$, with the help of which it is possible to study the basis of "equivalental" models (EMs) by analog with equalization function in known models. The connection of functions $\tilde{E}(\xi,\eta)$ with correlation functions was demonstrated (Krasilenko et.al., 1997,b). Normalized equivalence of two matrices $A = \{a_{ij}\}_{I \times J}$ and $B = \{b_{ij}\}_{I \times J} \in [0,1]^{I \times J}$ is determined in the following way (Krasilenko et.al., 1995, e; Krasilenko et.al., 1997,b): $A \underset{n}{\sim} B = \sum_{i=1}^{I}\sum_{j=1}^{J}\frac{(a_{ij} \sim b_{ij})}{I \cdot J}$, and correspondingly normalized non-equivalence: $A \underset{n}{\not\sim} B = \sum_{i=1}^{I}\sum_{j=1}^{J}\frac{\left(a_{ij} \not\sim b_{ij}\right)}{I \times J}$. It should be noted, if $\mathbf{1}=\left[1\right]_{I\times J}$ (matrix of units) then $A \underset{n}{\sim} \mathbf{1} = a_m$ (mean arithmetic value), and $A \underset{n}{\not\sim} \mathbf{1} = A \underset{n}{\sim} \mathbf{0} = 1 - a_m = \overline{a}_m$ (complementary mean arithmetic value). Operations $\left(\tilde{n}\right)$ and $\left(\overset{\not\sim}{n}\right)$ are measures of similarity (equivalence) and difference (non- equivalence, distance) of matrices, which are connected with. Hamming distance, in particular (Krasilenko et.al., 1997,b). Thus, by components operations $(\sim)$ and $(\not\sim)$ of scalar NBL are generalized on matrix case, and NBL logic becomes matrix NBL, i. e. (MNBL). The peculiarity of above considered measures or criteria on their base is that they are invariative to scale change (range) of input components of vectors, matrix, to the change of signal polarity, to the choice of coding type (unipolar of bipolar) to the change of constant component (simultaneous shift of all components by level) belong to the same range [-D, D], are normalized and interconnected (Krasilenko et al., 1997,a; Krasilenko et al. , 2001, a; Krasilenko et.al., 1997,b). Without loosing of community we can consider carrying set (for scalar case) $C_u = \left[0,1\right]$ and $C_u^N = \left[0,1\right]^N$ (for N-dimension (vector, matrix) case). Normalized equivalence $(\underset{n}{\sim})$ and nonequivalence $(\underset{n}{\not\sim})$ are more general new complementary metrics in matrix space R. In particular, $(\underset{n}{\not\sim})$ is a normalized metric distance $d_1\left(\mathbf{A},\mathbf{B}\right)/ m \times n$, and for A and B $\in$ $\{0,1\}^N$ it turns into normalized distance of Hamming $d_n\left(\mathbf{A},\mathbf{B}\right)/ N$. The variants of operations of equivalence and nonequivalence depend on different types of operations of t-norms and s-norms used in them and integrated operations of crossing and joining up in fuzzy logic. Depending on type, variants of equivalent algebra (EA) (Krasilenko et al., 1997,a), as a new algebro-logical instrument for creation of equivalental theory of NNAM based on matrix NBL. We should note the fact that the usage of above-mentioned operations ($\sim$) and ($\not\sim$) allowed to create on their base "equivalence" models of artificial networks (ANN) and corresponding equivalence paradigm and theory, that allowed to describe processes in ANN, explain their dynamics, substantiate architectural realizations and technical solutions, that permitted to recognize even space-invariant and strongly correlated images at considerable increase (up to (2.5-4.0)N! at experiments) of the volume of associative memory (Krasilenko et.al., 2001, f; Krasilenko et.al., 2001, g). The contribution to the development of neuromodules is the enlargement of scalar NBL (generalized operations of equivalence) and mathematical models of NN on vector-matrix and matrix-tensor cases, and introduction of spatial normalized equivalence functions of two 2D-images, operations of "auto-equivalency" and "non-equivalency" (Krasilenko et al. , 2001, a; Krasilenko et.al., 2001, f; Krasilenko et.al., 2001, g; Krasilenko et.al., 2002, f).

2.2 Fuzzy set theory and fundamental operation of fuzzy logic

In common case in order to determine the fuzzy set the refection of universal set of objects ^C in the section [0, 1] is used: $\mu_A\left(x\right)$: ^C $\rightarrow \left[0,1\right]$, determining for each $x \in$ ^C its degree of belonging to fuzzy set A. Thus, the fuzzy set
can be determined as the set of pairs $\left(x,\mu_A\left(x\right)\right)$, where $x \in$ ^C , $\mu_A\left(x\right)$ - belonging function. Thus fuzzy subsets can be considered as an extension of traditional crisp set, of in other

words, the classical crisp set is a special case of a fuzzy set (Zhiging Wen, 1996). Assuming there are two fuzzy subsets $A=\{x,\mu_A(x)\}$ and $B=\{x,\mu_B(x)\}$, $x\in{}^\wedge C$ some common operations are defined as follows (Klir & Folger , 1998; Schmucker, 1989):

1. Complement with $\mu_{\bar{A}}(x)=1-\mu_A(x)$,
2. Union of fuzzy subsets A and B $\mu_{A\cup B}(x)=\max(\mu_A(x),\mu_B(x))\in{}^\wedge C=[0,1]$,
3. Intersection $\mu_{A\cap B}(x)=\min(\mu_A(x),\mu_B(x))$,
4. Equalization: $A=B$, $\mu_A(x)=\mu_B(x)$,
5. Symmetric difference $\mu_{A\Delta B}(x)=|\mu_A(x)-\mu_B(x)|$ $\forall x\in{}^\wedge C$.

2.3 The family of equivalence (non- equivalence) operations

Let us introduce new generalized operation of the its type of equivalence (non-equivalence) operation, written in the following form:

$$^{t,s}E'(a,b)=(atb)s(a_ntb_n)=(atb)s(\overline{a}t\overline{b}), \tag{7}$$

and as *t*-norm and *s*-norm any of their variants can be used (Kuzmin et.al., 1992; Krasilenko et al., 2002, a). The analysis of the whole spectrum of all possible $^{t,s}E'(a,b)$ shows, that the following operations are the most interesting in case of certain *t*- and *s*-norms:

$^{\bullet,+}E'(a,b)=a\cdot b+\overline{a}\cdot\overline{b}=a\dot{\sim}b$, (known operation see formula (5))

$^{\bullet,\dot{+}}E'(a,b)=a\cdot b+\overline{a}\cdot\overline{b}-a\cdot b\cdot\overline{a}\cdot\overline{b}=(a\dot{\sim}b)-(a\cdot b)\cdot(\overline{a}\cdot\overline{b})$, (new operation)

$^{\bullet,\vee}E'(a,b)=a\cdot b\vee\overline{a}\cdot\overline{b}=\max(a\cdot b,\overline{a}\cdot\overline{b})$, (new operation)

$^{\wedge,+}E'(a,b)=\min(a,b)+\min(\overline{a},\overline{b})=a\overset{+}{\sim}b$, (see formula (3) known operation)

$^{\wedge,\vee}E'(a,b)=a\eqsim b=(a\wedge b)\vee(\overline{a}\wedge\overline{b})$, (see formula (1) known operation)

$$^{\wedge,\dot{+}}E'(a,b)=(a\wedge b)\dot{+}(\overline{a}\wedge\overline{b})=\min(a,b)+\min(\overline{a},\overline{b})-\min(a,b)\cdot\min(\overline{a},\overline{b})=$$
$$=\left(a\overset{+}{\sim}b\right)-\min(a,b)\cdot\min(\overline{a},\overline{b});$$

It should be noted, that for the case of taking the complement by one the variables, these operations will have the following form:

$^{\bullet,+}E'(a,\overline{b})=a\cdot\overline{b}+\overline{a}\cdot b=a\dot{\nsim}b={}^{\bullet,+}E'(\overline{a},b)$, (see formula (6))

$^{\bullet,\dot{+}}E'(a,\overline{b})=a\cdot\overline{b}+\overline{a}\cdot b-(a\cdot\overline{b})\cdot(\overline{a}\cdot b)=(a\dot{\nsim}b)-(a\cdot b)\cdot(\overline{a}\cdot\overline{b})$,

$^{\bullet,\vee}E'(a,\overline{b})=a\cdot\overline{b}\vee\overline{a}\cdot b=\max(a\cdot\overline{b},\overline{a}\cdot b)$, $^{\wedge,+}E'(a,\overline{b})=(a\wedge\overline{b})+(\overline{a}\wedge b)$,

$^{\wedge,\vee}E'(a,\overline{b})=(a\wedge\overline{b})\vee(\overline{a}\wedge b)=a\not\eqsim b$, (see formula (2) known operation)

$$^{\wedge,\dot{+}}E'(a,\overline{b})=(a\wedge\overline{b})+(\overline{a}\wedge b)-\min(a,\overline{b})\cdot\min(\overline{a},b)={}^{\wedge,+}E'(a,\overline{b})-\min(a,\overline{b})\cdot\min(\overline{a},b)$$

Introducing new generalized operation of equivalence of II type (non-equivalence) we write in the following form: ${}^{s,t}E''(a,b)=(asb)t(\overline{a}s\overline{b})$ or taking into consideration the law of De Morgan so:

$$ {}^{s,t}E''(a,b)=\left((asb)_n s(\overline{a}s\overline{b})_n\right)_n=\left((\overline{a}t\overline{b})s(atb)\right)_n=\left({}^{t,s}E'(a,b)\right)_n=\overline{{}^{t,s}E'(a,b)} \quad (8) $$

we will obtain the connection between operations of I and II type. That is why, the II type of operations can be called the operation "non-equivalence" of the I type and designate it as: ${}^{t,s}NE'(a,b)={}^{s,t}E''(a,b)=\left({}^{t,s}E'(a,b)\right)_n$ .Thus, formula (7) and (8) which is analogous to formula (7) determine new generalized operations of comparison (determination of equivalence or non-equivalence).

2.4 The short browse mathematical equivalence models of neural networks

The weighing coefficients of synapse connections matrix of equivalence models are determined through the normalized equivalence of *f*, namely (Krasilenko et al. , 2001, a; Krasilenko et al., 1997,a):

$$ T_{ij}^{0}=\frac{1}{M}\sum_{m=1}^{M}\left(S_m^i\sim S_m^j\right)=f\left(\vec{S}^i,\vec{S}^j\right) $$

where S_m^i, S_m^j are proper values of *i*-th and *j*-th neuron of *m*-th standard learning pattern, $\vec{S}^i$ and $\vec{S}^j$ are vectors from all *i*-th or *j*-th components of all set from **M** vectors, whether $T_{ij}^{\beta}=\frac{1}{M}\sum_{M=1}^{M}\left(|\beta_m|\sim S_m^i\sim S_m^j\right)$ where β_m is vectors equivalence coefficient, and also it is the normalized equivalence of f vectors, that $\beta_m=f\left(\vec{X},\vec{S}^m\right)=\frac{1}{N}\sum_{i=1}^{N}(S_i^m\sim X_i^{in})$. Formulas for the neurons initial signals calculation in the equivalence models also taken to determination of the normalized equivalence of *f*, namely: $X_j^{out}(t+1)=\varphi_j\left[f(\vec{T}_j^{0/\beta},\vec{X}_{(t)}^{in})\right]$ where $\vec{T}_j^{0/\beta}$ is a *j*-th vector-column from a matrix $\mathbf{T}^{0/\beta}$, and $\vec{X}^{in}$ is an input vector. For initial vector X_j^{out} component calculation it is possible to take advantage of formula from works (Krasilenko et al., 1997,a; Krasilenko et.al., 1996; Krasilenko et.al., 1997,b) which is taken to finding of the normalized equivalence: $X_j^{out}=\varphi_j[f(\vec{R}^j\sim\vec{\beta})]$ where $\vec{R}^j$ is a vector from the *j*-th components of all **M** learning vectors $\vec{S}_{m\in 1\div M}$, and as a vector $\vec{\beta}$ with **M** dimension it is possible to use $\vec{\beta}$, $\vec{\beta}^{\alpha}$, and $\vec{\beta}^{\text{nonlinear}}$, thus $f\left(\vec{R}^j,\vec{\beta}\right)=\frac{1}{M}\sum_{m=1}^{M}(\vec{R}^j\sim\vec{\beta})$.

In addition, it is possible to show many other formulas, which are used in the equivalence models paradigm and based on calculations of the normalized equivalence or nonequivalence of vectors or matrices (Krasilenko et.al., 2008, a; Krasilenko et.al., 2009, a; Krasilenko et.al., 2002, f). The analysis of these formulas shows that in most cases one of vectors from which calculated *f*, there is a component with binary values, that $x_i\in\{0,1\}$, but not $[0,1]$. It considerably simplifies realization of such f operations, as all of equivalence types, namely ($\overline{\sim}$), ($\dot{\sim}$), ($\overset{\times}{\sim}$), are taken to one. Mathematical formulas for calculation of f and $\overline{f}$, or

$\tilde{e}(\vec{x},\vec{w})$ and $\widetilde{ne}(\vec{x},\vec{w})$ are showed in paper (Krasilenko et.al., 2009, c) . But in most general case optoelectronic complementally dual neuron (equivalentor/nonequivalentor), including the normalized components of vectors $\vec{x}$ and $\vec{w}$, has analog homopolar encoded components, $x_i \in [0,1]$ and $w_i \in [0,1]$. We will mark also, that in work (Krasilenko et.al.,2008, b) it is showed that set of operations, normalized equivalence and nonequivalence is the functional complete system of continuous logic functions. Therefore f realizing from vector information is very actual. As follows from works, for example from paper (Krasilenko et.al., 2002, f), for the calculation of mean value (expected value) component of vector, it is necessary to calculate a normalized equivalence f from vector $\vec{A}$ and vector $\vec{1}$ with single components: $a_m = f[\vec{A},\vec{1}]$. Also $\overline{a}_m = f[\vec{A},\vec{0}] = \overline{f}[\vec{A},\vec{1}]$,where $\overline{f}$ –normalized nonequivalence. We will designate f various types of erations ($\approx$), ($\overset{+}{\sim}$), ($\overset{\vee}{\sim}$) or their generalizations (Krasilenko et al., 2002, a); ${}^{t,s}E'(a,b) = (atb)s(\overline{a}t\overline{b})$ by character ($\sim$) for simplicity.

3. Designing and modelling of multifunctional units of neural (continuous) logic

3.1 Structural-functional design of universal elements for neural (continuous) logic

Now let us discuss continuous logic (CL), modifying the approach, suggested in papers (Krasilenko et al., 1995, a; Shimbirev, 1990, a), but to make the understanding of the problem easier, we will consider here the scalar case for n (an example for two) arguments and for the realization of a random continuous-logic function, where their complements participate. We assume x_1, x_2 to be the arguments and their complements $\overline{x_1} = 1 - x_1$, $\overline{x_2} = 1 - x_2$, taking into account their dependence. The number of cube partitioning $[0,1]^n$ into areas or number of situations of mutual location of arguments in this case is $H = 2^n n!$ areas (if n=2 the number of the areas is 8), but not $(2n)!$ that is seen from the Fig. 2a, where the square $[0,1]^2$ partitioning is shown. In Fig. 2a the number of each area coincides with the number of situation indicated above, for which the relations between arguments in the given situation and values of the function $f(x_1,x_2) = (\overline{x_1} \wedge x_2) \vee (x_1 \wedge \overline{x_2})$ for each area are shown in the table, presented in Fig. 2b. At all points in the same area the function value, not in case of const "0" and "1", equals one of either the arguments or the complements. Fig. 2c shows values of $\widetilde{EQ}$ function (notation from paper (Krasilenko et al., 1997, a)), which is calculated in the following way: $\widetilde{EQ}\,(x_1,x_2) = x_1 \wedge x_2 \vee (\overline{x_1} \wedge \overline{x_2})$. Fig. 2d shows relations between x_2 and $\overline{x_1}$ for area I (the shaded area) and area II. Fig. 2e shows pairs, setting order relations between pairs, the connections in the pairs are shown by dotted and solid lines. Each situation of known subcode is definitely determined by setting of order relations between pairs and the number of such pairs equals 4 for n=2 (see Fig. 2e), in general case $n + 2C_n^2 = n^2$. Table (Fig. 2b) shows the states of comparators $K_1 \div K_4$ signals, comparing by pair the order relations in Fig. 2e. The usage of decoder, converting 4-digit binary codes at the output of comparators in 8-digit one-positional code (this is the difference between our approach from the approach, known from paper (Shimbirev, 1990, a), and in general case n^2-digit code of comparators into ($H = 2^n 2!$)-digit one-positional code, allows by means of formation of H adjusting signals y_h from the set of $2n$ arranged variables $\{x_1, \dots x_n, \overline{x_1}, \dots \overline{x_n}\}$ to select for each situation one of $2n$ arguments (direct or its complement). In our modified variant, as it is seen from Fig. 2,c,d the value of $\widetilde{EQ}$ function ($\widetilde{EQ} \in \{x_1, x_2, \overline{x_1}, \overline{x_2}\}$) determines one of the regions {7, 2, 3, 5} in the area I or one of the regions {4, 6, 8, 1} in the

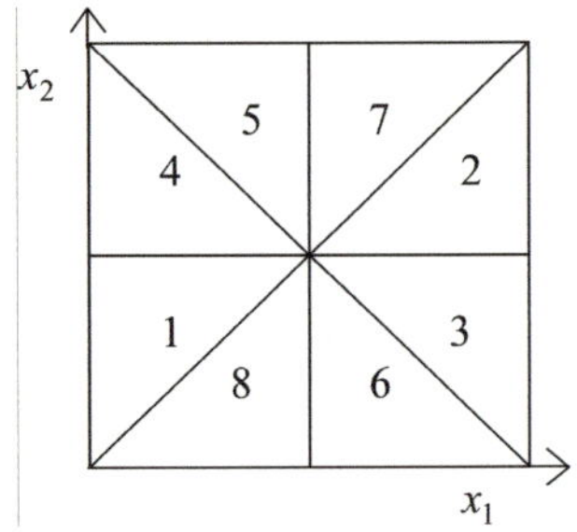

Fig. 2a. The number of each area coincides with the number of situation indicated above, for which the relations between arguments in the given situation and values of the function

Region	Situation	Situation	$f(x_1,x_2)$	Comparators outputs			
				1	2	3	4
1	$0 \le x_1 \le x_2 \le 1/2$	$x_1 \le x_2 \le \overline{x_2} \le \overline{x_1}$	x_2	1	0	1	0
2	$0 \le \overline{x_1} \le \overline{x_2} \le 0.5$	$\overline{x_1} \le \overline{x_2} \le x_2 \le x_1$	$\overline{x_2}$	0	1	0	1
3	$0 \le \overline{x_1} \le x_2 \le 0.5$	$\overline{x_1} \le x_2 \le \overline{x_2} \le x_1$	$\overline{x_2}$	0	1	1	1
4	$0 \le x_1 \le \overline{x_2} \le 0.5$	$x_1 \le \overline{x_2} \le x_2 \le \overline{x_1}$	x_2	1	0	0	0
5	$0 \le \overline{x_2} \le x_1 \le 0.5$	$\overline{x_2} \le x_1 \le \overline{x_1} \le x_2$	$\overline{x_1}$	1	1	0	0
6	$0 \le x_2 \le \overline{x_1} \le 0.5$	$x_2 \le \overline{x_1} \le x_1 \le \overline{x_2}$	x_1	0	0	1	1
7	$0 \le \overline{x_2} \le \overline{x_1} \le 0.5$	$\overline{x_2} \le \overline{x_1} \le x_1 \le x_2$	$\overline{x_1}$	1	1	0	1
8	$0 \le x_2 \le x_1 \le 0.5$	$x_2 \le x_1 \le \overline{x_1} \le \overline{x_2}$	x_1	0	0	1	0

Fig. 2b. Table shows the states of comparators $K_1 \div K_4$ signals, comparing by pair the order relations in Fig. 2e

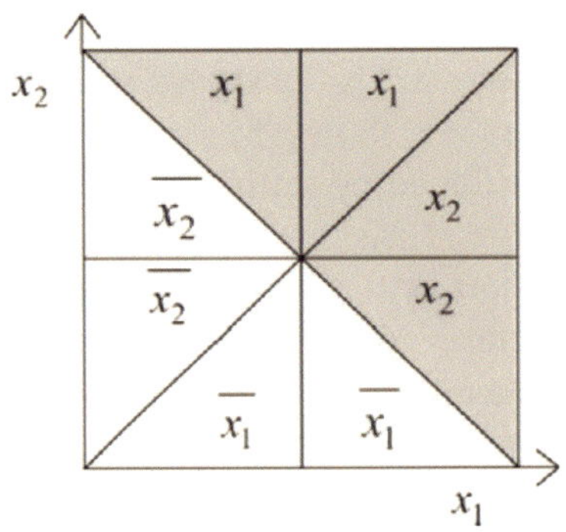

Fig. 2c. Values of $\overset{\vee}{\tilde{EQ}}$ function, which is calculated in the following way:

$$\overset{\vee}{\tilde{EQ}}\ (x_1,x_2) = x_1 \wedge x_2 \vee (\overline{x_1} \wedge \overline{x_2})$$

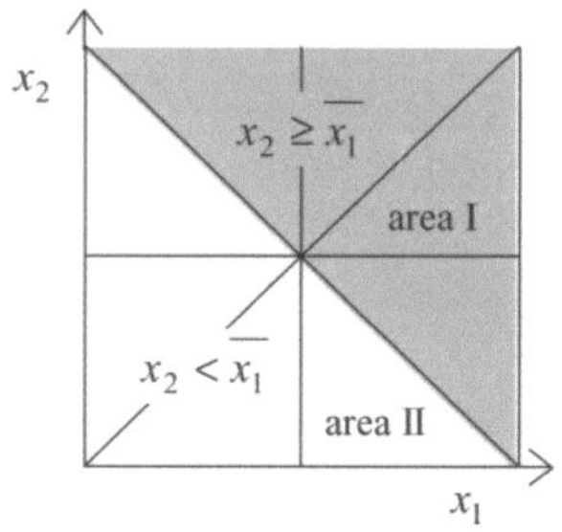

Fig. 2d. Shows relations between x_2 and $\overline{x_1}$ for area I (the shaded area) and area II

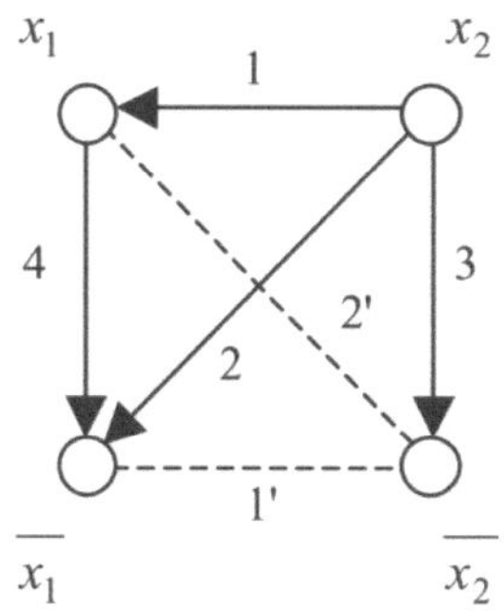

Fig. 2e. Shows pairs, setting order relations between pairs, the connections in the pairs are shown by dotted and solid lines

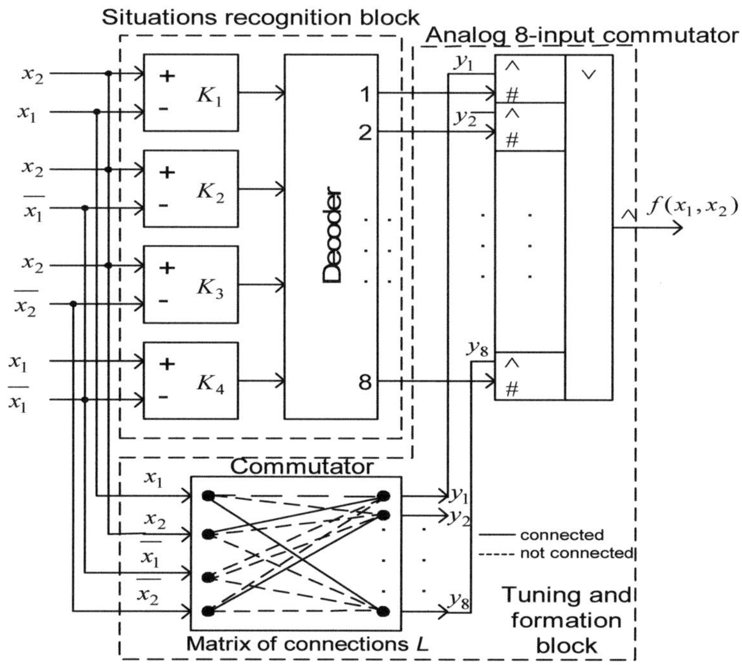

Fig. 3a. The universal continuous-logic neural element with programmable tuning

$\mathbf{L} =$

	y_1	y_2	y_3	y_4	y_5	y_6	y_7	y_8
x_1	0	0	0	0	0	1	0	1
x_2	1	0	0	1	0	0	0	0
$\overline{x_1}$	0	0	0	0	1	0	1	0
$\overline{x_2}$	0	1	1	0	0	0	0	0

Fig. 3b. Tuning matrix for the function $f(x_1,x_2)$ for the example, shown in Table 1

$\mathbf{L}' =$

	y_1	y_2	y_3	y_4	y_5	y_6	y_7	y_8
x_1	0	0	0	0	0	1	0	0
x_2	1	0	0	1	0	0	0	0
$\overline{x_1}$	0	0	0	0	1	0	1	0
$\overline{x_2}$	1	1	1	0	0	0	0	0

Fig. 3c. The tuning matrix *L'* shown on the figure differs from *L* in the first situation the constant $"1" = x_2 + \overline{x_2}$ is at the output, and in the 8-th situation "0" is at the output

area II. The scheme of the universal continuous logic element will have the view, shown in Fig. 3a and adjusting matrix is shown in Fig. 3b. By multiplying inputs vector $\{x_1,x_2,...x_n,\overline{x_1},\overline{x_2},...\overline{x_n}\}^{tr}$ by this matrix *L*, the adjusting vector $Y=\{y_1,...y_H\}^{tr}$ is formed. The tuning matrix *L'* shown in Fig. 3c differs from *L* in the first situation the constant $"1" = x_2 + \overline{x_2}$ is at the output, and in the 8-th situation "0" is at the output. Thus, we have synthesized universal continuous-logic element (UCLE), its possible realization for matrix case we construct in the next sections. It should be noted, that the situations recognition block might include elements, which realize only neuro-fuzzy logic operations and can be considered as micro-level hetoro-associative neural network with analog inputs and digital outputs.

3.2 Time-pulse coding (TPC) universal elements of neuro-biologic

Let us return to the problem of presentation and coding of continuous signals. It is seen from Fig. 3a, that signal x_2 must be sent to 3 comparators and to 8 synapse-connections in adjusting commutator, thus it become clear, that coding of optic power arguments within the scheme is not effective. Having chosen both in matrix multivalued logic (MML) universal picture element (UPE) and matrix twovalued logic (MTL) UPE time-pulse coding, the scheme of UCLE can be converted into the form, shown in Fig. 4. It is taken into account that RS-triggers by means of sending pulses from the PPM output to these triggers can act as comparator of time delays, as it has been shown in paper (Krasilenko et al., 1995, a). The analysis of the scheme Fig. 4a shows that since time-pulse signals are formed at the circuit output, then phase-pulse photoconverters with complementary outputs are needed only for input of initial information; as a matter of fact, processor part can operate with already formed at previous step time-pulse coding results (CL functions of arguments t_{out}, $\overline{t_{out}}$) which are again sent at the input of this or the next array of such universal neural elements.

That is why, if we form, as it is seen in the scheme of Fig. 4b, two outputs: $f(x_1,x_2) \Rightarrow t_{out}(t_1,t_2)$ and $\overline{f}(x_1,x_2) \Rightarrow \overline{t_{out}}(t_1,t_2)$ or in general case several functional outputs, then pulse phase modulator (PPM) must have direct and complementary optic outputs.

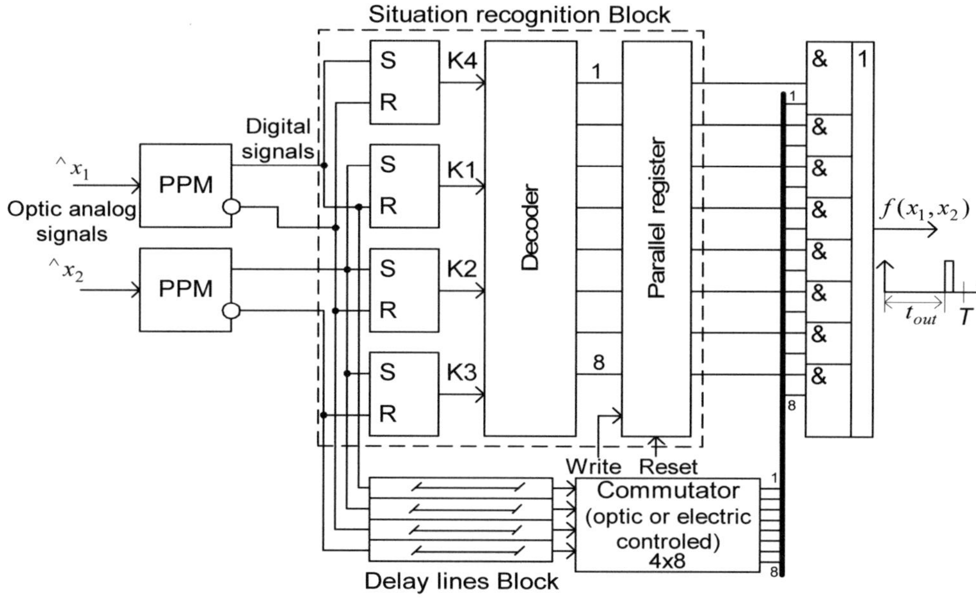

Fig. 4a. The universal continuous-logic neural element

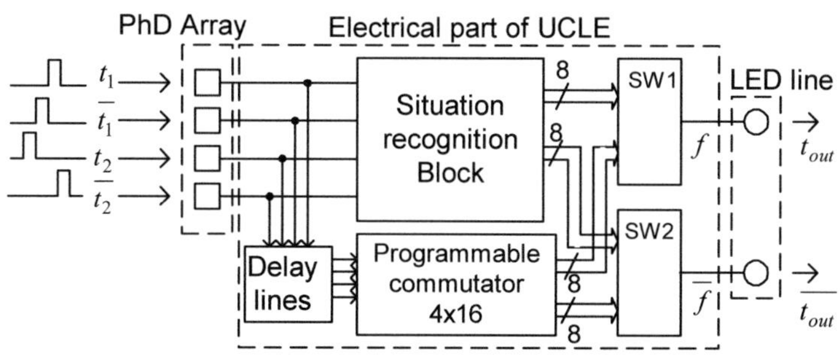

Fig. 4b. Scheme's variant with electronic unit of digital signals delay - two outputs: $f(x_1,x_2) \Rightarrow t_{out}(t_1,t_2)$ and $\overline{f}(x_1,x_2) \Rightarrow \overline{t_{out}}(t_1,t_2)$

In this case, as in variant with electronic unit of digital signals delay (Fig. 4b) as well as in variant of optic delay unit (better for picture organization of matrix logic) universal continuous logic element will be completely digital, operating with two level signals, tuning and information inputs are realized on photosensitive elements, distinguishing only two levels of optic power. It is not difficult to create an array of such digital elements. The electrical digital part of the UCLE can be realized on programmable logic devices (PLD). Simulation results of the scheme Fig. 4a by Altera MAX-PLUS II v.10 software are shown in Fig. 5. The scheme successfully fits into device EPM7032LC44-6 from MAX7000 family. The simulation results are shown for the realization of 16-valued functions (Fig. 2b). The *L*

matrix of connections (Fig. 3d) is used. At the cycle 1 at input pins we have the set of phase-modulated pulses that corresponds to the situation 1 and so at the output pin (Out_f) at the cycle 2 we have a pulse that corresponds to x_2 (see table in Fig. 2b) and so on. For 100 distinguishable time positions and 10ns input pulse width the processing time will be 1μs and can be less if faster PLD is used. Hence, the electronic part of the UCLE does not limit the processing speed in contrast to PPM, which, as it is shown below, is a slower device.
Thus, proceeding from the above-mentioned study of picture type universal elements with programmable tuning of continuous and discrete logics it is necessary to develop arrays of time-pulse, and especially pulse-phase photoconverters if time-pulse coding is used and comparators of analog optic signals if amplitude coding for continuous logic is used.

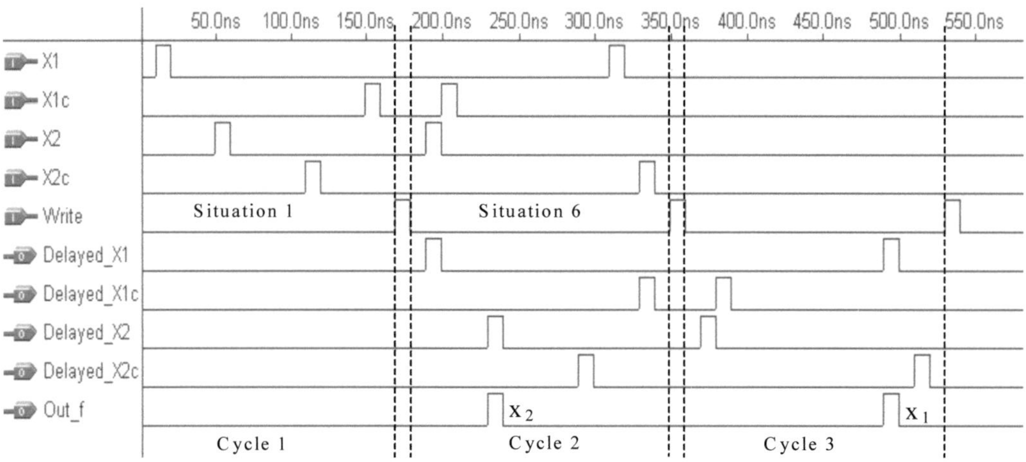

Fig. 5. Simulation results of digital part of UCLE (Fig. 4a)

3.3 Picture continuous logic elements (PCLE)

Fig. 6 shows the structural diagram of picture neural element (PNE) for computation of all basic matrix-continuous-logic (MCL) operations in matrix quasiboolean algebra **C=((A,B),∨, ∧ -)**(Krasilenko et al., 1995, a) for which in any set of MCL arguments matrix continuous logic function (MCLF) **F** takes the value of a subregion of one of the arguments or its supplement. PE of MTL, performing MTL operations over matrix temporal functions $O^i_t(t)$ (in point of fact two-valued 2D-operands) realize MCLF over continuous logic variables (CLV) O^i_t. The time-pulse coding of a grayscale picture is shown in Fig. 7. As it is seen from Fig. 8 at each point of picture output of PNE, MCL can be performed over continuous logic variables (CLV) $O^1_{ijT},\ldots O^{n=2}_{ijT}$, presented by $t^1_{ij},\ldots t^n_{ij}$ durations of time pulse signals, during each interval T one of the following operations of CL : min(a,b), max(a,b), mod(a-b), $\overline{\mathrm{mod}(a-b)}$, complementary $\overline{a}=1-a$, equivalence, etc. (Krasilenko et al., 1995, a; Levin , 1990, a; Krasilenko, 1988,b). The duration of MTL formed at the output, and as a result of PNE, signal $f^{NE}_{ij}(t)=f^{NE}_{ij}\left(O^1_{ij}(t),O^2_{ij}(t)\right)$, is CL function of input binary temporal variables durations. Thus, as it is seen from Fig. 8, almost all basic operations of continuous logic, neuro-fuzzy logic, that are shown in work (Krasilenko et al., 2002, a), can be realized with the help of a time-pulse coding of variables $\mathbf{X}_1,\ldots,\mathbf{X}_n$ and universal picture element (UPE) of array two-level logic (ATL). But for that PWM PT are needed. It is not needed to form contrast-conversion (complementary operand) image for analog picture optic inputs if PWMs PT have complementary outputs.

Thus, we have shown, that the basic unit of components considered above is width- pulse or phase-pulse matrix (multichannel) photoconverters, and main characteristics of basic macrounits of future advanced systems of 2D-data processing optical computers and neural networks greatly depend on the parameters of above mentioned components, especially, such parameters as linearity, accuracy and transformation rate.

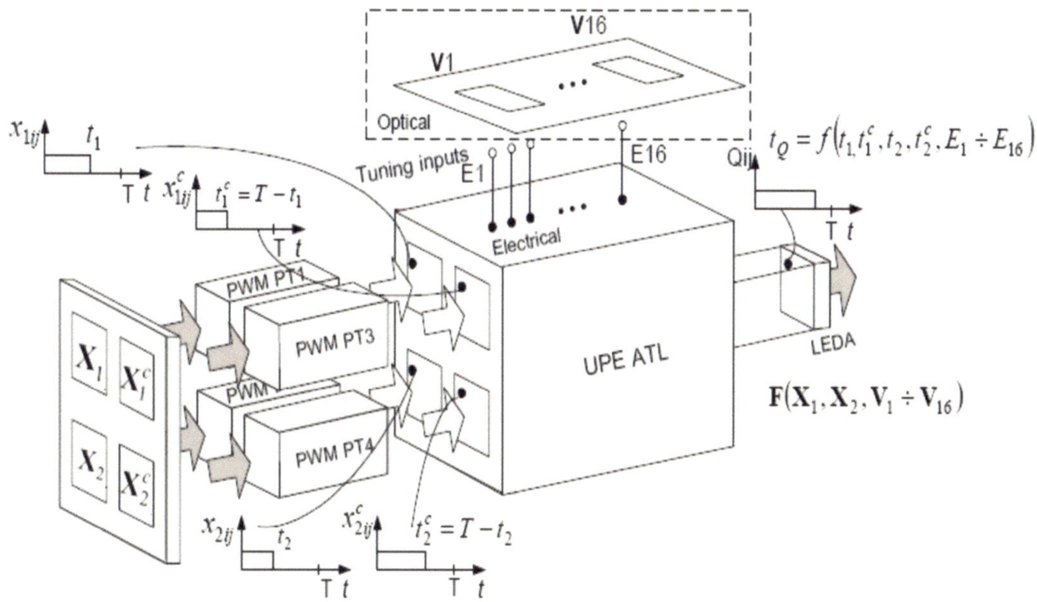

Fig. 6. PNE ANL based on PWM PT with programmable tuning

P =

	0	1	2	3	4	5	6	7	8	9
0	0	1	2	3	4	5	6	7	8	0
1	1	0	3	4	5	6	7	8	0	10
2	2	3	0	5	6	7	8	0	10	11
3	3	4	5	0	7	8	0	10	11	12
4	4	5	6	7	0	0	10	11	12	13
5	5	6	7	8	0	0	11	12	13	14
6	6	7	8	0	10	11	0	13	14	15
7	7	8	0	10	11	12	12	0	15	15
8	8	0	10	11	12	13	13	14	0	15
9	0	10	11	12	13	14	14	15	15	0

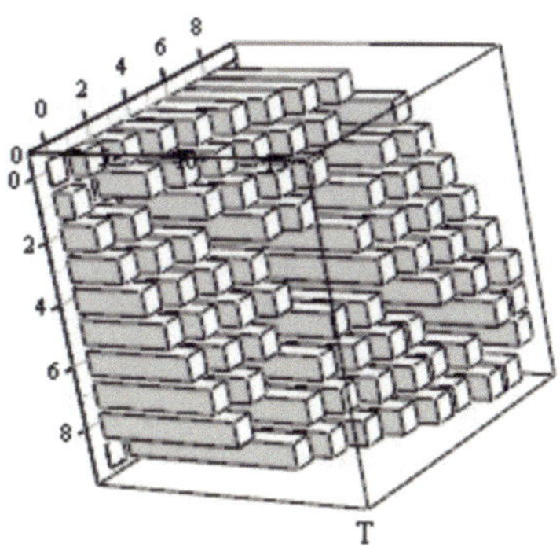

a) b) c)

Fig. 7. Time-pulse coding of a grayscale image: a) input grayscale image; b) matrix of light intensities; c) output time-pulse code

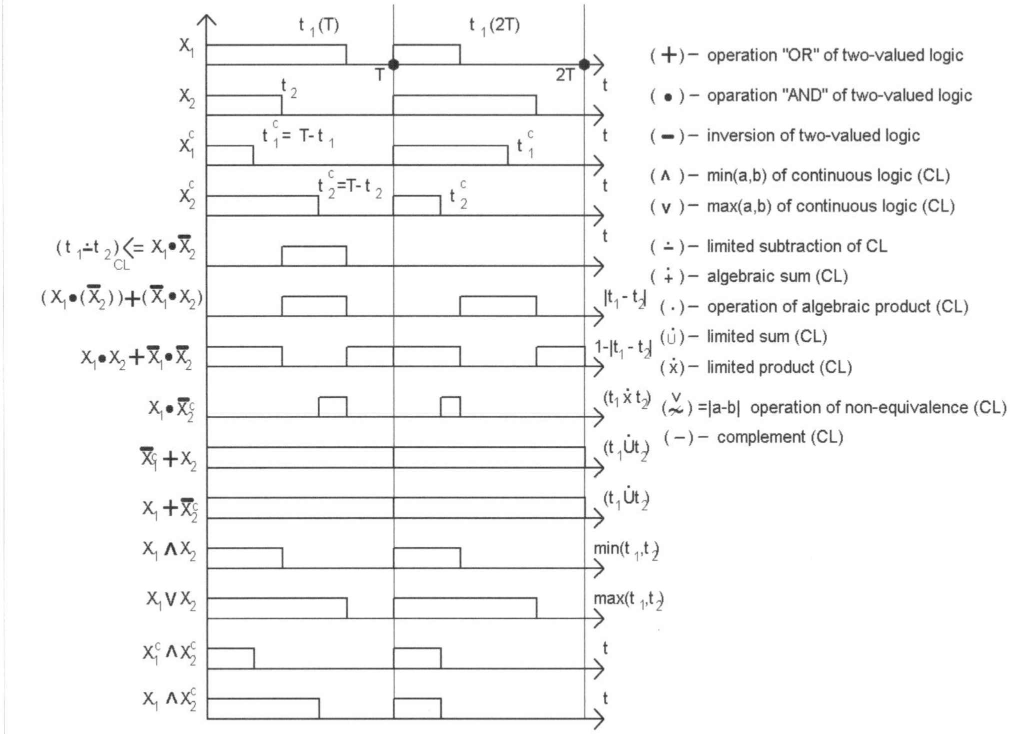

Fig. 8. Time diagrams of CL operation fulfillment by means of time CL variables

3.4 Time-pulse coding photoconverters

3.4.1 Optic signals comparators

The suggested time-pulse coding photoconverters are realized on the basis of optic signals comparators, that is why first we will consider the principle of construction and operation of optic radiation power comparator with potential output. The circuit of suggested comparator with potential output is shown in Fig 9 and consist of two photodiodes, intended for transportation of power of input optic radiation into corresponding value of input current, and of current comparator, realized on two current mirrors, on CMOS-transistors. Power of input optic radiation P1 and P2 are transformed into corresponding input currents I1 and I2 by means of photodiodes VD1, VD2. Further these currents pass to the input of the first current mirror, realized on transistors VT1, VT3 and second current mirror, realized on transistors VT2, VT4. At the output of comparator (Out), the voltage of logic "1" is formed, if current I1>I2 (power P1>P2), and voltage of logic "0", if current I1<I2 (power P1<P2).

The results of this circuit simulation by means of OrCad PSpice software are shown in Fig 10. 1.5μm CMOS-transistors technology (AMIS 1.50) is used, the parameters of transistors models are taken from the website of MOSIS company (www.mosis.org). For transistors, the length L and width W of channel is: for NMOS L=1.6μm, W=4μm; for PMOS L=1.6μm, W=8μm. The model of CMOS-photodiode is used (Swe & Yeo, 2001). The range of input currents is 100nA...100μA. Supply voltage is 3...15V. Time delay of voltage setting at the output of comparator is in inverse proportion to input currents difference. In case of

Δ=10μA time delay is t_d≈40ns, in case of Δ=1μA, t_d≈100ns. Fig. 11 shows the transfer characteristics of comparator for two values of input current I2=20μA and I2=60uA for input current I1 changes from 0 to 100μA. The transfer characteristics of the comparator can be improved if we put a key-inverter (Fig. 12) at the output. In this case, we obtain transfer characteristics that are shown in Fig. 13. At the output we have voltage of logic level "1", if current 11<12 and output voltage of logic level "0" if current I1>I2. The voltage of logic "1" corresponds to supply voltage. The accuracy of comparison is 0.5nA, if input currents are 100nA, and 40nA, if input currents are 10μA, i.e. the relative error is <0.5%. If we put a driver circuit (Ahadian & Fonstad, 1998) and a light-emitting diode or a semiconductor laser at the output, then we obtain the comparator with optic output (Fig. 12).

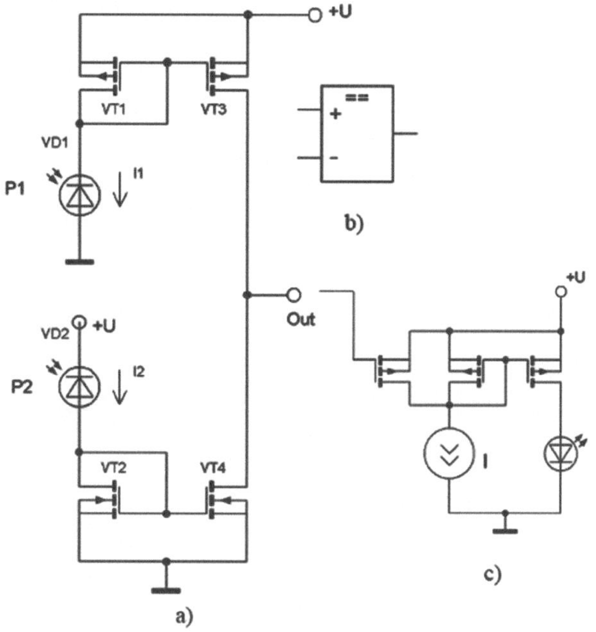

Fig. 9. Comparator of photocurrent with potential output (a), conventional graphical symbol (b), circuit of optical output (c)

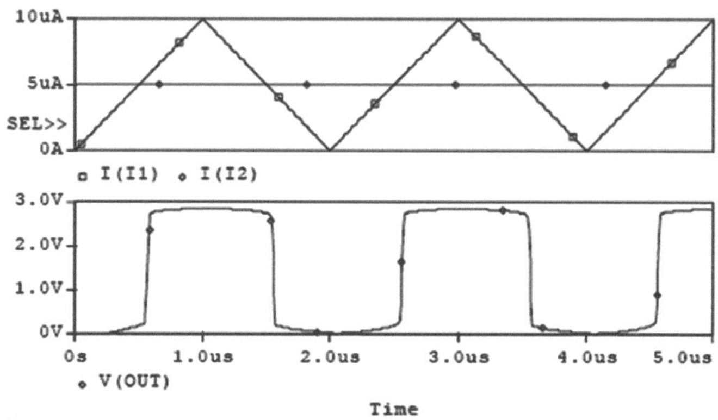

Fig. 10. Time diagram of comparator operation

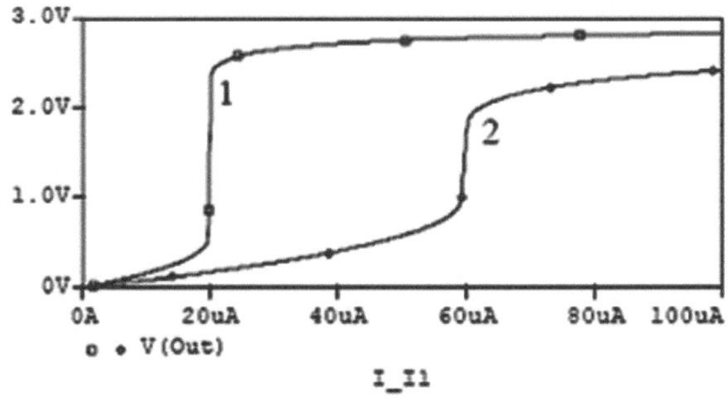

Fig. 11. Transfer characteristic of the comparator: 1 – threshold current I2=20μA; 2 – threshold current I2=60μA

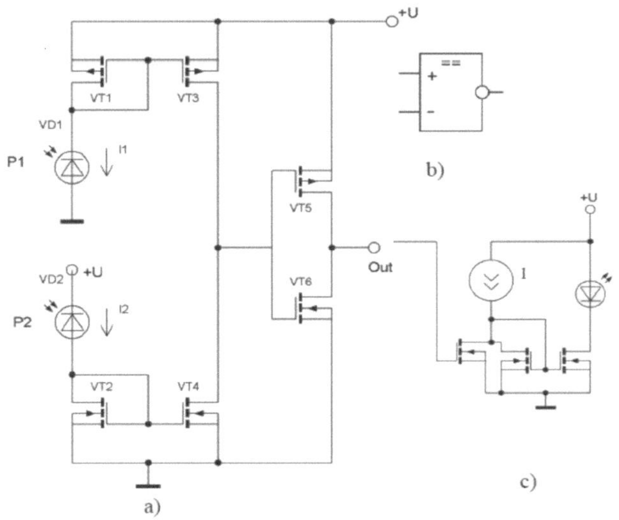

Fig. 12. Circuit of photocurrent comparator with inverter at the output (a), conventional graphical symbol (b), optical output circuit (c)

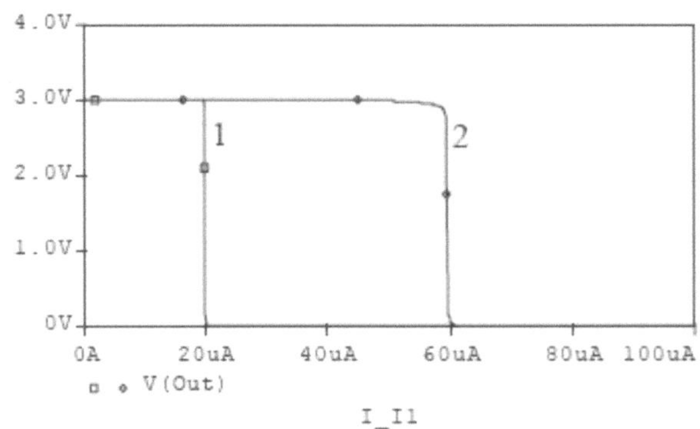

Fig. 13. Transfer characteristic of the comparator: 1 – threshold current I2=20μA; 2 – threshold current I2=60μA

3.4.2 Optoelectronic PWM

Having replaced photodiode VD1 in the circuit, present in Fig 12, by a saw-type current generator, we obtain width-pulse modulator. Voltage pulse duration on its output is directly proportional to the value of the input current I2. The results of simulation of the given circuit for three values of input current I2=(1μA, 4μA, 8μA) are shown in Fig 14. Pulse duration delay of output voltage for value of the input current I2=1μA is 350ns, that leads to rather big relative error at small currents. Fig 15 shows the diagram of relative error of transformation for the range of input current 1.....100μA, and period of transformation is 100μs. In order to decrease the error it is necessary to use transistors that are faster or increase the conversion coefficient. If the transistors of TSMC 0.25μm technology are used the delay of output voltage pulse duration is 180ns at the input current I2=1μA, that reduces the relative error to 18% at small input currents. If we increase the conversion period to 1ms than the relative error can be reduced to 7% at small inputs currents and be less than 1% at input currents more than 5μA. The transition on GaAs technology allows to increase the conversion speed and scale of integration. Power consumed by modulator is 155μW if the value of the input current is 1μA, and 605μW if the value of the input current is 100μA, and supply voltage is 3V. The application of 0.25μm technology transistors allows decreasing the supply voltage to 1.5 V, and hence reducing the consumed power. The average consumed power by one cell is about 0.5mW, and it is about 1mW with LED. So the average consumed power by 1000 of such cell is about 1W that allows to carry out their integration in arrays of 32×32, 64x64 elements and more.

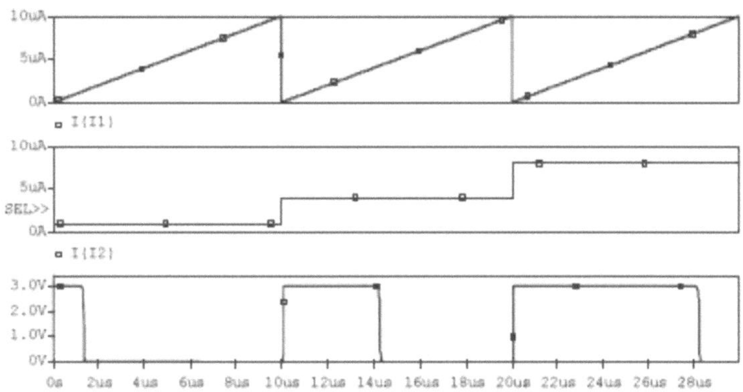

Fig. 14. Time diagrams of PWM operation

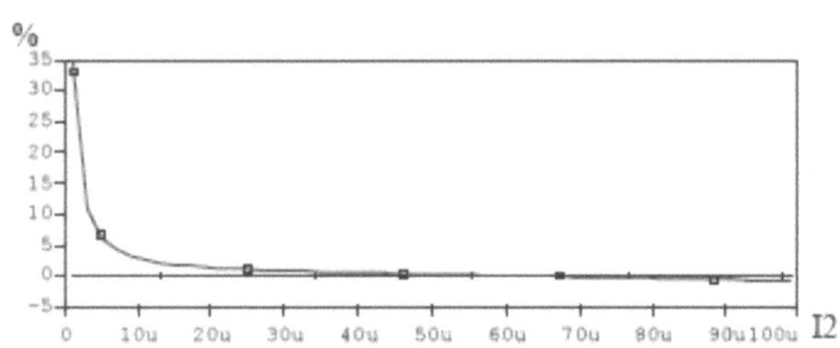

Fig. 15. Relative error of current I2 to output voltage pulse-width conversion for 100μs period

Earlier we have considered PWM with optic inputs based on λ-diodes (Krasilenko et.al., 2003). That modulator provides non-linearity of not more than 0.1% at transformation period of 1ms in the range of input currents 1uA…1mA and satisfactory parameters at frequencies up to 1MHz. But the drawback of the circuit is the realization of λ-diodes on field transistors with controlling p-n junction, that are not used in standard digital CMOS IC technology, and the necessity to use reset current signals with two times more amplitude than the λ- diodes threshold.

3.4.3 PWM with complementary outputs

In order to construct a various devices of continuous, neural, multilevel and other logics it is often needed to have direct and complementary signals. For this purpose we suggest to consider the operation of circuit of width-pulse modulator with complementary time-matched outputs (Fig 16). The sum of durations of voltage pulses at the complementary and direct output equals conversion period *T*. The availability of the complementary output expands functional possibilities of UPE ANL. The circuit is obtained on the base of the circuit presented in Fig 12. The formation of inverse input current is realized on current mirrors on VT2-VT5 transistors by means of subtraction from base current I3 of input current I2. The resulting current arrives at the input of comparator, realized on VT1, VT8-VT10 transistors, at the output of which the pulse of complementary time-matching voltage is formed. Fig 17 presents the results of circuit operation simulation. Current I1 of saw-shaped form with the period of 100μs and amplitude of 100μs. Reference current I3=100μA. Input optic flow is sent to the photodiode VD3, where it is transformed into input current I2, in our case of simulation the current I2=20μA. At the direct output Out1 the voltage pulse with duration of 20μs is formed, voltage pulse with duration of 80μs is formed at complementary output Out2.

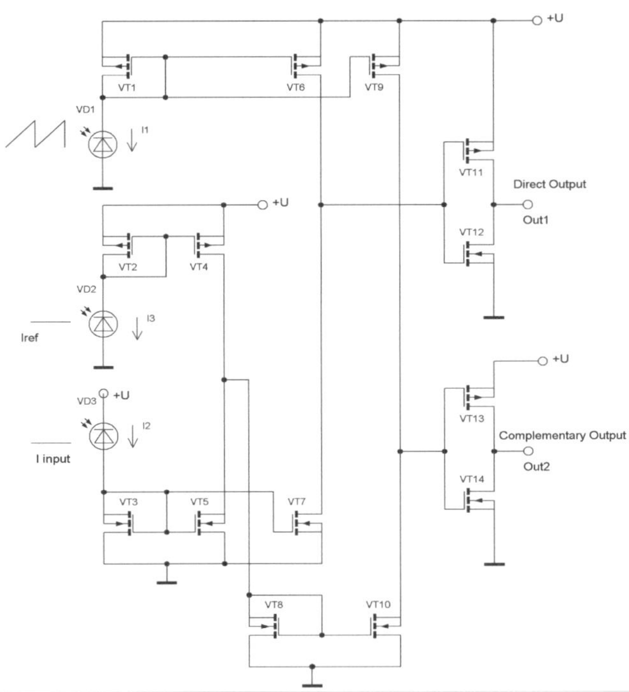

Fig. 16. Circuit of PWM with complementary outputs

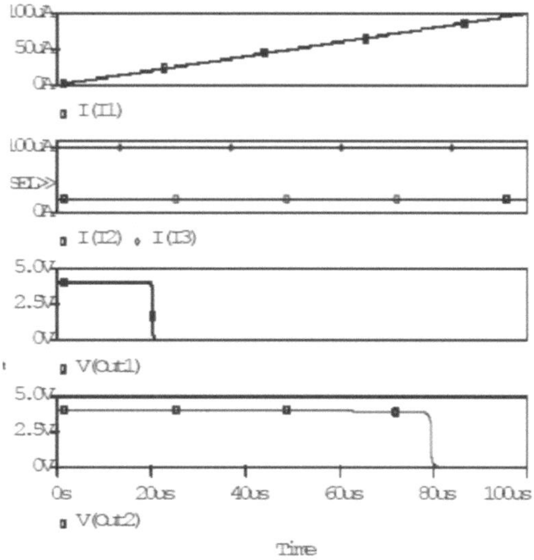

Fig. 17. Time diagram of operation of PWM with complementary outputs (I_{input}=20μA)

3.4.4 Designs and simulation of optoelectronic pulse-phase modulators

There are different ways to build PPM on different devices, particularly for optoelectronic applications (Krasilenko et al., 1995, a; Krasilenko et.al , 1995, c; Krasilenko et.al., 2003; Krasilenko et.al., 2001, h). We consider realization of PPM on CMOS transistor current mirrors and CMOS digital elements. The operation of phase-pulse photoconverter can be based on the two-threshold current comparator, at which output the voltage of logic "1" is formed, if the value of input signal is within the range of two thresholds values and the circuit consists of two current comparators, realized on current mirrors, and XOR gate. We considered current comparator on basis of current mirrors in work (Krasilenko et.al, 2004, b). PPM can be built in several ways. The first way assumes that saw-shaped currents of the same amplitude and period, but biased relatively each other by certain phase, are threshold currents. The drawback of the given approach is the necessity to use two saw-shaped current generators of high linearity. The linearity of the saw-shaped current generator in considered modulators will mainly determine the linearity of the modulators. The second approach is to use one generator of saw-shaped current $I1$, which will serve as the input current for two-threshold comparator. Input optic signal is converted by photodiode VD1 into the input photocurrent, which serves as the first threshold current $I2$, from which the current $I3$ of the second threshold is obtained by means of adding of small current ΔI, $I3=I2+\Delta I$ (Fig. 18).
The simulation results of the circuit by OrCad PSpice software are shown in Fig. 19. AMIS 1.5 μm technology CMOS transistors with minimal channel length and width (L=1.6μm, W=4μm) are used. The saw-shaped current $I1$ of 10μA amplitude and 10μs period arrives at the input of the first and second comparators, realized on VT4-VT10 transistors. The current ΔI=20nA and correspondingly duration of output voltage pulses equals 20ns. The range of input current is 100nA...10μA that corresponds to input optic power of 200nW...20μW if the photodiode responsivity is 0.5A/W. The PPM has ramp conversion law. As shown from simulation results, there is a big relative conversion error at small input currents because of

output pulse time delay. To reduce the error faster CMOS transistors or greater conversion period or input bias current must be used. The PPM power consumption is 83μW for 3V supply and does not depends on the period value (for values 10μs, 100μs, 1ms). Another variant of PPM on pulse-width modulator (PWM) is shown in Fig. 20. The PWM is realized on current comparator on transistors VT1-VT4, which form up and down current mirrors. Inverters on transistors VT7-VT10 form time-delay line that ensures 30ns delay if C1=1pF. When a saw-shaped current value becomes greater than an input current value the logic "1" voltage (3V for 3V supply voltage) is established at output Out1 and in 30ns at Out2. These signals come to XOR gate that extracts the output phase modulated pulse signal. Simulation results of the circuit for two values of input photocurrent, *I*2=1μA and 8μA are shown in Fig. 21. To reduce the relative conversion error at small input currents the bias current of 5μA is used. The range of input currents is 100nA…10μA. The full conversion period is 12μs. The relative conversion error is 0.1÷1% with taking into account the constant delay time. The power consumption is 83μW for 3V supply. To reduce input current values a current amplifier based on current mirrors at the circuit input can be used. For example, if the current amplification gain is 10 then the range of input currents will be 10nA…1μA that allows to operate with smaller input optic power (20nW…2μW for 0.5A/W photodiode responsivity). The same way can be used to construct PPM on PWM based on λ-devices, which can work with satisfactory parameters at frequencies up to 1MHz (Krasilenko et.al., 2003). The small power consumption of the considered above PPM enables successfully their integration in 2D-array with size of 128x128 elements and more and productivity equals 1…10Giga continuous logic operations per sec. The considered PPMs are realized on CMOS transistors so an array of such elements can easily be integrated with CMOS digital processing block.

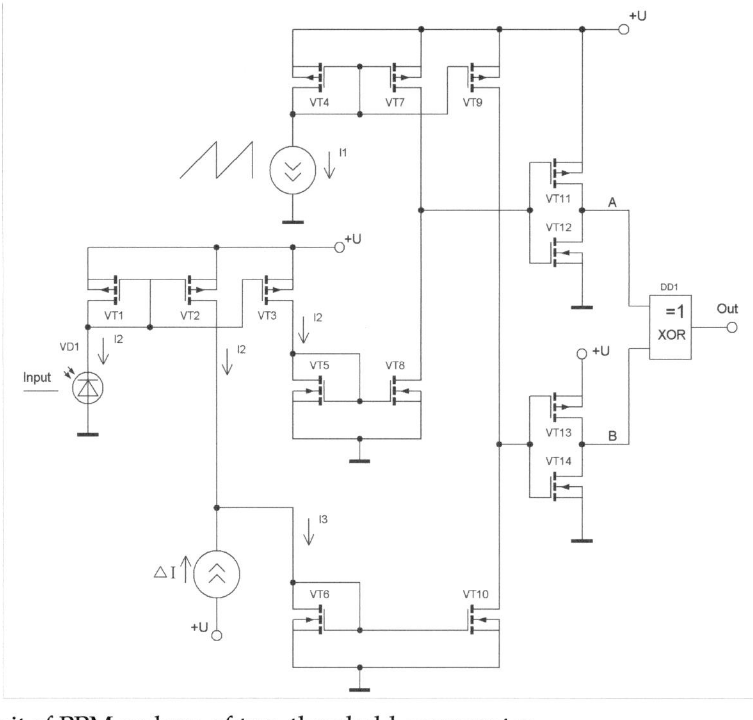

Fig. 18. Circuit of PPM on base of two-threshold comparator

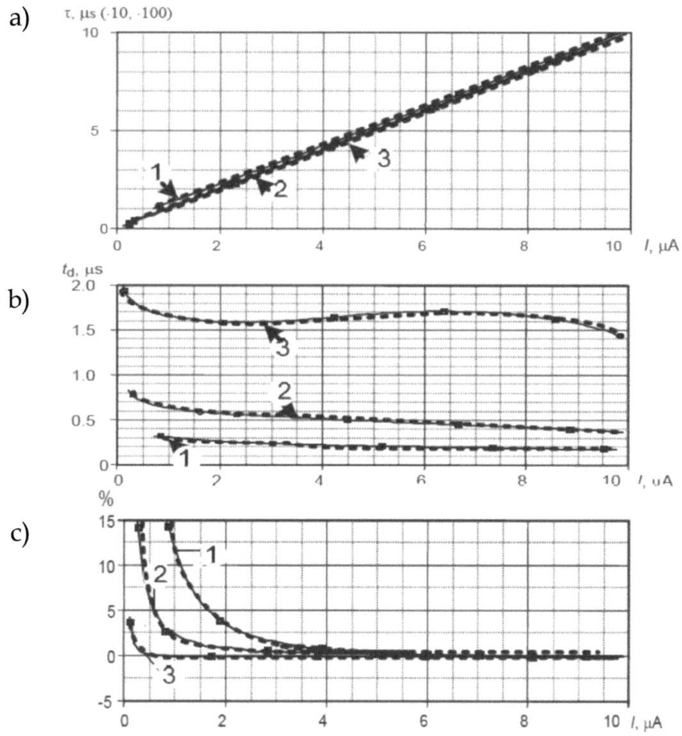

Fig. 19. Dependences of output pulse phase (a), time delay (absolute conversion error) (b) and relative conversion error (c) on input current for three values of conversion period: 1) 10μs, 2) 100μs, 3) 1ms

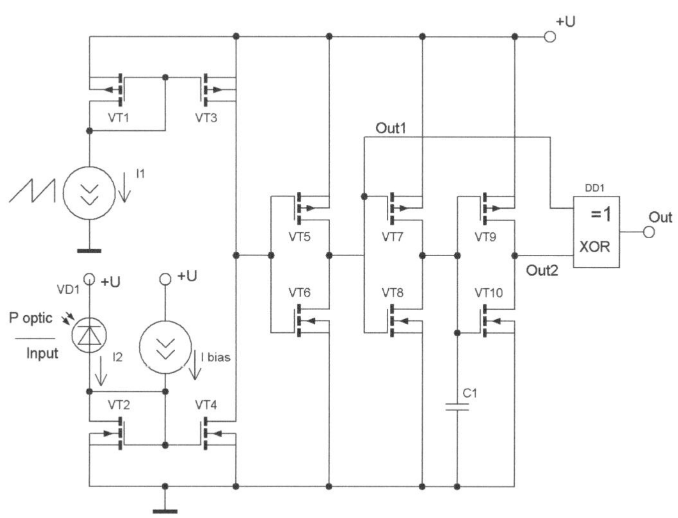

Fig. 20. PPM on base of PWM

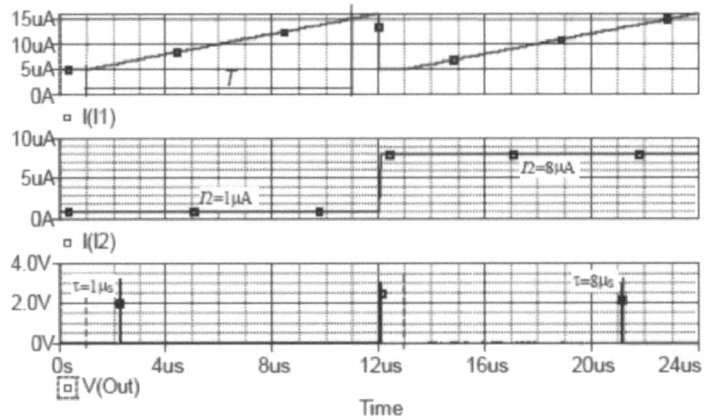

Fig. 21. Simulation results of PPM (Fig. 20)

3.5 Optoelectronic photocurrent logical elements on the basis of current mirror and comparators

3.5.1 Ground of necessity for universal photoreceiver logical elements for scalar-relation vector processors

The use of the mentioned above generalizing operations, among which it is possible to allocate such operations as min, max, restricted product, algebraic product, restricted subtraction and others of analog logics allows to determine not only difference between vectors, but also to discover any other relations between them, including the relations between matrixes. The research and demonstration of design results of optoelectronic scalar-relation vector processors (SRVP) with time-pulse coding are shown in work (Krasilenko et.al., 2005, a). Because the application of the standard bistable electronic logical element technology produces some known problems it is more promising a compromise between optic and electronic. Since at photoreceivers outputs we have current signals it is better to use current logic elements for the signals processing. And so the purpose of the work is design and simulation of current logical elements. At the same time a more wide set of operations ${}^{t,s}e_i$ is probable in such SRVP after elementwise processing over result vector $\vec{E}=\left\{{}^{t,s}e_1,{}^{t,s}e_2,\ldots,{}^{t,s}e_n\right\}$. Therefore in a general view the SRVP can be described by the following model:

$$\Psi_{SRVP}\left(\vec{A},\vec{B}\right)=f^2\left(\vec{E}\right)=f^2\left(e_1\left(a_1,b_1\right),\ldots,e_n\left(a_n,b_n\right)\right)=f^2\left(f_1^1\left(a_1,b_1\right),\ldots,f_n^1\left(a_n,b_n\right)\right), \qquad (9)$$

where f^2 - one of a set of possible functions mapping a vector $\vec{f}^1=\left\{f_1^1,\ldots,f_n^1\right\}$ into Ψ, where Ψ, $f_i\in M=\left[0,1\right]$, i.e. $\left[0,1\right]\times\left[0,1\right]\times\ldots\left[0,1\right]\rightarrow\left[0,1\right]$, and f_i^1 - one of a set of possible functions mapping a vector $\left(a_i,b_i,1-a_i,\bar{b}_i\right)\rightarrow f_i$.

The scalar-relation vector processors considered in the given paper can be applied for decision of problems on the fuzzy logic basis. For such scalar-relation vector processors in each concrete case it is possible to select those or others *s*-norm and *t*-norm, but then the processor will execute only particular procedure. Such approach is perspective a little. Therefore we want to allot our scalar-relation vector processors by universality (or at least quasiuniversality), therefore requirements to such processors the following: changing type of operations at algorithmic steps (f^1 at the first step, f^2 at the second step), we can ensure execution of required relations.

The structural diagrams of the scalar-relation vector processor are shown in Fig.22. It consists of a group of elements ULE_1-ULE_n (Fig.22a), which implement f_i^1 operations, and element of second stage, which implements function f^2 (see formula (9)). And, the type of *t*-norm operation (f_i^1) can vary with the help of special adjusting signals $\vec{C}_1 - \vec{C}_m$, and the type of *s*-norm operation (f^2) can vary too with the help of special adjusting signals $\vec{G}_1 - \vec{G}_p$. At the input of the scalar-relation vector processor the operands like vectors $A=\{a_i\}$ and $B=\{b_i\}$, $\left(i=\overline{1,n}\right)$, and also their complements $A^c=\{a^c{}_i\}$ and $B^c=\{b^c{}_i\}$ come. In structural diagram shown in Fig.22b output signals from universal logical element of two-valued logic (ULE TVL) come to a converter (normalizer) that realizes signals temporal integrations. The time diagram that explains the scalar-relation vector processor operation is shown in Fig.23.

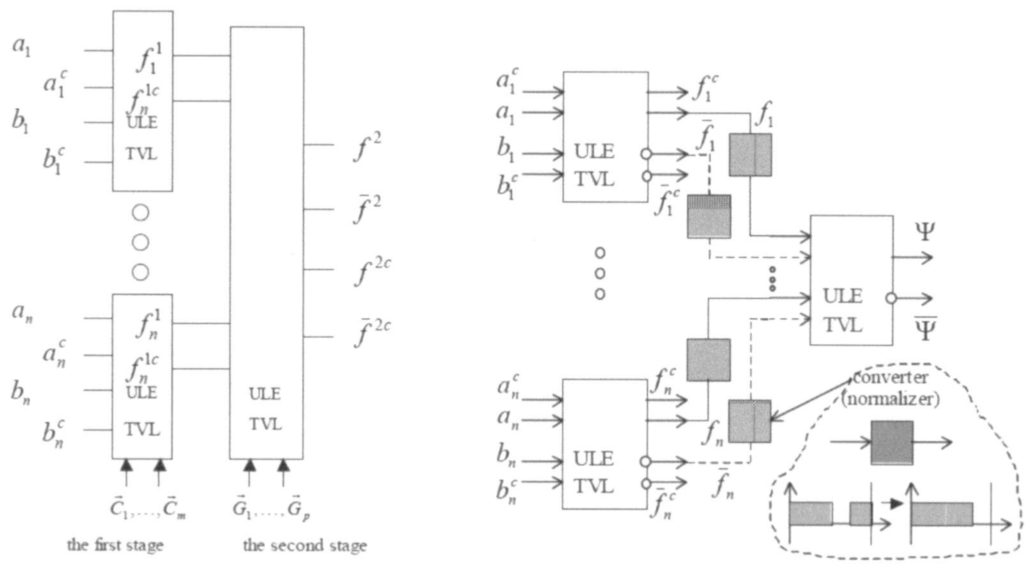

a) b)

Fig. 22. The structural diagrams of the scalar-relation vector processor

A variant of two-valued logic universal logical element realization on the basis of current mirrors comparators (current mirror - CM) is shown in Fig.24. The ULE TVL is, as a matter of fact, a quasiuniversal element of time-pulse analogous logic. Changing the output elements functions by the help of adjusting signals C_{1i} - C_{4i} it is possible to evaluate simultaneously some functions. This circuit is, as a matter of fact, an analog-to-digital converter of parallel type. The time-pulse coding is selected for operands representation (Krasilenko et al., 1995, a), because it allows, using universal logic gates of two-valued logic, to implement the majority of continuous logic functions. Therefore at such coding our problem becomes simpler, as we can, tuning two-valued logic universal logic gates to any of possible two-valued operations with the help of signals C_{1i} - C_{4i}, implement the majority of continuous logic operations and neuro-fuzzy generalizing logic, such as different *s*-norm and *t*-norm.

3.5.2 Designs and simulation results of the logical elements on the basis of current mirrors

Circuits that realize NOT, AND, OR functions with optical inputs-outputs are needed in many cases for max-min image processing. The circuit of an inverter on the bases of current mirrors comparator and its simulation results by PSpice OrCad are shown in Fig.25. The inverter has current (optical) input and current (optical if a proper LED driver will be connected) output. AMIS 1.5μm CMOS transistors MOSIS models were used. The supply voltage is 3V. The circuit shown in Fig.26 can realize both AND and OR functions depending on the reference current value. Both cases simulation results are shown in Fig. 26. Input optical signals and corresponding input currents are discrete and equal to 10μA in the considered case. The number of inputs can be up to 50 and more that is the advantage of the circuit. If the reference current is less than one input discrete than the circuit execute OR function. If the reference current is more than the sum of n-1 input discretes than the circuit executes AND function. If an additional current mirror is connected at the circuit output then we receive the current output. These circuits are based on current mirrors comparator and so their performance depends on the comparators and current mirrors performance. The comparator was considered in our work (Krasilenko et.al, 2004, b) and has the following characteristics: realized on 1.5μm CMOS transistors; the input currents range - 100nA...100μA; the supply voltage - 3...15V; the relative error is less than 0.5%; the output voltage time delays is 10…100ns. The using of more advanced CMOS technologies (0.35μm, 0.13μm and so on) can significantly improve the performance, in particular the operation speed. Time-pulse converters on CMOS current mirrors comparators and on λ-diodes were considered in our previous works (Krasilenko et.al, 2004, b; Krasilenko et.al., 2003). AND, OR and OR-NOT function circuits with current output and its simulation results are shown in fig. 27, 28, 29.

The conception of construction of the family of the offered optoelectronic photocurrent logical elements (OPLE) consists in the use of a few current mirrors realized on 1.5μm technology CMOS transistors. Connection of such mirrors with photodetectors and reference current generators allows also to create circuits realizing the limited difference operations of continuous current logic and current comparators with potential or current outputs. Presence of four - ten transistors, one - two photodetectors makes the offered circuits quite compact and allows their integration in 1D and 2D arrays. The factor of ramification at outputs and the factor of unification at inputs of the offered elements are greatly increased. Further we will consider simulation results of the circuits realizing different Boolean operations including AND, OR, OR-NOT, and others. Thus we will consider two variants of the circuits: 1- circuits with potential outputs; 2-circuits with logical current inputs and current outputs.The simulation results on the 1.5μm technology CMOS transistors showed that the level of logical unit can change from 1 uA to 10 uA for low-power consumption variants and from 10uA to 100uA for high-speed variants. Signals delays, values of fronts and cutoffs at operation with impulse logical signals with 1uA logical unit are not exceed 70-140ns and at operation with impulse logical signals with 100uA logical unit are no more than 4-6ns and the consumption power is 200-400uW.

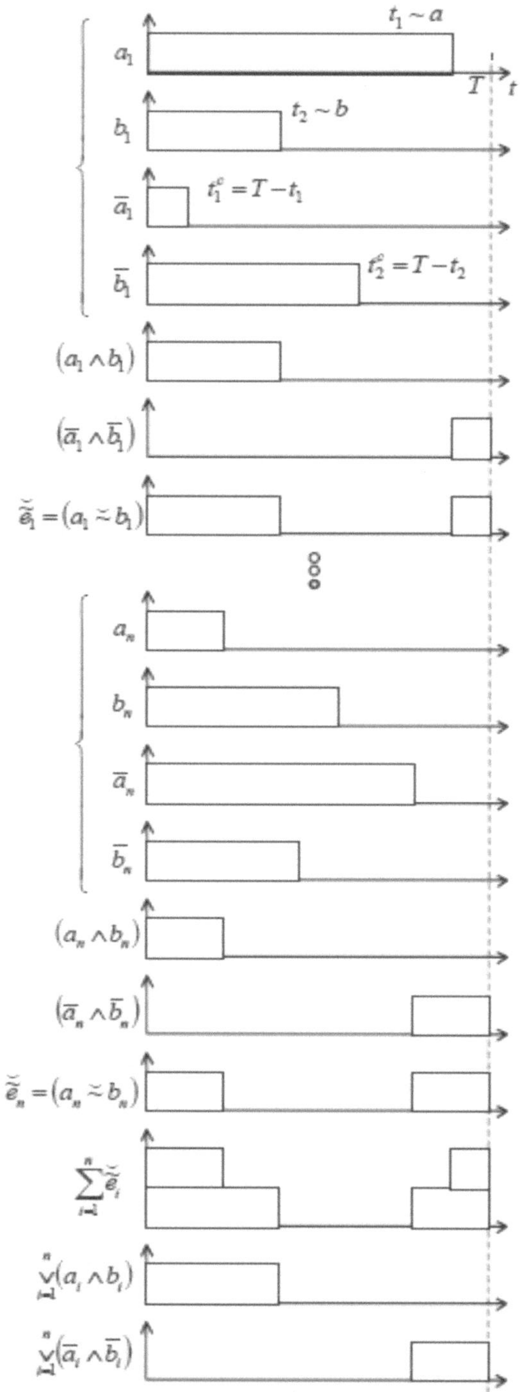

Fig. 23. The time diagram of the scalar-relation vector processor operation

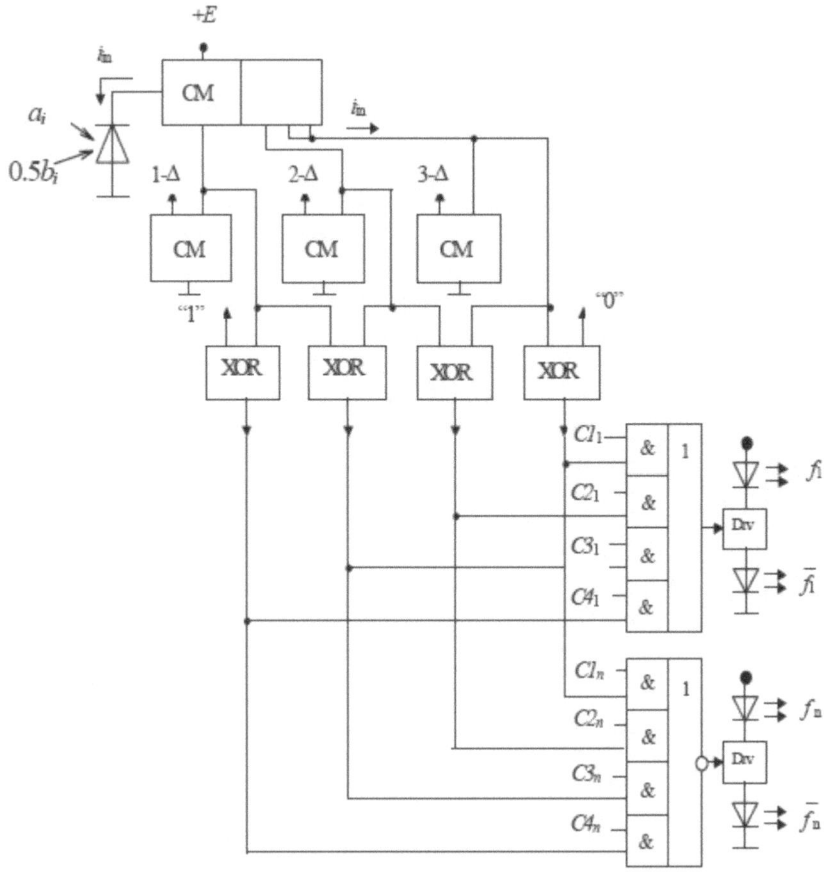

Fig. 24. The universal logical element of two-valued logic

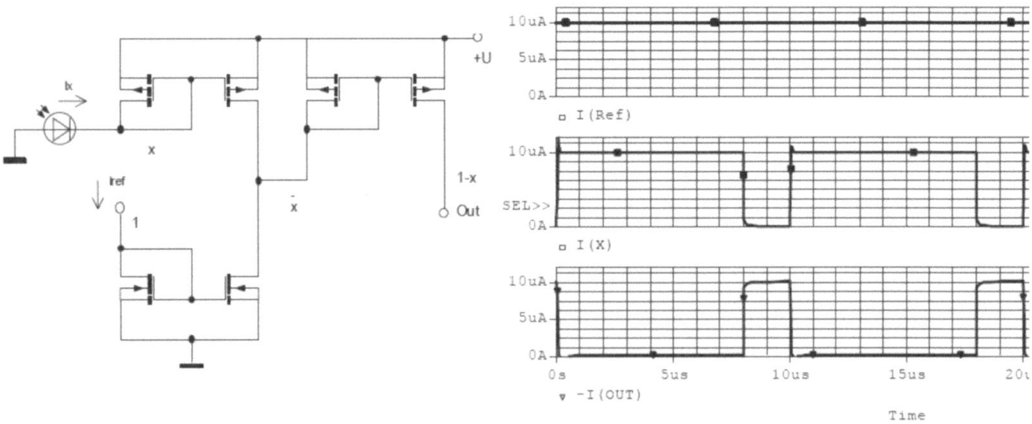

Fig. 25. Inverter circuit on current mirrors and its simulation results

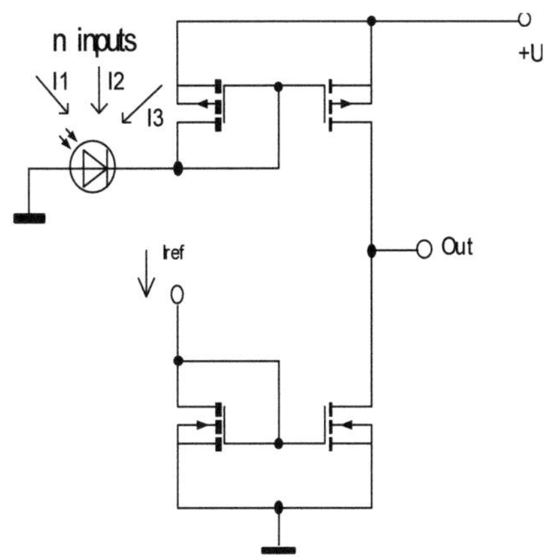

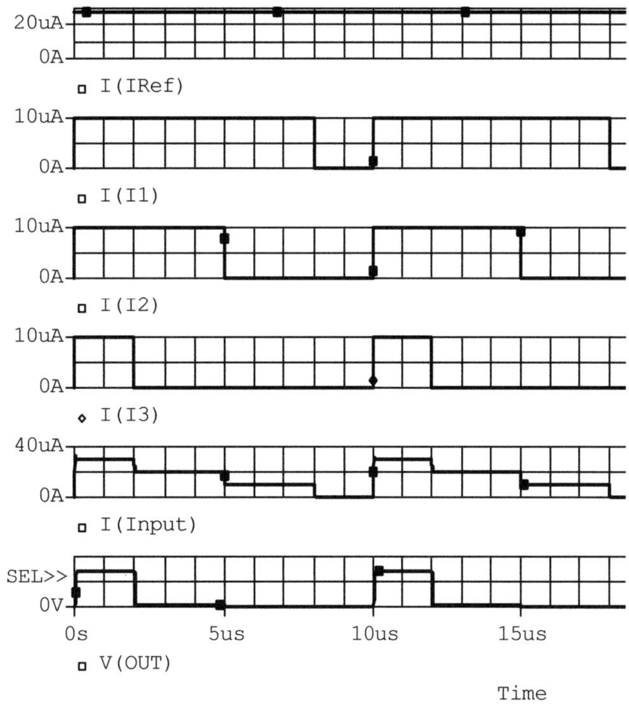

Time diagram of realization of AND function

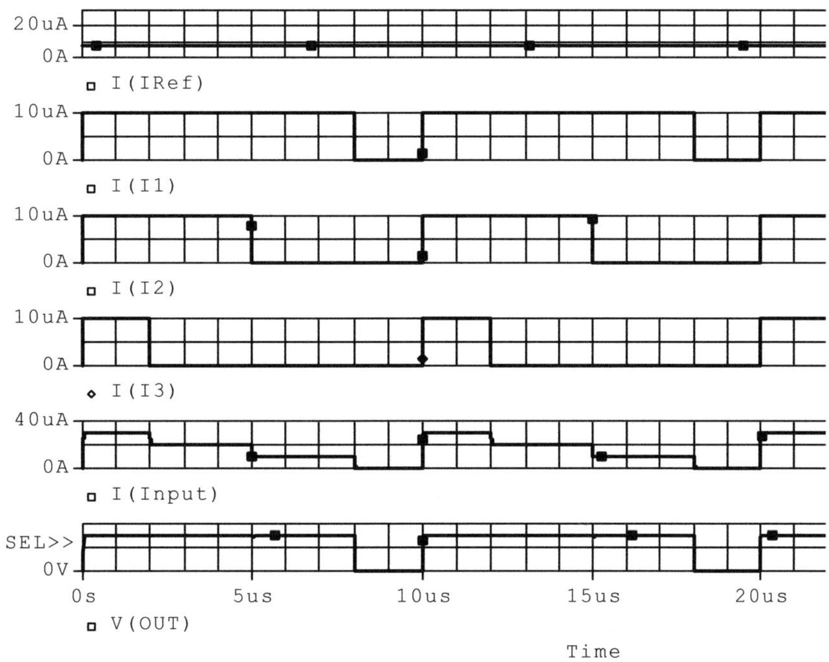

Time diagram of realization of OR function

Fig. 26. AND (OR) circuit with potential output on current mirrors and its simulation results

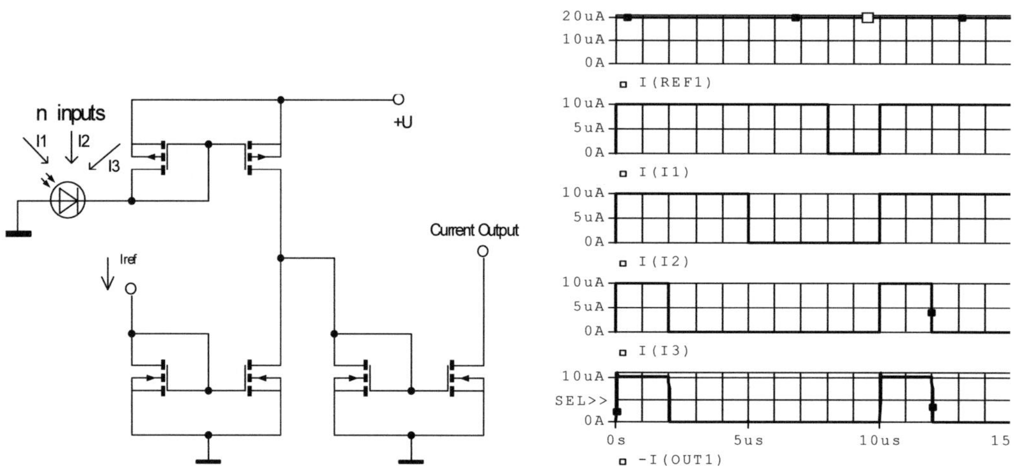

Fig. 27. AND function circuit with current output and its simulation results

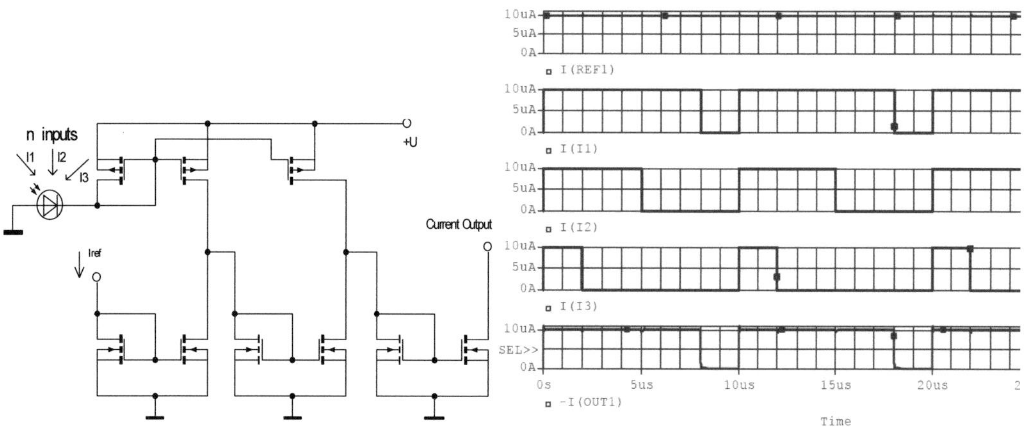

Fig. 28. OR function circuit with current output and its simulation results

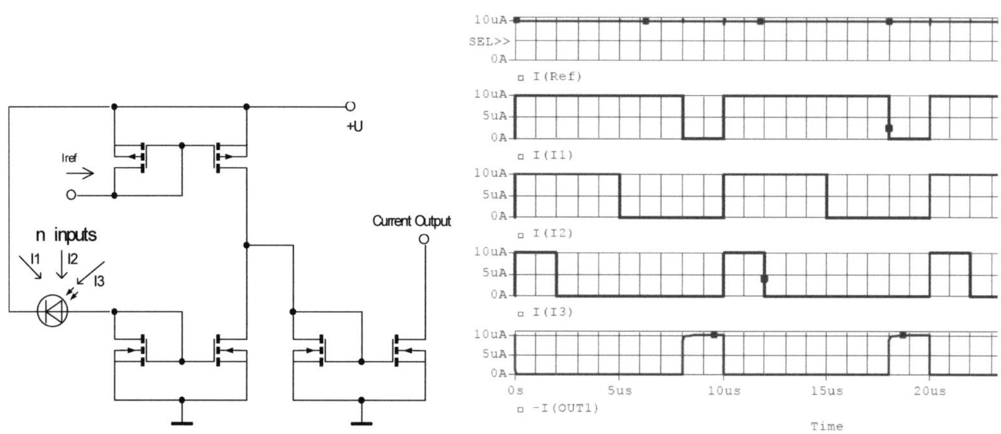

Fig. 29. OR-NOT function circuit with current output and its simulation results

3.6 Spatial-time integrators and equivalence neural elements on their basis

3.6.1 Spatial-time integrators

Realization of integrators on current mirrors on CMOS-transistors is perspective due to their circuit simplicity and compatibility with the well developed CMOS technology of modern integral digital microcircuits. An integrator of the time type consists of two current mirrors (Fig. 30). Transistors VT1 and VT2 are intended for integration (algebraic accumulation) of input signal; VT3, VT4 - for result of integration memorizing. An input optical signal is transformed in photocurrent with photodiode, and the signal controls keys S1 and S2.Then the input signal is on, the key is on. If the signal is off, the key if off. Integration is carried out due to charge of capacities C1 and C2, which hold a value of output current to the end of transformation cycle. Thus the output current value is determined by the sum of input signal pulse durations during the period of integration (Fig. 30,b). For example in Fig. 30 value of input signal pulse duration at the first period of integration is 50μs, at the second period - 100μs, and at the third - 150μs. The output current is proportional to the overall input signal pulse durations, so after the third period the output current value is 5+10+15=30μA. Thus the circuit carried out integration of input signals during three periods. The integration circuit has relative error 1…2 %. Input signals photocurrent range 1-100μA, power consumption 100…500 μW, signal period 0.1…1ms, output current delays - less 1μs Maximal amount of periods, during which the circuit can carry out integration of input signals, is limited to the maximal possible output current. If value of maximal output current is 100μA the maximal amount of periods of integration for this circuit is 10 for 10μA maximal current input signals.

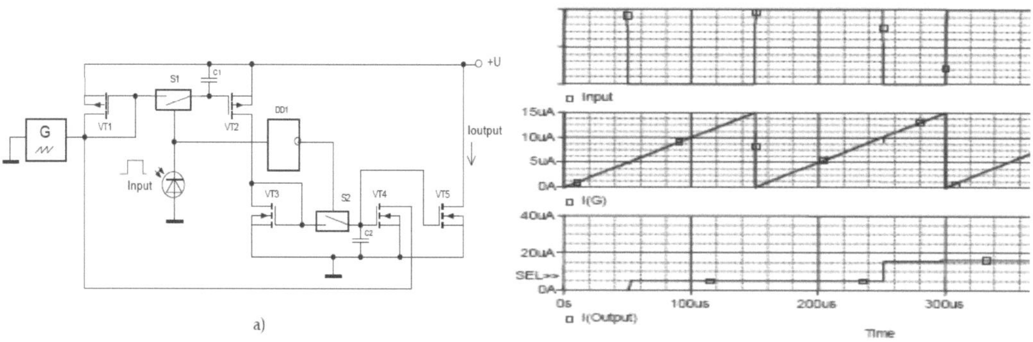

Fig. 30. Integrator circuit (a), time diagrams (b)

In the same way it is simple to provide the spatial integration of optical signals (Fig. 31). Input optical signals are converted in photocurrents with photodiodes and the signals control corresponding keys S1 of selection and storage devices on current mirrors. Accumulative signal of integration is written in the output device at the end of period by the control signal of generator G2. Thus the output current is determined by duration of input signals from every optical channel. For example in Fig. 31 there are three optical channels and input optical signals are 50μs, 20μs, 30μs accordingly. The output current after the first period is determined by the pulse duration sum of input signals and it is equal to 5+2+3=10μA. Maximal amount of input signals is determined by the maximal possible output current. The considered circuits can also perform time integration during some periods, so it performs spatial-time integration. Circuit simplicity and small power consumption allows to create arrays of such integrators with size 10x10 and more. The range of input photocurrent is 0.1-10μA, transformation time (period) 100-1000μs, supply voltage 1.5V, average consumption power 150μW per channel.

3.6.2 Neural elements circuits for equivalence and non- equivalence functions

The family of new functions "equivalence" and "non-equivalence" of neuro-fuzzy logic, which we have elaborated on the based of such generalized operations of fuzzy-logic as fuzzy negation, t-norm and s-norm are shown in paper (Krasilenko et al., 2002, a). The functions are prospective for neural networks creation with selforganization and artificial intelligence properties. New generalized operation of equivalence (non-equivalence) operation can be written in the following form: ${}^{t,s}E'(a,b)=(atb)s(a_n tb_n)=(atb)s(\overline{a}t\overline{b})$, and *t*-norm and *s*-norm of any of their variants can be used. The analysis of the whole spectrum of all possible ${}^{t,s}E'(a,b)$ shows, that the following operations are the most interesting in case of certain *t*- and *s*-norms (Krasilenko et al., 2002, a). New generalized operation of equivalence of II type (non-equivalence) can be written in the following form: ${}^{s,t}E''(a,b)=(asb)t(\overline{a}s\overline{b})$, or taking into consideration the law of De Morgan so: ${}^{s,t}E''(a,b)=\left((asb)_n s(\overline{a}s\overline{b})_n\right)_n=\left((\overline{a}t\overline{b})s(atb)\right)_n=\left({}^{t,s}E'(a,b)\right)_n=\overline{{}^{t,s}E'(a,b)}$

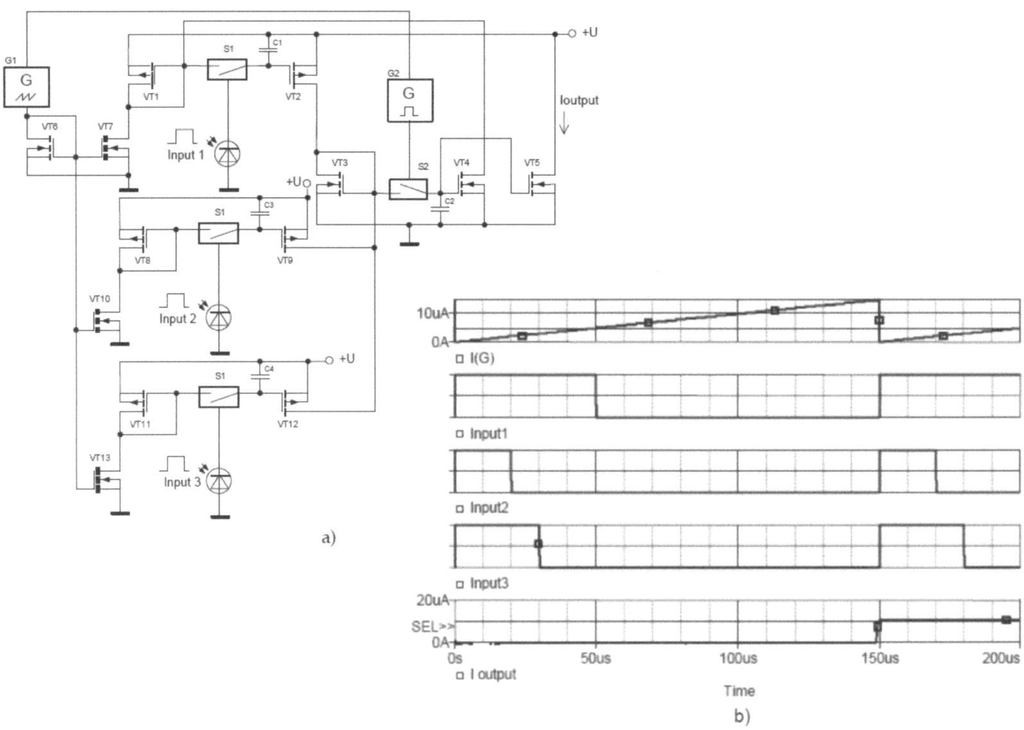

Fig. 31. Spatial-time integrator (a), time diagrams (b)

The II type of operations can be called the operation "non-equivalence" of the I type and designate it as:

$$^{t,s}NE'(a,b) = {}^{s,t}E''(a,b) = \left({}^{t,s}E'(a,b)\right)_n .$$

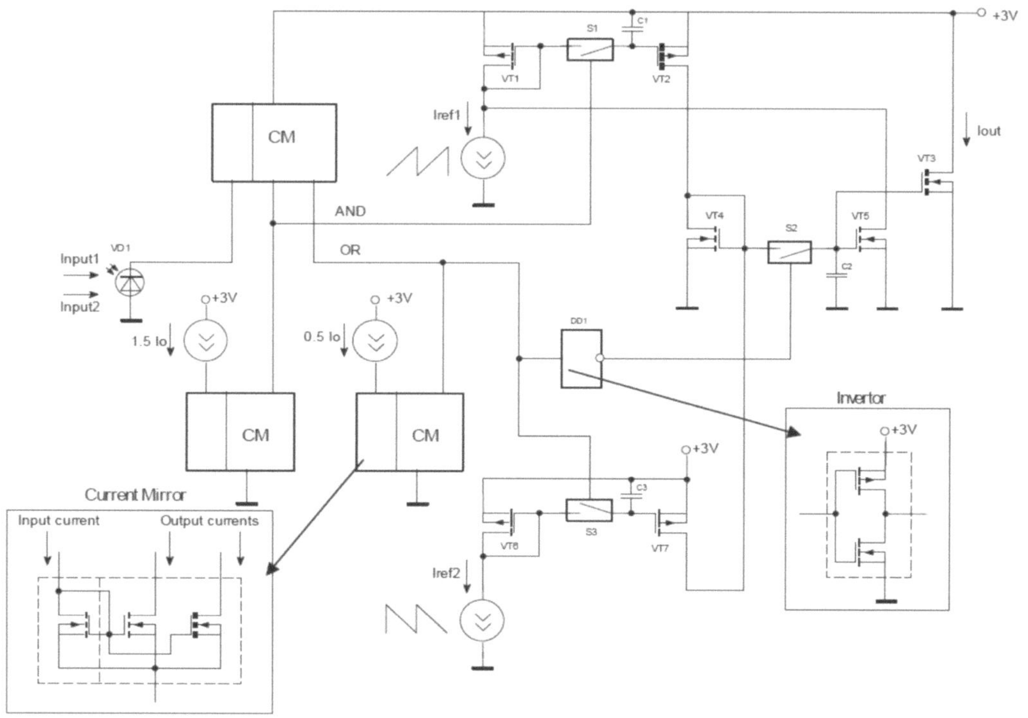

Fig. 32. Equivalence (non-equivalence) function circuit

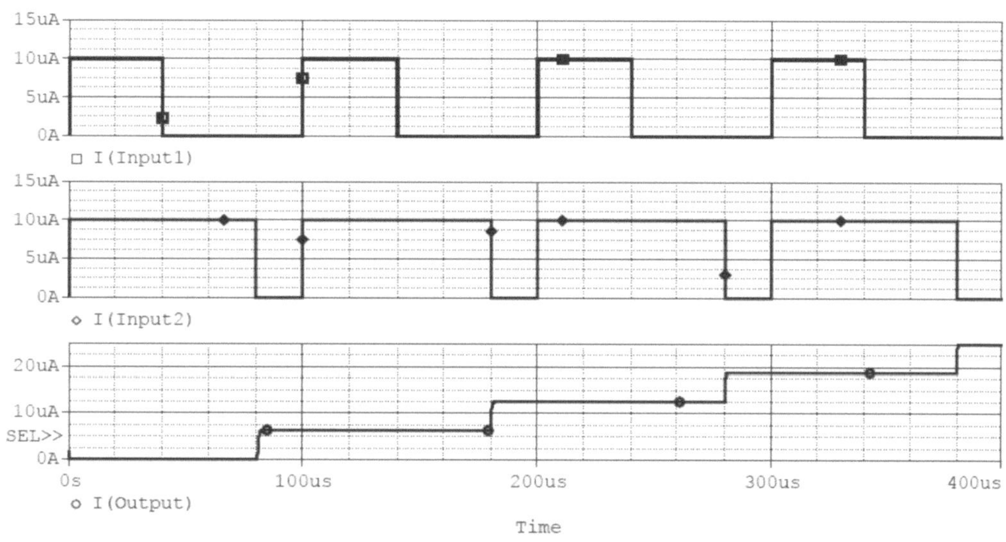

Fig. 33. Time diagrams of the equivalence circuit operation

Thus, the functions determine new generalized operations of comparison (determination of equivalence or non-equivalence).
The equivalence (non-equivalence) functions can be realized on current mirrors and the proposed integrators. Output current value Iout is the equivalence of two input optical signals pulse durations (Fig. 32). In the circuit CMOS current mirrors (CM) with one and two outputs are used. For example in Fig. 33 input signals pulse durations 40μs and 80μs (the equivalence of the input signals is 60μs), period 100μs, Io=10μA, output current after the first period Iout=6uA, after the third period Iout=18μA (Fig. 33), relative error about 3%, as more than one integrators and current mirrors are used and their errors are summed.
To obtain non-equivalence function circuit the reference saw-tooth current generators Iref1 and Iref 2 must be interchanged. For above mentioned input signals the non-equivalence of input signals is 40μs (it is the time then signals are different), so the output current after the first period Iout=4μA, after the third period Iout=12μA.

3.6.3 Weighted and non-linear integration for neural elements circuits for equivalence and non- equivalence functions

The algebraic spatio-temporal weighted integration of optical signals can be realized by the integrator with variable slope angle of saw-tooth current generator G (Fig. 34). For example in Fig. 35, input signal pulse duration 50μs, period 100μs. The saw-tooth current generator G determines integration weight. If maximal saw-tooth current is 10uA than transformation weight is 0.1 μA/μs. So after the first period output current is 5μA. At the second period maximal saw-tooth current is 10μA that corresponds to transformation weight of 0.15 μA/μs. So the transformation weight is in 1.5 times increased. After the second period the output current is 5+7.5=12.5μA. At the third period maximal saw-tooth current is 20μA that corresponds to transformation weight of 0.2 μA/μs. So after the third period the output current is 5+7.5+10=22.5μA. The considered circuit performs only time weighted integration of one input signals. But the circuit in Fig. 31 can perform spatio-temporal weighted integration of multichannel optical input signals if variable angle of slope of saw-tooth current generator G1 is used. The circuits can be used for neural networks adaptations.

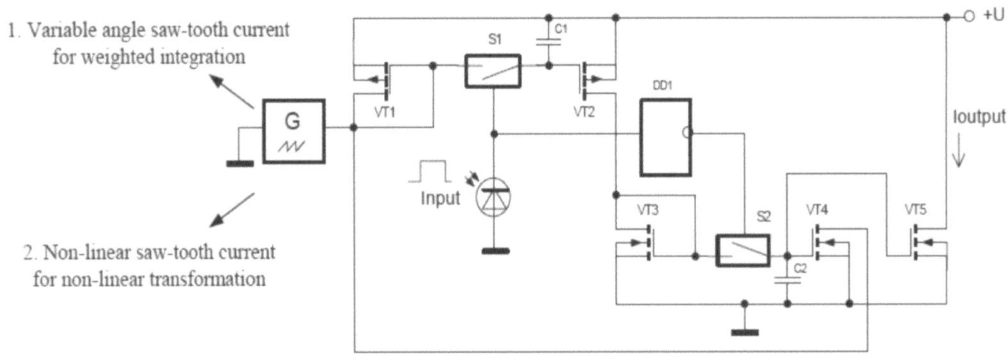

Fig. 34. Weighted or non-linear integrator

Non-linear signal transformation can be realized with the integrator circuit Fig. 34 if a non-linear saw-tooth current generator is used. For example in Fig. 36 input signal pulse durations are 20μs, 50μs, 80μs, and period 100μs. An exponential function for saw-tooth current generator G is used (Fig.36). So we get exponential transformation of input signals.

After the first period the output current is 7.5μA, after the second period the output current is 7.5+9.5=17μA, and after the third period the output current is 7.5+9.5+10=27μA. Thus in spite of identical input signals there are different corresponding output current values. The circuit can be used for non-linear neural networks activation functions. The non-linear transformation signals function is determined by a non-linear function of saw-tooth current generator G.

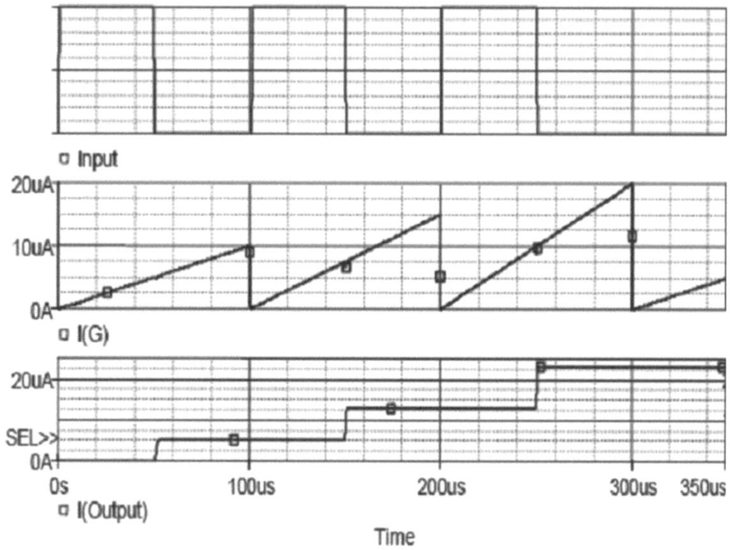

Fig. 35. Time diagrams of weighted integration

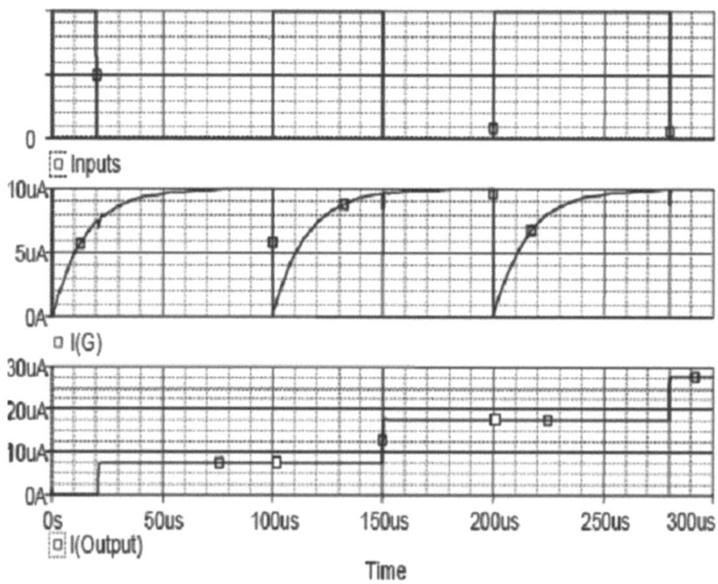

Fig. 36. Time diagrams of non-linear signal transformation

4. Conslusion

We consider design and hardware realizations of optoelectronic logical elements of two-valued logic with current inputs and current outputs on the basis of CMOS current mirrors. In the chapter we consider the whole family of the offered circuits, show the simulation results and possible prospects of application of the circuits in particular for time-pulse coding for multivalued, continuous, neuro-fuzzy and matrix logics. We consider neurons design and simulation results of multichannel spatio-time algebraic accumulation - integration of optical signals. Advantages for nonlinear transformation and integration are shown. The integrator-neuron is based on CMOS current mirrors and comparators. The performance: consumable power - 10...500 μW, signal period- 0.1...5μs, input optical signals power - 0.2... 20 μW; time delays - less 0.01-1μs, the number of input optical signals - 2...64 and more, integration time - 10...100 of signal periods, accuracy or integration error - about 1%-10%. A biologically motivated concept and time pulse encoding principles of continuous logic photocurrent mirrors and sample-storage devices with pulse-width photoconverters have allowed us to design generalized structures for realization of a family of normalized linear vector operations "equivalence"-"nonequivalence".

5. References

Ahadian J.F., Fonstad C.G., Epitaxy on electronics technology for monolithic optoelectronic integration: foundations, development and status, *Opt. Eng.*, 37(12), pp. 3161-3174, 1998.

Antamonov Y.G., Principles of neurodynamics, *Naukova dumka*, Kiev, 1974.

Awwal A.A., Iftekharuddin K.M., Computer arithmetic for optical computing, (Special Section), *Opt. Eng.*, Vol. 38(3), pp. 400-513, 1999.

Bardachenko In. F., Kolesnickiy of O. K., Vasileckiy S., Timer neuron elements and structures. UNIVERSUM-Winnitsa, 2004. - 126p.

Baukens V., Thienpont H., Goulet A. , et al, 2D parallel optical data input for arrays of optoelectronic transceiver elements. *Optical processing and computing* , pp15-16, October 1997.

Berger Christoph, Collings Neil, Gehriger Daniel, Recurrent Optical Neural Network for the Study of Pattern Dynamics, *Proc. SPIE*, Vol. 3402, pp.233-244, 1997.

Freeman James and Skapura D.M., Neural Networks: Algorithms applications and programming techniques, Addison-Wesley Publishing Company, 1992.

Guilfoyle P.S., McCallum D.S., High-speed low-energy digital optical processors, *Opt. Eng.*, 35(2), pp. 436-442, 1996.

Huang Kung-Shiuh, Keith Tenkins B., Sawchuk Alexander A.. Image algebra representation of parallel optical binary arithmetic , *Aplied Optics*.- 1989.- Vol. 28, N6. - P.1263-1278.

Huang H., Itoh M., Yatagai T., Optical scalable parallel modified signed-digit algorithms for large-scale array addition and multiplication using digit-decomposition-plane representation, *Opt. Eng.*, Vol. 38(3), pp. 432-440, 1999.

Klir G.I. and Folger T.A., Fuzzy sets, Uncertainty and Information, Prentice-Hall, Englowood Cliffs, NI, 1998.

a, Krasilenko V.G. etc., Structure of optoelectronic processors of parallel type for image processing, *New electronic installations and devices*, Moscow, 1988.

b, Krasilenko V.G., Optoelectronic structures in information-measuring systems for image processing: Dissertation for the degree of cand. of tech. science. - Vinnitsa, 1988.

a, Krasilenko V.G. et al., Author certificate of USSR N1668984, Arrangement for logical image processing ,*Bulletin of inventions*, N26, 1991.

b, Krasilenko V.G., Kolesnitsky O.K., Dubchak V.N., Creation principles and circuitry construction question of multichannel arrangements and systems for parallel image analysis and processing, *Proc. 1-st All-Union Meeting on Pattern recognition and image analysis*, Minsk, Part 3, pp. 83-87, 1991.

Krasilenko V.G., Dubthak V.N., Author certificate of USSR N1711140, Optoelectronic arithmetic device, *Bulletin of inventions*, N5, 1992.

a, Krasilenko V.G. et al, Creation opportunities of optoelectronic continuous logic neural elements, which are universal circuitry macrobasis of optical neural networks, *SPIE*, Vol. 2647, pp. 208 - 217, 1995.

c, Krasilenko V.G. et al., Lines of optoelectronic neural elements with optical inputs/outputs based on BISPIN-devices for optical neural networks, *Proc. SPIE*, Vol. 2647, pp. 264 - 272, 1995.

e, Krasilenko V.G., Boguhvalsky A.K., Volosovich A.E., The simulation of information processing algorithms in optical neural networks in training and recognition models, *Proc. SPIE*, Vol.2647, pp.307 - 312, 1995.

Krasilenko V. G., Bogakhvalskiy A.K., Magas A.T.. Equivalental Models of Neural Networks and Their Effective Optoelectronic Implementations Based on Matrix Multivalued Elements. *Proc. SPIE*, Vol. 3055, pp. 127-136, 1996.

a, Krasilenko V.G. et al., Continuos logic equivalental models of Hamming neural network architectures with adaptive correlated weighting, *Proc. SPIE*, Vol.3402, pp.398-408, 1997.

b, Krasilenko V. G. et. al. Applications of Nonlinear Correlation Functions and Equivalence Models in Advanced Neuronets. *Proc. SPIE*, Vol. 3317, pp. 211-222, 1997.

Krasilenko V.G., Magas A.T., Fundamentals of design of multifunctional devices of matrix multiciphered logic with fast programmed adjusting, *Measuring and computer technique in technological processes (MCTTP)* , №4,pp.113-121,1999.

a, Krasilenko V.G., Nikolskyy A.I. et al., Demonstration of efficiency of equivalence models of neuron networks with adaptive-correlation tuning of matrix of connections // MCTTP - 2000. - №4.- p. 199-122.

b, Krasilenko V.G., Nikolskyy A.I. et al, Gnoseological approach to search of most general functional model of neuron, Collected research works, №7 (2000) according to the results of 7-th STC "Measuring and computer technique in technological processes", Khmelnitsky, MCTTP, pp. 23 -27, 2000.

a, Krasilenko V.G. Nikolskyy A.I. et al., Optical pattern recognition algorithms on neural-logic equivalental models and demonstration of their prospective ness and possible implementations, *SPIE*, Vol. 4387, pp. 247 - 260, 2001.

c, Krasilenko V. G. and etc. Organization and design of computing structures of matrix-quaternion sign-digit arifmetic, *Measuring and Computer Technique in Technological Processes*, N1, pp. 146-150, 2001.

d, Krasilenko V.G., Nikolskyy A.I et al., Some newest results and achievements of domestic neurobiologic. // *Collection of scientific works*. 8 NTK - 2001.- №8. - p. 320-324.

e, Krasilenko V.G., Nikolskyy A.I. et al., . Research of dynamic processes in neural networks by equivalent system power functions, // *Collection of sciences. labours* №8 2001, VNTK VOTTP-2001, p. 325-330.

f, Krasilenko V.G., Nikolskyy A.I., Ogorodnik K.V., Some advanced results and achievements of domestic neuro-information science, *Proc. the 8-th STC "Measuring and computer devices in technological processes"*, №8 (2001), pp. 320 - 324, 2001.

g, Krasilenko V.G., Nikolskyy A.I., Parashuk A.V., Research of dynamic processes in neural networks with help of system energy equivalence functions, *Proc. the 8-th STC "Measuring and computer devices in technological processes"*, №8 (2001), pp. 325 - 330, 2001.

h, Krasilenko V.G. et al., An approach to design of matrix one-digit processor of quaternary signed-digit arithmetic, *Proc. the 8-th STC "Measuring and computer devices in technological processes"*, №8 (2001), pp.106-111.

a, Krasilenko V.G. et al., The family of neuro-fuzzy logics, their optoelectronic realization and applications, *SPIE* Vol. 4732, pp.106-120, 2002.

b, Krasilenko V.G., Nikolskyy A.I., Pavlov S.N.. The associative 2D memorys based on metric-tenzor acvivalentel models, // *Radio electronics. Informatics. Management.* №2(8), 2002.-- p.45-54.

d, Krasilenko V.G., Nikolskyy A.I. et al., Patent of Ukraine №51813, Node of matrix for comparison of images, *LI* Vol. №12, 2002.

e, Krasilenko V.G., Nikolskyy A.I., Kurdybaxa S.V., Lischenko C.V., Ogorodnik K.V., Designing and research of circuitry for realization of the generalized equivalence (nonequivalence) of neurobiologic operations, //Measuring and calculable technique in technological processes, Collection of scientific labours 9 RTC Vol. №9. - 2002. - T. 1. - pp. 160-165.

f, Krasilenko V.G., Nikolskyy A.I., Bilynska M.Y., Pastushenko A.L., Prospects of liquid crystal structures application in instrumental realizations of neural network matrix-tensor equivalence models (MTEM), *Proc. SPIE*, Vol. 4938, p. 212-222, 2002.

Krasilenko V.G., Nikolsky A.I., Lazarev A.A., Optoelectronic triggers based on λ-devices as advanced components for optical computing arrays, *Proc. SPIE*, Vol. 5104, 2003.

a, Krasilenko V. G., Nikolskyy A. I., Lazarev A. A., Sholohov V. I., The concept of biologically motivated time-pulse information processing for design and construction of multifunctional devices of neural logic // *Proc. SPIE*, Vol. 5421, 2004, pp.183-194.

b, Krasilenko V.G., Nikolsky A.I., Lazarev A.A., Michalnichenko N.N. Smart time-pulse coding photoconverters as basic components 2D-array logic devices for advanced neural networks and optical computers // *Proc. SPIE*, Vol. 5439, 2004, pp.198-209.

a, Krasilenko V.G., Bardachenko V.F., Nikolsky A.I., Lazarev A.A., Kolesnitsky O.I. Design of optoelectronic scalar-relation vector processors with time-pulse coding // *Proc. SPIE*, Vol. 5813, 2005, pp. 333-341.

b, Krasilenko V.G., Nikolskyy A.I., Lazarev A.A., Pavlov S.N. Design and applications of a family of optoelectronic photocurrent logical elements on the basis of current mirror and comparators // *Proc. SPIE*, Vol. 5948, 2005, pp. 426-435.

a, Krasilenko V.G., Yackovska R.O., Yackovskiy V.. «Equivalent model of geteroasociativno memory and design results», // Scientifically methodical collection of materials 8 NPK VSEI UU - 2008.-p. 116-118 b, Krasilenko V.G., Yackovskiy V.I., Yackovska R.O. Leading to and design of functional plenitude of the system of the rationed equivalence (nonequivalence) functions of continuous logic. Materials of Fourth Scientifically practical conference the «Scientific days»-- 2008 T.16. *Technologies*. Sofiya. p.76-78.

c, Krasilenko V.G., Nikolsky A. I., Lazarev A. A., Lazareva M. V., Design of time-pulse coded optoelectronic neuronal elements for nonlinear transformation and integration // *Pгoc. SPIE*, Vol. 6974. - 2008. - pp. 286-297.

a, Krasilenko V.G., Yackovska R.O., Yackovskiy V.I.. «Research and design of equivalent model of getero associative storage», // Scientifically materials of the 9-th conference in VSEI of University Ukraine-2009.-p.59-61.

b, Krasilenko V.G., Nikolskyy A.I., Lazarev A.A. Perfection of charts for realization of the generalized equivalence (nonequivalence) of neurobiologic operations, // Announcer of the Khmel'nickogo national university.-2009.-- №2. *Engineerings sciences*. - p. 174-178.

c, Krasilenko V.G., Nikolsky A.I., Lobodzinska R.F. Design of neurophysiologically motivated structures of time-pulse coded neurons. *Proc. SPIE*, 2009. - Vol. 7343. - P. 341-352.

Krasilenko Vladimir, Bardachenko Vitaliy, Nikolsky Alexander, Lazarev Alexander, Programmed optoelectronic time-pulse coded relational processor as base element for sorting neural networks, // Independent Component Analyses, Wavelets, Unsupervised Nano-Biomimetic Sensors, and Neural Networks V, Harold H. Szu; Jack Agee, Editors, 657610, *Proceedings SPIE* Vol. 6576. p. 597 - 610.

Kuzmin V.B., Travkin S.I., Theory of fuzzy sets in control tasks and organization principles of fuzzy processors (survey of foreign literature), *Automatica and telemechanica*, №11, pp. 3 - 36, 1992.

Leondes G.T., Algorithms and architectures, Vol. 1, *Neural network systems, techniques and applications*, Academik press, USA, 1998.

Levshin V.L., Biocybernetical optoelectronic devices of automatical recognition images, *Mashinostroenie*, Moscow, 1987.

a, Levin V.I., Continuous logic, its generalization and application, Automatica and telemechanica, N8, pp. 3-22, 1990.

Madani K. and Chebira A., Hybrid neural-based decision level fusion architecture: application to road traffic collision avoidance, *Opt.Eng*. 37(2), pp. 370 - 377, 1998.

Mori Masahiko, Toyohiko Yatagai, Optical learning neural networks with two dimensional structure, *Proc. SPIE*, Vol. 3402, pp. 226-232, 1997.

N.V. Posin, Modeling of neural structures, *Nauka*, Moscow, 1970.

Rudenko O.G., Bodyanskiy E. V. Artificial neural networks, Kharkov:. SMIT, 2005. - 408p.

Schmucker K.I., Fuzzy sets, Natural langwage computations and risk analysts, 2 nd ed. Computer science press, Rockville, MD, 1989.

Shimbirev P.N., Hybrid continuouse-logic arrangements, Energoizdat, Moscow, 1990.

Shirman D. et al., Methods of radio-location recognition and their simulation, *Foreign radio electronics*, № 11, pp.3 - 65, 1996.

Swe T.N. and Yeo K.S., An Accurate Photodiode Model for DC and High Frequency SPICE Circuit Simulation, *Modeling and Simulation of Microsystems*, pp. 362-365, 2001.

Volgin L.N., Predicative algebra of choice and its modifications. Vol. 4. Tallinn, Valgus, pp. 64 - 104, 1986.

Zadeh L.A., Fuzzy sets, *Inform Control*, Vol.8, pp.338 - 353, 1965.
Zadeh L., The concept of a linguistic variable and its applications to approximate reasoning, *Information sciences*, V. 8, pp. 199 - 249, V. 9, pp. 43 - 80, 1975.
Wen Zhiging, Yeh Pochi, Yang Xiangyang, Fuzzy neural network for invariant optical pattern recognition, Opt. Eng. 35 (8), pp.2188-2195,1996.

Optical and Electrical Spectrum Analysis of Optoelectronic Devices

Ning Hua Zhu, Wei Chen, Wei Li, Li Xian Wang,
Xiao Qiong Qi and Bang Hong Zhang
Institute of Semiconductors, Chinese Academy of Sciences
P. R. China

1. Introduction

1.1 Introduction of optical and electrical spectrum analysis

Electrical/optical (E/O) and optical/electrical (O/E) devices constitute a cascaded network as shown in Fig. 1.1. Four scattering parameters (S_{11}, S_{12}, S_{21} and S_{22}) of the cascaded network can be easily obtained using a calibrated microwave vector network analyzer (VNA). However, to characterize the individual E/O or O/E device in such a cascaded network, a calibrated O/E or E/O device is required. Even though, other characteristics in both optical and electrical domains can not be extracted from these four scattering parameters. Optical and electrical spectrum analyses with the combination of the VNA sweeping techniques can provide simple and effective approaches to characterize high-speed E/O and O/E devices.

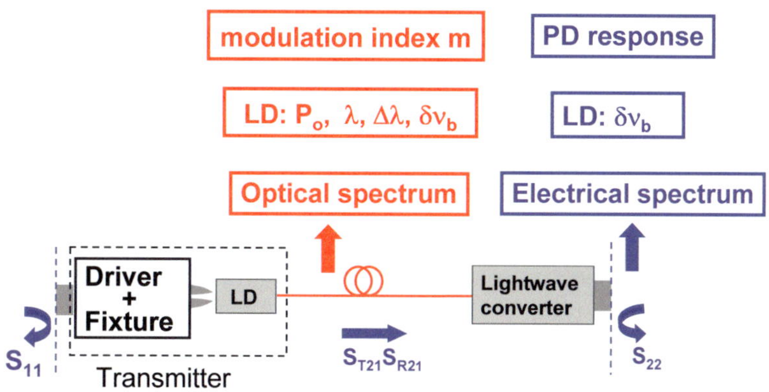

Fig. 1.1. Schematic diagram of optical and electrical spectrum analyses for characterizing optoelectronic devices

From Fig. 1.1, it can be seen that the optical spectrum emitted from the light transmitter and the electrical spectra at the electrical output port of the lightwave converter can be measured by using an optical spectrum analyzer (OSA) and an electrical spectrum analyzer (ESA), respectively. Many important characteristics of optoelectronic devices can be obtained from these spectra. For instance, optical power, wavelength, mode interval and linewidth can be extracted from the optical spectra. Other properties, which are usually measured in the electrical domain, can also be obtained, such as frequency responses of broadband optical external modulators (Yan et al., 2005) and directly modulated semiconductor lasers (Henery et al., 1989; Bjerkan et al., 1996), chirp parameters of light transmitters (Villafranca et al., 2005) and optical injection properties of semiconductor lasers (Simpson et al., 1997). Since optical receiver is not required in the optical spectrum analysis, there is no upper limit in the measurement of frequency responses. Moreover, this technique is simple to implement because only one OSA is used in the measurement. Nevertheless, the resolution of an OSA is usually several tens of picometers, which limits the application of this technique in frequency response measurement at lower modulation frequency.

Characteristics of optoelectronic devices, such as frequency responses of high-speed photodetectors (Zhu et al., 2006), tuning characteristics of tunable lasers, time-resolved optical spectrum (Yu & O'Dowd, 2002), chirp parameters and fibre dispersion coefficients (Kowalski et al., 1999), can be precisely measured by means of electrical spectrum analysis. Compared to the optical spectrum analysis, electrical spectrum analysis is more accurate in analysis of optical spectral properties. For instance, in the optical spectral linewidth measurement, a resolution of 1 mHz can be achieved theoretically (Zhu et al., 2007).

According to the above discussions, optical and electrical spectrum analyses extend the applications in optoelectronic device characterization. Meanwhile, the combination of these two techniques would bring a new perspective in optoelectronic device designing and testing. This chapter is organized as follows. In Section 2, the techniques in characterizing optoelectronic devices in the electrical domain are introduced. In Section 3, the techniques in characterizing optoelectronic devices in the optical domain are introduced. In Section 4, the optical spectrum analysis is used to investigate the optical multi-beam injected semiconductor lasers, followed by the conclusion and prospect in Section 5.

2. Characterization based on electrical spectrum analysis

Along with the development of the used optical spectrum analyzer communication system, characterizing optoelectronic devices accurately and rapidly becomes more and more important. In the past decade, reasearchers have dedicated to develop techniques based on electrcial domain ananlysis because of its high resolution. In this sectin, we will introduce some of the recent progress of these techniques.

2.1 Spectral linewidth measurement

The widely used optical spectrum analyzer uses a scanning diffraction grating as frequency selective filter and its resolution is limited to tens of picoseconds (>1GHz), whereas, the linewidths of distributed feedback (DFB) laser and distributed Bragg reflector (DBR) laser are of the order of 10MHz and that of fiber laser is even narrower than 1 kHz. Since it is difficult to improve the resolution bandwidth of the optical spectrum analyzer, the linewidth measurement mainly takes advantage of optical heterodyne technique. The main idea is transferring the detect domain from optical frequency to microwave frequency and calculating the optical spectral 3 dB linewidth by measuring the electrical spectral linewidth and analyzing the spectral lineshape. Recently Agilent, Tektronix and R&S companies all have electrical spectrum analyzer with the resolution bandwidth lower than 10 Hz and the real-time electrical spectrum analyzer improves the resolution to 1 mHz. Theoretically, optical heterodyne method can be used to measure the laser linewidth of the order of millihertz.

In the following we will introduce several methods of linewidth measurement by using electrical spectrum analyzer.

1. Dual-beam optical heterodyne method

This method needs a stable and linewidth-known reference tunable laser. For two beams of Lorentzian lineshapes, the linewidth of their beat signal is the sum of that of the two optical spectra. For both Gaussian lineshapes, the squared linewidth of the beat signal is the sum of the squared linewidths of the two beams. So, if the linewidth of the reference laser is far narrower than that of the under test laser, the linewidth of the beat signal can be estimated to be that of the under test laser.

The dither between the reference laser and the under test laser limits the measurement accuracy.

2. Self-homodyne method

The under test lightwave is injected to a Mazh-Zehnder interferometer and split into two unequal arms. The beat signal is at zero frequency. If the delayed fiber line is much longer than the coherence length, the measured 3 dB linewidth of the self-homodyne signal is two times of that of the under test laser.

In this method, the amplitude noise and phase noise are mutually overlapped. The important linewidth information is, however, connected with the phase noise.

3. Self-heterodyne method

To distinguish the amplitude and phase noise spectra, it is better to shift the beat signal to none-zero frequency. The simplest way is to add an acousto-optical frequency shifter to one arm. Thus, the beat signal could be detected by a standard detector, but the inducing of the shifter will cause an optical loss.

In practice, frequency and polarization fluctuations of the under test laser will affect the linewidth measurement. So we can say that it is hard to accurately measure the laser linewidth when frequency and polarization jitters are serious.

2.2 Frequency response measurement for high-speed photodetectors

Wide-bandwidth photodetectors are essential devices for very high-speed optical transmission and signal processing systems. Many methods for measuring the frequency responses of photodetectors have been proposed, such as the swept frequency method (Humphreyst, 1989), the pulse spectrum analysis method (Shao & Gallawa, 1986), the interferometric FM sideband method (Ohishi et al., 1989) the optical heterodyne detection method (Hawkins et al., 1991), the intensity noise technique (Baney et al., 1994).

2.2.1 Improved self-heterodyne method

Among these methods, optical heterodyne method has drawn more and more attention of engineers due to its advantages. In the conventional optical heterodyne method, two lasers with the requirement of precise wavelength (less than 0.5 nm) and polarization mode matching are utilized. One of the lightwave sources is lasing at fixed wavelength and the other one is tuned continuously over the whole measurement range. However, the wavelength stability of both lasers is rigorous because any slight drifts in the operating condition of the lasers can cause unwanted frequency and power fluctuations which affect the accuracy of measurements of frequency responses. All of these problems can be avoided by using the self-heterodyne method (Hou et al., 1989).

The measurement setup is shown in Fig. 1.2 (Zhu et al., 2006). In the measurement, a 1.55μm four-section DBR laser is used. The two Bragg sections are shorted. The output wavelength is tuned by the current applied to the phase section and is measured by an OSA. With the application of a square wave voltage, the tunable DBR is operated in two states corresponding to high and low levels of a square-wave voltage. The low level of the square wave voltage is fixed, corresponding to a fixed optical wavelength. The high level of the square wave voltage is tuned, producing a tunable optical wavelength. Therefore, the laser switches periodically between two optical wavelengths.

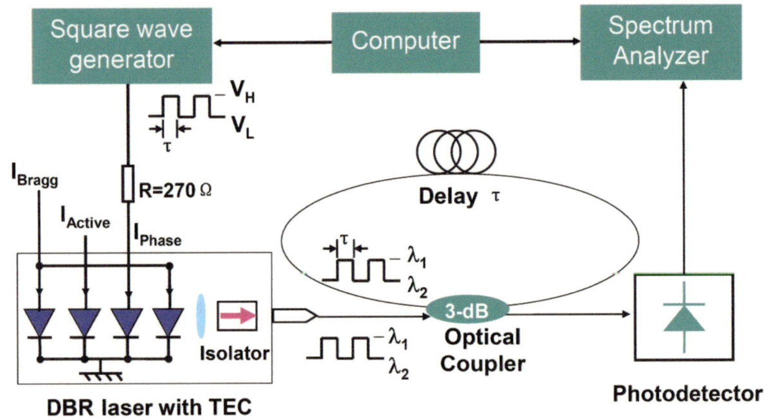

Fig. 1.2. Measurement setup of self-heterodyne method for characterizing frequency responses of photodetectors

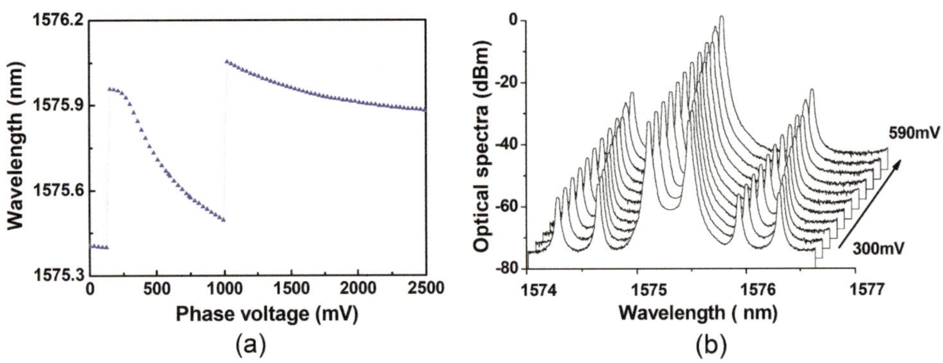

Fig. 1.3. (a) Measured optical wavelength and (b) spectra as functions of the phase voltages

Part of the optical power is delayed by half of the square wave voltage period after passing through the fiber ring interferometer. The delayed and undelayed light waves beat in the photodetector to generate a microwave signal. The center frequency of the beat signal equals to the wavelength difference between the two wavelength states. The tuning properties as functions of the phase voltage are shown in Fig. 1.3. It can be seen that the wavelength continuously decreases with the applied phase voltage, and the frequency sweeps from DC to tens of gigaherz. Fig. 1.3(b) shows the optical spectra when a square-wave voltage is applied, in which the high voltage level is fixed, and the low voltage level varies

The experimental results are shown in Fig. 1.4. The measured linewidths are between 30 MHz and 100 MHz. The trace of the beat signal peaks represents the measured frequency responses of the photodetector. However, this response still includes the influences of the fluctuations of the optical power and the linewidths of the beat signals.

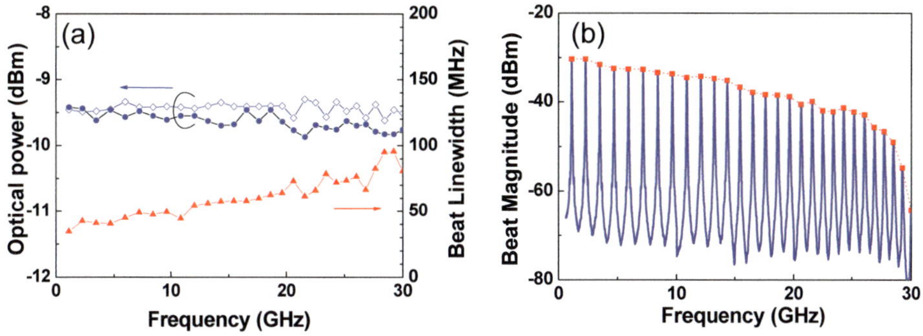

Fig. 1.4. (a) Measured linewidths of the beat signals and optical peak powers of two wavelengths. (b) Measured magnitudes of the beat signals, where V_L is fixed

Assuming the line shape is Lorentian, the peak power of the beat spectrum measured using an ESA can be expressed as (Zhu et al., 2006):

$$P_b(\nu_b) \propto (\frac{e\eta}{h\nu})^2 \cdot E_1^{\ 2} \cdot E_2^{\ 2} \cdot R_L \cdot F^2(\nu_b) \cdot \int_{\nu_b - B/2}^{\nu_b + B/2} \frac{\delta\nu_b}{2\pi\left\{(\nu-\nu_b)^2 + (\frac{\delta\nu_b}{2})^2\right\}} d\nu$$
$$\propto \frac{2}{\pi} \cdot (\frac{e\eta}{h\nu})^2 \cdot E_1^{\ 2} \cdot E_2^{\ 2} \cdot R_L \cdot F^2(\nu_b) \cdot \arctan\frac{B}{\delta\nu_b} \quad (1)$$

where h is the photon energy, e is the electron charge, η is quantum efficiency, E_1 and E_2 are the optical fields corresponding to the voltages V_H and V_L applied to the phase section, respectively. $F(\nu_b)$ is the frequency response coefficient of the photodetector, B and R_L are resolution bandwidth of the ESA and the input impedance, respectively. Therefore, the frequency response of the photodetector can be obtained as follows:

$$F(\nu_b) = C \cdot \left| \frac{P_b(\nu_b)}{E_1^{\ 2} E_2^{\ 2} \arctan(B / \delta\nu_b)} \right|^{\frac{1}{2}} \quad (2)$$

As the amplitudes of driving signal increase, the beat signal may not be a Lorentian line shape. In this case, the calibration should be made using the actual lineshape, and (2) becomes:

$$F(\nu_b) = C \cdot \left| \frac{P_b(\nu_b)}{E_1^{\ 2} E_2^{\ 2} \int_{\nu_b - B_0}^{\nu_b + B_0} S_b(\nu) d\nu} \right|^{\frac{1}{2}} \quad (3)$$

where B_0 and $S_b(\nu_b)$ is the 10-dB linewidth of the beat signal and the power spectrum of the beat signal respectively. From the measurement results, on can see that, the results obtained using both sweeping methods are similar after calibration, and agree well with the Agilent data.

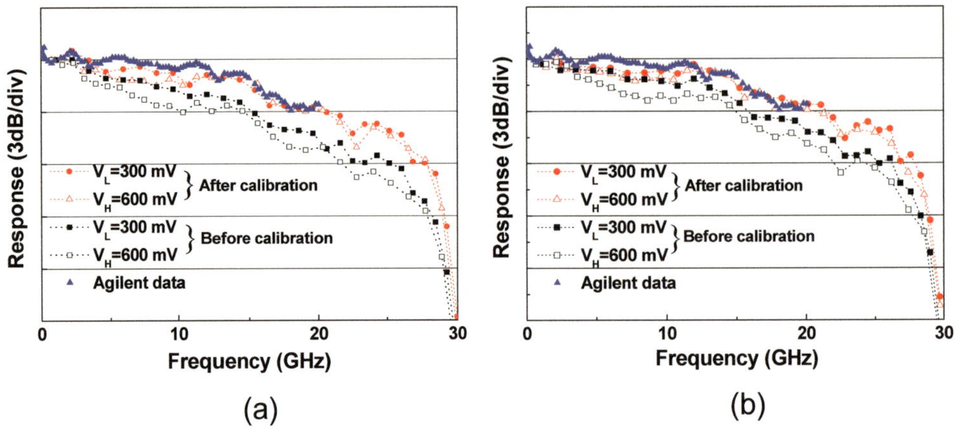

Fig. 1.5. Measured frequency responses of the photodetector from Agilent when the Bragg sections are (a) shorted and (b) biased at 12 mA

2.2.2 Harmonic analysis technique

In this section, the swept frequency method for measuring the frequency response of a PD based on harmonic analysis is introduced. The schematic diagram is shown in Fig. 1.6. The output of a laser is modulated by a swept frequency RF signal via a Mach-Zehnder $LiNbO_3$ modulator, and then detected by a photodetector. The 2nd-order harmonic of the RF signal contains information of the frequency responses and the nonlinearities of the RF source,optical modulator and PD. It has been shown that the small-signal power measuring technique can be used to measure the frequency response of the RF source and modulator (Huang & Zhu, 2006). Therefore, by deducting the known frequency responses and the nonlinearities of the RF source and modulator, the frequency response of the photodetector can be obtained (Zhang et al., 2009).

The optical output of an $LiNbO_3$ modulator can be written as:

$$\begin{aligned} P_{\text{out}} = \frac{1}{2} P_0 \{ & 1 + \cos\phi_0 J_0(F(\omega_{\text{m}})) \\ & + 2\cos\phi_0 \sum_{n=1}^{+\infty} (-1)^n J_{2n}(F(\omega_{\text{m}})) \cos(2n\omega_{\text{m}} t) \\ & + 2\sin\phi_0 \sum_{n=1}^{+\infty} (-1)^n J_{2n-1}(F(\omega_{\text{m}})) \cos[(2n-1)\omega_{\text{m}} t] \} \end{aligned} \quad (4)$$

where $F(\omega_m)$ is the frequency response of the RF source and modulator. The detected power of 2nd harmonic can be expressed as follows, in which $M_p(\omega_m)$ denotes the frequency response of the photodetector.

$$P_{2\omega_{\text{m}}} \propto J_2^2(F(\omega_{\text{m}})) M_{\text{p}}^2(2\omega_{\text{m}}) \propto F^4(\omega_{\text{m}}) M_{\text{p}}^2(2\omega_{\text{m}}) \quad (5)$$

When drive signal consists of a single tone at a low frequency ω_1 low enough that $M_p(\omega_1)=1$, the detected power of the signal at ω_1 is (Huang & Zhu, 2006):

$$P_{\omega_1} \propto i_{\omega_1}^2 \propto F^4(\omega_m) \quad (6)$$

Substituting (6) into (5), the frequency response of the PD can be easily obtained by normalizing the results. Fig. 1.7 shows the results when a 10 GHz modulator is employed. It can be seen that the data after calibration fit well with the reference data. We can also see the measurement range is doubled using harmonic analysis method compared to that using the conventional swept frequency technique.

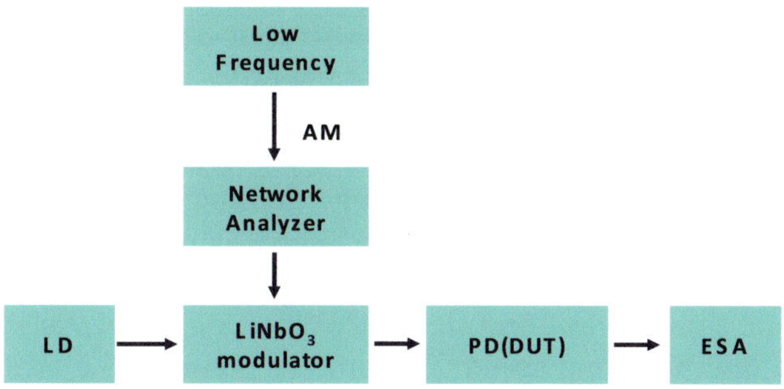

Fig. 1.6. Schematic diagram of the measurement setup

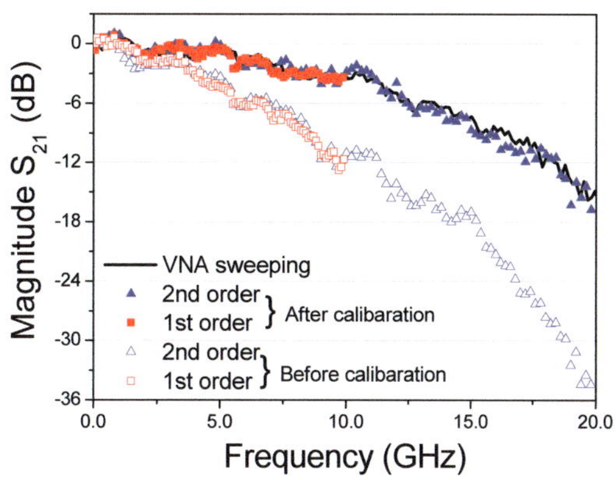

Fig. 1.7. Measurement results

2.3 Tuning characteristics of tunable lasers measurement

The lightwave wavelengths of semiconductor lasers vary with bias current and environment temperature. The central wavelength can be easily obtained from optical spectrum analyzer. With the bias current tuned, the wavelength variation of normal lasers is actually due to the change of junction temperature. It is the heat induced change in refractive index that leads to the changes in wavelength. This period is relatively slow and the response time is of the order of milliseconds. However, the lightwave wavelength of DBR tunable laser varies fast with phase section current, and the response time is of the order of nanoseconds. In practice, fast tunable semiconductor lasers are considered to be the essential devices in optical packet switching technique which is one of the key techniques of the next optical communication network. The prerequisite of such lasers is the ability of fast switching in the limitation of the order of nanometer. Thus, it is important to study the instantaneous properties of the tunable laser. On the one hand, by studying the instantaneous characteristics one can achieve the wavelength switching time and the optimum time for the wavelength-stable output. These parameters are important to judge whether a system meets requirement. On the other hand, linewidth broadening is also a dynamic process. The linewidth of the tunable laser in the tuning process is quite different from that in steady state. The linewidth broadening will induce dispersion which will deteriorate the transmission. Therefore, deeply studying the instantaneous characteristics of the tunable laser and the influence factors is critical to the assessment and improvement of device properties.

The resolution of optical spectrum analyzer is of the order of 0.01 nm, which can not meet the requirement of measuring the dynamic process, whereas by combining optical and electrical spectrum analyses we can easily solve this problem. We introduce two methods to measure the tuning speed. One is using optical filter to filter out one wavelength. Then after detected by a photoelectric detector, the electric signal is observed using an oscilloscope. Another method is using a reference laser and taking advantage of optical heterodyne technique to generate a microwave beat signal. By detecting the beat signal's varying speed one can obtain the optical tuning speed. The time response of the beat signal is detected using "zero span" function of the electrical spectrum analyzer.

2.4 Time-resolved optical spectrum measurement

In many applications, such as the transmission of AM and FM optical signal, channel switch in FDM system, the transient characteristics of laser wavelength should be always taken into account. Researchers have paid much attention in developing techniques to measure the time-resolved optical spectra. One of them (Bergano, 1988) is based on an optical filter that is used as a discriminator. The absorption coefficient of the filter is a function of the optical wavelength, so that the variation of wavelength can be converted into intensity variation. In this technique, the pass band of the optical filter is relatively wide. Another method (Yu & O'Dowd, 2002) also uses an optical filter, but its pass band is narrow and only one wavelength can be selected. After passing through this filter, an oscilloscope is used to observe the intensity variation of the output. Commonly the signal under test is periodic, so that when the frequency centre of the optical filter is changed, optical signals with other frequencies can be measured. By combing all the data, the time-resolved optical spectrum can be obtained. The schematic setup of these two methods is shown in Fig. 1.8(a).

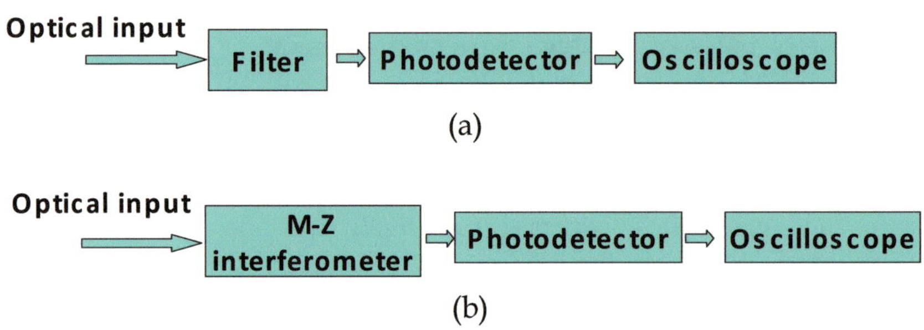

Fig. 1.8. Schematic setup for measuring time-resolved optical spectra

There is also a method that uses self-heterodyne technique (Joseph & Sadot, 2004), which measuring the time-resolved optical spectra in electrical domain. The input optical signal is divided into two branches, one of which is delayed. Then the beat signal of the two signals is measured, which is equivalent to obtain the autocorrelation function. After some calculation, the time-resolved optical spectrum can be deduced. The schematic setup is shown in Fig. 1.8(b).

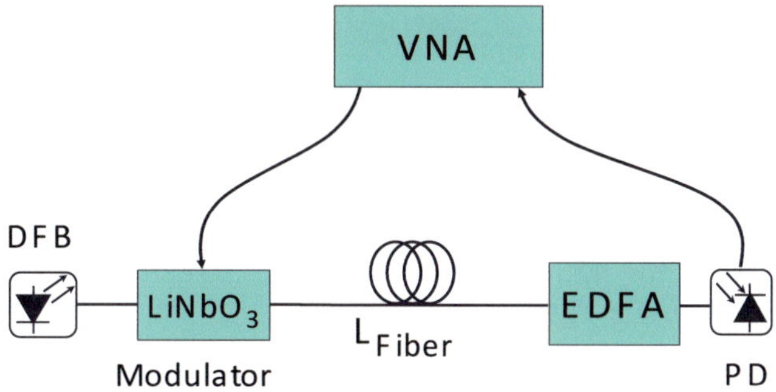

Fig. 1.9. Experiment setup for measuring the chirp parameter and fibre dispersion

2.5 Measurement of chirp parameter and fiber dispersion

Almost all of the laser sources, including directly modulated semiconductor laser, electroabsorption modulator, $LiNbO_3$ modulator, have the characteristic of frequency chirp. The chirp parameter indicates the relationship between the intensity variation and the phase variation:

$$\delta\phi = \frac{\alpha\delta\phi}{2I} \tag{7}$$

I is the optical intensity, Φ denotes the phase of the laser, a is the chirp parameter.
The chirp of the laser source together with the chromatic dispersion of single mode fibre (SMF), which is simply a variation in the speed of propagation of a lightwave with a wavelength, can severely affect the performance of optical communication systems based on fibre. Consequently, characterizing these two parameters accurately plays an important role in designing fibre based communication system.

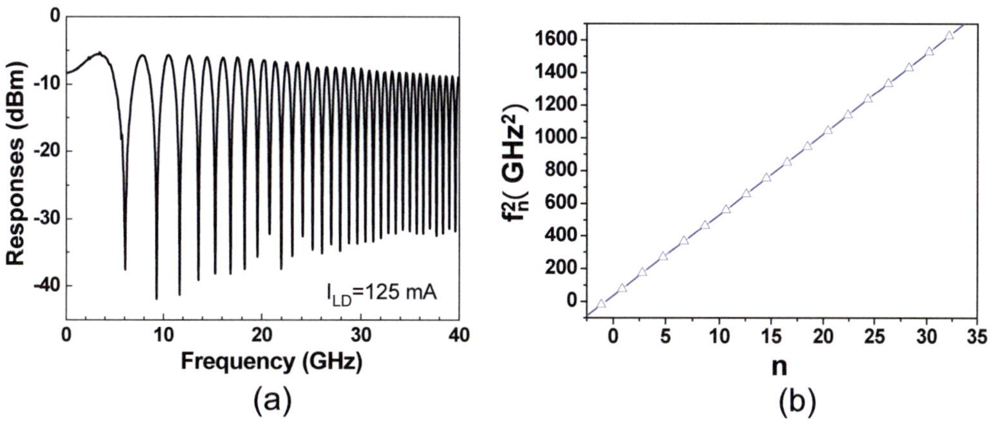

Fig. 1.10. (a) Calibrated fibre transfer function. (b) The measured results according to (9)

By using the sweep method based on VNA, the small-signal chirp parameter and the dispersion coefficient can be measured simultaneously (Kowalski et al., 1999). The fibre transfer function can be expressed as follows:

$$I_f \propto \left| \cos\left(\frac{\pi\lambda^2 DLf^2}{c} + \arctan\alpha\right) \right| \tag{8}$$

Firstly, the long fibre under test is replaced by a short length of fibre and the optical-to-optical calibration is implemented to remove the impact induced by the VNA, the modulator and the photodetector. Then the scattering parameter S_{21} of the long fibre is measured. The square of the frequency f_n according to the nth zero pole is proportional to the reciprocal of the dispersion coefficient:

$$f_n^2 L = \frac{c}{2D\lambda^2}\left(1 + 2n - \frac{2}{\pi}\arctan\alpha\right) \tag{9}$$

The measurement setup is shown in Fig. 1.9. The results are shown in Fig. 1.10. As shown in Fig. 1.10(b), we can obtain the fibre dispersion coefficient and the chirp parameter from the slope and the intercept respectively according to (9). This method is accurate and simple. The measurement uncertainty is about ±0.1.

3. Characterization based on optical spectrum analysis

In this section, we introduce the techniques for characterizing optoelectronic devices based on optical spectrum analysis, which can be easily carried out with simple experimental setups. The frequency responses of the lasers or modulators can be extracted from the measured optical spectrum.

3.1 Frequency responses of broadband modulators

We use sideband calculating method which is proposed by Auracher and Keil for the first time (Auracher & Keil, 1980), and developed by Shi and Willner *et al.* (Shi et al., 2003; Yan et al., 2005; Yan et al., 2003; Kawanishi et al., 2001; Oikawa et al., 2003), to obtain the frequency responses of modulators based on optical spectrum analysis. The sidebands produced by direct or external modulation can be observed by OSA. The frequency modulation index can be calculated from the intensities of the optical sidebands, and then the frequency response of the modulator can be obtained. For a Mach-Zehnder modulator, the output optical field can be written as:

$$E_o = \frac{E_i}{2} e^{j\omega_0 t}\left(e^{j\Phi_1} e^{j\eta_1 U_m \cos\omega_m t} + e^{j\Phi_2} e^{j\eta_2 U_m \cos\omega_m t}\right) \tag{10}$$

where $U_m cos\omega_m t$ is the applied voltage, ω_0 is the frequency of the lightwave, and Φ_1 、Φ_2, η_1 、η_2 are the inherent phase shifts and the phase modulation indices in two arms, respectively.

Assuming that the incident optical field is equally split into two beams, and the applied electrical fields in the two arms are with the same phase, the above equations can be expanded into Fourier series by Bessel function as follows:

$$E_o = \frac{E_i}{2} e^{j\omega_0 t} \sum_{n=-\infty}^{\infty} j^n \left[e^{j\Phi_1} J_n(\eta_1 U_m) + e^{j\Phi_2} J_n(\eta_2 U_m) \right] \times e^{jn\omega_m t} \tag{11}$$

From (11), we can see that the output optical field includes not only the carrier component ω_0, but also the infinite sidebands components $\omega_0+n\omega_m$. The optical intensity of each sideband can be expressed as:

$$I(\omega_0 + n\omega_m) = \frac{I_i}{4}\left[J_n^2(\eta_1 U_m) + J_n^2(\eta_2 U_m) + 2J_n(\eta_1 U_m)J_n(\eta_2 U_m)\cos(\Phi_1 - \Phi_2) \right] \tag{12}$$

For M-Z interference intensity modulator, we have $\eta_1 = -\eta_2 = \eta$, and (12) can be rewritten as:

$$I(\omega_0 + n\omega_m) = \frac{I_i}{2} J_n^2(\eta U_m)\left[1 + (-1)^n \cos(\Phi_1 - \Phi_2) \right] \tag{13}$$

Under small-signal modulation, the Bessel function can be expanded as Taylor series. That is:

$$J_1(\eta U_m) \approx \frac{\eta U_m}{2}, \; J_2(\eta U_m) \approx \frac{(\eta U_m)^2}{8} \tag{14}$$

The modulation index can be obtained from the ratio between the intensities of the 1st and 2nd harmonic components,

$$R_{2,1}(\eta) = \frac{J_2^2(\eta U_m)}{J_1^2(\eta U_m)} = \frac{I(\omega_0 \pm 2\omega_m)}{I(\omega_0 \pm \omega_m)} = \frac{1}{16}(\eta U_m)^2 \tag{15}$$

These two intensities can be obtained from the measured optical spectra.

3.2 Frequency response of directly modulated lasers

Here we introduce the optical spectrum analysis method for frequency response measuring of the directly modulated laser in detail. When a single-mode laser is modulated at ω_m, the electrical field can be written as (Henery et al., 1989; Bjerkan et al., 1996)

$$E(t) = E_0\sqrt{1 + m\cos(\omega_m t)}e^{i[\omega_0 t + \beta\sin(\omega_m t + \theta)]} \tag{16}$$

where E_0 is the optical intensity without modulation, m is the intensity modulation index, ω_m is the angular frequency, β is the frequency modulation index, θ is the phase difference between frequency modulation and intensity modulation.

Assuming that the intensity modulation index m<<1, the electrical intensity can be expanded as:

$$E(t) = E_0 \sum_{p=-\infty}^{+\infty} e^{i[\omega_0 t + p(\omega_m t + \theta)]} \times \left\{ J_p(\beta) + \frac{m}{4}\left[J_{p-1}(\beta)e^{-i\theta} + J_{p+1}(\beta)e^{i\theta} \right] \right\} \tag{17}$$

Hence the ratio between the intensities of 1st sideband and the carrier can be written as:

$$I_{+1} / I_0 = J_1^2(\beta)(1 + \frac{m}{\beta}\cos\theta) / J_0^2(\beta) \tag{18}$$

$$I_{-1} / I_0 = J_1^2(\beta)(1 - \frac{m}{\beta}\cos\theta) / J_0^2(\beta) \quad (19)$$

Then, we have

$$J_1^2(\beta) / J_0^2(\beta) = (I_{+1} + I_{-1}) / (2 \times I_0) \quad (20)$$

When the frequency modulation index β<<1, the equation above can be further simplified as:

$$J_1^2(\beta) / J_0^2(\beta) \approx 0.25\beta^2 \quad (21)$$

It has been shown that (21) is accurate enough when β<0.6. Since the optical power of the carrier and sidebands can be easily observed from the optical spectrum, the frequency modulation index β can be solved from the above equation. Once β is determined, intensity modulation index m can be obtained from the following equation (Schimpe et al., 1986),

$$m(\omega_m) = \frac{2\beta(\omega_m)}{\alpha\sqrt{1 + \left(\frac{\omega_g}{\omega_m}\right)^2}} \quad (22)$$

where a is the chirp parameter, namely linewidth enhancement factor, and ω_g is the characteristic angular frequency. For a given laser, ω_g usually varies in a small range. Under small-signal modulation, a is a constant. For example, for a typical 1.55μm InGaAsP DFB laser, a is between 2 to 3, and $f_g(\omega_g/2\pi)$ is about 2.1 GHz (Petermann 1991). Hence, we can get the frequency response of the laser from the measured optical intensities of the carrier and sidebands.

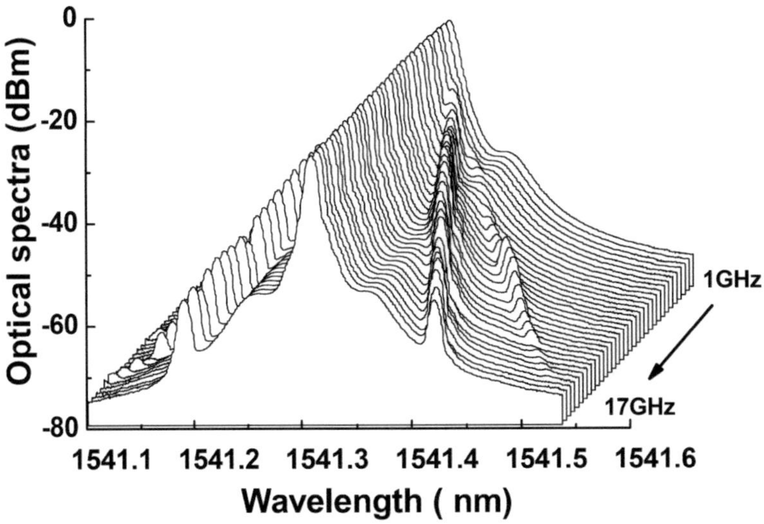

Fig. 1.11. Optical spectra at different modulation frequencies

Fig. 1.11 shows the measured optical spectra. Since the amplitude of the modulation signal is far lower than the dynamic range of the laser, it can be considered as the case of small signal modulation.

The frequency modulation index β solved from the optical spectra under different modulation frequencies is depicted in Fig. 1.12(a).

The amplitude modulation index calculated from (22) is plotted in Fig. 1.12(b). The frequency response obtained using sweeping method is also given in Fig. 1.12(b) for comparison, from which we can see that a good agreement has been obtained.

For conventional sweeping method using a VNA, a high speed photodetector is required, whereas, the optical spectrum analysis method is simple, and there is no need to use a calibrated phtodetector.

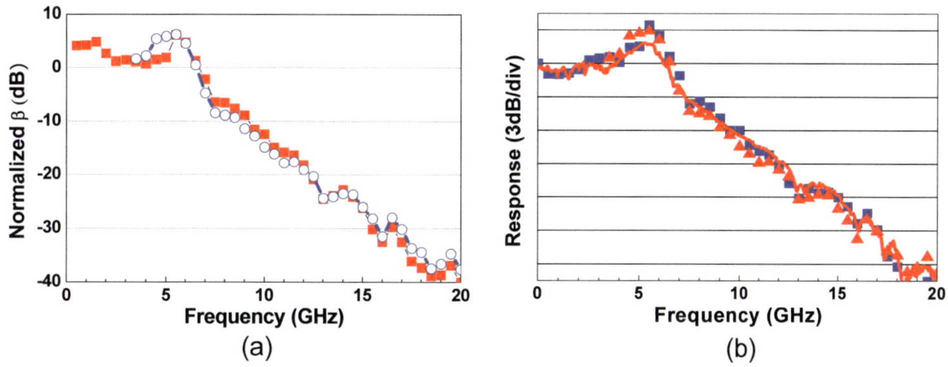

Fig. 1.12. (a)Normalized FM-modulation index and (b) frequency responses of the DFB laser

3.3 Chirp parameters of directly modulated lasers

The modulated optical spectrum of a laser comprises a series of sidebands with their frequencies at $\omega=\omega_0+n\omega_m$, where n is an integer. As shown in Fig. 1.11, as the semiconductor laser's modulation frequency increases, the wavelength detuning between its optical sidebands increases.

In the case of small signal modulation, the optical power of the main peak, 1st order sideband and 2nd order sideband are determined by the following equations (Shi et al., 2003; Kruger et al., 1995),

$$I_0 = E_0^2[\mathrm{J}_0^2(\beta) + m^2\mathrm{J}_1^2(\beta)] \tag{23}$$

$$I_1 = E_0^2[\mathrm{J}_1^2(\beta) + (m^2 / 4)(\mathrm{J}_2(\beta) - \mathrm{J}_0(\beta))^2] \tag{24}$$

$$I_2 = E_0^2[\mathrm{J}_2^2(\beta) + (m^2 / 4)(\mathrm{J}_3(\beta) - \mathrm{J}_1(\beta))^2] \tag{25}$$

where E_0 is the amplitude of the optical field; β is the frequency modulation index; $J_n(\beta)$ is the first kind Bessel function. The sideband powers are the average values of the upper and lower sidebands, that is, $I_1= (I_{+1}+I_{-1})/2$; $I_2= (I_{+2}+I_{-2})/2$.

According to (23)-(25), FM and IM modulation indices can be obtained by solving the following equations,

$$\xi = \frac{I_1}{I_0} = \frac{J_1^2(\beta) + (m^2/4)[J_2(\beta) - J_0(\beta)]^2}{J_0^2(\beta) + m^2 J_1^2(\beta)} \tag{26a}$$

$$\zeta = \frac{I_1^2}{I_0 I_2} = \frac{[J_1^2(\beta) + (m^2/4)(J_2(\beta) - J_0(\beta))^2]^2}{(J_0^2(\beta) + m^2 J_1^2(\beta))[J_2^2(\beta) + (m^2/4)(J_3(\beta) - J_1(\beta))^2]} \tag{26b}$$

For directly modulated laser, the relation between FM and IM indices can be expressed as (Schimpe et al., 1986)

$$\frac{\beta}{m} = \frac{\alpha}{2}\sqrt{1 + (\frac{\omega_g}{\omega_m})^2} \tag{27}$$

where α is the chirp parameter, ω_g is the characteristic angular frequency. For a high-speed semiconductor laser, the typical value of ω_g is 0~3GHz*2п. If the modulation frequency $\omega_m >> \omega_g$, (27) can be simplified as:

$$\alpha = 2\beta / m \tag{28}$$

When $\omega_m > 5\omega_g$, the (27) has an error less than 2%. Using (26), the chirp parameter can be obtained by measuring the optical powers of the sidebands. However, (26) does not have analytic solutions, so the graphic solution is introduced.

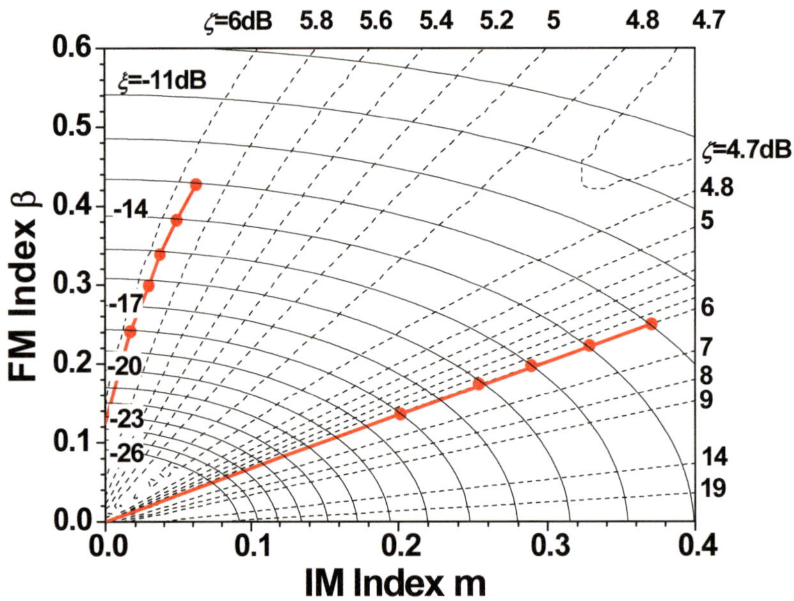

Fig. 1.13. Contour diagram for the graphic solution of FM and IM indices, where the solid line indicates ξ and the dashed line indicates ζ

Fig. 1.13 shows the contour diagram of ξ and ζ, with FM and IM indices as two orthogonal axes. The value of ξ increases when the IM index increases. A straight line passing through the origin can be obtained by measuring ξ and ζ at different modulation powers, which are defined as the RF power applied to the laser. According to (28), the slope of the straight line is $\alpha/2$ at high modulation frequencies. Hence, the chirp parameter a of the laser can be extracted from the graph. As shown in Fig. 1.13, there might be two solutions for a single measurement. Considering there is a linear relationship between β and m according to (28), the straight line which passes through the origin indicates the right solution.

For lasers with high chirp parameters, the straight line appears in the upper left part of the contour diagram. While for lasers with low chirp parameters, the line locates in the lower right part of the diagram. According to (26) and Fig. 1.13, the modulation power should be large enough to ensure the second sideband of the laser in the dynamic range of the OSA. Besides, the measurement should be in the case of small-signal modulation, so ξ should be lower than -10dB. As a result, the modulation power should meet the above two conditions.

Although our analysis is extended via the single longitudinal mode lasers, this method can be also used in characterizing multi-mode lasers with large mode spacing. For Fabry-Perot (FP) laser, the mode spacing is usually larger than 100GHz. Therefore, this method might be suitable for FP lasers.

As the development of the techniques of the optical spectrum measurement, the methods for characterizing optoelectronic devices using optical spectrum analysis are also continuously developing. It becomes a simple auxiliary method in the Lab to measure the frequency response and chirp parameter by optical spectrum analysis.

4. Optical spectrum analysis of the optical multi-beam injected semiconductor lasers

External optical injection of semiconductor lasers changes the dynamic characteristics of the injected laser cavity and leads to many phenomena, such as injection locking, push-pull effects, chaos, and period doubling (Simpson et al., 1997). In previous works, different kinds of semiconductor optical injected lasers have been investigated (Hui et al., 1991; Chang et al., 2003; Zhang et al., 2007). However, the investigations were mainly concerned about single-beam injected lasers. As to the authors' knowledge, little work has been done to study the multi-beam injected lasers. In this section, we investigate the optical spectral characteristics and frequency responses of an FP laser subject to the external light injection from another FP laser (Zhu et al., 2008) and frequency-pushing effect in a dual-beam injection system (Li et al., 2010).

4.1 Enhanced modulation bandwidth of a Fabry-Perot semiconductor laser subject to light injection from another Fabry-Perot laser

Fig.1.14 illustrates the typical optical spectra of the slave and injected master lasers. The FP mode spacings of the slave and the master FP lasers are 200 and 203 GHz, respectively. Therefore, the wavelength detuning is different for all FP modes. From Fig. 1.14, it can be seen that the mode spacing difference between the main FP modes of the two lasers are 9, 12, and 15GHz, respectively. It has been shown that the frequency response of an injection locked laser depends on both the injection optical power and wavelength detuning (Chrostowski, 2006). Therefore, we intuitively assume that a flat frequency response can be obtained after multimode light injection when the wavelength detunings and injection optical power are different for all FP modes.

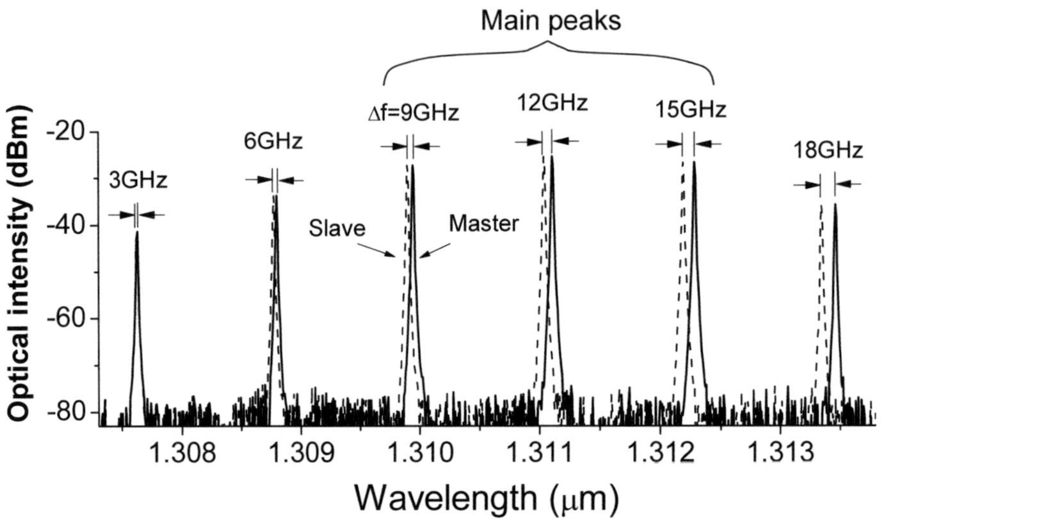

Fig. 1.14. Ideal optical spectra of the master and slave FP lasers. The mode spacing difference is 3 GHz

The optical injection operation efficiency of the two FP lasers is not always optimal. To achieve the best modulation bandwidth enhancement, we investigate the optimal conditions of the two FP lasers through the following experiments.

a. Wavelength matching conditions

The wavelength detuning of the FP modes can be adjusted by changing the bias current of the master and the slave lasers. Fig. 1.15 gives the experimental results when the mode spacing difference is 4 GHz. From the measured optical spectra shown in Fig. 1.15(a) and the inserted figures, it can be seen that for the main peaks the wavelength detuning are 25.5, 21.5, 17.5, 13.5, 9.5, and 5.5 GHz, respectively. Therefore, the wavelength detunings vary in a wide frequency range for different FP modes. From the measured frequency responses given in Fig. 1.15(b), it can be seen that both the 3-dB modulation bandwidth and the relaxation oscillation frequency increase by 10 GHz, and a very flat frequency response can be achieved within 18 GHz after optical injection.

Fig. 1.16 shows the results when the mode spacings of the slave and master lasers are almost identical. The mode spacings of the master and slave lasers are 178.68 and 178.82 GHz. Since the mode spacing difference of two FP lasers is only 0.14 GHz, the wavelength detunings for different FP modes vary in a narrow frequency range. From Fig. 1.16(a), it can be seen that the optical spectra of all main peaks are very similar. Therefore, the frequency response improvement can happen in a narrow frequency range, even for higher injection optical power. As shown in Fig. 1.16(b), the resonant frequency increases by 2 GHz after optical injection.

b. Central wavelength overlap

The central wavelength overlap can be adjusted by changing the bias current of the master laser. In the measurement, an EDFA and an attenuator are used to keep the injection optical power unchanged when adjusting the bias current of the master laser. From Fig. 1.17(a) it can be seen that the central wavelength are fully overlapped when the master laser is biased at 31.5 mA. The measured frequency responses (Fig. 1.17(b)) show that the resonance frequency increases by 4GHz. However, when the bias current of the master laser increases to 57 mA, the central wavelengths become mismatched. In this case, the 3-dB modulation bandwidth increases slightly.

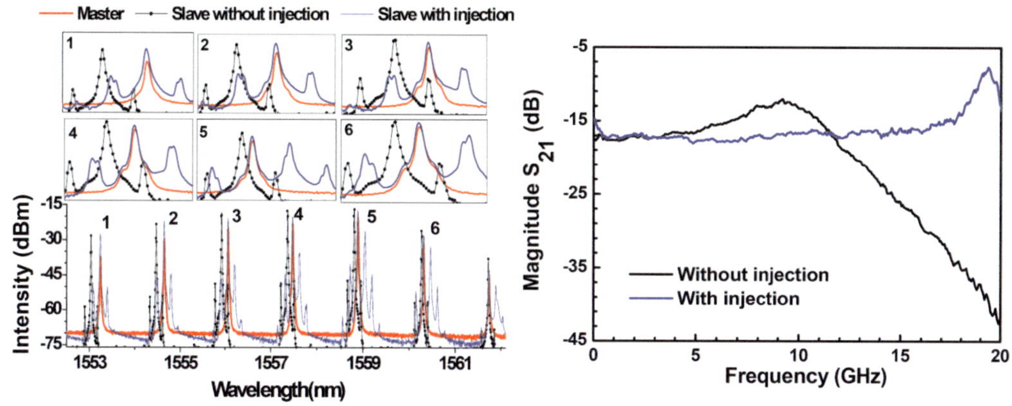

(a) (b)

Fig. 1.15. (a) Measured optical spectra at the modulation frequency of 17 GHz. The insets show the higher resolution spectra. (b) Frequency responses of the slave laser, where the injection optical power is 8.3 mW, and the mode spacing difference is 4 GHz

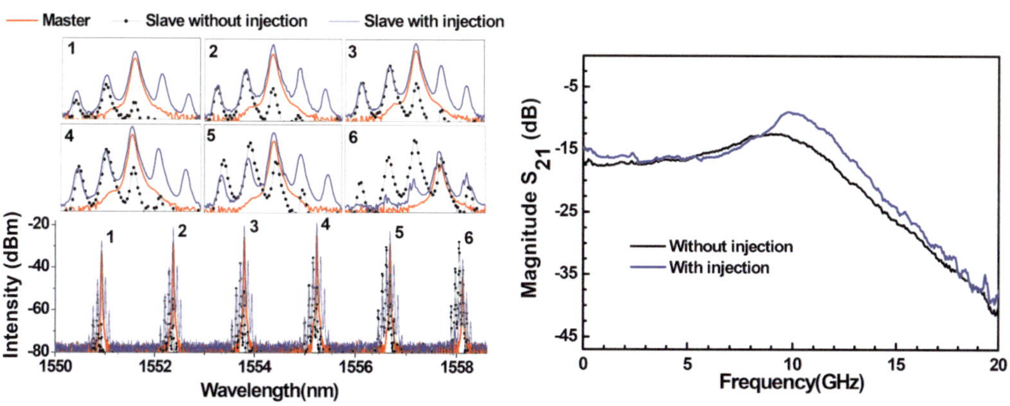

(a) (b)

Fig. 1.16. (a) Measured optical spectra at the modulation frequency of 10 GHz. The insets show the higher resolution spectra. (b) Frequency response of the slave laser at different wavelength detuning, where the injection optical power is 3.6 mW, and the mode spacing difference is 0.14 GHz

In addition to the above two main factors, the total optical injection power of the master laser and FP mode numbers can also affect the injection operation efficiency (Zhu et al., 2008). The optimal wavelength matching is mainly twofold: first, the close central wavelengths and more FP modes, and second, the mode spacing difference of several gigahertz.

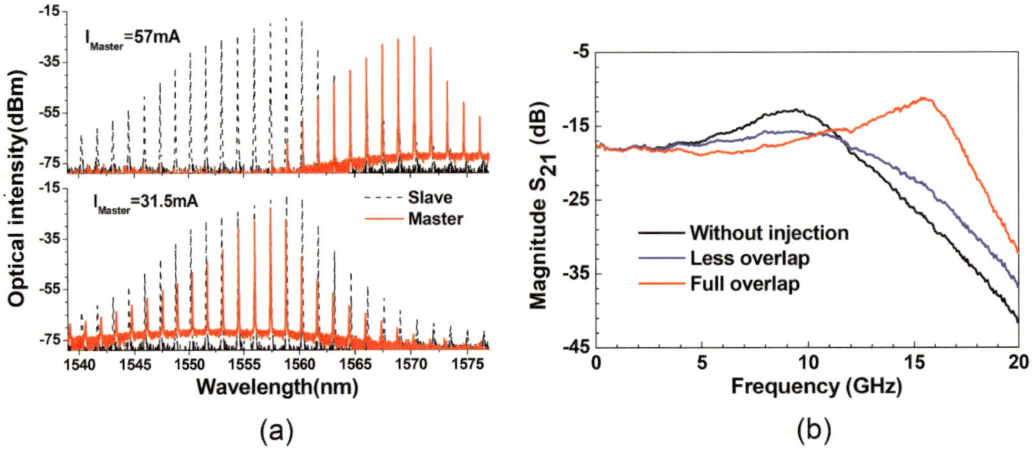

Fig. 1.17. (a) Measured optical spectra of the master laser biased at 57 and 31.5 mA, compared with the measured spectrum of the slave laser. (b) Corresponding frequency responses of the slave laser, where the injection optical power is 3.9 mW, and the mode spacing difference is 4 GHz

4.2 Frequency-pushing effect in single-mode diode laser subject to external dual-beam injection

A frequency-pushing effect can occur for semiconductor laser at positive detuning frequencies and higher injection ratios. The degree of pushing increases as the detuning frequency of the injection ratio increases or the external signal decreases (Simpson et al., 1997). Frequency convertor based on nearly degenerate four-wave mixing (NDFWM) in injection-locked semiconductor laser has been demonstrated for frequency conversion in optical communication systems, where two injection lasers are used. NDFWM in an injection-locked semiconductor laser has been studied in detail by Li and Petermann (Li & Petermann, 1993). The NDFWM pump signal is provided by a semiconductor laser frequency-locked through optical injection of an external signal, and another external injection signal acts as the probe signal. Therefore, a conjugate signal is produced in the NDFWM process.

The influence of the frequency-pushing effect on dual-beam injection system can be observed intuitively in Fig. 1.18. Fig. 1.18(a) shows the free-running spectrum consisting of a single peak and weak relaxation resonance sidebands. The negative detuning frequency between the slave and the injection light (M1) is fixed out of the locking range as shown in Fig. 1.18(b). Then, another light (M2) is simultaneously injected at positive detuning frequency. As shown in Fig. 1.18(c), the spectrum consists of a slave signal and two regenerative amplified injected signals M1 and M2, and their detuning frequencies are both out of the locking range. For optical injection at positive detuning frequency, M2 pushes the slave signal to the negative detuning frequency. While the detuning frequency between M2 and the slave decreases, the degree of the frequency-pushing increases. Therefore, at a critical point, M1 falls into the locking range of the slave laser, i.e. the slave signal is locked to M1 as shown in Fig. 1.18(d). A converted signal C1 is observed at the red side of M1. As M2 continuously moves towards M1, the amplitude of C1 increases at first and then decreases as shown in Fig. 1.18(d)-(f). This means that the conversion efficiency has maximum (Li & Petermann, 1993). In Fig. 1.18(g), the slave light jumps nearby of its free-running frequency when M2 gets close to M1 because M1 and M2 are both out of the locking range again.

To get a deep insight of this process, we explain it on the viewpoint of medium-gain. Firstly we consider a single-beam injection system. Since the external injection light enhances the stimulated emission inside the cavity, the carrier population gets depleted faster, which results in a reduction of the gain and the carrier density in the laser cavity. This gain reduction induced by external light injection (blue line) is illustrated in Fig. 1.19 by showing the gain spectrum and the loss spectrum. Furthermore, this gain reduction through *ß* (the linewidth enhancement factor) changes the slave laser cavity frequency. The amount of cavity resonance shift is: $\Delta\lambda_{cav}$ $(N)= -(\lambda_0^2/2\pi c)$ $\beta g \Delta N/2$ (Zhao et al., 2006), where g is the differential gain, N is the carrier density and c is the speed of light. Since ΔN is always negative, the cavity resonance is red-shifted and $\Delta\lambda_{cav}$ (N) is increased as the injection power increased.

Under the optical injection at positive detuning frequencies, the cavity resonance shift exhibits a frequency-pushing effect, which can be related to the medium-gain spectrum of the slave laser. For an injection wavelength far from the cavity resonance, the medium gain is small, and the amount of stimulated emission (regenerative amplified master light) caused by injection light is also small. Thus, the cavity resonance shift as well as the carrier density reduction is small. The injection light experiences more cavity gain as it getting closer to the cavity resonance under a fixed injection power, leading to a further reduced carrier density and a further red-shifted cavity resonance. In this process, the regenerative amplified master light is amplified by the cavity gain and the cavity resonance mode is suppressed. Finally, the slave laser gain is smaller than its threshold value, the master laser compensates this reduction and the slave laser is lasting at the master wavelength when it is injection locked. This process is illustrated in Fig. 1.19(a)-(d).

For optical injection at negative detuning frequencies (Fig. 1.19(e)), the cavity resonance shift exhibits a frequency pulling effect towards the injected light. As the frequency detuning between the maser and slave light is decreased at a critical point (Fig. 1.19(f)), the cavity resonance shifts automatically and continually to the frequency of the injection light until injection locking. It can be explained as follows: at the critical point (the margin of the locking range), the negative detuning injection light pulls the cavity resonance to the red side by a little amount, thus, the injection light can experience more cavity gain, which results in a further reducing of the carrier density and a more red-shifted cavity resonance. The red-shifted cavity resonance makes the injection light experience even more cavity gain. This process repeats automatically until the slave laser gain is smaller than its threshold value, leading to injection locking as shown in Fig. 1.19(f)-(h). The physically process in Fig. 1.19 is important in understanding the frequency-pushing effect in dual masters-slave system.

The gain and optical spectra in Fig. 1.20(a)-(f) correspond to the process in Fig. 1.18(b)-(g). The slave laser gets sufficient push when M2 moves closer (Fig. 1.20(b)), as explained in Fig. 1.19(a)-(c), and then it falls into the locking range of M1. Thus, the slave light is locked to the frequency of M1 just like the process illustrated in Fig. 1.19(f)-(h). NDFWM between the M2and injection-locked slave accounts for the converted signal C1 and C2 as shown in Fig. 1.20(c). When the converted signal C1 coincides with the red-shifted cavity resonance (amplified spontaneous emission), C1 is amplified by the cavity resonance as shown in Fig.

1.20(d). This explains the maximum of conversion efficiency from Fig. 1.18(d) to Fig. 1.18(f). In Fig. 1.20(f), the injection signals are out of the locking range of the slave laser. The interaction between the injected lights and the slave gain is very weak. Thus, the slave laser is lasing near its free-running frequency.

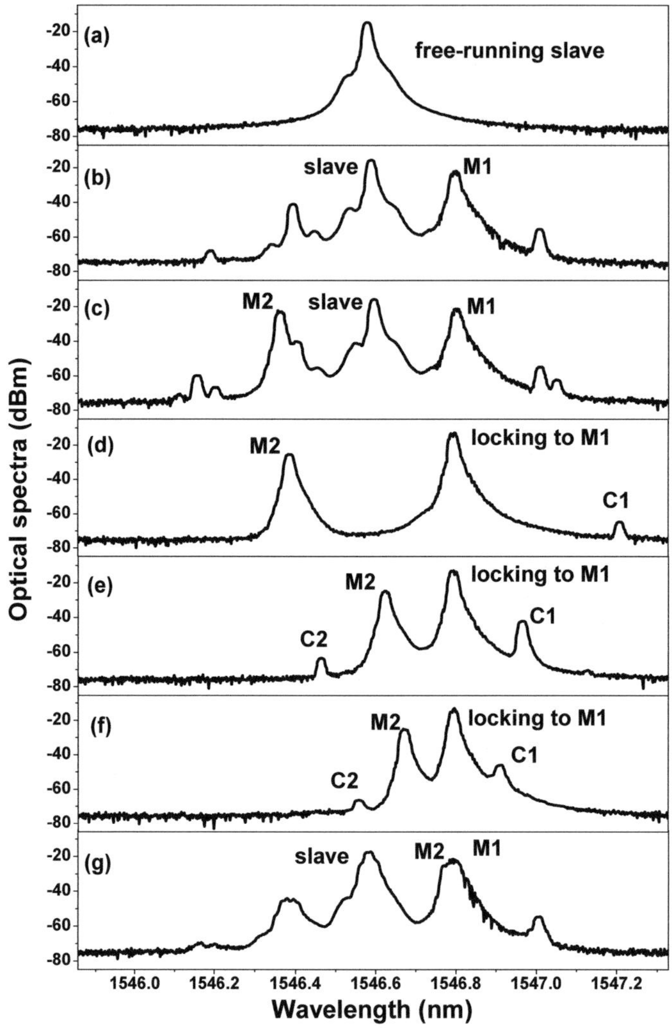

Fig. 1.18. Measured optical spectra in the process: M2 moves towards the slave signal from positive detuning frequency with a fixed optical power (-15.37 dB) while M1 is fixed at an operating point (-13.73 dB, -12.59 GHz)

The further influence of the frequency-pushing effect on dual-beam injection system is shown in Fig. 1.21(a). Firstly, the conventional stable locking diagram is generated in a single master-beam injection system. The locking boundary is marked by open circles and dashed curve as shown in Fig. 1.21(a). Next, another injection signal from M2 injects simultaneously at a fixed operating point (-10.12 dB, 12.08 GHz). A pushed locking range is observed in this case due to the frequency-pushing effect, whose boundary is indicated by red open triangles and dashed curve. The direction of pushing is indicated by a blue arrow, i.e. towards negative detuning frequency and low injection ratio. This is mainly because the injection of M2 pushes the slave frequency to the negative detuning frequency and suppresses its optical power by decreasing the slave gain.

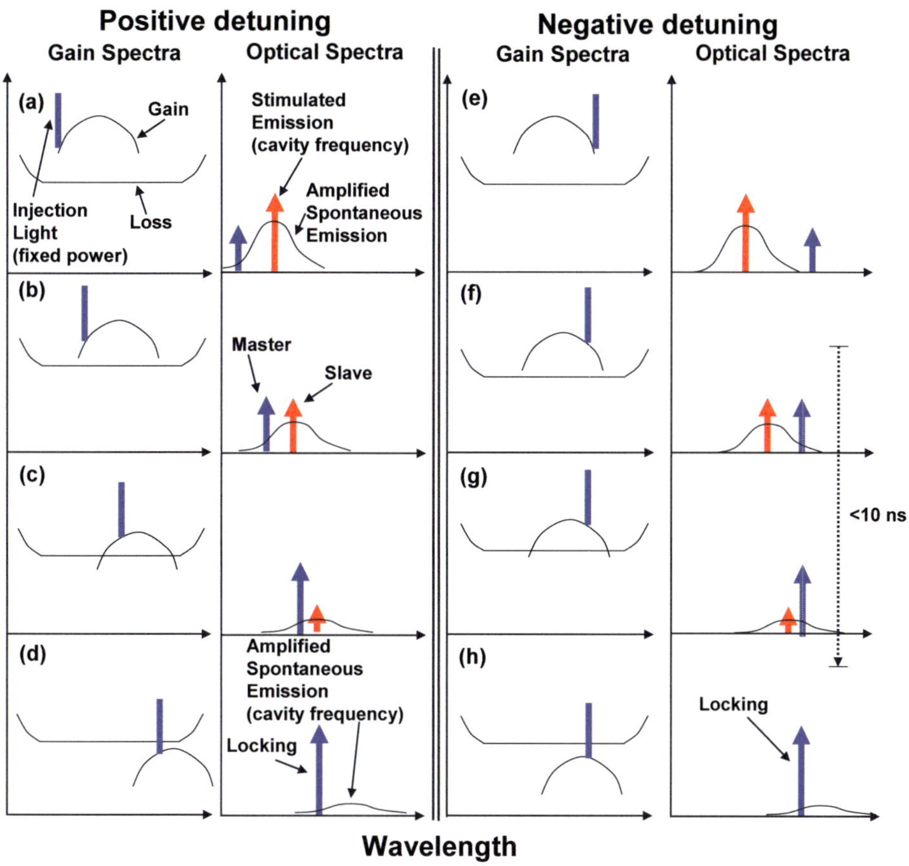

Fig. 1.19. Schematic showing the processes of frequency pushing and injection locking by medium-gain and optical spectra changing at positive and negative detuning frequencies with a fixed injection power in single-beam injection system

Fig. 1.21(b) gives the locking bandwidth as a function of power injection ratio under single-beam injection (curve a) and dual-beam injection (curve b). It is obvious that the locking bandwidth varies with the power injection ratio. The variation of the locking bandwidth before and after injecting another beam is also shown in Fig. 1.21(b) (curve c= curve b-curve a). The horizontal dashed grey line as a reference indicates the identical locking bandwidths before and after injecting another beam. Under dual-beam injection, it can be seen that the frequency-pushing effect results in an increased locking bandwidth at most injection ratios except that around -24 dB. One can see that in Fig. 1.21(a) the negative detuning is limited to about 5 GHz under moderate injection ratio in single-beam injection system. In the injection-locked NDFWM process, the pump signal is provided by the slave laser which is injection locked to a master laser. The injection of the probe signal with moderate or high injection ratio can extend the negative detuning locking boundary, which allows a larger detuning frequency and lower injection ratio between the slave and M1.

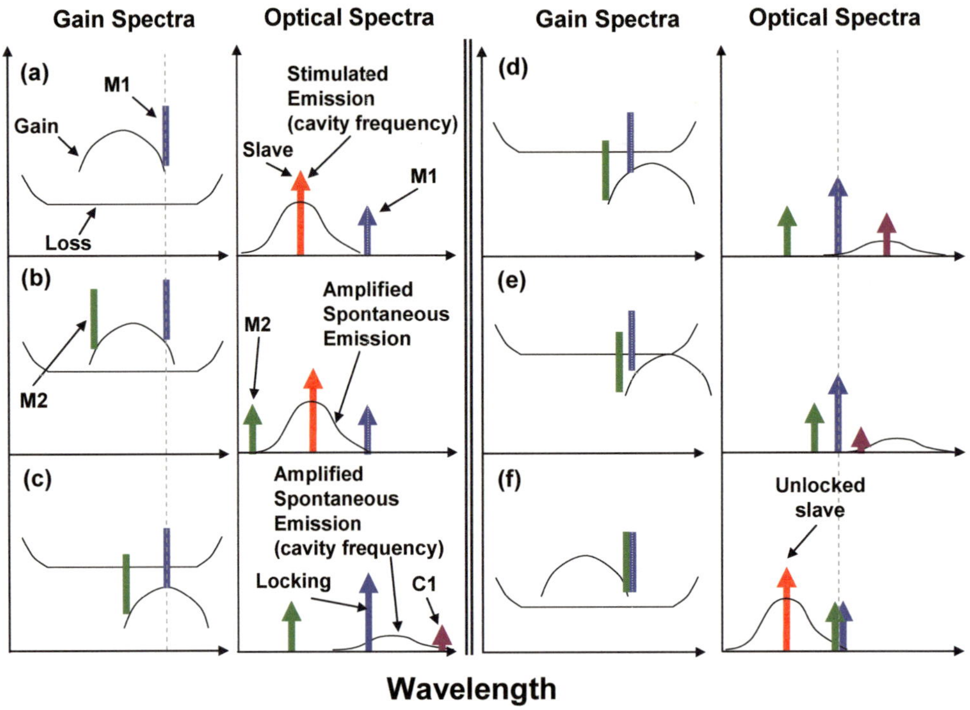

Fig. 1.20. Schematic showing the processes in Fig. 1.18 by medium-gain model

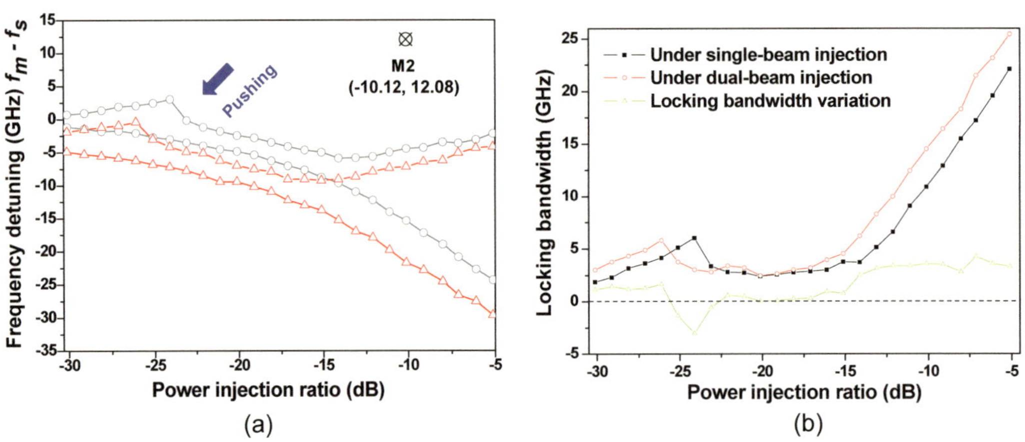

Fig. 1.21. (a) Experimental mapping of the stable locking range. The boundary of locking range in single-beam injection system is indicated by open circles and dash curve. The red open triangles and dash curve indicate the boundary of the pushed locking range by a fixed M2 (-10.12 dB, 12.08 GHz). (b) Locking bandwidth as a function of power injection ratio

5. The prospect of optical and electrical spectra analyses

In the previous sections, optoelectronic device characterization in the electrical and optical domain has been extensively investigated. In this section we make a comparison between these two techniques and discuss their potential applications. Some conclusions could be made as follows:

1. The optical spectrum contains rich information of characteristics of optoelectronic devices. For example, the junction temperature can be obtained by measuring the output spectrum of a semiconductor laser driven by short pulses due to the close relationship between the output wavelength and the junction temperature. The frequency response of an optoelectronic device can also be estimated by analyzing the intensities of the optical spectrum sidebands. One of the main advantages of this technique is that there is no upper limit in measurement bandwidth compared to the methods based on electrical spectrum analysis, since no lightwave receiver is needed in the measurement. As the development of the spectrometry measurement techniques, the optical spectrum analysis has become a promising measurement technique in many microwave photonics labs over the world.
2. Although optical spectrum analysis has many advantages, its accuracy can not meet the requirements in some special applications due to the limited resolution of the OSA. The typical resolution is around several tens of picometers. By using the optical heterodyne technique, the optical spectra could be converted to intermediate frequencies in the electrical domain. Thus, the resolution of the optical spectrum analysis can be dramatically improved. For instance, the measured average linewidth of wave trains is narrower than 1 mHz, and the frequency response of an photodetector with ultra-wide bandwidth can also be precisely measured. It is also possible to analyze the real-time spectra of all types of wideband lightwave sources with the developed techniques.
3. Both the optical and electrical spectrum analyses play an important role in the optical generation of microwave techniques. A microwave signal can be generated in the optical domain by using optical heterodyning, in which two lightwaves with different wavelengths beat in a nonlinear optoelectronic device. This technique is able to generate a microwave signal with ultra high frequency (e.g. more than 100GHz). Further more, the optical generated microwave signal can be distributed via optical fiber, so that it could greatly simplify the equipment requirement in applications such as the wireless communication system.

The amazing advantage of microwave photonics lies in the combination of the optical domain and electrical domain. There are still a lot of issues to be further investigated. For example, it is difficult to measure the optical spectrum with high resolution at a wide optical spectrum range simultaneously with good stability. In other words, the measurement resolution, spectral range and the system stability are difficult to be guaranteed simultaneously. The main factors, which have significant influence on the measurement results, are polarization state, phase and intensity noises of the light and so on. In summary, the combination of the analyses in optical and electrical domains is very promising for the measurement of optoelectronic devices, which is increasingly improved on the stage of rapid development.

6. References

Auracher F. & Keil R. (1980). Method for measuring the RF modulation characteristics of Mach-Zehnder-type modulators. *Appl. Phys. Lett.*, Vol.36, No. 8, 626-629, ISSN 0003-6951.

Baney, D. M.; Sorin, W. V.; & Newton, S. A. (1994). High-frequency photodiode characterization using a filtered intensity noise technique. *IEEE Photon. Technol. Lett.*, Vol. 6, No.10, (October, 1994) 1258-1260, ISSN 1041-1135

Bergano, N. S. (1988). Wavelength discriminator method for measuring dynamic chiro in DFB lasers. *Electron. Lett.*, Vol. 24, No. 20, (September, 1988) 1296-1297, ISSN 0013-5194

Bjerkan L., Royset A., Hafskjaer L. & Myhre D. (1996). Measurements of Laser Parameters for Simulation of High Speed Fiberoptic Systems. *J. Lightwave Technol.*, 1996, Vol. 14, No. 5, 839-850, ISSN 0733-8724.

Chang, C.-H.; Chrostowski, L. & Chang-Hasnain, C. J. (2003). Injection locking of VCSELs. *IEEE J. Quantum Electron.*, Vol. 9, No. 5, (2003) 1386-1393, ISSN 1077-260X.

Chrostowski, L.; Zhao, X.; Chang-Hasnain C. J.; Shau, R.; Ortsiefer, M. & Amann M.-C. (2006). 50-GHz optically injection-locked 1.55-μm VCSELs. *IEEE Photon. Technol. Lett.*, Vol. 18, No. 2, (Jan. 2006) 1063-1066, ISSN 1041-1135.

G. Agrawal. (1985). Power spectrum of directly modulated single-mode semiconductor lasers: Chirp-induced fine structure. *J. Quantum Electron.*, Vol. 21, No. 6, 680-686, ISSN 0018-9197.

Hawkins, R. T.; Jones, M. D.; Pepper, S. H. & Goll, J. H. (1991). Comparison of fast photodetector response measurements by optical heterodyne and pulse response techniques. *J. Lightwave Technol.*, Vol. 9, No. 10, (October, 1991) 1289-1294, ISSN 1289-1294

Henery E., Chusseau L. & Lourtioz J. M. (1989). Frequency characterization of photo-detectors by fabry-perot interferometry of modulated semiconductor lasers. *Electron. Lett.*, Vol. 25, No. 1, 42-44, ISSN 0013-5194.

Hou, S.; Tucker, R. S. & Koch, T. L. (1989). High-speed photodetector characterisation by delayed self-heterodyne method. *Electron. Lett.*, Vol. 25, No. 24, (November, 1989) 1632-1634, ISSN 0013-5194

Huang, H. P.; Zhu, N. H. (2006). Small-signal power measuring technique for measuring the frequency response of electroabsorption modulators. *IEEE Photon. Technol. Lett.*, Vol. 18, No. 23, (December, 2006) 2451-2453, ISSN 1041-1135

Hui, R.; D'Ottavi, A.; Mecozzi, A. & Spano, P. (1991). Injection locking in distributed feedback semiconductor lasers. *IEEE J. Quantum Electron.*, Vol. 27, No. 6, (June 1991) 1688-1695, ISSN 0018-9197.

Humphreyst, D. (1989). Integrated-optic system for high-speed photodetector bandwidth measurements. *Electron. Lett.*, Vol. 25, No. 23, (November, 1989) 1555-1557, ISSN 0013-5194

Joseph, H. & Sadot, D. (2004). A novel self-heterodyne method for combined temporal and spectral high-resolution measurement of wavelength transients in tunable lasers. *IEEE Photon. Technol. Lett.*, Vol. 16, No. 8, (August, 2002) 1921-1923, ISSN 1041-1135

Kawanishi T., Kogo K., Oikawa S. & Izutsu M. (2001). Direct measurement of chirp parameters of high-speed Mach-Zehnder-type optical modulators. Opt. Commun., 195:399-404, ISSN 0030-4018.

Kowalski B., Debeau J. & Boittin R. A. (1999). Simple and novel method for measuring the chirp parameter of an intensity modulated light source. *IEEE Photon. Technol. Lett.*, Vol. 11, No. 6, 700-702, ISSN 1041-1135.

Kruger U. & Kruger K. (1995). Simultaneous Measurement of the Linewidth, Linewidth Enhancement Factor and FM and AM Response of a Semiconductor Laser. *J. Lightwave Technol.*, Vol. 13, No. 4, 592-597, ISSN 0733-8724.

Li, L. & Petermann, K. (1993). Small-signal analysis of THz optical-frequency conversion in an injection-locked semiconductor laser. *IEEE J. Quantum Electron.*, Vol. 29, No. 12, (Dec. 1993) 2988-2994, ISSN 0018-9197.

Li, W.; Zhu, N. H.; Wang, L. X.; Ke, J. H.; Chen, S. F.; Qi, X. Q.; Zhang, B. H. &Xie, L. et al (2010). Frequency-Pushing Effect in Single-Mode Diode Laser Subject to External Dual-Beam Injection. *IEEE J. Quantum Electron.*, Vol. 46, No. 5, (May. 2010) 796-803, ISSN 0018-9197.

Ohishi, T.; Ohtsuka, K.; Matsui, T. & Ogata, H. (1989). Frequency characterisation of photodetectors by Fabry-Perot interferometry of modulated semiconductor lasers. *Electron. Lett.*, Vol. 25, No. 1, (January, 1989) 42-44, ISSN 0013-5194

Oikawa S., Kawanishi T. & Izutsu M. (2003). Measurement of chirp parameters and half-wave voltage of Mach-Zehnder-type modulators by using a small signal operation. IEEE Photon. Technology Lett., Vol. 15, 682-684, ISSN 1041-1135.

Petermann K. (1991). *Laser diode modulation and noise*. Kluwer Academic Publishers, ISBN 90-277-2672-8.

Schimpe R., Bowers J. E. & Koch T. L. (1986). Characterization of frequency response of 1.5 μm InGaAsP DFB laser diode and InGaAs pin photodiode by heterodyne measurement technique. *Electron. Lett.*, Vol. 22, No. 9, 453-454, ISSN 0013-5194.

Shao, Y. & Gallawa, R. L. (1986). Fiber bandwidth measurement using pulse spectrum analysis. *Appl. Opt.*, Vol. 25, No. 7, (April, 1986) 1069-1071, ISSN 1559-128X

Shi Y., Yan L. S., & Willner A. E. (2003). High-speed electrooptic modulator characterization using optical spectrum analysis. *J. Lightwave Technol.*, Vol. 21, No. 10, 2358-2367, ISSN 0733-8724.

Simpson, T. B.; Liu, J. M.; Huang, K. F. & Tai, K. (1997). Nonlinear dynamics induced by external optical injection in semiconductor lasers. *Quantum Semiclass. Opt.*, Vol. 9, (May, 1997) 765-784, ISSN 1355-5111.

Villafranca A., Lazaro J. A., Salinas I. &Garces I. (2005). Measurement of the linewidth enhancement factor in DFB lasers using a high-resolution optical spectrum analyzer. Photon. *Technol. Lett.*, 2005, Vol. 17, No. 11, 2268-2270, ISSN 1041-1135.

Yan L. S., Shi Y., & Willner A E. (2003). Chirp Measurement of Electro-Optic Modulators Using Simple Optical Spectrum Analysis. *OFC' 03*, paper MF58, ISBN 1-55752-746-6.

Yan L. S., Shi Y., & Willner A E. (2005). Graphical Solution for RF Half-Wave Voltage and Chirp Parameter of Electro-Optic Modulators Using Optical Spectrum Analysis. *IEEE Photon. Technol. Lett.*, Vol. 17, No. 7, 1486-1488, ISSN 1041-1135.

Yu, Y. & O'Dowd, R. (2002). Interpretation of wavelength switching effects of widely tunable lasers. *IEEE Photon. Technol. Lett.*, Vol. 14, No. 10, (October, 2002) 1397-1399, ISSN 1041-1135

Zhang, B. H.; Zhu, N. H.; Han, H.; Ke, J. H.; Zhang, H. G.; Ren, M.; Li, W. & Xie, L. (2009). Development of Swept Frequency Method for Measuring Frequency Response of Photodetectors Based on Harmonic Analysis. *IEEE Photon. Technol. Lett.*, Vol. 21, No. 7, (April, 2009) 459-461, ISSN 1041-1135

Zhang T., Zhu N. H., Zhang B. H. & Zhang X. (2007). Measurement of chirp parameter and modulation index of a semiconductor laser based on optical spectrum analysis. *IEEE Photon. Technol. Lett.*, Vol. 19, No. 4, 227-229, ISSN 1041-1135.

Zhang, X. ; Zhu, N. H. ; Xie, L. & Feng, B. X. (2007). A stabilized and tunable single-frequency erbium-doped fiber ring laser employing external injection locking. *J. Lightwave Technol.*, Vol. 25, No. 4, (Apr. 2007) 1027-1033, ISSN 0733-8724.

Zhao, X. ; Change-Hasnain, C. J. Hofmann, W. & Amann, M.-C. (2006). Modulation efficiency enhancement of 1.55-μm injection-locked VCSELs. *Proceeding of IEEE 20th int. Semiconductor laser conf.* pp. 125-126, New York.

Zhu, N. H.; Li, W.; Wen, J. M.; Han, W.; Chen, W. & Xie, L. (2008). Enhanced modulation bandwidth of a Fabry-Perot semiconductor laser subject to light injection from another Fabry-Perot laser. *IEEE J.Quantum Electron.*, Vol. 44, No. 6, (June 2008) 528-535, ISSN 0018-9197.

Zhu, N. H.; Wen, J. M.; San, H. S.; Huang, H. P.; Zhao, L. J. & Wang, W. (2006). Improved self-heterodyne method for measuring frequency responses of photodetectors. *IEEE Quantum Electron.*, Vol. 42, No. 3, (March, 2006) 241-248, ISSN 0018-9197

24

Bistable Photoconduction in Semiconductors

Stefano Lagomarsino
Department of Physics - University of Florence
Italy

1. Introduction

Bistable conduction in semiconductors, supported or not by an external source of ionizing radiation, has long been related to instabilities emerging in ionization-recombination processes. Under certain circumstances, a same applied voltage or a same irradiation level can produce two different current intensities, depending on the history of the biasing-irradiation settings, due to different conditions of equilibrium between the population levels of the impurity centers. The populations of the conduction and the valence band, along with those of the impurity inter-bandgap levels, depend in a complex way on the equilibrium between ionization, produced by irradiation and/or impact ionization by hot carriers, and recombination, mainly in deep impurity recombination centers. Since late fifties, negative conductivity regimes in doped semiconductors, exposed or not to ionizing irradiation, has revealed that more than one equilibrium point between ionization and recombination processes can hold in certain doping conditions, as evidenced by the typical S-shape of the current-voltage characteristics of the conductor (Khosla et al., 1973; Koenig, 1958). Several mechanisms has been invoked to explain the bistable behavior of semiconductors, all involving some kind of avalance-related phenomenon: screening of the impurity charge by free carriers (Crandal, 1970; Kim & Yung, 2004; Stobbe et al., 1994; Yoffa, 1981) and excitation followed by ionization of metastable states of dopants (Brandl & Prettl, 1991; Schöll, 1982), are probably the most fortunate among them, but absolutely not the unique ones; dissipation effects (Golik et al., 1990; Koepp & Urbelis, 1968), hole-assisted autocatalysis of electrons (Schöll, 1979), autocatalytic generation of excitons (Landsberg & Pimpale, 1976) have served as well to the modelling of some kind of negative conductivity regime. Moreover, other works report on possible instability mechanisms due to field dependent effects other than autocatalytic charge generation, namely field-enhanced trapping (Bonch-Bruevich et al., 1975) and Poole-Frenkel emission (Frenkel, 1938; Reggiani & Mitin, 1989). All these *ways* toward bistability suffer a common drawback from being all originated by electric-field related phenomena: thermodynamic considerations (Aoki, 2001; Schöll, 1987) prescribe that once the bistable regime has been reached, spatial patternization of high charge-density regions occurs spontaneously, manifesting itself in drops (Bel'kov et al., 1999) or much more often in filaments (Hirschinger et al., 2000). As a matter of fact, even though, at the beginning of these studies, application of bistable conduction was foreseen in switching devices and information storage (see e.g. the "cryosar" (Mcwhorter & Rediker, 1959; Menoret, 1968)), uncontrollable spatial patterns lead to the emergence of deterministic chaos, resulting in unreliable switching(K.Aoki & K.Yamamoto, 1989).

In this chapter, I report on a theoretical research about if and how a recombination bistability can occur supported *only* by light irradiance, in presence of arbitrarily small, or also null, electric field, in a way that this kind of bistability can be reliably employed in optoelectronic devices (Lagomarsino, 2007; 2008). In section 2 a very general result about stability of dynamical systems is stated, in order to determine conditions for photoconductive bistability. In section 3 some well known models of recombination bistability are resumed and studied with the aid of the stability criterion stated in section 2. The same criterion, in section 4, is employed to determine the two mechanisms which can assure bistability driven by light irradiance alone, with zero applied electric field. In section 5 and 6 these two mechanisms are studied in detail, and sufficient algebraic conditions on the recombination coefficients are found assuring photoconductive bistability in suitably doped semiconductors. Finally, in section 7 data from current literature are employed to study one of the centers which are the best candidates to exhibit a bistable recombination behavior driven by light irradiance in germanium.

2. A necessary condition for dynamical instability

The occurrence of multistability in a dynamical system implies its phase-space to be partitioned in domains which are mutually inaccessible, called basins of attraction. If two representative points, at a given instant of time, are arbitrarily close to each other, but belong to different, contiguous basins of attraction, the evolution of the two points will carry them, after a sufficient amount of time, sensitively apart from each other. Let a homogeneous system, subject to ionization-recombination processes, be adequately described by the populations of the atomic levels, in a way that the representative points of its phase space are given by the values of the populations themselves $N_1, N_2, \ldots, N_k, \cdots \equiv \{N_i\}$. If $\{N_i(t)\}$ is a trajectory of the system in the phase space, and $\{\delta N_i(t)\}$ is a perturbation, the existence of two or more basins of attraction can be excluded stating that, for each $\{N_i(0)\}$, a positive number ϵ exists such that, if $|\{\delta N_i(0)\}| < \epsilon$ then $\lim_{t\to\infty} |\{\delta N_i(t)\}| = 0$ (asymptotic stability). Now, let the rate equations of the system be the following:

$$\dot{N}_i = f_i(\{N_k\}). \tag{1}$$

The perturbation evolves in time according to the equations

$$\delta \dot{N}_i = \sum_j \boldsymbol{T}_{ij} \delta N_j + o(\{\delta N_k\}) \text{ with } \boldsymbol{T}_{ij} = \frac{\partial f_i}{\partial N_j}. \tag{2}$$

It is worth noting that from the neutrality condition $\sum_i N_i(t) = \text{const.}$ two statements about the matrix $\boldsymbol{T}_{ij}$ and the perturbation δN_i follow straightforwardly, namely

$$\sum_i \boldsymbol{T}_{ij} = \frac{\partial}{\partial N_j} \frac{d}{dt} \sum_i N_i = 0, \tag{3}$$

and

$$\sum_i \delta N_i = 0. \tag{4}$$

Under these last assumptions, we will state some sufficient conditions on the Jacobian matrix $\boldsymbol{T}_{ij}$ under which any perturbation $\{\delta N_i(t)\}$ vanishes for time approaching infinity. Preliminarily, we assume that no rearrangement of the indexes exists such that $\boldsymbol{T}_{ij} = 0$ for all the $i \leq k$ and $j > k$, for a convenient k (irreducibility for permutations). In this way we exclude that the system can be divided into two or more non-interacting parts. In this case, the following, equivalent statements hold

- Non-negativity of the off-diagonal entries of $\boldsymbol{T}_{ij}$, implies stability of the system.
- Dynamical instability requires as a necessary condition the negative sign of some off-diagonal entry of $\boldsymbol{T}_{ij}$.

A first step to realize the truth of the two propositions derives from consideration of the matrix $\boldsymbol{T}_{ij}$ under the light of the first Gershgorin theorem(Todd, 1962), concerning the limitation of the eigenvalues of a (generally complex) matrix $\boldsymbol{T}_{ij}$. The theorem affirms that every eigenvalue of the matrix belongs to some of the disks in the complex plain having centers in $\boldsymbol{T}_{jj}$ and radii $r_j \equiv \sum_{i \neq j} \left| \boldsymbol{T}_{ij} \right|$. Due to the eq.3 and to the non-negativity of the off-diagonal entries, it follows that each $\boldsymbol{T}_{jj}$ is surely non-positive, and that $r_j = -\boldsymbol{T}_{jj}$. Consequently (see also fig.1), all of the eigenvalues of the matrix have a real part either negative or null.

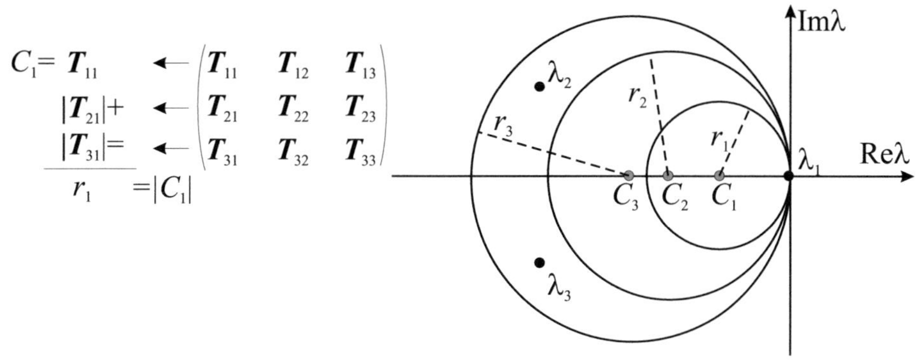

Fig. 1. Scheme of a 3×3 Jacobian matrix $\boldsymbol{T}$ with its eingenvalues and the relative Gershgorin disks in the complex plane.

Now, it is well known that the negative real part of *all* the characteristic exponents of the system could be a sufficient condition for dynamic stability (Minorsky, 1962). Nevertheless, eq. 3 implies singularity of the matrix $\boldsymbol{T}_{ij}$, then at least one of the roots of the characteristic equation is null, and an asymptotic divergence of the rate equations cannot be excluded at this point, due to the higher order terms in eq.2. The key for the solution is to notice that the space S spanned by the points $\{\delta N_i\}$ satisfying condition 4 is closed with respect to the matrix $\boldsymbol{T}_{ij}$, as can easily be seen by application of eq.3. Therefore, it will be sufficient to test the sign of the eigenvalues of $\boldsymbol{T}^S$: the restriction to the space S of the linear operator defined by $\boldsymbol{T}_{ij}$. To this aim, we recall two conclusions concerning sign-symmetric and irreducible matrices, reached by Hearon(Hearon, 1963) and more recently surveyed in the context of general M-matrix theory (see chapter 6 of Berman & Plemmons (1994)):

1. An eigenvector of the zero eigenvalue exists, such that all its components are positive.
2. The multiplicity of the zero characteristic exponent is one.

For statement 1, the eigenspace S_0 corresponding to the zero-eigenvector is not included in the space S spanned by the solutions of eq. 2, because its vectors do not satisfy condition 4.

Thus, S and S_0 both being closed with respect to the matrix $\boldsymbol{T}_{ij}$, a similarity transformation A_{ij} exists such that

$$A^{-1}\boldsymbol{T}A = \left(\begin{array}{c|c} 0 & 0 \\ \hline 0 & \boldsymbol{T}^S \end{array}\right); \quad (5)$$

where $\boldsymbol{T}^S$ operates in the space S.

For statement 2, only one of the roots of the characteristic equation of $\boldsymbol{T}_{ij}$ vanishes, so that the eigenvalues of $\boldsymbol{T}^S$ are all non-zero, and equal the non-null eigenvalues of $\boldsymbol{T}_{ij}$, which have negative real part. Since $\boldsymbol{T}^S$ is the restriction of $\boldsymbol{T}_{ij}$ to the space spanned by the solutions of system 2, the negative real part of its eigenvalues implies asymptotic stability. Of course, the negativity of some off-diagonal entry of the Jacobian matrix is not a sufficient, but only a necessary condition for instability; nevertheless it provides a very simple tool to greatly restrict the field of investigation and to identify all the relevant phenomena inducing bistability in ionization-recombination processes.

3. Avalance-related bistabilities in photoconductors

In this section we consider two well known cases of bistable photo-conduction triggered by impact ionization of impurities. Even if our principal aim in this chapter is the study of field-independent bistabilities, the study of two simpler cases provides useful illustrations of the necessary criterion for instability introduced in the previous section. Moreover, it will be useful as an introduction to some basic concepts largely employed in the following.

3.1 Screenig effect in impact ionization of one-level impurities

Let us consider a deep donor doped, partially compensated semiconductor, exposed to a sub-bandgap radiation which is able to ionize the donor centers (extrinsic photoconduction), and subject to an electric field strong enough to sustain impact ionization of impurities. Assuming n and n_t to be the concentration of the electrons in the conduction band and in the impurity centers, and being N_D and N_A the concentration of donor and acceptor centers, respectively, the rate equations of the system are the following (see also fig 2 (left) for a schematics of the transition rates):

$$\dot{n} = (A_I n + A_R)\, n_t - B_T n\,(N_D - n_t) \quad \text{and} \quad \dot{n}_t = -\dot{n}$$
$$\text{with} \quad n + n_t = N_D - N_A. \quad (6)$$

In eq.6, A_R is a generation term proportional to the radiation irradiance, B_T is a recombination

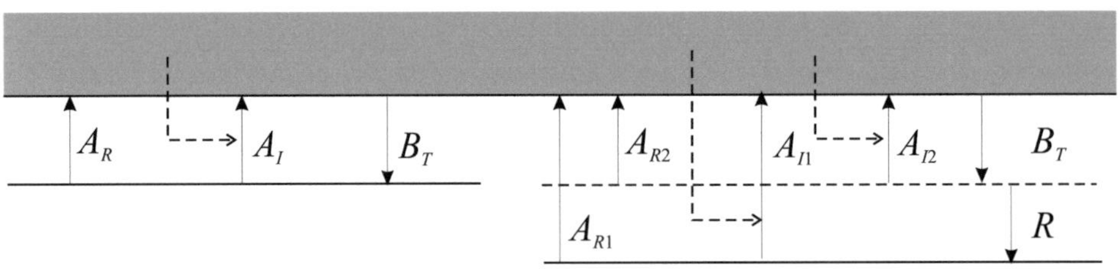

Fig. 2. Diagram of the transitions for a one-level and a two-levels impurity.

coefficient, and A_I is the impact ionization coefficient, depending on the electric field and on the shielding of charged scatterers produced by the free electrons themselves, and given by Aoki (2001):

$$A_I = A_I^0 (E) + \sigma n. \tag{7}$$

At equilibrium, when $\dot{n} = \dot{n}_t = 0$, equation 6 reduces, in the adimensional variable $x = \frac{n}{N_D - N_A}$, to:

$$x^3 - \alpha x^2 + \beta x - \gamma = 0, \tag{8}$$

with the coefficients α, β and γ given by:

$$\alpha = 1 - \frac{A_I^0 + B_T}{\sigma (N_D - N_A)}, \quad \beta = \frac{A_R}{\sigma (N_D - N_A)^2} - \frac{A_I^0 - B_T \frac{N_A}{N_D - N_A}}{\sigma (N_D - N_A)}, \quad \gamma = \frac{A_R}{\sigma (N_D - N_A)^2}. \tag{9}$$

In order to occur bistability, equation 8 has to admit three real and positive solutions, which is true if and only if the following inequalities hold:

$$\begin{aligned} &\left(\alpha^3 - \frac{9}{2}\alpha\beta + \frac{27}{2}\gamma\right)^2 < \left(\alpha^2 - 3\beta\right)^3, \quad \text{for reality,} \\ &\alpha > 0, \quad \beta > 0, \quad \gamma > 0, \quad \text{for positivity.} \end{aligned} \tag{10}$$

The points (α, β, γ) satisfying the inequalities 10, identify a region Ω in the three-dimensional space which is represented in figure 2, along with two sections ω_γ and ω_α parallel to the $\alpha\beta$ and to the $\beta\gamma$ plains, respectively. The geometric locus χ of the cusp points C_γ (or C_α) is given by the curve whose equations are $\beta = \frac{1}{3}\alpha^2$ and $\gamma = \frac{1}{27}\alpha^3$.
In order for the system to undergo bistable transitions, the curve described in the $\alpha\beta\gamma$ space by the representative point of the system, while varying the radiation irradiance term A_R, has to cross through the bistability region Ω. Since in this case the curve is simply a straight line r parallel to the $\beta\gamma$ plane and starting, for $A_R = 0$, from a point P_0 on the $\alpha\beta$ plain, it is sufficient to impose that P_0 doesn't belong to the region Ω and the cusp point $C_{\alpha_0} = \left(\frac{1}{3}\alpha_0^2, \frac{1}{27}\alpha_0^3\right)$ lays under the straight line r in the plane $\beta\gamma$ (see also fig.3b (left)). Simple though annoying algebra shows that these two conditions hold if the ionization and recombination coefficients satisfy the following inequalities:

$$\begin{aligned} &\left(\frac{A_I^0 + B_T}{\sigma (N_D - N_A)}\right)^2 + \frac{2A_I^0}{\sigma (N_D - N_A)} - \frac{N_D + N_A}{N_D - N_A} \frac{2B_T}{\sigma (N_D - N_A)} + 1 < 0, \\ &\left(1 - \frac{A_I^0 + B_T}{\sigma (N_D - N_A)}\right)^2 \left(8 + \frac{A_I^0 + B_T}{\sigma (N_D - N_A)}\right) > 27 \frac{B_T \frac{N_A}{N_D - N_A} - A_I^0}{\sigma (N_D - N_A)}. \end{aligned} \tag{11}$$

In figure 4 the area in the $A_I^0 B_T$ plane satisfying rel.11 is shown for several values of the ratio N_D / N_A. It is quite evident, from inspection of the figure, that once known the screening and the recombination coefficients, the bandgap can be tailored, by suitable choice of N_D and N_A, in order to manifest photoconductive bistability.
Let us consider, now, the Jacobian matrix $\boldsymbol{T}_{nn_t}$ of the system 6:

$$\boldsymbol{T} = \begin{pmatrix} n_t \frac{\partial}{\partial n} (A_I n) - B_T (N_D - n_t) & A_I n + A_R + B_T n \\ -n_t \frac{\partial}{\partial n} (A_I n) + B_T (N_D - n_t) & -A_I n - A_R - B_T n \end{pmatrix}; \tag{12}$$

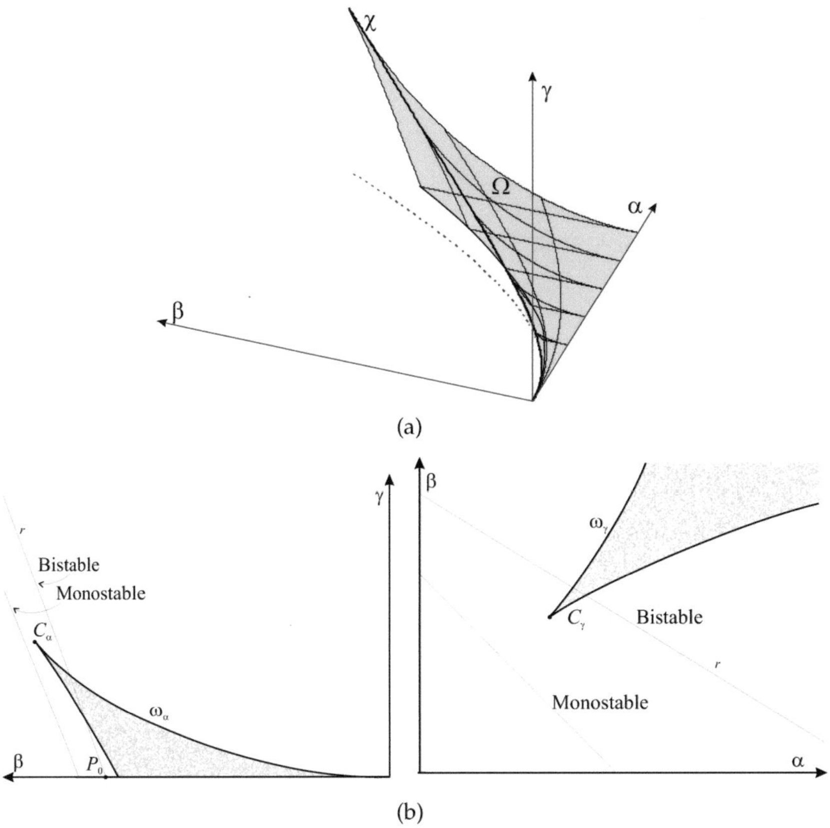

Fig. 3. (a) Diagram of the Ω region in the $\alpha\beta\gamma$ space, whose points correspond to three states of equilibrium, two stable and one instable; χ is the geometric locus of the cusp points. (b-left) Section of the Ω region in a plane with $\alpha = const$. (b-right) Section of the Ω region in a plane with $\gamma = const$.

according to the stability criterion of the non-negativity of the off-diagonal terms, stated in the preceding section, the observation of the Jacobian matrix simply gives the indication that, in order for the bistability to occur, the impact ionization term A_I has to be nonnull. Compared with eq. 11 this seems a quite poor and qualitative information, but we will verify how, in more complex situations, with an arbitrary number of energy levels, for instance, this kind of insight will be of unvaluable usefullness.

3.2 impact excitation and ionization of metastable states

A widely adopted model of impact ionization introduced by Schöll (1987), suggests that the impact ionization processes of impurity multilevels can induce bistable conduction in semiconductors. Here we extend the model to the photo-ionization of the impurity levels and neglect thermal excitation and ionization of the deep donor centers. Given n, n_0 and n_1 the population of the conduction band, the ground and the excited level of the donor centers,

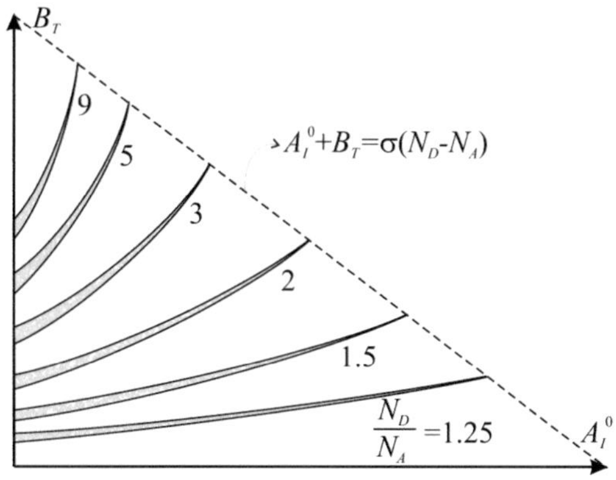

Fig. 4. In grey, the areas in the $A_I B_T$ plane which correspond to bistable situations, for different values of the constant N_D/N_A.

respectively, the rate equations of the system are:

$$\begin{aligned}
\dot{n} &= (A_{I1}n + A_{R1})\, n_1 + (A_{I2}n + A_{R2})\, n_2 - B_T n\,(N_D - n_1 - n_2),\\
\dot{n}_2 &= -\,(A_{I2}n + A_{R2})\, n_2 + B_T n\,(N_D - n_1 - n_2) - R n_2,\\
\dot{n}_1 &= -\,(A_{I1}n + A_{R1})\, n_1 + R n_2,\\
&\text{with} \quad n + n_1 + n_2 = N_D.
\end{aligned} \tag{13}$$

Given $x = n/N_D$, and considering that the ratio of the generation terms $A_{R2}/A_{R1} \equiv \rho$ the solving equation at the equilibrium is again of the type of eq.8, with the coefficients α, β and γ equalling:

$$\begin{aligned}
\alpha &= \frac{N_D A_{I1} A_{I2} - R\,(A_{I1} + B_T) - A_{R1}\,(\rho A_{I1} + A_{I2} + B_T)}{N_D A_{I1}\,(A_{I2} + B_T)},\\
\beta &= \frac{\rho A_{R1}^2 + A_{R1}\,(R - \rho A_{I1} N_D - A_{I2} N_D) - A_{I1} R N_D}{N_D^2 A_{I1}\,(A_{I2} + B_T)},\\
\gamma &= \frac{\rho A_{R1}^2 + A_{R1} R}{N_D^2 A_{I1}\,(A_{I2} + B_T)}.
\end{aligned} \tag{14}$$

In this case, the curve describing the system in the $\alpha\beta\gamma$ space is of the second order in the generation term, and the algebraic discussion of its intersection with the region Ω is considerably more complex.
The stability criterion of the non-negativity of the off-diagonal entries of the Jacobian matrix of the system can give useful insight in the physics of the processes. These have the following values:

$$\begin{aligned}
&\boldsymbol{T}_{nn_1} = A_{I1}n + A_{R1} + B_T n, \quad \boldsymbol{T}_{nn_2} = A_{I2}n + A_{R2} + B_T n,\\
&\boldsymbol{T}_{n_1 n} = -n_1 \frac{\partial}{\partial n}(A_{I1}n), \quad \boldsymbol{T}_{n_1 n_2} = R,\\
&\boldsymbol{T}_{n_2 n} = B_T\,(N_D - n_1 - n_2) - n_2 \frac{\partial}{\partial n}(A_{I2}n), \quad \boldsymbol{T}_{n_2 n_1} = -B_T n,
\end{aligned} \tag{15}$$

As expected, also for the Schöll model, the non-negativity of the off diagonal entries, along with the stability of the system, are not guaranteed if the impact ionization coefficients $A_{I1,2}$ are nonnull. Moreover, one of the non-diagonal terms is negative independently on the ionization mechanism, only due to the new recombination channel given by the metastable state, determining the term $-B_T n$ in eqs.15. This seems to open the way to some kind of different instability mechanism, independent on the electric field and determined only by the competition of different recombination channels. The next section will be devoted to the throughout investigation of all the possible mechanisms which can break the sign-simmetry of the Jacobian matrix of a sistem subject to ionization-recombination processes. The analysis will evidence the existence of two possible ways toward instability in photoconductors with zero electric field.

4. Dynamical instability in ionization-recombination reactions

If r_{ik} represents the number of electrons that, in the unit time, change their state from k to i, the functions f_i in eq.1 have the form

$$f_i = \sum_{k \neq i} (r_{ik} - r_{ki}) \tag{16}$$

The transition rates r_{ik}, in their turn, depend on the occupation numbers N_l in one of the following two ways:

$$\text{excitation-relaxation} \qquad r_{ik} = R_{ik}(\{N_l\})\, N_k^{\mathrm{A}} \tag{17}$$

$$\text{ionization-recombination}\; r_{ik} = S_{ik}(\{N_l\})\, N_k^{\mathrm{A}} N_i^{\mathrm{F}} \tag{18}$$

where N_k^{A} is the number of electrons in the state k which are available to make a transition to another state, while N_i^{F} is the number of free places in the state i. Transition probabilities S_{ik} and R_{ik} may depend or not on the occupation numbers $\{N_l\}$, according to the nature of the reactions involved. For example, in impact ionization and in Auger recombination, S_{ik} depends on the number of free electrons. N_k^{A} and N_i^{F}, on the other hand, depend on the nature of the state labelled by the index. Figure 5 gives a schematics of the various cases involved in neutral and singly ionized atoms (left side of the figure), multiply ionized atoms (right side) and free states (top of the figure).

- *Neutral or singly ionized atoms.* Since all the valence electrons can be excited or extracted from the atoms, the number N_k^{A} of electrons available to make a transition is simply N_k. On the other hand, the number of "free places" for an electron to recombine is given by the number of ions, then

$$N_k^{\mathrm{A}} = N_k$$
$$N_i^{\mathrm{F}} = g_i - \sum_l \Delta_{il} N_l, \tag{19}$$

where g_i is the number of atoms of the specie labelled by i, while Δ_{il} is equal to one only if i and l refer to states of a same atomic specie, otherwise is zero (notice that the index labels both the atomic specie and the energy level).

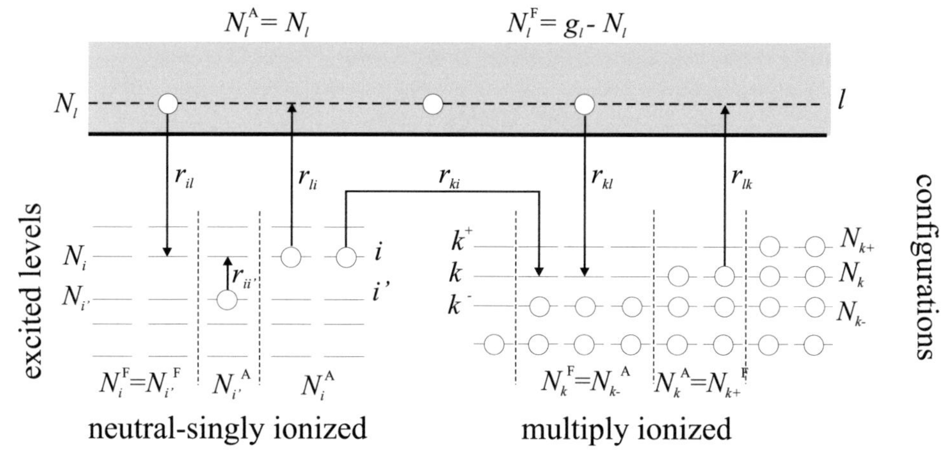

Fig. 5. A schematics of the transitions described by the model represented by eq. 17 and 18. On the left side, a set of atoms of a given specie is represented, each with its level diagram, three of whom are neutral and two singly ionized. On the right side, a set of neutral, singly or multiply ionized atoms are represented. In this case, the excited levels of each configuration are disregarded, in the assumption that their population is ruled by lattice or free charges temperature. The upper side represents the unbound state.

- *multiply ionized atoms.* Let the indexes k^-, k and k^+ indicate three adjacent energy levels in a given atomic configuration. An electron in the state k can make a transition in low energy processes only if no other electrons occupy the state k^+ in the same atom. Analogously, a free state k can be reached by an unbound electron only if the state k^- is occupied in its turn. Consequently, we have

$$\begin{aligned} N_k^{\mathrm{A}} &= N_k - N_{k^+} \\ N_i^{\mathrm{F}} &= N_{i^-} - N_i \end{aligned} \tag{20}$$

- *Free states.* If g_i is the degeneration of an unbound state, we can obviously state

$$\begin{aligned} N_k^{\mathrm{A}} &= N_k \\ N_i^{\mathrm{F}} &= g_i - N_i \end{aligned} \tag{21}$$

  It is worth noting that, if we formally state $N_{k^+} = 0$ and $N_{i^-} = g_i$, the present case reduces to the previous one, which will be useful in the following discussion.

Once the expressions for the transition probabilities r_{ik} are determined by eq.17-18, the entries of the matrix $\boldsymbol{T}_{ij}$ are calculated from their derivatives with respect to the occupation numbers, as given in eq.2 , and can be then expressed as a sum of two contributions:

$$\boldsymbol{T}_{ij} = \boldsymbol{T}_{ij}^{\mathrm{nc}} + \boldsymbol{T}_{ij}^{\mathrm{c}} \quad \text{with}$$

$$\boldsymbol{T}_{ij}^{\mathrm{nc}} = \sum_{k \neq i} \left(S_{ik} \frac{\partial N_k^{\mathrm{A}} N_i^{\mathrm{F}}}{\partial N_j} - S_{ki} \frac{\partial N_i^{\mathrm{A}} N_k^{\mathrm{F}}}{\partial N_j} \right) + \left(R_{ik} \frac{\partial N_k^{\mathrm{A}}}{\partial N_j} - R_{ki} \frac{\partial N_i^{\mathrm{A}}}{\partial N_j} \right) \quad \text{and}$$

$$\boldsymbol{T}_{ij}^{\mathrm{c}} = \sum_{k \neq i} \left(\frac{\partial S_{ik}}{\partial N_j} N_k^{\mathrm{A}} N_i^{\mathrm{F}} - \frac{\partial S_{ki}}{\partial N_j} N_i^{\mathrm{A}} N_k^{\mathrm{F}} \right) + \left(\frac{\partial R_{ik}}{\partial N_j} N_k^{\mathrm{A}} - \frac{\partial R_{ki}}{\partial N_j} N_i^{\mathrm{A}} \right). \tag{22}$$

The apices "nc" and "c" refer respectively to "non catalytic" and "catalytic" processes. In fact, $\boldsymbol{T}_{ij}^{c}$ is non-zero only if the transition probabilities S_{ik} and R_{ik} depend explicitly on $\{N_l\}$, that is, if the transition of an electron from state k to i is catalyzed or somewhat influenced by the presence of electrons in some specific quantum state. If this is not the case, only the $\boldsymbol{T}_{ij}^{nc}$ (non-catalytic) term is non zero. In this last case, the equations of the system 1 are of the second degree in $\{N_i\}$. Indeed, non linearity of the second order has been known to be a cause of instability in dynamical systems since the very beginning of chaos studies (Lorentz, 1963; Roessler, 1976). Nevertheless, almost without exceptions, only catalytic processes have been considered as possible mechanisms leading to instability in semiconductors. In the following we will study non-catalytic processes, the only ones which can hold in absence of an externally applied electric field.

- *Neutral or singly ionized atoms.* The rate equations assume the form (from eq.16,19,22)

$$\dot{N}_i = \sum_{l \neq i} \left\{ \left[S_{il} \left(g_i - \sum_k \Delta_{ik} N_k \right) + R_{il} \right] N_l \right\} - \left\{ \sum_{l \neq i} \left[S_{li} \left(g_l - \sum_k \Delta_{lk} N_k \right) + R_{li} \right] \right\} N_i \tag{23}$$

while the off-diagonal entries of the matrix $\boldsymbol{T}_{ij}$ are

$$\boldsymbol{T}_{ij} = \boldsymbol{T}_{ij}^{nc} = S_{ij} \left(g_i - \sum_l \Delta_{il} N_l \right) + S_{ji} N_i + R_{ij} - \Delta_{ij} S_{ij} N_j + \sum_{l \neq i,j} \left(\Delta_{lj} S_{ki} N_i - \Delta_{ij} S_{il} N_l \right) \tag{24}$$

Observing eq. 24, we notice that the only way for the off-diagonal entries to assume a negative value is to admit transitions involving several excited states of a same atomic specie, corresponding to the terms $-\Delta_{ij} S_{ik} N_k$ and $-\Delta_{ij} S_{ij} N_j$. These transitions may be free carrier capture or emission, or also electron exchange by tunneling between impurities in solid state plasmas. Even in these cases, however, the negative contributions of the expression can be overwhelmed if the exchange contributions are lesser than the collisional or radiative rates R_{ij}, which are always positive. In conclusion, instability, if any, requires electron exchange with metastable states and low excitation-relaxation rates.

- *Multiply ionized atoms* Sufficiently high collisional rates involve stability in case of single ionization, as just shown. This is not necessarily true for multiple ionization. If the collisional rates are high enough to termalize the distribution of the excited levels of each electronic configuration, the transition probabilities S_{ij} between configurations depend on temperature, and the rate equations can be written in the following form (from eq.16,19,22):

$$\dot{N}_i = \sum_{k \neq i} \left[S_{ik} \left(N_{i^-} - N_i \right) \left(N_k - N_{k^+} \right) - S_{ki} \left(N_{k^-} - N_k \right) \left(N_i - N_{i^+} \right) \right] \tag{25}$$

This equation describes not only electronic exchange between singly or multiply ionized atoms, but also ionization or recombination toward or from an unbound state k, formally stating $N_{k^+} = 0$ and $N_{k^-} = g_k$. For $i \neq j, j^{\pm}$, the off-diagonal terms of the Jacobian matrix $\boldsymbol{T}_{ij}$ assume the simple form:

$$\boldsymbol{T}_{ij} = \boldsymbol{T}_{ij}^{nc} = \left(S_{ji} - S_{j^+i} \right) \left(N_i - N_{i^+} \right) + \left(S_{ij} - S_{ij^-} \right) \left(N_{i^-} - N_i \right) \tag{26}$$

Since N_i and N_{i^-} are always not less than N_{i^+} and N_i, respectively, sufficient conditions for non-negativity of eq.26, and than for the stability of the dynamic system, are: $S_{ij} \geq S_{ij^-}$ and

$S_{ji} \geq S_{j+i}$ for every i and j. That is, non-negativity of $\boldsymbol{T}_{ij}$ is assured if any further ionization of less ionized atoms is more favored than the more ionized ones, and recombination of the more ionized atoms is more probable compared with the less ionized. This is effectively true for shallow states of the band gap of semiconductors, but for deeper states the opposite can hold, due to the high energy transfer required by the recombination in deep centers, which make the multiphonon or radiative transition less likely than the phonon-cascade trapping by the shallower levels. Infact, several deep recombination centers in low-bandgap semiconductors manifested this behavior since the very beginning of the studies on limiting charge life-time centers (Wertheim, 1959). Particularly, in the last years an extensive thermal spectroscopy study revealed that this behavior is quite common for transition metals in semiconductors (Grillenberger, 2004; Knack et al., 2002; Sachse et al., 2000; Shiraishi et al., 1999; Yarykin et al., 1999).

In conclusion, we have shown that dynamical instability in ionization-recombination processes, in absence of electric fields sustaining catalytic carriers generation, requires the occurring of one of the following conditions:

- electronic exchange with metastable states with low excitation-relaxation rates;
- inversion of transition probability in multiply ionized atoms.

In the next sections, sufficient conditions on the recombination coefficients will be given for the occurring of bistability in each of the two cases.

5. Bistability by meta-stable states of recombination centers

We will study the equilibrium, under over-bandgap irradiation, of a semiconductor in which recombination is assured by deep donor centers (density M) admitting a metastable excited state. Since the general case would require discussion of a 5th-degree equation, we restrict to the case of a semiconductor strongly doped (density $N >> M$) with shallow donor or acceptor impurities, that, as shall be seen, substantially reduces the algebraic complexity of the problem. The rate equations of the system (see fig. 5, left) for a schematics of the transition rates) are the following:

$$\begin{aligned} \dot{n} &= -\left(C_{1n} + C_{2n}\right) n \left(M - n_1 - n_2\right) + g, \\ \dot{n}_2 &= C_{2n} n \left(M - n_1 - n_2\right) - C_{2p} p n_2 - R_{21} n_2, \\ \dot{n}_1 &= C_{1n} n \left(M - n_1 - n_2\right) - C_{1p} p n_2 + R_{21} n_2, \\ \dot{p} &= -\left(C_{1p} n_1 + C_{2p} n_2\right) p + g; \\ &\text{with} \quad n + n_1 + n_2 - M \pm N = p. \end{aligned} \tag{27}$$

Let us express the constant and variable quantities of the system 27 in term of the adimensional parameters:

$$\begin{aligned} &x = \frac{p}{N}, y = \frac{n}{N}, y_i = \frac{n_i}{N}, \\ &R_i = \frac{C_{ip}}{C_{in}}, \rho = \frac{R_{21}}{N\sqrt{C_{1n}C_{2n}}}, K_i = \frac{C_{in}}{\sqrt{C_{1n}C_{2n}}}, G = \frac{g}{NM\sqrt{C_{1n}C_{2n}}}. \end{aligned} \tag{28}$$

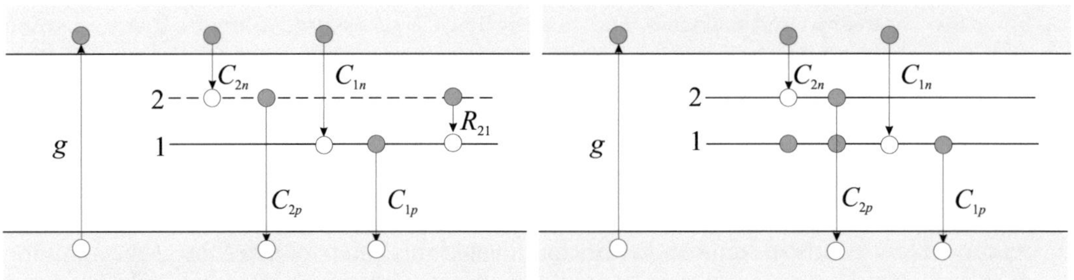

Fig. 6. The levels diagram of a semiconductor doped with a deep level having; (left) a metastable excited state or (right) two different ionized states.

From the second and the third equation of system 27, we can calculate y_i:

$$y_1 = \frac{R_2xy + (K_1 + K_2)\,\rho y}{R_1R_2x^2 + (R_1 + R_2)\,xy + (K_1 + K_2)\,\rho y + \rho K_1R_1x},$$
$$y_2 = \frac{R_1xy}{R_1R_2x^2 + (R_1 + R_2)\,xy + (K_1 + K_2)\,\rho y + \rho K_1R_1x}. \tag{29}$$

Substituting eqs.29 in the first and the last equation of system 27, we obtain a system of two equations in the x and y unknowns:

$$\frac{(K_1 + K_2)\left(R_1R_2x^2y + \rho K_1R_1x\right)}{R_1R_2x^2 + (R_1 + R_2)\,xy + (K_1 + K_2)\,\rho y + \rho K_1R_1x} = G,$$
$$y - x \mp 1 - \frac{M}{N}\frac{R_1R_2x^2 + \rho K_1R_1x}{R_1R_2x^2 + (R_1 + R_2)\,xy + (K_1 + K_2)\,\rho y + \rho K_1R_1x} = 0. \tag{30}$$

As announced, in the general case the solving equation of the system is of the fifth grade, but if $\frac{M}{N} << 1$ the second equation is linear, and the system reduces to a third degree equation of the type $x^3 - \alpha x^2 + \beta x - \gamma$ (equation 8), with

$$\alpha = G\frac{R_1R_2 + R_1 + R_2}{(K_1 + K_2)\,R_1R_2} \mp 1,$$
$$\beta = \left[1 - GK_1\left(1 + \frac{(K_1 + K_2)\,K_1\,(\rho K_2 \pm 1) \pm (R_1 + R_2)}{\rho K_1R_1}\right)\right]\frac{1}{(K_1 + K_2)\,K_2R_1R_2},$$
$$\gamma = \pm G\frac{\rho}{R_1R_2}. \tag{31}$$

From a first inspection of the third definition of system 31 and of the inequalities 10, p-doping results to be incompatible with bistability if recombination is assured by deep donors, because in this case γ would be negative. On the other hand, in case of n-doping, the representative points $(\alpha\,(G), \beta\,(G), \gamma\,(G))$ of the system in the $\alpha\beta\gamma$ space belong to a straight half-line starting from the $\alpha\beta$ plain and lying on the semi-space with $\gamma \geq 0$. In this case, bistability can hold if the straight line passes under the cusp line χ (see fig. 3b, left). This happens if the following necessary and sufficient condition holds:

$$\gamma\,(G_0) < \frac{1}{27}\alpha\,(G_0)^3 \quad \text{with} \quad \beta\,(G_0) = \frac{1}{3}\alpha\,(G_0)^2. \tag{32}$$

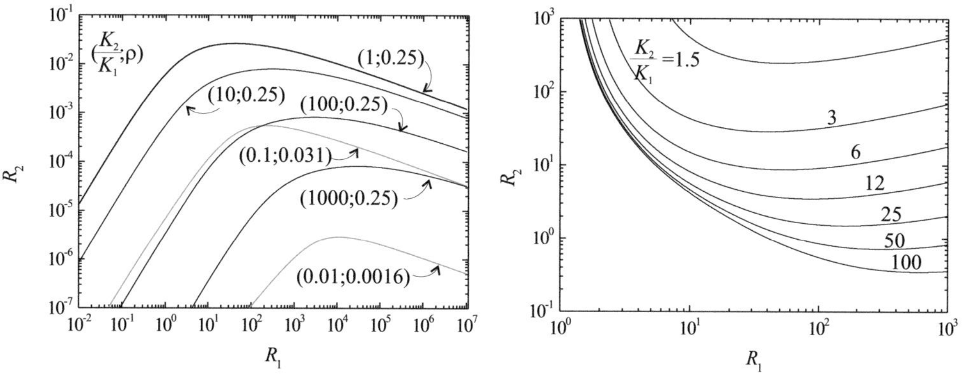

Fig. 7. (left) bistability by excited levels: lines which define the bistable region of the plane R_1R_2 for different values of $\frac{K_2}{K_1}$ and for the value of ρ which make maximum the area of the bistable region itself. (right) bistability by multiply charged centers: lines which define the bistable region of the plane R_1R_2 for different values of $\frac{K_2}{K_1}$.

It is found that the inequality 32 is satisfied if, for given values of $\frac{K_2}{K_1} = \sqrt{\frac{C_{2n}}{C_{1n}}}$ and $\rho = \frac{R_{21}}{N\sqrt{C_{1n}C_{2n}}}$, the points ($R_1$,$R_2$), belong to the area below a given line. Since for a given recombination center the parameter ρ is inversely proportional to the doping density N, the bistable zone below the line can be made maximum by tuning the parameter ρ. In figure 7 (left), several lines are represented with $0.01 \leq \frac{K_2}{K_1} \leq 1000$, with the values of ρ corresponding to the maximum area. From their inspection, bistability results to be possible practically for all the values of R_1, if R_2 is sufficiently small and/or $\frac{K_2}{K_1}$ is not too small or large respect to the unit. As a matter of fact, capture of an electron in the excited state by a positive center seems much more likely than capture of a hole by the neutral excited state, especially if the latter is closer to the conduction band than to the valence one, so that $R_2 = \frac{C_{2p}}{C_{2n}}$ is likely very small. In conclusion, bistable photoconduction in semiconductors, if recombination is assured by deep centers with metastable states, appears likely to be feasible by suitable choice of the doping parameters M and N and of the irradiation conditions.

6. Bistability by multiply ionized recombination centers

Photoconductive instability is possible, in case of multiply charged recombination centers, if the capture of a carrier by a neutral center is less likely than trapping by the center which have the same charge of the carrier itself, in spite of columbic repulsion (see section 4). This seemingly unlikely condition happens to be verified by several transition metal impurities in germanium, silicon, silicon carbide (Grillenberger, 2004; Knack et al., 2002; Sachse et al., 2000; Shiraishi et al., 1999; Wertheim, 1959; Yarykin et al., 1999), both in case of doubly and triply ionized centers. A treatment of photoconduction in semiconductor in which recombination is assured by multiply ionized centers can be found in ref.(Lagomarsino, 2008), where cases of double or triple donors or acceptors, donor-acceptors, double donor-acceptors, ecc. are studied in a general perspective. Here we restrict to the case of double acceptor impurities, which happens to comprehend the more interesting practical case. The rate equations of the system, in this case, are the following (see also fig.5 right):

$$\begin{aligned}
\dot{n} &= g - C_{2n}n\left(n_1 - n_2\right) - C_{1n}n\left(M - n_1\right),\\
\dot{n}_2 &= C_{2n}n\left(n_1 - n_2\right) - C_{2p}pn_2,\\
\dot{n}_1 &= C_{1n}n\left(M - n_1\right) - C_{1p}p\left(n_1 - n_2\right),\\
\dot{p} &= g - C_{2p}pn_2 - C_{1p}p\left(n_1 - n_2\right),\\
n &+ n_1 + n_2 - p - N = 0
\end{aligned} \tag{33}$$

We will consider all the parameters and the unknown quantities expressed in terms of the following, non-dimensional parameters:

$$\begin{aligned}
x &= \frac{p}{n},\ y = \frac{n}{N},\ y_i = \frac{n_i}{M}\ R_i = \frac{C_{ip}}{C_{in}},\\
K_i &= \frac{C_{in}}{\left(C_{1n}C_{2n}\right)^{1/2}},\ G = \frac{g}{NM\left(C_{1n}C_{2n}\right)^{1/2}}.
\end{aligned} \tag{34}$$

As a consequence, the second and the third of Eqs. 33 give, at equilibrium ($\dot{n}_i = 0$), the following conditions:

$$y_1 = \frac{1 + R_2x}{1 + R_2x + R_1R_2x^2},\quad y_2 = \frac{1}{1 + R_2x + R_1R_2x^2}. \tag{35}$$

Their substitution in the first and the last of eqs.33 gives:

$$\begin{aligned}
y &= G\frac{1 + R_2x + R_1R_2x^2}{K_2R_2x + K_1R_1R_2x^2},\\
y &= \frac{1}{1 - x}\left(1 - \frac{M}{N}\frac{2 + R_2x}{1 + R_2x + R_1R_2x^2}\right).
\end{aligned} \tag{36}$$

Although the solving equation of the system, in the general case, is of the fifth degree in the unknown x, if the concentration of deep centers is much lower than that of the dopants ($M << N$), the second of the Eqs.36 reduces to:$y = \frac{1}{1-x}$, and the solving equation reduces to one of the third degree of kind $x^3 - \alpha x^2 + \beta x - \gamma = 0$ (eq.8), with

$$\alpha = 1 - \frac{1}{R_1} - \frac{K_1}{G},\quad \beta = \frac{1}{R_1}\left(\frac{K_2}{G} + \frac{1}{R_2} - 1\right),\quad \gamma = \frac{1}{R_1R_2}. \tag{37}$$

In order to find conditions for bistability, one has to state which values of the parameters K_i and R_i make the coefficients α, β, γ to satisfy the inequalities 10 for some positive value of the control parameter G. Since G is not contained in γ, we can consider, for a given value of γ, the 2-dimensional set ω_γ of the couples (α, β) which satisfies the inequalities 10. The shape of ω_γ is shown in fig.3b (right), with a cusp point $C(\gamma)$ whose coordinates are

$$C\left(\gamma\right) = \left(\alpha_\gamma, \beta_\gamma\right) = \left(3\gamma^{\frac{1}{3}}, 3\gamma^{\frac{2}{3}}\right). \tag{38}$$

By elimination of G in the first two equations of 37, we find that the coefficients α and β define a straight line r with a negative angular coefficient in the plane $\alpha\beta$. This line crosses the region ω_γ if and only if the cusp point C_γ lies below it, that means

$$\beta_\gamma + \frac{K_2}{K_1 R_1}\alpha_\gamma < \frac{1}{R_1}\left[\frac{K_2}{K_1}\left(1-\frac{1}{R_1}\right)+\frac{1}{R_2}-1\right]. \tag{39}$$

From this expression, taking into account expressions 38 and 37, it turns out that recombination bistability is possible if and only if the following inequality holds:

$$3\left(\frac{K_1}{K_2}\left(\frac{R_1}{R_2^2}\right)^{\frac{1}{3}}+\left(\frac{1}{R_1 R_2}\right)^{\frac{1}{3}}\right)+\frac{K_1}{K_2}\left(1-\frac{1}{R_2}\right)+\frac{1}{R_1}<1 \tag{40}$$

This inequality gives a necessary and sufficient condition for having photoconductive multistability driven by the generation factor g. In the three-dimensional space-parameters given by R_1,R_2 and $\frac{K_2}{K_1}$, the region satisfying the inequality 40 is divided into two distinct branches. The first branch lies in the region with $\frac{K_2}{K_1} > 1$ and $R_1 > 1$, the second one in the "slice" $0 < R_2 < 1$. Nevertheless, only the first branch has a physical meaning, because the values of the parameters corresponding to the second branch give bistability only if $G < 0$. The projection of the first branch on the $R_1 R_2$ plane is shown in Fig.7 (right), for several values of $\frac{K_2}{K_1}$.

From the inspection of fig.7 we can set the rule of thumb according to which bistability occur if:

1. $\frac{K_2}{K_1}$ is large enough, and in any case greater than 1. That is, the charged center capture another carrier of the same charge more easily than the neutral one in spite of columbic repulsion.
2. R_1, R_2 are sufficiently large too, and R_1 is in any case greater than 1. That is, the columbic capture cross-section of the charged centers for the opposite-charged carriers is higher than that for the carriers of the same polarity (which is generally true);

condition 1 has been stated in section 4 as a necessary condition for dynamical instability referred to an arbitrary number of states of charge. Relation 40 also gives a sufficient condition for doubly charged centers.

It will be useful for the following considerations to infer the order of magnitude of the generation factor g and the time scale τ involved in the bistable transition. From Eqs. 38 and 37, we have, for the value G_γ at the cusp point $C(\gamma)$:

$$G_\gamma = K_1\left(1-\frac{1}{R_1}-\frac{3}{(R_1 R_2)^{\frac{1}{3}}}\right)^{-1}, \tag{41}$$

which gives, in the limit $R_{1,2} >> 1$ (condition 2):

$$G_\gamma \approx K_1 \text{ thus } g_\gamma \approx NMC_{1n}. \tag{42}$$

The time scale τ can be obtained expressing Eq.33 in term of the adimensional parameters, which gives:

$$\tau\frac{dy}{dt} = G - \sum_{i=1}^{2} K_i y\left(y_{i-1}-y_i\right), \quad \text{with} \quad \tau = \frac{1}{M\left(C_{1n}C_{2n}\right)^{1/2}}. \tag{43}$$

7. One example of possible photoconductive bistable behavior in germanium

In order to design a practical bistable photoconductive device, an accurate determination of the recombination center's trapping parameters is necessary in two respects: first, the values of the capture coefficients of a center having either multiple states of charge or excited metastable levels, define if the center is or not a candidate to exhibit multistability; second, the same values determine which are the more suitable values of the doping densities in order to obtain bistability for the desired radiation irradiances and with the appropriate response time.

Even though several deep centers having metastable electronic states are known, especially in large bandgap semiconductors (Ganichev et al., 1997; Ghosh et al., 1973; Jelezko & Wrachtrup, 2006; Pagonis et al., 2010), there is no plenty of detailed studies about the capture cross-section of both the ground and the excited state. For instance, for native EL2 centers in gallium arsenide, undoubtely one of the most studied (Delaye & Sugg, 1994; Lin et al., 2006; Sudzius et al., 1999; Vincent et al., 1982), $R_1 = C_{1p}/C_{1n}$ (see section 5) is known to vary in the range $10^{-3} - 1$ from the ambient to cryogenic temperatures , the dependence of its recovery rate R_{21} (having an activation energy of about 0.3 eV) on temperature is also reported, but the capture rates of the metastable level are only known to be much smaller than those of the ground level (Delaye & Sugg, 1994), without any indication on the value of their ratio.

In comparison, the capture rates of muptiply charged recombination centers have been studied in more detail, and several transition metal impurities in silicon, germanium and silicon carbide have exhibited a favourable ratio of the higher level cross section to the lower level one ($K_2/K_1 > 1$, see section 6) (Grillenberger, 2004; Knack et al., 2002; Sachse et al., 2000; Shiraishi et al., 1999; Wertheim, 1959; Yarykin et al., 1999). Here we present a study on a center which is likely the most deeply and the longest studied of these, namely Nickel in Germanium. As a matter of fact, the conclusions reached about the capture capture rates of this deep double acceptor center are not unanimous (see tab. 1). To my knowledge, there are three works offering a quite complete picture of the recombination coefficients, based on different experimental tecniques, along with a number of studies giving one or two among the four capture rates. Klassen et al. (1961) obtains $\frac{K_2}{K_1} = 0.14 < 1$ by generation-recombination noise measurements, but Wertheim (1959), Tseng & Li (1972), and Eliseev & Kalashnikov (1963) obtain $\frac{K_2}{K_1} = 6 > 1$, with photoconductive methods. Wertheim and Tseng give three of the four coefficients necessary for the calculation of rel.40 (K_1, K_2, and R_2). Wertheim obtains $20 < R_2 < 80$, which could give bistability, according to rel.40 if $R_1 > 5$, that is: $\sigma_{1p} > 5 \times 10^{-16}\text{cm}^2$. This inequality is largely satisfied by the values of σ_{1p} independently obtained by all the other reserchers (see tab. 1) at sufficiently low temperatures. Tseng and Li, on the other hand, give a value of $R_2 \approx 5$, wich is slightly too small to permit bistability, also with very large values of R_1. In conclusion, there is conclusive evidence, by fotoconductive measurements, that the necessary condition $K_2 > K_1$ is satisfied for Nickel in Germanium; moreover, there are suggestions that also the sufficient condition of rel. 40 is satisfied for this center, at least at low temperatures ($\approx 100K$).

A numerical simulation of the behavior of n-doped germanium, lightly compensated with Nickel, has been carried out, solving eqs. 33. The parameters were those given by Wertheim (1959), along with $\sigma_{1p} \equiv \sigma_{p-/0} = 10^{-13}\text{cm}^2$, which is a conservative estimate based on the values of tab. 1. The doping centers and the recombination centers concentrations were, respectively, $N = 10^{16}\text{cm}^{-3}$, $M = 5 \times 10^{14}\text{cm}^{-3}$. Figure 8 (left) represent the hysteresis cycle of the valence band population, as a function of the generation factor, and Fig.8 (right) represents the response of the valence band population to pulses above and below the bistable

Author method	$\sigma_{n-/0}$ $\sigma_{n--/-}$ (10^{-16}cm^2)	Temperature	$\sigma_{p-/0}$ $\sigma_{p--/-}$ (10^{-16}cm^2)	Temperature
(Wertheim, 1959)	0.96	indep.		
PC	5	indep.	100 – 400	lowers with T
(Tseng & Li, 1972)	1.4	300K		
PC	6.8	300K	40	300K
(Klassen et al., 1961)	17		350	$\sigma_{200K} \times T^{-1.5}$
GR Noise	2.3	300K	> 1000	300 K
(Eliseev & Kalashnikov, 1963)	0.5	300K indep.		
PEM, PC, PC decay	3	300K indep.		
(Kalashnikov & Tissen, 1959)				
PC, PC decay	20	150-300K	200	150-300K
(Kotina et al., 1987)			8800	95-110K
DLTS	180	105-125K		
(Rupprecht, 1961)			7000	$\sigma_{100K} \times T^{-(0.5-1)}$
Pulsed Field				

Table 1. Capture cross sections for Nickel centers in Germanium, as measured by several authors. All the authors but Klaassen which measured $\sigma_{n-/0}$ and $\sigma_{n--/-}$ give a ratio $\frac{\sigma_{n--/-}}{\sigma_{n-/0}}$ of about 6, compatible with bistability.

interval. For the values of the parameters given above, bistability occurs for generation factors $g \approx 7.5 \times 10^{21}\text{cm}^{-3}\text{s}^{-1}$ which are obtained, for an over band-gap radiation with a penetration lenght of 100μm, at a power density of about 8W/cm^2. Transitions ON-OFF and OFF-ON occurs with pulses of at least 2μs of duration. The transition time and the power density, according to rel. 43 and 42, scale as $\frac{1}{M}$ and $N \times M$, respectively, which could give a slight freedom of choice in the bandgap engineering to achieve a bistable photoconducting behavior.

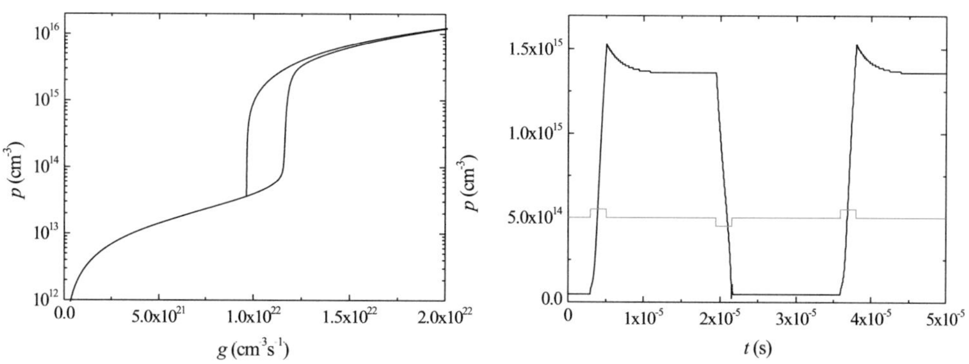

Fig. 8. (left) Hysteresis cycle of the valence band population, as a function of the generation factor, for n-doped germanium lightly compensated by nickel centers (parameters in section 7). (right) Responce of the valence band population to pulses above and below the bistable interval.

If a photoconduction experiment is performed in the current regime (Bass et al., 1970), with a ballast resistance between the sample and a DC power supply, the bistability effect should be detected also with an electric field as low as 1Vcm^{-1}, measuring the voltage drop across the photoconductor. This electric field intensity is well below the limit for impact ionization of impurities in germanium (Kahn et al., 1992), and could not be mistaken with any avalance-related bistability effect.

8. Conclusions

Instability in photoconduction is a long recognized issue, whose study, both experimental and theoretical, has been grounded on the basis of avalance-related phenomena. This approach has been fruitful in term of comprehension of an amount of phenomena involved: spatial patternization, temporal behavior including chaotic oscillations in semiconductors has been deeply studied and accounted for. Nevertheless, other routes toward instability or multistability, potentially more reliable for the applications, has been neglected, in a way that none of the possible instability mechanisms in ionization-recombination processes, apart from those involving impact ionization, has been explored before the work exposed in this chapter. We have realized that, even without the intervention of any electric field, instability in photoconduction processes can occur due to competition between mutually interacting recombination channels, namely in case of deep recombination centers with metastable excited states and in the presence of multiple ionization of recombination impurities. Moreover, the variational analysis of the problem showed that these are the only two cases which admit recombination instability in absence of avalance-related processes. Sufficient, close algebraic conditions on the recombination coefficients has been stated assuring bistabilty in this two cases, and some deep centers which are candidate to exhibit recombination bistability has been identified. Experimental work is worth doing, on one hand to obtain a deeper characterization of the recombination coefficients of the centers which are potential candidate to assure photoconductive bistability, on the other, to test the effective ability to trigger instability of the centers which, at the moment, seems to be the most promising candidates.

9. References

Aoki, K. (2001). *Nonlinear Dinamics and Chaos in Semiconductors*, Institute of Physics Publishing, Bristol and Philadelphia.

Bass, F., Bochov, V. & Gurevich, Y. G. (1970). *Sov. Phys. JEPT* 31: 972.

Bel'kov, V. V., Hirschinger, J., Novak, V., F.-J.Niedernostheide, Ganichev, S. & Prettl, W. (1999). *Nature* 397, 4 February 1999: 398.

Berman, A. & Plemmons, R. (1994). *Nonnegative Matrices in the Mathematical Sciences*, Society for Industrial and Applied Mathematics, Phyladelphia.

Bonch-Bruevich, V. L., Zvyagin, I. P. & Mironov, A. (1975). *Domain Electrical Instabilities in Semiconductors*, Consultants Bureau, New York.

Brandl, A. & Prettl, W. (1991). *Phys. Rev. Lett.* 66: 3044.

Crandal, R. (1970). *Phys. Rev.* 1 (2): 730–738.

Delaye, P. & Sugg, B. (1994). Temperature enhancement of the photorefractive effect in gaas due to the metastable state of the el2 defect, *Phys. Rev. B* 50: 16973.

Eliseev, P. & Kalashnikov, S. (1963). *Sov. Phys. Solid State* 5: 233.

Frenkel, J. (1938). *Phys. Rev.* 54: 647.

Ganichev, S. D., Raab, W., Zepezauer, E., Prettl, W. & Yassievich, I. N. (1997). Storage of electrons in shallow donor excited states of gap:te, *Phys. Rev. B* 55: 9243.

Ghosh, A. K., Lauer, R. & Jr., R. A. (1973). Photoconductivity and photoluminescence kinetics associated with the luminescence center in rutile (tio2), *Phys. Rev. B* 8: 4842.

Golik, L., Gutman, M. M. & Pakseev, V. E. (1990). Nonlinear oscillations and chaos in cds single crystals under temperature-electric instability conditions, *phys. stat. sol. (b)* 199(162).

Grillenberger (2004). *Phys. Rev. B* 70: 205209.

Hearon, J. (1963). *Ann NY Acad Sci* 108: 36.

Hirschinger, J., Niedernostheide, F.-J., W.Prettl & V.Novàk (2000). *Phys. Rev. B* 61: 1952.

Jelezko, F. & Wrachtrup, J. (2006). phys. stat. sol. (a), *Unknown Journal* 203: 3207.

Kahn, A. M., Mar, D. J. & Westervelt, R. M. (1992). Spatial measurements near the instability threshold in ultrapure ge, *Phys. Rev. B* 45: 8342.

Kalashnikov, S. G. & Tissen, K. P. (1959). Unknown title, *Sov. Phys. Solid State* 1: 1603.

K.Aoki & K.Yamamoto (1989). *Appl. Phys. A* 48: 111.

Khosla, R., Fisher, J. & Burkey, B. (1973). *Phys. Rev. B* 7 (6): 2551–2564.

Kim, C.-G. & Yung, Y.-D. (2004). *J. Appl. Phys.* 96 (1): 913.

Klassen, F., Blok, J. & Booy, H. (1961). *Physica* 27: 48.

Knack, S., Weber, J., Lemke, H. & Riemann, H. (2002). *Phys. Rev. B* 65: 165203.

Koenig, S. (1958). *Phys. Rev.* 110: 986–988.

Koepp, R. & Urbelis, A. (1968). Low frequency photocurrent oscillations in cds crystals with thermally induced negative differential conductivity, *Phys. stat. sol.* 30: K5.

Kotina, I., Kuryatkov, V., Novikov, S. & Pirozhkova, I. (1987). Unknown title, *Sov. Phys. Semicond.* 21: 635.

Lagomarsino, S. (2007). Instability in nonautocatalytic ionization-recombination processes, *Phys. Rev. B* 75: 024305.

Lagomarsino, S. (2008). Optoelectronic bistability effect in semiconductors, *Phys. Rev. B* 78: 125201.

Landsberg, P. & Pimpale, A. (1976). *J. Phys. C.* 9: 1243–1251.

Lin, A. L., Omelianovski, E. & Bube, R. H. (2006). Photoelectronic properties of high-resistivity gaas:o, *Journal of Applied physics* 47(1852).

Lorentz, E. (1963). *Journal of the atmospheric sciences* 20: 130–141.

Mcwhorter, A. & Rediker, R. (1959). *Proceedings of the IRE* 47: 1207–1213.

Menoret, M. (1968). *U.S. PATENT* 3404383.

Minorsky, N. (1962). *Nonlinear Oscillations*, D. Van Nostrand Company, Inc., Princeton, New Jersey.

Pagonis, V., C.Ankjærgaard, A.S.Murray, M.Jain, R.Chen, J.Lawless & S.Greilich (2010). Modelling the thermal quenching mechanism in quartz based on time-resolved optically stimulated luminescence, *Journal of luminescence* 130: 902.

Reggiani, L. & Mitin, V. (1989). *Rev. Nuov. Cim.* 12: 1.

Roessler, O. (1976). *Phys. Lett.* 57A: 397.

Rupprecht, G. (1961). Unknown title, *J. Phys. Chem. Solids* 22: 255.

Sachse, J.-U., Weber, J. & Lemke, H. (2000). *Phys. Rev. B* 61: 1924.

Schöll, E. (1979). *Proc. R. Soc. Lond. A.* 365: 511–521.

Schöll, E. (1982). *Z. Phys. B* 46: 23–30.

Schöll, E. (1987). *Nonequilibrium Phase Transitions in Semiconductors - Self-organization Induced by Generation and Recombination Processes*, Springer, Berlin.

Shiraishi, M., Sachse, J.-U., Lemke, H. & Weber, J. (1999). *Material Science and Engineering* B58: 130.

Stobbe, M., Redmer, R. & Schattke, W. (1994). *Phys. Rev. B* 49 (7): 4494.

Sudzius, M., Bastys, A. & Jarasiunas, K. (1999). Optical nonlinearities at transient quenching of el2 defect at room temperature, *Optics Communications* 170: 149.

Todd, J. (1962). *Survey of Numerical Analysis*, McGraw-Hill Book Company, Inc., New York.

Tseng, H. F. & Li, S. S. (1972). *Phys. Rev. B* 6 (8): 3066.

Vincent, G., Bois, D. & Chantre, A. (1982). Photoelectric memory effect in gaas, *Journal of Applied Physics* 53: 3643.

Wertheim, G. (1959). *Phys. Rev.* 115: 37.

Yarykin, N., Sachse, J.-U., Lemke, H. & Weber, J. (1999). *Phys. Rev. B* 59: 5551.

Yoffa, E. (1981). *Phys. Rev. B* 23 (4): 1909.

Part 6

Laser Devices and Methods

Electromechanical 3D Optoelectronic Scanners: Resolution Constraints and Possible Ways of Improvement

Oleg Sergiyenko[1], Vera Tyrsa[2], Luís C. Basaca-Preciado[1], Julio C. Rodríguez-Quiñones[1], Wilmar Hernández[4], Juan I. Nieto-Hipólito[1], Moisés Rivas Lopez[1] and Oleg Starostenko[3]
[1]*Autonomous University of Baja California, Mexicali-Ensenada,*
[2]*Polytechnic University of Baja California, Mexicali,*
[3]*Universidad de Las Americas, Puebla,*
[4]*Polytechnic University of Madrid,*
[1,2,3]*Mexico*
[4]*Spain*

1. Introduction

Non-Contact optoelectronic 3D measurement is a rapidly growing field. Three-Dimensional Non-Contact Measurement Technologies are very common for research due to multiple practical applications expecting for its benefits. Many fields are using in any way 3D measurements or shape recognition, some of them there are vision assisted assembly in various branches of industry, autonomous mobile robots navigation, structural health monitoring, micro surfaces inspections, precise automated surgery, etc.

In this chapter it is expedient to mention and briefly cross-compare the following emerging technologies for 3D measurements: laser scanners, lasers based on conoscopy holography technology and 3D cameras.

Laser scanners: Most contemporary non-contact 3D measurement devices are based on laser range scanning. The simplest devices (Fischer, 2007) are based on the laser triangulation technique. This is an active stereoscopic technique in which the distance of the object is computed by means of a directional light source and a video camera. The CCD camera's 2D array captures the image of surface profile and digitizes all data points along the laser disadvantage of this method is that a single camera collects only a small percentage of the reflected energy. The amount of the collected energy can be drastically increased by trapping the entire reflection cone, thus significantly increasing the precision and reliability of the measurements.

Lasers based on Conoscopic Holography technology: Conoscopic Holography is a simple implementation of a particular type of polarized light interference process based on crystal optics. In the basic interference set-up, a point of light is projected onto a diffuse object. This point creates a light point, which diffuses light in every direction. In a conoscopic system, a complete solid angle of the diffused light is analyzed by the system. The measurement process retrieves the distance of the light point from a fixed reference plane. The problem inherent in laser scanning is its relatively low measurement speed, though it is faster than traditional contact Coordinate measurements machines (CMM's).

a)

b)

Fig. 1. A scanner laser (a) and a 3D camera (b)

3D cameras: 3D photography is based on reconstructing 3D data from 2D images, taken from different points of view (stereo- graphic) the basic problem with this approach is the correspondence problem. In 3D cameras a pattern is projected on the object and the same pattern points are identified on each image. This approach is much more efficient, since it does not require marking specific points, and it can produce a very large number of measurements in one shot of camera.

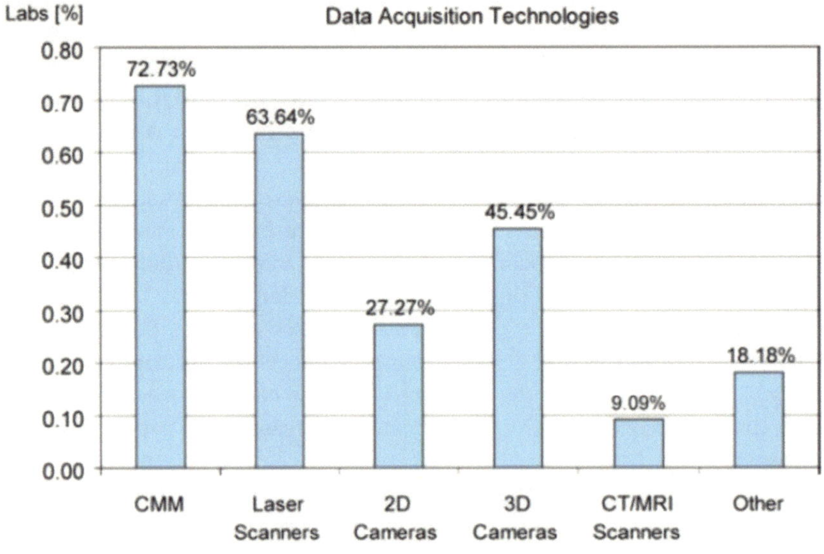

Fig. 2. Distribution of 3D scanning technologies among 23 engineering universities and research centers across Europe according to (Fischer, 2007)

The typical distribution of 3D scanning technologies research among engineering universities and research centers is given on Fig.2. CMM is still numerous due to non-innovative routines in industry with traditional investments for research. But this technology is contact and applicable for very limited area of tasks. So, the future is for laser scanning and 3D camera vision.

In fact, specialists already get a consensus regard to most evidences of these two branches. Cameras are winning in a point of less energy consuming and a relatively longer range of action. But laser scanning systems always win at resolution, accuracy and data processing time.

Basing in this, it is easy to show that for various tasks (such as: vision assisted assembly, robot navigation in densely cluttered environment, structural health monitoring, etc.) the laser scanner is an optimal way to obtain quality 3D information about objects in nearest surrounding in higher resolution.

But the majority of modern laser scanners have a certain part of rotating/moving electromechanical components, which are closely related with such negative phenomenon like mechanical vibrations, friction and wear, mechanical delays, etc.

Present paper scope is precisely to research what are the most prospective ways to increase the electromechanical laser scanners resolution and robustness. With this aim let us deeper study most typical approaches to laser scanner construction, as one of the most promising technical vision system (TVS).

2. Typical laser scanner constructions and their constraints

According to the recent literature electromechanical laser scanners have sufficiently variety in its constructions. At the same time, they have enough similarities and common components in its general structure. Let will see some of the typical constructions of electromechanical laser scanners for to have the possibility systemize such constructions and define their common advantages and lacks.

An obvious optimization of the measurement system (Wulf & Wagner, 2003) is to take the scanning method that is most suitable for the application. However, taking a 2D laser scanner with 180° scanning range and a servo drive it results in a number of possible combinations of scan planes and rotation axis to get a 3D scan. This section describes four of these combinations. We have named the scanning methods as *pitching scan, rolling scan, yawing scan* and *yawing scan top*. The *pitching scan* (Fig. 3a) has a horizontal scan plane and is pitching up and down. This method is for example used in (Surmann et al., 2001) and (Hähnel & Burgard, 2002). A method that is newly introduced here is the *rolling scan* (Fig. 3b). This scan is rotating around the center of the scanner, with the advantage of only one focus point in front of the sensor. The *yawing scan* (Fig. 3c) and the *yawing scan top* (Fig. 3d) has a vertical scan plane and is rotating around the upright z-axis. This method is used e.g. by (Wulf, O. & Wagner, B. (2003).

Let us compare constructions and advantages/disadvantages of the different laser scanner constructions according to the next literature sources: (Son et al., 2002), (Wulf & Wagner, 2003), (Nüchter, 2007), (Nüchter, 2008), (Wulf et al., 2004), (Surmann, 2003), (Surmann et al., 2001), (Hähnel & Burgard, 2002), (Blais et al., 1988), (Blais et al., 2000), (Beraldin et al., 2000), (Andersen et al., 2006), (Laurin et al., 1996), (Blais et al., 1991), (Klöör et al., 1993), (Vandapel et al., 2004), (Montemerlo & Thrun, 2004), (Pagnottelli et al., 2005), (Sergiyenko et al., 2006).

a)

b)

c)

d)

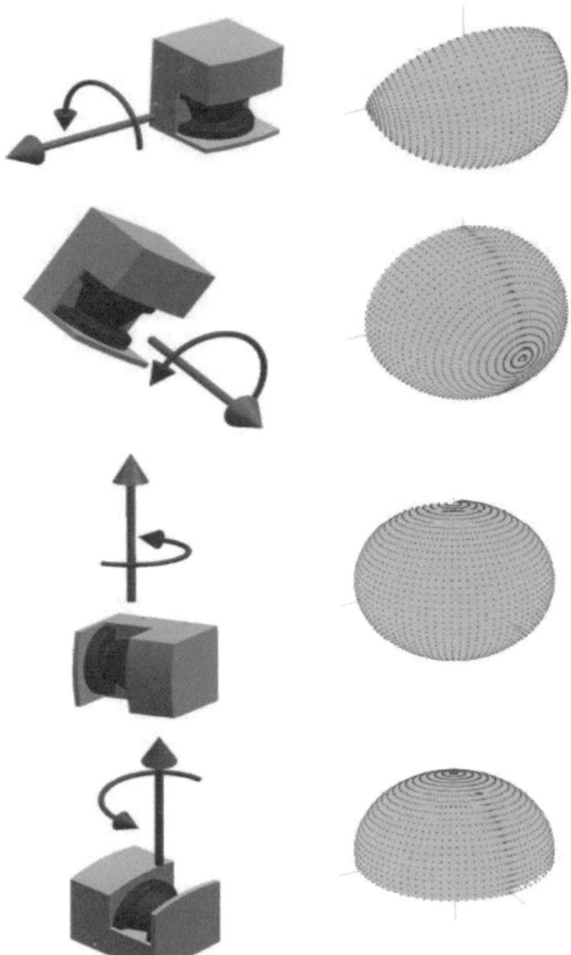

Fig. 3. The *pitching scan* (a), the *rolling scan* (b), the *yawing scan* (c) and the *yawing scan top* (d) (Wulf & Wagner, 2003)

2.1 Automated laser scanning system for reverse engineering and inspection

The mechanism of the 3D laser scanner used in this research (Son et al., 2002) is illustrated in Fig. 4. A laser stripe is projected onto a surface and the reflected beam is detected by CCD cameras. Through image processing and triangulation method, three-dimensional coordinates are acquired. The laser probe is mounted on a three-axis transport mechanism and moves along the scan path that consists of a series of predetermined line segments. It also rotates in two directions.

When the laser scanner captures an image, the system automatically finds an optical focus and keeps the standoff distance. The length of laser stripe and the stand-off distance cannot be changed by an operator. Since a laser scanner consists of optical sensors and mechanical moving parts, various constraints must be satisfied when measuring a certain point on a part (Fig. 5). The goal of this section is to generate an optimal scan plan that satisfies the following major constraints (Zussman et al., 1994):

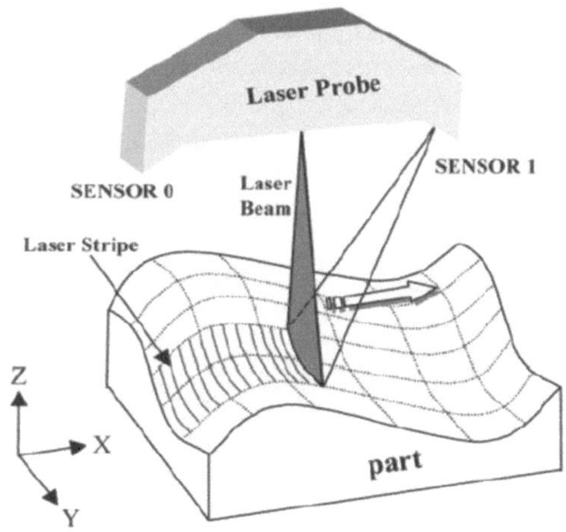

Fig. 4. Laser scanning mechanism

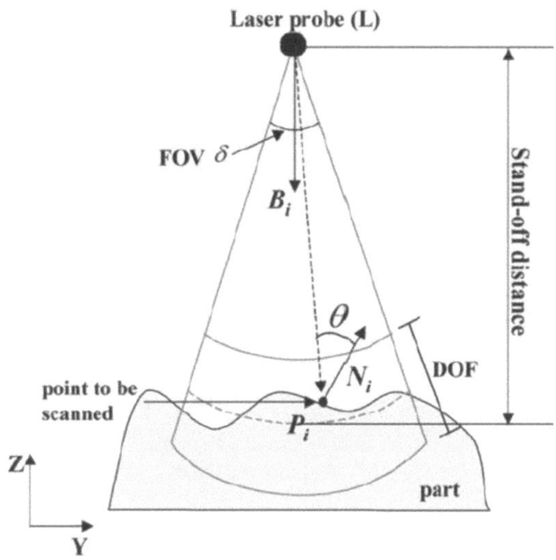

Fig. 5. Constraints for laser scanning

1. View angle: the angle between the incident laser beam and the surface normal of a point being measured should be less than the maximum view angle γ

$$d_i \bullet N_i \geq \cos(\gamma), \tag{1}$$

where

$$d_i = \frac{L - P_i}{|L - P_i|}. \tag{2}$$

2. FOV: the measured point should be located within the length of a laser stripe

$$\left(-d_i\right)\bullet B_i \geq \cos\left(\frac{\delta}{2}\right), \tag{3}$$

where δ is the FOV angle

3. DOF: the measured point should be within a specified range of distance from the laser source

$$l_{STAND} - \frac{l_{DOF}}{2} \leq \left|L - P_i\right| \leq l_{STAND} + \frac{l_{DOF}}{2}, \tag{4}$$

where l_{STAND} and l_{DOF} denotes stand-off distance and DOF length.

4. Occlusion: the incident beam as well as the reflected beam must not interfere with the part itself.
5. The laser probe should travel along a path that is collision-free.
6. If the part is shiny or transparent, preprocessing is required such as spraying.

2.2 Range error analysis of an integrated time-of-flight, triangulation and photogrammetric 3D laser scanning system

Laser scanner model (Blais et al., 2000). Models must be able to relate design parameters, laboratory measurable, and operational performance. Figure 6 illustrates the major subsystems of an active electro-optical system: projector sources and collimating optics, deflection mechanism, collecting optics, detector, signal conditioning and processing, and final output. The collecting optics images the radiation onto the detector. In the example of Figure 7, the scanner optically moves the detector's instantaneous field-of-view (IFOV) across the total field-of-view (FOV) to produce an output voltage (signal) proportional to the local scene intensity (produced by ambient light conditions) and the laser light reflected back from a reflective surface.

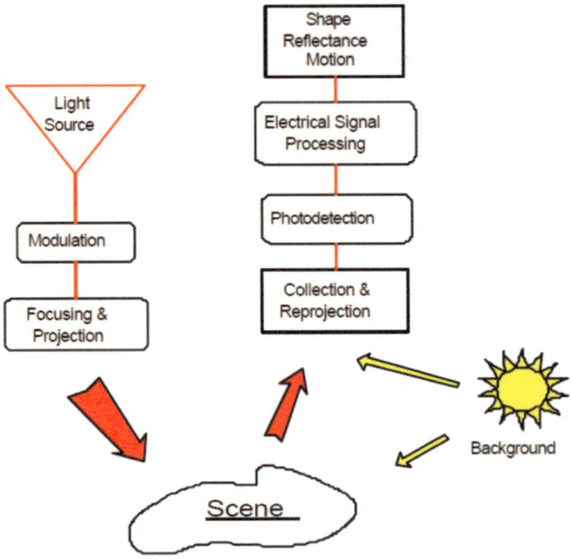

Fig. 6. Generic sensor operation applied to active electro-optical systems

The detector is at the heart of the electro-optical system because it converts the scene radiation (reflected flux) into a measurable electrical signal. Amplification and signal processing creates a signal in which voltage differences represent scene intensity differences due to various objects in the field-of-view.

The majority of electro-optical quality discussions are centered on resolution and sensitivity evaluation. System sensitivity deals with the smallest signal that can be detected. It is usually taken as the signal that produces a signal-to-noise ratio of unity at the system output. Sensitivity is dependent upon the light-gathering properties of the optical system, the responsivity of the detector, the noise of the system and, for this application, the background flux. It is independent of resolution.

In the case of metrology, resolution is not sufficient and stability and accuracy must also be considered. Resolution has been in use so long that it is thought to be something fundamental, which uniquely determines system performances and is often confused with accuracy. It is often specified by a variety of sometimes-unrelated metrics such as the Airy disk angular size, the detector subtense, or the sampling frequency. Resolution does not usually include the effect of system noise.

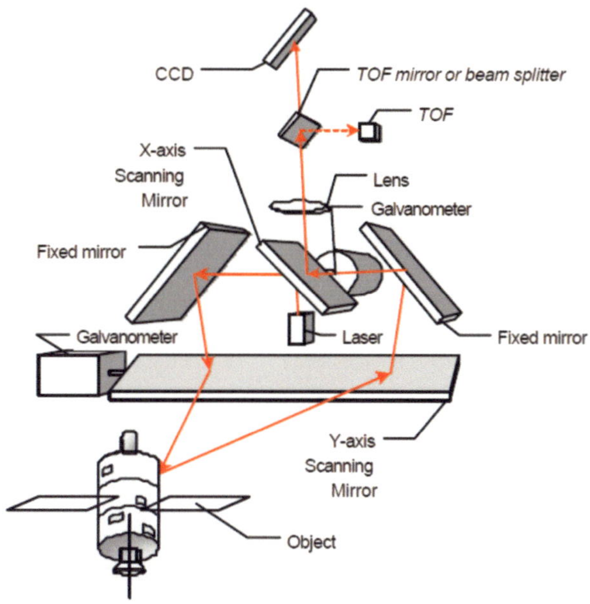

Fig. 7. Schematic representation of the auto-synchronized geometry

2.3 The auto-synchronized laser scanner

Figure 7 shows a photograph of the prototype of the auto-synchronized laser scanner developed for this demonstration. The scanner uses a variation of the auto-synchronized triangulation range sensor based on one galvanometer (Blais et al., 1988). The system comprises two orthogonally mounted scanning mirrors and a linear discrete-response photosensitive position device (e.g. linear CCD) used for short to medium range measurement as triangulation (Beraldin et al., 2000). An optional avalanche photo-diode-based Time-of-Flight (LIDAR) ranging module is used for longer-range measurement

(Laurin et al., 1996). Only resolved targets using TOF are considered for this application. Laser illumination is provided using a laser source coupled to a single-mode fiber, either pulsed (TOF mode) or CW (triangulation mode). The laser scanner operates at a relatively eye safe wavelength of 1.5 mm (compared to visible laser wavelengths).

The basic concept of auto-synchronization is that the projection of the light spot is synchronized with its detection as illustrated in Figure 7. The instantaneous field of view (IFOV) of the position sensor follows the spot as it scans the scene. Therefore, an external optical perturbation can potentially interfere with the detection only when it intersects the instantaneous field of view (IFOV) of the scanner. At this level, electronic signal processing is used to filter these false readings to obtain correct 3-D measurement (Blais et al., 1991). With synchronization, the total field of view (FOV) of the scanner is related to the scanning angles of the galvanometers and mirrors as opposed to a conventional camera-based triangulation. Here the field of view and image resolution are intimately linked (Blais et al., 1988); a large field of view produces a lower pixel resolution. In summary, the instantaneous field-of-view of the scanner plays a major role in the system sensitivity analysis.

2.4 Range measurement

Figure 7 shows the optical geometry of the auto-synchronized laser scanner. The laser scanner system can measure range information for each voxel (3-D pixel) in the scene using two modes of operation: (1) triangulation as illustrated in Figure 8, and (2) time-of-flight shown in Figure 9. It is beyond the scope of this paper to discuss the details of operation of the scanner and the exact mathematical model. This information is available from previous publications where the scanner is operated in imaging mode (Blais et al., 1988; Beraldin et al., 1993; Beraldin et al., 2000). Here, we will use the simplified models illustrated in Figure 10 to model range measurement and to associate object pose estimation obtained using video camera models shown in Figures 10 and 11 and techniques discussed in section 4.

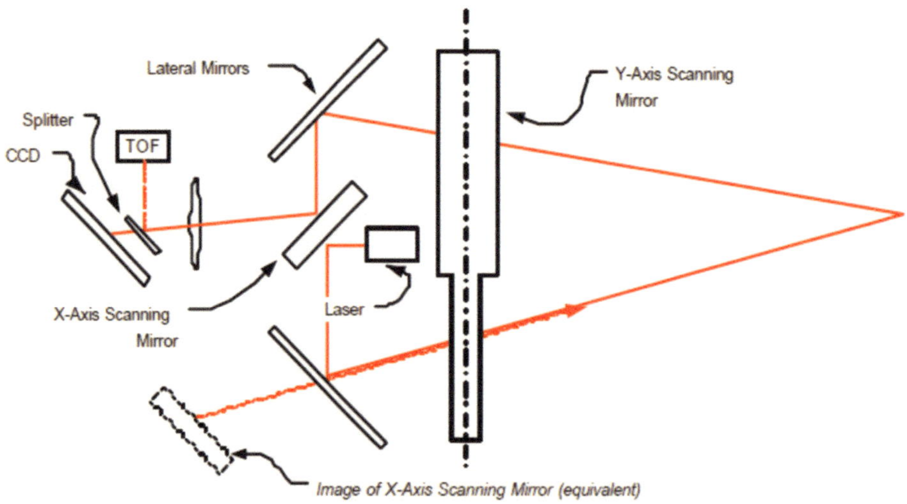

Fig. 8. Simplified geometry of the laser scanner for the triangulation mode

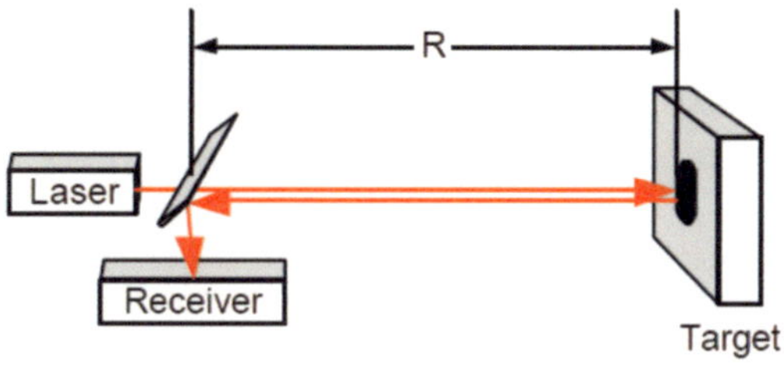

Fig. 9. Schematic of the time-of-flight principle

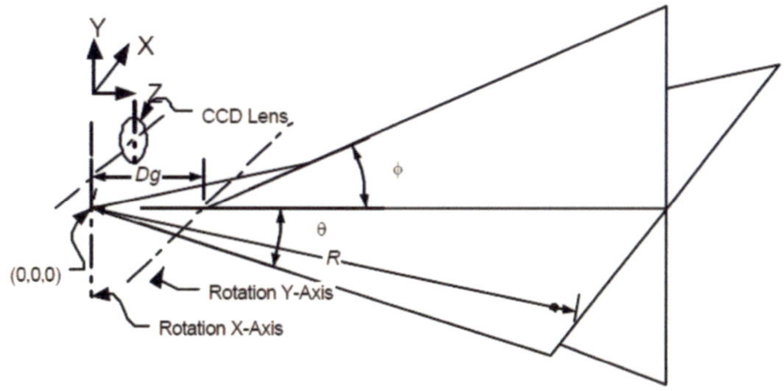

Fig. 10. Simplified geometrical model of the laser scanner showing the effect of astigmatism between the X and Y scanning axis

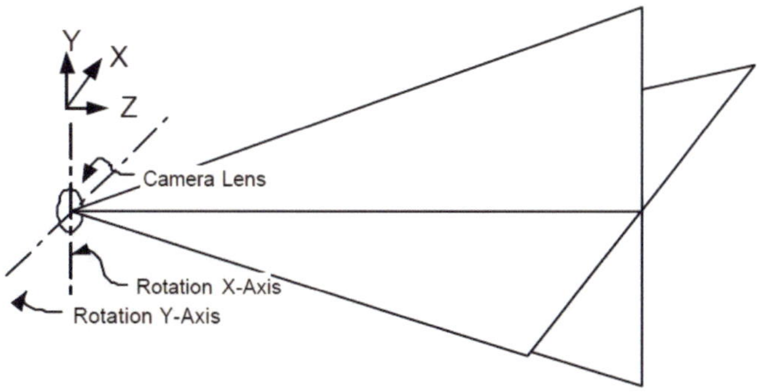

Fig. 11. Simplified geometrical model of a simple camera lens system used by conventional camera and photogrammetric methods

From (Blais et al., 1988), knowing that $R=z/\cos(\theta)$, and from Figure 10, range R can be calculated either using triangulation methods or TOF. The simplified aberrations free model is presented here. For triangulation, range is given by

$$R_{Trian} = \frac{f \cdot d}{p}\cos(\theta) + d\sin(\theta) \tag{5}$$

where f is the focal length of the lens, d is the triangulation base, θ is the deflection angle following the x-axis, and p is the position of the imaged laser spot of the position sensor (see (Blais et al., 1988) for details). For the TOF method of Figure 9, range is simply obtained based on the speed of light c and the propagation delay τ of a laser pulse:

$$R_{TOF} = c\frac{\tau}{2} \tag{6}$$

From Figure 10, the x-y-z coordinates of a point are

$$\begin{bmatrix} x \\ y \\ z \end{bmatrix} = R \cdot \begin{bmatrix} \sin(\theta) \\ (\cos(\theta) - \psi)\sin(\theta) \\ (1-\cos(\theta))\psi + \cos(\theta)\cos(\theta) \end{bmatrix} \tag{7}$$

where θ and φ are the deflection angles, and $\psi = Dg/R$ where Dg is the separation between the two scanning axis shown in Figure 10. Range R is obtained using either R_{Trian} or R_{TOF} depending on the operating mode of the scanner. Because Dg<<R, error propagation calculations (in triangulation mode) can be approximated by

$$\Delta R_{Trian} \approx \frac{R^2}{f \cdot d} \Delta p \tag{8}$$

$$\begin{bmatrix} \Delta x \\ \Delta y \\ \Delta z \end{bmatrix}^2 = \begin{bmatrix} \sin(\theta) \\ \cos(\theta)\cdot\sin(\varphi) \\ \cos(\theta)\cos(\varphi) \end{bmatrix}^2 \left(\frac{R^2}{f \cdot d}\Delta p\right)^2 + \\ + \begin{bmatrix} \cos(\theta) \\ -\sin(\theta)\cdot\sin(\varphi) \\ -\sin(\theta)\cos(\varphi) \end{bmatrix}^2 R^2 \cdot \Delta\theta^2 + \begin{bmatrix} 0 \\ \cos(\varphi) \\ -\sin(\varphi) \end{bmatrix}^2 R^2\cos^2(\theta)\cdot\Delta\varphi^2 \tag{9}$$

where Δp is the uncertainty associated with the laser spot measurement.

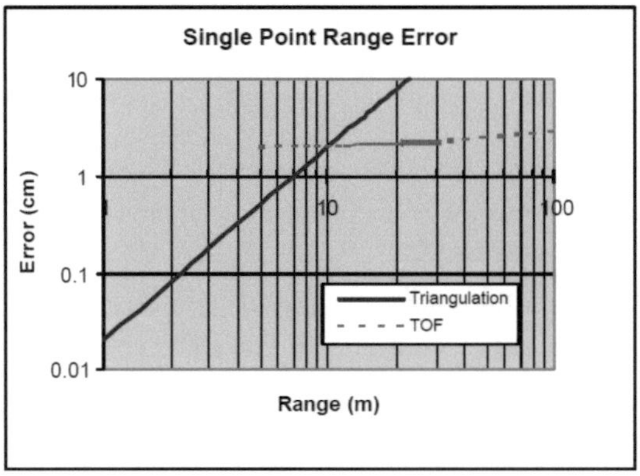

Fig. 12. Range error accuracy of the Laser Scanner System

Figure 12 shows range error measured with the scanner in triangulation (notice the R^2 dependence of the error), and from the manufacturer specifications for the TOF mode of operation (notice the almost flat error over range). Other typical system parameters for the prototype used in Figure 7 are maximum deflection angles of 0.5 rad (30 deg) and angular errors of 50-100 μrad, depending on the target array. From equation 9 and Figure 12, the total system error, for medium to long range, is mostly contributed by range error measurement ΔR, i.e., Δp the uncertainty associated with the laser spot measurement.

2.5 Traversable terrain classification for outdoor autonomous robots using single 2D laser scans

Autonomous navigation by mobile robots in unstructured or semi-structured outdoor environments presents a considerable challenge (Andersen et al., 2006). Adequately solving this challenge would allow robotic applications within industries such as agriculture, mining and logging. To achieve the level of autonomy required for such operation, a robot must be able to perceive and interpret the environment in a meaningful way. Limitations in current sensing technology coupled with a dynamically changing unknown environment, and difficulties in modeling the interaction between robot and terrain all make this task difficult.

The sensors used for this task could include vision, ultrasound, radar and laser scanner, both as combined solutions and each sensor type used individually. Multi sensor solutions would be expected to provide superior results but computational efforts and requirements of fast real-time classification make it interesting to explore what would be achievable with a single sensor.

Current work in the area tends to focus on using 3D laser scanners, vision or a combination of 3Dlaser scanners and vision. In vision solutions (Bertozzi & Broggi, 1997) argues that this problem can be divided into two sub-problems: Lane following and obstacle detection, and describes a stereo vision based solution for both. Edge detection in vision systems is one of the possibilities to identify road borders and had some success already in 1986 as described in (Wallace et al., 1986), color segmentation is used in (Fernandez & Price, 2005) for tracking of dirt roads, road types similar to the ones used in this paper. A method for road following using vision and neural network was developed by (Jochem et al., 1993). A combined stereo vision and 2D laser scanner solution for outdoor obstacle avoidance is presented in (Pagnottelli et al., 2005). Using 3D laser scanner solutions have been proposed by (Vandapel et al., 2004) by transforming point clouds into linear features, surfaces, and scatter. These were classified using a Bayesian filter based on a manually classified training set. Identification of navigable terrain using a 3D laser scanner by checking if all height measurements in the vicinity of a range reading had less than a few centimeters deviation is described in (Montemerlo & Thrun, 2004). An algorithm that distinguished compressible grass (which is traversable) from obstacles such as rocks using spatial coherence techniques with an omni-directional single line laser was developed by (Macedo et al., 2000). A method of traversing a rock field based on three metrics extracted from stereovision was likewise developed by (Wettergreen et al., 2005). A tactile and vibration-based approach to terrain classification was proposed by (Legnemma et al., 2004). A method for detection and tracking the vertical edges of the curbstones bordering the road, using a 2D laser scanner, was developed by (Wijesoma et al., 2004). Detection of borders or obstacles using laser scanners are often used both indoors and in populated outdoors environments, and is the favored method when the purpose includes map building, as in (Guivant et al., 2001; Klöör et al., 1993).

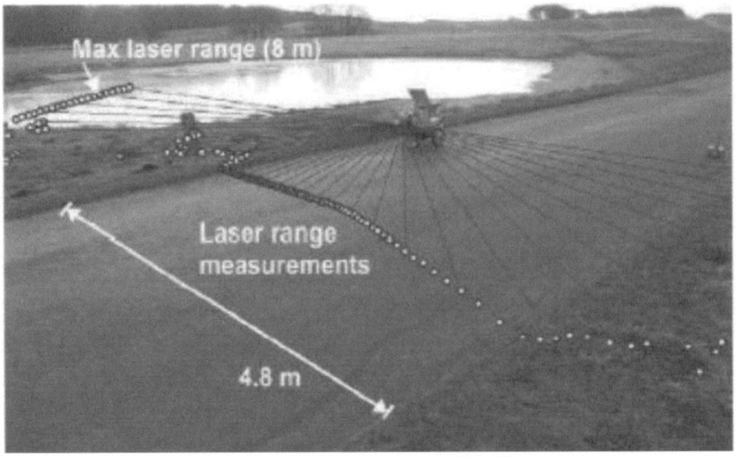

Fig. 13. The robot uses a laser scanner to detect the terrain in front of the robot

Fig. 14. The robot used in the experiments has a laser scanner at a height of 41 cm looking down in an angle of $\theta_L = 9°$. The road is therefore detected at 2.6 m. The position of the last scan (scan n − 1) is shown to illustrate the largest undetected obstacle. The unseen object is not detected in scan n − 1 and is too short to be seen by scan n

2.6 Constructing a 3D laser range finder

Now we will consider a three-dimensional laser range finder and its design.

The presented 3D laser range finder is built on the basis of a 2D range finder by extension with a mount and a servomotor. The 2D laser range finder is attached to the mount so that it can be rotated. The rotation axis is horizontal. A standard low-cost servomotor is connected on the left side (fig. 15).

It is an alternative approach if we have to rotate the range finder around the vertical axis. In (Surmann at al. 2001) it is only discussed the approach based on a horizontal rotation, but all presented algorithms can be used in the same way. The differences between both approaches are the orientation of the apex angle and the effects of dynamic objects moving through the scene, e.g. persons. Using vertical scanning, a perturbation either appears with low probability within a few scans making them useless for further data processing, or does not appear at all. The first approach on the other hand shows perturbations throughout the whole scene, but these data can still be used for robot navigation and object detection.

Fig. 15. A 3D laser range finder for autonomous mobile robots. The servo is mounted at the left side. A camera on top is used to get texture images for the realistic appearances of a 3D scene (Surmann at al. 2001)

The given setup determines an intrinsic order of the acquired data. The data coming from the 2D laser range finder is ordered anticlockwise. In addition the 2D scans (scanned planes) are ordered due to the rotation. A digital camera for texture mapping is mounted on top of the 3D laser range finder. While rotating, several photos are taken so that the texture mapping can be applied to a larger area. The 3D laser range finder uses only standard interfaces of a computer. The servomotor is directly connected to the parallel port, the 2D laser range finder to the serial port and the camera is connected to an USB port. Nowadays, every computer (esp. laptops) does have these interfaces and the built 3D laser range finder can therefore easily be used on mobile platforms. The mount and the servo are inexpensive and no special electronic devices are used, so the price of the whole system mainly depends on the used 2D laser range finder.

3. Typical constraints of laser scanners constructions

As evident from above analysis, in the different laser scanner constructions by its nature are involved the next typical constraints and limitations:

- a limited geometry of the field-of-view with non-uniformly distributed uncertainty of the sinus law application (Zussman et al., 1994), (Montemerlo & Thrun, 2004), (Sergiyenko et al. [a]; 2009), (Sergiyenko et al. [b]; 2009).
- non-uniform spatial distribution of the scanning tool - laser ray - along the variable striking distance (Guivant et al., 2001; Klöör et al., 1993) , (Surmann at al. 2001);
- not balanced torque of the rotating electromechanical system; wear process of friction surfaces and rocking of electromotor shafts (Surmann at al. 2001), (Pagnottelli et al., 2005);
- unwanted quaziperiodical noise of the complex nature in a photodetecting circuit (Básaca [a] et al., 2010);
- limited range of action due to non-lineal photodetected signal attenuation (Sergiyenko et al. a; 2009), (Sergiyenko O.Yu.; 2010).

The most prospective ways of the mentioned constraints decrease is a subject under consideration it the further sections.

4. Limitations decrease of the field-of-view geometry design

As we can observe from the several different laser scanner constructions mentioned above, they have a common point: for general compactness of device the emitter and receiver are located very closely to each other. But it is well known that for small angles a trigonometric function of sinus/tangens are approximately equal to the angle because of linearization of the trigonometric functions (truncation of their Taylor series). In our task these functions are the basic for used mathematical formalism. Hence, it causes non uniform resolution of the theoretical method inside a full field-of-view of the system. The same reason causes a problem of resolution at the angles close to 90°.
Experimental proofs of such non-uniform uncertainty distribution are shown in (Rivas Lopez et al. [a], 2008), (Rivas Lopez et al. [b], 2008), (Rivas Lopez et al. , 2010), (Sergiyenko et al. [a]; 2009). See Fig. 16.

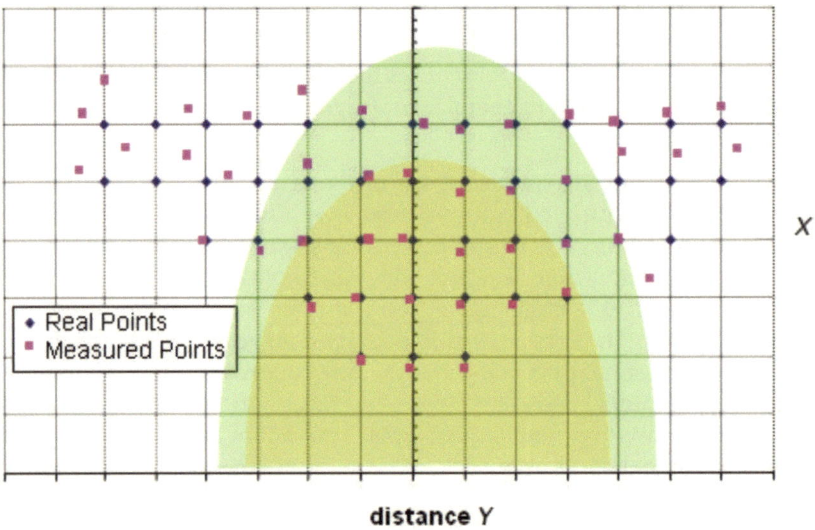

Fig. 16. Experimental graphic of point's measurement in 2D at variable obstacle positioning. TVS field-of-view and "accuracy zones", according to (Rivas Lopez et al. [a], 2008)

The accuracy of coordinates measurement is not uniform (see Fig. 16) in all TVS field-of-view, but in the olive- and green-zone correspondingly it is not more than 1% and 4% out of level of confidence (Rivas Lopez et al. [a], 2008), (Rivas Lopez et al. [b], 2008). Usually, modern regular step drives are operated with average velocity of 1 KHz. It means that we can obtain coordinates at least of 1000 points per second, each X, Y, and Z with metrological accuracy and defined uncertainties. This is a point to apply this TVS as input data sensory system.
Additionally, it is obviously that a triangulation base of 1m (used in mentioned publications) not just permits an optimal design of the optoelectronic system field-of-view, but also permits increase a computation speed for used trigonometric mathematical formalism because multiplication for 1 is an empty operation.

5. Geometrical centre search of the laser spot projection

The laser ray as the scanning tool of any laser scanner unfortunately has a non-uniform spatial distribution along the variable striking distance. This circumstance unfortunately causes that a shape and size of the laser spot projection on the scanning surface is significantly variable. It is shown in (Sergiyenko et al. [a]; 2009), (Rivas Lopez et al., 2010), and on the Fig.17.

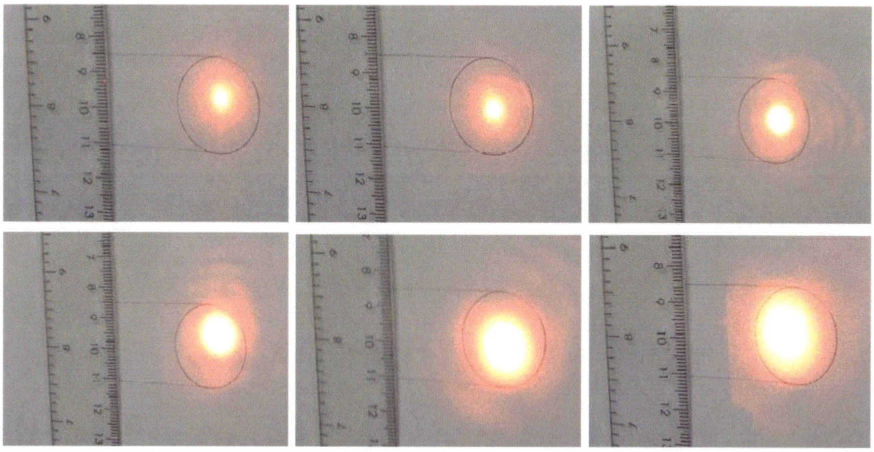

a)

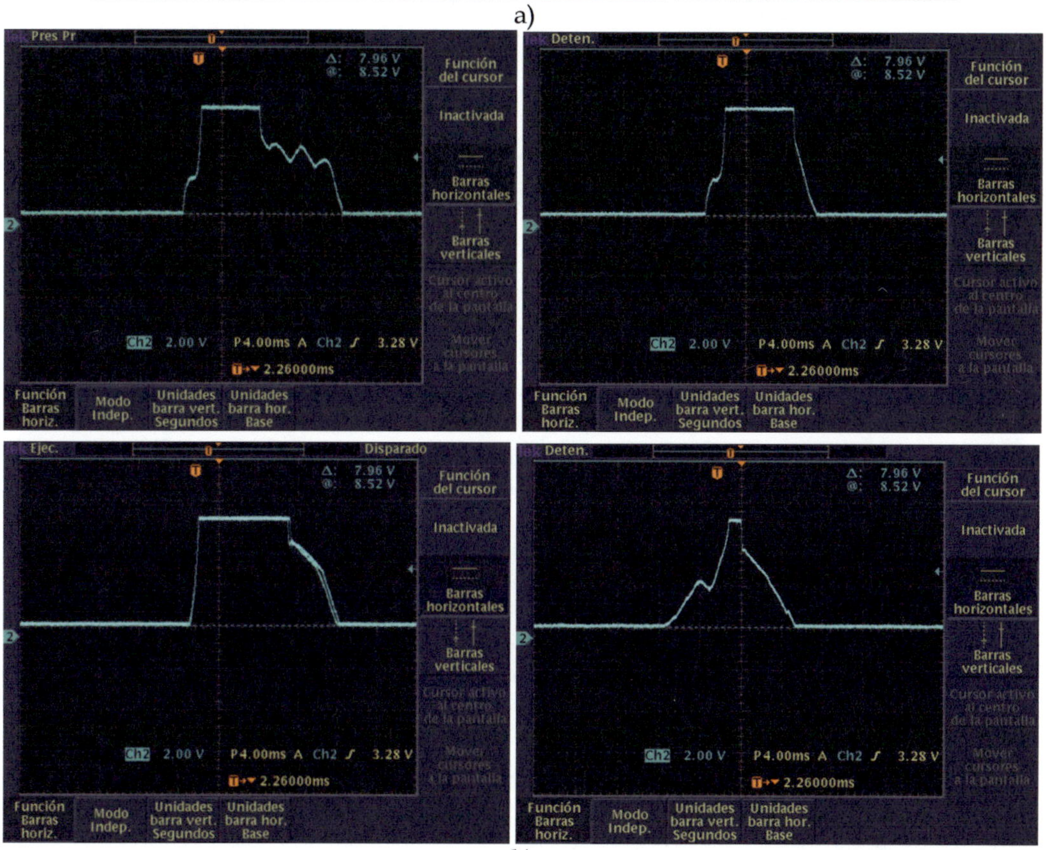

b)

Fig. 17. General view of the stop-signal at the different distances and highlighted spot sizes: a) spot sizes variation; b) stop-signal screenshot

It causes that spatial resolution of any method based on the spatial position estimation by such signals registration has a strong source of uncertainty.
The possible way for this disadvantage decrease is based on the theoretical method and special circuit of the signal energetic center search introduced in (Sergiyenko et al. [a]; 2009), (Rivas Lopez et al., 2010).

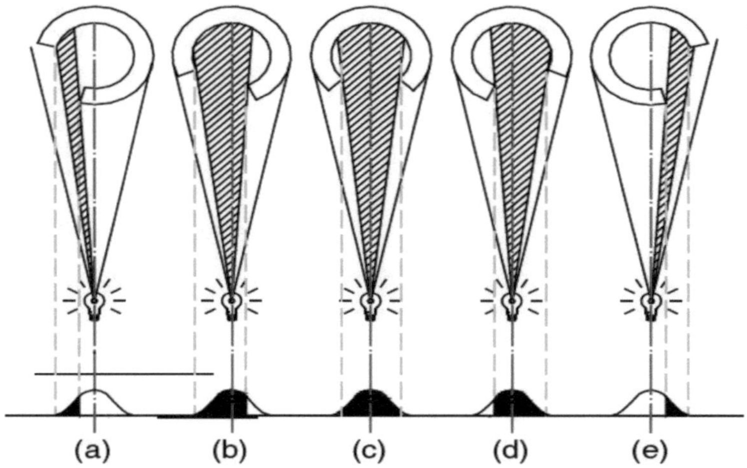

Fig. 18. The principle of noisy electrical stop-pulse formation during rotational scanning

The triangulation mathematic framework is still strongly theoretical. Because in the geometrical scheme we considering a pure geometrical objects, i.e. the straight line is precisely straight, without any curvature, have only length, all another sizes are zero; points have no any sizes and its location is characterized only by three Cartesian coordinates in the space.
Practically it is different (Rivas Lopez et al. , 2010). The optical ray it is a cone in general or even more complex shape depending on medium properties. The vertex of this cone also isn't a point, but distributed in some small spatial scope. The divergence inside this cone it is caused by many natural factors. It causes that practically the stop-pulse in photo receiver is not a short pulse of the standard form, but it is formed as shown on Fig. 18 (a-e).
It growth (Fig. 18, a) and falling down (Fig. 18, e), and fluctuating around its maximum area in figs. 18 (b-d). Taking to account the presence of some natural noise (as rule, white Gaussian noise) in this process, we can conclude that finally we need to operate the single electric signal of non-regular shape presented on Fig.19, a. Moreover, the shape and the width of this pulse are closely related to scanning velocity. But the scanning velocity in practical case is not a constant also. In this case the best problem solution is adequate signal processing. And our contribution is based on simple idea. What is it the red point on the Fig. 19, a? In the point of view of electric signals area on the graphic limited by signal curve is the energy transferred by this electric signal. So, this red point with certain small uncertainty is the signal energetic center. It is essential to note that position of the energetic center coincides for both pulses on the Fig. 19, a: ideal theoretical square pulse, and its real noisy performance. Much more essential is to note in this case, that the energetic center position uncertainty the less depends on noise, than becomes more the pulse area. In our task is the same like less scanning velocity.

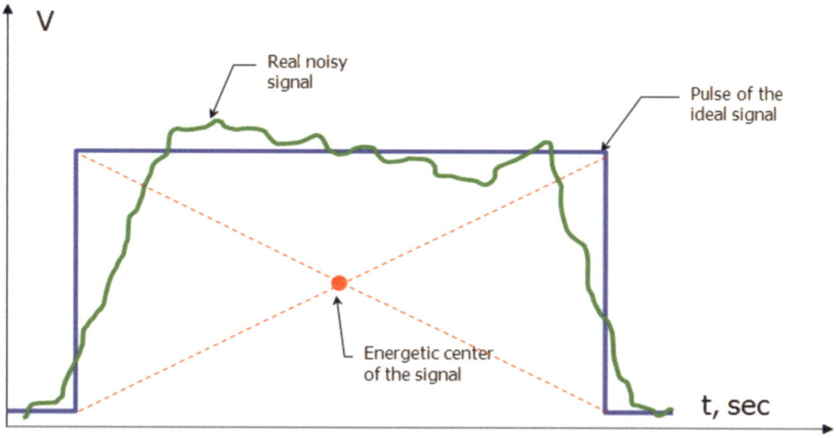

a)

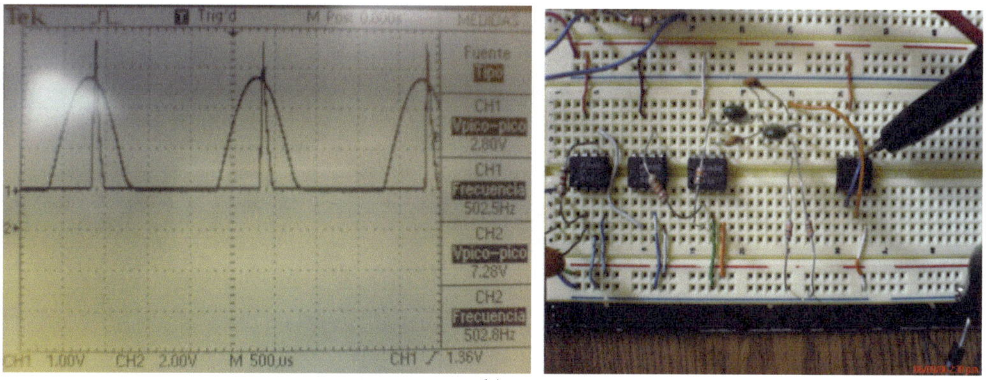

b)

Fig. 19. Method of the signal energetic center search: a) general purpose; b) differentiator circuit & functioning screenshot

From the other hand, as we consider this pulse as the optic blur equivalent registered by photosensor, it is the center of optical ray spatial cone. Such way we can detect in noisy electric signal the truth position of the unique straight line which belongs the same time the center of light source (active target) and the center of photoreceiver.

It is possible to provide practically using another strong and simple rule: the function (signal curve) maximum it is always where its derivative is zero. In another words, after finding the zero-cross position of first derivate function of the registered signal, we can find a real position of the EB in our coordinate system.

The practical realization of described operation it is possible in a several ways. However, in our primary experimentation we realize this operation using standard differentiator circuit (Fig.19, b). On the operation screenshot on Fig. 19, b it is clear evident that it is completely possible in a real time scale to obtain an electrical mark of the real light source position. Of cause it is a certain delay between real maximum position and its mark by pin-pulse (Fig. 19, b) because of differentiator circuit operating time. But it is strongly evident that it has a constant character and can be eliminated on the processing stage by simple correction factor.

6. Optimal design of laser scanner mechanical part

Serious problems of scanner normal functioning and sources of spatial resolution uncertainty are caused by not balanced torque of the rotating electromechanical system; mechanical movements delay; wear process of friction surfaces; rocking of electromotor shafts, etc.

The optimal design of the mechanical part is given in (Básaca [a,b] et al., 2010), it permits significant decrease of the most important electromechanical lacks of the previous prototype presented in (Sergiyenko et al. [a - c]; 2009), (Rivas Lopez et al. [a,b], 2008), (Rivas Lopez et al. , 2010).

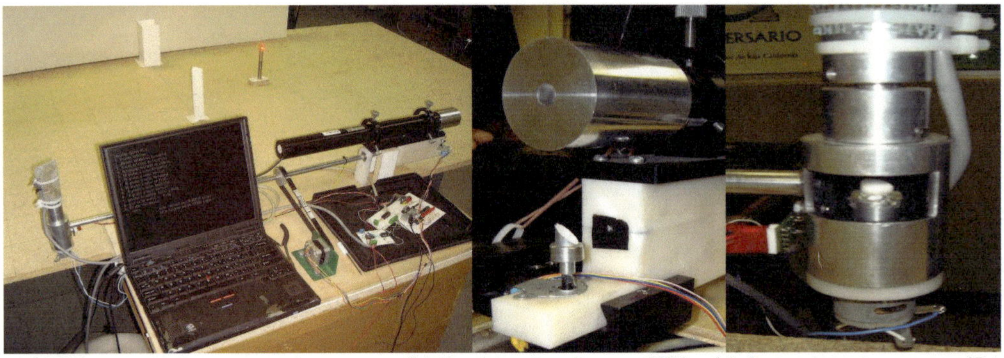

(a) Prototype I system view (b) Positioning Laser (PL) (c) Scanning Aperture (SA)

Fig. 20. Prototype I

(a) Full Size Prototype II system view (b) Positioning Laser (PL) (c) Scanning Aperture (SA)

Fig. 21. Prototype II

Figure 20 shows Prototype I, this system was the first we built to demonstrate our method and although testing and experimentation were successful, we detected uncertainty points (See Fig. 17) that could be eliminated with an improved mechanical design.

Prototype II shown in Fig. 21, a, has several advantages in comparison to its predecessor; the most important is that in the new design most of the components, including the Laser are installed inside a cylindrical tube, thus placing the center of gravity in the center of the tube and providing an easier and more precise way of rotating the tube than the band used in Prototype I (see Fig. 20, a).

In other words this improvement decreases the TVS own torque which provide us with several advantages over Prototype I such as:

- The possibility of using an electric motor rated with lower torque, voltage and current per phase consumption.

- As a result, full system power consumption is reduced.
- Extended battery life which is essential for mobile applications.

Moreover, cross-comparing Fig. 20, b and Fig. 21, b and the pair Fig. 20, c - Fig. 21, c, in the constructions of PL and SA, the following improvements are presented.

The older Positioning Laser design used a fixture with a 45° mirror attached to the laser to redirect the laser beam in a 90° angle to a second 45° mirror which is attached directly to a stepper motor (Fig. 20, b). This motor is controlled to redirect the laser beam orthogonally to scan the area in front of the mobile robot.

On the other hand, the new PL design (Fig. 21, b) is based on the same general principle but has mechanical differences. E.g. the new PL has no fixture attached directly to the laser; instead of attaching the 45° mirror, part of the inside surface of the cylindrical tube is machined with mirror finish, with this, the possibility of uncertainty due to mirror displacement or improperly installed fixture is eliminated. It is the way to decrease the uncertainty deviation on Fig.17.

Another uncertainty source is that the more far away the surface marked by the laser, the greater the laser beam or spot diameter becomes. Fig 20, b shows that the laser beam travels a certain distance and is redirected by a mirror on two occasions, due to this travel, at the PL output the laser beam diameter is already greater than the originally emitted laser beam diameter.

Hence, another improvement is that the distances between the laser and the mirrors, i.e. the laser beam travel was reduced from centimeters to millimeters, by doing this the laser beam diameter in the PL output equals to the beam diameter emitted from the laser output, maintaining the lowest beam diameter possible within the TVS, therefore theoretically increasing the TVS range. Analysis of laser spot diameter variation is given in (Básaca [a,b] et al., 2010), it shows that this circumstance plays a significant role in total uncertainty distribution, and this part of general design is critical for total uncertainty decrease.

The Scanning Aperture design is maintained from Prototype I to II (Fig 20, c and Fig 21, c), the only difference is that the new SA is smaller in size and is installed within the cylindrical tube, however it's important to mention that the most critical point with the SA is to maintain it aligned in the same plane as PL in order to be able to create the dynamic triangle more solid or fixed in mechanical meaning, i.e. only this design guarantee that scanning ray and scanning plane are meeting exactly in the same plane triangle.

7. Noise filter for photoreceiver circuit

Experimental studies of electromechanical laser scanner prototype as the rule shows the presence of noise of the complex nature in a photodetecting circuit. It is caused by cross-action of two main components: mechanical vibration and, properly, optical noise.

The typical procedure of the filter design for such kind of noise, for different scanning velocities, will be introduced in full text of this paper basing on the material of publication in (Básaca [a] et al., 2010).

One of the most challenging points of normal TVS functioning is the presence of typical input noise mixed with the "stop-signal" train in the form of the screenshot on Fig. 22. This noise can be filtered by specially designed circuit (Fig. 23) with the aim of guaranteed detection of true position of the scanning ray reflected points, in other words the guaranteed localization of the real "stop-pulses".

As in every system, noise is present here too in the signal acquired from the photo receiver, after studying the signal we observed that the noise reaches maximum amplitude of 120 mV (see Fig. 22) with frequencies that vary between 400 Hz and 20 KHz. To eliminate this noise a Butterworth third order low pass filter (-60 dB per decade) was designed and implemented, see Fig. 23. The filter's output connects to a voltage level detector with a reference voltage of 120mV to reject the noise amplitude, whenever an obstacle is detected by the photo receiver a spike in the signal is detected, the voltage of this spike varies depending on the distance of the detected obstacle but is always greater than 120mV from the noise amplitude. The output of the voltage level detector is a signal of 5Vdc when an obstacle is detected and remains at 0Vdc when there is no obstacle, in other words providing us with a 0-5Vdc square signal which indicate us whether an object is present or not.

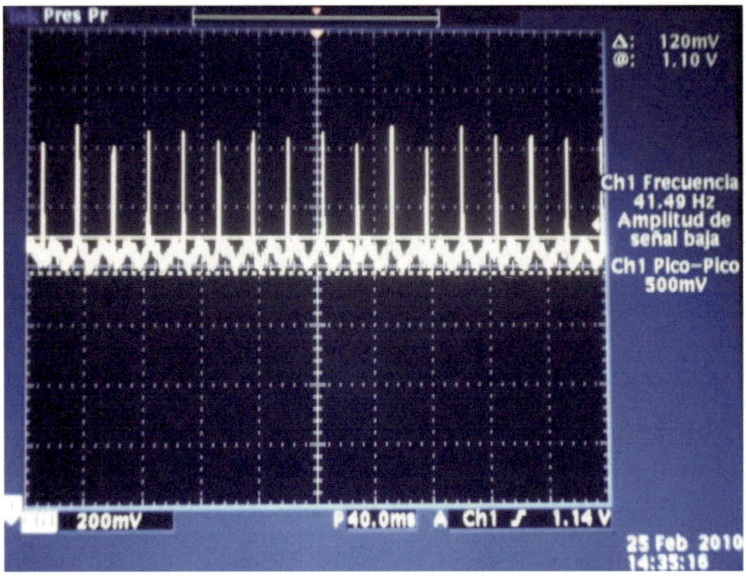

Fig. 22. Typical experimental noise voltage and frequency

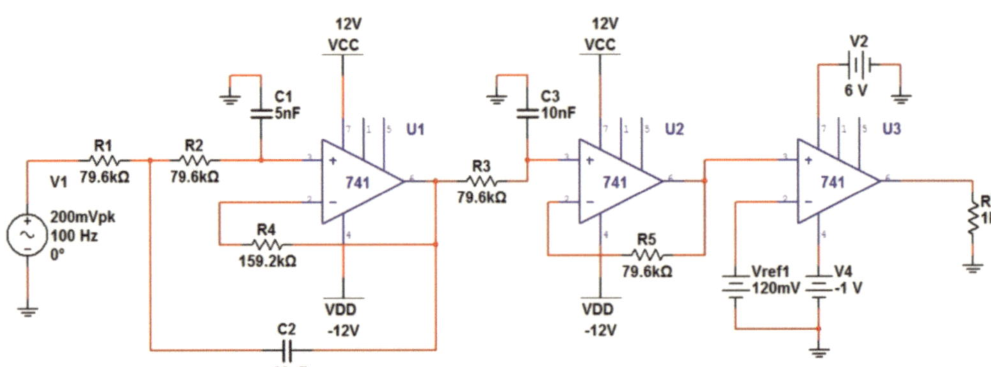

Fig. 23. Butterworth third order low pass filter (-60 dB/decade) and voltage level detector

The Butterworth type filter was chosen over other types such as Chebyshev filters due to Butterworth's more linear phase response and flat frequency response in our passband according to (Coughlin & Driscoll, 1999). Our filter was designed as follows:

$$\omega_c = 200;\ C_3 = 10nF \tag{10}$$

where ω_c is the cutoff frequency, the value assigned is 200 Hz in order to let pass the frequency band between 0 and 200 Hz, letting pass the signal of a detected obstacle and attenuating all the frequencies higher than 200 Hz.

$$C_1 = \frac{1}{2}C_3 = 5nF \tag{11}$$

$$C_2 = 2C_3 = 20nF \tag{12}$$

$$R = \frac{1}{\omega_c C_3} = \frac{1}{(6.25)(200)\left(5x10^{-9}\right)} = 79.617k \tag{13}$$

$$R_1 = R_2 = R_3 = R = 79.617k\Omega \tag{14}$$

$$R_{f1} = 2R = 159.235k\Omega \tag{15}$$

$$R_{f2} = R = 79.617k\Omega \tag{16}$$

The filter with -60 dB/decade attenuation is achieved by cascading a -40 dB/decade filter and a -20 dB/decade filter. The total closed loop gain is the multiplication of each filter's gain. See Fig. 23 for the frequency response of the circuit shown on Fig. 23, (Coughlin & Driscoll, 1999).

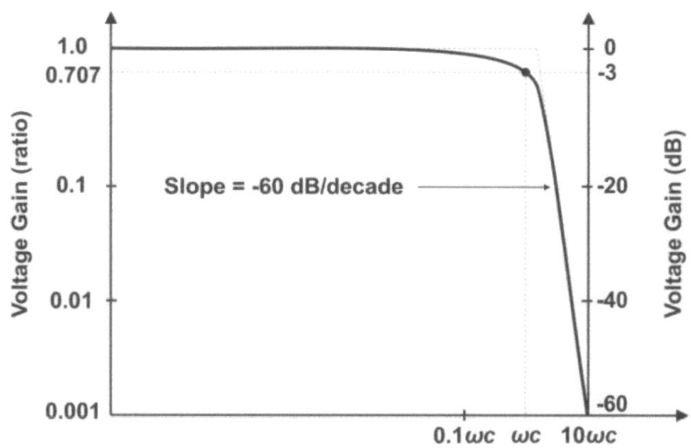

Fig. 24. Frequency response for Butterworth third order low pass filter (-60 dB/decade)

7.1 Simulation

The circuit presented on Fig. 23 was simulated with software from National Instruments, NI Multisim 10. Fig. 25, a shows a caption of a Simulated Tektronix Oscilloscope with three

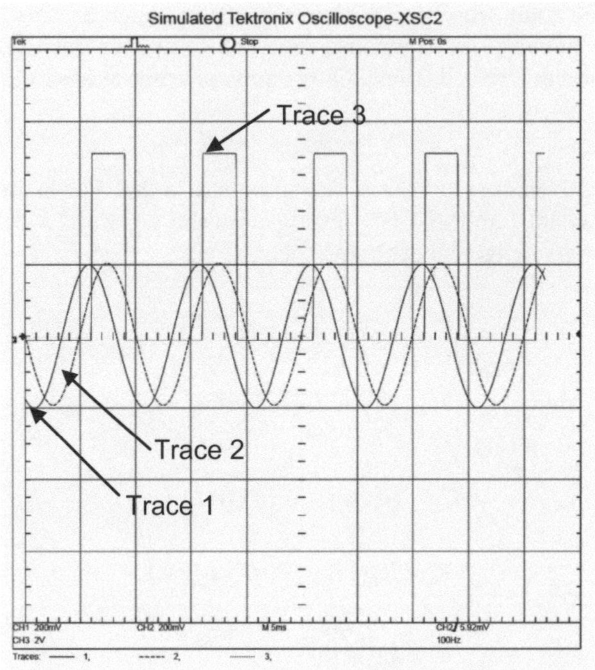

(a) Input signal is 200mV at 100 Hz; output is a 0-5Vdc square signal

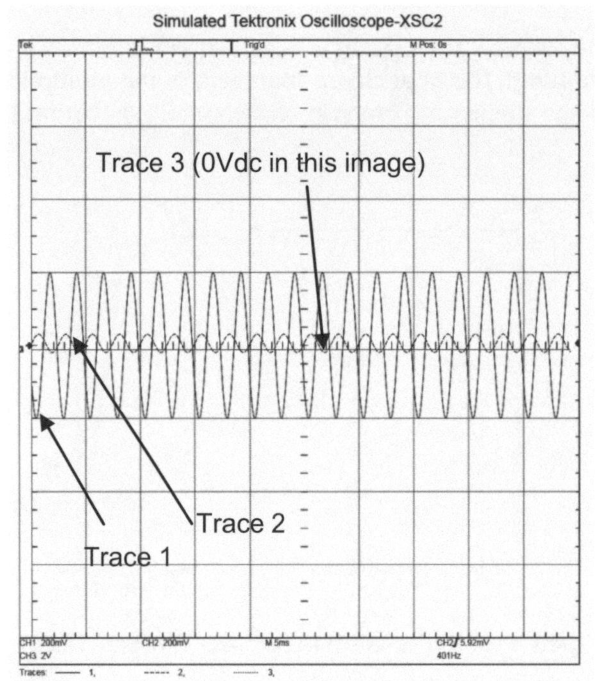

(b) Input is 200mV at 400 Hz; output is attenuated due to input frequency being higher than cutoff frequency (200 Hz), regardless of the input voltage.

Fig. 25. Low pass filter and voltage level detector output (circuit on Fig. 23) signals, trace 1 is the circuit input signal that simulates the photo receiver output, this is the signal that is filtered and amplified, for this simulation the input signal is 200mV at 100 Hz. Trace 2 is the filter output, as shown below, the signal has the same amplitude as the original as long as its frequency is in the passband region (below 200 Hz), therefore it's not attenuated, it also has a delay of 8ms which is acceptable. And the last, trace 3, is the voltage level detector output, this signal will be 5Vdc if the signal voltage coming from the filter is greater than Vref (120mVdc), meaning an obstacle has been detected, and it remains 0Vdc if there is no obstacle in sight.

Fig. 25, b shows the case when the input signal is 200mV at 400 Hz, this frequency is in the stopband region (higher than 200 Hz), for this reason the trace 2 signal which is the filter output is attenuated to 35mV that for this purpose equals 0V. Therefore trace 3 remains at 0Vdc.

Thus, our simulation demonstrates that the designed filter can attenuate the experimentally detected undesired noise in TVS prototype above 400Hz.

7.2 Slower scanners for application in medical surgery

At the more slow scanning, which permits increase the resolution in spite of operating time, it was experimentally obtained the noise of same character, but with another typical values. Such scanning velocity decrease it is expedient, for example, for medical scanners application (Rodriguez et al., 2009).

The dead zone circuit DZC Fig. 28 cut any signal that is greater than 200mV; this eliminates any noise left over after applying the filter of -40db. The detected signal corresponding to the laser is greater than 400mV which allows the cut rise. In experimentation is carried out that useful signal is berried in noise, wish have in our case a typical form presented on Fig. 25. With the aim to delete this noise in this paper design DZC under consideration. After DZC we make a saturation of the signal and amplify to 5 Volts, thereby obtain a processable signal to detect the moment of triangulation. It should be mentioned that the period of existence of the triangle can be measured in the order of a millisecond, but this time is sufficient to calculate the necessary angles.

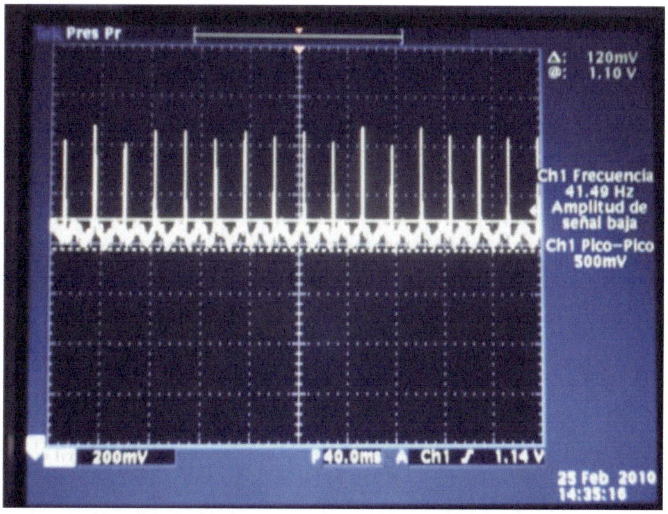

Fig. 26. 200 mv Noise, the peaks are the useful signal

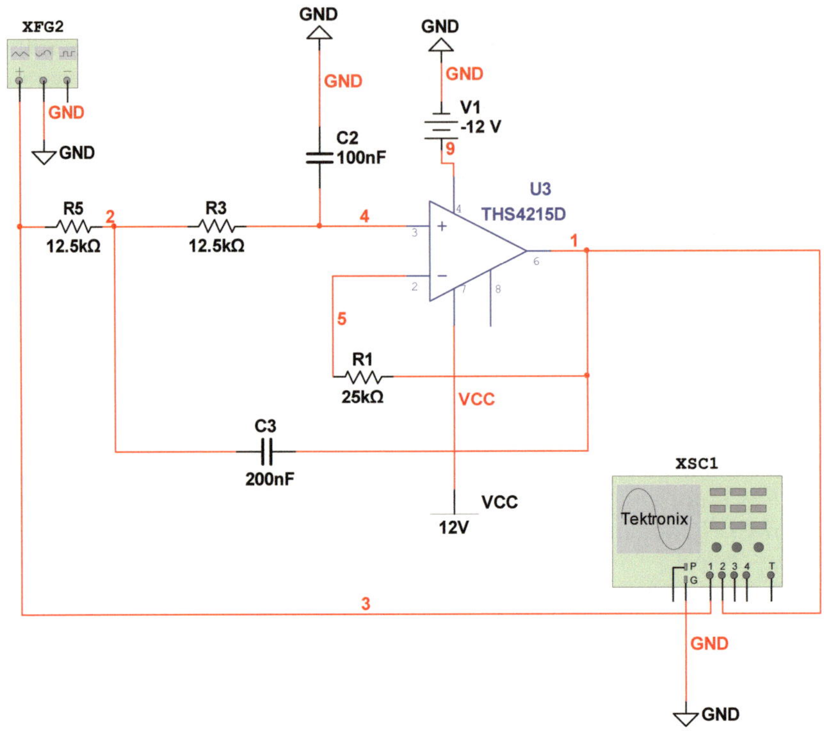

Fig. 27. Low pass Filter with -40db and fast attenuation response at 10ωc

The filter produces an attenuation of -40db/decade, after the cut-off the input magnitude decreases 40db when ω increases to a value of 10ωc Equation 18. The operational amplifier is connected so as to obtain unitary dc gain. It includes resistance R1 to the dc deviation. Since the operational amplifier circuit is a unitary gain amplifier, the voltage on C1 equals the output voltage (Coughlin & Driscoll, 1999, pp 294 - 297). The design procedure is simplified too if R3 and R5 are equal. Then follow the following equations:
Initial data

$$f_c = 90Hz \quad C_2 = 100nf \quad R_3 = R_5 = R \tag{17}$$

Cut frequency expressed in Radians/ Second

$$\omega_c = (2\pi)(f_c) \tag{18}$$

Capacitor 3 Value

$$C_3 = (2)(C_2) \tag{19}$$

Voltage Gain of the close loop

$$A_{CL} = \frac{1}{1 + j\omega RC} \quad A_{CL} = \frac{1}{1 + j1} = \frac{1}{\sqrt{2}\angle 45^{o}} \quad |A_{CL}| = \frac{1}{\sqrt{2}} = .707 \tag{20}$$

R Value using -40db as parameter

$$R = \frac{|A_{CL}|}{(\omega_c)(C_1)} \tag{21}$$

R1 Value:

$$R_1 = 2R \tag{22}$$

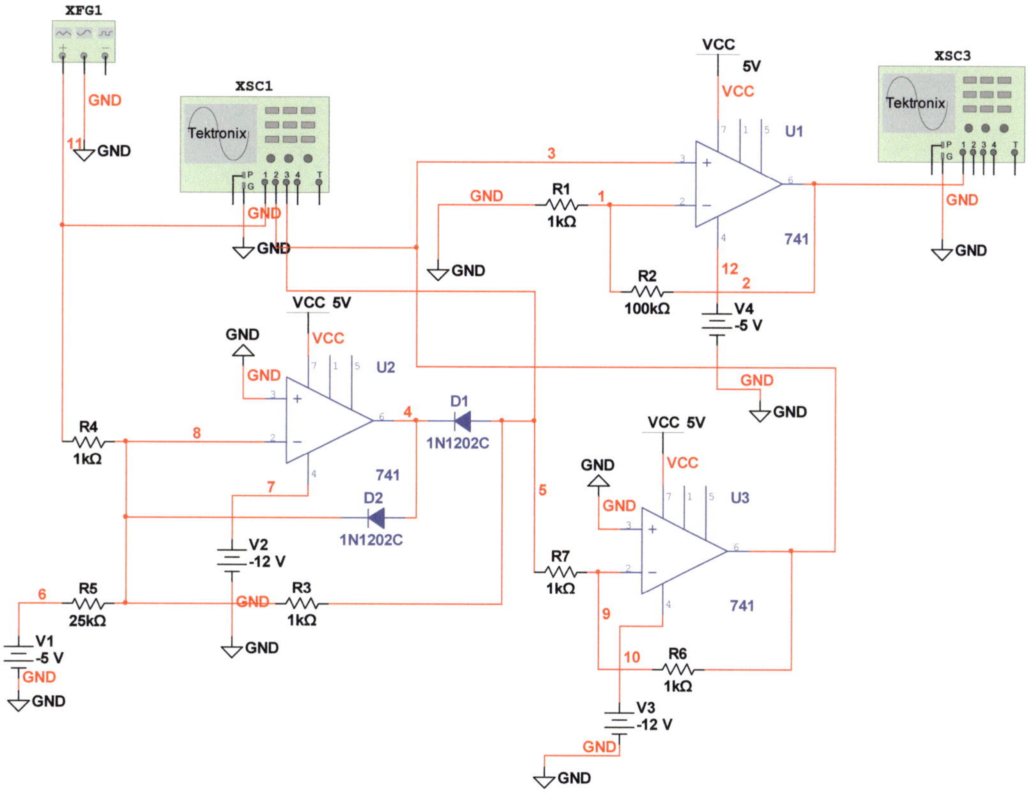

Fig. 28. Dead Zone to complement application filter circuit with saturation amplifier

Through a regulated supply voltage V1 and a resistance mR (R5) sets the reference voltage Vref. This is calculated from Equation 24. As will be shown the negative value of Vref is what defines the dead zone. The diode D2 always leads XFG1 values and set the "node 5 =-XFG1 - Vref". Whenever XFG1 exceeds the value of -Vref = 200mV, the output node 3 lets you know the amount XFG1 exceeds the value-Vref. There is a dead zone when there are values XFG1 below -Vref. (Coughlin & Driscoll, 1999, pp 200 - 204).
Initial data

$$R = R1 = R3 = R4 = R6 = R7 = 1k \tag{23}$$

Reference Voltage in the dead zone

$$V_{ref} = \frac{V_1}{m} \quad \text{(Node 6)} \tag{24}$$

Gain to be amplified and detect the signal as 0 volts or 5 volts (0 or 1 logic)

$$Gain = \frac{R2}{R1} \tag{25}$$

R5 Value

$$R5 = (m)(R) \tag{26}$$

In Fig. 30 are presented the results of simulation of the designed filter functioning. As shown Fig. 30 without a DZC the signal is not detectable (As in Fig. 26) and after DZC implementation it is eliminated undesired noise and the useful signal is clear end detectable.

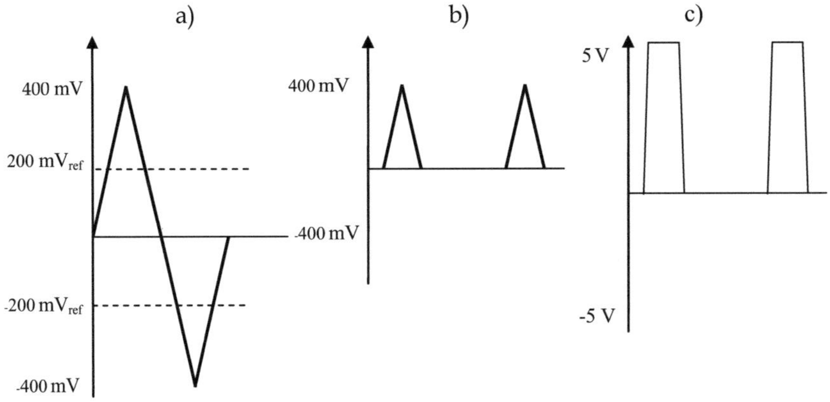

Fig. 29. a) dead zone circuit, b) Inverse Follow Circuit, c) Saturation amplifier

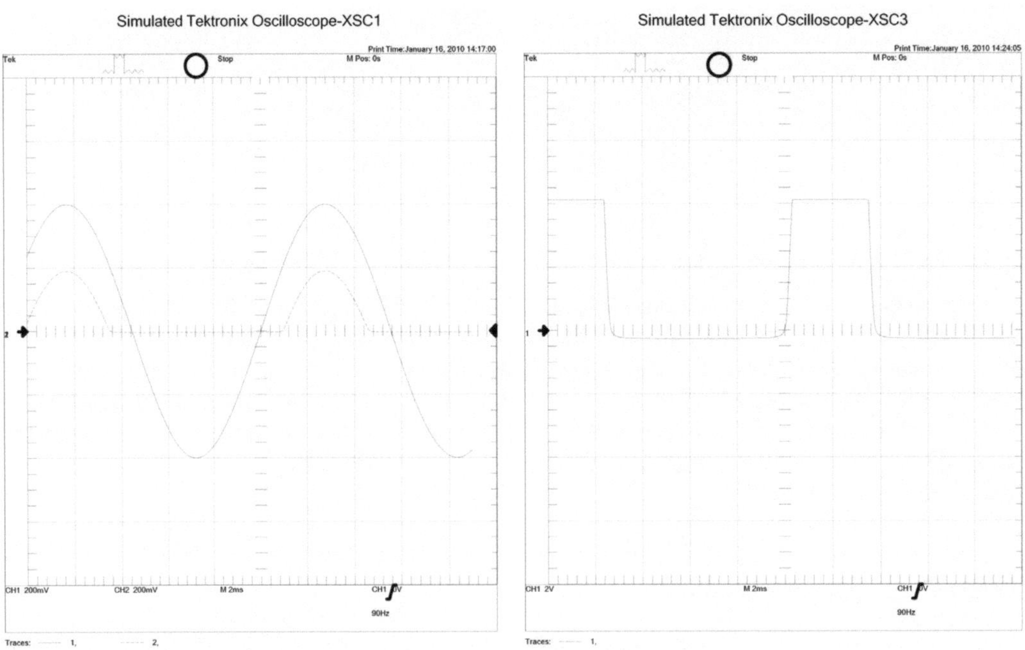

Fig. 30. Filter Output in 90 Hz Signal a) Signal as Fig. 25 without DZC b) Output Signal after DZC

8. Robust detection of the weak reflected signals

In (Sergiyenko et al. [a]; 2009), (Sergiyenko et al. [b]; 2009), (Sergiyenko et al. [c]; 2009), (Sergiyenko et al.; 2008), (Rivas Lopez et al. [a], 2008), (Rivas Lopez et al. [b], 2008), (Rivas Lopez et al. , 2010) it is shown various times that practically the most strong constraint of electromechanical laser scanner use is the signal attenuation (voltage decrease) in a photodetecting circuit. It is shown on Fig. 31:

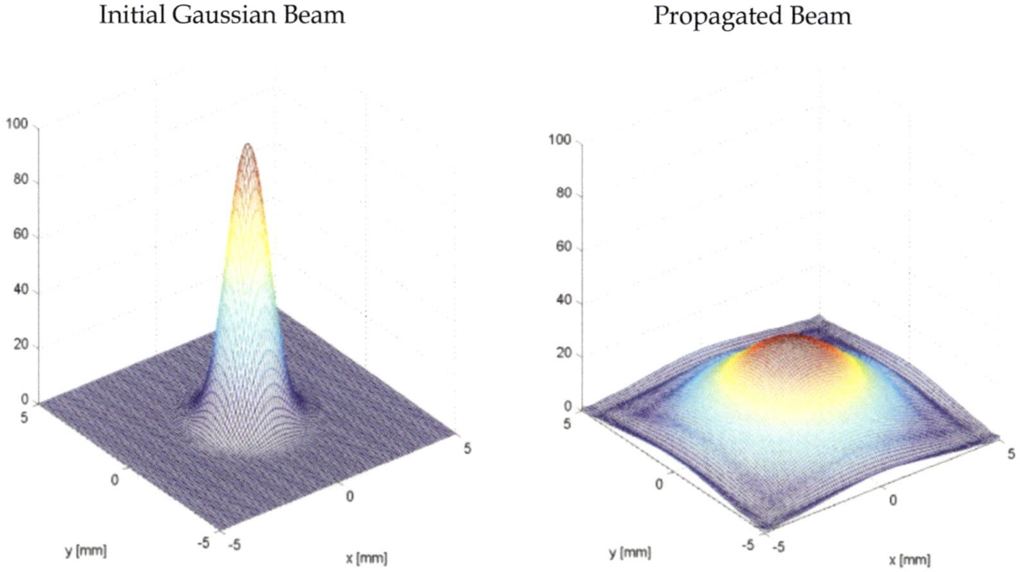

a) b)

Fig. 31. a) Power distribution pattern for 0 meter from laser source. b) Power distribution pattern for 15 meters from laser source

The possible way for this disadvantage decrease is based on the theoretical method of robust photometer circuit introduced in (Hernandez, 2007), (Hernandez [a], 2008), (Sergiyenko et al. [a]; 2009), (Hernandez [b], 2008).

In the scientific literature (Hernandez [a], 2008) on photometer circuits operational amplifiers are frequently used for photodiode monitoring. Nevertheless, most of the operational amplifier connections used in this kind of application are based on basic current-to-voltage converter circuits.

Figure 32 shows the basic circuit, in which D represents the photodiode and R_f is the negative feedback resistor used to convert the photocurrent into an output voltage linearly related to the light energy.

One of the clear advantages of the circuit shown in Fig. 32 is that it is of easy implementation. Nevertheless, on the other hand, as the diode is not forming part of the feedback loop, there is no way to compensate for disturbances, variations in temperature, structure and unstructured uncertainties in the photodiode, noise, and so on.

Therefore, it is important to place the sensor in a feedback loop able to deal with the above-mentioned problems. Figure 33 shows an example of a photometer circuit with the diode placed in the feedback loop (Hernandez, 2007), (Hernandez [a], 2008), (Hernandez [b], 2008):

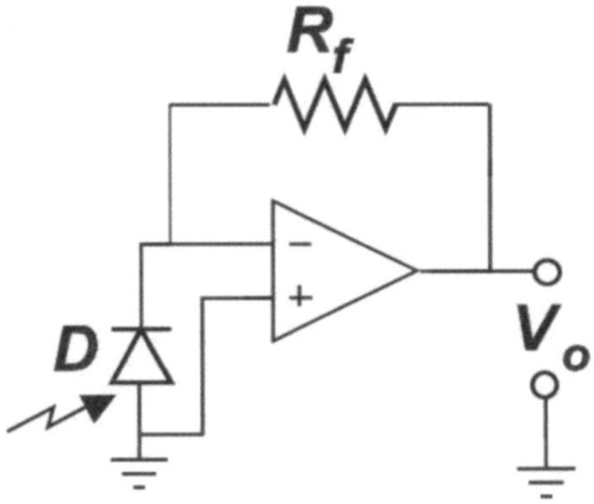

Fig. 32. Current-to-voltage converter circuit

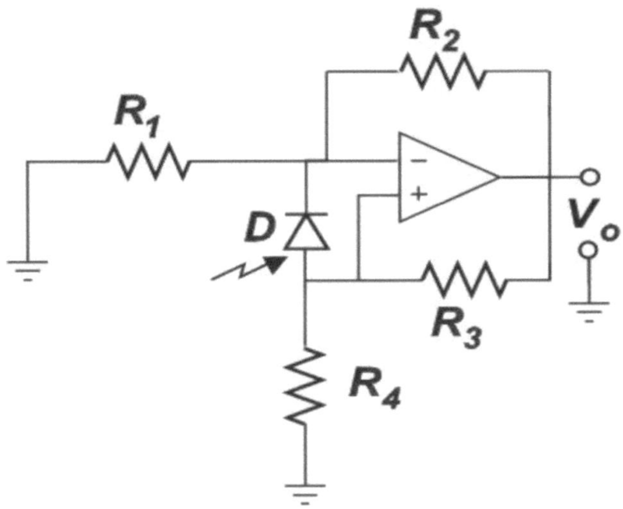

Fig. 33. Feedback photometer circuit

However, it is important to point out that in spite of the fact that the circuit shown in Fig. 33 is also of easy implementation, due to the fact that it works with positive feedback compensation it can become unstable if the values of the resistors are not chosen properly.
In (Hernandez [b], 2008), an in-depth analysis of such a circuit is carried out taking into consideration operational amplifier parameters such as the input resistance, input capacitance, open-loop gain and gain bandwidth product. Furthermore, the input-output transfer function analysis of the circuit shown in Fig.33 is carried out from the linear system theory point of view, considering that the operational amplifier is ideal.
Finally, considering the photodiode model shown in Fig. 34 (in which C_J is the junction capacitance, I_P is the light generated photocurrent, R_{SH} is the shunt resistance, R_S is the series resistance, R_L is the external load resistance and V_O is the output voltage), in (Hernandez, 2007), (Hernandez [a], 2008), (Hernandez [b], 2008) a procedure for the measurement uncertainty estimation of the circuit shown in Fig. 35 is presented.

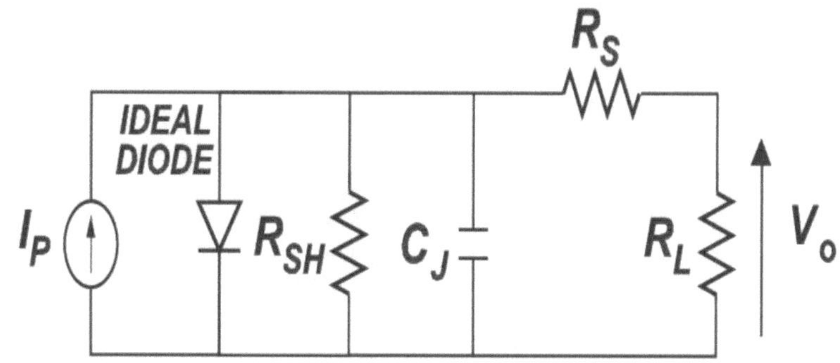

Fig. 34. Simplified version of the equivalent circuit for a photodiode

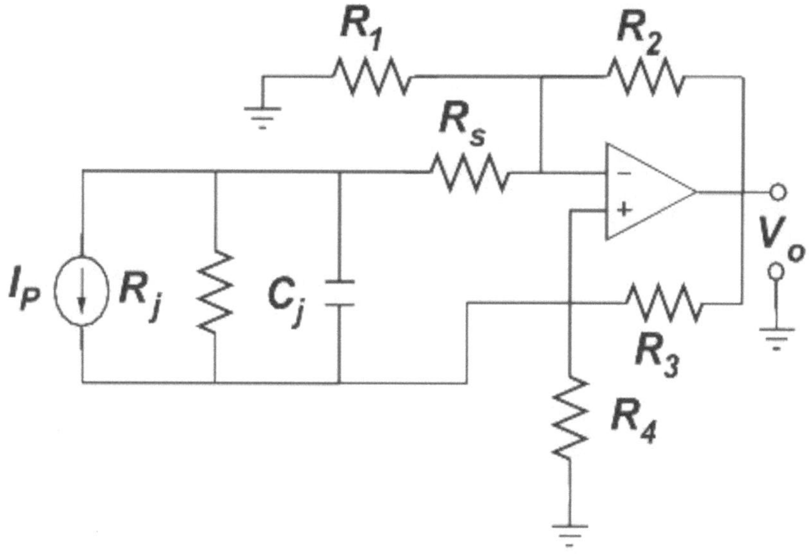

Fig. 35. Feedback photometer circuit with equivalent photodiode circuit

9. Conclusions

Finally, we can get the conclusion that possible grade of 3D scanners resolution improvements by combined application of all the mentioned actions will be efficient.

- Laser scanning systems always give a more resolution, accuracy and the better data processing time over other known Machine Vision systems. This is a reason for future research of the 3D electromechanical laser scanners for practical application matched for it in the best way. For best resolution in spatial domain it is expedient as shown above to use some additional technologies.
- Laser scanning systems always give a worse solution than other known Machine Vision systems in a part of energy consuming, medium optical conditions dependence and

range of action. So, Laser scanning systems designers always must to take in mind this fact, and to provide additional corrective measures for better system application.

- Due to non-uniform uncertainty distribution in instantaneous field-of-view (IFOV), as well in full field-of-view, it desirable the optimal geometrical design of the optoelectronic system field-of-view strictly linked to the certain practical application (i.e. static monitoring of large civil engineering structures; fast dynamic monitoring of the certain sector in indoor or outdoor autonomous robot navigation; medical scanning of the biological object with reduced movement activity; visual control of surfaces in automated assembly; etc.). In other words, the spatial sector under inspection must to be located in a smallest error sector of total TVS field-of-view. This requirement in general coincides with the basic theoretical concepts of the electro-optical system design (Fig. 5.1 on p.63 in (Wyatt, C.L., 1991)). It is expedient to note, that optimal geometrical design also permits sometimes increase a computation speed for used trigonometric mathematical formalism (for example, because multiplication or division for 1 is an empty operation).
- The use of the theoretical method and special circuit of the signal energetic center search permits us eliminate totally one of the most complex sources of uncertainty. It is uncertainty caused by irregular form and variable size of the projected light spot. It is especially important note that the variable size of the projected light spot has the non-linear relation to the distance between scanner and object surface. The uses of our theoretical method permits completely exclude this complex dependence and to establish the uncertainty rate only as a rigorous function of internal circuit parameters.
- The optimal design of the TVS mechanical part permits significant decrease of undesirable axial play, torque and wear. Mentioned above actions implementation gives a possibility to possibility of using an electric motor rated with lower torque, voltage and current per phase consumption for full system power consumption reduction and extended battery life which is essential for mobile applications.
- The typical challenging point of normal TVS functioning is the presence of input noise. As shown in the present paper, it is always necessary to realize an electronic filtration of such noise and parameters of such noise are mostly dependant on scanning velocity. Unfortunately, for this particular task only experimental test of electromechanical laser scanner prototype shows the presence of noise of the complex nature in a photodetecting circuit and its physical characteristics. It is caused by cross-action of mechanical vibration and, properly, optical noise. However, the special filter design for different scanning velocities is very similar and regular in its procedure, as it was shown in examples described in subsections 7.1 and 7.2.
- Theoretical method of robust photometer circuit introduced on photometer circuits operational amplifiers used for photodiode monitoring can permits to detect a weak signal which never been detected before in electromechanical scanners.

10. Acknowledgements

This work has been partially supported by the Autonomous University of Baja California, Mexico (projects No. 2352, in 2003-2004 and N2386, in 2005-2006) and the Ministry of Science and Innovation (MICINN) of Spain in 2009-2010 under the research project TEC2007-63121, and the Universidad Politecnica de Madrid. The authors would like to thank to various international congresses organizers, where the different aspects of the present work were presented, for fruitful discussions and valuable remarks. The authors dedicate this work to the grateful memory of Dr. Valentin Yevstaf'yevich Tyrsa; and would like to thank to Prof. Christofer Druzgalsky and Jason McDowell (Electrical Engineering Department of California State University, Long-Beach, USA) for their invaluable assistance with English language. We appreciate also Dr. Daniel Hernandez-Balbuena's invaluable assistance in a part of discussion about uncertainty of the repetitive cyclic measurements.

11. References

Andersen, J. C.; Blas, M. R.; Ravn, O.; Andersen, N. A. & Blanke, M. (2006). Traversable terrain classification for outdoor autonomous robots using single 2D laser scans. *Integrated Computer-Aided Engineering*, Vol. 13, No. 3, 2006, 223-232, ISSN 10692509.

Básaca, L.C. a; Rodríguez, J. C.; Sergiyenko, O.; Tyrsa, V.V.; Hernández, W.; Nieto Hipólito, J. I. & Starostenko, O. (2010). 3D Laser Scanning Vision System for Autonomous Robot Navigation, *Proceedings of IEEE International Symposium on Industrial Electronics (ISIE-2010)*, Bari, Italy, July 4-7, 2010, pp.1773-1779, ISBN 978-1-4244-6390-9

Básaca, L.C. b; Rodríguez, J. C.; Sergiyenko, O.; Tyrsa, V.V.; Hernandez, W.; Nieto Hipolito, J.I. & Starostenko, O. (2010). Resolution improvement of Dynamic Triangulation method for 3D Vision System in Robot Navigation task, *Proceedings of 36th Annual Conference of the IEEE Industrial Electronics Society (IECON-2010)*, Glendale, Arizona, USA, November 7-10, 2010, pp. 2880-2885. ISBN 978-1-4244-5225-5, ISSN 1553-572X

Beraldin, J.-A.; Blais, F.; Rioux, M.; Cournoyer, L.; Laurin, D. & MacLean, S.G. (2000) Eye-safe digital 3D sensing for space applications, *Optical Engineering*, Vol. 39, No, 1, January, 2000, 196-211, ISSN 0091-3286.

Beraldin, J.-A.; El-Hakim, S.F. & Cournoyer, L. (1993). Practical Range Camera Calibration, *Videometrics II Π(Proceedings Volume)*, Sabry F. El-Hakim, pp. 21-31, Proceedings of SPIE Vol. 2067, ISBN 9780819413321.

Bertozzi, M.; Broggi, A. (1997). Vision-based vehicle guidance, *Computer*, Vol. 30, No. 7, July, 1997, 49–55, ISSN 0018-9162.

Blais, F.; Beraldin, J.-A. & El-Hakim, S.F. (2000). Range Error Analysis of an Integrated Time-of-Flight, Triangulation, and Photogrammetry 3D Laser Scanning System. *Proceedings of SPIE Vol. 4035 AeroSense 2000*, pp. 236-247, ISBN 0-8194-3661-5, Orlando, FL, USA, April, 2000, SPIE, Bellingham, Washington.

Blais, F.; Rioux, M. & Beraldin, J.-A. (1988). Practical Considerations for a Design of a High Precision 3-D Laser Scanner System, *Proc. Soc. Photo-Opt. Instrum. Eng.*, Vol. 959, 1988, 225-246, ISSN 0361-0748.

Blais, F.; Rioux, M. & MacLean, S.G. (1991). Intelligent, Variable Resolution Laser Scanner for the Space Vision System, *Acquisition, Tracking, and Pointing V (Proceedings Volume)*, Michael K. Masten & Larry A. Stockum, pp. 473-479, Proceedings of SPIE Vol. 1482, ISBN 9780819405913.

Coughlin, R. F.; Driscoll, F. F. (1999). Operational Amplifier with diodes, in *Operational Amplifier and lineal integrate circuits*, Prentice Hall, Mexico DF, ISBN 970-17-0297-0, pp 294 - 297.

Fernandez, D. & Price, A. (2005). Visual detection and tracking of poorly structured dirt roads, *Proceedings of 12th International Conference on Advanced Robotics 2005. ICAR*

'05, pp. 553–560, ISBN 0-7803-9178-0, Seattle Sheraton Hotel & Towers, July, 2005, IEEE, Seattle.

Fischer, A. (2007). Digital Processing and Fusion of 3D Data from Emerging Non-Contact 3D Measurement Technologies, In: *The future of product development*, Frank-Lothar Krause, (Ed.), pp. 387-396, Springer, ISBN: 978-3-540-69819-7, Berlin, Germany.

Guivant, J. E.; Masson, F. R. & Nebot, E. M. (2001). Optimization of the simultaneous localization and map-building algorithm for real-time implementation, *IEEE Transaction on Robotics and Automation*, Vol. 17, No. 3, June 2001, 242–257, The University of Sidney, Australia, ISSN 1042-296X.

Hähnel, D; Schulz, D. & Burgard, W. Map Building with Mobile Robots in Populated Environments. *Proceedings of IEEE/RSJ International Conference on intelligent robots and systems,* pp.496-501, ISBN 0-7803-7398-7, Lausanne , SUISSE, 30 September - 4 October 2002.

Hernandez, D.; Sergiyenko, O.; Tyrsa, V.; Burtseva, L. & Rivas López, M. (2009). Signal frequency measurement by rational approximations, *Measurement*, vol.42, issue 1, Elsevier, January 2009, p. 136-144. ISSN: 0263-2241

Hernandez, W. (2007). Robustness and noise Voltage Analysis in Two Photometer Circuits, *IEEE Sensors Journal*, vol. 7, 2007, USA, pp. 1668-1674. ISSN 1530-437X

Hernandez [a], W. (2008). Linear Robust Photometer Circuit, *Sensors and Actuators A*, vol. 141, Elsevier, 2008, p.447-453, ISSN 0924-4247

Hernandez [b], W. (2008). Performance Analysis of a Robust Photometer Circuit, *IEEE Transactions on Circuits and Systems-II,* vol. 55, No.12, December, 2008, p. 106-110, ISSN: 1549-7747

Jochem, T.; Pomereau, D. & Thorpe, C. (1993). Maniac: a next generation neurally based autonomous road follower, in F. Groen, S. Hirose and C. Thorpe, eds, *Proceedings of the International Conference Intelligent Autonomous Systems IAS-3*, pp. 592–601, ISBN 90-5199-122-3, Pittsburgh, PA, USA, February 1993, IOS Press, Pittsburgh.

Klöör, P.L.; Lundquist, P.; Ohlsson, P.; Nygrds, J. & Wernersson, A. (1993) Change detection in natural scenes using laser range measurements from a mobile robot, *Proceedings of 1st IFAC International Workshop on Intelligent Autonomous Vehicles*, pp. 71–76, ISBN 0080422233, Southampton, UK, April, 1993, Pergamon Press.

Laurin, D.; Blais, F.; Beraldin, J.-A. & Cournoyer, L. (1996). An eye-safe Imaging and Tracking laser scanner system for space Applications, *Laser Radar Technology and Applications (Proceedings Volume)*, Gary W. Kamerman (Ed.), pp. 168-177, Proceedings of SPIE Vol. 2748, ISBN 9780819421296.

Legnemma, K.; Brooks, C. & Dubowsky, S. (2004). Visual, tactile, and vibration-based terrain analysis for planetary rovers, *Proceedings of IEEE Aerospace Conference 2004*, Vol. 2, March, 2004, 841–848, ISSN 1095-323X.

Macedo, J.; Matthies, L. & Manduchi, R. (2000) Ladar-based discrimination of grass from obstacles for autonomous navigation, *Proceedings of Experimental Robotics VII ISER 2000*, pp. 111–120, ISBN 3-540-42104-1, Waikiki, Hawaii, December, 2000, Springer, Waikiki.

Montemerlo, M. & Thrun, S. (2004). A multi-resolution pyramid for outdoor robot terrain perception, *Proceedings of the Nineteenth National Conference on Artificial Intelligence AAAI-04*, pp 464–469, ISBN 978-0-262-51183-4, San Jose, California, at the San Jose Convention Center, July, 2004, AAAI Press, San Jose, California.

Nüchter, A. (2007). 6D SLAM - 3D Mapping Outdoor Environments, in: Quantitative Performance Evaluation of Robotic and Intelligent Systems, *Journal of Field Robotics*, Vol. 24, No. 8-9, November, 2008, pp. 699-722, ISSN 1556-4959

Nüchter, A. (2008). Towards semantic maps for mobile robots, *Robotics and Autonomous Systems, Elsevier,* Vol. 56, No. 11, November, 2008, pp. 915-926, ISSN 0921-889.

Pagnottelli, S.; Taraglio, S.; Valigi, P. & Zanela, A. (2005). Visual and laser sensory data fusion for outdoor robot localisation and navigation, *Proceedings of 12th International Conference on Advanced Robotics 2005. ICAR '05,* 171–177, ISBN 0-7803-9178-0, Seattle; Washington, July 2005, IEEE, Seattle.

Rivas Lopez [a], M.; Sergiyenko, O. & Tyrsa, V. (2008). Machine vision: approaches and limitations, In: *Computer vision,* Xiong Zhihui, (Ed.), pp. 395-428. Editorial: IN-TECH, ISBN 978-953-7619-21-3, Vienna, Austria.

Rivas Lopez [b], M.; Sergiyenko, O.; Aguirre, M.; Devia, L.F.; Tyrsa, V.; & Rendón, I. (2008). Spatial data acquisition by laser scanning for robot or SHM task. *Proceedings of IEEE-IES International Symposium on Industrial Electronics (ISIE-2008),* p.1458-1463, ISBN 978-1-4244-1666-0, Cambridge, United Kingdom, 30 June -2 July 2008, IEEEXplore, Cambridge.

Rivas Lopez, M.; Sergiyenko, O.; Tyrsa, V.; Hernández Perdomo, W.; Hernández Balbuena, D.; Devia Cruz, L.; Burtseva, L.; & Nieto Hipólito J. I. (2010). Optoelectronic method for structural health monitoring. *International Journal of Structural Health Monitoring,* Vol.9, No.2, March, 2010, SAGE Publications, pp.105-120, ISSN 1475-9217

Rodríguez, J.C., Sergiyenko, O.; Básaca, L.C. (2009). Medición de parámetros biométricos específicos en el rostro humano por medio de barrido láser dinámico (in Spanish). *Proceedings of Congreso Nacional de Estudiantes de Posgrado del Instituto de Ingeniería,* UABC Mexicali B.C, November, 25-27, 2009, 6p., ISBN: 978-607-7753-33-9.

Sergiyenko, O.; Burtseva, L.; Bravo, M.; Rendón, I. & Tyrsa V. (2006). Scanning vision System for Mobile Vehicle Navigation. *Proceedings of IEEE-LEOS Multiconference on Electronics and Photonics (MEP-2006),* p.178-181, ISBN 1-4244-0627-7, Guanajuato, Mexico, 7-10 November, 2006, IEEEXplore, Guanajuato.

Sergiyenko, O.; Tyrsa, V.; Hernandez-Balbuena, D.; Rivas López, M.; Rendón López, I. & Devia Cruz, L.F. (2008). Precise Optical Scanning for practical multiapplications. *Proceedings of IEEE-34th Annual Conference of IEEE Industrial Electronics (IECON'08),* p.1656-1661, ISBN 978-1-4244-1766-7, Florida, USA, 10-13 November, 2008, IEEEXplore, Orlando.

Sergiyenko, O. [a]; Hernandez, W.; Tyrsa, V.; Devia Cruz, L.; Starostenko, O. & Pena-Cabrera, M. (2009). Remote Sensor for Spatial Measurements by Using Optical Scanning, *Sensors,* 9(7), July, 2009, MDPI, Basel, Switzerland, pp. 5477-5492. ISSN 1424-8220.

Sergiyenko, O. Y. [b]; Tyrsa, V. V.; Devia, L. F.; Hernandez, W.; Starostenko, O. & Rivas Lopez M. (2009). Dynamic Laser Scanning method for Mobile Robot Navigation. *Proceedings of ICROS-SICE International Joint Conference (ICCAS-SICE 2009),* p. 4884-4889, ISBN: 978-4-907764-33-3, Fukuoka, Japan, August, 18-21, 2009, IEEEXplore, Fukuoka.

Sergiyenko, O. Yu. [c]; Hernandez, W.; Tyrsa, V. V. & Hernández-Balbuena, D. (2009). Precise Optical Scanning for multiuse. *Proceedings of IEEE-35th Annual Conference of IEEE Industrial Electronics (IECON'09),* pp.3399-3404. ISBN 978-1-4244-4649-0/09, 3-5 November, 2009, Porto, Portugal, IEEEXplore, Porto.

Sergiyenko, O. Yu. (2010) Optoelectronic System for Mobile Robot Navigation. Springer/Allerton Press, Inc., *Journal Optoelectronics, Instrumentation and Data Processing*, Vol. 46, No. 5, October, 2010, pp.414-428 – ISSN 8756-6990

Son, S.; Park, H. & Lee, K. H. (2002). Automated laser scanning system for reverse engineering and inspection. *International Journal of machine tools and manufacture*. Pergamon Press, Great Britain, Vol. 42; No. 8, 2002, p. 889-897, ISSN 0890-6955

Surmann, H.; Lingemann, K.; Nüchter, A.; & Hertzberg, J. (2001). A 3D laser range finder for autonomous mobile robots. *Proceedings of the 32nd ISR (International Symposium on Robotics)*, pp. 153 - 158, 19-21 April 2001.

Surmann, H. (2003), An autonomous mobile robot with a 3D laser range finder for 3D exploration and digitalization of indoor environments, *Journal Robotics and Autonomous Systems (JRAS), Elsevier Science*, Volume 45, No. 3-4, December, 2003, pages 181 – 198, ISSN 0921-8890.

Vandapel, N.; Huber, D.F.; Kapuria, A. & Hebert, M. (2004). Natural terrain classification using 3-d ladar data, Proceedings *of IEEE International Conference on Robotics and Automation 2004. IEEE ICRA '04*, Vol. 5, April-May, 2004, 5117–5122, ISSN 1050-4729.

Wallace, R.; Matsuzaki, K.; Goto, Y.; Crisman, J.; Webb, J. & Kanade, T. (1986). Progress in robot road-following, *Proceedings of IEEE International Conference on Robotics and Automation 1986*, Vol. 3, April, 1986, 1615–1621, ISSN 0882-4967.

Wettergreen, D.; Tompkins, P.; Urmson, C.; Wagner, M. & Whittaker, W. (2005) Sun-synchronous robotic exploration: Technical description and field experimentation, *International Journal of Robotics Research*, Vol. 24, No. 1, January, 2005, 3–30, ISSN 0278-3649.

Wijesoma, W.; Kodagoda, K. & Balasuriya, A. (2004). Road-boundary detection and tracking using ladar sensing, *IEEE Transactions on Robotics and Automation*, Vol. 20, No. 3, June, 2004, 456–464, ISSN 1042-296X.

Wulf, O.; Kai, O; Christensen, Henrik I.; Wagner, B (2004), 2D Mapping of Cluttered Indoor Environments by Means of 3D Perception, *International Conference on Robotics & Automation*, ISBN : 0-7803-8232-3, New Orleans, LA, April 2004, IEEE, New Orleans

Wulf, O. & Wagner, B. (2003), Fast 3D Scanning Methods for Laser Measurement Systems, *Proceedings of International Conference on Control Systems and Computer Science*, 2003, p. 312-317, ISBN: 973-8449-18-9 Bucharest, Romania, July 2003.

Wyatt, C.L., (1991). Electro-optical system design: For information Processing .McGraw-Hill, ISBN 0-07-072184-X, New York, 1991, 343p.

Zussman, E.; Schuler, H. & Seliger, G. (1994). Analysis of the geometrical feature detectability constraints for laser-scanner sensor planning. *The International Journal of Advanced Manufacturing Technology*, Vol. 9, No. 1, January, 1994, p. 56–64, ISSN 1433-3015.

26

Employment of Pulsed-Laser Deposition for Optoelectronic Device Fabrication

Ullrich Bruno
Air Force Research Laboratory, Materials & Manufacturing Directorate, Wright Patterson AFB, OH 45433-7707 USA

1. Introduction

Thin-film preparation and its controlled mastery - specifically of semiconductors - became imperative for modern devices including all kind of applications such as electronics, optoelectronics, photonics, and superconductivity. Many methods and their technological applications have been explored and studied during the last decades (George, 1992, Smith, 1995, Ullrich et al. 1988, Bouchenaki et al., 1991 a, Bouchenaki et al., 1991 b, Ullrich et al., 1992, Ullrich and Kobayashi, 1995): Vacuum evaporation, molecular beam epitaxy (MBE), spray pyrolysis, closed-space deposition, sputtering, and pulsed-laser deposition (PLD). The difference between the latter and the aforementioned methods is that the film deposition process takes place only by photons, which naturally do not effect, alter or contaminate the ambient conditions of the substrate, which is kept in vacuum (typically 10^{-6} torr$\approx1.3\times10^{-4}$ Pa). This feature puts PLD on top of the stoichiometry maintaining thin-film deposition methods. However, theory does not go along with reality all the time because the intrinsic atomic target features might influence the stoichiometry as well - for example PLD of CdS leads most of the time to slightly Cd enriched films. It is presumed that the heavier Cd atoms displace some of the S atoms from their designated target-to-substrate transfer path. This brings us to the basics of PLD - how does it work? The ablating light, which is provided by a pulsed laser, hits the substrate and, in case the convolution of laser fluence (i.e., the incident laser energy by illuminated area and pulse) and absorption is sufficiently high, material is ablated from the target. As an example, the deposition rates vs. fluence for different laser wavelengths of GaAs are shown in Fig. 1. The threshold fluence at 355 nm and 532 nm is at around 0.3 J/cm^2, whereas at 1064 nm, the ablation onset requires higher fluence of approximately 0.5 J/cm^2 due to weaker absorption of the infrared laser pulses. The qualitative appearance of the rates is the same at all wavelengths. Beyond threshold, the inset in Fig. 1 shows that the ablation rate exponentially increases with the fluence (F), i.e., $\propto\exp(kF)$, where k=6.0, 6.7, and 5.9, at 355 nm, 532 nm, and 1064 nm, respectively, followed by a fairly linear growth, which finally turns to a flat saturating rate of the ablated material. However, Fig. 2 shows that the deposition rate depends on the material - apparently the same fluence ablates more material from the ionically bonded II-VI compound CdS than from the covalently bonded III-V compound GaAs. The deposition rates have been recorded with the Sloan 200 monitor using a quartz crystal in the vacuum chamber.

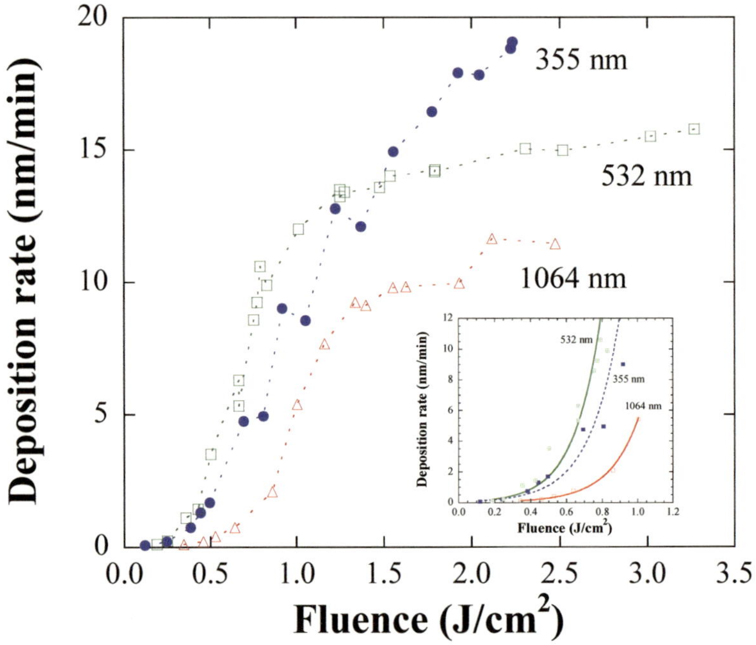

Fig. 1. GaAs deposition rates for the laser lines at 355 nm, 532 nm, and 1064 nm. The inset shows the threshold in detail

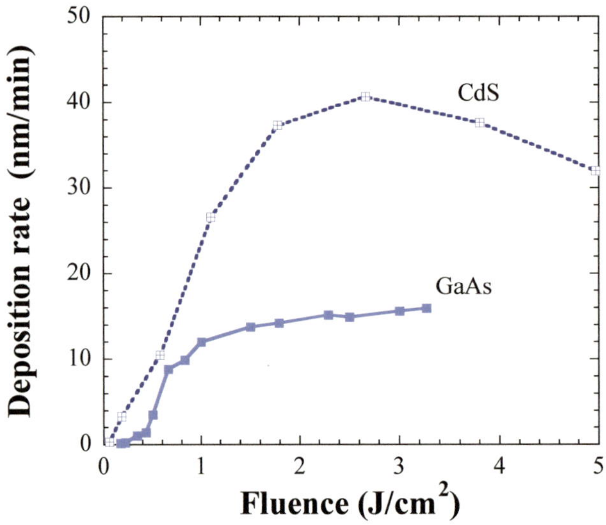

Fig. 2. Comparison of the GaAs and CdS deposition rates at 532 nm

The thin-film CdS sample was formed using the PLD setup employed to form the GaAs films (Ullrich et al., 2003). The CdS ablation was carried out with a Neodymium doped

Yttrium Aluminum Garnet (Nd:YAG) laser using the frequency doubled 532 nm line. We used the Spectron laser model SL456G with 6 ns pulse duration and 10 Hz repetition rate. The laser beam passed through a 9 mm aperture, a 532 nm interference filter, a 25 cm converging lens, and a calcium fluoride window of the vacuum chamber. During the PLD, the target rotated with 16 revolutions per minute in order to ablate the material as homogeneous as possible without local burns. The unheated substrate was mounted at distance of 6 cm parallel to the target, which was tilted 45° with respect to the impinging laser beam. The material was ablated for 10 minutes with a fluence of 1.65±0.12 J/cm^2 at $5.1x10^{-7}$ Torr, resulting in an approximately 400 nm thick CdS film. Figure 3 shows the optical images of the CdS plume for various fluences. Effective ablation, i.e., a visible bluish plume, starts around 0.64 J/cm^2, which is in reasonable agreement with expectations from the CdS deposition rate in Fig. 2.

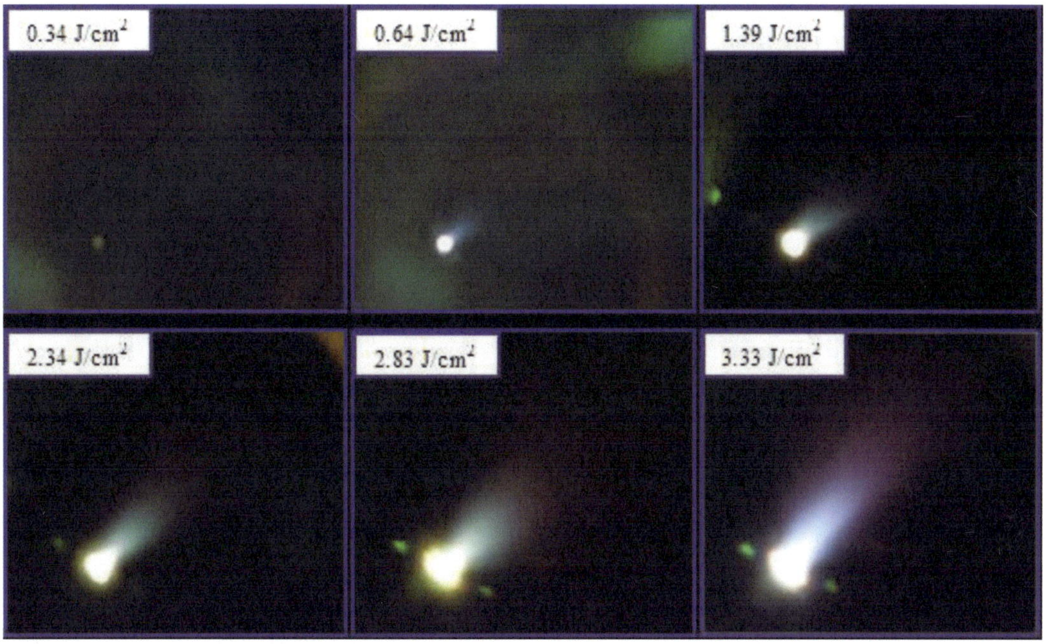

Fig. 3. CdS plume pictures at different laser fluences. Plasma is seen around 0.64 J/cm^2 indicating the threshold of the ablation fluence

Properly termed, the results in Figs. 1 and 2 refer to a blend between ablation and deposition rate. The latter is commonly accepted and therefore we use the nomenclature "deposition rate". Figure 2 reveals that the CdS deposition rate first increases with increase in fluence. However, around 1.90 J/cm^2 the deposition rate starts to saturate and after reaching the maximum at 2.68 J/cm^2, the deposition rate decreases because at higher fluences the denser and denser "particle cloud" and possibly ablated clusters absorb and scatter the incoming laser light (Acharya et al., 2010). Due to the use of relatively short laser pulses (ns pulse duration and faster), PLD can be apostrophized as "explosive material flash". The point is that the film formation is over before the material decomposes owing to the supersonic velocity (10^4-10^5 m/s) of the plume particles (Ezumi & Keitoku, 1993). For the distance of 6 cm, the deposition takes place within 600 ns-6 μs.

The unmatched flexibility of PLD is given by the ease to modify the external parameters, which are the laser line and fluence used to ablate the material. The table below reveals the considerable influence of the ultraviolet (UV) and infrared (IR) laser lines employed on the resultant CdS film features (Ullrich et al., 2001, Ullrich, 2007):

UV (355 nm, 5 ns, 10 Hz)	IR (1064 nm, 180 μs, 10 Hz)
Extremely photosensitive	Low photosensitivity
Dark conductivity $\sigma \leq 0.01$ $(\Omega\ \text{cm})^{-1}$	Dark conductivity $\sigma \approx 1$ $(\Omega\ \text{cm})^{-1}$
Dark carrier concentration $n \approx 10^{16}$ cm^{-3}	Dark carrier concentration $n \approx 10^{18}$ cm^{-3}
c-axis orientation either perpendicular or parallel with respect to the substrate - tunable dichroism	*c*-axis orientation perpendicular with respect to the substrate
Weak emission	Visible strong green emission
Lasing not observed	Lasing due to microcavities

Table 1. Thin-film CdS properties formed at 355 nm and 1064 nm

Work is still ongoing to clarify the orgin of the considerable difference of the dark carrier concentration between the CdS samples (Ullrich, 2007). Most striking is the capability of PLD to control the *c*-axis orientation of II-VI compounds (Sakai et al. 1998). Irradiating the target with 355 nm and a fluence of about 2 J/cm^2, the CdS film deposited on glass exhibits a perpendicularly oriented *c*-axis with respect to the glass surface, whereas a fluence beyond 3 J/cm^2 results in CdS films with a *c*-axis parallel to the glass surface. The origin of the *c*-axis turn underlies the fluence dependent plume composition (Sakai et al. 1998, Ullrich 2007). Fluences around 1 J/cm^2 consist mainly of atoms, while a higher fluence increases the ablated cluster density. Those clusters break loose from the target along the cleavage direction; arriving at the substrate surface the clusters arrange themselves in the most probable position, which is the flat cleavage surface. The cleavage line runs parallel to the *c*-axis and, therefore the film orientation is parallel to the glass surface.

In summary, effective absorption of the target at the ablation laser line transfers the energy of the laser beam into the target. The excited volume remains fairly small under these circumstances (the longitudinal dimensions are about ≤0.1-1 μm). The absorbed energy locally increases the temperature of the target and ablation takes place. The most important PLD parameter is the laser fluence, which is given by,

$$F = P \times l / (R \times A), \tag{1}$$

where P is the impinging laser power, l are the transmission losses mainly caused by the window of the vacuum system (typically 15%), R is the repetition rate (10 Hz), and A is the area of the laser spot (here about 0.0157 cm^2) on the target. With a power of 184 mW a fluence of 1 J/cm^2 is achieved. Further details about the PLD system used and important parameters are published in the Dissertation of Artur Erlacher (Erlacher, 2005). The capability of changing material properties by simply changing the fluence and/or the ablation wavelength makes PLD a versatile deposition method ideally suited for prototyping. Furthermore, PLD is widely independent of substrate qualities and matching conditions (Ullrich et al., 1997). The following items show device related applications of PLD formed thin films.

2. Hybrid logic device formation

With regard to information transmission, all-optical logic operations are very appealing to technological applications since electron based signal processing has certain limitations due to charge storage in capacitors, crosstalk and charge loss or attenuation. However, optical computing - a phrase frequently used in the mid-1980s - remains more or less a fiction and there was disillusion already 20 years ago (Gibbs & Khitrova, 1990) that raised serious doubts regarding the impact non-linear photonics would have on computing. We believe that this development is mainly caused by the circumstance that photonics lacks the inherent ease based on the overwhelmingly perfect mastery of Si (and also GaAs and related compounds), which enables nearly effortless solutions for complex electronic and optoelectronic devices. Hence, it is not a competition between photons and electrons but a struggle between a non-existent technology on an industrial level with the established Si/GaAs industry.

The presented results promote PLD formed glass/GaAs interfaces as effective photonic switch media. All-optical and hybrid red laser reflection modulations have been achieved by altering the electronic state of the interface with a green laser. The concept enables the flip from negative to positive logics by solely applying an electric field. Furthermore, reflective femtosecond operations of the interface are demonstrated with pump-probe experiments.

Manipulating light with light seems to be an impossible task since photons are gauge bosons and as such not influenced by themselves or by electromagnetic means. Nevertheless, during the 1970's and 1980's, notable research efforts revealed ways to make optical computing a reality (Hajto & Janossy, 1983, Gibbs, 1985, Wherrett & Hutchings, 1993). Since then, many switching concepts, various computer architectures, and core materials have been investigated for light driven data transmission (Erlacher et al. 2006, and references therein). Although there are currently various commercially available optical switches, such as the mirror-based Micro-Electro-Mechanical Systems (MEMS) (Shepard, 2001), an overall satisfactory optical switch fabric has not yet been established. In spite of the unsolved switching issue, optical fibers conquer the world of data transportation. Their unmatched transmission speed (currently terabit/s) is capable of bypassing the looming interconnect bottleneck of electric connections. Consequently, as the unit costs keep coming down, fiber to the home (FFTH) might be upon us (Bates, 2001).

Neither necessarily linked to concepts described in the "classical" literature (Hajto & Janossy, 1983, Gibbs, 1985, Wherrett & Hutchings, 1993) nor going distinctly along with currently discussed optical techniques in the literature (Papadimitriou, 2007), we introduced all-optical operations via laser crossing (LC) in thin-film semiconductors on glass (Erlacher and Ullrich, 2004, Liyanage et al., 2009). The main advantaging difference between LC and other concepts is its low-cost realization potential due to extreme straightforwardness of the basic principle. Here, we focus on the employment of the glass/GaAs interface as active LC medium for all-optical applications and hybrid computing architectures.

The schematic of the experiment is shown in Fig. 4. Red (633 nm) and green (530 nm) continuous wavelength (cw) emissions of a 10 mW He-Ne laser and 20 mW Ar-Kr laser, respectively, were crossed in one spot at the glass/GaAs interface opposite to the opening of Al contacts, which have been evaporated on the film surface. The gap between the contacts was about 1 mm and the impinging laser intensities were kept below 10 W cm^{-2}. The arrangement of the laser rays permitted the measurement of reflection and transmission of the red laser as a function of the on-off sequence of the green laser with the additional option to apply an electric field across the illuminated film area. The reflected and transmitted signals of the He-Ne laser were measured with Si photodiodes, which were connected to amperemeters. The on-off cycling of the Ar-Kr laser beam was achieved by means of an optical shutter.

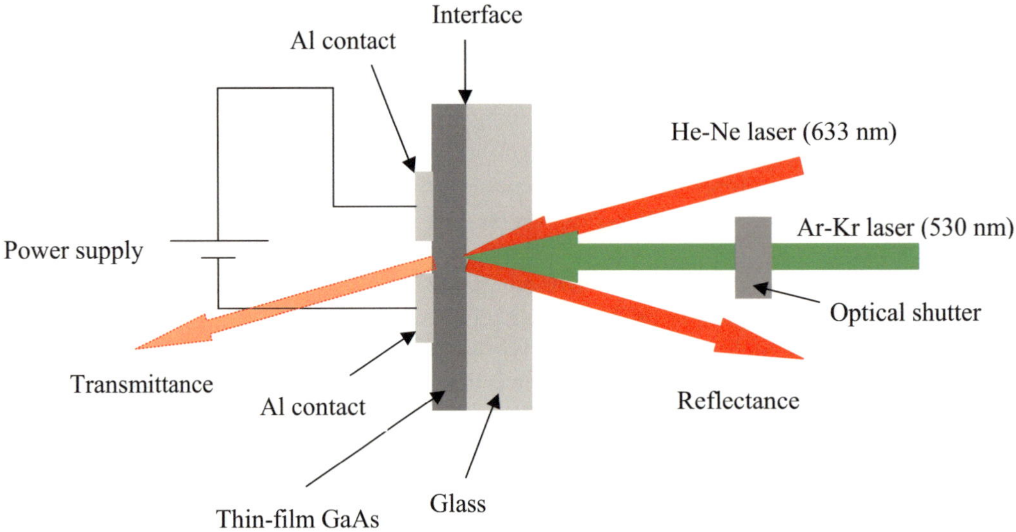

Fig. 4. Arrangement of reflective LC at the glass/GaAs interface. Red and green emissions from a He-Ne and Ar-Kr laser, respectively, are crossed at one point oppositely to the gap between two Al contacts on the film surface (US Patent No. 7,826,058)

The GaAs film has been formed with PLD in vacuum (10^{-3} Pa) by employing the fundamental emission (1064 nm) of the Nd:YAG laser (6 ns, 10 Hz). The target was GaAs:Zn (doping concentration 10^{19} cm^{-3}) and the film was ablated with a fluence of about 0.8 J cm^{-2} to be 0.5 μm thick. The distance between target and fused silica glass substrate, which was not heated during the deposition, was 6 cm. It is of interest to briefly discuss the complex texture and unusual high-ohmic behavior of PLD GaAs. We performed x-ray analysis, Raman measurements, and atomic force microscopy (Erlacher et al. 2004 & 2005). Based on these results, the films are of multi-phase nature consisting of microcrystallites, amorphous parts, and nano-sized zincblende migrations. Hall measurements showed acceptor concentration as low as 10^{13} cm^{-3}, i.e., six orders of magnitude below the target value and one might conclude that the Zn doping of the target vanishes somewhere during PLD. However, secondary ion mass spectroscopy (SIMS) showed that the Zn concentration in the films is maintained but due to the peculiar texture the impurities are hardly electrically active. This observation explains the measured resistance of $>10^9$ Ω between the two evaporated Al contacts. From this point of view, low-temperature PLD GaAs on glass refers to dielectric material rather than to typical semiconductors.

It is worthwhile to note that the motivation for the interface investigations came from photoluminescence (PL) measurements rather than from previous switching experiments. Comparing the PL excited at the surface and through the glass substrate, we recognized that PLD CdS on glass exhibits an increased absorption at the glass/CdS interface with respect to the rest of the film (Ullrich & Schroeder a, 2001, Ullrich and Schroeder b, 2001). Hence, we presumed that the same might be true for PLD GaAs on glass and enhanced LC modulation amplitudes should be possible by using the rear reflection of the film. Indeed, in the following it is demonstrated that the assumption was right.

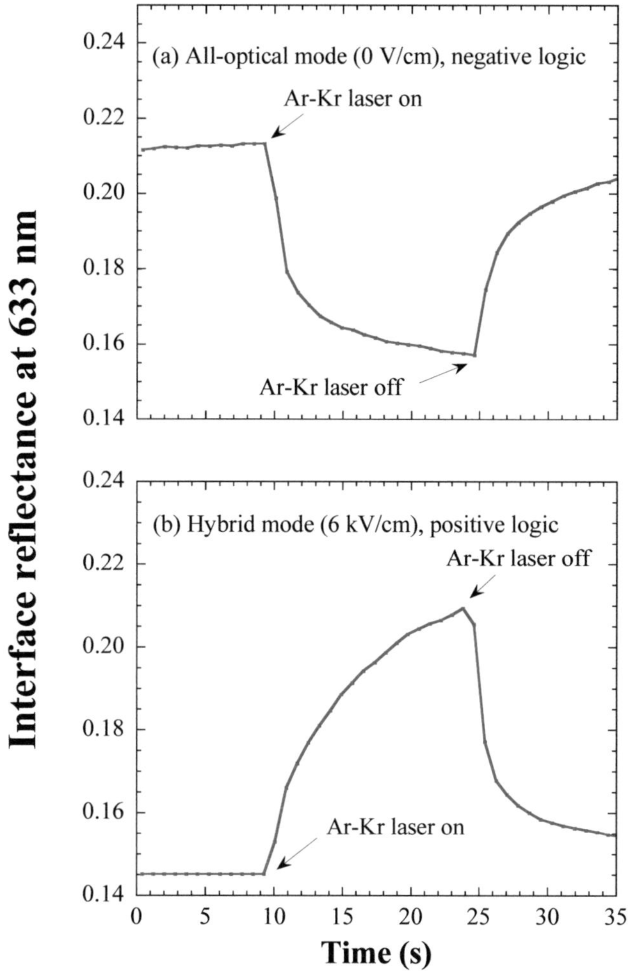

Fig. 5. Modulation of the red laser reflectance at the interface due to the on-off sequence of the green laser: (a) is the intrinsic all-optical response and (b) is the hybrid mode measured with an applied electric field of 6 kV/cm

Figure 5 shows the modulated He-Ne laser reflection caused by the impinging Ar-Kr laser beam. The measurements have been performed at room temperature (a) without and (b) with applied electric field of 6 kV/cm. The achieved modulation amplitudes are up to 35% and, notably, both responses are almost perfectly inverted. According to our chosen definition, the modulation flip refers to the cross-over between negative (all-optical) and positive (hybrid) logics. To invert logic operations by simply applying an electric field is a novel aspect in the world of digitizing devices. We are not aware of another switch fabric with such a useful capability.

According to Fig. 5, around 15% and more of the red laser light was reflected and for the reflected portion of the green laser we found 30%. Thus, with consideration of the losses in the lenses used to adjust the laser beams, the overall optical power dissipated in the film did not exceed 20 mW. It was demonstrated in a previous publication (Ullrich et al., 2004) that a moderate laser power like that only marginally warms up the GaAs sample and does not cause noticeable photothermal effects, i.e., the observation in Fig. 5 (a) is driven by the alteration of the local electronic state of the film. The same is true for the hybrid operation. Due to the extreme resistance of the sample, the dissipated electrical power under illumination did not exceed 10^{-4} W and, therefore, is certainly not the origin of modulation flip in Fig. 5 (b) with respect to the all-optical mode in Fig. 5 (a). The inversion of the modulation is rather caused by the reduction of the electron number available for absorption transitions, i.e., with growing electric field more and more electrons join the circulation of the photocurrent rather than to maintain the matrix element of the absorption coefficient due to recombination into the valence band. As a consequence, the sample becomes more transparent and the modulation sequence is inverted in Fig. 5 (b) with respect to Fig. 5 (a). Figure 6 reveals the dependence of the modulation amplitude on the applied electric field of the reflected and transmitted signal of the He-Ne laser. Positive and negative signs of the modulation amplitude correspond to positive and negative logic.

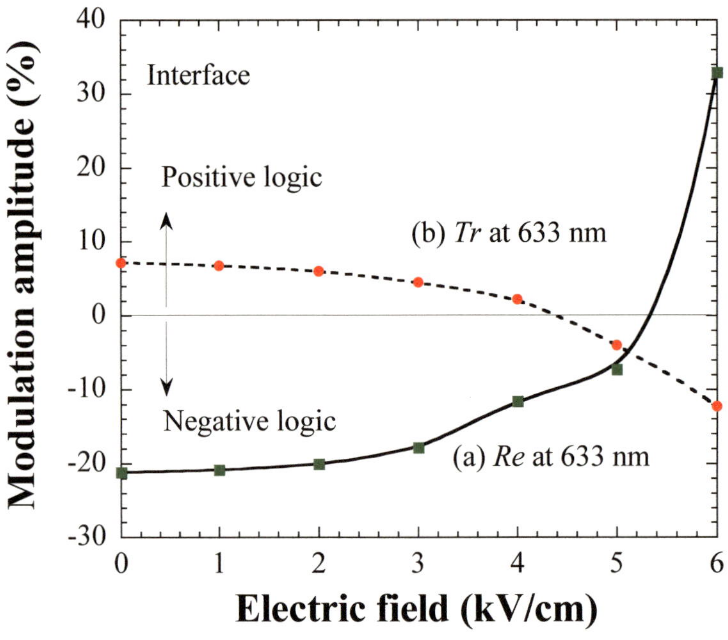

Fig. 6. The modulation amplitude vs. electric field of (a) the reflectance (*Re*) and (b) transmittance (*Tr*) of the He-Ne laser measured at the glass/GaAs interface. Positive and negative signs refer to positive and negative logic, respectively. The solid and broken lines are guides for the eyes

For comparison, Fig. 7 shows the corresponding LC results for the thin-film, i.e., the film side was excited as in previous works (Erlacher and Ullrich, 2004, Ullrich et al. 2004). The film reflection exhibits almost no modulation, whereas the maximum of the transmission modulation is comparable with the results in Fig. 6, while, as suspected from the PL results (Ullrich & Schroeder a, 2001, Ullrich and Schroeder b, 2001), at the interface, more absorption transitions are excited with the same amount of impinging photons than at the film surface itself. Furthermore, from the viewpoint of device engineering, with the employment of the reflected beam effective fan-out is easier to achieve than with the much weaker transmitted signals, which were approximately 0.8% and 0.2%, for the red and green laser, respectively.

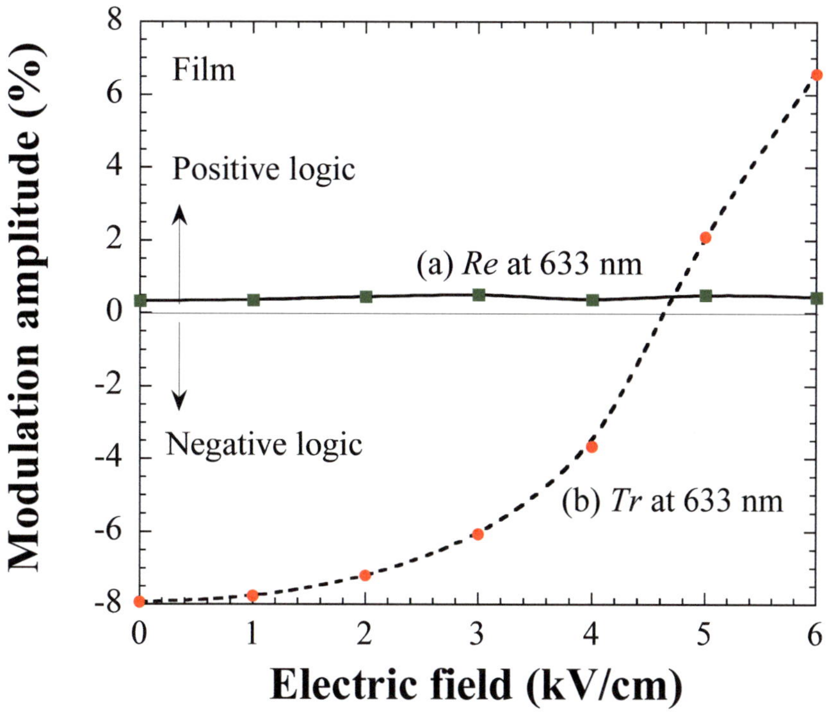

Fig. 7. Corresponding results to Fig. 6 measured at the film side

Equivalently to Fig. 4, pump-probe experiments have been performed at room temperature by replacing the green cw laser irradiation with green (532 nm) ultrafast (<100 fs) laser pump pulses with a repetition rate of 1 kHz, and, instead with the He-Ne laser, the interface was probed with white light continuum. The reflected spectrum was guided in a fiber spectrometer. A typical result of the reflection kinetics for 633 nm is shown in Fig. 8, which depicts the ratio of the reflection without pump over the reflection with pump (Re_0/Re). Turn-on and recovery time are extremely fast resulting in a kinetics profile with a full width at half maximum (FWHM) of about 600 fs. Hence, in contrast to the thin-film itself whose recovery time exceeded more than 10 ps (Ullrich et al., 2004), metastable states with unusual short lifetimes in the fs range are generated at the glass/GaAs interface. It is worth mentioning that our temporally resolved reflection alteration takes place on a similar time scale as the transmission kinetics of CdTe nanocrystals (Padilha, 2005) and our observed recovery time can be compared with that in Bragg-spaced quantum well structures (Johnston, 2005).

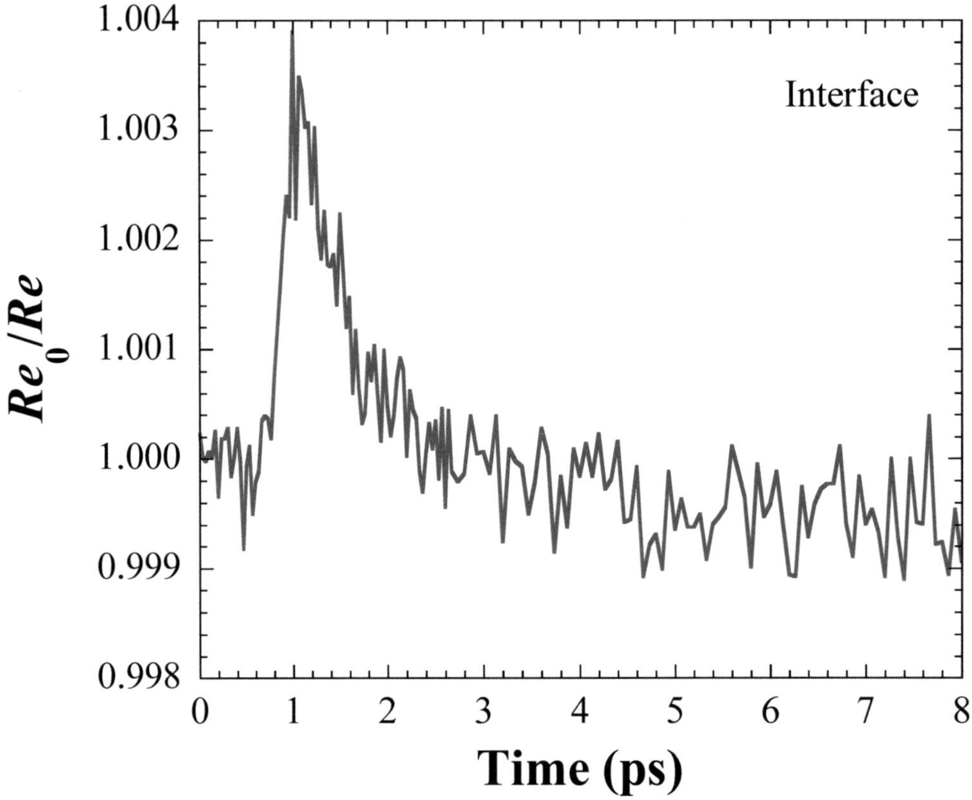

Fig. 8. Ultrafast reflection kinetics of the glass/GaAs interface at 633 nm. The ordinate shows the signal without pump pulse (Re_0) over the signal with the pump pulse (*Re*)

3. Hybrid photosensitive device formation

The need for operative optoelectronic devices composed of various semiconductors is based on technological requirements as well as economical arguments, while in many cases simultaneous improvement of both branches is anticipated. In particular, operative hetero-pairing of GaAs and Si has been pursued for quite some time owing to its considerable technological importance for efficient and low-cost photovoltaic and optoelectronic devices (Alberts et al., 1994). However, GaAs-on-Si formations are not a straightforward task. The mismatch of thermal expansion coefficients (60%) and lattice constants (4%) causes stress in the film during deposition and the formation of defect states at the film substrate interface, respectively. Besides improvement of the electrical and structural properties of the GaAs/Si heterostructure can be achieved using a ZnSe (Lee et al., 1994) or $SrTiO_3$ (internet, 2001) buffer layer, direct growth and deposition were pursued by using a wide range of methods since several decades with varying success and satisfaction. For example, the direct growth of GaAs on Si employing MBE was studied (Usui et al., 2006).
Concerning the hetero-pairing of PLD GaAs with Si, previous research mainly focused on rectifying properties and bias dependent alternating photocurrent (APC) of *p*-GaAs on *n*-Si

(Ullrich and Erlacher a, 2005, Ullrich and Erlacher b, 2005). Here, stoichiometry, surface morphology, crystallographic properties, and quality of rectification of PLD formed *n*-GaAs on *p*-Si heterostructures were studied. In addition, the bias dependence of APC spectra and direct photocurrent (DPC) spectra is demonstrated.
The deposition of GaAs onto Si was carried out with the second harmonic emission at 532 nm of the *Q*-switched Nd:YAG laser (6 ns, 10 Hz). The substrate was boron (B) doped prime grade 250-300 µm thick *p*-type Si (100) wafer and the target was tellurium (Te) doped *n*-type GaAs (100) with a doping concentration of 1.3-3.4×10^{18} cm^{-3}. As usual, the rotating (16 rpm) GaAs target was kept at 45° with respect to the impinging laser beam. The rotation was performed in order to ablate the material homogeneously. Parallel to the ablation area of the target, the substrate was mounted 6 cm above the target. The deposition was carried out in vacuum of 6.7×10^{-5} Pa without heating the substrate in order to avoid interface stress and deformation during cooling. The target was exposed to a laser fluence of 0.98 ±0.1 J/cm^2 for one hour resulting in a GaAs film with a thickness of about 600 nm. In order to perform APC and DPC measurements, electrical contacts were made with evaporated aluminum (Al) to the surface and by silver (Ag) paste to the rear of the sample.
Stoichiometry, crystallographic properties, and surface quality of the GaAs film were investigated by electron probe micro-analyzer (EPMA) technique, x-ray diffraction, and scanning electron microscopy (SEM), respectively. The current-voltage (*I-V*) characteristic, of the heterojunction was measured in the dark by sweeping the applied bias between the Al and Ag contacts with a programmable source meter. The APC spectra were measured with lock-in technique by exciting the sample surface with chopped (107 Hz) light, while steady-state illumination and a pico-amperemeter were used to record the DPC responses. The sample was irradiated with a 200 W halogen lamp, whose emission was monochromatically dispersed with a ½ m monochromator. Excitation light was passed through a long pass filter and the photon energy was varied from 1.1 to 1.80 eV and the biases were applied using a programmable power supply. The PC was measured vertically through the device using contacts 0 and 1 or laterally employing contacts 1 and 2 as schematically shown in Fig. 9. All the PC spectra shown have been corrected with a calibrated Si photodiode in order to express the results in terms of responsivity (A/W). All measurements were carried out at room temperature.
The EPMA results show Ga and As concentrations of 48.80 (at. %) and 51.20 (at. %), respectively. Hence, low-temperature PLD forms quasi-stoichiometric GaAs films on Si with a slight As surplus. The slight deviation of stoichiometry reflects the trend known from MBE formed GaAs, i.e., the growth of GaAs at low substrate temperatures results in GaAs with excess As incorporated into the GaAs matrix (Ma et al., 2004), which is amorphous as the x-ray result in Fig. 10 reveals. The only GaAs related peak observed is the fairly faint feature at 27.25 degree indicating that the film texture is predominately of amorphous nature embedding small portions of crystallites. We should stress that previous x-ray investigations on *n*-GaAs/*p*-Si did not show peaks, presumably due to the extremely thin film. The SEM image, (Acharya et al., 2009), shows droplet-like micron-sized clusters. With selectively performed Raman measurements, we found that these clusters are single crystalline GaAs, which are not grown but directly transferred from the target during PLD (Erlacher et al. 2006). Since these droplets increase the surface area, the formed hetero-pairing has the potential to be used as chemical sensors and chemical reaction catalysts.

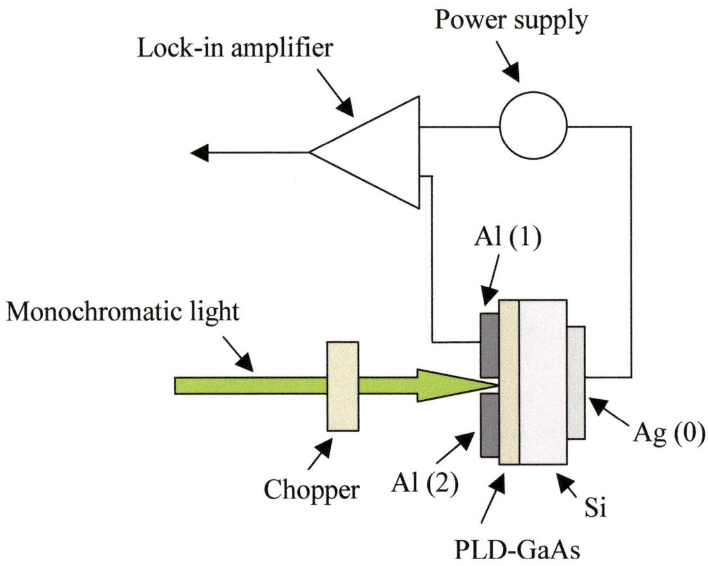

Fig. 9. Experimental setup used for the APC measurements. DPC was measured without chopper and the lock-in amplifier was replaced with a pico-amperemeter

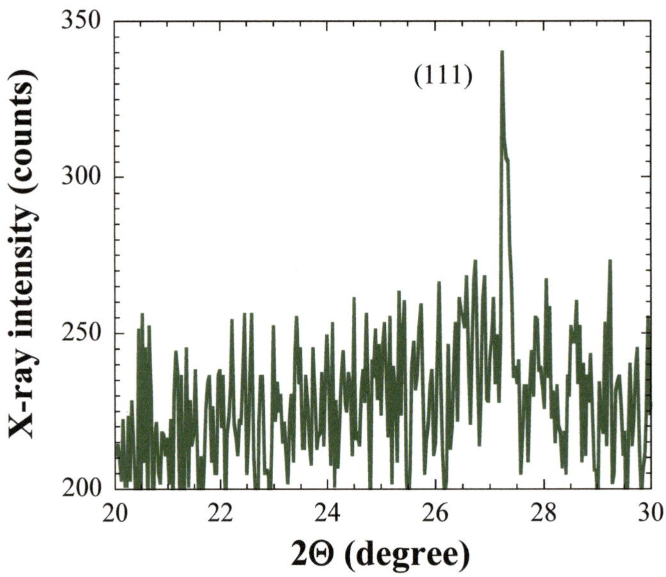

Fig. 10. X-ray pattern of the GaAs film

The dark *I-V* characteristic of the *n*-GaAs/*p*-Si device is shown in Fig. 11. The device showed rectification and the symbols represent the measurements, while the solid line visualizes close fit of the forwardly biased *I-V* curve using the common diode equation (Fahrenbruch &Bube, 1983),

$$I = I_s\left(\exp\left[\frac{eV_b}{\beta kT}\right]-1\right), \qquad (2)$$

where, I_s is the saturation latching current, e is the elementary charge, V_b is the applied bias, β is a factor taking into account deviations from the ideal diode characteristic, and kT (=25 meV) is the thermal energy. The fit delivered the following values: I_s=2.01×10^{-7} A and β=35.2. The applicability of Eq. (2) is quite notable since it did not apply for *p*-GaAs/*n*-Si hetero-structures (Ullrich and Erlacher a, 2005), which required a modified version of Eq. (2) in order to fit the forward current.

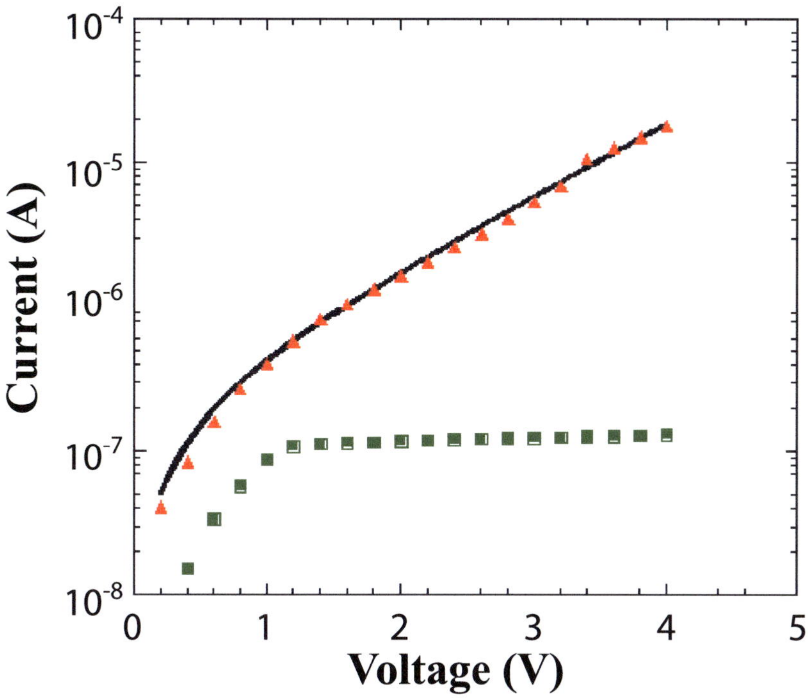

Fig. 11. *I-V* characteristic of the junction measured using contacts 1-0 in Fig. 9. The symbols represent the measurements and the solid line the fit using Eq. (2)

The doping of the target and its reestablishment in the deposited film was investigated. Comparing the results of SIMS and Hall measurements for *p*-GaAs/*n*-Si samples, we established that the physical concentration of doping atoms in the film closely corresponds to the target level but, due to the amorphous film texture, the concentration of electronically active dopants in the film is several orders of magnitude below the initial target value (Ullrich et al. 2008). Since the texture of *n*-GaAs deposited on *p*-Si is amorphous as well, we conclude that the improved diode features of the current sample is most likely caused by a comparably more abrupt interface with less recombination centers and charge alignments, allowing the exponential dependence on the bias.

The APC responsivity across the sample was measured by using contact combination 1-0 and 1-2 for various biases shown in Figs. 12 (a) and 12 (b). We first discuss the results in Fig. 12 (a). Negative bias signs mean that contact 0 (*p*-Si) was connected to the negative terminal of the power supply, corresponding to backward polarity, whereas positive signs indicate forward bias direction. Figure 12 (a) reveals all the features of an operating photodiode: a) The hetero-structure is sensitive to light without bias, b) increase in APC under reverse bias points to enlargement of the depletion layer, while c) a forward bias shrinks the depletion layer making the structure less light sensitive. In fact, the APC is completely bleached at a bias of +2 V on a linear scale.

Comparing the results shown here with the spectra of *p*-GaAs/*n*-Si photodiodes investigated in detail in Ref. Ullrich & Erlacher b, 2005, the most notable difference is the spectral shape of the intrinsic APC shown in Fig. 13. We should stress that despite the inverted doping, both samples in Fig. 13 have been formed under the same conditions. The *n*-GaAs/*p*-Si hetero-pairing exhibits APC in the absorption range of GaAs and Si, while the responsivity of the *p*-GaAs/*n*-Si is clearly dominated by the absorption in the Si substrate with trifling contributions of the GaAs film. The GaAs film contributes to the photocurrent if the depletion width in the GaAs film is comparable to that in the Si substrate, or, at least the diffusion length of the minority carriers in the GaAs film is long enough to reach the depletion region in Si (Ullrich et al., 1999). Obviously, this condition is not fulfilled for the *p*-GaAs/*n*-Si sample. Furthermore, forward bias induced APC bleaching is much less effective for the *p*-GaAs/*n*-Si structures and, while the energy position of the APC maxima in Fig. 12 (a) does not depend on bias polarity, the APC peak of *p*-GaAs/*n*-Si samples undergoes a shift of ≈100 meV by flipping the bias from +1 V to –1 V. These observations - in combination with the result revealed in Fig. 11 - clearly point to the fact that low-temperature PLD of *n*-GaAs onto *p*-Si forms the more defined junction device with clearly improved operational features.

The APC responsivities achieved with contacts 1 and 2 for different biases are shown in Fig. 12 (b). The sign notation refers to a polarity change rather than to a specific bias direction since we measure along the GaAs layer. The APC grows with increase in bias in either direction as expected from a common photoconducting material. Ideally, the photoresponse of the sample ought to be symmetric, i.e., the same APC spectrum is expected for either polarity. The difference in the spectra is most likely caused by inequalities of the Al electrodes. Notably, the sample showed responsivity at 0 V, like a diode though the current was not measured across the junction. The reason is discussed below.

The DPC responsivity spectra of the sample using contacts 1 and 0 are shown in Fig. 14 (a). The results correspond to these in Fig. 12 (a) besides the forwardly biased sample did not expose DPC, while a fairly weak DPC at 0 V took place. The latter is negative because in the measuring configuration employed the positive terminal of the amperemeter was connected with the *n*-side of the sample. The spectra in Fig. 14 (b) measured with the contacts 1 and 2 expose the features of a semiconducting layer like in Fig. 12 (b), showing a similar broken symmetry. However, in contrast to the APC measurements, the DPC shows only a very weak (<50 μA/W) signal for 0 V. Hence, the DPC spectra reveal a behavior much closer to expectation than the zero biased APC measurements. The answer to this apparent contradiction lies in the fact that the APC at 0 V measures the light dependence of the charging current of the device capacity rather than its responsivity, and, as a consequence, the APC spectra for 0 V barely depend on the contact configuration.

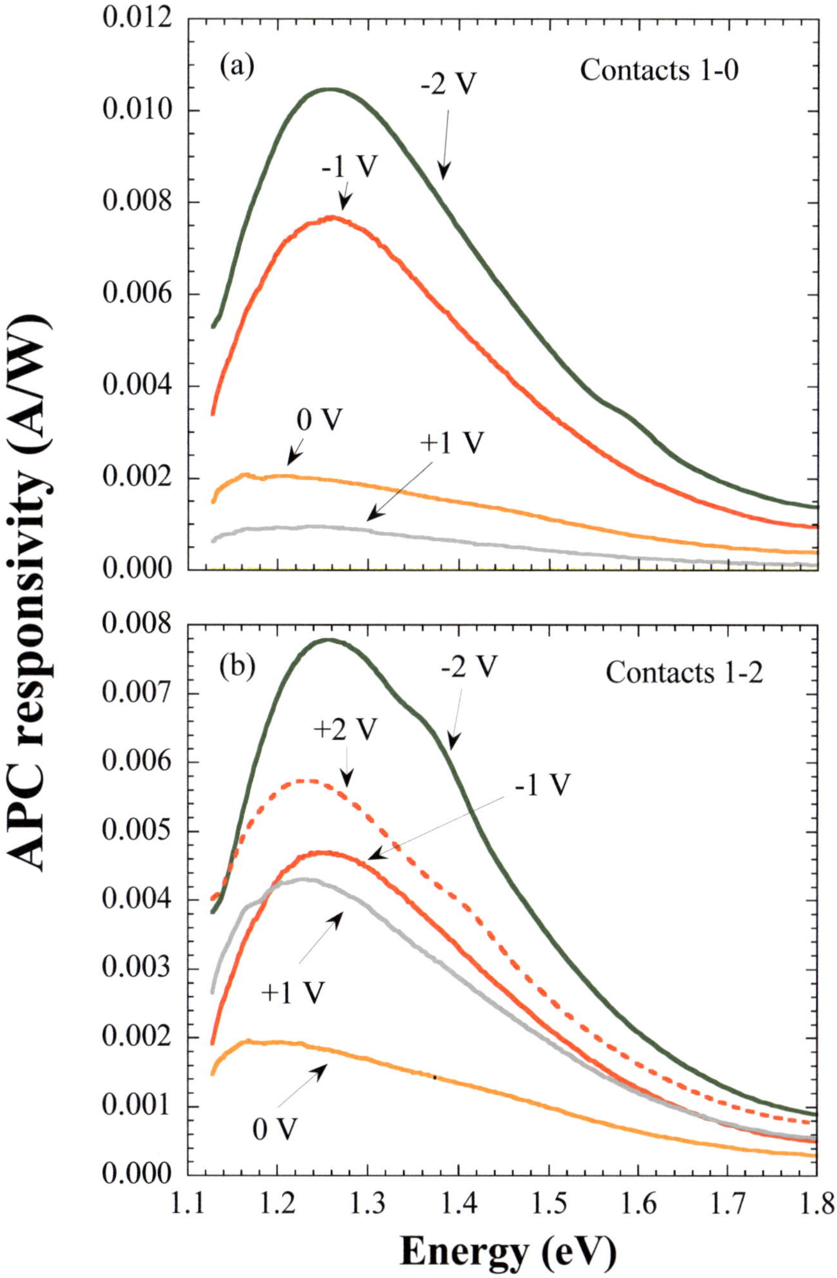

Fig. 12. APC spectra measured (a) across the junction and (b) along the GaAs film employing the contacts 1 and 0, and 1 and 2, respectively

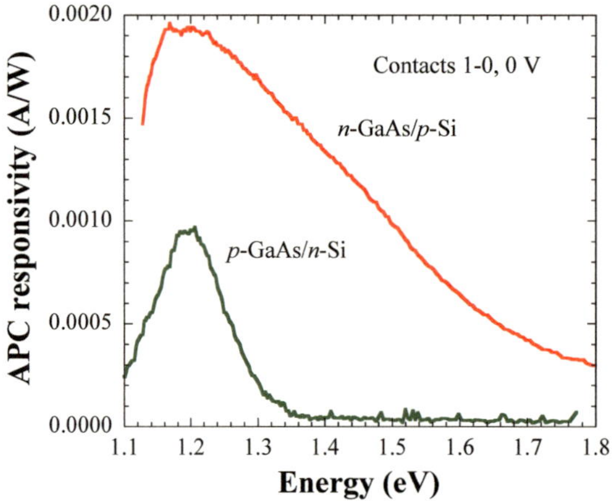

Fig. 13. Comparison of the APC responsivity at 0 V for *n*-GaAs/*p*-Si in Fig. 12 (a) and for *p*-GaAs/*n*-Si taken from Fig. 4 in Ref. Ullrich & Erlacher b (2005)

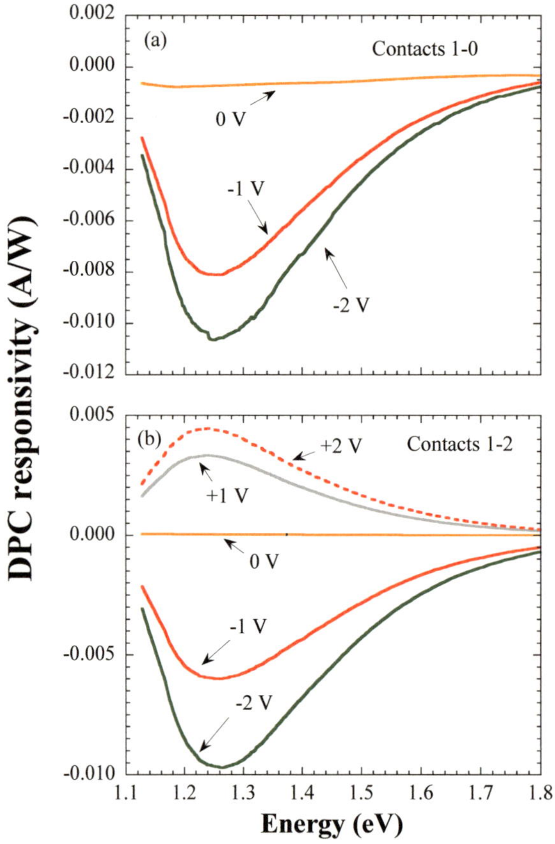

Fig. 14. DPC spectra (a) measured across the junction and (b) along the GaAs film employing the contacts 1 and 0, and 1 and 2, respectively

Concluding, we have presented a detailed study of thin-film *n*-GaAs deposited onto *p*-Si using nanosecond laser ablation. The results demonstrate that low-temperature PLD delivers stoichiometric GaAs films, which have some crystalline sections embedded in an amorphous matrix. The appearance of the film surface is very similar to GaAs/glass samples, i.e., smooth with some spherical micron-sized droplets on top. It is clearly worthwhile to stress that the achieved optoelectronic results here are quite different from previous investigations on *p*-GaAs/*n*-Si samples. Most striking is that the deposition of *n*-GaAs on counter-doped Si results in a *pn*-junction, which is accurately described by the common diode equation. The opposite material combination, however, exhibits a peculiar, but nevertheless rectifying, *I-V* characteristic. The responsivity features go along with the *I-V* curves: Contrasting the results in Ullrich B. & Erlacher A. b (2005), the junction spectra did not shift when a reverse bias was applied, pointing to a depletion region with a defined uniform band gap, and, at a fairly low bias of 2 V, the APC was drastically bleached. The APC results were confirmed with DPC investigations. In this context, by measuring the responsivity along the GaAs film, the work further stresses that intrinsic features (meaning intrinsic solely from the electronic viewpoint) measured with lock-in technique do not necessarily correspond to the *intrinsic ground state* of the device investigated.

4. Summery and future developments

Formation of thin-film device structures with low-temperature PLD is a very active, appealing and extremely versatile applied research field. PLD is in particular well suited for the production of operative optoelectronic/photonic device prototypes such as the GaAs-on-glass merger and GaAs/Si photodiodes.

Low-temperature PLD opens another door of fruitful research: The formation of predominantly amorphous thin films, which reveal quantum-confined features (Acharya et al., 2007). Ongoing work on nanostructured CdS films on flexible polyethylene terephthalate (PET) substrates formed with nanosecond low-temperature PLD is a promising alternative to fragile, rigid device structures. The film consists of intrinsically formed self-assembled ~ 100 nm grains speckled with CdS 10-20 nm quantum dots. The films are stoichiometric and show blue shifted photocurrent spectra due to quantum confinement. The nanocrystal size of about 6 nm calculated from the photocurrent peak shift agrees reasonably well with the texture analysis. Similar results have been achieved currently with ZnTe.

Due to the complex and therefore electronically sensitive texture, the hybrid merger of high-absorptive materials such as GaAs or CdS on plastic is an ideal host for logic devices based on LC using common visible laser sources. Promising preliminary results including optoelectronic and photonic operations have been demonstrated and discussed here. Figure 15 shows one of the advanced results at present under investigation. By superposition of red (632.8 nm) and blue (488.0 nm) cw laser irradiation in the same spot of the GaAs thin-film surface, i.e., by additional irradiation with the red laser line, the reflected spot (a) of the blue line changed to a ring-like pattern (b). The effect can be used as photonic laser steering without externally provided mechanical means. The effect is most likely caused by the formation of a photothermal lens due to mechanical deformation of the plastic substrate caused by the warming of the GaAs top layer. Eventually, future photonic data transfer technologies will employ laser pattern alterations and laser steering by non-mechanical means (Liyanage et al., 2009).

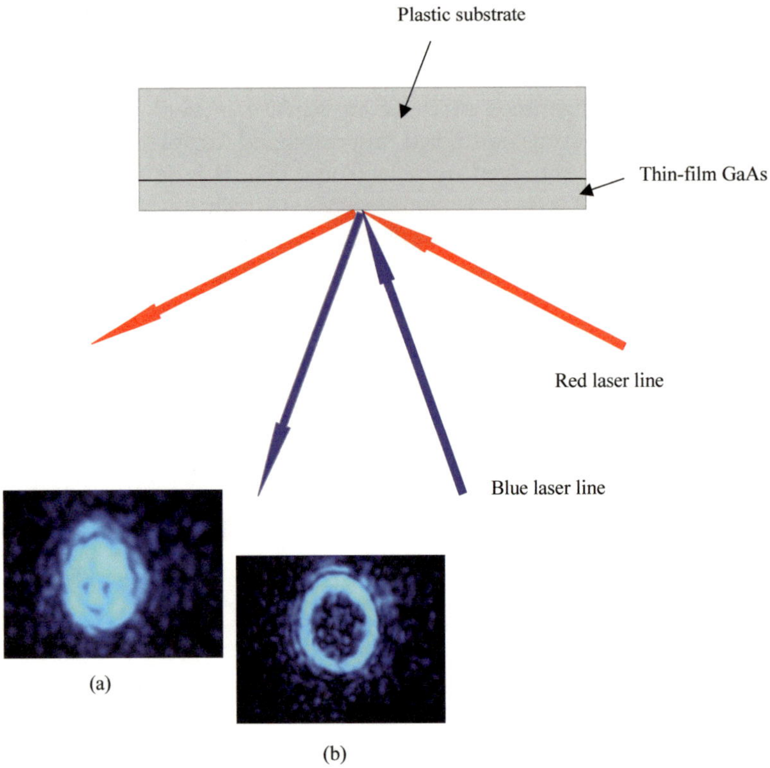

Fig. 15. Laser steering by superposition of two laser lines using a GaAs/plastic hetero-pair

5. References

George J. (1992). *Preparation of Thin Films,* Marcel Dekker, ISBN 0-8247-8196-1, New York

Smith D. L. (1995). *Thin-film Deposition, Principles & Practice*. MacGraw-Hill Inc., ISBN 0-07-058502-4, New York

Ullrich B., Zhang C. & v. Klitzing K. (1988). Quantum-confined subband transitions of GaAs sawtooth doping superlattices, *Appl. Phys. Lett.*, Vol. 54, Issue 12, 1133-1135

Bouchenaki C. a, Ullrich B. & Zielinger J. P. (1991). Luminescence investigations performed on differently prepared thin CdS layers. *J. Luminescence,* Vol. 48&49, Part 2, 649-654

Bouchenaki C. b, Ullrich B., Zielinger J. P., Cong H. N. & Chartier P. (1991). Preparation, characterization and bistable photoconduction of thin CdS layers. *J. Opt. Soc. Am. B*, Vol. 8, Issue 3, 691-700

Ullrich B., Kulac I. & Pint H. (1992). Photocurrent in thin $YBa_2Cu_3O_6$ films on sapphire. *Jap. J. Appl. Phys.*, Vol. 31, No. 7A, L856-L859

Ullrich B. & Kobayashi T. (1995). A new architecture for optical data processing devices by the application of bistability in luminescence. *Optical Review*, Vol. 2, No. 1, 43-46

Ullrich B., Erlacher A., Yano S., Schroeder R., Gerasimov T. G. & Haugan H. J. (2003). Preparation of thin-film GaAs on glass by pulsed-laser deposition. *Proceedings of SPIE* 4977, pp. 180-187, Photonics West, San Jose

Acharya K. P., Mahalingam K., & Ullrich B. (2010). Structural, compositional, and optoelectronic properties of thin-film CdS on p-GaAs prepared by pulsed-laser deposition. *Thin Solid Films*, Vol. 518, Issue 7, 1784-1787

Ezumi H. & Keitoku S. (1993). Influence of pulse width on CdS film prepared by YAG laser ablation. *Jap. J. Appl. Phys*. Vol. 32, Issue 4R, 1783-1786

Ullrich B. (2007). Thin-film CdS formed with pulsed-laser deposition towards optical and hybrid device applications. *J. Mater. Sci.: Mater. Electron*, Vol. 18, 1105-1108

Sakai H., Tamaru T., Sumomogi T., Ezumi H. & Ullrich B. (1998). Crystal direction of CdS thin films produced by laser ablation. *Jap. J. Appl. Phys*. Vol. 37, 4149-4153

Erlacher A. (2005). Pulsed-laser deposition and characterization of hetero-paired thin-film GaAs. Dissertation BGSU. *Source ProQuest DAI-B 66/10*, p. 5459, Apr. 2006.

Ullrich B., Sakai H., Dushkina N. M., Ezumi H., Keitoku S. & Kobayashi T. (1997). The effect of the substrate on the emission features of thin CdS films formed by laser ablation. *Mater. Sci. Eng. B* Vol. 47, Issue 2, 187-189

Gibbs H. M. & Khitrova G. (1990). Introduction, in Nonlinear Photonics, Gibbs H. M., Khitrova G. & Peyghambarian N. (Eds.), 1-3, Springer, ISBN-10 0387521941, Berlin

Hajto J. & Janossy I. (1983). Optical Bistability observed in amorphous semiconductor films. *Phil. Mag. B* Vol. 47, 347-366

Gibbs H. M. (1985). *Optical Bistability: Controlling light with light*, Academic Press, ISBN-10 0122819411, San Diego.

Wherrett B. S. & Hutchings D. C. (1993). Optical bistability, In: *Nonlinear Optics in Signal Processing*, Eason R. W., Miller A. (Ed.), 145-189, Chapman & Hall, 0 412 39560 6, London

Erlacher A., Danilov E. O. & Ullrich B. (2006). Introduction of glass/GaAs interfaces for all-optical and hybrid switch fabrics. *Semicond. Sci. Technol*. Vol. 21, No. 9, 1331-1334

Shepard S. (2001). *Optical Networking Crash Course*, McGraw Hill, ISBN 0-07-137208-3, New York

Bates R. J. (2001). *Optical Switching and Networking Handbook*, MacGraw Hill, ISBN 0-07-137356-X, New York

Papadimitriou G. I, Papazoglou C. & Pomportsis A. S. *Optical Switching*, Wiley & Sons, ISBN-10 0-471-68596-8, Hoboken

Erlacher A. & Ullrich B. (2004). All-optical digitizing of laser transmission through thin-film GaAs on glass. *Semicond. Sci. Technol*. Vol. 19, L9-L12

Liyanage C., Acharya K. & Ullrich B. (2009). Photonic digitizing and pattern alteration with flexible CdS and GaAs film surfaces. *J. Opt. Soc. B*, Vol. 26, Issue 2, 254-258

Erlacher A., Ambrico M.. Capozzi V., Augelli V., Jaeger H. & Ullrich B. (2004). X-ray, absorption and photocurrent properties of thin-film GaAs on glass formed by pulsed-laser deposition. *Semicond. Sci. Technol*. 19 No. 11, 1322-1324

Erlacher A., Ullrich B., Komarova E. Y., Jaeger H., Haugan H. J. & Brown G. J. (2006). Texture and surface analysis of thin-film GaAs on glass formed by pulsed-laser deposition. *J. Non-Cryst. Solids*, Vol. 352, Issue 2, 193-196

Ullrich B. a & Schroeder R. (2001). Green single- and two-photon gap emission of thin film CdS formed by infrared pulsed-laser deposition on glass. *IEEE J. Quantum Electron*. Vol. 37, No. 10, 1363-1367

Ullrich B. b & Schroeder R. (2001). Intrinsic gap emission and its geometry dependence of thin-film CdS excited by two-photon absorption. *Semicond. Sci. Technol*. 16, No. 12, L89-L92

Ullrich B., Erlacher A. & Danilov E. O. (2004). Switch performance and electronic nature of photonic laser digitizing through thin GaAs films on glass. *Semicond. Sci. Technol.* Vol. 19, No. 12, L111-L114

Padilha L. A., Neves A. A. R., Rodriguez E., Cesar C. L., Barbosa L. C. & Brito Cruz C. H. (2005).Ultrafast switching with CdTe nanocrystals in a glass matrix. *Appl. Phys. Lett.* Vol. 86, 161111 (3 pages)

Johnston W. J., Yildirim M., Prineas J. P., Smirl A. L., Gibbs H. M. & Khitrova G. (2005). All-optical spin-dependent polarization switching in Bragg-spaced quantum well structures. *Appl. Phys. Lett.* Vol. 87, 101113 (3 pages)

Alberts V., Neethling J. H. & Leitch A. W. (1994). Correlation between structural, optical, and electrical properties of GaAs grown on (001) Si. *J. Appl. Phys.*, Vol. 75, Issue 11, 7258-7265

Lee M. K., Horng R. H., Wuu D. S. & Chen P. C. (1991). Improvements in the heteroepitaxy of GaAs on Si by incorporating a ZnSe buffer layer. *Appl. Phys. Lett.* Vol., 59, Issue 2, 207-209

Internet (2001), Motorola grows GaAS on Si

http://www.thinfilmmfg.com/Noteworthy/Noteworthy01/GaAs20Sept01.htm

Usui H., Ishiji K., Yasuda H. & Mori H. (2006). Growth process and nanostructure of crystalline GaAs on Si (110) surface perpared by molecular beam epitaxy. *J. Cryst. Growth*, Vol . 289, Issue 1, 76-80

Ullrich B. & Erlacher A. a (2005). Rectification and intrinsic photocurrent of GaAs/Si photodiodes formed with pulsed-laser deposition at 1064 nm. *Appl. Phys Lett.*, Vol. 87, 151115 (3 pages)

Ullrich B. & Erlacher A. b (2005). Photosensitive hetero-pairing of *p*-GaAs/*n*-Si by pulsed-laser deposition. *J. Phys. D: Appl. Phys.*, Vol. 38, No. 22, 4048-4052

Ma K., Urata R., Miller D. A. B. & Harris J. S. Jr. (2004). Low-temperature growth of GaAs on Si used for ultrafast photoconductive switches. *IEEE J. Quantum Electron.*, Vol. 40, No. 6, 800-804

Acharya K. P., Khatri H. & Ullrich B. (2009). Physical characterization of *n*-GaAs on *p*-Si formed by low-temperature pulsed-laser deposition. (2009). *J. Appl. Phys.* Vol. 105, 103111 (5 pages)

Fahrenbruch A. L. & Bube R. H. (1983). *Fundamentals of Solar Cells: Photovoltaic Solar Energy Conversion*, Academic Press, ISBN 0122476808, New York

Ullrich B., Erlacher A., Smith H. E., Mitchel W. C. & Brown G. J. (2008). Electronic properties of *p*-GaAs deposited on *n*-Si with pulsed laser deposition. *J. Phys.: Condens. Matter* Vol. 20, 175217 (3 pages)

Ullrich B., Loeher T., Segawa Y. & Kobayashi T. (1999). The influence of hydrogen passivation of silicon on the photocurrent of CdS/Si heterodiodes. *Mater. Sci. Eng.* Vol. 65, Issue 3, 150-152

Acharya K. P., Skuza J. R., Lukaszew R. A., Liyanage C. & Ullrich B. (2007). CdS thin films formed on flexible plastic substrates by pulsed-laser deposition. *J. Phys.: Condens.* Matter, Vol. 19, No. 19, 196221 (5 pages)

27

Optical Spectral Structure and Frequency Coherence

Ning Hua Zhu, Wei Li, Jian Hong Ke, Hong Guang Zhang,
Jiang Wei Man and Jian Guo Liu
Institute of Semiconductors, Chinese Academy of Sciences
P. R. China

1. Introduction

To account for the various phenomena of light, two theories have been proposed: the corpuscular and the undulatory. The former assumes that the light is a stream of corpuscles, namely photons, discrete photons carrying packets of energy and momentum. The undulatory theory, on the other hand, requires that light consists of a series of wave trains (Michelson, 1927). Each wave train is followed by another which has a random change in phase (Mathieu, 1975). A single wave train is made up of monochromatic components, i.e. the wave train is polychromatic. Although wave trains are supposed not to be strictly monochromatic, experimental demonstration is extremely difficult and has not been reported to date (Diitchburn, 1963). In addition, many efforts have been made to measure the laser coherence length which is regarded as the length of the wave train. Among the methods for measuring the coherence length of lasers (Geng et al., 2005; Ryabukho et. al., 2005 ; Wheeler et al., 2003), Michelson interferometer-based method (Ryabukho et al., 2005) is the one widely used in the past. However, these methods suffer from mechanic vibration, thermal and acoustic fluctuations, and beam divergence, and errors in the observation of the spatial coherence are difficult to eliminate. For lightwave from a real laser source, the wave trains are neither identical nor of simple form (Born & Wolf, 1999). Unfortunately, a complete description of other properties, such as linewidth, intensity profile, and frequency spacings among wave trains, has not been explicitly given due to resolution limitations in both measuring techniques and instruments.

Understanding the spectral structure of semiconductor laser is a fundamental issue. The spectral analysis, especially for the fine spectral structure, reveals the important properties of semiconductor laser, such as mode characteristics, atom emission behavior, high-frequency performances, and coherence features. Spectral lines of light are broadened by various processes. For semiconductor lasers, Henry's model (Henry, 1982) can be used to explain linewidth broadening of the laser by two mechanisms: 1) the instantaneous phase change caused by the spontaneous emission results in the linewidth broadening of the light; 2) the instantaneous fluctuations of the field intensity through linewidth enhancement parameter a results in a delayed phase change, which further broadens the linewidth. However, this model also can not describe the properties of wave trains mentioned above distinctly. Indeed, it implies that the spectral linewidth we observed, in fact, results from the

rapid shift of some narrower linewidth elements. This rapid shift derives from the above two fluctuation mechanisms through a series of spontaneous emission events.
In this chapter, we use an interference technique based on the division of frequency to characterize the wave trains of semiconductor lasers, and present a spectral structure model in the frequency-time domain. The assumptions for the model are demonstrated through a series of experiments. It has also been proved that this interference technique has a resolving power of 10^{17} in optical spectral analysis compared to 10^5 of Michelson-interferometer-based methods. At the same time, the powerful resolving ability of this technique enables us to show that the spectral linewidth of the wave train is narrower than 1 mHz. Based on the spectral structure model, a new concept of frequency coherence is proposed, which is different from the conventionally used concepts, such as temporal, spatial and spectral coherence. Moreover, the elimination of the frequency coherence between two lightwaves is also discussed. Frequency coherent lightwaves generation is achieved by a self-injected distributed Bragg reflector (DBR) laser. In addition, monolithically integrated photonic microwave source based the frequency coherent lightwaves generation is experimentally demonstrated.

2. Optical spectral structure model

The commonly used method for spectral analysis has been two-beam or multiple-beam interference. In general, there are two methods to obtain beams from a lightwave source: division of amplitude and division of wave-front (Born & Wolf, 1999). The frequency-division scheme shown in Fig. 2.1(a) is used to measure the linewidth of the wave train. In the following, the spectral linewidth, frequency spacings, and intensity profile among wave trains are investigated through a series of experiments (Zhu et al., 2007).

a. Spectral linewidth and Frequency spacings

Fig. 2.1(a) illustrates the frequency-division scheme, which allows us to perform optical heterodyning at nonzero frequency. This makes it possible to measure the linewidth of wave train with a high-resolution electrical spectrum analyzer instead of observing the fringe visibility.

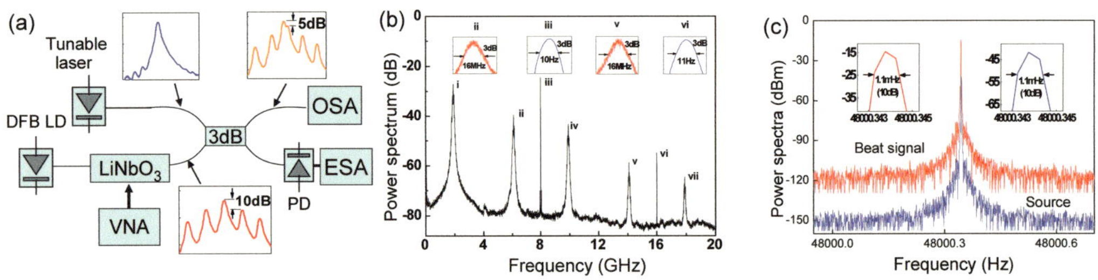

Fig. 2.1. Spectral linewidth analysis of lightwave source and wave trains. (a) Experimental setup. (b) Measured power spectrum of the beat note. Insets show the higher resolution spectra. (c) Measured hyperfine spectrum compared with the power spectrum of the waveform generator

The results (Fig. 2.1(b)) show that the linewidth of the beat note between the lightwaves from different sources is 16 MHz, corresponding to the linewidth of the distributed feedback laser diode (DFB LD). The beat note at the modulation frequency is about 10 Hz, which is much narrower than the linewidth of the lightwave source. This narrow beat note comes from the interference between the coherence wave trains in the carrier and the sidebands of the intensity modulated lightwave. To precisely estimate the linewidth of the wave trains, a high-resolution spectrum analyzer and a pure electrical source are used. Fig. 2.1(c) shows that the measured 10-dB linewidths of both the beat note and the electrical source are about 1.1 mHz, and no obvious broadening is observed for the beat note. Hence, the linewidth of the wave train should be less than 1 mHz. The above experiment also verifies the assumption that the wave trains emitting simultaneously have random frequency spacings. If this is not the case, the beat note at the modulation frequency cannot be in such a narrow frequency range. From Fig. 2.1(a) and (b), it can be seen that the optical power of the reference signal is 5 dB higher than that of the carrier in the signal channel. But peak ii (beating between the reference signal and the lower sideband) is 15 dB lower than peak iii (beating between carrier and two sidebands). The 17-dB (5+15-3=17) discrepancy is due to the beat note between the wave trains in the carrier and the corresponding wave trains in the two sidebands which are always superposed at the modulation frequency.

b. Intensity profile and duration

The experimental setup shown in Fig. 2.2(a) is used to investigate the interference between lightwaves from the same laser source when a path difference exists. The optical intensity of the reference signal is 15 dB higher than that of the carrier in the signal channel. When the fiber interferometer is made asymmetric by inserting an optical fibre into the reference arm, the beat peak decreases and the noise level increases with path difference. It is because some wave trains in the reference channel, which have length shorter than the path difference, become incoherent with the corresponding wave trains in the sidebands in the signal channel. Interference among these incoherent wave trains produces beat notes whose frequencies spread around the modulation frequency randomly, and leads to an increase in the noise level. This experiment implies that the wave trains have different spatial or temporal lengths.

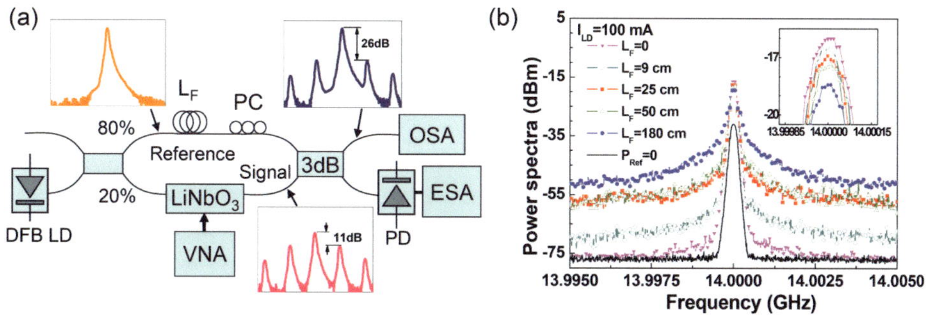

Fig. 2.2. Investigation of interference between lightwaves from one laser source. (a) Experimental setup. PC, polarization controller; LF is the fiber length difference between two arms of the M-Z interferometer. (b) Measured power spectra at various fiber length differences. The curve "P_{ref}=0" indicates the spectrum without reference signal injection. The inset shows the higher resolution spectrum

In order to estimate the length of the wave trains we propose a filtered M-Z interferometer scheme as shown in Fig. 2.3(a). This all-fibre experimental setup is superior to the Michelson interferometer because there is no problem of beam divergence. For lightwave from a real source, atom emission is irregularly modified by the disturbance from its neighbors, and the duration of wave trains will vary randomly in a certain range. The fluctuation of the beat note is the reflection of the essential pattern of atom emission. The results (Fig. 2.3(b)) show that the beat notes at different bias currents have a dip at the fiber length difference of 2.2 m, and a peak at 4.6 m. Another small peak at 9.2 m can also be observed. The dip and peak values increase with bias current, but the locations change slightly. The conventional ideal models presented in the past literatures (Diitchburn, 1963 ; Born & Wolf, 1999) are unable to explain this phenomenon. If all wave trains are identical and of simple form, the measured beat magnitude will decrease linearly with path difference. Based on the atom emission law (Mandel & Wolf, 1995), we can assume that the rising and decay times are not identical, and before a wave train vanishes it seeds another wave train with the same frequency. Therefore, the probability of occurrence for the two or more joint wave trains with the same frequency is high. Subsequent wave trains may have different frequencies, but the probability is higher for those having closer frequency. When the path difference is half the wave train length, the beat note is at its minimal value. When the path difference is about the wave train length, the beat note reaches its second peak. Therefore, the average length of wave trains is estimated to be about 4.6 m.

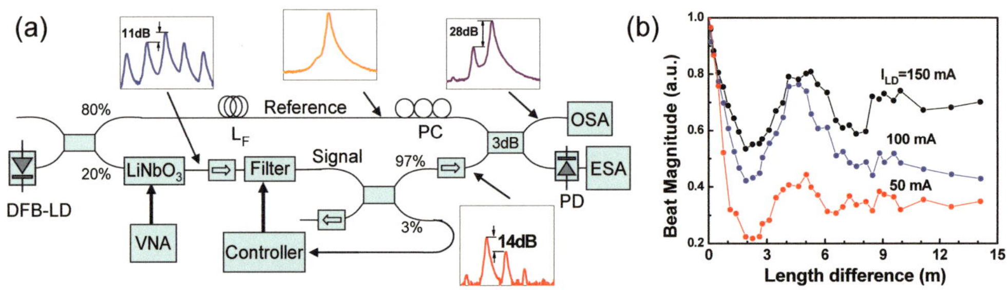

Fig. 2.3. Estimation of coherence length. (a) Experimental setup with the measured optical spectra. A narrow band (2GHz) optical filter is added to the setup (Fig. 2.3(a)) to remove the carrier signal in the signal channel. Three optical isolators are used for the filter to be locked to one of the first-order sidebands. (b) Measured magnitude of the beat notes at different fiber length differences when the DFB laser is biased at 50, 100, and 150 mA, respectively

The resolving power of the Michelson interferometer is limited by the beam divergence. Frequency modulated continuous wave measurement has been used to estimate the coherence length of over 210 km in air using an all-fibre experimental setup (Geng et al., 2005). For the same reason, the interferometer (Fig. 2.3(a)) is stable and can achieve a relatively large path difference without losing resolving power if a zero-dispersion fiber is used as the delay line.

Here we give an alternative method for determining precisely the average coherence length. Fig. 2.4(a) shows the typical setup for measuring the small-signal frequency response, from which the fibre dispersion and the chirp parameter of light emitter can be obtained (Devaux et al., 1993). We found that the setup can also be used to determine the length of the wave trains. In this case, the long optical fiber functions as an "interferometer" and its path difference between the carrier and the sidebands depends on the modulation frequency and fiber dispersion. The optical path difference at the pth peak frequency f_p can be expressed as

$$\Delta L_p = \frac{c}{f_p}\left[p - \frac{1}{2}\,\frac{f_{D2}^2 - 3f_{D1}^2}{f_{D2}^2 - f_{D1}^2}\right],\quad p\text{=0, 1, 2, …} \tag{1}$$

where c is the light speed in free space, and f_{D1} and f_{D2} are the first and second dip frequencies. It is interesting to notice that ΔL_p depends only on the peak and dip frequencies, and it is not necessary to determine the physical length of the fiber and its dispersion coefficient. Since large dispersion is desirable, the long fiber can be replaced by dispersion compensation fiber or chirped fiber grating.

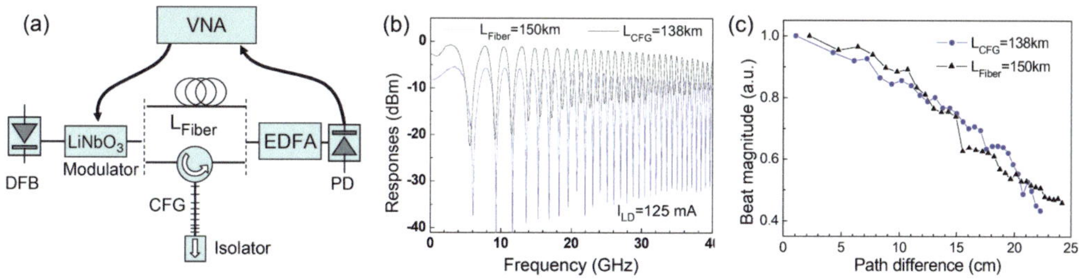

Fig. 2.4. Estimation of coherence length. (a) Experimental setup, in which a long optical fiber or chirped fiber grating (CFG) can be used to achieve a path difference for the lightwaves with different frequencies. (b) Measured frequency responses using a 150-km G652 optical fiber and two chirped fiber gratings for compensating the dispersion of 81+57km G652 fiber when the DFB laser is biased at 125 mA. The measurements are made for 50 times and the curves show the average values. (c) Normalized magnitudes of the corresponding beat note peaks

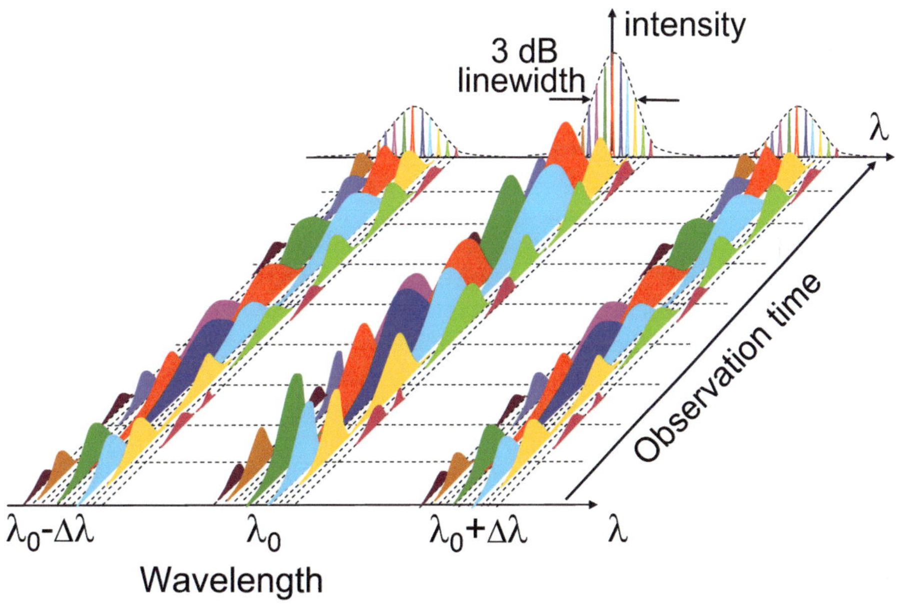

Fig. 2.5. Optical spectral structure model of a semiconductor laser after intensity modulation

The typical frequency responses using a long optical fiber and two chirped fiber gratings are plotted in Fig. 2.4(b), and the corresponding variations of the beat intensity with path difference are shown in Fig. 2.4(c). When the path difference of the "two arms" reaches 20 cm, the average beat note decreases by 3 dB. During the observation time, the changes in modulation frequency and the total fiber length can be neglected since the path difference is less than 25 cm after 150 km propagation. The interference is extremely stable, and the signal to noise ratio can be significantly improved because the wavelength and the polarization of the carrier and the sidebands changes in the same direction, and the two waves propagate in the same medium. Thus, this scheme is simple and suitable for the spectral analysis of a laser with relatively short coherence length.

Based on the above experimental results, we propose an optical spectral structure model as shown in Fig. 2.5, where only the carrier and the first sidebands are shown for simplicity. This model can be summarized as follows: 1) the spectrum of a laser consists of a large number of wave trains. Wave train is not monochromatic and its spectral linewidth is extremely narrow (less than 1 mHz); 2) wave trains are neither identical nor of simple form, and they have variable lengths; 3) wave trains emitting simultaneously will have random frequency spacings, and a wave train can seed another wave train with the same frequency.

3. Frequency coherence

This section presents a new concept of frequency coherence to describe the field correlations between two lightwaves with different frequencies. This concept is different from other well-known coherence concepts, such as temporal coherence, spatial coherence, polarization coherence, quantum coherence, and spectral coherence. Spatial coherence and temporal coherence describe the correlations between beams at different space points and different moments, respectively. Our frequency coherence describes the field correlation between two lightwaves in the frequency-time domain. This model gives a straightforward illustration of optical spectral structure, which is helpful for understanding the spectral structure of semiconductor laser.

3.1 Basic concept of optical coherence

In the broadest sense, optical coherence theory is concerned with the statistical description of the fluctuations of optical fields. Interference is a typical phenomenon that reveals correlation between light beams. Temporal and spatial coherence have been extensively studied in the past come of Michelson and Young's interference experiments, respectively. In a typical interferometer, a beam of lightwave is split into two beams, and the two light beams are recombined together with different delay times. The two beams are perfectly coherent when the lengths of the two paths are identical. For a certain delay difference, the degree of coherence depends on the linewidth and wavelength stability of the light beam. Spatial coherence describes the correlation between signals at different points in space. Temporal coherence describes the correlation between signals observed at different moments. There are other concepts on coherence in accordance with different physical parameters, such as polarization coherence, quantum coherence, and spectral coherence. Spectral correlation, which is not so widely used as temporal and spatial coherence, describes the correlation that exists between the spectral components at a given frequency in the light oscillations at two points in a stationary optical field (Mandel & Wolf, 1976).

3.2 Description in three-dimensional space

The coherence properties of two beams should be described in three-dimensional spaces. Take coherence in space-time domain for example, the three dimensions are distance, time delay, and amplitude. The complex degree of coherence is traditionally used to characterize correlations in stationary fields. It is defined as normalized cross-correlation function of the optical fields at two points. In the following, we briefly introduce the theory of the complex degree of coherence in space-time domain (Mandel & Wolf, 1965).

Suppose that $V(r_1, t)$ and $V(r_2, t)$ are the analytic signal representations of the light oscillations at two points with position vectors r_1 and r_2, the complex degree of coherence can be expressed as

$$\gamma(r_1,r_2,\tau)=\Gamma(r_1,r_2,\tau)/[I(r_1)I(r_2)]^{1/2} \tag{2}$$

where $\Gamma(r_1,r_2,\tau)=<V^*(r_1,t)V(r_2,t+\tau)>$ is the mutual coherence of the light, and $I(r)=\Gamma(r,r,0)$ is the average intensity of the light. For all possible values, $0\leq\gamma(r_1,r_2,\tau)\leq1$.

If the light is quasimonochromatic, the visibility of fringes at position $P(r)$ on the interference screen is

$$\upsilon(r)=\frac{I_{max}(r)-I_{min}(r)}{I_{max}(r)+I_{min}(r)}=\left|\gamma(r_1,r_2,\tau_{12})\right| \tag{3}$$

which means that $|\gamma|$ is a measure of the sharpness of the interference fringes. The complex degree of spectral coherence in the frequency-time domain (Mandel & Wolf, 1976) is given by

$$\mu(r_1,r_2,\nu)=W(r_1,r_2,\nu)/[W(r_1,r_1,\nu)W(r_2,r_2,\nu)]^{1/2} \tag{4}$$

Where $W(r_1,r_2,\nu)$ is the cross-spectral density function (also known as the cross power spectrum) of the two optical fields. Therefore, the coherence properties in the space-time domain depend on position and on the delay time, and the coherence properties in the space-frequency domain depend on position and on the frequency of the light.

3.3 Concept of frequency coherence

It is desirable to give general description of the field correlations between two lightwaves with different frequencies, and to investigate the related phenomena and their applications. Optical heterodyne technique using two lightwaves with different wavelengths has been widely used to generate microwave and millimeter waves. The spectral characteristics and correlations of the two lightwaves are critical in obtaining a stable and narrow-linewidth microwave signal. Referring to the concept of spatial and temporal coherence in the space-time domain and the spectral coherence in the space-frequency domain, we introduce a new concept of frequency coherence in the frequency-time domain, which describes the field correlation between two lightwaves with different frequencies at a given moment.

Given that two arbitrary beams a and b overlapped in wavelength, the optical fields at optical frequencies ω_1 and ω_2 can be expressed as

$$E_1(\omega_1,t)=[E_A f_a(\omega_1,t)e^{-j\varphi_a}+E_B f_b(\omega_1,t)e^{-j\varphi_b}]e^{-j\omega_1 t} \tag{5}$$

$$E_2(\omega_2,t)=[E_A f_a(\omega_2,t)e^{-j\varphi_a}+E_B f_b(\omega_2,t)e^{-j\varphi_b}]e^{-j\omega_2 t} \tag{6}$$

E_A and E_B are the maximal optical fields of beams *a* and *b*, respectively. $f_a(\omega,t)$ and $f_b(\omega,t)$ are the normalized power spectrum profiles and meet

$$\int_0^{+\infty} f_x^2(\omega,t)d\omega=1, \qquad x=a,b \tag{7}$$

The photocurrent could be written as

$$\begin{aligned} i(t) \propto & \frac{1}{2}E_A^2 f_a^2(\omega_1,t)+\frac{1}{2}E_B^2 f_b^2(\omega_1,t)+\cos\phi E_A E_B f_a(\omega_1,t)f_b(\omega_1,t)\cos\Delta\varphi \\ & +\frac{1}{2}E_A^2 f_a^2(\omega_2,t)+\frac{1}{2}E_B^2 f_b^2(\omega_2,t)+\cos\phi E_A E_B f_a(\omega_2,t)f_b(\omega_2,t)\cos\Delta\varphi \\ & +F(\omega_m)E_A^2 f_a(\omega_1,t)f_a(\omega_2,t)\cos(\omega_m t)+F(\omega_m)E_B^2 f_b(\omega_1,t)f_b(\omega_2,t)\cos(\omega_m t) \\ & +\cos\phi F(\omega_m)E_A E_B f_a(\omega_1,t)f_b(\omega_2,t)\cos(\omega_m t+\Delta\varphi) \\ & +\cos\phi F(\omega_m)E_A E_B f_b(\omega_1,t)f_a(\omega_2,t)\cos(\omega_m t-\Delta\varphi) \end{aligned} \tag{8}$$

where ϕ is the angle between the polarization directions of the two beams, $F(\omega_b)$ is the frequency response coefficient of the photodetector, and $\Delta\varphi=\varphi_a-\varphi_b$ is the phase difference between the two beams. The first six terms in (8) represent the DC beat notes. The 7th and 8th terms indicate homodyne signals. Intensity noise exists at all frequencies simultaneously for wide spectral optical source, and a simple method for the calibration of wide bandwidth photoreceiver has been established (Eichen & Silletti, 1987). The 9th and 10th terms indicate heterodyne signals. Collecting all the optical current components at frequency ω_m from the beat notes of the two beams, we have

$$\begin{aligned} i_{\omega_m}(t) \propto & F(\omega_m)E_A^2\cos(\omega_m t)\int_0^{\infty} f_a(\omega_1,t)f_a(\omega_2,t)d\omega_1 \\ & +F(\omega_m)E_B^2\cos(\omega_m t)\int_0^{\infty} f_b(\omega_1,t)f_b(\omega_2,t)d\omega_1 \\ & +\cos\phi F(\omega_m)E_A E_B\cos(\omega_m t+\Delta\varphi)\int_0^{\infty} f_a(\omega_1,t)f_b(\omega_2,t)d\omega_1 \\ & +\cos\phi F(\omega_m)E_A E_B\cos(\omega_m t-\Delta\varphi)\int_0^{\infty} f_b(\omega_1,t)f_a(\omega_2,t)d\omega_1 \end{aligned} \tag{9}$$

When two beams do not overlap, i.e., $f_a(\omega_2,t)=0$, and $f_b(\omega_1,t)=0$, (9) reduces to

$$i_{\omega_m}(t) \propto \cos\phi F(\omega_m)E_A E_B\cos(\omega_m t+\Delta\varphi)\int_0^{\infty} f_a(\omega_1,t)f_b(\omega_2,t)d\omega_1 \tag{10}$$

When the light beams are optical carrier and the sidebands produced by optical intensity modulation, ω_m becomes the modulation frequency. The optical spectral distribution in the frequency-time domain is shown in Fig. 2.5, and only the carrier and the first sidebands are shown for simplicity. Based on the atom emission law and the time response of ultrafast photonic crystal laser (Yariv, 1997), we can assume that the rising and decay times are not identical. From our understanding of optical spectral distribution, the spectrum of a laser consists of a large number of wave trains. It has been shown that the spectral linewidth of wave trains is narrower than 1 mHz, and the linewidth of DFB laser used is ~ 16 MHz. Also, the simultaneous emitted wave trains have random frequency spacings (Zhu et al., 2007). Therefore, the probability of occurrence of two wave trains from the same lightwave source with a frequency spacing $\omega_m=\omega_2-\omega_1$ is rather low. Thus, we can assume that

$$\int_0^{+\infty} f_x(\omega,t) f_x(\omega+\omega_m,t)d\omega << 1, \qquad \omega_m \neq 0 \tag{11}$$

From the definition of the complex degree of coherence in the space-time and space-frequency domains, the degree of frequency coherence of two beams in the frequency-time domain can be expressed as

$$\gamma(\omega_1,\omega_2,t) = \int_0^{+\infty} f_a(\omega_1,t) f_b(\omega_2,t)d\omega, \tag{12}$$

According to the definition of the power spectrum profile, $0 \leq |\gamma(\omega_1,\omega_2,t)| \leq 1$ for all possible values of γ.

The wave trains in the carrier and the corresponding wave trains in the sidebands appear simultaneously, and their frequency intervals exactly equal the modulation frequency. Since the carrier and the sidebands have identical intensity profile, polarization and phase, i. e., $f_b(\omega_2,t)=f_b(\omega_1+\omega_m,t)=f_a(\omega_1,t)$, and considering the optical spectrum properties indicated by (7) and (11), we have $\gamma(\omega_1,\omega_2,t)=1$. Thus, (9) can be approximated as

$$i_{\omega_m}(t) \propto F(\omega_m)E_A E_B \cos(\omega_m t + \Delta\varphi) \tag{13}$$

It means that the linewidth of the beat note at the modulation frequency is independent of the optical spectral profile of the beams. In this case, the beat note at the modulation frequency is much stronger than the beat notes at other frequencies which become random noise. The carrier and its sidebands produced by intensity modulation are perfectly coherent if there is no time delay. As the path difference increases the beat note at the modulation frequency will decrease and the noise level will increase (Zhu et al., 2007). The carrier and the sidebands may then become partially coherent.

It can be concluded that for the coherent beams the maximum intensity of the mixed beams may exceed the sum of the intensities of the beams (Born & Wolf, 1999), and the beat note between the two beams would have a linewidth much narrower than the sum of the beam linewidths.

4. Experimental analysis of frequency coherence properties

In the above sections, we have presented the new hyperfine spectral structure of semiconductor lasers and the theory of frequency coherence. In this section, frequency coherence properties of different optical lightwaves are experimentally investigated using optical heterodyne technique. The results indicate that the optical carrier and its intensity modulated sidebands are perfectly coherent and the longitudinal modes of an FP laser are partially coherent (Zhu et al., 2009).

4.1 Modulated optical spectrum: perfectly coherent

For the beams which are coherent, the maximum intensity of the mixed beams may exceed the sum of the intensities of the beams, and the linewidth of the beat note between the two beams would be much narrower than the sum of the beam linewidths. These discrepancies in intensity and linewidth are the key features in describing the coherence properties of lightwaves. The following experiments were designed to analyze the optical and electrical spectra of the beams and the corresponding beat notes.

Fig. 2.6 is the measurement setup. In the first experiment, five lightwave sources, which have different spectral widths, are used, i.e., an amplified spontaneous emission (ASE) lightwave source (an erbium-doped fiber amplifier without optical input signal), the same ASE source together with a 28GHz optical filter and a 5GHz optical filter, an DFB laser with a 16 MHz linewidth, and a tunable laser with a linewidth narrower than 100 kHz. These lightwave sources are all modulated by a 15-GHz signal from a vector network analyzer (VNA) through a $LiNbO_3$ modulator with a bandwidth of 40-GHz. An optical fiber coupler is used to split the output into two waves, and the optical spectrum is measured by an optical spectrum analyzer, and a electrical spectrum analyzer (ESA) and a high-speed photodetector is used to measure the power spectrum. The spectrum is recorded using the "Max Hold" function of the ESA for several seconds. The peak power of the beat note is kept the same for different lightwave sources.

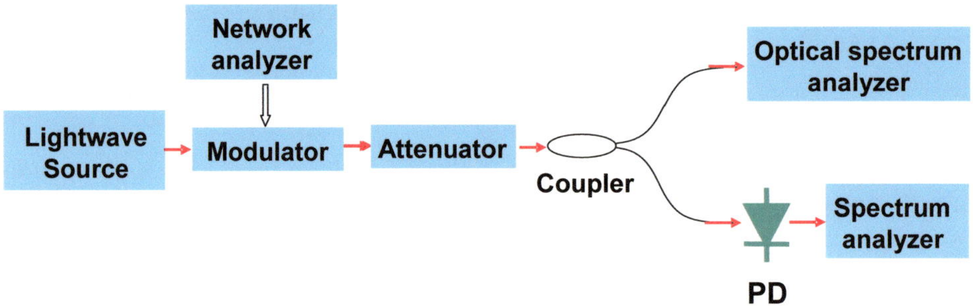

Fig. 2.6. Experiment setup for testing the coherence properties of modulated optical spectrum

Fig. 2.7 shows the measured optical spectra and the corresponding power spectra. For different lightwave sources whose spectrum widths vary from 100 kHz to 5 THz, the measured power spectra are almost identical, and the 6-dB linewidths of the beat notes are about 154 kHz, which depends on the parameters of the instruments, such as the frequency span and the resolution bandwidth of the spectrum analyzer. In the next experiment it will be shown that the beat note between the carrier and the sidebands is extremely narrow.

In the second experiment the measurement setup is the same, except a tunable laser is connected to the other input port of the coupler and acted as a reference lightwave source, and its wavelength is 2 GHz lower than the central wavelength of a DFB laser which is the lightwave source. The modulation frequency is set to be 8 GHz.

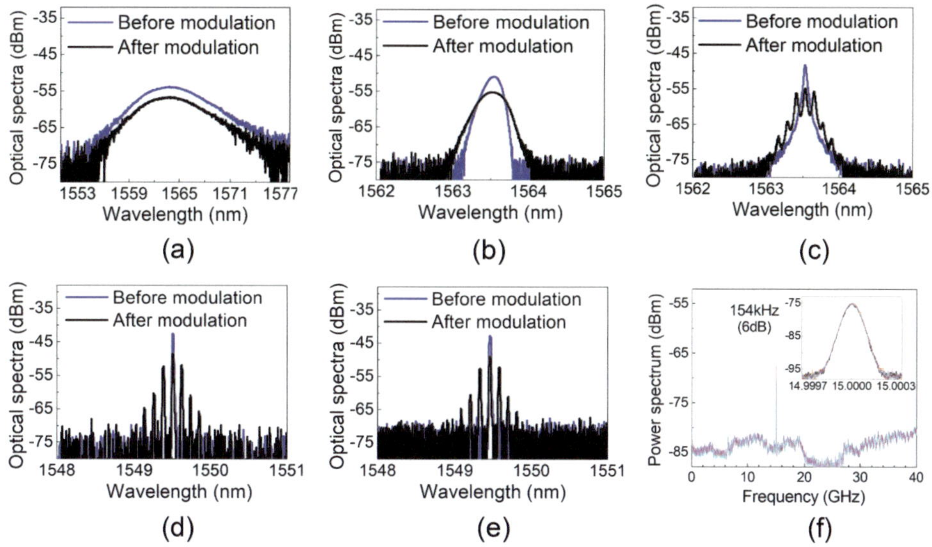

Fig. 2.7. Measured optical spectra (a-e) and the corresponding power spectra (f)

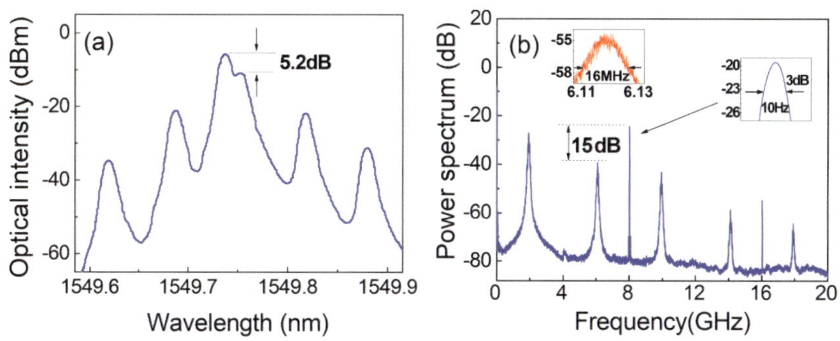

Fig. 2.8. Measured optical and electrical spectra, when the lightwave sources used was a DFB laser, the reference tunable lightwave was from a tunable laser, and the modulation frequency was 8GHz. The inset shows the higher resolution spectrum at the modulation frequency

The optical and electrical spectra are shown in Fig. 2.8. The signal at 6.12 GHz denotes the beat note between the first left sideband and the reference signal. The linewidth of the beat note is 16 MHz and can be considered as the linewidth of the DFB laser since the linewidth of the tunable laser is much narrower, which is less than 100 kHz. The beat note at 8GHz is between the two first sidebands and the carrier. Its linewidth is only 10 Hz.

From Fig. 2.8 it shows that the optical power of the reference signal is 5.2dB higher than that of the carrier of the DFB laser. However, the beat note (at 6GHz) between the carrier and the reference signal is 15dB lower than that of the beat note at 8 GHz. There is a discrepancy of 17.2-dB (15+5.2−3=17.2), assuming that the intensities of the two sidebands are equal and the photodetector has the same frequency responses at 6 and 8 GHz.

The experiments above show that the beat note between the carrier and the sidebands is extremely narrow, which is about 10 Hz, and its intensity is 17.2 dB higher than that of the beat notes between the lightwaves emitted from different lightwave sources. Therefore, the carrier and its sidebands are perfectly coherent.

4.2 Longitudinal modes of FP laser: partially coherent

The mode spacing of a FP laser usually exceeds 100GHz. Thus, the conventional optical heterodyne technique can not be directly used to investigate the frequency coherence properties of an FP laser. To overcome this problem, a measurement setup similar to that shown in Fig. 2.6 is used. The lightwave source is a FP laser diode. After modulation, the high-order sidebands of the adjacent FP modes become close and fall into the operation frequency range of the ESA. An Erbium-doped optical fiber amplifier (EDFA) is used to raise the optical power after intensity modulation. Fig. 2.9(a) shows the measured optical spectra before and after intensity modulation at 30GHz. From this figure it can be seen that the frequency difference between the higher second sideband of FP mode 01 and the lower second sideband of FP mode 02 is about 18 GHz, whose 6-dB linewidth of the beat note is about 7.1 MHz.

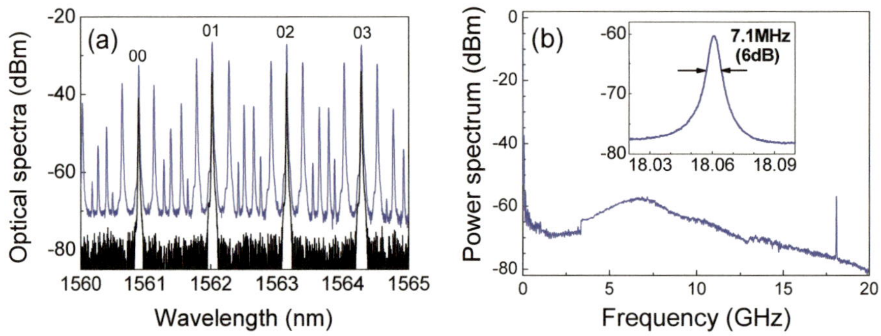

Fig. 2.9. (a) Measured optical spectra of a Fabry-Perot laser before (solid line) and after (dashed line) modulation, where the modulation frequency is 30 GHz. (b) Corresponding spectra, where inset shows the higher resolution spectrum

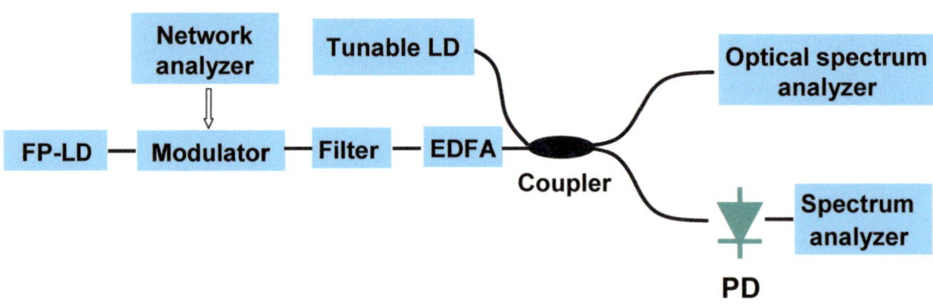

Fig. 2.10. Experiment setup for analysing the frequency coherence properties of longitudinal modes of an FP laser using an optical filter

Another experiment is implemented as shown in Fig. 2.10. A tunable laser is the reference lightwave source, and its wavelength is between the lower second sideband of FP mode 02 and the higher second sideband of FP mode 01. Other optical modes are removed by a 28-GHz bandpass optical filter, and only the shifted second sidebands of FP modes 01 and 02 are considered.

As shown in Fig. 2.11(b), the 6-dB linewidth of the beat note (18.06 GHz) between the second sidebands of FP modes 01 and 02 is about 7.2 MHz. The spectral widths of the beat notes (8.3 and 9.76 GHz) between the reference lightwave and the shifted second sidebands of FP modes 01 and 02 are about 130 MHz. From Fig. 2.11(a) one can see that the optical power of the reference signal is 5dB higher than that of the higher second sideband of FP mode 01. However, the beat note between the lower second sideband of FP mode 02 and the reference signal is 7 dB lower than the beat note between the second sidebands of the FP modes 01 and 02. A discrepancy of 12dB exits. The above results indicate that the FP modes of the laser are coherent. However, the FP modes are not perfectly coherent, if not, the linewidth of the beat note would be extremely narrow. The linewidth of the beat note between the two second sidebands of the FP modes is about 7.2 MHz. This means that the longitudinal modes of the FP laser are partially coherent.

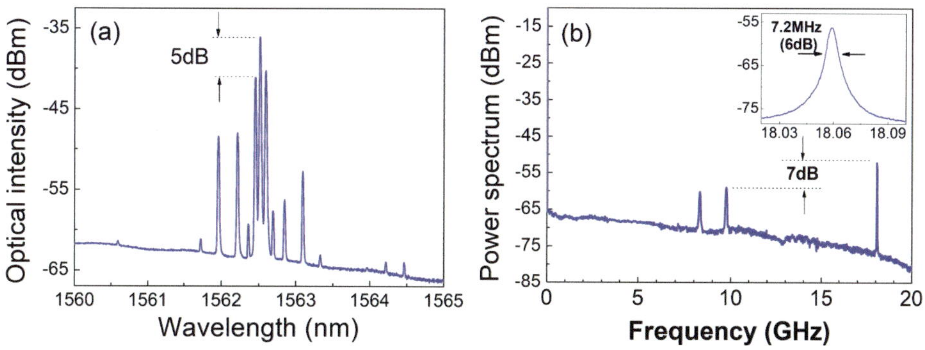

Fig. 2.11. (a) Measured optical spectra when the FP laser is modulated at 30GHz, an optical filter is used. (b) Corresponding spectra, where inset shows the higher resolution spectrum

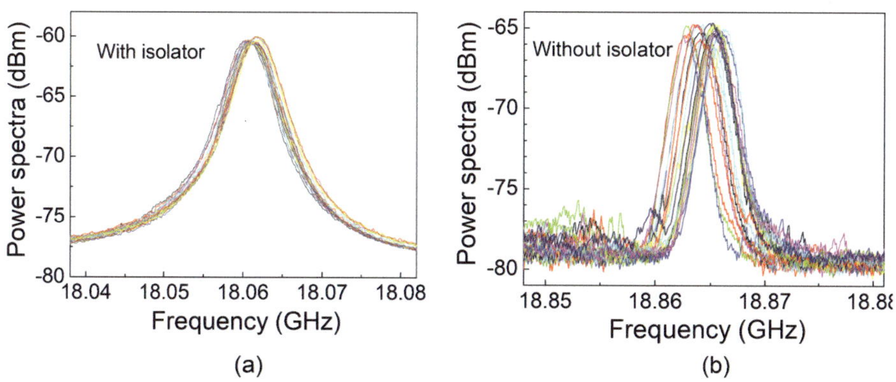

Fig. 2.12. Higher resolution spectra of the laser with (a) and without (b) optical isolator measured for 20 times

The higher resolution spectrum of the optical beat note is measured for 20 times and is shown in Fig. 2.12(a), from which it can be seen that the measured beat note is quite stable. If optical filter is not used, beating may occur between the second sidebands of all neighbouring FP modes. However, from Fig. 2.11(b) and 2.9(b) one can conclude that the mode spacings and the linewidths of all the adjacent FP modes are almost identical. This coincides with the statement that the modes of FP laser are locked by mutual injection (Sato et al., 2001). However, FP modes are only partially coherent, because the mode locking is based on four-wave mixing, and all modes originate from at least two modes.

Similar experiments are carried out using a FP laser without an optical isolator. The threshold current of the laser is about 10 mA and the bias current is 60 mA. The modulation frequency is 40 GHz. The results are given in Fig. 2.12(b), 2.13, and 2.14.

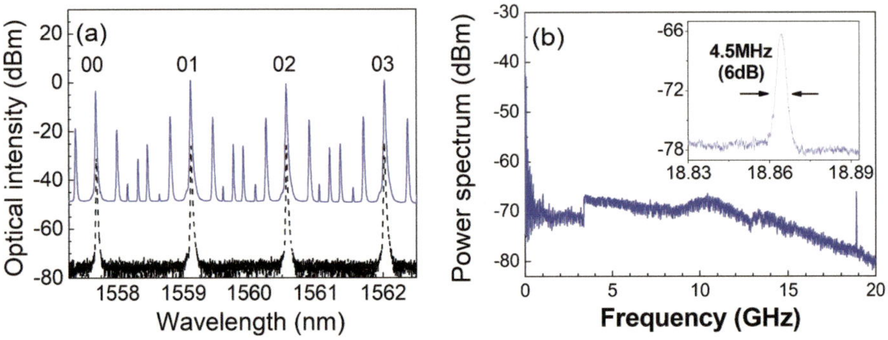

Fig. 2.13. (a) Measured optical spectra when the FP laser without an optical isolator is modulated at 30GHz. (b) Corresponding spectra, where inset shows the higher resolution spectrum

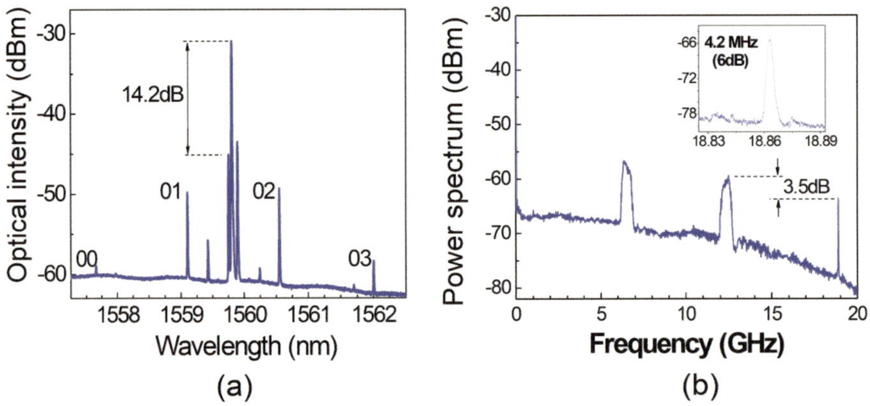

Fig. 2.14. (a) Measured optical spectra when the FP laser without an optical isolator is modulated at 30GHz, an optical filter is used. (b) Corresponding spectra, where inset shows the higher resolution spectrum

Normally the optical wavelengths of the FP laser may be affected by the ambient temperature and bias current. It has been shown that when the chip temperature varies by 1°C the FP modes of the laser will change by 0.1 nm, corresponding to a frequency change of 12.5GHz at 1.55μm (Zhu et al., 2006). Although a Peltier cooler can be used to control the temperature, it is impossible to maintain the chip temperature within 0.0001°C. From Fig. 2.14 one can see that the FP modes shift is about 610 MHz. However, from Fig. 2.12(b) it can be observed that the changes of the FP mode spacing are within 4 MHz. This clearly shows that the longitudinal mode spacing of the FP laser is relatively fixed, although the wavelengths of the FP modes change with temperature. Therefore, the beat note between the shifted second sidebands is much more stable.

Comparing Fig. 2.12(a) and 2.12(b) it follows that the wavelength of FP laser becomes more stable when an optical isolator is used. Wavelength stability plays an important role in the generation of a stable and narrow-linewidth microwave signal.

From the measured optical and electrical spectra shown in Fig. 2.14, it can be seen that the optical power of the reference signal is 14.2dB higher than that of the higher second sideband of FP mode 01. The beat note between the reference signal and the lower second sideband of FP mode 02 is only 3.5 dB higher than the one between the higher second sideband of FP mode 01 and the lower second sideband of FP mode 02. This discrepancy (10.7dB) and 4.5 MHz linewidth of the beat note between the two sidebands also indicate that the FP modes of the FP laser are partially coherent.

5. Elimination of frequency coherence

In the previous sections, we have theoretically and experimentally demonstrated the concept of frequency coherence. The degree of frequency conherence mainly depends on the spectral characteristics and correlations of the two lightwaves. Although two lightwaves with highly frequency coherence are desirable for generating pure microwave or millimeter wave signals, it is also needed to eliminate the frequency coherence of two lightwaves in some cases, for example, the linewidth measurement of the lasers. In this section, the elimination of the frequency coherence of two lightwaves is investigated experimentally and the phenomena observed in the experiments are explained using the three-dimension optical spectral structure model.

5.1 Three-dimension model of optical spectrum

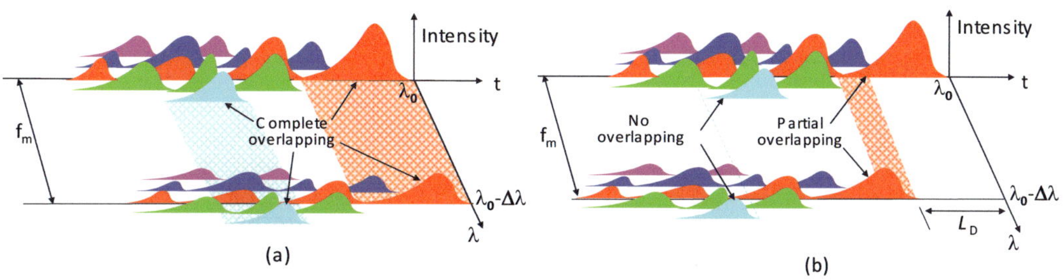

Fig. 2.15. Configuration of optical spectral structure in frequency-time domain after intensity modulation (a) without and (b) with the time delay

Based on optical spectral structure in the frequency-time domain, the spectrum of the light beam consists of a large mount of wave trains around the center wavelength, which have distinct features just as the followings (Zhu et al., 2007).

a. In the frequency domain, it is not strictly monochromatic and the spectral linewidth is much narrower than 1 mHz, corresponding to a wavelength range of 10^{-23} m at 1.55 μm.
b. The wave trains are emitted simultaneously with random frequency spacings. A wave train is able to seed another wave train with the same frequency, and the probability of

occurrence of two or more joint wave trains with the same frequency is rather high. The subsequent wave trains may be with different frequencies, but the probability is much lower.

c. In the time domain, the spatial and temporal intensity profiles of wave trains are neither identical nor of simple form. The length of wave trains has a large variable range. The intensity profile and average duration mainly depend on the laser structure, the bias condition and the optical cavity.

When a light beam is modulated, any wave train in the carrier and the corresponding wave train in the sidebands may appear simultaneously with perfect frequency coherence, and their frequency interval is exactly identical to the modulation frequency. Fig. 2.15 gives the optical spectral structure of the carrier and only one first sideband in the time-frequency domain for simplicity. All the beat notes between wave train pairs in the sideband and carrier are always superposed at the modulation frequency. The beat note at the modulation frequency is much stronger than beat notes at other frequencies, which will become random noise originated from the vicinity of the center wavelength.

If there is a time delay between these two beams, from the optical spectral structure shown in Fig. 2.15, one can see that the corresponding wave trains in carrier and sidebands having coherence lengths longer than the delay time are partially overlapping in time domain. That means these trains are partially frequency coherent, and only part of the beat notes between these wave train pairs are superposed at the modulation frequency. For the wave trains with lengths shorter than the delay time, there is no overlapping in the time domain. The corresponding wave trains in the carrier and sidebands have no coherence and do not beat. If the delay time between sidebands and the carrier is long enough and longer than all wave trains, the two beams become completely incoherent. In this case the measured linewidth of beat note is so called spectral linewidth of light beams.

5.2 Dependence of frequency coherence on delay time

Referring to the Wiener-Khinchin theory (Richter et al., 1986), the optical spectral structure consists of incoherent and coherent parts. We will give the formulation description (Zhu et al. 2010). If the delay time between sideband and the carrier is τ_0, the total optical field can be expressed as:

$$E(t) = E_0 \exp[j(\omega_0 t + \varphi(t))] + \beta E_0 \exp\{j[(\omega_0 + \omega_m)(t + \tau_0) + \varphi(t + \tau_0)]\} \tag{14}$$

where E_0 is the amplitude of optical field, β is a real factor accounting for the amplitude ratio between two fields, ω_0 is the angular frequency of laser beam, ω_m is the modulation frequency, the phase section$\varphi(t+\tau_0)$ and $\varphi(t)$ introduce the phase jitter which is assumed to be a Gaussian distribution. After the necessary formula derivation, the power spectrum $S(\omega)$ can be expressed as:

$$S(\omega) = f_\delta \delta(\omega - \omega_{\mathrm{m}}) + f_L \frac{1}{1 + \left[(\omega - \omega_{\mathrm{m}}) / S_0\right]^2}, \tag{15a}$$

where

$$f_\delta = 2\beta^2 E_0^{\ 4} \exp(-S_0 \tau_0), \tag{15b}$$

$$f_L = 4\beta^2 E_0^4 \left\{ 1 - \exp(-S_0\tau_0) \left[\cos((\omega-\omega_m)\tau_0) + \frac{S_0}{\omega-\omega_m} \sin((\omega-\omega_m)\tau_0) \right] \right\}. \qquad (15c)$$

Here only the white noise S_0 originated from atom spontaneous radiation is included. In (15a), the first term is a δ function at modulation frequency, and is the beat note between the coherent wave trains of light beams when the average coherence length is longer than the delay length. It can be concluded that its intensity will decrease with the increase of the delay time. The second term is the beat note between incoherent wave trains which is broadened by the phase random noise. It has a quasi-Lorentzian profile with a weight factor f_L located at the modulation frequency. From (15c) it can be seen that the amplitude of quasi-Lorentzian profile will be proportional with the delay time.

Actually, when the delay fiber is long enough, both the white noise and the $1/f$ noise component are included, and the $1/f$ noise component gives a similar Gaussian profile. In this case, the two light beams will become incoherent and the δ-peak disappears. The power spectrum will become a Voigt lineshape, which is the convolution of the Gaussian profile and Lorentzian profile. The disappearance of the δ-peak can be regarded as the criterion of coherence elimination and the lineshape broadening at the moment can be used as the optical linewidth.

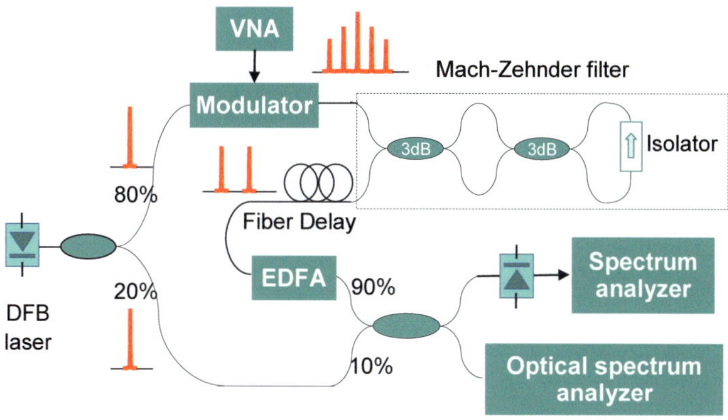

Fig. 2.16. Measurement scheme for lineshape analysis using delayed optical self-heterodyne method

Up to date, lots of methods for analyzing lineshape and linewidth have been established in the past two decades (Chan, 2007; Dawson et al., 1992; Ludvigsen et al., 1998; Richter et al., 1986; Signoret et al., 2001; Signoret et al., 2004; Zhu et al., 2010). The method widely used for analyzing lineshape is delayed self-heterodyne technique. The frequency fluctuations or optical phase of the laser source under test can be converted into intensity variations by an asymmetric Mach-Zehnder type interferometer. Enough fiber delay, which is longer than coherence length, can make no overlapping in time domain between the carrier and sideband, and the frequency coherence between them can be eliminated. Fig. 2.16 illustrates the measurement setup for carrying out the lineshape analysis of beat note between the optical carrier and shifted sidebands based on delayed optical heterodyning in the experiment. A VNA and a $LiNbO_3$ modulator were used to modulate the light beams from the DFB laser and a Mach-Zehnder filter was used to separate the sidebands from the modulated lightwaves.

It has been shown that no matter what optical sources are used, the beat note between first sidebands and the carrier has an extremely narrow linewidth if there is no time delay between them (Zhu et al., 2009). With the increase of the delay line, the coherence between the carrier and the delayed first sidebands will be reduced. Fig. 2.17 shows the measured power spectra of the DFB laser using different delay lines. The power ratios between the δ-peak and the Lorentzian component with different delay lengths are summarized in Table I. It can be seen that the δ-peak decreases and the amplitude of the quasi-Lorentzian profile originated from the noise increases with the increased delay line. When the delay fiber is over 10 km, δ-peak disappears and the coherence between the carrier in the reference channel and the delayed first sidebands has been eliminated completely.

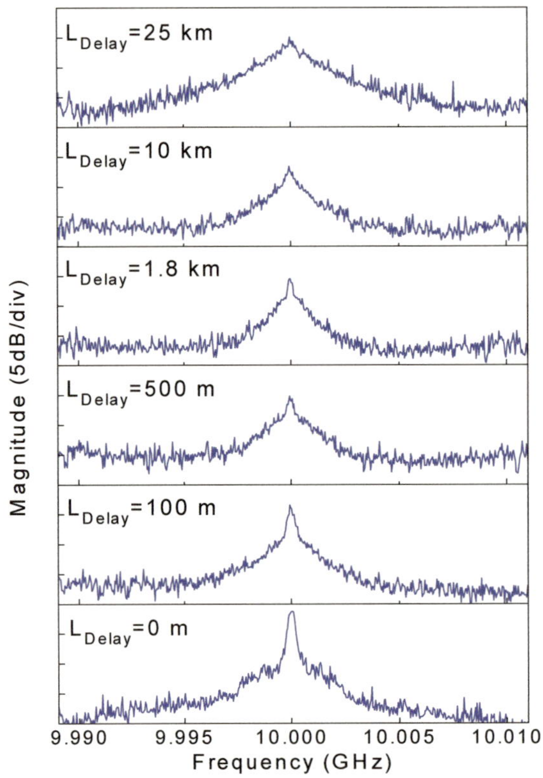

Fig. 2.17. Measured power spectra of DFB laser with fiber delay changing from 0 m to 25 km

A long delay line over coherence length of the two light beams is required to completely eliminate frequency coherence. However, long delay fiber will introduce a high insertion loss. Although an EDFA can be used to compensate the optical power level, the amplifier introduces noise and leads to a poor signal-to-noise ratio. In order to improve the previous experiment, a recirculating scheme as shown in Fig. 2.18 is proposed. In experiment, the first sidebands can be reamplified through the circulating loop. This increases the length of the delay line efficiently. It must be mentioned that this technique is not suitable for measuring linewidth since the output optical signal contains different circulation orders. The method proposed by M. Han (Chen et al., 2006; Han et al., 2005) is more suitable method for linewidth measurement.

Delay length (km)	0	0.1	0.5	1.8	10
Power ratio (dB)	7.3	4.6	3.0	2.7	0.8
10-dBlinewidth (MHz)	1.3	3.2	5.0	5.5	5.7

Table 1. Power ratio between δ-peak and Lorentzian component, and 10-dB linewidth

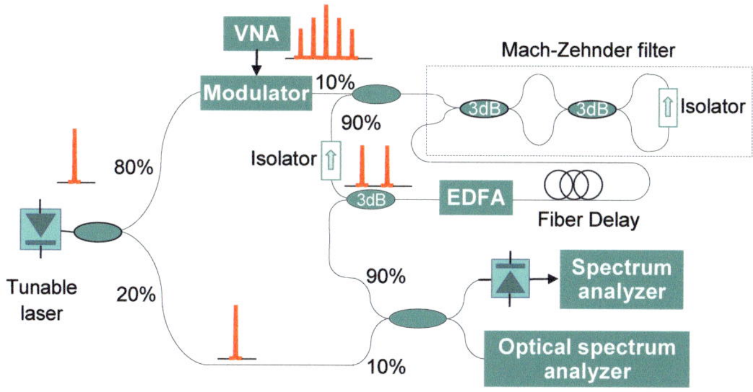

Fig. 2.18. Experimental configuration of optical self-heterodyne scheme with fiber delay loop for lineshape analysis

The peak at the modulation frequency gets wider when the recirculating scheme in Fig. 2.18 is used. For this scheme, the output optical signal comprises higher circulation-order sidebands. Thus, the beat may occur between the carrier and sidebands with a delay time in a wide time period, in which the wavelength may shift. That means that the carrier and the delayed sidebands are launched out from the lightwave source at different time. In this relatively long delay time, the optical wavelength may shift due to the instability of laser source. Consequently, broadening of the measured power spectra reveals wavelength stability of the lightwave source in the observation period. Therefore, the measured optical spectral linewidth depends on the observation time due to the instability of laser.

6. Narrow-linewidth microwave generation

Optical generation of frequency-tunable, narrow-linewidth, and stable microwave signals is desirable for many applications such as in radar, wireless communications, and satellite communication systems. Conventionally, a microwave signal can be generated in the optical domain using optical heterodyning, in which two optical waves of different wavelengths beat at a photodetector (PD). An electrical beat note is then generated at a PD, and its frequency depends on the wavelength spacing of the two optical waves (Gliese et al., 1998). This method is capable of generating microwave and millimeter wave signals. The only frequncy limit is the bandwidth of the PD. However, the beating of two optical waves from two independent optical sources would lead to a microwave with unstable frequency and high phase noise since there is no frequency coherence between them.

In the previous sections, it has been shown that the linewidth of the generated microwave signal depends only on the frequency coherence properties of the two lightwaves, not on the spectral linewidths of the individual light beams. To generate a pure microwave signal, two optical waves used for heterodyning must be highly frequency coherent(Zhu et al., 2009). This section presents two typical approaches to obtaining two frequency coherent lightwaves. One is to make the light correlated to the light emitted in the past time from the same active region. Another way is using two correlated lightwave sources, such as two lasers with mutual injection or two monolithically integrated lasers.

6.1 Microwave generation using a self-injected DBR laser

It has been mentioned that the linewidth of the beat note between two light beams depends on the frequency coherence, not on the spectral linewidths of the two beams. Therefore, we can determine the frequency coherence of two light beams from the linewidth of their beat note.

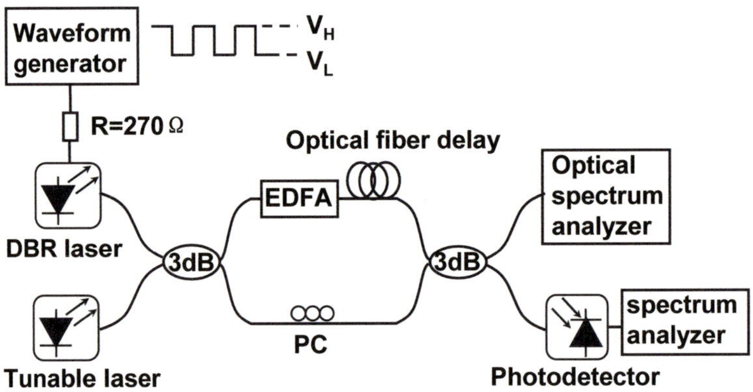

Fig. 2.19. Experimental setup

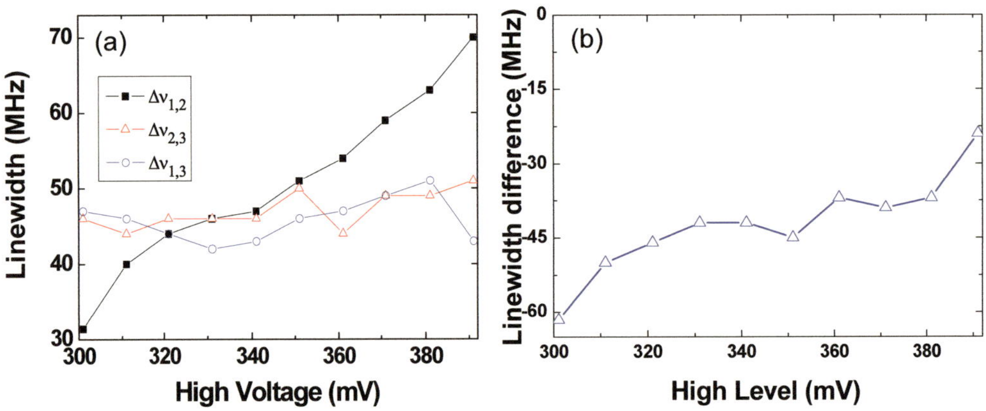

Fig. 2.20. (a) Linewidth measurement when fixing V_L at 300 mV and tuning V_H from 301 mV to 391 mV, and (b) Linewidths difference: $\Delta\nu_{1,2}-(\Delta\nu_{2,3}+\Delta\nu_{1,3})$, where $\Delta\nu_{1,2}$, $\Delta\nu_{2,3}$, and $\Delta\nu_{1,3}$ are 6-dB linewidths of beat notes between λ_1 and λ_2, λ_2 and λ_3, and λ_1 and λ_3, respectively

In this system, a square-waveform voltage generated from a waveform generator (Agilent 33250A) was applied to the phase section of the DBR tunable laser. Two beams at different wavelengths λ_1 and λ_2 were generated corresponding to the high voltage V_H and low voltage V_L of the waveform, respectively. Another lightwave λ_3 from a narrow-linewidth tunable source (Agilent 8164B) was used as a reference. These lightwaves were launched into a Mach-Zehnder interferometer and then arrived at an 18-GHz photodetector (Agilent 11982A). The spectra of their beat notes were measured by an electrical spectrum analyzer (ESA, Advantest R3182).

The optical fiber was 500 m long corresponding to a delay time of 2.5 μs, and the period of square waveform was set to be 5 μs to achieve a steady and high efficient optical heterodyne (Zhu et al., 2006). We fixed the low level V_L of the square voltage at 300 mV, and tuned the high level V_H from 301 mV to 391 mV at a step of 10 mV. In the measurement, the peak power of lightwave at λ_3 was adjusted to be equal to those of lightwaves at λ_1 and λ_2. The measured 6-dB linewidths of beat notes and the results are given in Fig. 2.20(a). $\Delta\nu_{1,2}$, $\Delta\nu_{2,3}$, and $\Delta\nu_{1,3}$ denote 6-dB linewidths of beat notes between λ_1 and λ_2, λ_2 and λ_3, and λ_1 and λ_3, respectively.

From Fig. 2.20 one can see that the measured linewidths $\Delta\nu_{1,3}$ and $\Delta\nu_{2,3}$ are between 40 and 50 MHz and do not change much in the whole tuning range. $\Delta\nu_{1,2}$ is obviously narrower than the sum of $\Delta\nu_{2,3}$ and $\Delta\nu_{1,3}$ when V_H is 301 mV and increases with the increment of V_H. Linewidth of the optical lightwave from DBR laser may broaden with the increase of phase section voltage, but in our experiment, the tuning range is rather small (~90 mV) during which the broadening effect does not that apparent as shown in Fig. 2.20 and can be ignored. Additionally, the square-wave modulation to phase section of DBR laser may cause jitter of the emitted lightwaves and thus affect the linewidth measurement. However, in this experiment, we only care about the comparison between the linewidth of the beat note $\Delta\nu_{1,2}$ and the sum of linewidths of the two lightwaves ($\Delta\nu_1+\Delta\nu_2$), not the absolute linewidth value. Thus, we plotted the difference between them in Fig. 2.20(b). The jitter can be eliminated after making the subtraction. Since linewidth of the lightwave λ_3 from the narrow-linewidth tunable laser is narrower than 100 kHz, $\Delta\nu_3$ can be neglected. The difference can be written as,

$$\Delta\nu_{1,2}-(\Delta\nu_{1,3}+\Delta\nu_{2,3})\approx\Delta\nu_{1,2}-(\Delta\nu_1+\Delta\nu_2) \quad (16)$$

As shown in Fig. 2.20(b), in the tuning range, the linewidth of beat note between the ligtwaves at λ_1 and λ_2 is apparently narrower than the sum of that of the two lightwaves. The difference between them is diminishing as the increase of V_H. Although the curve is not smooth due to the accumulated measurement error, the trend is obvious.

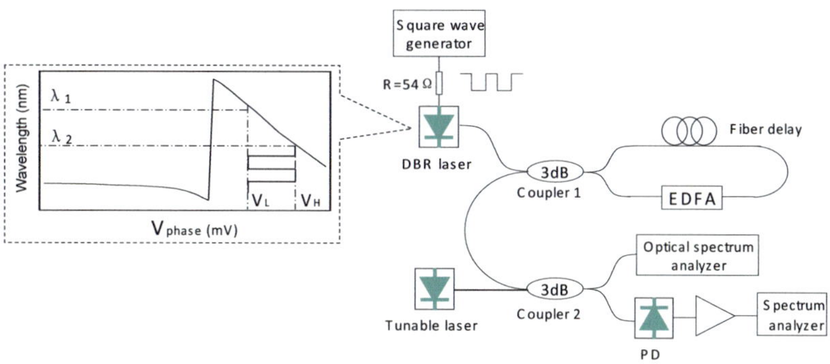

Fig. 2.21. Delayed optical self-injection system

These results reveal that the two lightwaves at λ_1 and λ_2 from the same active region of DBR laser are partially frequency coherent when voltage difference between V_H and V_L is smaller (i.e., the wavelengths of the two lightwaves are close), and the coherence of the two light beams becomes weaker as the increase of the amplitude of the square-waveform voltage.

The frequency coherence between the two lightwaves is weak because they are generated asynchronously. By a method of delayed optical self-injection, the lightwave λ_1 and λ_2 can be generated synchronously. The experimental setup of delayed optical self-injection is shown in Fig. 2.21.

In this system, the laser source is a sampled-grating distributed Bragg reflector (SG-DBR) laser without isolator. A square-wave voltage is applied to the phase section of the laser source. As shown in the insert figure, two lightwaves at different wavelengths λ_1 and λ_2 are generated alternately corresponding to the higher voltage V_H and the lower voltage V_L. After a fiber delay the lightwaves are injected back to the laser. The delay time is exactly equal to the half period of the square-wave voltage. Thus, when the laser oscillated at λ_1 (or λ_2), the injected lightwave is at λ_2 (or λ_1). Then the DBR laser can generate two lightwaves at both λ_1 and λ_2 simultaneously.

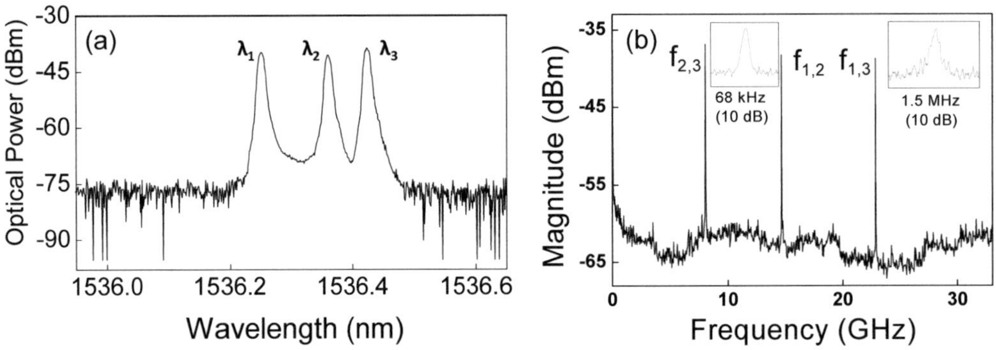

Fig. 2.22. (a)Optical spectra of lightwaves and (b) electrical spectra of their beat notes, where $f_{1,2}$, $f_{2,3}$ and $f_{1,3}$ denote the beat notes generated by the lightwaves λ_1 and λ_2, λ_2 and λ_3, and λ_1 and λ_3 respectively

The optical spectra of the lightwaves are shown in Fig. 2.22(a). The electrical spectra of their beat notes are described in Fig. 2.22(b). $f_{1,2}$, $f_{2,3}$ and $f_{1,3}$ denote the beat notes generated by the lightwaves λ_1 and λ_2, λ_2 and λ_3, and λ_1 and λ_3 respectively. The insert figures are the refined electrical spectra of $f_{1,2}$ and $f_{1,3}$. The 10-dB linewidths of beat notes $f_{1,2}$ and $f_{1,3}$ are 68 kHz and 1.5 MHz, respectively, which are much narrower than that of the lightwaves. There are two reasons for the linewidth reduction. For one thing, the fiber ring used for the optical feedback configures an external cavity. The cavity narrows the linewidth of lightwaves and suppresses the noise of the laser. For another, by optical self-injection, the two lightwaves λ_1 and λ_2 are generated synchronously. The frequency coherence between the two lightwaves is enhanced, which further narrows the linewidths of the beat notes.

The method of the delayed optical self-injection makes the active region oscillate at two different wavelength simultaneously, which strengthens the frequency coherence between the two lightwaves. This method provides us an effective way to generate the microwave signals with narrow linewidth and low phase noise.

6.2 Monolithically integrated microwave source

Fig. 2.23 shows the experimental setup for microwave signal generation based on an EML, which emits one light beam. Another light beam from a narrow-linewidth tunable laser is injected into the electro-absorption modulator (EAM) through an optical circulator. These two beams mixed in the modulator. It has been observed that a reversely biased EAM can be utilized as high-frequency pohotodetector to generate beat signal (Wood et al., 1986; Westbrook et al., 1996). The frequency of the generated signal exactly depends on the wavelength difference, and the power can be expressed as:

$$P_{\text{Microwave}} = \frac{1}{4}(mP_{\text{opt}}R)^2 R_{\text{d}} \tag{17}$$

where m is the modulation depth, P_{opt} is the optical power coupled into EAM, R is the DC responsibility, and R_d is the load impedance.

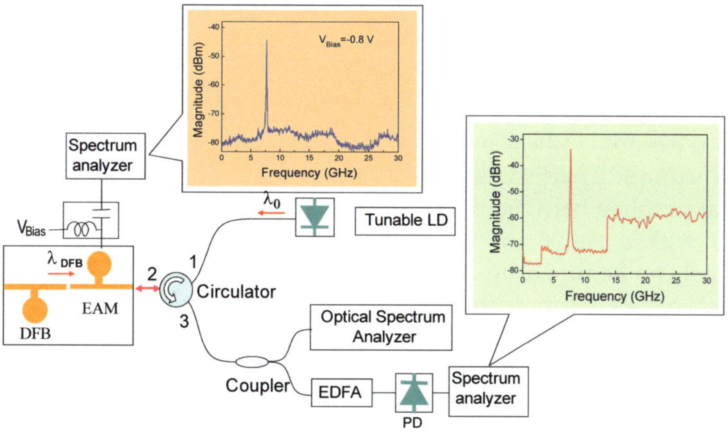

Fig. 2.23. Microwave generation by EML subject to optical injection

As shown in Fig. 2.23, the EAM is biased through a bias Tee and the generated microwave signal from the EAM was measured by an ESA. The output from port 3 of the optical circulator consists of the lightwaves from the DFB laser and the tunable laser. The mixed lightwaves are split into two waves by a coupler. One of them is launched into an optical spectrum analyzer (OSA). Another is received by a high-speed photodetector and the electrical spectra are measured by another ESA. In this way, the spectrum of the microwave signals generated in both the high-speed photodetector and the EAM can be measured simultaneously, as shown in the insets of Fig. 2.23. It has been shown that the frequency of the microwave generated in the EAM can be tuned when the wavelength of the DFB laser is fixed and the wavelength of the narrow-linewidth tunable laser is changed. On the other hand, the wavelength of the DFB laser can be shifted by adjusting the bias voltage of the EAM due to adiabatic chirp. When the isolation resistance between the EAM and the integrated DFB laser is not large enough, the laser working current will vary with the reverse bias voltage. This results in the laser wavelength shift, and can be used for fine tuning of the frequency of the generated microwave signal.

Based on the above work, a tunable monolithic microwave source is proposed (Zhu et al., 2009). The schematic diagram of the chip and the experiment setup is shown in Fig. 2.24. The structures of the modulator and lasers are similar to those of the devices shown in Fig. 2.23. In this scheme, the wavelengths of the two DFB lasers can be tuned by adjusting their bias currents respectively. The lightwaves emitted from both DFB lasers are injected into the

EAM and mixed with each other to generate a microwave signal. A lensed fiber is used to monitor the change of the optical wavelength.

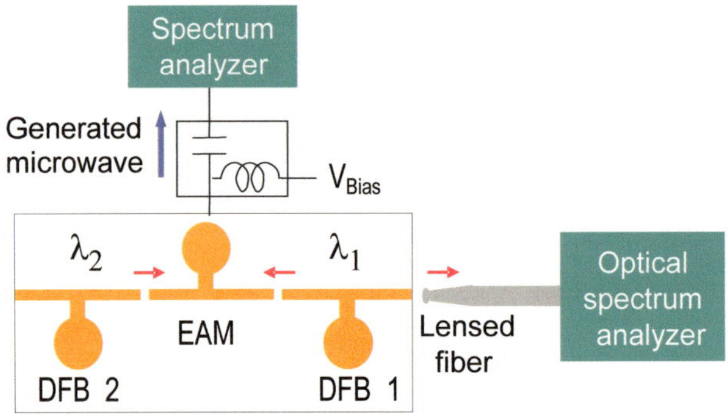

Fig. 2.24. Microwave signal generation using an EAM integrated in between two DFB lasers

Fig. 2.25 shows the output optical and electrical spectra. From Fig. 2.25(a), it can be seen that the four-wave mixing still exists when the optical wavelength difference is over 30 GHz due to strong optical coupling between the two lasers. In Fig. 2.25(b), there is a sharp peak at the beat frequency with a signal to noise ratio of 24-dB. The frequency of the generated microwave signal can be tuned by changing the bias currents of the DFB lasers. Here, the modulator in this device has three functions:

1. Generate microwave signal.
2. Tune the frequency of the generated microwave.
3. Control the intensity of the generated microwave.

The results show that an EAM integrated in between two DFB lasers can be used as a monolithic integrated microwave source.

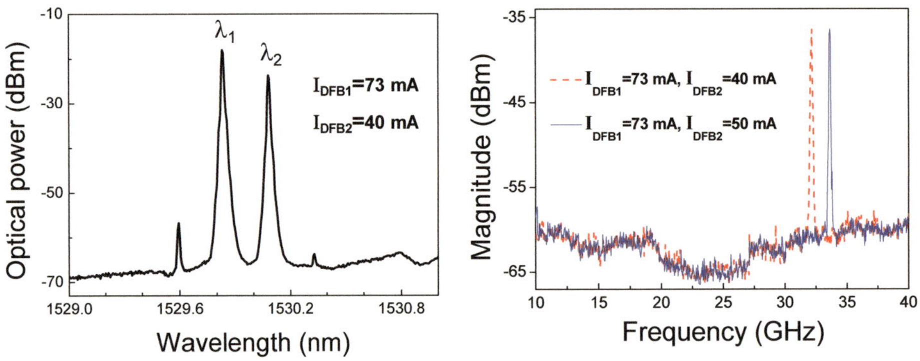

Fig. 2.25. (a) Optical spectrum and (b) corresponding electrical spectrum (dashed line). The electrical spectrum after adjusting the bias current of the DFB laser 2 is also included

7. References

A. A. Michelson (1927). *Studies in Optics*, University of Chicago Press, Chicago

Born, M. & Wolf, E. (1999). *Principles of Optics (7th ed)*, Cambridge University Press, ISBN 0 521 642221, New York

Chan, E. H. W. (2007). Coherence-free optical delay line signal processor. *Electron. Lett.*, Vol. 43, No. 17, (Aug. 2007) 947-948, ISSN 0013-5194.

Chen, X. P.; Han, M.; Zhu, Y. Z.; Dong, B. & Wang, A. B. (2006). Implementation of a loss-compensated recirculating delayed self-heterodyne interferometer for ultranarrow laser linewidth measurement. *Appl. Opt.*, Vol. 45, No.29, (Oct. 2006) 7712-7717, ISSN 0003-6935.

Dawson, J. W.; Park, N. & Vahala, K. J. (1992). An improved delayed self-heteodyne interferometer for linewidth measurements. *IEEE Photon. Technol. Lett.*, Vol. 4, No. 9, (Sept. 1992) 1063-1066, ISSN 1041-1135.

Devaux, F.; Sorel, Y. & Kerdiles, J. F. (1993). Simple measurement of fiber dispersion and of chirp parameter of intensity modulated light emitter. J. Lightwave Technol., Vol. 11, No. 12, (Dec. 1993) 1937-1940, ISSN : 0733-8724.

Diitchburn, R. W. (1963). *Light*, Blackie & Son Ltd., Glasgow.

Eichen, E. & Silletti, A. (1987). Bandwidth measurements of ultrahigh-frequency optical detectors using the interferometric FM sideband technique. *J. Lightwave Technol.*, Vol. 5, pp. 1377-1381

Gliese, U.; Nielsen, T. N.; Nørskov, S. & Stubkjær, K. E. (2005). Narrow linewidth fiber laser for 100-km optical frequency domain reflectometry. *IEEE Photon. Technol. Lett.*, Vol. 17, No. 9, (Sept. 2005) 1827-1829, ISSN 1041-1135.

Geng, J.; Spiegelberg, C. & Jiang, S. (2005). Narrow linewidth fiber laser for 100-km optical frequency domain reflectometry. *IEEE Photon. Technol. Lett.*, Vol. 17, No. 9, (Sept. 2005) 1827-1829, ISSN 1041-1135.

Han, M. & Wang, A. (2005). Analysis of a loss-compensated recirculating delayed self-heterodyne interferometer for laser linewidth measurement. *Appl. Phys. B*, Vol. 81, (Jul. 2005) 53-58, ISSN 0946-2171.

Henry, C. H. (1982) Theory of the linewidth of semiconductor lasers. *IEEE J.Quantum Electron.*, Vol. 18, No. 2, (May 2009) 259-264, ISSN 0018-9197.

Ke, J.; Zhu, N.; Zhang, H.; Man, J.; Zhao, L.; Chen, W.; Wang, X.; Liu, Y.; Yuan, H.; Xie, L. & Wang, W. (2009). Frequency and Temporal Coherence Properties of Distributed Bragg Reflector Laser. *Microwave Opt. Technol. Lett.*, Vol. 52, No. 4, (Aprial, 2010)822-825, ISSN 0895-2477

Ludvigsen, H.; Tossavainen, M. & Kaivola, M. (1998). Laser linewidth measurements using self-homodyne detection with short delay. *Opt. Commun.*, Vol. 155, No. 5, (Oct. 1998) 180-186, ISSN 0030-4018.

Mandel, L. & Wolf, E. (1965). Coherence Properties of Optical Fields, *Rev. Mod. Phys.*, Vol. 37, No. 2, pp. 231-287

Mandel, L. & Wolf, E. (1976). Spectral coherence and the concept of cross-spectral purity. *J. Opt. Soc. Am.*, Vol. 66, No. 6, (Jun. 1976) pp. 529-535

Mandel, L. & Wolf, E. (1995) *Optical Coherence and Quantum Optics*, Cambridge University.

Mathieu, J. P. (1975). *Optics*, Oxford, 0080171575, New York.

Richter, L. E.; Mandelberg, H. I.; Kruger, M. S. & Mcgerath, P. A. (1986). Linewidth determination from self-heterodyne measurements with subcoherence delay times. *IEEE J. Quantum Electron.*, Vol. QE-22, No. 11, (Nov. 1986) 2070-2074, ISSN 0018-9197.

Ryabukho, V.; Lyakin, D. & Lobachev, M. (2005). Longitudinal pure spatial coherence of a light field with wide frequency and angular spectra. *Opt. Lett.*, Vol. 30, No. 3, (Feb. 2005) 224-226, ISSN 0146-9592.

Sato, K. (2001). 100 GHz optical pulse generation using Fabry-Perot laser under continuous wave operation. *Electron. Lett.*, Vol. 37, No. 12, (June, 2001) 763-764, ISSN 0013-5194

Signoret, P.; Marin, F.; Viciani, S.; Belleville, G.; Myara, M. & Tourrenc, J. P. (2001). 3.6 MHz linewidth 1.55μm monomode vertical-cavity surface-emitting laser. *IEEE Photon. Technol. Lett.*, Vol. 13, No. 4, (Apr. 2001) 269-271, ISSN 1041-1135.

Signoret, P.; Myara, M.; Tourrenc, J. –P.; Orsal, B.; Monier, M. –H.; Jacquet, J.; Leboudec, P. & Marin, F. (2004). Bragg section effects on linewidth and lineshape in 1.55-μm DBR tunable laser diodes. *IEEE Photon. Technol. Lett.*, Vol. 16, No. 6, (Jun. 2004) 1429-1431, ISSN 1041-1135.

Sun, C.; Zhu, J.; Liu, Y. & Yu P. (1996). Dual-function electroabsorption waveguide modulator/detector for optoelectronic transceiver applications. *IEEE Photon. Technol. Lett.*, Vol. 8, No.11, (November, 1996)1540-1542, ISSN 1041-1135

Wheeler, C. R.; Ramsier, R. D. &Henriksen, P. N. (2003). An investigation of the temporal coherence length of ligh. *Eur. J. Phys.*, Vol. 24, No. 4, (2003) 443-450, ISSN 0143-0807.

Wood, T.; Carr, E.; Kasper, B.; Linke, R. & Burrus, C. (1986). Bidirectional fibre-optical transmission using a multiple-quantum-well (MQW) modulator/detector. *Electron. Lett.*, Vol. 22, No. 10, (March, 1986)528-529, ISSN 0013-5194

Yariv, A. (1997). *Optical Electronics in Modern Communications,* Oxford University Press, ISBN-13: 9780195106268, New York

Zhang, H.; Zhu, N.; Man, J.; Ke, J.; Zhang, B.; Han, W.; Chen, W.; Wang, X.; Xie, L.; Zhao, L. & Wang, W. (2009). Narrow-linewidth microwave generation using a self-injected DBR laser diode. *IEEE Photon. Technol. Lett.*, Vol.21, No.15, (August, 2009)1045-1047, ISSN 1041-1135

Zhu, N. H.; Wen, J. M.; Chen, W. & Xie, L. (2007). Hyperfine spectral structure of semiconductor lasers. *Physical Review A*, Vol. 76, (Dec. 2007) 063821, ISSN 1050-2947.

Zhu, N. H.; Li, W.; Wang, L. X.; Chen. S. F. & Ke, J. H. (2009). Study on frequency coherence properties of light beams. *IEEE J. Quantum Electron.*, Vol. 45, No. 5, (May. 2009) 514-522, ISSN 0018-9197.

Zhu, N. H.; Man, J. W.; Zhang, H. G. & Ke, J. H. et al (2010). Lineshape analysis of the beat note between optical carrier and delayed sidebands. *IEEE J. Quantum Electron.*, Vol. 46, No. 3, (Mar. 2010) 347-353, ISSN 0018-9197.

Zhu, N. H.; Wen, J. M.; Song, H. P.; Zhang, S. J. & Xie, L. (2006). Measurement of small-signal and large-signal responses of packaged laser modules at high temperature. *Opt. Quantum Electron.*, Vol. 38, (December, 2006) 1245-1257, ISSN 0306-8919

Zhu, N.; Zhang, H.; Man, J.; Zhu, H.; Ke, J.; Liu, Y.; Wang, X.; Yuan, H.; Xie, L. & Wang, W. (2009). Microwave generation in an electro-absorption modulator integrated with a DFB laser subject to optical injection. *Optical Express*, Vol. 17, No. 24, (November, 2009) 22114-22123, ISSN 1094-4087

Part 7

Optical Communications

28

Optoelectronic Chaotic Circuits

M.P. Hanias, H.E. Nistazakis and G.S. Tombras
Department of Electronics, Computers, Telecommunications and Control, Faculty of Physics, National and Kapodistrian University of Athens, Athens, 15784 Greece

1. Introduction

Deterministic chaotic systems exhibit great efficiency, since they react sensitively to small perturbations and thus it is feasible to be controlled and produce signals with small amounts of energy. Moreover, these systems, have the potential for great information-bearing capacity, since the complex signals and the variety of the produced states offers more possibilities for compact conventional encoding of information signals. One of the major application of deterministic chaotic systems is the field of optical communications (Argyris et al., 2005). The significant advantage chaotic optoelectronic systems is that there are not the restrictions of the standard spectra of sinusoidal frequency bands and thus, the number of available communication channels could be larger than the ones of linear systems and the only limit by the ability of receivers to distinguish between different chaotic states (Rizomiliotis et al., 2010). Due to the significant benefits which can offer this scientific area the research direction of dynamical chaos is gradually moving towards practical applications and thus, there is a growing interest for chaotic signal generation sources. In this frame, various circuits have been proposed, among which active chaotic oscillators are preferably considered due to their relative simplicity and energy efficiency. Such a circuit may be externally triggered, i.e. externally driven to chaotic oscillation and it can typically consist of only one active and a few passive components (Argyris et al., 2010). In this respect, it is reasonable to expect that optoelectronic elements, such as a light emitting diode (LED) and optocoupler devices, will provide for the nonlinear characteristics required towards chaos.

The optoelectronic circuits of the chaotic oscillators must be as simply as they can so in order to be possible to synchronize them. Chaos, in this case, is created by the way that the components of the circuits are connected in order to operate in a non linear manner and not by the specific characteristic that each item has (Romeira et al., 2009). The advantage of such an operation is that in this way we avoid the inherent and unpredictable instabilities in the operation of optical devices (i.e. impurities concentration, rate of carrier recombination, crystal defects, e.t.c.).

In this chapter, we present simple chaotic circuits that can be utilized as chaotic signal generation, sources, main transmitters, as well as nonlinear observer-based main receivers for chaos based communication systems. These circuits are classified as Resistor-Inductor-LED optoelectronic chaotic circuits, single optocoupling device and optoelectronic Simulation of the Duffing-Holmes Equation.

The operational characteristics of these circuits are studied, by simulations, using circuit simulation software (e.g. Multisim). Following that, we investigate the influence of various circuit parameters to the complexity of the so generated strange attractors. From the obtained -calculated and recorded- time series, we estimate, with non-linear analysis, the invariant parameters, as correlation, embedding dimension, Kolmogorov entropy and Lyapunov exponents, of the corresponding strange attractors as function of the control parameters.

2. RL-LED optoelectronic circuits

2.1 Circuit's description

A non autonomous chaotic circuit driven RL-LED circuit (Hanias et al., 2008) is shown in Figure 1. It consists of a series connection of an ac-voltage source, a linear resistor R_1, a linear inductor L_1 and a typical LED. The value of R_1 100 Ω and inside the circuits has been placed in series with the LED. In the circuit's input has been applied a sinusoidal voltage with amplitude V_1 as applied through an inductor L_1 with value 47mH. The simulated circuit operation is monitored by checking the voltage value across resistor R_1. In Figure 2 is shown the, obtained by the simulation, chaotic time series at the output for input signal's amplitude $V_{1,rms}$ = 7V and frequency f=10 KHz.

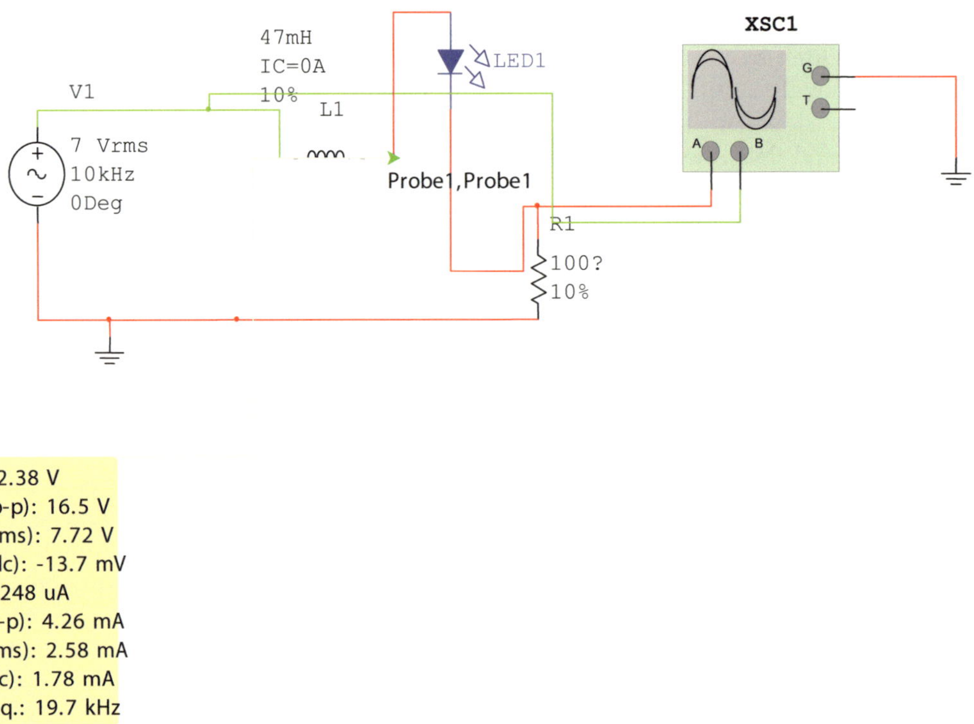

Fig. 1. RL-LED chaotic circuit in Multisim circuits simulation software

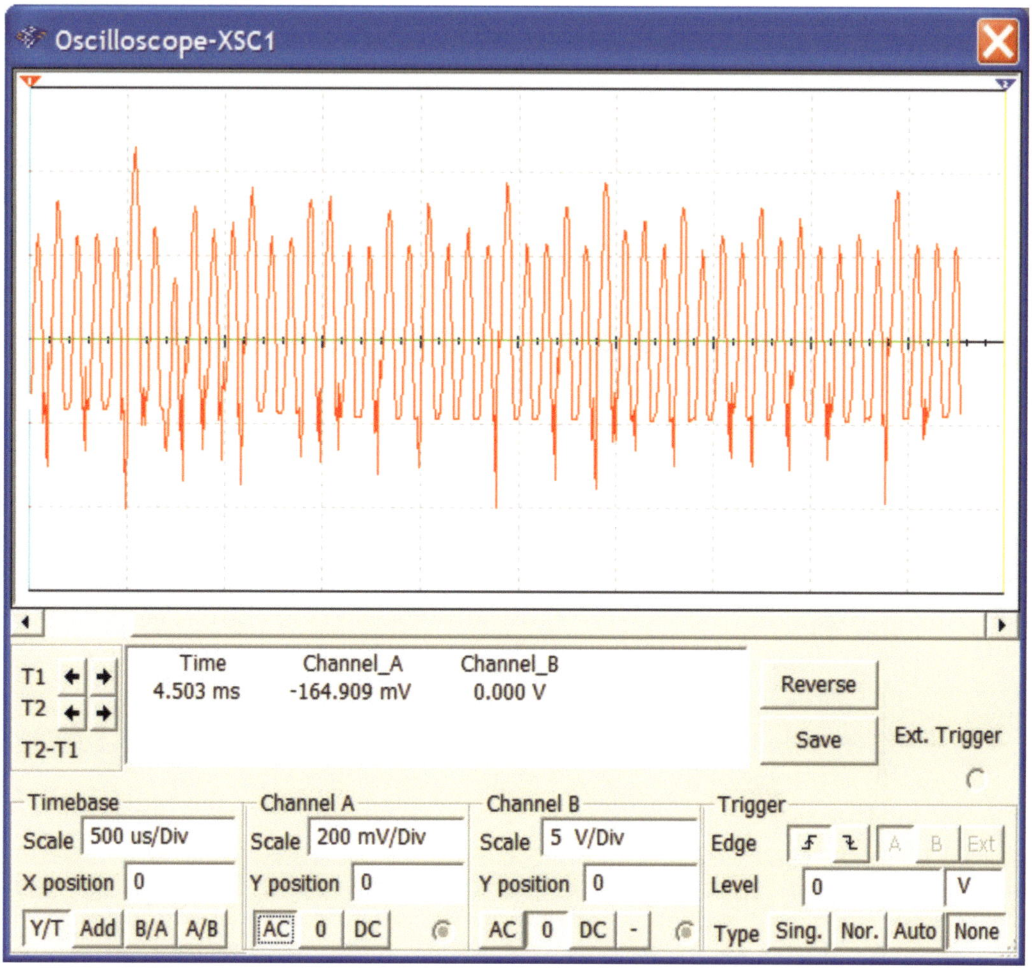

Fig. 2. Chaotic signal *V=V(t)* across resistor R_1 for the RL-LED circuit of Figure 1

2.2 Non-linear analysis

Next, we proceed to the analysis of the obtained chaotic time series following the method proposed by Grassberger and Procaccia (Grassberger & Procaccia, 1983) and successfully applied in similar cases (Hanias & Anagnostopoulos, 1993). Additionally, according the Takens theory (Takens, 1981), the measured time series can be used to reconstruct the original phase space. At first, we calculate the correlation integral *C(r)* for the simulated output signal, for $lim(r)=0$ and $lim(N)=\infty$ (*N* represents the number of the corresponding time series points), as defined by Kantz and Schreiber (Kanz & Schreiber, 1997):

$$C(r) = \frac{1}{N_{pairs}} \sum_{\substack{l=1,\\ j=l+W}}^{N} H\left(r - \left\|\vec{X}_l - \vec{X}_j\right\|\right) \tag{1}$$

where *W* is the Theiler window (Kanz and Schreiber, 1997), *H* is the Heaviside function, and N_{pairs} is given as:

$$N_{pairs} = \frac{2}{(N-m+1)(N-m+W+1)} \qquad (2)$$

with m being the embedding dimension. It is clear that the summation in (1) counts the number of pairs $\left(\vec{X}_l, \vec{X}_j\right)$ for which the distance, i.e. the Euclidean norm $\left\|\vec{X}_l - \vec{X}_j\right\|$ is less than r in an m dimensional Euclidean space. Here, the number of the experimental points is N=10896, while, considering the m dimensional space, each vector $\vec{X}_l$ will be given as,

$$\vec{X}_l = \{V(t_i), V(t_i+\tau), V(t_i+2\tau), \ldots, V(t_i+(m-1)\tau)\} \qquad (3)$$

and represents a point of the m dimensional phase space in which the attractor is embedded each time. In equation (3), τ is the delay time factor determined by the first minimum of the mutual information function $I(\tau)$ and defined as $\tau=\ell\Delta t$ with ℓ=1,2,...,N where Δt=6.25μs is the sample rate. As shown in Figure 3, in our case, the mutual information function $I(\tau)$ exhibits a local minimum at τ=5 time steps and thus we consider this value as the optimum one.

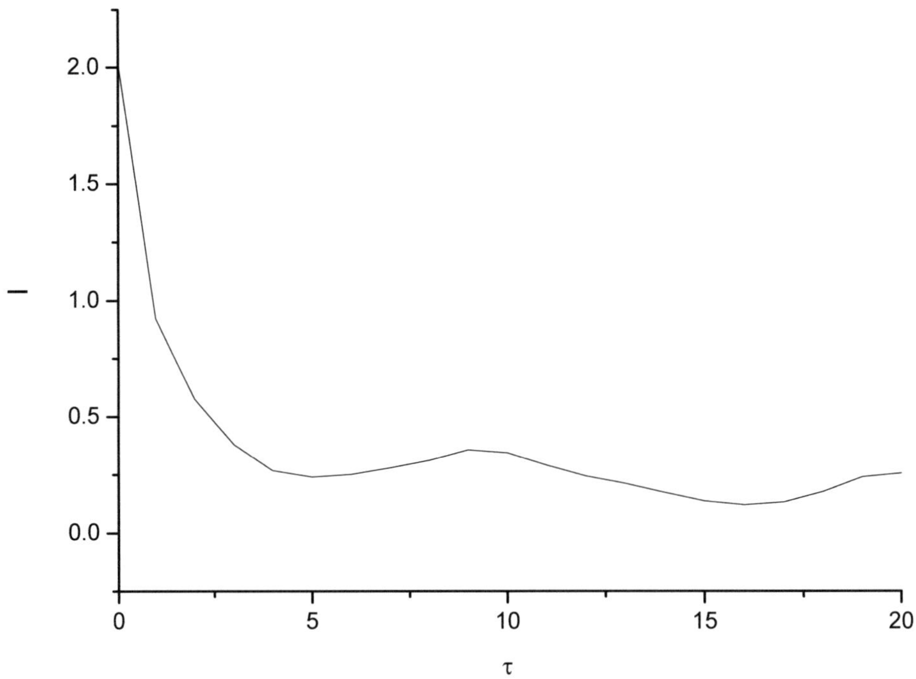

Fig. 3. Average mutual information $I(\tau)$ versus delay time τ

Next, we investigate the parameter W which is the Theiler window. As Theiler pointed out if temporally correlated points are not neglected, spuriously low dimension estimate may be obtained (Stelter & Pfingsten, 1991). However, since there is no concrete rule of how to choose *W*, it may take the first zero-crossing value of the correlation function *CR(τ)* (Kanz & Schreiber, 1997), as suggested by Kantz and Schreiber (Kanz & Schreiber, 1997). This means that we can use the correlation length as a starting value for *W*. As shown in Figure 4, the correlation length is equal to 5 and thus, *W*= 5 time lags. Figure 4 also depicts a strong correlation between the data indicating the way past states affect the system's current state. Hence, we can use these values for phase space reconstruction.

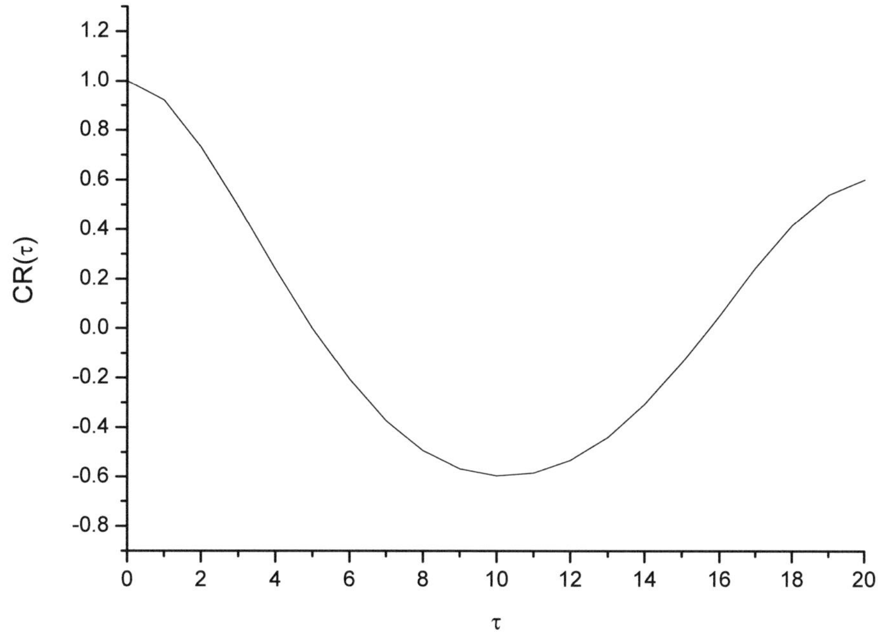

Fig. 4. Correlation function *CR(τ)* versus delay time *τ*

It has been proven (Grassberger & Procaccia, 1983) that if the attractor is a strange one, the correlation integral will be proportional to r^{v}, where v is a measure of the attractor's dimension called correlation dimension. By definition, the correlation integral *C(r)* is the limit of correlation sum of equation (1) and is numerically calculated as a function of *r*, from equation (1), for embedding dimensions m=1,...,10. Figure 5, depicts the relation between the logarithms of correlation integral *C(r)* and *r* for different embedding dimensions *m*. As seen in Figure 6, the slopes v of the lower linear parts of these *log-log* curves provide all necessary information for characterizing the attractor. Then, in Figure 7, the corresponding average slopes v are given as functions of the embedding dimension *m*, indicating that for high values of *m*, v tends to saturate at the non integer value of v=2.23. For this value, the minimum embedding dimension can be 3 (Kanz and Schreiber, 1997) and thus the minimum embedding dimension of the attractor for one to one embedding will be equal to 3.

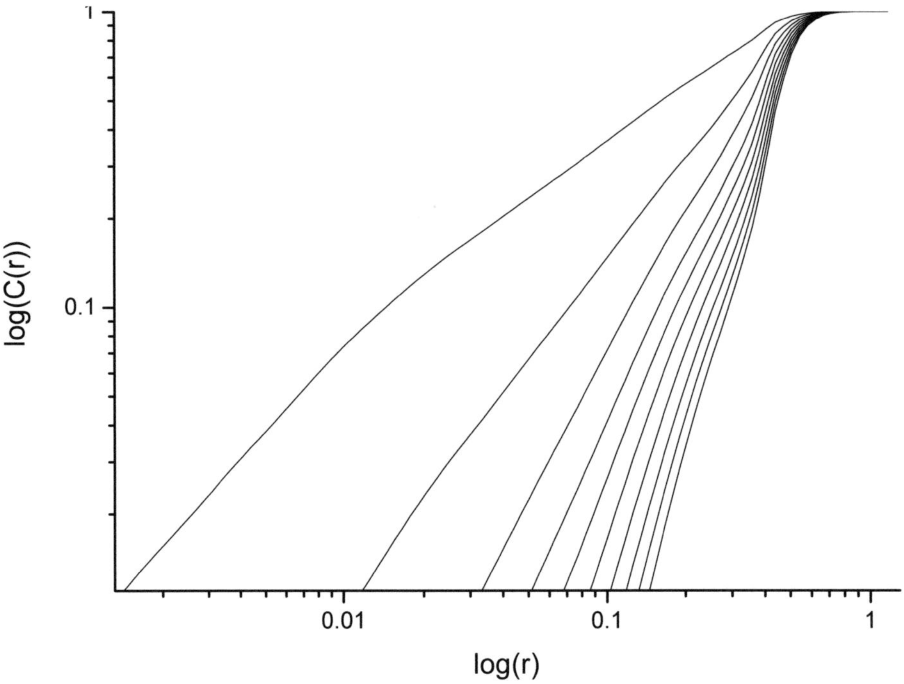

Fig. 5. Relation between *log(C(r))* and *log(r)* for different embedding dimensions *m*

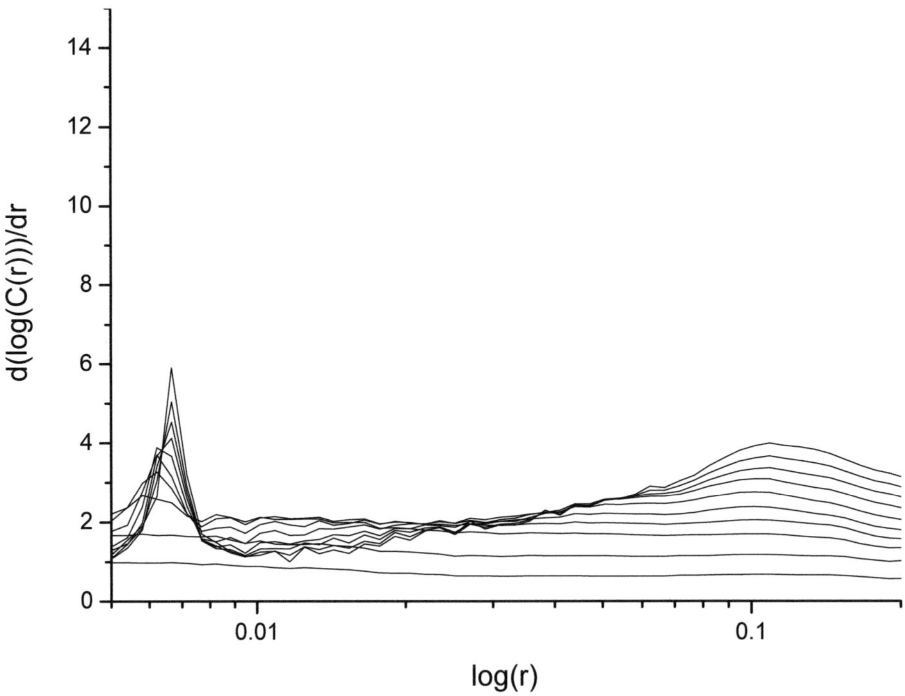

Fig. 6. The corresponding slopes and scaling region of Figure 5

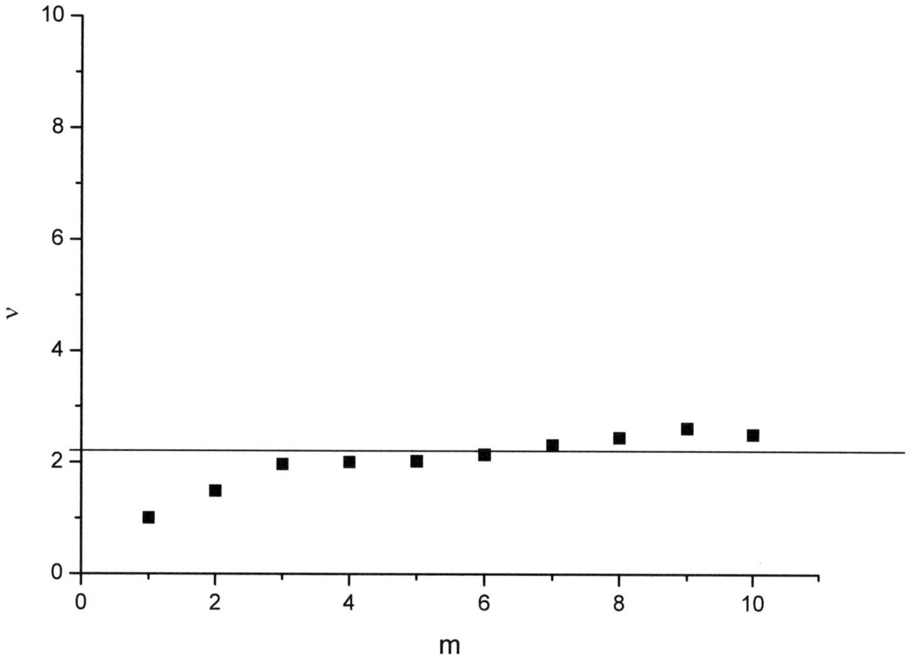

Fig. 7. Correlation dimension v versus embedding dimension m

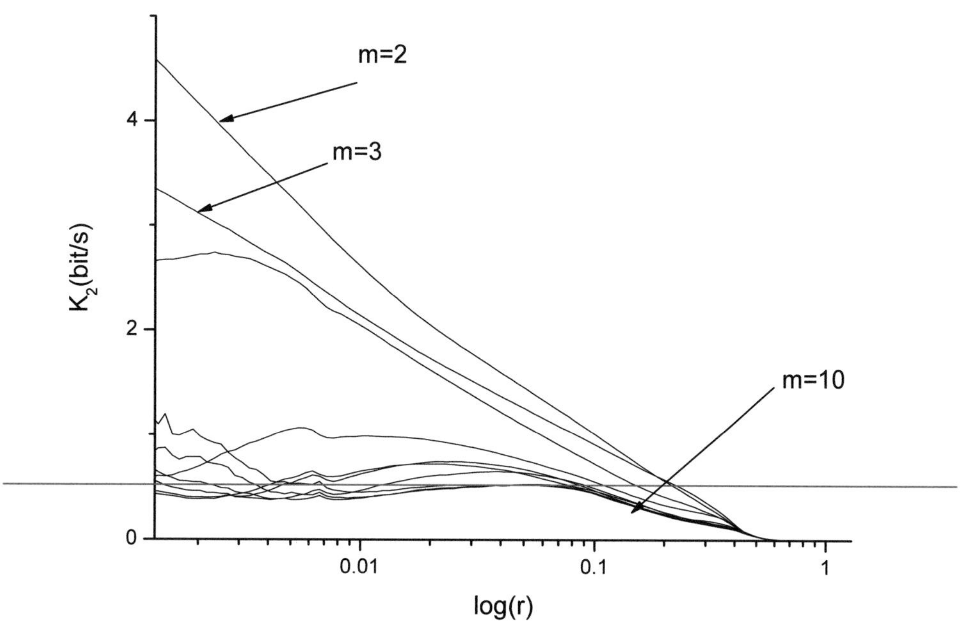

Fig. 8. Kolmogorov entropy versus *log(r)* for embedding dimensions m=2,...10

Following the above, in order to get accurate measurements of the strength of the chaos present in the oscillations of the simulated output signal, we introduce the Kolmogorov entropy. According to Kanz and Schreiber (Kanz & Schreiber, 1997), the method followed so far, also leads to an estimation of the Kolmogorov entropy, i.e. the correlation integral *C(r)* scales with the embedding dimension *m*, since:

$$C(r) \sim e^{-m\tau_d K_2} \tag{4}$$

where K_2 represents a lower bound to the Kolomogorov entropy. In figure 8 is shown the relation between K_2 and the logarithm of *r* for different embedding dimensions *m*. It is clear from figure 8 that around K_2=0.52 bit/s the trajectories appear a "plateau", a red line marks the region, which indicates that there is a steady loose of information at a constant rate given by K_2.

3. Single chaotic optocoupling device

3.1 Description of the circuit

There is a growing interest for non-autonomous chaotic signal generation circuits. Such a circuit may be externally driven and it can typically consist of only one active and a few passive components. Here, we consider a particularly simple circuit based on a single optocoupling device. Its complete layout is shown in figure 9 and it consists of a 4N25 optocoupler, in a typical common emitter configuration, along with an emitter degeneration resistor R_2=3.3KΩ, a collector resistor *R*=33Ω, and a DC power supply V_{SS}=12V. The circuit is driven by an input sinusoidal voltage $\upsilon_{IN}(t)$ with amplitude V_{in}=13V and frequency *f*=800 Hz, which is applied through an inductor *L*=1mH connected in series to the driver LED and a resistor R_1, the value of which, as we will see, plays a crucial role in the circuit's operation and the generation of chaotic voltage time series across R_1 and R_2. In this respect, we use the MultiSim circuit simulation software in order to examine the complete circuit operation by monitoring the voltages $\upsilon_D(t)$ and $\upsilon_E(t)$ across R_1 and R_2, respectively.

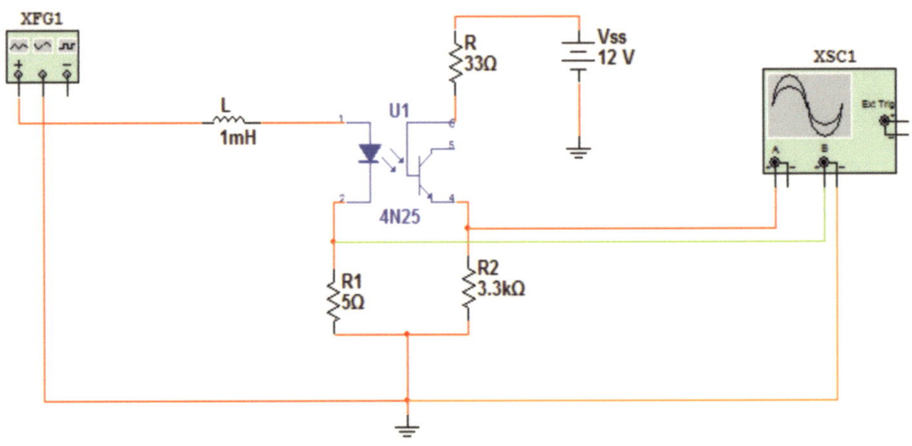

Fig. 9. The optocoupling circuit and its MultiSim simulation environment

Here, we must note that the considered circuit layout resembles the resistor-inductor-diode (RLD) and the resistor-inductor-transistor (RLT) circuits, whose chaotic operation details have been presented and discussed in (Hanias & Tombras, 2009). More specifically, following the conclusions derived in (Hanias & Tombras, 2009), we use various values for R_1 in order to achieve chaotic operation in both, if it is possible, the input LED loop and the emitter output loop, i.e. across R_1 and R_2, respectively. Then, as mentioned above, the simulated circuit operation is monitored by checking the voltages $\upsilon_D(t)$, across R_1, and $\upsilon_E(t)$, across R_2, since both of these voltage signals depend on the chosen value of the LED resistor R_D.

3.2 Simulation results

For the considered circuit operation simulation, we use $\upsilon_{IN}(t)=V_{in}\sin(2\pi ft)$ with V_{in}=13 Volts and f=800 Hz. It is known that, chaotic operation may be generated under various operation conditions and parameters' values. In this chapter, we choose to keep V_{in} and f constant, while varying the value of R_1. Our goal is to first examine whether chaos can be achieve for a specific R_1 value and then to examine whether variation of that value of R_1 may strengthen, weaken, or even destroy the achieved chaos, by returning the circuit to its typically expected operation.

Under these conditions and after some try-and-error selections for the value of R_1, we conclude that if R_1 is set equal to 5Ω then the circuit will exhibit a fully chaotic operation, meaning that both voltage signals $\upsilon_D(t)$ across R_1, and $\upsilon_E(t)$ across R_2, are chaotic. This is readily shown in figure 10. Then, using this as starting point, we see that an increase of R_1 weakens the already obtained chaotic signals and this continues up to a value of R_1=500Ω, for which a weak chaotic signal can still be seen across R_1, but not across R_2. This is shown in figure 11. Finally, further increase of R_1 leads to the total destruction of the remaining chaotic signal across R_1. This is shown in figure 12 where, for R_1=1KΩ, both voltage signals $\upsilon_D(t)$ and $\upsilon_E(t)$ are not chaotic.

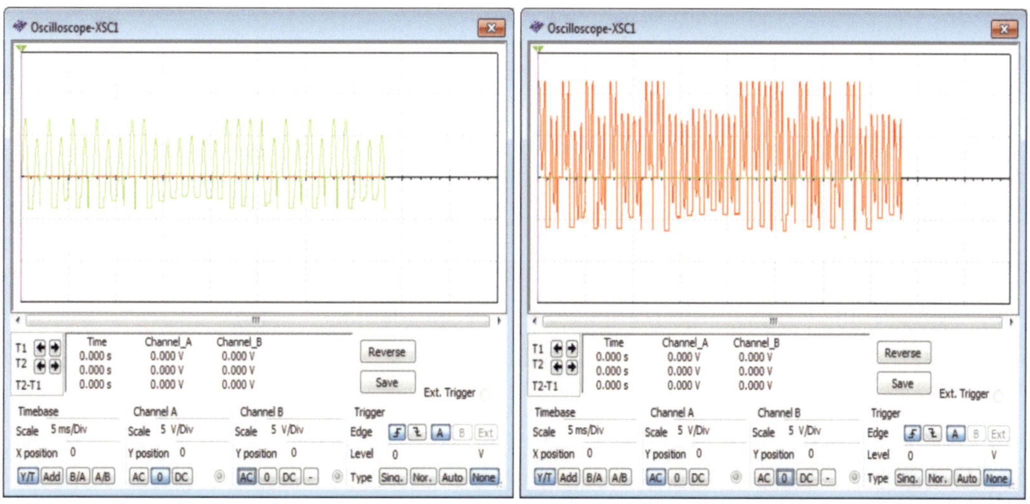

(a) (b)

Fig. 10. For R_1=5Ω: both voltage signals, (a) $\upsilon_D(t)$ across R_1 and (b) $\upsilon_E(t)$ across R_2, are chaotic

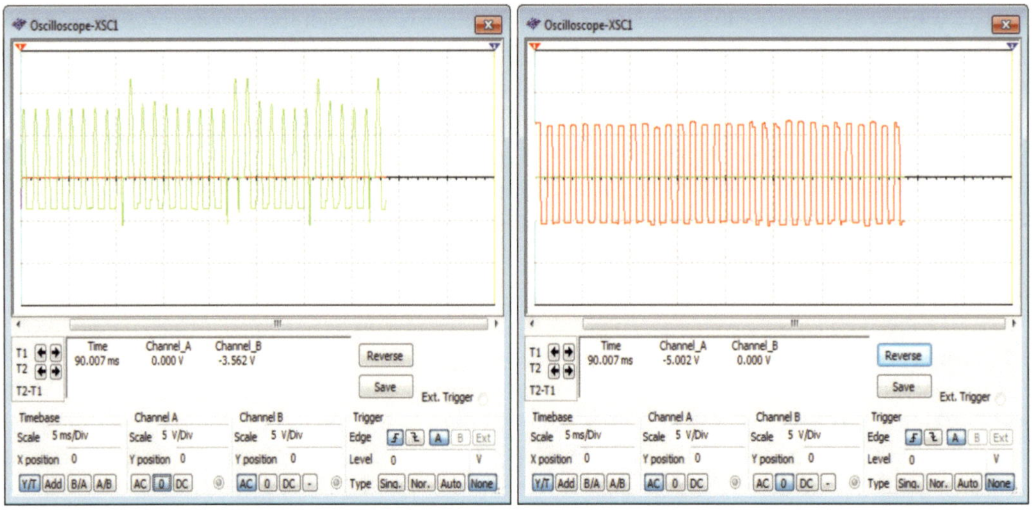

(a) (b)

Fig. 11. For R_1=500Ω, (a) $\upsilon_D(t)$ is a weak chaotic signal across R_1 and (b) $\upsilon_E(t)$, across R_2, is non chaotic

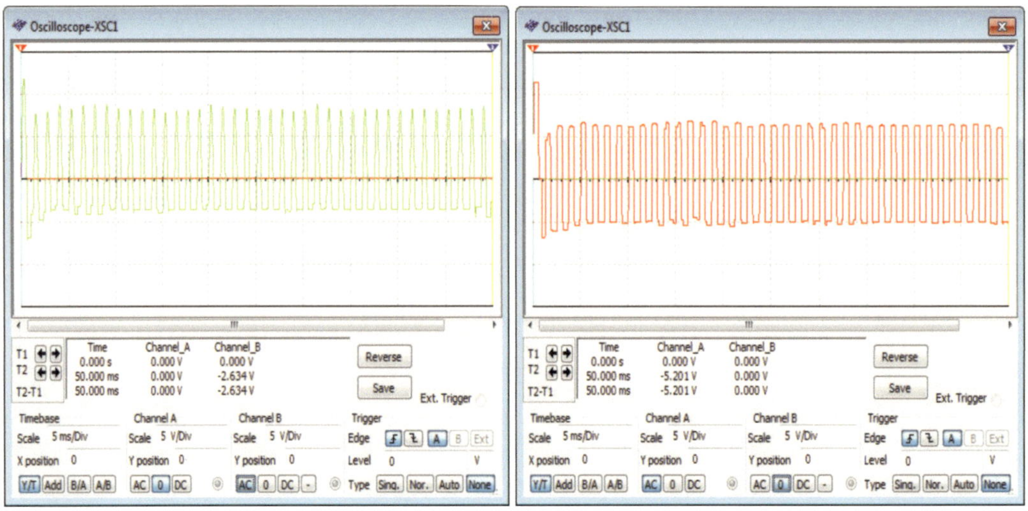

(a) (b)

Fig. 12. For R_1=1KΩ, both voltage signals (a) $\upsilon_D(t)$ across R_1 and (b) $\upsilon_E(t)$ across R_2, are non chaotic

3.3 Nonlinear analysis

In this section, we proceed to the analysis of the obtained chaotic signals time series when R_1=5Ω. Using our data, with value of R_1=5Ω we construct a vector $\vec{X}_i$, i=1..N, where N=5000 data values in the m dimensional phase space given in the following form:

$$\vec{X}_i = \left\{ \upsilon_i, \upsilon_{i-\tau}, \upsilon_{i-2\tau}, \ldots \upsilon_{i-(m-1)\tau} \right\} \quad (5)$$

This vector, represents a point of the m dimensional phase space in which the attractor is embedded each time, where τ is the time delay τ=iΔt and Δt=0.1μs is the sample rate. The element υ_i represents a value of the examined scalar time series in time, i.e. here a voltage value υ_*, across R_1 or R_2, corresponds to the i-th component of the time series. The use of this method reduces phase space reconstruction to a problem of proper determination of suitable values of m and τ. The choice of these values is not always simple, especially when there is no additional information about the original system and the only source of data is a simple sequence of scalar values as acquired from that system.

The dimension, where a time delay reconstruction of the phase space provides for the necessary number of coordinates to unfold the dynamics from overlaps on itself caused by projection is called embedding dimension m. Using the average mutual information, we can then obtain τ as being less associated with a linear point of view, and thus, more suitable for dealing with nonlinear problems. The average mutual information $I(\tau)$, expresses the amount of information (in bits), which may be extracted from the value in time υ_i about the value in time $\upsilon_{i+\tau}$. As optimum τ, suitable for the phase space reconstruction, the position of the first minimum of I(τ) is usually used. In this case τ =77 time steps for chaotic signal across emitter resistor R_2 and τ=38 steps for chaotic signal across LED resistor R_1 as shown in figure 13.

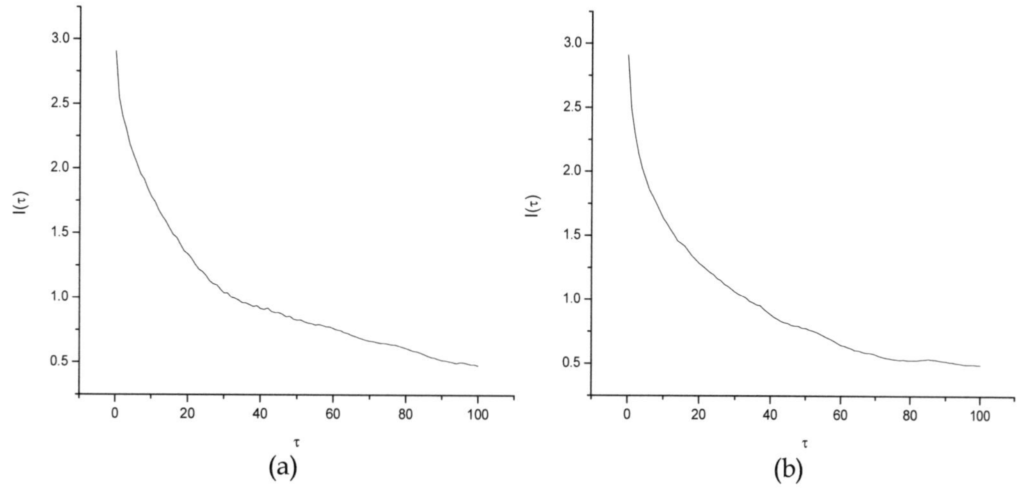

Fig. 13. Mutual information I versus time delay τ for both chaotic signals across (a) R_1=5Ω and (b) R_2

Next, we use the method of False Nearest Neighbors (FNN), [Hanias et al., 2010], in order to estimate the minimum embedding dimension. This method is based on the fact that when the embedding dimension is too small, the trajectory in the phase space will cross itself. Hence, if we are in position to detect these crossings, we may decide whether the used m is large enough for correct reconstruction of the original phase space [i.e. when no intersections occur] or not. If, however, intersections are present for a given m, then the embedding dimension must be considered too small and it must be increased by one at least. Then, again, we test the eventual presence of self-crossings.

The practical realization of the described method is based on testing the neighboring points in the m-dimensional phase space. Typically, we take a certain amount of points in the phase space and find the nearest neighbor to each of them. Then we compute distances for all these pairs and also their distances in (m+1)-dimensional phase space. The rate of these distances is given by

$$P = \frac{\left\| \vec{X}_i(m+1) - \vec{X}_{n(i)}(m+1) \right\|}{\left\| \vec{X}_i(m) - \vec{X}_{n(i)}(m) \right\|} \tag{6}$$

where $\vec{X}_i(m)$ represents the reconstructed vector as described in (1), belonging to the i-th point in the m-dimensional phase space and index $n(i)$ denotes the nearest neighbor to the i-th point. If P is greater than some value P_{max}, we call this pair of points false nearest neighbors (i.e. neighbors, which arise from trajectory self-intersection and not from the closeness in the original phase space. The dimension m will then be found when the false nearest neighbors percentage falls below some limit, typically set to 1%, [Kennel et al.,1992]. Thus, by choosing P_{max}=10, we finally calculate the quantity m. The so obtained results are shown in figure 14 indicating that the application of the FNN method yields a minimum embedding equal to value 5 for both chaotic signals.

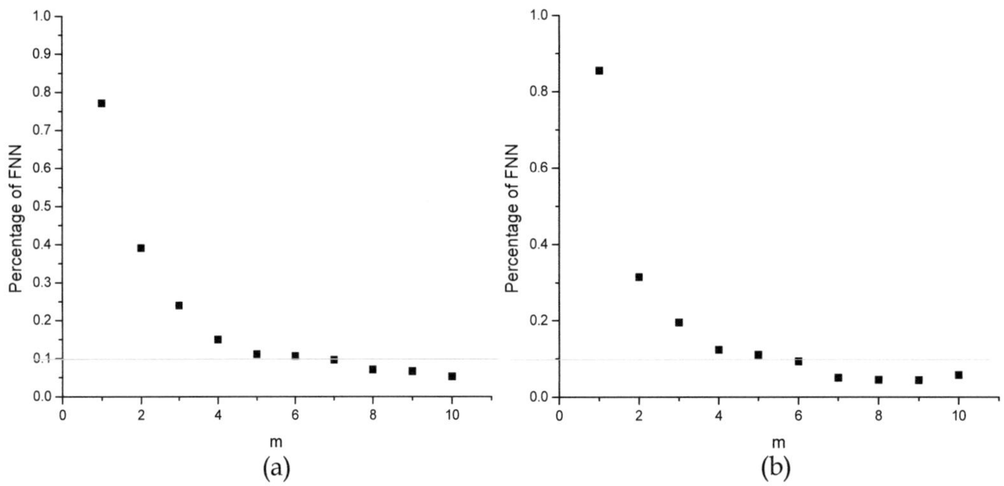

Fig. 14. False nearest neighbor ratio as a function of the embedding dimension. The false nearest neighbors become negligible after m = 5 for the chaotic signals (a) across LED resistor R_1=5Ω and (b) across emitter resistor R_2

4. Optoelectronic simulation of the Duffing-Holmes equation

4.1 Description of the circuit

It is well known that the extremely simple analogue electrical circuit dedicated for simulation the Duffing-Holmes (DH) equation (Holmes, 1979). Our approach is to modify the circuit E (Tamaseviciute et al., 2008), by adding a LED component and finding the various operation conditions and parameters' values that enhanced or not the chaotic operation of LED component. The suggested circuit is shown in figure 15. It is an externally driven damped RLC oscillator with all elements linear. The nonlinearity is involved by the positive feedback loop consisting of the resistor R_3 and two diodes D_1 and D_2. The operational amplifier OA plays the role of both, the buffer for the external sinusoidal force and the amplifying stage for the positive nonlinear feedback. The electrical circuit resembles the Young-Silva oscillator (Lai et al., 2005), but is essentially simpler. It includes a single operational amplifier, two diodes, and four resistors only, in contrast to the Young-Silva circuit containing four operational amplifiers, four diodes, and nine resistors

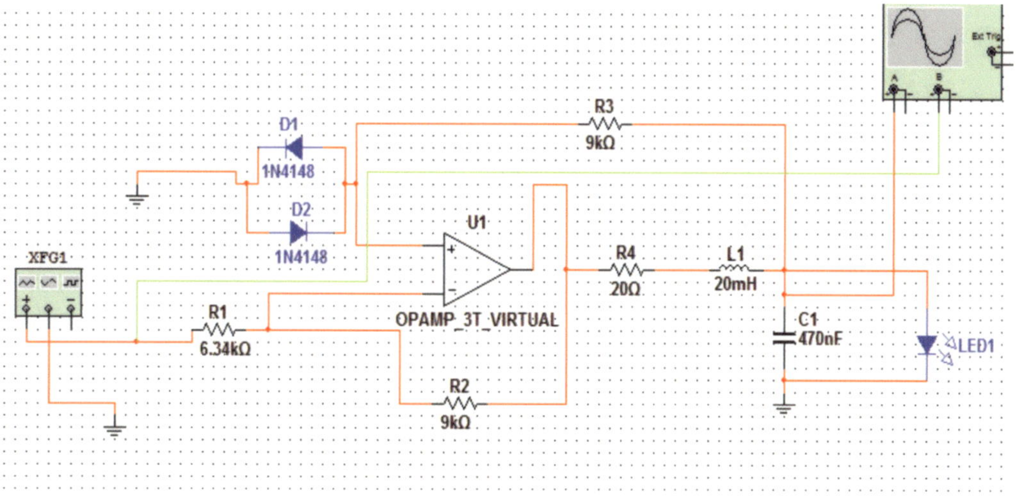

Fig. 15. The considered optoelectronic Duffing-Holmes circuit and its MultiSim simulation environment

The circuit in figure 15 has been simulated using the "Electronics Workbench Professional" package which is based on the SPICE software. The following circuit element values have been used: L=20mH, C=470nF, R_1=6.34KΩ, R_2=9KΩ, R_3=9KΩ, R_4=20Ω. The operational amplifier OA is virtual one but the circuit works the same way with a LM741 type chip, the diodes D_1 and D_2 are general-purpose 1N4148 or similar type silicon devices. Simulation results are shown in figure 16 and figures 17–18. The circuit is driven by an input sinusoidal voltage with amplitude V_{in}=200mV and frequency f=1.5kHz

4.2 Simulation results

It is known that, chaotic operation may be generated under various operation conditions and parameters' values. In this work, we keep V_{in} and f constant, while varying the value of R_1. Our aim is to first examine whether chaos can be achieve for a specific R_1 value and then to examine whether variation of that value R_1 may strengthen, weaken, or even destroy the achieved chaos, by returning the circuit to its typically expected operation. Under these conditions and after some try-and-error selections for the value of R_1, we conclude that if R_1 is set equal to 5.76KΩ then the circuit will exhibit a periodic operation. This is readily shown in figure 16. Then, using this as a starting point, we see that an increase of R_1 the circuit will exhibit a fully chaotic operation. This is shown in figure 17 for R=6KΩ and in figure 18 for R=6.81KΩ.

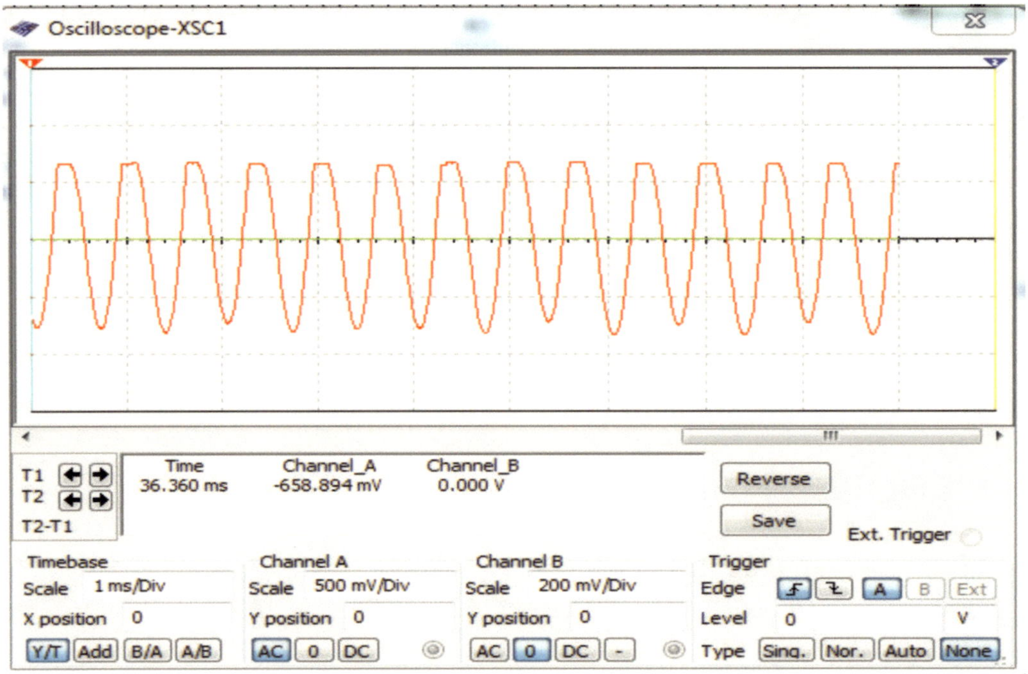

Fig. 16. For R_1=5.76KΩ the voltage signal and LED response are periodic

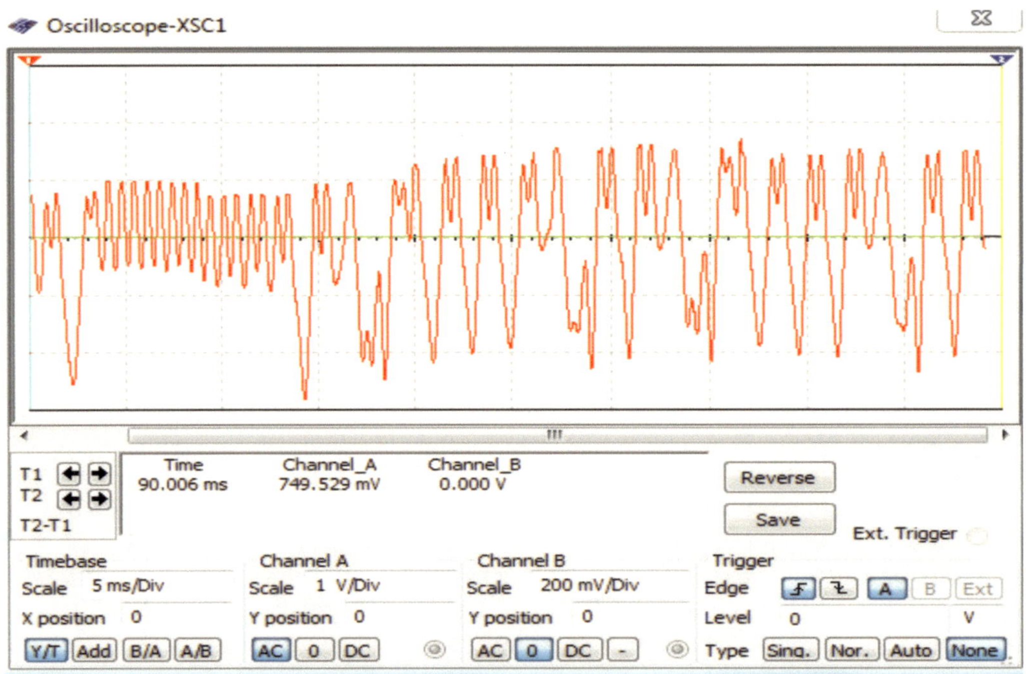

Fig. 17. For R_1=6KΩ, the voltage signal and LED response are chaotic

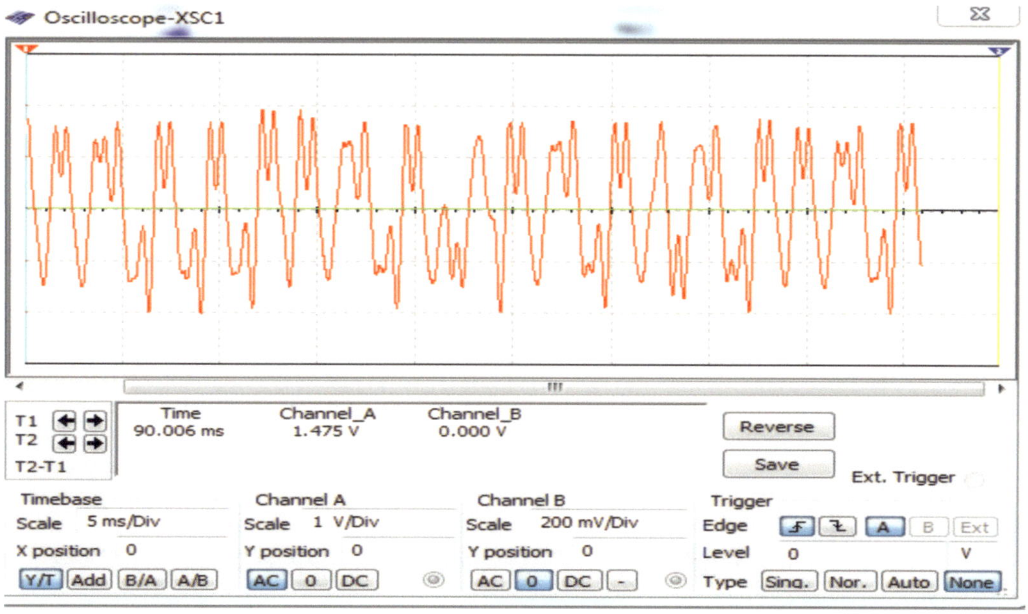

Fig. 18. For R_1=6.81KΩ the voltage signal and LED response are chaotic

4.3 Nonlinear analysis

We apply the analysis as we did in section 2.2. For R_1=6.8 kΩ we examine the signal across the LED while the LED is in operation mode The signal is shown in figure 18. We record N=13454 points with sample rate Δt=6.66μs. As shown in figure 19, the mutual information function exhibits a local minimum at τ=26 time steps and, thus, we shall consider τ=26 as the optimum delay time.

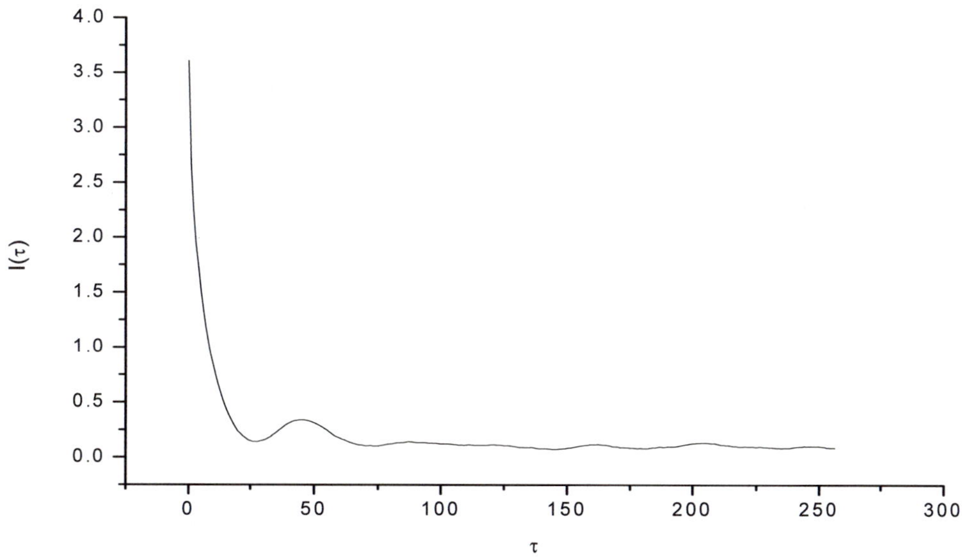

Fig. 19. Average Mutual Information $I(\tau)$ versus time delay τ

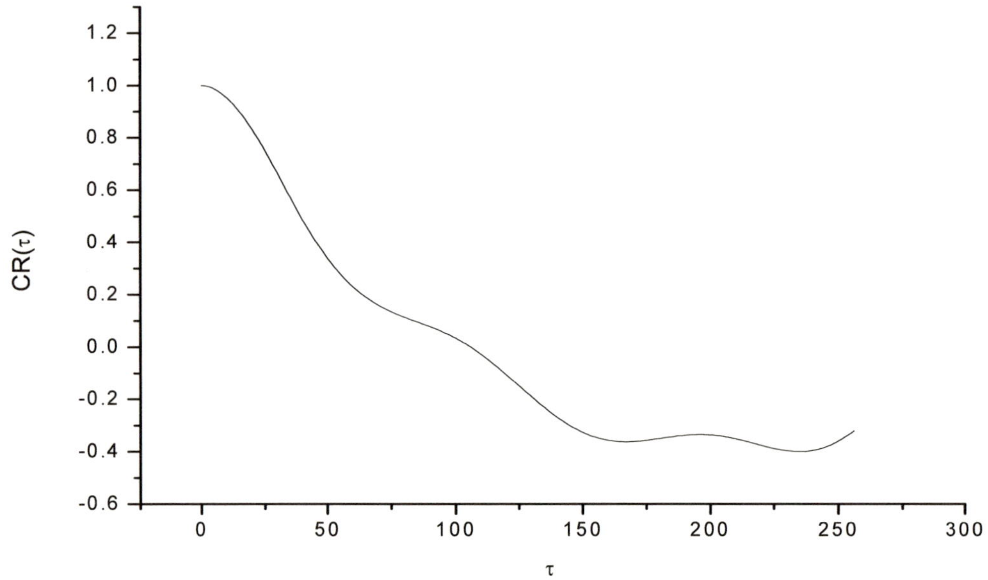

Fig. 20. Correlation function *CR(τ)* versus time delay τ

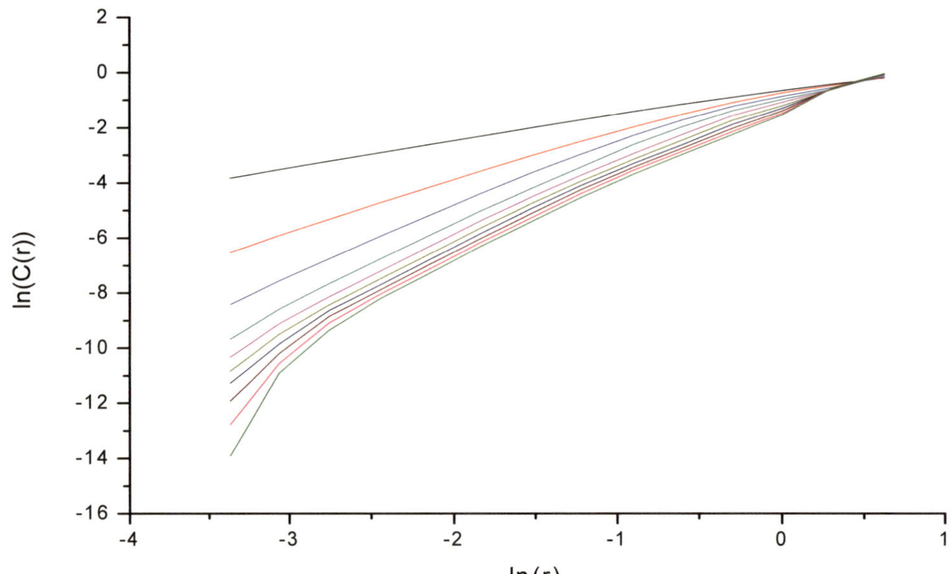

Fig. 21. Relation between *ln(C(r))* and *ln(r)* for different embedding dimensions *m*

Next we calculate the autocorrelation function as we did in section 2.2 As shown in figure 20, the first zero crossing is at τ=107, then the correlation length is equal to 107 and thus, *W*= 107 time lags. Hence, we can use these values for phase space reconstruction.

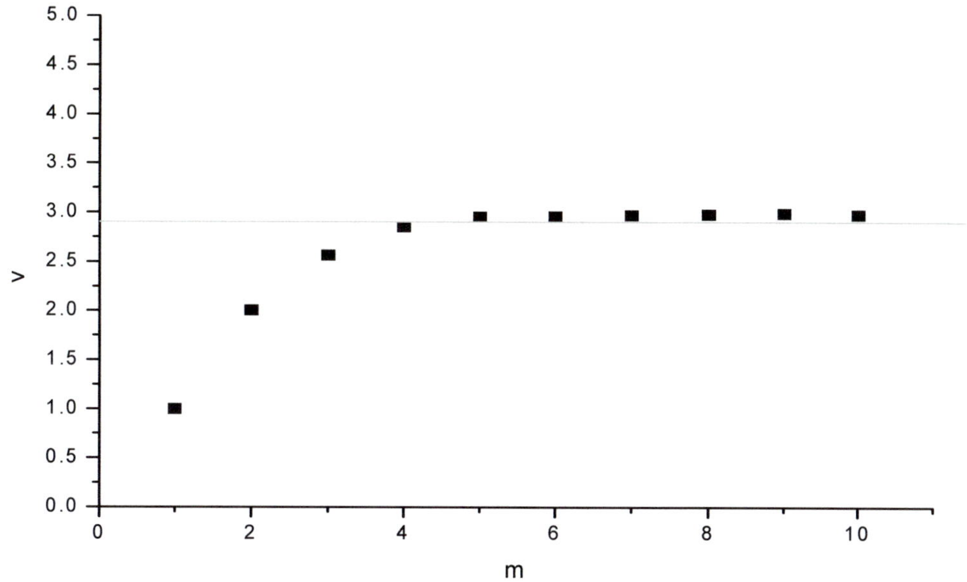

Fig. 22. Correlation dimension v versus embedding dimension m

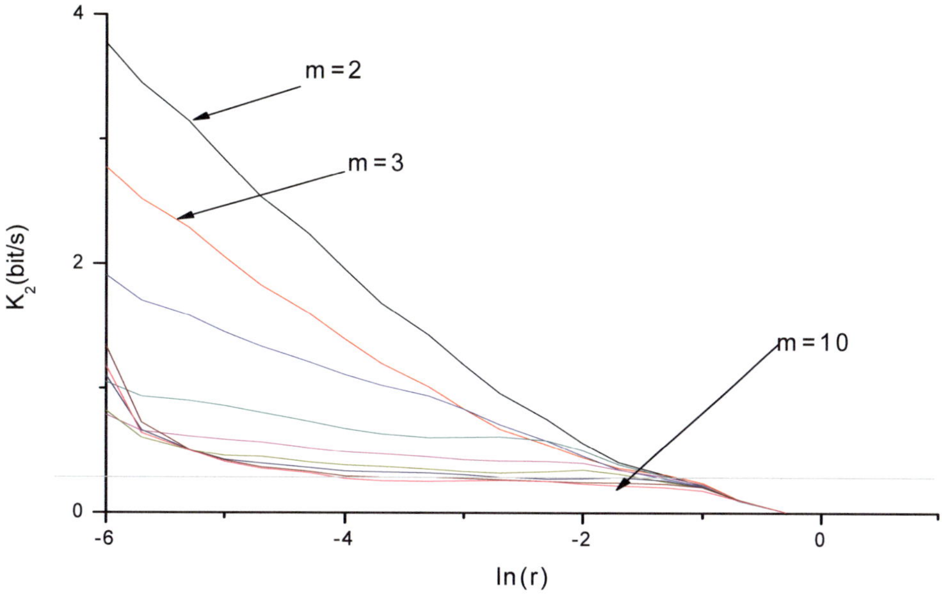

Fig. 23. Kolmogorov entropy versus $ln(r)$ for embedding dimensions m=2,...10

Figure 21 depicts the relation between the natural logarithms of correlation integral $C(r)$ and r for different embedding dimensions m. In figure 22, the corresponding average slopes v are given as a function of the embedding dimension m indicating that for high values of m, v tends to saturate at the non integer value of v=2.9. For this value of v, the minimum embedding dimension can be m_{min}=3 and thus, the minimum embedding dimension of the attractor for one to one embedding will be equal to 3.

The Kolmogorov entropy is calculates as in section 2.2. Figure 23 shows the relation between K_2 and the logarithm of r for different embedding dimensions m, while the plateau, indicates that K_2=0.281 bit/s, meaning that there is a steady loose of information at a constant rate given by K_2.

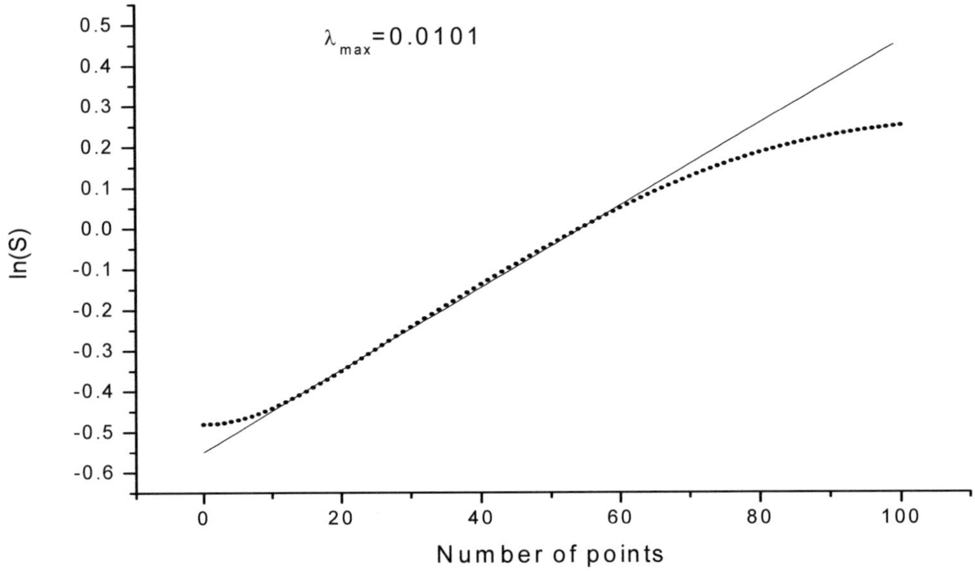

Fig. 24. Estimation of the largest Lyapunov exponent after the Rosenstein method. The portion of the curve used for the least-squares line fits the starting point up to the saturation point. The straight line, represents the resulting least-squares line fit with slope of 0.00101 for m=3

In order to determine the strength of the observed chaotic behaviour, we calculate the largest Lyapunov exponent, which measures the divergence of nearby trajectories. As the system evolves, the sum of a series of attractor point values (in each dimension) will either converge or diverge. Lyapunov exponents measure this convergence/divergence rate in each dimension and a chaotic system will exhibit trajectory divergence in at least one dimension. Thus, a positive Lyapunov exponent is a strong indicator of chaos (Rosenstein et.al., 1993). The method used in calculating the largest Lyapunov exponent is based on averaging the local divergence rates or the local Lyapunov exponents. Rosenstein in (Rosenstein et.al., 1993) proposed a new method to calculate the largest Lyapunov exponent from an observed times series. Hence, after reconstructing the phase space using suitable values for τ and m, we can compute the logarithm of the average distance of a point X_{n0} in phase space with respect to all points X_n in its r-neighbourhood. This is repeated for N number of points along the orbit so as to calculate an average quantity S known as the stretching factor and given by

$$S = \frac{1}{N}\sum_{n0=1}^{N} ln\left[\frac{1}{\left|u_{X_{n0}}\right|}\sum\left|X_{n0} - X_n\right|\right] \tag{7}$$

Where $u_{X_{n0}}$ is the number of neighbours found around point X_{n0}. In the case of chaotic dynamics, a plot of the stretching factor S against the number of points N will yield a curve with a linear increase at the beginning, followed by an almost flat region. The slope of the linear portion of the first part of this curve gives an estimate of the largest Lyapunov exponent. The value of r is taken as the data interval is divided by 200 and again, in order to avoid temporal correlations we use the Theiler window of 107. The magnitude ln(S) versus the number of points N shows the expected linear increase and then after a flat region. This is clearly shown in figure 9, where the slope value that corresponds to the largest Lyapunov exponent is obtained after the least-squares line fit and is found to be equal to λ_{max}=0.0101 for m=3. Moreover, the dependence of λ_{max} on the embedding dimension m is shown in figure 24.

5. Conclusion

From the previous analysis, it is clear, that simple optoelectronic circuits can easily constructed and also be easily controlled via a control parameter as is the value of Resistor or the frequency of voltage source. For RL -LED circuits the result is that the scaling behaviour of the correlation integral and the saturation of correlation dimension ν with increasing embedding dimensions m reflect low dimensionality. The strange attractor that governs the phenomenon has a correlation dimension υ=2.23 stretching and folding in a 3 dimension phase space. Thus, the number of degrees of freedom of the whole domain structure is limited at 3 and these results in the low value of the correlation dimension. The LED exposes chaotic behaviour even if it works in its operation point. In this work, the obtained simulation results indicate that the proposed circuit can be used to generate chaotic signal, in a light emitting manner, useful in code and decode applications. The Optoelectronic simulation of the Duffing-Holmes Equation is governed by a strange attractor that stretching and folding in a 3 dimension phase space too.

The higher value of its correlation dimension 2.9 instead of 2.23 of RL -LED circuit shows that this circuit is better for secure transmission of information using chaos while the loss of information is K_2=0.281bit/s instead of 0.52bit/s of RL -LED circuit.

For the optocoupler circuit the chaos is generated easily. It is clear that this simple externally triggered optoelectronic circuit can be used in order first to generate chaotic voltage signals and then to control the obtained chaotic signals by varying specific circuit parameters, for example, the value of a specific component. In the considered optoelectronic circuit, the crucial parameter is the input loop resistor R_1, the value of which leads to the generation of a chaotic voltage signal $\upsilon_D(t)$ across the resistor R_1. However, this chaotic voltage reflects a chaotic current through R_1 and thus, through the optocoupler's input driving LED. Hence, the so driven phototransistor will generate a chaotic voltage signal $\upsilon_E(t)$ across emitter resistor R_2, according to the strength of the driving chaotic light signal as controlled by the value of the input loop resistor R_1.

6. References

Argyris, A.; Syvridis, D.; Larger, L.; Annovazzi-Lodi, V.; Colet, P.; Fischer, I.; García-Ojalvo, J. & Shore, K.A.;(2005). Chaos-based communications at high bit rates using commercial fibre-optic links, *Nature*, vol. 438, no. 7066, pp. 343-346.

Argyris, A.; Grivas, E.; Hamacher, M.; Bogris, A. & Syvridis, D. (2010). Chaos-on-a-chip secures data transmission in optical fiber links, *OPTICS EXPRESS,* Vol. 18, No. 5 ,pp 5188-5198.

Grassberger, P & Procaccia, I. (1983). Characterization of Strange Attractors, *Phys. Rev Lett.,* vol. 50, pp. 346-349.

Hanias, M.P. & Anagnostopoulos, J.A.N. (1993). Negative-differential-resistance effects in the TlGaTe2 ternary semiconductor, *Phys. Rev B.,* vol. 47, pp. 4261 -4267.

Hanias, M.P.; Magafas, L. & Kalomoiros, J. (2008). Non- linear Analysis in IN RL-LED optoelectronic circuit, *Optoelectronics and advanced materials – Rapid Communications,* vol. 2, no. 2, pp. 126 - 129.

Hanias, M.P. & Tombras, G.S. (2009). Time series analysis in single transistor chaotic circuit, *Chaos, Solitons and Fractals,* vol. 40, pp.246–256.

Hanias, M.P. & Tombras G.S. (2009). Time series crossprediction in single transistor chaotic circuit, *Chaos, Solitons and Fractals,* vol. 41, pp.1167-1173.

Hanias, M. P.; Giannis, L. I. & Tombras, G. S. (2010). Chaotic operation by a single transistor circuit in the reverse active region, *Chaos,* vol. 20, pp. 0131051–0131057.

Holmes, P. (1979). A nonlinear oscillator with a strange attractor, *Phil. Trans. Roy. Soc. London A,* vol. 292, pp. 419–448.

Kennel M.B.; Brown R. & Abarbanel H.D.I. (1992). Determining embedding dimension for phase-space reconstruction using a geometrical construction, *Phys. Rev. A,* vol. 45, pp. 3403-34011.

Lai,Y-C.; Kandangath,A.; Krishnamoorthy,S.; Gaudet, J.A & de Moura, P.S. (2005). Inducing chaos by resonant perturbations: theory and experiment, *Phys. Rev. Lett.,* vol. 94, pp. 214101-4.

Rizomiliotis, P.; Bogris, A. & Syvridis, D. (2010). Message Origin Authentication and Integrity Protection in Chaos-Based Optical Communication, *IEEE Journal of Quantum electronics,* vol. 46, no. 3, pp. 377-383.

Romeira,B.; Figueiredo, J.M. L.; Ironside,C.N. & Slight,T. (2009). Chaotic Dynamics in Resonant Tunnelling Optoelectronic Voltage Controlled Oscillators, IEEE Photonics Technology Letters, vol. 21, no. 24, pp.1819-1821.

Rosenstein, M.T.; Collins, J.J. & Deluca C.J. (1993). A practical method for calculating largest Lyapunovexponents, *Physica D,* vol. 65, pp.117-134.

Stelter, P & Pfingsten, T. (1991). Calculation of the fractal dimension via the correlation integral, *Chaos Solitons& Fractals,* vol 1, iss. 3, pp. 273-280.

Takens, F. (1980). Detecting strange attractors in turbulence, *Dynamic Systems and Turbulence,* Warwick, 1980, *Lecture Notes in Mathematics,* vol. 898, edited by D. A. Rand and L. S. Young_Springer, Berlin, 1981, pp. 366–381.

Tamaseviciute, E.; Tamasevicius, A.; Mykolaitis, G., Bumeliene,S. & Lindberg, E. (2008). Analogue Electrical Circuit for Simulation of the Duffing-Holmes Equation, *Nonlinear Analysis: Modelling and Control,* vol. 13, no. 2, pp. 241–252.

Optoelectronic Feedback in Semiconductor Light Sources: Optimization of Network Components for Synchronization

Sora F. Abdalah[1], Marzena Ciszak[2], Francesco Marino[3], Kais Al-Naimee[4], Riccardo Meucci[5] and F. Tito Arecchi[6]
[1,2,4,5,6]*CNR-Isituto Nazionale di Ottica, Florence*
[1]*High Institute of Telecommunications and Post, Al Salihiya, Baghdad, Iraq*
[3,6]*Physics Department, University of Florence*
[4]*Physics Department, College of Science, University of Baghdad*
[1,2,3,4,5,6]*Italy*
[1,4]*Iraq*

1. Introduction

Synchronization phenomena have been the subject of intense studies since their first observation by Huygens in pendulum clocks. The subsequent discovery of deterministic chaos introduced a new kind of an oscillating system, a chaotic oscillator. In the late 1980s researchers turned their attention to the synchronization properties of chaotic systems (1–4), occurring when two, or more, chaotic oscillators are coupled, or, in case of the unidirectional synchronization, when a chaotic oscillator (master) drives another chaotic oscillator (slave). Despite their unpredictable nature, whose most remarkable feature is the exponential divergence of trajectories starting with infinitely close initial conditions, synchronization of chaotic systems is a phenomenon well established experimentally and reasonably understood theoretically.
The increasing interest in chaotic synchronization is motivated by its potential applications. Clear understanding of these phenomena and the dynamical mechanisms behind them opens new opportunities for applications in different fields of science. In biology, for instance, a challenging problem is to understand how a group of cells or functional units, each displaying complicated behavior, can interact with one another to produce a coherent response on a higher organizational level. In secure communications, a message can be hidden in the output of a chaotic system during transmission and can be recovered by using a copy of the original system, synchronized to the first (5). In this context, the investigation has been focused on chaotic optoelectonic devices.
Within the variety of dynamical regimes observed in semiconductor devices, chaotically spiking attractors play a special role. Typical time traces consist of large pulses separated by irregular time intervals in which the system displays small-amplitude chaotic oscillations. These erratic –though fully deterministic– sequences of pulses, mimic the dynamics experimentally observed in neurons (6) and many hypothesis have been done on their role in neural information processing (see (7) and references therein). Therefore, experiments

based on chaotically spiking units coupled in a network might be useful to simulate and model neuronal dynamics during a perceptual task. To this end, light-emitting diodes (LEDs) appear as ideal candidates since they allow the realization of a miniaturized chip of units optoelectronically coupled and an easier implementation in networks with various coupling topology.

In the following we report and characterize synchronization phenomena in such optoelectronic networks. In particular, we focus our attention on the experimental design, introducing a control bias voltage which allows to simplify the coupling scheme and to reduce the influence of small parameter mismatch in the network units.

2. Generation of chaotic optical pulses

The dynamics of LEDs can be typically described in terms of two coupled variables (intensity and carrier density) evolving with very different characteristic time-scales. The introduction of a third degree of freedom (and a third time scale) describing the AC-feedback loop, leads to a three-dimensional slow-fast system, displaying complicated bifurcation sequences arising from the multiple time-scale competition between optical intensity, carriers and the feedback nonlinear filter function. A similar scenario has been recently observed in semiconductor lasers with optoelectronic feedback (8). Here, we consider a closed-loop optical system, consisting of a LED with AC-coupled nonlinear optoelectronic feedback (see Fig. 1). The

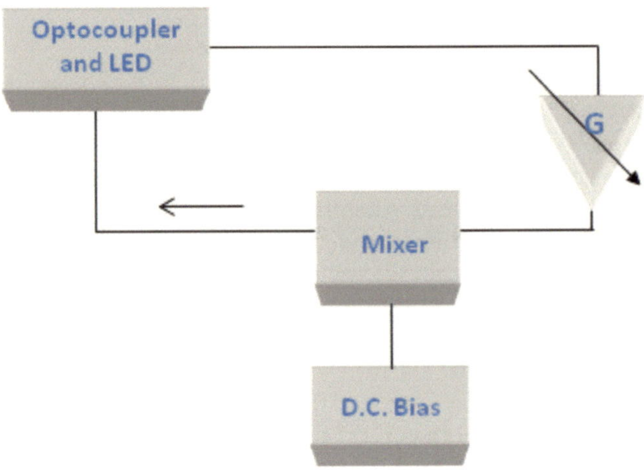

Fig. 1. Sketch of the experimental setup for LED.

output light is sent to a photodetector producing a current proportional to the optical intensity. The corresponding signal is sent to a variable gain amplifier characterized by a nonlinear transfer function of the form $f(w) = Aw/(1 + sw)$, where A is the amplifier gain and s a saturation coefficient, and then fed back to the injection current of the LED. The feedback strength is determined by the amplifier gain, while its high-pass frequency cut-off can be varied (between 1 Hz and 100 KHz) by means of a tunable high-pass filter. External signals or control bias can be added to the LED pumping current. Several of these units are eventually coupled through the detector photocurrent to form a network.

Fixed both the feedback gain and frequency cut-off and increasing the dc-pumping current, we observe the dynamical sequence shown in Fig. 2. In the upper panel, corresponding to the lowest current, the detected optical power is stable. As the current is delicately increased a transition to a chaotically spiking regime is observed, where large intensity pulses are separated by irregular time intervals in which the system displays small-amplitude chaotic oscillations. As shown in Ref. (13) the chaotic spiking regime can be understood in terms of excitability of a chaotic attractor, where the small chaotic background spontaneously triggers excitable spikes in an erratic but deterministic sequence. The time-scale of these dynamics, much slower respect to typical carriers time scales ($\sim$ 1 ns), is fully determined by the high-pass filter in the feedback loop (8). Although similar to the well known Shilnikov homoclinic chaos observed in various physical systems (9–12), here the attractor structure is substantially different since an exact homoclinic connection to a saddle-focus does not occur. Further increase of the current makes the firing rate higher until a periodic regime is eventually reached (lower panel in Fig. 2). As the large pulses, these oscillations can be decomposed into a sequence of periods of slow motion, near extrema, separated by faster relaxations between them. This behavior (relaxation oscillations) is typical in multiple time-scales nonlinear systems. The same dynamical sequence can be obtained as the pumping current is kept constant and the amplifier gain is changed.

3. Synchronization in optical network

3.1 Symmetric coupling with control bias

Interactions between the constituents of physical or biological systems occur due to the existence of different types of connections: global, local, unidirectional or bidirectional. Besides of synchronization due to a direct coupling the systems may be synchronized by interaction via a common medium where the units interact through the exchange of some substances as occurs in many biological and chemical systems. In one dimension, the unidirectionally coupled systems (Fig.3 (a) and (c)) are defined in the following way:

$$\dot{\mathbf{x}}_i = \mathbf{f}(\mathbf{x}_i) + \mathbf{K}(\mathbf{x}_{i-1} - \mathbf{x}_i) \tag{1}$$

while bidirectionally (or nearest-neigbour) coupled systems (Fig.3 (b) and (d)) are defined as follows:

$$\dot{\mathbf{x}}_i = \mathbf{f}(\mathbf{x}_i) + \mathbf{K}((\mathbf{x}_{i-1} - \mathbf{x}_i) + (\mathbf{x}_{i+1} - \mathbf{x}_i)) \tag{2}$$

Depending on the boundary conditions the system form an one-dimensional array (open boundary conditions, see Fig. 3 (a) and (b)) or a ring (closed boundary conditions, see Fig. 3 (c) and (d)). The synchronization in the systems described in Eqs. 1 and 2 can occur for particular values of coupling parameters defined by matrix $\mathbf{K}$, when the difference between variables $\Delta x_{ij} = x_i - x_j$ vanishes. The difference variable Δx_{ij} multiplied by coefficient matrix $\mathbf{K}$ plays a role of a corrective signal which pushes the systems to the desired state of synchronization. In higher than one dimensions systems can be coupled globally (see Fig.3 (e)):

$$\dot{\mathbf{x}}_i = \mathbf{f}(\mathbf{x}_i) + \frac{\mathbf{K}}{N-1}\sum_{j \neq i}(\mathbf{x}_j - \mathbf{x}_i) \tag{3}$$

The coupling scheme defined in Eq. 3 can be replaced by a simpler one:

$$\dot{\mathbf{x}}_i = \mathbf{f}(\mathbf{x}_i) + \mathbf{K}_1\sum_{j<i}(\mathbf{x}_j - \mathbf{x}_i) + \mathbf{K}_2\sum_{j>i}(\mathbf{x}_j - \mathbf{x}_i + \mathbf{B}_i) \tag{4}$$

where $\mathbf{B}_i$ is a bias voltage, meanwhile $\mathbf{K}_1$ and $\mathbf{K}_2$ are the fixed-valued coupling matrices. The optimization aspects of the coupling scheme presented in Eq. 4 respect to the coupling

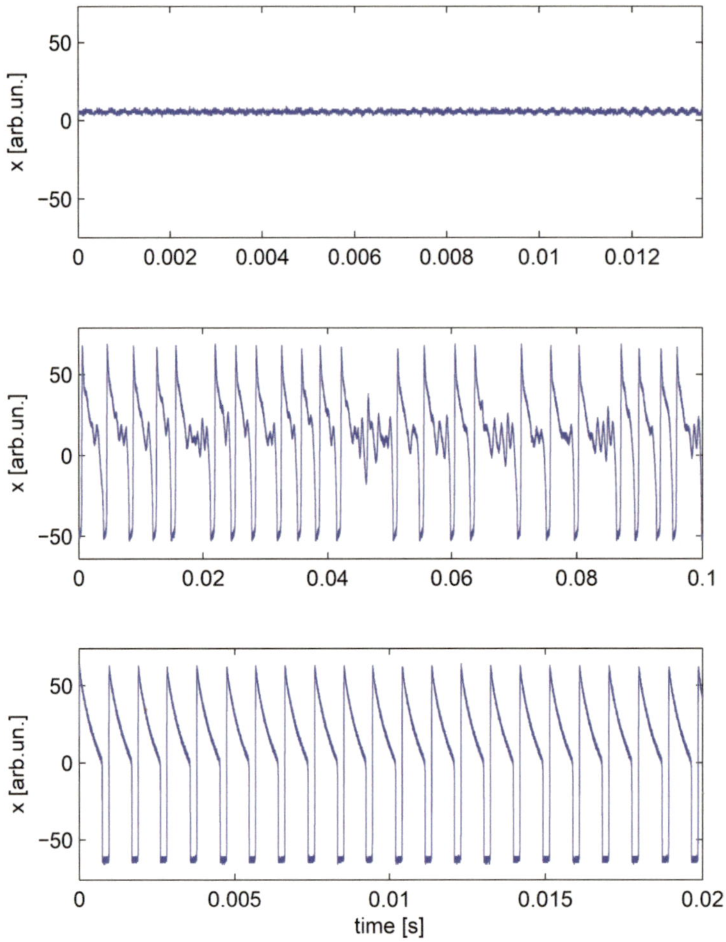

Fig. 2. Transition from a stationary steady state to chaotic spiking and eventually periodic self-oscillations as the dc-pumping current is varied in LED with optoelectronic feedback.

described in Eq.3 is explained below. Let us assume that the units are coupled only through a one system variable, i.e. a one observable in experiments, thus we consider $\mathbf{x}_k = x_k$. Then the total number of differential amplifiers used to get an input signal difference $\Delta x_{ji} = (x_j - x_i)$ which feeds the system x_i can be described by the binomial coefficient $E = 2\binom{N}{2} = \frac{N!}{(N-2)!}$, where N is the number of units in the network. It is however possible to decrease the number of elements by using the same signal difference Δx_{ji} as an input to a system x_j. In order to maintain the stability of synchronized solutions, in the second case, an additional bias voltage B_j is applied to the system x_j, so that it compensates the negative sign of the expression $-\Delta x_{ij}$ and allows to mimic an addition of the standard difference term Δx_{ij} required for the system x_j. For example, as shown in Fig. 4 in the case of $N = 4$, the number of elements reduces

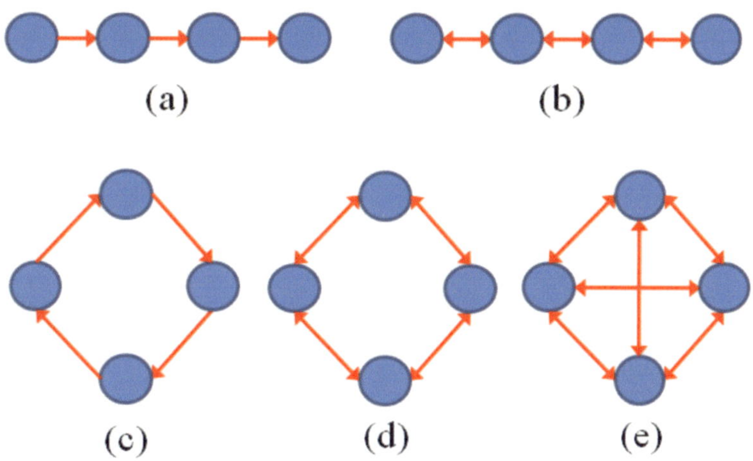

Fig. 3. Schematic presentation of different types of couplings (a) unidirectional, (b) bidirectional, (c) unidirectional in a ring, (d) bidirectional in a ring and (e) global.

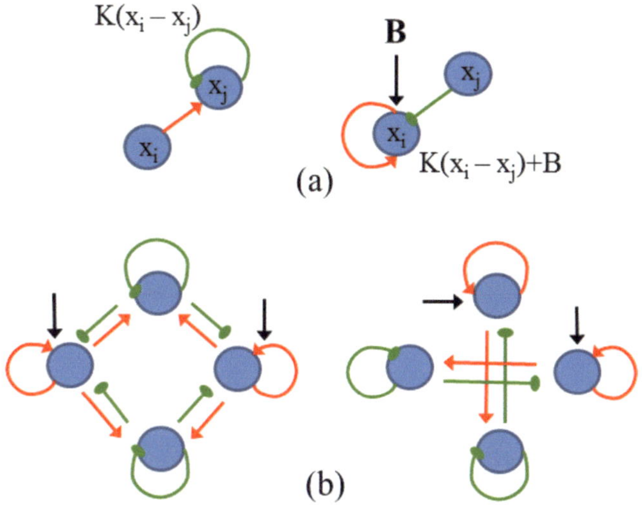

Fig. 4. Schematic presentation of the global coupling scheme with bias B: (a) compensation of the difference sign by B; (b) number of elements needed in the case of $N = 4$ units.

twice, from $E = 12$ to $E = 6$. The general value of elements E for N systems then becomes $E = \binom{N}{2}$.

Experimental results provided for 2 coupled systems, revealed that the transition to synchronization is not continuous as the control parameter, in our case the bias strength B, is varying. To describe quantitatively these abrupt changes we characterize the degree of order in the system by means of entropy S (see Ref. (14)). It is calculated from the distribution of the response times t_r in the time series. When the coupling is zero, this distribution is flat, i.e., the information on a site gives no information on the other ones. Increasing the coupling, we observe the birth of peaks for fixed time differences, due to the time correlation between

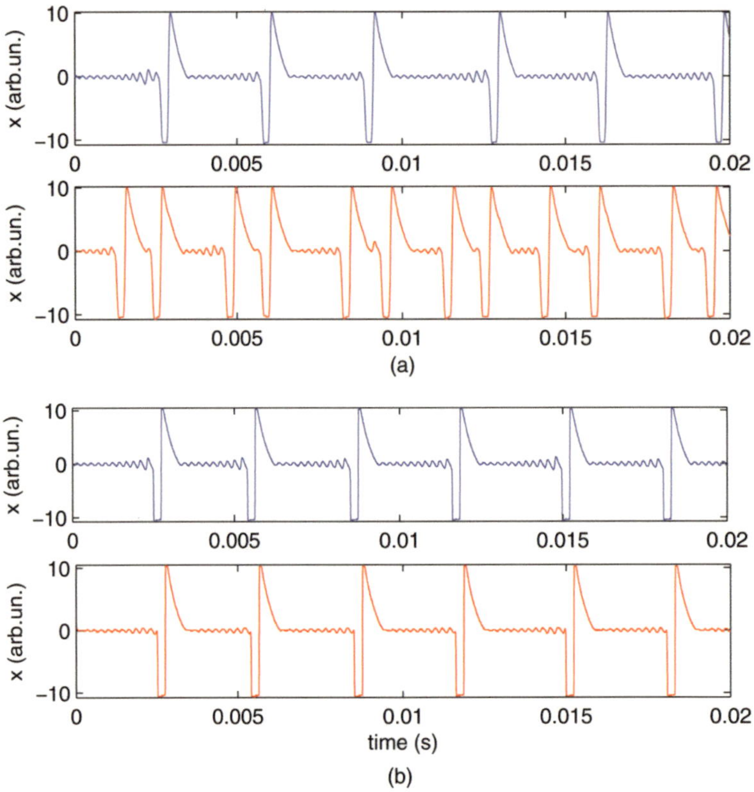

Fig. 5. Time series for 2 bidirectionally coupled LEDs in (a) desynchronized and (b) synchronized state.

spikes at adjacent sites. The entropy S is defined as follows:

$$S = -\sum_{t_r} p(t_r) \ln p(t_r) \tag{5}$$

where $p(t_r)$ is the probability distribution of t_r. The time series for two LED coupled with bias bidirectionally are shown in Fig. 5. Note, that in the case of N=2 systems, the implementation of Eq. 4 requires only one control bias B. The corresponding probability distributions are plotted in Fig. 6 (a). It can be seen that as the control bias is changing, the probability distribution becomes sharper giving rise to a low entropy values as reported in Fig. 6 (b).

3.2 Asymmetric coupling with control bias

The synchronization can be also obtained by constructing another types of couplings, not necessarily symmetric. One example, realized experimentally for 3 systems is based on a network with a doubly calculated difference $[x_3 - (x_2 - x_1)] + B$, which then feeds all 3 systems. Also in this case, the modulation of a control bias B enables to reach a synchronized state between all oscillators. Another type of asymmetric coupling has been applied to 6 LEDs. In this case it was defined as follows: $[(x_2 - x_1) + (x_4 - x_3) + (x_6 - x_5)] + B$, and again, used to feed all the six systems. The modulation of the control bias B enabled to reach synchronization between the systems as shown in Fig. 7. It is worth to note however, that the

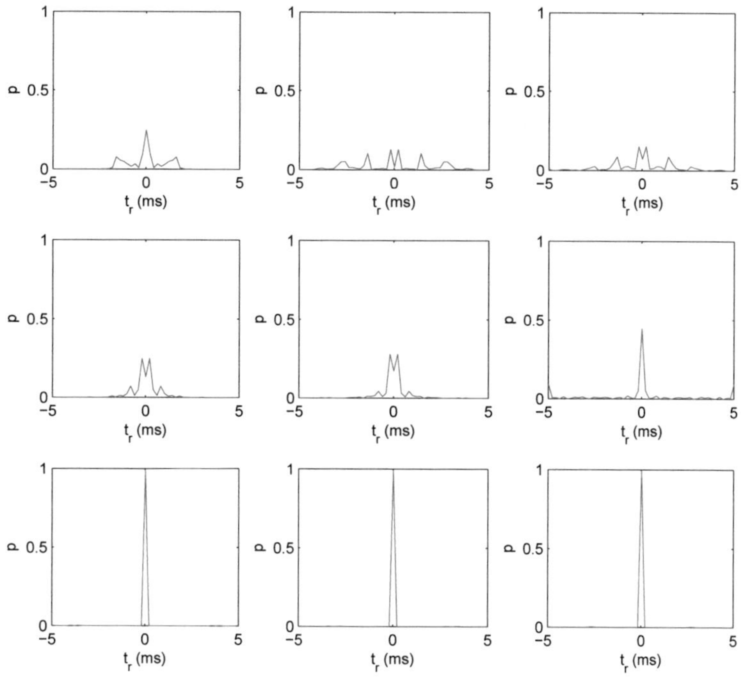

(a)

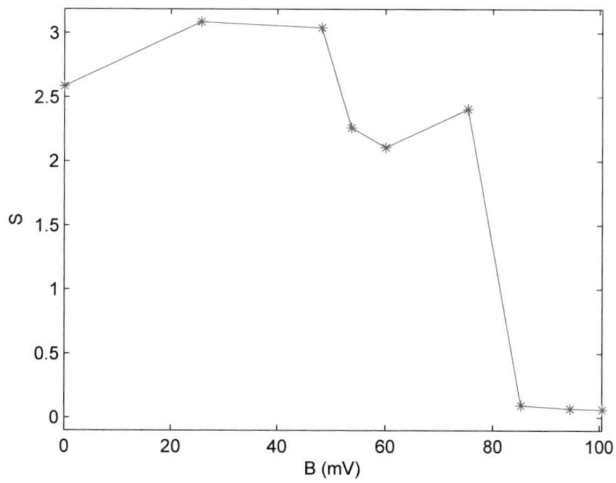

(b)

Fig. 6. (a) Distributions of the response times t_r and (b) corresponding entropy S versus bias B strength for two bidirectionally coupled LEDs.

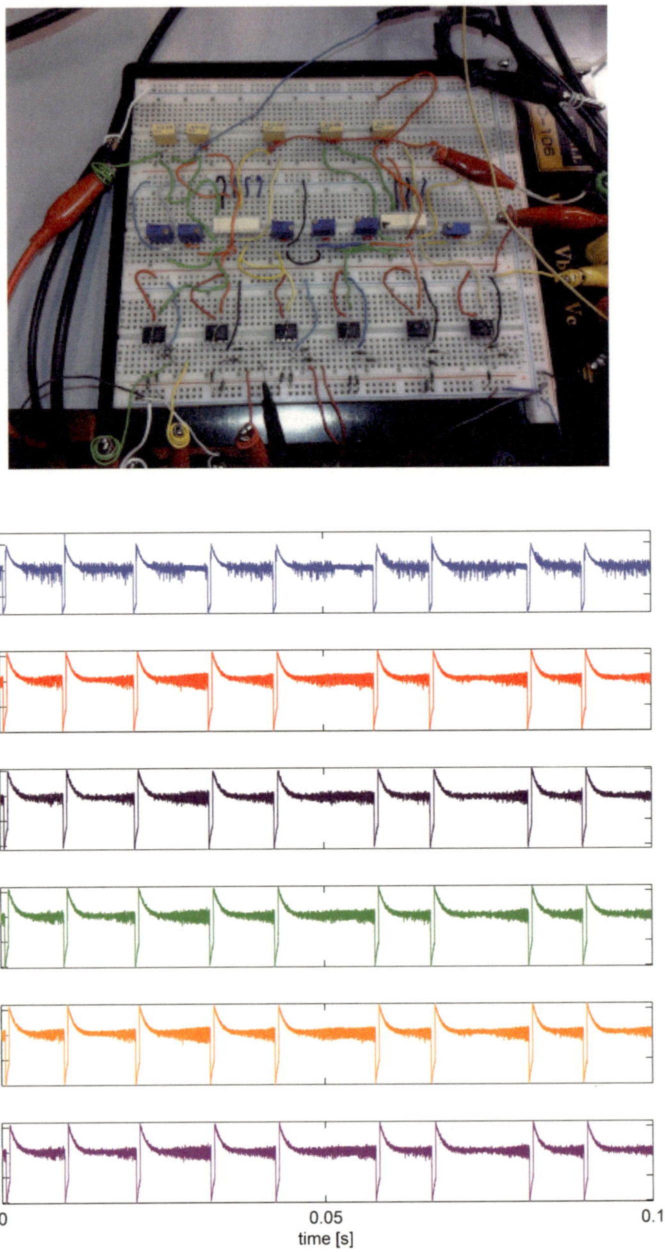

Fig. 7. Synchronization of six asymmetrically coupled chaotic LEDs.

synchronization occurs in high amplitude spikes, meanwhile the chaotic background of each oscillator is unsynchronized and follows its trajectory on its own chaotic attractor.

4. Conclusions

We presented the experimental results for the synchronization of chaotic optical network. The number of differential amplifiers has been reduced through adding of a bias voltage to the chaotic units. The adjustment of the single units dynamics, in order to reach a synchronized state within the all network, is a different kind of controlling the synchronization phenomenon as reported so far. This approach allows to control the parameter mismatch between the coupled units, what usually occurs in the experimental setups. Here, the collective interaction of the optical units has been reached through adjustment of the system states whereas the coupling strength is kept fixed. Moreover, we have realized experimental configurations for particular cases of asymmetric coupling in the case of three and six coupled oscillators. The synchronized state has been analyzed in terms of the probability distributions of the response times, measured between all interacting units. From the probability distributions the entropy has been evaluated what revealed the sharp decrease in its value as the bias voltage was modulated. An experimental observation of the transition to synchronization in an optical network is an important advance which can contribute not only to technological applications but also to experimental neuroscience. The latter problem, concerning the design of a miniaturized optical and functional neural network, will be the subject of our future study.

5. References

[1] Yamada, T. & Fujisaka, H. (1983). Stability Theory of Synchronized Motion in Coupled-Oscillator Systems. II. *Prog. Theor. Phys.* Vol. 70, pp. 1240-1248.
[2] Yamada, T. & Fujisaka, H. (1984). Stability Theory of Synchronized Motion in Coupled-Oscillator Systems. III. *Prog. Theor. Phys.* Vol. 72, pp. 885-894.
[3] Afraimovich, V. S.; Verichev, N. N. & Rabinovich, M. I. (1986). Stochastic synchronization of oscillations in dissipative systems. *Izv. Vys. Uch. Zav., Radiofizika* Vol. 29, pp. 1050Ű1060.
[4] Pecora, L. M. & Carroll T. L. (1990). Synchronization in chaotic systems. *Phys. Rev. Lett.* Vol. 64, pp. 821-824.
[5] Cuomo, K. M. & Oppenheim A. V. (1993). Circuit implementation of synchronized chaos with applications to communications. *Phys. Rev. Lett.* Vol. 71, pp. 65-68.
[6] Dafilis, M. P.; Liley, D. T. J. & Cadusch, P. J. (2001). Robust chaos in a model of the electroencephalogram: Implications for brain dynamics. *Chaos* Vol. 11, pp. 474-478.
[7] Arecchi, F. T. (2004). Chaotic neuron dynamics, synchronization and feature binding. *Physica A* Vol. 338, pp. 218Ű237.
[8] Al-Naimee, K.; Marino, F.; Ciszak, M.; Meucci, R.; Arecchi, F. T. (2009). Chaotic spiking and incomplete homoclinic scenarios in semiconductor lasers with optoelectronic feedback. *New J. Phys.* Vol. 11, pp. 073022.
[9] Marino, F.; Marin, F.; Balle, S. & Piro, O. (2007). Chaotically Spiking Canards in an Excitable System with 2D Inertial Fast Manifolds. *Phys. Rev. Lett.* Vol. 98, pp. 074104.
[10] Albahadily, F. N.; Ringland, J. & Schell, M. (1989). Mixed-mode oscillations in an electrochemical system. I. A Farey sequence which does not occur on a torus. *J. Chem. Phys.* Vol. 90, pp. 813.

[11] Petrov, V.; Scott, S. K. & Showalter, K. (1992). Mixed-mode oscillations in chemical systems. *J. Chem. Phys.* Vol. 97, pp. 6191-6198.

[12] Koper, M. T. M.; Gaspard, P. & Sluyters, J. H. (1992). Mixed-mode oscillations and incomplete homoclinic scenarios to a saddle focus in the indium/thiocyanate electrochemical oscillator. *J. Chem. Phys.* Vol. 97, pp. 8250.

[13] Al-Naimee, K.; Marino, F.; Ciszak, M.; Abdalah, S.F.; Meucci, R. & Arecchi F. T. (2010). Excitability of periodic and chaotic attractors in semiconductor lasers with optoelectronic feedback. *Eur. Phys. J. D* Vol. 58, pp. 187Ű189.

[14] Ciszak, M.; Montina, A. & Arecchi, F. T. (2008). Sharp versus smooth synchronization transition in locally coupled oscillators. *Phys. Rev. E* Vol. 78, pp. 016202.